EPILEPSY

The Intersection of Neurosciences, Biology, Mathematics, Engineering, and Physics

EPILEPSY

The Intersection of Neurosciences, Biology, Mathematics, Engineering, and Physics

EDITED BY
Ivan Osorio • Hitten P. Zaveri
Mark G. Frei • Susan Arthurs

CRC Press
Taylor & Francis Group
Boca Raton London New York

CRC Press is an imprint of the
Taylor & Francis Group, an **informa** business

MATLAB® is a trademark of The MathWorks, Inc. and is used with permission. The MathWorks does not warrant the accuracy of the text or exercises in this book. This book's use or discussion of MATLAB® software or related products does not constitute endorsement or sponsorship by The MathWorks of a particular pedagogical approach or particular use of the MATLAB® software.

CRC Press
Taylor & Francis Group
6000 Broken Sound Parkway NW, Suite 300
Boca Raton, FL 33487-2742

© 2011 by Taylor and Francis Group, LLC
CRC Press is an imprint of Taylor & Francis Group, an Informa business

No claim to original U.S. Government works

Printed in the United States of America on acid-free paper
10 9 8 7 6 5 4 3 2 1

International Standard Book Number: 978-1-4398-3885-3 (Hardback)

Library of Congress Cataloging-in-Publication Data

Epilepsy : the intersection of neurosciences, biology, mathematics, engineering and physics / [edited by] Ivan Osorio ... [et al.].
    p. ; cm.
    Includes bibliographical references and index.
    Summary: "Integrating the studies of epilepsy, neurosciences, computational neurosciences, mathematics, physics, engineering, and medicine, this volume provides the first means to a clear, structured, and enriching multidisciplinary perspective for developing students, researchers, and clinicians in the field of epilepsy. The text gives up to date information on the state of current research, possible future research, and clinical directions in the fields of epilepsy and seizure dynamics. Clear illustrations that illustrate the patterns of brain electrophysiology and behavioral manifestations are provided to aid comprehension. The authors include extensive references in each chapter to enhance further study"--Provided by publisher.
    ISBN 978-1-4398-3885-3 (hardback)
    1. Epilepsy. I. Osorio, Ivan.
    [DNLM: 1. Epilepsy--physiopathology. 2. Interdisciplinary Communication. 3. Predictive Value of Tests. 4. Seizures--prevention & control. WL 385]

    RC372.E67276 2011
    616.8'53--dc22   2011006202

Visit the Taylor & Francis Web site at
http://www.taylorandfrancis.com

and the CRC Press Web site at
http://www.crcpress.com

This book is dedicated to all epilepsy patients and their caregivers who carry the burden of this disease and for whom we work to provide better treatments. We also acknowledge all of the researchers from the many disciplines and from around the world who tirelessly focus on epilepsy in general and seizure prediction in particular.

Contents

SECTION I   Foundations of Epilepsy

SECTION II   Foundations of Engineering, Math, and Physics

SECTION III   The Challenge of Prediction

SECTION IV   The State of Seizure Prediction: Seizure Prediction and Detection

SECTION V   *The State of Seizure Prediction: Seizure Generation*

SECTION VI   *The State of Seizure Prediction: Seizure Control*

SECTION VII   The State of Seizure Prediction: Technology

SECTION VIII   Nocturnal Frontal Lobe Epilepsy: A Paradigm for Seizure Prediction?

Preface

Epilepsy: The Intersection of Neurosciences, Biology, Mathematics, Engineering, and Physics conflates the didactic sessions, keynote lectures, and scientific presentations delivered at the Fourth International Workshop on Seizure Prediction held in Kansas City, Missouri in the spring of 2009 and at the Sanibel Symposium (Nocturnal Frontal Lobe Epilepsy: An Interdisciplinary Perspective) convened on Sanibel Island, Florida in the winter of 2010. The complexity and diversity of the subject matter contained in this book coupled with the contributors' worthy efforts to meet the inherent challenge to make its contents accessible to researchers with disparate scientific backgrounds, makes this book a useful source for those outside the field as well those inside. Students of the dynamical behavior of seizures and those aspiring to predict their occurrence with a worthwhile degree of clinical utility and implement timely therapeutic interventions may be its greatest beneficiaries.

Of all neurological disorders, epilepsy demands of investigators the broadest and deepest knowledge of dynamical, control, and system theories, knowledge that cannot be amassed without possessing a certain level of sophistication in relevant areas of physics, mathematics, and engineering. A small but growing number of epileptologists realize that substantive progress in their field of endeavor will not materialize without transcending conventional approaches and the insularity of their efforts. Contributions from mathematicians, physicists, engineers, computer scientists, and other disciplines have been welcomed and are beginning to bear fruit. This book, albeit imperfectly, attempts to both capture and enrich the burgeoning interdisciplinary synergism in the nascent field of dynamical epileptology by narrowing the inescapable cultural chasm that commonly fragments multidisciplinary efforts. To address this innate risk, the meetings and the book's contents have been organized around five themes: 1. Foundations of Epilepsy (Chapters 1–5) introduces nonphysicians to language and concepts necessary to establish a meaningful dialog with epileptologists; 2. Foundations of Engineering, Math, and Physics (Chapters 6–12) expands the fund of knowledge of physicians into areas such as dynamical theory and signal processing without which the synthesis of concepts and ideas into testable hypotheses would be onerous and likely barren; 3. Challenge of Prediction (Chapter 13) delves into the issue of how to assess the degree of predictability from a system's behavior and the techniques required for fulfillment of this aim, by mining of knowledge from fields devoted to the investigation of aperiodic paroxysmal phenomena, such as earthquakes, so as to lay the proper foundations for dynamical epileptology and foster much needed growth efficiently; 4. The State of Seizure Prediction, Generation and Control (Chapters 14–31) provides an update of advancements in our understanding of the spatiotemporal behavior of seizures and of the mechanisms of epileptogenesis and ictogenesis as well as of seizure control and ancillary technology; and 5. Nocturnal Frontal Lobe Epilepsy: A Paradigm for Seizure Prediction? (Chapters 32–38) calls attention to an ignored syndrome, nocturnal frontal lobe epilepsy, whose proclivity for seizures to occur during a certain sleep stage may facilitate the task of identifying a precursory state, if existent, while markedly narrowing the search space.

The editors recognize and express their gratitude to those who generously put to pen their insights into a spinous but intellectually stimulating field of study and to the federal and private organizations and individuals* that provided the means to foster coalescence of seemingly disparate topics, concepts, and viewpoints into printed matter.

---

* See Acknowledgments for a complete list of sponsors.

MATLAB® is a registered trademark of The MathWorks, Inc. For product information, please contact:

The MathWorks, Inc.
3 Apple Hill Drive
Natick, MA 01760-2098 USA
Tel: 508-647-7000
Fax: 508-647-7001
E-mail: info@mathworks.com
Web: www.mathworks.com

Acknowledgments

Funding for Fourth International Workshop on Seizure Prediction and the Sanibel Symposium (Nocturnal Frontal Lobe Epilepsy: An Interdisciplinary Perspective) was provided by the following government agencies, foundations, companies, university and hospital partners, and individuals: National Institutes of Health[*] (Grant No. R13NS065535 from the National Institute of Neurological Disorders and Stroke, Office of Rare Diseases Research and National Institute of Child Health and Human Development), Alliance for Epilepsy Research, UCB, Cyberonics, Deutsche Gesellschaft für Epileptologie, NeuroVista, American Epilepsy Society, CURE, University of Kansas Medical Center, Children's Mercy Hospitals and Clinics, Honeywell–Kansas City Plant, Ad-Tech, Cardinal Health, Medtronic, DIXI, Boulevard Brewing Co., and Mary Shaw Branton, Don Alexander, and Frank and Helen Wewers.

---

[*] The views expressed in this publication do not necessarily represent the official views of the National Institutes of Health, NINDS, ORDR, or NICHD and do not necessarily reflect the official policies of the Department of Health and Human Services; nor does mention by trade names, commercial practices, or organizations imply endorsement by the U.S. Government.

Editors

Ivan Osorio— Ivan Osorio, a physician scientist and epileptologist, is professor of neurology at the University of Kansas Medical Center and the R. Adams Institute of Analytical Biochemistry. He is a Distinguished Visiting Neuroscience Professor at the Claremont Colleges' Joint Science Department and a visiting scholar at Harvey Mudd's Department of Mathematics. He serves on the editorial boards of several journals and on NINDS' study sections.

His research interests are the characterization of the spatiotemporal behavior of seizures, the development of intelligent devices for automated seizure detection, warning, and control, quantitative real-time assessment of therapeutic efficacy and real-time optimization, and the application of control and systems theory to the study and treatment of epilepsy.

Hitten Zaveri—Hitten P. Zaveri is an associate research scientist in the Department of Neurology at Yale University. Dr. Zaveri has received academic training in electrical engineering (BSE, MSE), computer engineering (BSE), and biomedical engineering (MS, PhD) from the University of Michigan in Ann Arbor and postdoctoral training in epilepsy and neurology from Yale University. He is the director of the Yale University Computational Neurophysiology Laboratory. His research interests lie at the intersection of neuroscience, engineering, and mathematics.

Mark G. Frei—Mark G. Frei is the managing director and technical director of Flint Hills Scientific, L.L.C. (FHS), a development stage research and development company in Lawrence, Kansas that Dr. Frei cofounded in 1995. He specializes in real-time quantitative analysis, filtering, identification and control of complex systems and signals, and in algorithm development for intelligent medical devices. He is an inventor on more than 30 patents, including several involving epileptic seizure detection. Prior to joining FHS, Dr. Frei was a postdoctoral fellow at the University of Kansas, in the Kansas Institute for Theoretical and Computational Science, the Comprehensive Epilepsy Center, and the Department of Mathematics. He received his PhD in mathematics from the University of Kansas with research specialties in the fields of modeling, prediction, and adaptive control of complex systems, and has authored or coauthored several scientific articles. He received his MS in applied mathematics/electrical engineering from the University of Southern California and BA in mathematics from the University of California Los Angeles. Prior to his doctoral training at the University of Kansas, Dr. Frei was a member of the technical staff of TRW, Inc.

Susan Arthurs—Susan Arthurs is a former United Airlines pilot who lost her career to epilepsy. In 1996, she cofounded the Alliance for Epilepsy Research, a 501(c)(3) charitable organization, and is currently executive director and secretary of the board. She is director and producer of the 2009 award-winning documentary, "It Is Epilepsy: The Challenges and Promises of Automated Seizure Control," which was created to increase public awareness of epilepsy and provide hope for those who have epilepsy in their lives. From 2000 to 2004, Susan was an adjunct instructor then assistant professor in the Department of Aviation at the University of Central Missouri. She also developed and is marketing the "Take Flight!" series of books with interactive CDs that contain ideas and resources for aviation and space activities for Grades K through 12. Susan has a BS degree from Penn State University in secondary mathematics education and a MS from the University of Central Missouri in aviation safety.

Contributors

L. Aires
Faculty of Sciences and Technology
University of Coimbra
Coimbra, Portugal

C. Alvarado-Rojas
Centre de Recherche de l'Institut du Cerveau
  et de la Moelle épinière
Hôpital de la Pitié-Salpêtrière
Paris, France

Ralph G. Andrzejak
Department of Information and
  Communication Technology
Universitat Pompeu Fabra
Barcelona, Spain

Daniel Bertrand
Department of Neuroscience
Medical Faculty
University of Geneva
Geneva, Switzerland

Stephan Bialonski
Department of Epileptology
and
Helmholtz Institute for Radiation and
  Nuclear Physics
and
Interdisciplinary Center for Complex Systems
University of Bonn
Bonn, Germany

Mark R. Bower
Mayo Systems Electrophysiology Laboratory
Division of Epilepsy and Section of
  Electroencephalography
and
Department of Neurology
Mayo Clinic
Rochester, Minnesota

Benjamin H. Brinkmann
Mayo Systems Electrophysiology Laboratory
Division of Epilepsy and Section of
  Electroencephalography
and
Department of Neurology
Mayo Clinic
Rochester, Minnesota

Molly N. Brown
Vanderbilt University
School of Medicine
Nashville, Tennessee

Anthony N. Burkitt
Department of Electrical and Electronic
  Engineering
The University of Melbourne
Victoria, Australia

Paul R. Carney
Departments of Pediatrics, Neurology, and
  Neuroscience
and
J. Crayton Pruitt Department of Biomedical
  Engineering
McKnight Brain Institute
University of Florida
Gainesville, Florida

Daniel Chicharro
Department of Information and
  Communication Technology
Universitat Pompeu Fabra
Barcelona, Spain

Jui-Hong Chien
Optima Neuroscience, Inc.
Alachua, Florida
and
J. Crayton Pruitt Family Department of
  Biomedical Engineering
University of Florida
Gainesville, Florida

R. P. Costa
Faculty of Sciences and Technology
University of Coimbra
Coimbra, Portugal

B. Crépon
Centre de Recherche de l'Institut du Cerveau et
    de la Moelle épinière
Hôpital de la Pitié-Salpêtrière
and
Epilepsy Unit, Groupe Hospitalier
    Pitié-Salpêtrière Assistance Publique
Hôpitaux de Paris
Paris, France

Al W. de Weerd
Department of Clinical Neurophysiology and
    Sleep Centre
Dokter Denekampweg
Zwolle, Netherlands

B. Direito
Faculty of Sciences and Technology
University of Coimbra
Coimbra, Portugal

William Ditto
School of Biological and Health Systems
    Engineering
Arizona State University
Tempe, Arizona

A. Dourado
Faculty of Sciences and Technology
University of Coimbra
Coimbra, Portugal

Tyler S. Durazzo
Yale University
New Haven, Connecticut

Christian E. Elger
Department of Epileptology
University of Bonn
Bonn, Germany

Hinnerk Feldwisch-Drentrup
Bernstein Center for Computational
    Neuroscience
University of Freiburg
Freiburg, Germany

Piotr J. Franaszczuk
Department of Neurology
Johns Hopkins University
School of Medicine
Baltimore, Maryland

Mark G. Frei
Flint Hills Scientific, L.L.C.
Lawrence, Kansas

Dennis Glüsenkamp
Department of Epileptology
and
Helmholtz Institute for Radiation and Nuclear
    Physics
University of Bonn
Bonn, Germany

Jeffrey H. Goodman
Department of Developmental Neurobiology
New York State Institute for Basic Research
Staten Island, New York
and
Department of Physiology and Pharmacology
State University of New York Downstate
    Medical Center
Brooklyn, New York

David B. Grayden
Department of Electrical and Electronic
    Engineering
University of Melbourne
Victoria, Australia

John V. Guttag
Electrical Engineering and Computer Science
    Department
Massachusetts Institute of Technology
Cambridge, Massachusetts

Wytske A. Hofstra
Department of Clinical Neurophysiology and
    Sleep Centre
Dokter Denekampweg
Zwolle, Netherlands

G. Huberfeld
Centre de Recherche de l'Institut du Cerveau
    et de la Moelle épinière
Hôpital de la Pitié-Salpêtrière
and
Epilepsy Unit, Groupe Hospitalier
    Pitié-Salpêtrière Assistance Publique
Hôpitaux de Paris
Paris, France

Dong-Uk Hwang
Korea Institute of Oriental Medicine
Daejeon, Republic of Korea

Leon Iasemidis
Harrington Department of Bioengineering and
    Electrical Engineering
Arizona State University
Tempe, Arizona
and
Department of Neurology
Mayo Clinic
Phoenix, Arizona

Pedro P. Irazoqui
Weldon School of Biomedical Engineering
Purdue University
West Lafayette, Indiana

M. Jachan
Albert-Ludwigs-University
Freiburg, Germany

John G. R. Jefferys
Neuronal Networks Group
School of Clinical and Experimental Medicine
University of Birmingham
Birmingham, United Kingdom

Premysl Jiruska
Neuronal Networks Group
School of Clinical and Experimental Medicine
University of Birmingham
Birmingham, United Kingdom

Bharat S. Joshi
Department of Electrical and Computer
    Engineering
University of North Carolina
Charlotte, North Carolina

Karin Jurkat-Rott
Institute of Applied Physiology
Ulm University
Ulm, Germany

Stiliyan N. Kalitzin
Medical Physics Department
Epilepsy Institute of the Netherlands
Heemstede, Netherlands

Ryan T. Kern
Optima Neuroscience, Inc.
Alachua, Florida

Dieter Krug
Department of Epileptology
and
Helmholtz Institute for Radiation and
    Nuclear Physics
University of Bonn
Bonn, Germany

Levin Kuhlmann
Department of Electrical and Electronic
    Engineering
University of Melbourne
Victoria, Australia

Themis R. Kyriakides
Departments of Pathology and Biomedical
    Engineering
Yale University
New Haven, Connecticut

Bruce Lanning
Thin Film Technologies
ITN Energy Systems
Littleton, Colorado

M. Le Van Quyen
Centre de Recherche de l'Institut du Cerveau
    et de la Moelle épinière
Hôpital de la Pitié-Salpêtrière
Paris, France

Frank Lehmann-Horn
Institute of Applied Physiology
Ulm University
Ulm, Germany

Klaus Lehnertz
Department of Epileptology
and
Helmholtz Institute for Radiation and
    Nuclear Physics
and
Interdisciplinary Center for Complex Systems
University of Bonn
Bonn, Germany

Holger Lerche
Department of Neurology and Epileptology
Hertie Institute of Clinical Brain Research
University of Tübingen
Tübingen, Germany

Ronald P. Lesser
Department of Neurology and Neurosurgery
Johns Hopkins University
School of Medicine
Baltimore, Maryland

Fernando Lopes da Silva
Swammerdam Institute of Neuroscience
University of Amsterdam
Amsterdam, Netherlands

Snezana Maljevic
Department of Neurology and Epileptology
Hertie Institute of Clinical Brain Research
University of Tübingen
Tübingen, Germany

Iven M. Y. Mareels
Department of Electrical and Electronic
    Engineering
University of Melbourne
Victoria, Australia

Gregory C. Mathews
Vanderbilt University
School of Medicine
Nashville, Tennessee

John G. Milton
W. M. Keck Science Center
Claremont Colleges
Claremont, California

Florian Mormann
Department of Epileptology
University of Bonn
Bonn, Germany

Gholam K. Motamedi
Department of Neurology
Georgetown University School of Medicine
Washington, DC

V. Navarro
Centre de Recherche de l'Institut du Cerveau
    et de la Moelle épinière
Hôpital de la Pitié-Salpêtrière
and
Epilepsy Unit, Groupe Hospitalier
    Pitié-Salpêtrière Assistance Publique
Hôpitaux de Paris
Paris, France

S. Nikolopoulos
Centre de Recherche de l'Institut du Cerveau et
    de la Moelle épinière
Hôpital de la Pitié-Salpêtrière
Paris, France

Elma O'Sullivan-Greene
Department of Electrical and Electronic
    Engineering
University of Melbourne
Victoria, Australia
and
National Information and Communications
    Technology Australia
Victorian Research Lab
Melbourne, Australia

Ivan Osorio
Department of Neurology
University of Kansas Medical Center
Kansas City, Missouri

Trudy Pang
Neurology Department
and
Clinical Investigator Training Program
Beth Israel Deaconess Medical Center
and
Harvard/MIT Health Sciences and Technology
  in collaboration with Pfizer Inc. and Merck
  and Company
Boston, Massachusetts

Panos M. Pardalos
Industrial and Systems Engineering Department
University of Florida
Gainesville, Florida

Austin R. Quan
Harvey Mudd College
Claremont, California

Shriram Raghunathan
Weldon School of Biomedical Engineering
Purdue University
West Lafayette, Indiana

Pooja Rajdev
Weldon School of Biomedical Engineering
Purdue University
West Lafayette, Indiana

Alexander Rothkegel
Department of Epileptology
and
Helmholtz Institute for Radiation and Nuclear
  Physics
and
Interdisciplinary Center for Complex Systems
University of Bonn
Bonn, Germany

Shivkumar Sabesan
Department of Neurology Research
Barrow Neurological Institute
St. Joseph's Hospital and Medical Center
Phoenix, Arizona
and
Harrington Department of Bioengineering
Arizona State University
Tempe, Arizona

J. Chris Sackellares
Optima Neuroscience, Inc.
Alachua, Florida

Steven C. Schachter
Neurology Department
Beth Israel Deaconess Medical Center
Boston, Massachusetts

Bjoern Schelter
Center for Data Analysis and Modeling
University of Freiburg
Freiburg, Germany

Andreas Schulze-Bonhage
Epilepsy Center
University Hospital
Freiburg, Germany

Deng-Shan Shiau
Optima Neuroscience, Inc.
Alachua, Florida

Ali Shoeb
Massachusetts General Hospital
Boston, Massachusetts
and
Massachusetts Institute of Technology
Cambridge, Massachusetts

Joseph Sirven
Department of Neurology
Mayo Clinic
Phoenix, Arizona

Didier Sornette
Swiss Federal Institute of Technology Zurich
and
Swiss Finance Institute
Zurich, Switzerland

Dennis D. Spencer
Department of Neurosurgery
Yale University
New Haven, Connecticut

Matthäus Staniek
Department of Epileptology
and
Helmholtz Institute for Radiation and Nuclear
    Physics
University of Bonn
Bonn, Germany

Matt Stead
Mayo Systems Electrophysiology Laboratory
Division of Epilepsy and Section of
    Electroencephalography
and
Department of Neurology
Mayo Clinic
Rochester, Minnesota

Ortrud K. Steinlein
Institute of Human Genetics
University Hospital, Ludwig-
    Maximilians-University
Munich, Germany

Sridhar Sunderam
Center for Biomedical Engineering
University of Kentucky
Lexington, Kentucky

Sachin S. Talathi
J. Crayton Pruitt Department of Biomedical
    Engineering
McKnight Brain Institute
University of Florida
Gainesville, Florida

C. A. Teixeira
Faculty of Sciences and Technology
University of Coimbra
Coimbra, Portugal

Jens Timmer
Freiburg Institute for Advanced Studies
University of Freiburg
Freiburg, Germany

David M. Treiman
Department of Neurology
Barrow Neurological Institute
St. Joseph's Hospital and Medical Center
Phoenix, Arizona

Konstantinos Tsakalis
Department of Electrical Engineering
Arizona State University
Tempe, Arizona

M. Valderrama
Centre de Recherche de l'Institut du Cerveau et
    de la Moelle épinière
Hôpital de la Pitié-Salpêtrière
Paris, France

Taufik A. Valiante
Krembil Neuroscience Center
University Health Network
Toronto Western Hospital
Toronto, Ontario, Canada

Andrea Varsavsky
Department of Otolaryngology
The University of Melbourne
Victoria, Australia

Demetrios N. Velis
Department of Clinical Neurophysiology
and
Epilepsy Monitoring Unit
Epilepsy Institute of the Netherlands
Heemstede, Netherlands

Christopher Warren
Mayo Systems Electrophysiology Laboratory
Division of Epilepsy and Section of
    Electroencephalography
and
Department of Neurology
Mayo Clinic
Rochester, Minnesota

Yvonne Weber
Department of Neurology and Epileptology
Hertie Institute of Clinical Brain Research
University of Tübingen
Tübingen, Germany

Steven Weinstein
Pediatric Epilepsy Center
Weill Cornell Medical Center
New York Presbyterian Hospital
New York, New York

Richard Wennberg
University of Toronto
Toronto, Ontario, Canada

Gregory A. Worrell
Mayo Systems Electrophysiology Laboratory
Division of Epilepsy and Section of
    Electroencephalography
and
Department of Neurology
Mayo Clinic
Rochester, Minnesota

Hitten P. Zaveri
Department of Neurology
Yale University
New Haven, Connecticut

Section I

Foundations of Epilepsy

1 Neuroanatomy as Applicable to Epilepsy
Gross and Microscopic Anatomy/Histology

Taufik A. Valiante

CONTENTS

If the brain were simple enough for us to understand it, we would be too simple to understand it.

—Ken Hill

1.1 INTRODUCTION

This is a brief summary of brain anatomy in the context of epilepsy. It reviews the gross anatomical and some ultrastructural details as a primer for those with little or no clinical orientation. This is by no means exhaustive, and there are ample texts for those wishing more detail (Nieuwenhuys et al. 1988; Carpenter 1991; Sheppard 2004; Andersen et al. 2007; Miller and Cummings 2007). As was so appropriately stated during the Fourth International Workshop on Seizure Prediction (IWSP4) in Kansas City by Walter Freeman, "We don't have a language that describes brain function." We are left then with trying to describe the brain from a structural perspective and implying function from physiological experimentation on animals and humans. It might be equally important to mention that, along with not having a language to describe brain function, we have little way to measure brain function—whatever that might be—if this is in fact what we are doing with various technologies currently at our disposal.

I will describe the various lobes of the brain and what structural and ultrastructural details give rise to these anatomical demarcations. How various regions are wired together will be dealt with through discussion of white matter tracts. Functional subregions relevant to epilepsy and its manifestations will follow from the above lobar demarcations as well as from cellular variability throughout the brain associated with specialized functions of these cellularly distinct regions. There is a more generous discussion of the temporal lobe compared with other lobes, as it is the portion of the brain that has been extensively studied in regard to epilepsy and the region most frequented by neurosurgeons in attempts to ameliorate epilepsy.

1.2 GROSS ANATOMY

The human brain weighs, on average, 1350 grams; however, in its native state suspended in nutritious cerebrospinal fluid (CSF), it weighs only 50 grams. It is a highly ordered and exceedingly organized structure with the majority of the brain volume being composed of cells and cellular processes. Hence, the brain is largely composed of the internal components of cells, with only a very narrow space (extracellular space) separating cellular components of the brain. In fact, only about 4% of the human brain volume can be attributed to extracellular space. The remainder of the brain volume can be attributed to neurons and glia, the two primary cellular components of the brain. There are three primary subdivisions of the brain (see Figure 1.1): cerebrum, brain stem, and cerebellum. The last two are not usually implicated in epilepsy and will not be discussed.

The two hemispheres (see Figure 1.2) are separated by the longitudinal or interhemispheric fissure and are covered by a 2- to 5-mm layer of neurons that forms the cortex (like the bark of a tree) or gray matter. Underlying the gray matter is white matter, which represents the cabling of the brain and is composed of axons. Axons convey all or none events to other neurons through the generation of action potentials that are conveyed in a metabolically efficient manner over long distances. The termination of the axon at a synapse generates graded potentials in the postsynaptic cell that will, if large (sufficiently depolarized) enough, cause the postsynaptic neuron to generate an action potential. There are other gray matter components—the largest being the striatum, basal ganglia, and the thalamus—that lie within the core of the brain with interposed white matter. These gray matter structures are not arranged in layers, like the cortex, but form clusters of neurons called ganglia. The two hemispheres are interconnected by a very large white matter bundle called the corpus callosum, which will be discussed in more detail later.

The total surface area of the cortex measures approximately 2500 cm^2 (0.5 × 0.5 m), with only one third of it being visible on the exposed surface of the brain (Douglas et al. 2004). The remainder

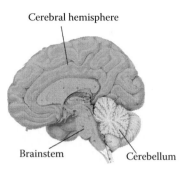

FIGURE 1.1   (See color insert.) The human brain. The cerebral hemispheres are paired (not shown), as are the cerebellar hemispheres. The brainstem is not a paired structure and is subdivided into the diencephalon, midbrain, pons, and medulla. Only the cerebral hemispheres and the thalamus are of concern in this chapter, as the other structures of the diencephalon have unclear roles in epilepsy. (Adapted from Nieuwenhuys, R. et al., *The Human Central Nervous System: A Synopsis and Atlas,* Springer-Verlag, New York, 1988. With permission; Carpenter, M. B., *Core Text of Neuroanatomy*, Williams & Wilkins, Baltimore, 1991.)

of the cortex is hidden within sulci (deep clefts) that are lined by cortex (Figure 1.2). The brain matter between these sulci are gyri. Sulci have a somewhat constant pattern between individuals, although this applies mainly to the larger sulci, which are called fissures. The variability of the gyral pattern makes inferences about function from gyral topology inconsistent. Given the laminar profile of the cortex, the electrophysiological manifestations of electrical fields generated at the top (crowns) of the gyri will be different from those generated in the walls of the sulci (Ebersole 2003). The engineering problems that arise from this are left to those more computationally adept than the author. The gyri can, in certain regions of the brain, be lumped together into functionally (and usually histologically) distinct regions called lobes. For example, the frontal lobe is demarcated from the parietal lobe and temporal lobe by the Rolandic fissure posteriorly and the Sylvian fissure

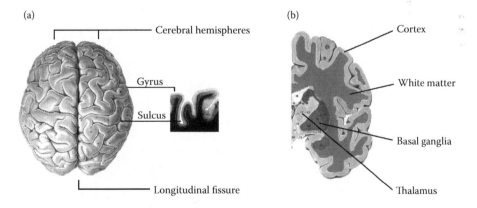

FIGURE 1.2   (See color insert.) The hemispheres and general organization. (a) The two hemispheres are separated by the longitudinal fissure. They are connected (although not visible) by the corpus callosum. The inset shows a cross section through the brain and the underlying structure and the relationship of gyri and sulci and how they are manifest on the surface of the brain. This can also be appreciated in (b). (b) Overall organization of the hemisphere. The cortex overlies the white matter that inspisates itself between the deep gray matter structures of the brain—the largest of these deep gray matter structures being the striatum, basal ganglia, and thalamus. (Adapted from Nieuwenhuys, R. et al., *The Human Central Nervous System: A Synopsis and Atlas,* Springer-Verlag, New York, 1988. With permission; Carpenter, M. B., *Core Text of Neuroanatomy,* Williams & Wilkins, Baltimore, 1991. With permission.)

inferiorly, respectively. With these gross demarcations, the frontal lobe can be further partitioned based on the sulcal pattern into various gyri (Figure 1.6).

1.3   CYTOARCHITECTURE

In the context of the brain, cytoarchitectonics refers to the study of the laminar organization of the cellular components of the brain. The neocortex has six layers with each mm³ containing approximately 50,000 neurons. It has been suggested that 1 mm² is the smallest area of cortex that can perform all the functions of a given cortical area (Douglas et al. 2004).

Broadly speaking, neurons can be subgrouped into those that have spines, and those that do not. A spiny neuron, unlike a nonspiny neuron, utilizes the neurotransmitter glutamate (excitatory neurotransmitter), whereas nonspiny neurons utilize gamma amino butyric acid (GABA) (Douglas et al. 2004). Excitation refers to the process of bringing the membrane potential of a given neuron to a more positive potential, and is also referred to as depolarization. Inhibition is the converse of excitation and is also referred to as hyperpolarization. In electrophysiology parlance, and the convention that currents are carried by positive charges, depolarization is a result of an increase in inward current or a decrease in outward current.

As mentioned above, the neocortex has six layers of neurons, and the cellular composition of these layers varies across different brain regions. The various layers and their names are illustrated in Figure 1.3. Broadly speaking, there appear to be three different patterns of cellular constituents, giving rise to cytoarchitecturally distinct types of cortex: (1) koniocortex or granular cortex of sensory areas, which contains many granule and stellate cells; (2) agranular cortex of motor and premotor regions, which is characterized by fewer stellate cells and more pyramidal neurons; and (3) eulaminate or homotypical cortex, which includes much of the association cortices (a cortex that is not primary motor or primary sensory). An example of a cytoarchitecturally distinct region is the primary motor area, also referred to as the motor strip, which is characterized by large pyramidal neurons in layers III and V and almost complete loss of layer 4 (Figure 1.4). In contrast to this primary motor area, the primary visual cortex, which is a primary sensory area, is characterized by a very prominent layer 4

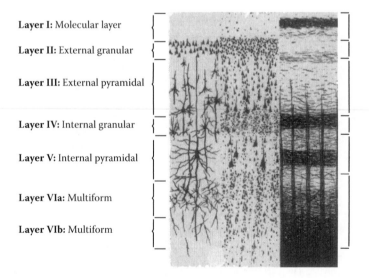

Layer I: Molecular layer

Layer II: External granular

Layer III: External pyramidal

Layer IV: Internal granular

Layer V: Internal pyramidal

Layer VIa: Multiform

Layer VIb: Multiform

FIGURE 1.3 Laminar structure of the cortex. A histological section through the human brain displaying the various layers. (From Nieuwenhuys, R. et al., *The Human Central Nervous System: A Synopsis and Atlas,* Springer-Verlag, New York, 1988; Carpenter, M. B., *Core Text of Neuroanatomy,* Williams & Wilkins, Baltimore, 1991. With permission.)

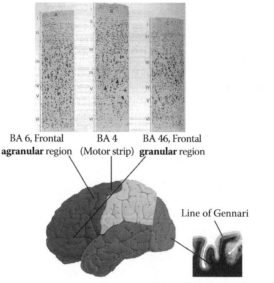

BA 6, Frontal   BA 4   BA 46, Frontal
agranular region (Motor strip) **granular** region

Line of Gennari

BA 17, Primary visual cortex

FIGURE 1.4 **(See color insert.)** Regional variability of cortical layering. Three different functional regions are depicted, and their variability in cortical layering. BA 6, or the premotor area, is just one sulcus anterior to the motor strip (primary motor area, BA 6) but lacks the large Betz cells of layer V that are characteristic of the primary motor area. BA 4 has a paucity of inputs from sensory and associational areas and thus has a less conspicuous layer IV, which is more prominent in the BA 46 (dorsolateral prefrontal cortex). In BA 17 (in the occipital lobe), note the very prominent black line (line of Gennari) noted in a gross specimen (not a histological specimen) that is visible without the aid of magnification. This line corresponds to the very prominent layer IV, where the massive thalamic input terminates. The numbers on the hemisphere do not correspond to Brodmann areas. (From Nieuwenhuys, R. et al., *The Human Central Nervous System: A Synopsis and Atlas,* Springer-Verlag, New York, 1988. With permission; Carpenter, M. B., *Core Text of Neuroanatomy,* Williams & Wilkins, Baltimore, 1991. With permission.)

(due to the massive thalamic input from the lateral geniculate nucleus) that is visible with the unaided eye (Figure 1.4). In all areas of the brain, layer IV is where input from the thalamus arrives.

The primary motor and primary visual areas are but two examples of 52 cytoarchitecturally distinct regions of the brain described by Brodmann, a German neurologist (Brodmann 1909). These areas also bear a close relationship to functionally distinct brain regions despite being characterized solely by their cytoarchitectural features. For example, the primary motor area is also known as Brodmann area 4, or BA 4, and the primary visual area is BA 17. This structural regional description will be useful when describing functional specialization within the brain.

1.4 LOBES OF THE BRAIN

In this section, I review six lobes of the brain (Figure 1.5), one of which is not truly a lobe, as it is a combination of varied structures and is, thus, a synthetic lobe. Enumerated, these are the frontal, parietal, occipital, temporal, insular, and limbic lobes (synthetic).

1.4.1 FRONTAL LOBE

The frontal lobe is demarcated anatomically from the other frontal lobes by the longitudinal fissure medially and from the parietal lobe posteriorly by the Rolandic fissure. It is the largest of the lobes

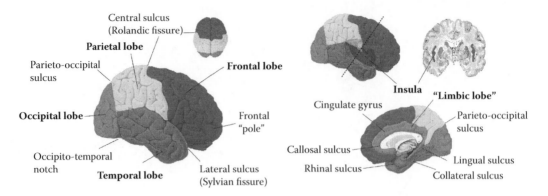

FIGURE 1.5   (See color insert.) Lobes of the brain. Various colors demarcate the different lobes of the brain. Some lobes are demarcated from other lobes by sulci and fissures (deep sulci), whereas others may have rather indistinct margins. The limbic lobe is a conglomerate of a number of disparate, although functionally related, gyri and subcortical structures. It is thus referred to as a synthetic lobe. The insula is shown hidden under the temporal and frontal gyri in a cutaway. The coronal section of the brain is taken along the dotted line. (From Nieuwenhuys, R. et al., *The Human Central Nervous System: A Synopsis and Atlas,* Springer-Verlag, New York, 1988. With permission; Carpenter, M. B., *Core Text of Neuroanatomy*, Williams & Wilkins, Baltimore, 1991. With permission.)

of the brain comprising one third of the hemispheric surface. From an epilepsy surgery perspective, it is the second most common lobe in which epilepsy surgery is performed.

1.4.1.1   Inferior Frontal Gyrus

The three gyri on the lateral surface of the frontal lobe are numbered from 1 to 3 and enumerated as F1, F2, and F3 (the F indicating frontal) (Figure 1.6a and 1.6d). These are also referred to as the superior, middle, and inferior frontal gyri, respectively. The inferior frontal gyrus, or F3, is functionally unique, as it "contains Broca's area," a region within the dominant hemisphere that is important for the production of speech (see Figure 1.10). Hemispheric dominance refers to the hemisphere in which verbal functions reside, including the reception and transmission of language, which is usually on the left side of the brain. Broca's area is comprised of two of the three parts (pars) of the inferior frontal gyrus: the pars triangularis and the pars opercularis (Figure 1.6a). There is a third part of F3 termed the pars orbitalis, which is usually not involved in speech generation. It was by studying the brain of an individual with damage to this area that Broca deduced this region of the brain to be involved in speech production. At the time of this discovery, there was ongoing debate as to whether the bumps on an individual's head could be used to localize brain function, an area of study called phrenology. To this Broca remarked, "I thought that if ever there were a science of phrenology, this was the phrenology of the convolutions, and not the phrenology of bumps" (Broca 1861). The disorder of speech generated by damage to this area is termed a nonfluent aphasia, as the individual's speech generation is impaired.

1.4.1.2   Precentral Gyrus

The precentral gyrus, also known as the motor strip, has already been mentioned. It resides just anterior to the Rolandic fissure (central sulcus), being the most posterior gyrus of the frontal lobe (Figure 1.6b). This is the primary motor area, the major output center to bulbar (brainstem) and spinal motor neurons that are involved in generating voluntary motor activity. The organization of the motor strip is somatotopically organized, which means that neighboring neurons within the motor strip control neighboring body parts and, thus, the physical organization of the motor strip is similar to the physical organization of the human body (Penfield and Rasmussen 1950). This functional specialization of the motor strip is mirrored by its cytoarchitectural organization, being characterized

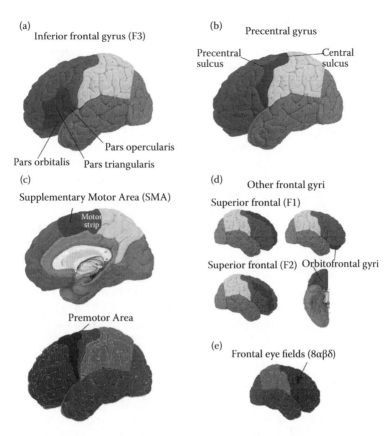

(a) Inferior frontal gyrus (F3)

Pars opercularis

Pars orbitalis   Pars triangularis

(b) Precentral gyrus

Precentral sulcus   Central sulcus

(c) Supplementary Motor Area (SMA)

Motor strip

Premotor Area

(d) Other frontal gyri

Superior frontal (F1)

Superior frontal (F2)   Orbitofrontal gyri

(e) Frontal eye fields (8αβδ)

FIGURE 1.6   (See color insert.) Various frontal gyri and functional regions within the frontal lobe. (a) The inferior frontal gyrus and its various divisions are shown as a darker blue. Pars orbitalis and pars triangularis comprise Broca's area. (b) Precentral gyrus, also known as the motor strip, is shown in dark blue and is demarcated anteriorly by the precentral sulcus and posteriorly by the central sulcus (also known as the Rolandic fissure). (c) The supplementary motor area (SMA) and its continuation on the lateral surface of the hemisphere as the premotor area. (d) Other frontal gyri, superior (F1), middle (F2), and orbitofrontal. (e) Frontal eye fields, BA 8αβδ, involved in voluntary movements of the eyes. (From Nieuwenhuys, R. et al., *The Human Central Nervous System: A Synopsis and Atlas,* Springer-Verlag, New York, 1988. With permission; Carpenter, M. B., *Core Text of Neuroanatomy,* Williams & Wilkins, Baltimore, 1991.)

by large layer V pyramidal neurons called Betz cells. During surgery, this region can be stimulated with current, giving rise to motor responses that aid in its identification (Penfield and Rasmussen 1950). The movements that result from stimulating this area are usually simple rudimentary movements of the limbs, hands, and face. It has the lowest threshold for generating an electrophysiological response of any other region of the brain. The output of this region is via the corticospinal tract, a white matter projection system that will be described later.

At this point, it is likely obvious to the reader that there are specific areas of the brain that appear to be dedicated to specific functions. This was not always so obvious, and it was the pioneering work of two German physicians, Fritsch and Hitzig, who in 1870 showed not only that, "One part of the convexity of the cerebrum of the dog is motor (this expression is used in the sense of Shiff), another part is not motor," but the brain is also electrically excitable (Fristsch and Hitzig 1870). In their experiments that were carried out in the home of one of the physicians' mothers, the exposed cerebrum of the dog was shown to be excitable by stimulation with electrical current via bipolar electrodes and observation of motor activity in the limbs of the animal. The localization of cerebral function was refined in animals by Ferrier (1876) and then brought into the realm of the human brain by Wilder Penfield. Penfield was able to map the human brain in conscious patients during surgery for epilepsy by electrically stimulating the brain and having the patient explain their feelings or observing their movements. Through this, he was able to essentially map out important cerebral functions including sensory, motor, speech, vision, hearing, memory, sensory perceptions, and dreams (Penfield and Rasmussen 1950).

1.4.1.3   Medial Frontal Gyrus: Supplementary Motor Area

The supplementary motor area (SMA), which was identified by Penfield, is in the medial aspect of the hemisphere, just anterior to the primary motor area (Figure 1.6c). When stimulated, it produces motor activity different from that obtained from stimulation of the motor strip in that it creates synergistic movements of a large number of muscle groups on the side opposite to the stimulation. This area of the brain is likely involved in the programming of complex motor movements, the initiation of movements, and the coordination of bimanual activity. This region is also somatotopically organized. Removal of this area results in a transient paralysis of the opposite (contralateral) side of the body and, if the dominant SMA is removed, there is transient mutism as well.

1.4.1.4   Other Frontal Gyri

The other frontal gyri, F2, F3, and orbital gyri, do not have somatotopically and uniquely localized functions (Figure 1.6d). What appears to be the case for these gyri is that portions of them combine into functionally and cytoarchitecturally discrete regions and are better demarcated by Brodmann areas than gyral anatomy.

1.4.1.5   Premotor Area

The premotor area is a continuation of the SMA, but on the lateral aspect of the hemisphere (see Figure 1.6c). This region is referred to as BA 6 and lies in the precentral gyrus, spanning a number of the frontal gyri, anterior to the motor strip, which is BA 4. Cytoarchitecturally, it is referred to as agranular cortex, since it lacks layer 4. This region is involved in sensorially guided voluntary movements (Geschwind and Iacoboni 2007). Neurons in this region are also activated by visual, auditory, and somotosensory stimuli. Electrophysiologically, this region has a higher threshold than the motor strip, and cytoarchitecturally it resembles the primary motor area, except that it does not contain Betz cells.

1.4.1.6   Frontal Eye Fields

This small region of the frontal lobe in BA $8\alpha\beta\delta$ is thought to be involved in voluntary eye movements in the absence of visual stimuli; for example, the initiation of purposeful rapid eye movements toward a target in the visual field—also known as saccades (Figure 1.6e) (Boxer 2007). Stimulation of this area causes deviation of the eyes away from the side of stimulation. It is thus thought that this region generates the "versive" or eye movements away from the side where a seizure is occurring.

1.4.1.7   Clinical Manifestations

The clinical manifestations of seizures arising from the various frontal regions relate to either an excessive activation of this region during the seizure or underactivity during periods between seizures (interictal periods). For example, a seizure that begins in the motor strip will be accompanied by repetitive movement of the part of the body in which the excessive synchronous activity is occurring. Likewise, a seizure focus within the frontal lobe may manifest in the interictal period as changes in frontal lobe function related to the planning or sequencing of daily events—a cognitive process that falls under the rubric of executive function. Another example would be a seizure within the frontal lobe involving the frontal eye fields, which would cause the eyes to deviate to the side opposite to where the seizure is occurring—so-called versive movement of the eyes. Conversely, during the interictal period, one might observe saccadic smooth pursuit, where the eyes make small jumps as one tracks a moving object rather than smoothly following the tracked object.

1.4.1.8 Prefrontal Area

This is an exceedingly expansive region that is thought to involve such diverse functions as attention, awareness, personality, emotion, sensory perception, speech and language, memory, and executive function.

1.4.2 PARIETAL LOBE

1.4.2.1 Primary Sensory Area

Juxtaposed to the primary motor area in the frontal lobe, and just behind the Rolandic fissure, is the primary sensory area (Figure 1.7). Like the primary motor area, it is somatotopically organized and is involved in discriminative touch sensation. It is composed of three Brodmann areas: BA 3, 1, 2. Area 3a receives information from area 1a muscle afferents and BA 3b from cutaneous stimuli. Area 1a receives a combination of cutaneous and deep receptor input and area 2 receives information from stretch receptors (Carpenter 1991). It is beyond the scope of this chapter to detail the sensory system; however, this region is the termination of thalamic efferents (thalamic outputs) that send sensory stimuli to the brain. Information that arrives here includes fine touch and vibration sensation, and the position sense of the joints that is required for accurate guidance of limb movement. Some pain information arrives here, albeit pain sensation is primarily processed in other parts of the brain.

When this area of the brain is stimulated, for example, while surgery is being performed in and around this region, the individual will describe a sensation of tingling, numbness, or electricity. Pain is never experienced here during electrical stimulation of this area of the brain. Removal of this area can be functionally devastating for the individual. Although motor strength is preserved, the limb is extremely hard to control as feedback during voluntary activity is lost, and the limb will often flail about violently.

1.4.3 OCCIPITAL LOBE

The occipital lobe is the lobe of the brain that is primarily involved in visual perception (Figure 1.8). On the lateral surface of the brain, its demarcations from the temporal lobe are not distinct (Figure 1.5) as there is no sulcal boundary between these two lobes. Medially, however, it is clearly demarcated from the parietal lobe by the parieto-occipital sulcus.

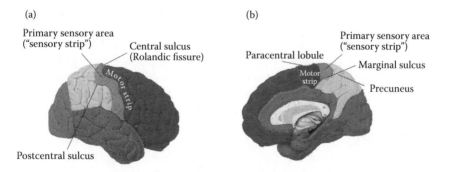

FIGURE 1.7 **(See color insert.)** Primary sensory area. (a) Lateral view of the primary sensory area in dark yellow, separated from the motor strip by the central sulcus. (b) It continues, as does the primary motor area on the medial aspect of the hemisphere. The motor strip plus the sensory area are together referred to as the paracentral lobule (outlined in orange). (From Nieuwenhuys, R. et al., *The Human Central Nervous System: A Synopsis and Atlas,* Springer-Verlag, New York, 1988. With permission; Carpenter, M. B., *Core Text of Neuroanatomy*, Williams & Wilkins, Baltimore, 1991.)

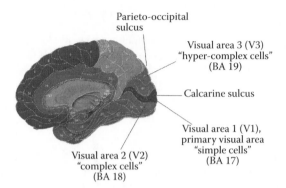

Parieto-occipital
sulcus

Visual area 3 (V3)
"hyper-complex cells"
(BA 19)

Calcarine sulcus

Visual area 1 (V1),
primary visual area
"simple cells"
(BA 17)

Visual area 2 (V2)
"complex cells"
(BA 18)

FIGURE 1.8   (See color insert.) Occipital lobe and vision. A medial view of the cerebral hemisphere with a superimposed map of Brodmann areas. The occipital lobe is demarcated from the parietal by the parieto-occipital sulcus. Concentrically arranged around the calcarine sulcus are visual areas of increasing functional complexity. These functionally distinct regions correspond to cytoarchitecturally distinct Brodmann areas. Information "radiates" out from the BA 17 to be processed further. This information is then sent forward (not shown) in two visual streams: the dorsal and ventral streams that subserve different aspects of visual perception. (From Nieuwenhuys, R. et al., *The Human Central Nervous System: A Synopsis and Atlas,* Springer-Verlag, New York, 1988. With permission; Carpenter, M. B., *Core Text of Neuroanatomy,* Williams & Wilkins, Baltimore, 1991. With permission.)

The occipital lobe is the first step in a complex circuit and represents the region to which the thalamus, specifically the lateral geniculate nucleus (LGN), projects. It is rather interesting that, in this area unlike in other areas of the brain, one begins to appreciate how complex representations are built of simpler representations (Kuffler et al. 1984). Simple cells in the primary visual area (BA 17) respond to oriented bars of light and will only respond to bars of light that are oriented at a specific angle on the retina. Visual area II, or the secondary visual area, is composed of complex cells that respond to a specifically oriented bar of light moving across the visual field. Even further refinements to detection are made in BA 18, where hypercomplex cells respond only to lines of a certain length that have a specific angle and a specific ratio of light-to-dark contrast.

The organization of the input in the primary visual cortex is retinotopic, which means that a specific region of the retina of the eye projects to a specific region of the brain and the physical relationships between different regions of the retina are maintained in the brain. It is important to note that each occipital lobe receives input from the contralateral visual field of each eye. Thus, the right hemifield (the visual field is split vertically) of each eye is transmitted to the left occipital lobe. Interestingly, the input from each eye remains segregated, resulting in the majority of the neurons in BA 17 being influenced independently by each eye. Thus, input from the two eyes remains segregated yet adjacent to each other, forming what are termed ocular dominance columns. Furthermore, neural assemblies that respond to a given orientation of a bar of light are organized adjacent to each other.

Damage to this region of the brain causes a homonymous hemianopsia, which is bilateral loss of vision on the side opposite the lesion in a vertical plane. Excessive activity in this region of the brain, either during a seizure or a migrainous attack, can result in both positive phenomena like flashing lights (scintillations), or negative phenomena like black spots (scotomata). The scotomata are thought to arise from regions of the brain that are undergoing a depolarizing blockage, whereas the scintillations arise from hypersynchronous activity within neuronal populations.

1.4.4   Temporal and Limbic Lobe

The temporal lobe is the most heterogeneous of the lobes with an isocortical (cortex containing six layers) mantle that hides both mesocortical (a less than six-layered cortex) and allocortical (a three-

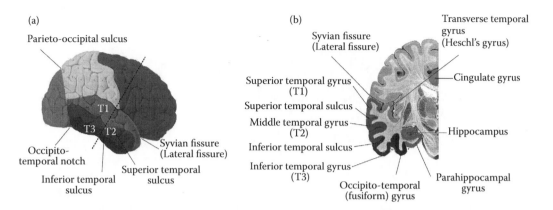

FIGURE 1.9 **(See color insert.)** The temporal lobe: (a) Superior (T1), middle (T2), and inferior (T3) temporal gyri are shown in varying shades of green. The posterior demarcation of the temporal lobe is indistinct from the parietal and occipital lobes. (b) A cross-section at the level shown by the dotted line in (a). The transverse temporal gyrus (Heschl's gyrus) and the basal temporal gyri (occipitotemporal and parahippocampal) now become visible, as well as the hippocampus. The hippocampus and the parahippocampal gyrus compose part of the limbic lobe, as well as the cingulate gyrus. (From Nieuwenhuys, R. et al., *The Human Central Nervous System: A Synopsis and Atlas*, Springer-Verlag, New York, 1988. With permission; Carpenter, M. B., *Core Text of Neuroanatomy*, Williams & Wilkins, Baltimore, 1991.)

layered cortex) regions of the temporal lobe. Both functionally and from an epilepsy perspective, the temporal neocortex is distinguished from the mesial (deep) temporal lobe structures that compose the "limbic lobe" (Figure 1.5). The temporal lobe is clearly demarcated from the frontal lobe by the Sylvian fissure, but posteriorly, its demarcation from the parietal and occipital lobe is somewhat arbitrary (see Figure 1.9).

The temporal neocortex exists both on the lateral aspect and base of the temporal lobe on the floor of the middle cranial fossa. Laterally, the exposed temporal neocortex consists of the superior, middle, and inferior temporal gyri: T1, T2, and T3, respectively. At the base of the temporal lobe is the occipital temporal gyrus—which is the last of the neocortical gyri—that is then followed more medially (toward the center of the head) by the parahippocampal gyrus, which is part of the limbic lobe (Figure 1.9).

1.4.4.1 Language and the Temporal Lobe

Language is, in part, a function that is localized to the neocortex of the dominant hemisphere, which is usually the left hemisphere. Stimulation of the Broca's area (Figure 1.10) results in speech arrest, whereas stimulation within the temporal lobe and within the appropriate location—which varies from person to person (Ojemann et al. 1989)—results in an anomia, which is the inability to name things. Regions within the brain that are involved in naming appear to be close (in physical distance) to Wernicke's area (Figure 1.10). Damage to Wernicke's area results in what is called a receptive aphasia, where the individual is unable to understand what is being said to them. Thus, they would be unable to understand even simple verbal commands.

1.4.4.2 Hearing

The temporal lobe is also involved in hearing, a function closely allied to speech. The primary auditory area consists of areas 41, and 42 of Brodmann, which are contained within the transverse temporal gyri—also known as Heschl's gyrus (Figures 1.9 and 1.10). This region of the temporal lobe receives fibers from the medical geniculate nucleus (lateral geniculate is for vision), a portion of the thalamus. This region of the brain is tonotopically organized, meaning that different frequencies are represented in different regions of the temporal lobe. Information is then transferred via short

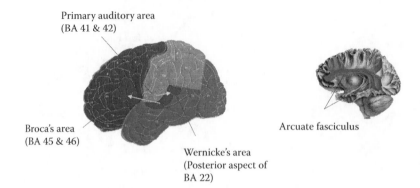

Primary auditory area
(BA 41 & 42)

Broca's area
(BA 45 & 46)

Wernicke's area
(Posterior aspect of
BA 22)

Arcuate fasciculus

FIGURE 1.10   (See color insert.) Speech and audition. The primary auditory area (BA 41 and BA 42) shown in orange. Through short association fibers (subcortical u-fibers) information is transferred from the primary auditory area to BA 22, which is the posterior aspect of the superior temporal gyrus (which includes Wernicke area), to Broca's area (BA 45 and BA 46) via the arcuate fasciculus (as shown). (From Nieuwenhuys, R. et al., *The Human Central Nervous System: A Synopsis and Atlas,* Springer-Verlag, New York, 1988. With permission; Carpenter, M. B., *Core Text of Neuroanatomy*, Williams & Wilkins, Baltimore, 1991. With permission.)

association fibers (described later) to Wernicke's area, which surrounds this region. Wernicke's area communicates with Broca's area in the generation of speech (Figure 1.10) via the arcuate fasciculus.

1.4.4.3   Memory

In the context of epilepsy and surgery for epilepsy, memory is an important aspect of temporal lobe function. Memory can be categorized into various types, with the mesial (or deep part of the temporal lobe) being involved in episodic and associational memory. We are all familiar with this type of memory, as memory for autobiographical events is one type of episodic memory. The two structures within the temporal lobe that are thought to be intimately involved in memory are the hippocampus and the parahippocampal gyrus and are the most common sites from where seizures are generated (Valiante 2009). These structures appear to be involved not only in storing memory (encoding) but also in recollection of memories (Moscovitch et al. 2005).

The medial temporal lobe containing the hippocampus and parahippocampal gyrus are the most common structures removed during temporal lobe surgery. Removal of these regions in the appropriately selected patient can result in an approximately 80% chance of seizure-freedom (McIntosh et al. 2001).

The hippocampus is well visualized on MRI (Figure 1.11), and with newer and more powerful magnets, more details are being described that help identify pathological changes within this structure (Howe et al.). The hippocampus can be divided into postcommissural (this is the portion within the ventricle—fluid filled space of the brain), supracommissural, and the precommissural hippocampus. The reference to "commissural" describes the anatomical relationship of the hippocampus to the corpus callosum, which will be described later. Only the postcommissural hippocampus (green part) can be removed (Figure 1.11). The hippocampus is a unique structure consisting of the paleocortex (the old cortex that has only three layers) with two interlocking c-shaped neuronal sheets, organized in three dimensions. The two sheets form the cornu ammonis (CA), which creates the various CA regions such as the CA1 region, and the dentate gyrus (Duvernoy 2005).

The hippocampus receives afferents from diverse regions of the brain, including all sensory regions, associational regions (regions which are not primarily motor or sensory), the other hippocampus, the hypothalamus (involved in homeostatic and autonomic functions), and fibers from the brainstem. Thus, the hippocampus is a surveillance system for all conscious and autonomic activities that sends projections back to the regions that project to it. An important output system of

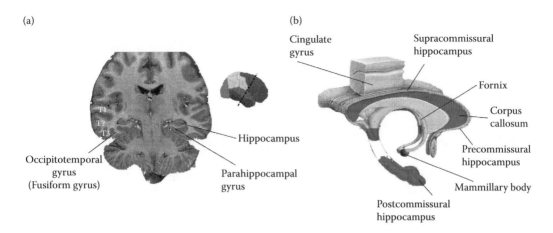

FIGURE 1.11 **(See color insert.)** The hippocampus. (a) Three Tesla magnetic resonance imagings (MRI) of the human brain shown with the hippocampus (shaded green), and temporal neocortical gyri labeled in white. The MRI section through the brain is taken along the dotted line in the inset. (b) The entire hippocampus, displaying is various components. The mammillary bodies receive output from the hippocampus via the fornix and then project to the anterior nucleus of the thalamus. The cingulate gyrus, shown above the supracommissural hippocampus is part of the limbic lobe. Only that portion of the hippocampus shaded in green can be removed during surgery. (From Nieuwenhuys, R. et al., *The Human Central Nervous System: A Synopsis and Atlas,* Springer-Verlag, New York, 1988. With permission; Carpenter, M. B., *Core Text of Neuroanatomy,* Williams & Wilkins, Baltimore, 1991.)

the hippocampus is the fornix, which consists of about 1 million axons that ultimately project to the anterior nucleus of the thalamus (via the mamillary bodies). Projections to the septal area then relay information back to hypothalamic and brainstem regions.

The circuitry of the hippocampus is among the most well studied of the brain, as it is intimately involved in seizure generation (Andersen et al. 2007). It is a convenient anatomical structure to study, as slices of this structure can generate spontaneous sustained activity. Furthermore, the typical preparation—the transverse hippocampal slice—preserves the majority of the functional connections, including important input circuits.

1.4.4.4 Amygdala

The amygdala is another structure that is removed during surgery for mesial temporal lobe epilepsy. It has been shown to have among the lowest seizure thresholds in the brain (Goddard et al. 1969). This means that the currents required to generate seizure-like activity following repetitive stimulation are lower in the amygdala than other brain regions, including the hippocampus and the parahippocampal gyrus. This region receives input from the olfactory tract (conveys smell information) and the hypothalamus. In its connection to the hippocampus, the amygdala's output is via the stria terminalis to the dorsal nucleus of the vagus and solitary nucleus, both of which are brainstem nuclei. The relevance of these connections becomes apparent when we consider the clinical manifestations of temporal lobe seizures.

1.4.4.5 Clinical Manifestations of Temporal Lobe Seizures

Temporal lobe seizures are often but not always heralded by an aura, a sensory or psychological experience that the individuals remember and that is usually followed by a loss of awareness. The aura is recalled since the patient still has the ability to encode it, whereas encoding does not occur during disturbances of consciousness and the individual remains amnestic (unable to remember) for the event. The aura is considered to be a seizure occurring in a focal area of the brain that is not so widespread as to impair consciousness. For example, although it is generally thought that a déjà vu experience is a glitch in the "matrix," given the involvement of the hippocampus in memory processes, it is not a stretch to consider a déjà vu experience as a result of aberrant activity within this structure. Likewise the association of smells with temporal seizures likely relates to activity within the corticomedial nuclei of the amygdala and the uncus. We Canadians have television commercials that commemorate events in Canadian history and among them is Wilder Penfield's mapping of the human brain that results in the patient re-experiencing the smell of "burnt toast" that is part of the patient's aura. However, the smells associated with auras of temporal lobe origin tend usually to be pungent, unidentifiable smells. Auras can consist of fear and anxiety, which suggest involvement of the central and basolateral nuclei of the amygdala. There can also be autonomic manifestations that relate to the output of the medial temporal lobe to the various brainstem nuclei, in particular the dorsal nucleus of the vagus. The vagus nerve innervates the heart and various abdominal organs. Individuals often describe a rising feeling in their stomach or a lump in their throat, which may be related to abnormal activity within the vagal nuclei.

Once the seizure activity spreads to involve areas of the brain that alter consciousness, the patient becomes unaware of their behaviors and does not encode the activity they are involved in. They are thus amnestic for this period, unable to recall what happened to them. This alteration of consciousness, usually termed loss of awareness or a dysconscious state, is different from unconsciousness. The individual with a loss of awareness is awake, and interacting with their environment, albeit in a most rudimentary way. This type of seizure is termed a complex partial seizure, with the term complex referring to the loss of awareness. An aura then is considered to be a simple partial seizure. The word partial in these terms refers to the fact that it is thought that the seizure is arising from a confined area of the brain, as opposed to a generalized seizure that is thought to involve both hemispheres of the brain simultaneously, usually from the onset of the seizure. During a complex partial seizure, the individual may be involved in automatic behavior like picking and fumbling with their clothes, orofacial automatisms (chewing movements of the mouth), and wandering (Tai et al. 2010). Following a seizure, the individual may be disoriented, feel extremely tired, have a headache, and may have symptoms of psychosis (paranoia, hallucinations, or delusions).

Involvement of neocortical structures can cause negative (loss of function) or positive (de novo sensory or perceptual experiences) manifestations, just as in the mesial temporal lobe. Thus, involvement of the neocortex can result in speech difficulties, auras that consist of hearing music or voices, and word finding problems.

1.4.5 INSULA

The insula has been considered part of the "limbic" lobe, but this assignment is certainly not consistent. It is hidden by the frontal and temporal lobes and sits deep to both these lobes. Visualization during surgery is made possible by dissecting the sylvian fissure. The insula receives sensory input from the thalamus and the amygdala. It contains the primary areas for taste, and in the dominant hemisphere has been shown to be critically involved in speech production (Dronkers 1996). It appears to be involved in the emotional experience of olfaction (smell), taste, visceral sensations, and autonomic inputs (Carpenter 1991). For example, it is thought that the insula is, along with the primary somatosensory area, partly responsible for the conscious perception of one's heart beat (Craig 2009).

Seizures that arise from the insula may appear to be, for all intent and purpose, both clinically and electrophysiologically like temporal lobe seizures. A unique aspect of seizures that arise from the insula is that they may be associated with hypersalivation, reflecting its connection to autonomic structures.

1.5 THALAMUS

The thalamus is a paired gray matter structure that sits at the top of the brainstem, deep to the cerebral cortex. It is situated on either side of the third ventricle. The ventricles are structures that produce and contain CSF. The thalamus sits in close proximity to the hypothalamus and is composed of a large number of nuclear groups, arranged in a complex three-dimensional structure.

This structure is part of the diencephalon, along with the epithalamus, hypothalamus, and subthalamus. The thalamus has reciprocal connection to the cortex. Projections to and from the cortex that initially go through the striatum form a number of distinct cortical–subcortical circuits that are thought to be involved in a spectrum of activities from releasing motor programs to conscious awareness. All sensory information except for olfactory information is relayed through the thalamus, ultimately to converge on their respective sensory areas. The idea of a "relay" is a rather simplistic mechanistic appellation and the thalamus would be more appropriately described as an integrator of widely disparate information destined for the cerebral cortex and coming from the cerebral cortex. Its rather extensive input–output relationship becomes apparent in Figure 1.12.

From a seizure perspective, the thalamocortical circuitry has been extensively studied (Steriade and Deschenes 1984). Unlike the focal epilepsies (as described elsewhere in this work), the thalamus has been implicated in the generation and propagation of generalized seizures. Generalized seizures have the electrophysiological hallmark of bilateral synchronous activity across the two hemispheres. Examples of this type of epilepsy are "absence" seizures, which are characterized by bilaterally synchronous activity at 3 Hz in both hemispheres. The thalamocortical circuitry, or cortical–subcortical circuits, that have been implicated in the generation of these types of seizures posited a novel mechanism that is largely dependent on inhibitory circuits (Kim et al. 1997).

The functional implications of seizures or damage to the thalamus are multiple. The thalamus is important in two aspects of consciousness: arousal and content. Arousal is the state of being awake, the intuitive understanding of having ones eyes open spontaneously. The aroused brain, however, has a specific electrophysiological signature characterized by an abundance of high-frequency activity, which appears on visual inspection (in general) to be aperiodic in nature (Ebersole and Pedley 2003). The unaroused (sleep or anesthesia to some extent) brain is usually characterized by an abundance of high-amplitude, low-frequency activity, decreased high-frequency activity, and certain stereotypical electrophysiological signatures that appear to occur in a seemingly periodic pattern (Ebersole and Pedley 2003). The transition from aroused to unaroused, for example, from wakefulness to sleep, is thought to fundamentally involve the thalamus and to be mediated by its specific connections to the ascending reticular activating system (ARAS). The ARAS is one of two

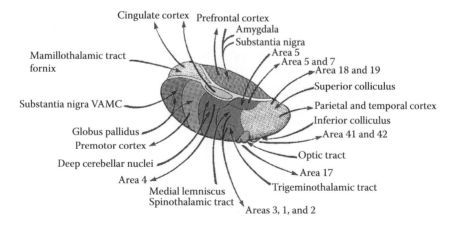

FIGURE 1.12   (See color insert.) Input-output of the thalamus. Arrows projecting toward the thalamus represent afferent (incoming) information flow, whereas arrows projecting away from the thalamus represent efferent (outgoing) information. The listed structures are either recipients or originators of thalamic informational flow. (From Nieuwenhuys, R. et al., *The Human Central Nervous System: A Synopsis and Atlas,* Springer-Verlag, New York, 1988; Carpenter, M. B., *Core Text of Neuroanatomy,* Williams & Wilkins, Baltimore, 1991. With permission.)

ascending systems encompassing an anatomically diffuse conglomeration of brainstem nuclei that sends projections to the rostral intralaminar nuclei of the thalamus. From here, one can envisage a sprinkler-type arrangement, with the intralaminar nuclei essentially "showering" the entire cerebral cortex with neurotransmitters that increase cortical excitability.

The "content" aspect of consciousness that is associated with our aroused state and sleep states as well—particularly rapid eye movement (REM) sleep—is functionally a result of the extensive cortical–subcortical circuitry, as mentioned above (Chow and Cummings 2007). It is this circuitry that has been implicated in the generation of widespread synchronization and desynchronization during physiological and pathological states (Llinas et al. 1998).

1.6   WHITE MATTER

The cerebral localization of function necessitates that functionally specialized regions of the brain be connected to each other so that information from disparate regions of the brain can be integrated. The term integration has been used before, and I will leave it to the reader's gestalt as to what is meant by this rather nebulous term. As mentioned above, we scarcely have the language or tools to describe what the brain does, but we know that it does something to make us aware of ourselves and our environment. Integration occurs through connectedness, and this connectedness can occur on different physical scales—short, medium, and long range—and is accomplished by white matter structures within the brain.

White matter, so described because it looks white in the real brain, lies below the neocortex and forms another "core" of the cerebral hemisphere (Figure 1.2b). It is interposed within each hemisphere between the cortex and the subcortical gray matter structures, as well between the subcortical structures and between the two hemispheres. It is composed of axons, the equivalent of electrical cables, of varying diameter and insulation thickness. Larger, more heavily insulated axons conduct faster and use less energy that smaller uninsulated fibers, since conduction along these larger axons is saltatory. These larger axons thus tend to be more efficient and, hence, compose the white matter core of the brain.

Wiring the brain comes at a price, both metabolically and space-wise. However, counter intuitively, each mm^3 of white matter contains 9 m of axons, whereas each cubic millimeter of gray matter contains 3 km of axons (Douglas et al. 2004). Thus, if one were to try to fuse 100 cortical areas together to avoid using white matter, the cortical volume would increase by 10-fold. Eliminating white matter would result in an increase in intra-areal volume that would far exceed the reduction in interareal axon volume, whereas connecting everything via white matter tracts would be metabolically exorbitant. Somewhere in between is a balance of these two types of architecture, and the brain seems to balance the two rather well.

White matter comes in three "flavors": projection fibers, association fibers, and commissural fibers. Projection fibers convey impulses to and from the cortex, association fibers convey impulses between cortical regions of the same hemisphere, and commissural fibers convey impulses between the two hemispheres.

1.6.1   Projection Fibers

Projection fibers are the "long-range" fibers within the brain that convey information either away from the cortex or toward it and do not terminate within another cortical area or arise from another cortical area. For example, fibers destined to move one's finger arise in BA 4, the motor strip, and terminate in the spinal cord, about 1 m away. The fibers that send information from the cortex (efferent fibers) arise in the deep layers of the cerebral cortex, and fibers going toward the cortex (afferent fibers) that terminate in more superficial layers of the cortex form a broad sheet of white matter called the corona radiata (Figure 1.13).

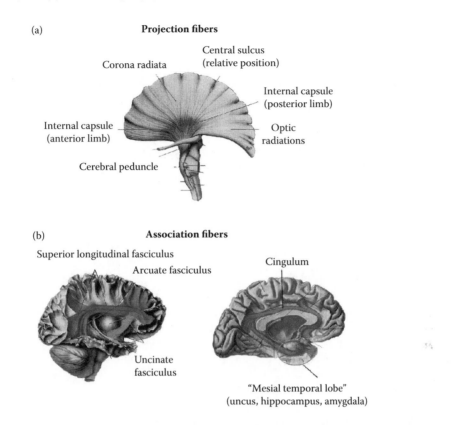

(a) **Projection fibers**

Central sulcus
(relative position)
Corona radiata

Internal capsule
(posterior limb)

Internal capsule
(anterior limb)

Optic
radiations

Cerebral peduncle

(b) **Association fibers**

Superior longitudinal fasciculus

Cingulum

Arcuate fasciculus

Uncinate
fasciculus

"Mesial temporal lobe"
(uncus, hippocampus, amygdala)

FIGURE 1.13 **(See color insert.)** White matter—projection and association fibers. (a) The projection fibers going to and away from the cortex form a large fan-like structure termed the corona radiata. Shown are fibers projecting away from the cortex (internal capsule) and others projecting toward the cortex like the optic radiations that are transmitting visual information to the occipital lobe (BA 17, see Figure 1.12). (b) Association fiber tracts, also known as fascicles, connect different cortical regions to one another (see text for details). (From Nieuwenhuys, R. et al., *The Human Central Nervous System: A Synopsis and Atlas,* Springer-Verlag, New York, 1988. With permission; Carpenter, M. B., *Core Text of Neuroanatomy,* Williams & Wilkins, Baltimore, 1991. With permission.)

1.6.2 ASSOCIATION FIBERS

These types of fibers interconnect cortical areas within the same hemisphere and are of two types: short and long. Short association fibers interconnect adjacent gyri, whereas long association fibers interconnect regions of the cortex within different lobes.

1.6.2.1 Long Association Fibers

There are four primary long association fiber systems termed "fascicles," which is another term for a bundle. Thus, a fascicle is composed of a large number of individual axons. The uncinate fasciculus interconnects the orbitofrontal gyri, which are those gyri that sit at the base of the frontal lobe (Figure 1.6d) with parts of the temporal lobe. Another frontal–temporal fascicle is the arcuate fasciculus that interconnects the frontal lobes to the more posterior portions of the temporal lobe. This fasciculus, among other functions, is thought to transmit information from Wernicke's area to Broca's area, two areas that are important in language. Damage to this fasciculus results in what has been termed "conduction aphasia"—referring to the conduction of information from Wernicke's area to Broca's area. The superior longitudinal fasciculus interconnects the frontal lobes to the

(a)

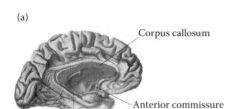

Corpus callosum

Anterior commissure

(b)

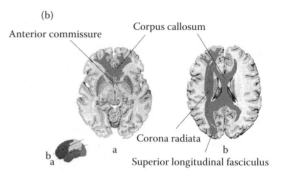

Anterior commissure

Corpus callosum

Corona radiata

Superior longitudinal fasciculus

FIGURE 1.14   (See color insert.) White matter—commissural fibers. (a) A medial view of the hemisphere showing the massive corpus callosum in dark green, and the smaller anterior commissure in lighter green. (b) Two cross-sections of the brain at different levels. Note the layering of the various types of white matter tracts. (From Nieuwenhuys, R. et al., *The Human Central Nervous System: A Synopsis and Atlas,* Springer-Verlag, New York, 1988. With permission; Carpenter, M. B., *Core Text of Neuroanatomy,* Williams & Wilkins, Baltimore, 1991. With permission.)

occipital and parietal lobes. These different fascicles are depicted in Figure 1.13b. The cingulum is the principal association fiber pathway on the medial aspect of the hemisphere contained within the cingulate gyrus (Figure 1.5). This fasciculus connects the medial frontal and parietal lobes with the parahippocampal gyrus and other mesial temporal lobe structures.

1.6.2.2   Commissural Fibers

Commissural fibers convey information between the two hemispheres. The corpus callosum is the largest of the "commissures," interconnecting vast areas of the two cerebral hemispheres, whereas the anterior commissure is much smaller and interconnects the middle and inferior temporal gyri of the two hemispheres and a small portion of the olfactory bulbs (Figure 1.14).

The corpus callosum reciprocally interconnects all of the lobes of the brain with their corresponding (homologous) lobe on the other side of the brain. It contains 300 million axons, which pales in comparison to the 10 billion neurons and 200 trillion synaptic connections within the cerebral hemispheres. It has been estimated that about 2%–3% of all the cortical neurons in the frontal lobe send projections to the contralateral hemisphere (Lamantia and Rakic 1990). Higher order associational areas (those areas most removed, from a functional perspective, from primary sensory areas) have the greatest density of connections, whereas primary sensory and motor areas have the least. The corpus callosum carries both excitatory and inhibitory connections and is thought to be involved in the integration and transmission of information between the two hemispheres. Thus, it has been postulated that seizure activity can, via the corpus callosum, be transmitted from one hemisphere to another. Based upon this assumption, sectioning of the corpus callosum is done in an effort to palliate individuals with a specific type of seizure, called atonic seizures. Sectioning of the corpus callosum, or a "corpus callosotomy," which means making a hole in the corpus callosum, is rarely curative (Engel 1993).

This homologous connectivity has been beautifully demonstrated in chronic models of epilepsy. In fact, the first chronic model of epilepsy involved placing alumina cream on one motor cortex of nonhuman primates and recording seizure-like activity from the other (contralateral or on the other side) motor cortex—a so-called mirror focus (Harris and Lockard 1981).

1.6.2.3   Short Association Fibers

Short association fibers interconnect adjacent gyri and are often referred to as subcortical U-fibers. These fibers form a "U" in the sense that they travel from one crown of a gyrus through the base of the sulcus and then back up to the crown of the adjacent gyrus (Figure 1.2).

1.6.3 CONNECTIVITY

It is thought that short-range fibers increase local computational ability, whereas long-range connections are required to bring distant regions together functionally. The balance achieved within the brain between these two types of connectedness has been described as having small-world network characteristics (Buzáski 2006). Small-world networks remain clustered through local connections but have devised a simple strategy to bring these clusters closer together through the reduction of path length. From seminal work on graph theory, it has been shown that by replacing 2% of the short-range connections with long-range connections, local architecture is robustly preserved, while the distance between any two clusters (for example, between two functionally specialized regions of the brain) is significantly reduced (Watts and Strogatz 1998). Thus, with only a small number of long-range connections, various regions of the brain are brought very close together in a metabolically efficient manner by being organized in a small-world network fashion. It is interesting to note, as mentioned above, that about 2%–3% of all cortical neurons in the frontal lobe send projections to the contralateral hemisphere.

REFERENCES

Andersen, P., R. G. Morris, D. G. Amaral, T. Bliss, and O'Keefe. 2007. *The hippocampus book*. New York: Oxford University Press.

Boxer, A. L. 2007. Principles of motor control by the frontal lobes as revealed by the study of voluntary eye movements. In *The human frontal lobes*. B. L. Miller and J. L. Cummings. New York: Guilford Press.

Broca, P. 1861. Remarques sur le siège de la faculté du language articulé, suives d'une observation d'aphémie (perte del la parole). *Bull. Soc. Anat.* 280:834–843.

Brodmann, K. 1909. *Vergleichenden Lokalisationslehre der Grosshirnrinde*. Leipzig: Johann Ambrosius Barth Verlag.

Buzáski, G. 2006. *Rhythms of the brain*. New York: Oxford University Press.

Carpenter, M. B. 1991. *Core text of neuroanatomy*. Baltimore: Williams & Wilkins.

Chow, T. W. and J. L. Cummings. 2007. Frontal–subcortical circuits. *The human frontal lobes*. B. L. Miller and J. L. Cummings. New York: Guilford Press.

Craig, A. D. 2009. How do you feel—now? The anterior insula and human awareness. *Nat. Rev. Neurosci.* 10(1):59–70.

Douglas, R., H. Markram, and K. Martin. 2004. Neocortex. In *The synaptic organization of the brain*. G. M. Sheppard, 719. Oxford: Oxford University Press.

Dronkers, N. F. 1996. A new brain region for coordinating speech articulation. *Nature* 384(6605):159–161.

Duvernoy, H. M. 2005. *The human hippocampus: Functional anatomy, vascularization, and serial sections with MRI*. New York: Springer-Verlag.

Ebersole, J. S. 2003. Cortical generators and EEG voltage fields. *Current practice of clinical electroencephalography*. J. S. Ebersole and T. A. Pedley. Philadelphia: Lippincott, Williams & Wilkins.

Ebersole, J. S. and T. A. Pedley. 2003. *Current practice of clinical electroencephalography*. Philadelphia: Lippincott, Williams & Wilkins.

Engel, J. J. 1993. *Surgical treatments of the epilepsies*. New York: Raven Press.

Ferrier, D. 1876. *The functions of the brain*. London: Smith, Elder.

Fristsch, G., and E. Hitzig. 1870. Ueber die elektrische Erregbarkeit des Grosshirns. *Arch. Anat. Physiol. Wiss. Med.* 300–332.

Geschwind, D. H., and M. Iacoboni. 2007. Structural and functional asymmetries of the human frontal lobes. *The human frontal lobes*. B. L. Miller and J. L. Cummings. New York: Guilford Press.

Goddard, G. V., D. C. McIntyre, and C. K. Leech. 1969. A permanent change in brain function resulting from daily electrical stimulation. *Exp. Neurol.* 25(3):295–330.

Harris, A. B., and J. S. Lockard. 1981. Absence of seizures or mirror foci in experimental epilepsy after excision of alumina and astrogliotic scar. *Epilepsia* 22(1):107–122.

Howe, K. L., D. Dimitri, C. Heyn, T.-R. Kiehl, D. Mikulis, and T. Valiante. 2010. Histologically confirmed hippocampal structural features revealed by 3T MR imaging: Potential to increase diagnostic specificity of mesial temporal sclerosis. *Am. J. Neuroradiol.* (epub ahead of print June 10, 2010).

Kim, U., M. V. Sanchez-Vives, and D. A. McCormick. 1997. Functional dynamics of GABAergic inhibition in the thalamus. *Science* 278(5335):130–134.

Kuffler, S. W., J. G. Nicholls, and A. R. Martin. 1984. *From neuron to brain*, 2nd ed. Sunderland, MA: Sinaur Associates.

Lamantia, A. S., and P. Rakic. 1990. Cytological and quantitative characteristics of four cerebral commissures in the rhesus monkey. *J. Comp. Neurol.* 291(4):520–537.

Llinas, R., U. Ribary, D. Contreras, and C. Pedroarena. 1998. The neuronal basis for consciousness. *Philos. Trans. R. Soc. Lond. B. Biol. Sci.* 353(1377):1841–1849.

McIntosh, A. M., S. J. Wilson, and S. F. Berkovic. 2001. Seizure outcome after temporal lobectomy: Current research practice and findings. *Epilepsia* 42(10):1288–1307.

Miller, B. L., and J. L. Cummings. 2007. *The human frontal lobes*. New York: Guilford Press.

Moscovitch, M., R. S. Rosenbaum, A. Gilboa, D. R. Addis, R. Westmacott, C. Grady, M. P. McAndrews, B. Levine, S. Black, G. Winocur, and L. Nadel. 2005. Functional neuroanatomy of remote episodic, semantic and spatial memory: A unified account based on multiple trace theory. *J. Anat.* 207(1):35–66.

Nieuwenhuys, R., J. Voogd, and C. van Huijzen. 1988. *The human central nervous system: A synopsis and atlas*. New York: Springer-Verlag.

Ojemann, G., J. Ojemann, E. Lettich, and M. Berger. 1989. Cortical language localization in left, dominant hemisphere. An electrical stimulation mapping investigation in 117 patients. *J. Neurosurg.* 71(3):316–326.

Penfield, W., and T. Rasmussen. 1950. *The cerebral cortex of man*. New York: Macmillan.

Sheppard, G. M. 2004. *The synaptic organization of the brain*. Oxford: Oxford University Press.

Steriade, M., and M. Deschenes. 1984. The thalamus as a neuronal oscillator. *Brain. Res.* 320(1):1–63.

Tai, P., S. Poochikian-Sarkissian, D. Andrade, T. Valiante, M. del Campo, and R. Wennberg. 2010. Postictal wandering is common after temporal lobe seizures. *Neurology* 74(11):932–933.

Valiante, T. A. 2009. Selective amygdalo-hippocampectomy. In *Textbook of stereotactic and functional neurosurgery*. A. Lozano, P. L. Gildenberg, and R. Tasker. Heidelberg, 2677–2714. New York: Springer-Verlag.

Watts, D. J., and S. H. Strogatz. 1998. Collective dynamics of 'small-world' networks. *Nature* 393(6684): 440–442.

2 Introduction to EEG for Nonepileptologists Working in Seizure Prediction and Dynamics

Richard Wennberg

CONTENTS

2.1 INTRODUCTION

The aim of this chapter is to provide a brief introduction to electroencephalography (EEG), specifically oriented toward nonclinicians working in the fields of seizure prediction and dynamical analysis. As such, only very limited attention will be given to the use and clinical interpretation of EEG in the diagnosis and treatment of the various epilepsy syndromes. For those interested in reading more about the clinical applications of EEG in epilepsy, the textbooks edited by Daly and Pedley (1990) and by Niedermeyer and Lopes da Silva (1993) are highly recommended. Additional information pertaining to the neurophysiology of electrogenesis and other aspects of EEG recording and interpretation can be found in the material used for the Fourth International Workshop on Seizure Prediction (IWSP4) didactic lecture that served as the basis for this book chapter (available at www.iwsp4.org/DidacticForNonMDs/Wennberg_Richard_EEG_for_NonEpileptologists.ppt). Perhaps the single most useful reference paper that could be recommended for this audience would be that of Gloor (1985) on neuronal generators, localization, and volume conductor theory in EEG.

2.2 GENERATION OF THE POTENTIALS RECORDED WITH EEG

The signal recorded with EEG is a representation of voltage fluctuations in space and time, the electrical potentials arising from summated excitatory and inhibitory postsynaptic potentials that are generated mainly by cortical pyramidal cells (Gloor 1985; Pedley and Traub 1990; Lopes da Silva and Van Rotterdam 1993; Speckmann and Elger 1993). "Synchronous" neural activity involving at least 6 cm² of cortex is considered necessary for detection with scalp EEG (Cooper et al. 1965), and pathological epileptiform potentials typically involve larger cortical areas extending over at least 10–20 cm² (Tao et al. 2005).

It is the parallel arrangement of pyramidal neurons in the cortical mantle, with their apical dendrites extending to the most superficial cortical layers, that gives rise to the effective "cortical dipole layer" that dominates the signal recorded with EEG (Gloor 1985). Sufficiently summated excitatory postsynaptic potentials (EPSPs) at the apical dendrites will cause inward current flow and a superficial cortical "sink"—if the area of synchronous excitation is large enough, the extracellular negative field resulting from the inward apical current flow will be detected by scalp EEG and depicted as an upward waveform (Pedley and Traub 1990; Speckmann and Elger 1993). The "source" corresponding to the simultaneous outward current flow will be situated deeper, near the pyramidal cell body layer, and thus not detected by the surface EEG recording. The superficial versus deep location of sources and sinks in the cortical dipole layer will reflect the predominant direction of current flow in the cortical dipole layer at a given point in time. Superficial sinks can arise from either summated apical EPSPs or summated deep-cell body layer inhibitory postsynaptic potentials (IPSPs), whereas superficial sources can result from either summated apical IPSPs or deep EPSPs (Pedley and Traub

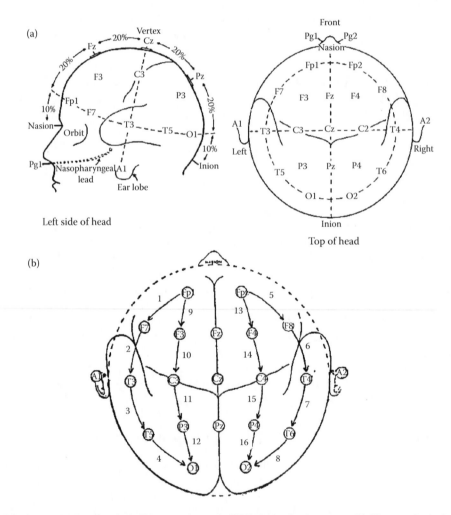

FIGURE 2.1 (a) International 10-20 system for scalp EEG electrode placement. (b) Sixteen-channel array of the anterior–posterior longitudinal ("double banana") montage.

1990; Speckmann and Elger 1993). It is the extracellular polarity of the superficial cortical field that is recorded with EEG, negative for a superficial sink, positive for a superficial source (Gloor 1985; Pedley and Traub 1990; Lopes da Silva and Van Rotterdam 1993; Speckmann and Elger 1993).

2.3 TYPES OF POTENTIALS RECORDED WITH EEG

In a broad sense, the potentials recorded with EEG can be classified into two types of phenomena: oscillations and transients. Oscillatory activity recorded with EEG is generated in the cortex and dependent to varying degrees on thalamocortical reciprocity (Steriade 1993). The recorded oscillations can be normal (alpha, beta, gamma, and mu rhythms, sleep spindles, and delta activity in sleep) or abnormal (seizures, burst-suppression). Oscillatory activity and, especially, seizure activity is, of course, the type of EEG signal of interest to this audience.

EEG sharp transients are usually of less interest for the purposes of seizure prediction and dynamical analyses (and can sometimes confound these analyses), although they are important for diagnostic purposes. Normal transients include a variety of sleep potentials (vertex waves, K-complexes, positive occipital sharp transients of sleep, benign epileptiform transients of sleep) as well as a number of noncerebral electrical potentials that are detected with EEG—eye blinks, cardiac impulses (EKG), and muscle activity (EMG). Abnormal EEG sharp transients often form the basis for the diagnosis of epilepsy and are referred to as interictal epileptiform potentials, identifiable by their particular morphological characteristics, which may be in the form of spikes, polyspikes, spike and wave

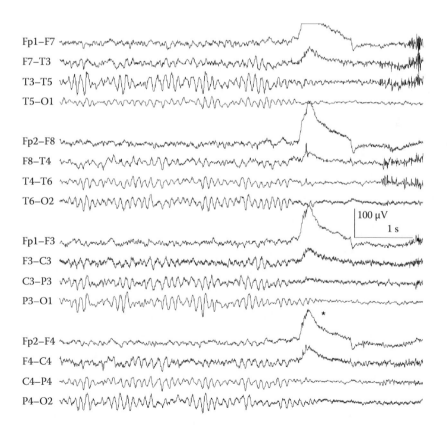

FIGURE 2.2 Subject is awake and resting. Normal posterior alpha rhythm disappears with eye opening (*). High-frequency activity at the end of the figure after eye opening is muscle artifact. Anterior–posterior bipolar montage. LFF 0.5 Hz, HFF 70 Hz in this and all other figures.

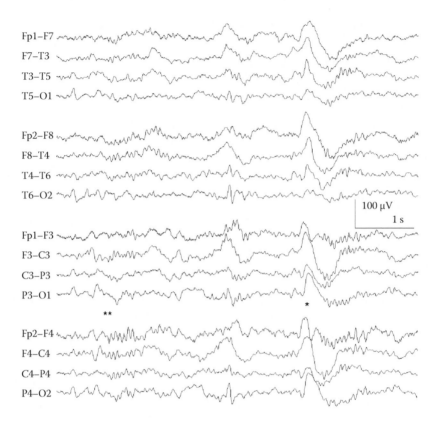

FIGURE 2.3   Stage II sleep. K-complex (*); sleep spindles (**). Anterior–posterior bipolar montage.

complexes, sharp waves, and sharp and slow wave discharges. Certain other nonepileptiform EEG transients, such as periodic complexes (lateralized and generalized) and triphasic waves can be indicators of various encephalopathies (Daly and Pedley 1990; Niedermeyer and Lopes da Silva 1993).

2.4   EEG RECORDING CONVENTIONS

There is a long-established convention for head measurement and EEG scalp electrode placement and labelling known as the international 10-20 system (Daly and Pedley 1990; Niedermeyer and Lopes da Silva 1993), with electrodes on the left assigned odd numbers, electrodes on the right assigned even numbers, and midline electrodes given the suffix "Z." Electrodes over frontal regions are denoted with the prefix "F," with "C," "T," "P," and "O" indicating electrode placement over central, temporal, parietal, and occipital regions, respectively (Figure 2.1).

EEG recordings are displayed for review and interpretation using different configurations of electrode combinations organized into "channels," with each channel containing two electrode inputs that are arranged into either *bipolar* or *referential* montages. All recorded EEG signals are a measure of the voltage differences at one electrode site relative to another electrode. By convention, polarity in clinical EEG is depicted as "negative up." Specifically, if the field potential recorded at the first electrode (input) of a channel is relatively more negative than the potential recorded at the same channel's second electrode input, then the difference is reflected as an upward waveform, and vice versa.

Bipolar montages are typically arranged in straight line chains of electrodes, where the second input of the first channel becomes the first input of the second channel, and so on, allowing for

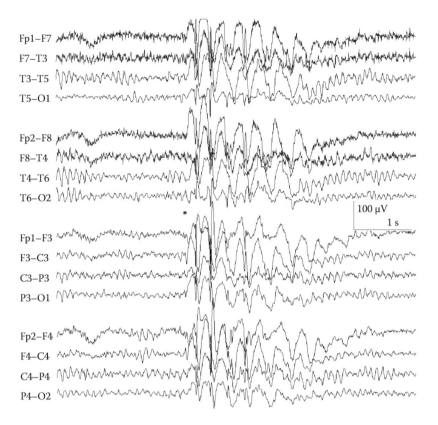

FIGURE 2.4 A burst of generalized 3-Hz spike and wave activity (*). Primary generalized epilepsy. Anterior–posterior bipolar montage.

localization of electrical field maxima and minima by so-called "phase reversal." Note that "phase" is used here in a different sense from that used in dynamical studies; a more accurate term would be "polarity reversal." Referential montages are actually also bipolar arrangements, except that the second input to each channel (referred to as the "reference electrode") is more distant in space from the first electrode position. It is usually situated somewhere on the scalp, occasionally on the mastoids or ear lobes or, less commonly, over the cervical spine or nose. In referential montages, electrical fields are localized by identification of amplitude maxima and minima at the recording electrode (first input) sites. These types of referential montages are sometimes also referred to as *monopolar* montages, which is not a truly accurate designation but is understood clinically. An ideal monopolar reference electrode would be electrically neutral to cerebral activity and uncontaminated by electrical artifact. Unfortunately, no such ideal electrode exists and all usable reference electrode positions have drawbacks. The reference electrode input can also be constructed from an average of some or all of the signals recorded from the other scalp electrodes. Examples including the common average reference, used extensively for EEG source localization and modelling of sharp transients, and the surface Laplacian, a more sophisticated local average reference that may be useful for decreasing the contribution of volume conducted distant electrical fields to the signal recorded from the electrode site of interest, the first input to the channel (Nunez et al. 1997).

Clinically, one of the most commonly used screening montages for scalp EEG review is the so-called anterior–posterior longitudinal bipolar (or "double banana") montage, which is shown at the bottom of Figure 2.1 and used to depict the scalp EEG examples in Figures 2.2 through 2.6.

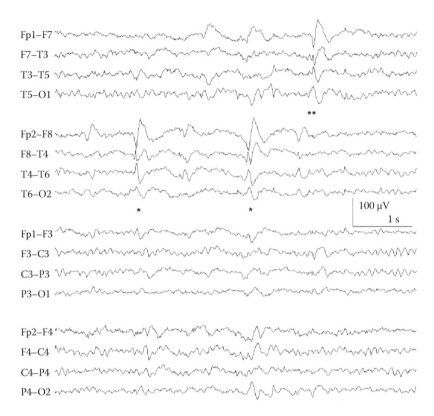

FIGURE 2.5   Bilateral temporal lobe (focal, partial) interictal epileptiform activity. Independent sharp and slow wave complexes over right (*) and left (**) anterior–mid-temporal regions. Temporal lobe epilepsy. Anterior–posterior bipolar montage.

2.5   EEG EXAMPLES

Figure 2.2 shows an example of the EEG of normal wakefulness. This can be contrasted with the EEG of normal (stage 2) sleep that is shown in Figure 2.3. The classical 3-Hz spike and wave activity of primary generalized epilepsy is shown in Figure 2.4. Figure 2.5 shows temporal lobe interictal epileptiform discharges (spikes) such as may be seen in temporal lobe epilepsy. Figure 2.6 shows part of an ongoing right temporal lobe seizure.

2.6   ARTIFACTS AND REFERENCE ELECTRODES

The EEG time series data that is available to nonepileptologists for analysis is typically first selected by clinical electroencephalographers after transformation of the data to some type of manipulable format (e.g., a large .txt file). Detailed consideration may not always be given to the issue of possible artifacts in the recording, and the issue of reference electrode choice is frequently not addressed. Indeed, at times mathematicians and physicists working with EEG data may not have been made aware by their clinical colleagues of the reference electrode used for a particular segment of data.

Although the issues related to artifacts and reference electrode contamination form an integral part of routine clinical EEG interpretation, they are really no less important for nonclinicians using EEG signal analysis to study seizure prediction and dynamics.

A good example of the importance of understanding artifacts is the recent demonstration that scalp EEG studies of beta and gamma oscillations in cognition over the past many years would

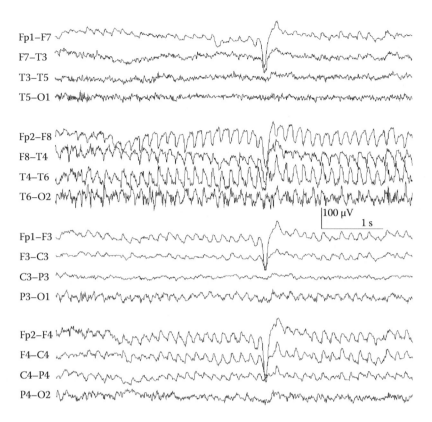

Fp1–F7
F7–T3
T3–T5
T5–O1

Fp2–F8
F8–T4
T4–T6
T6–O2

100 μV
1 s

Fp1–F3
F3–C3
C3–P3
P3–O1

Fp2–F4
F4–C4
C4–P4
P4–O2

FIGURE 2.6   Ictal EEG showing focal rhythmic seizure pattern localized to right temporal region ("equipotentiality" at F8–T4). Temporal lobe epilepsy. Anterior–posterior bipolar montage.

appear to have been analyzing mainly muscle artifact (Whitham et al. 2007, 2008), a finding not entirely surprising to some clinical electroencephalographers but one that throws into question decades of published research.

The effects of reference electrodes on measures of EEG correlation, coherency, or synchronization have been discussed in the past (Nunez et al. 1997; Zaveri et al. 2000) but are nevertheless often overlooked in the signal analysis literature (Guevara et al. 2005). The choice of reference electrode can have greater or lesser effects on classical measures of synchronization such as correlation or coherency analyses (Nunez et al. 1997; Zaveri et al. 2000). Modern techniques such as phase synchronization analysis are also affected by the choice of reference electrode, with the resulting amplitude variations influencing the synchrony measured and disrupting the intended benefit of the technique, which is designed to remove the effects of signal amplitude from the synchrony ("phase locking") measured (Guevara et al. 2005).

For a visual presentation of the effects of reference electrode choice (and referential vs. bipolar arrays), a single 10-second epoch of EEG is presented in Figure 2.7a through f in six different montages, including one bipolar montage and five different referential montages. A careful analysis of the different waveforms—especially the very evident, labelled transient physiological artifacts—with respect to their amplitude and polarity changes across the different montages will provide the interested reader with an excellent opportunity to begin to understand the process of visual EEG interpretation. Even a cursory comparison of the six different depictions of the same 10 s of EEG will make clearly evident the important effects of reference electrode choices. In Figure 2.8, graphic representations of 300 ms of EEG data points—selected from the same 10-second epoch, using the six different recording arrays shown in Figure 2.7 (one bipolar and five referential), the EEG

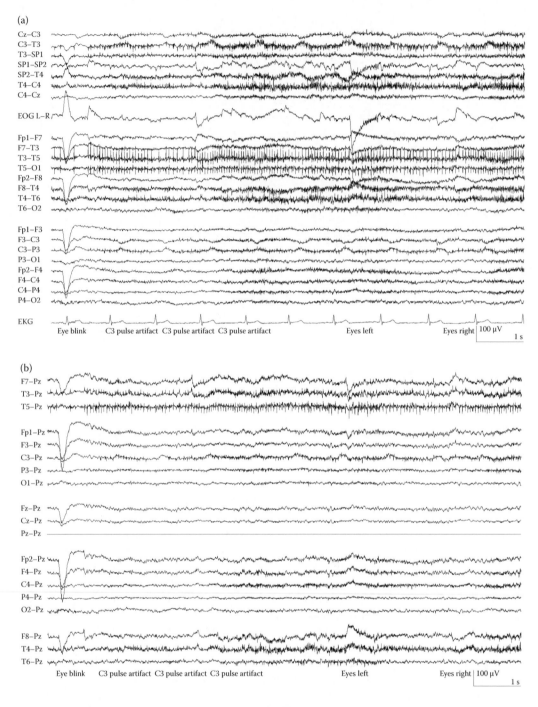

FIGURE 2.7 Ten seconds of normal interictal scalp EEG recording with subject awake and eyes open. Numerous different common physiological artifacts are evident, including an eye blink, horizontal eye movements, frontalis and temporalis EMG, lateral rectus EMG, and pulse artifacts. To demonstrate the effects of reference electrode choice, the same ten-second EEG epoch is depicted in (a) combined circular and anterior–posterior bipolar montage; (b) monopolar referential montage, reference = Pz.

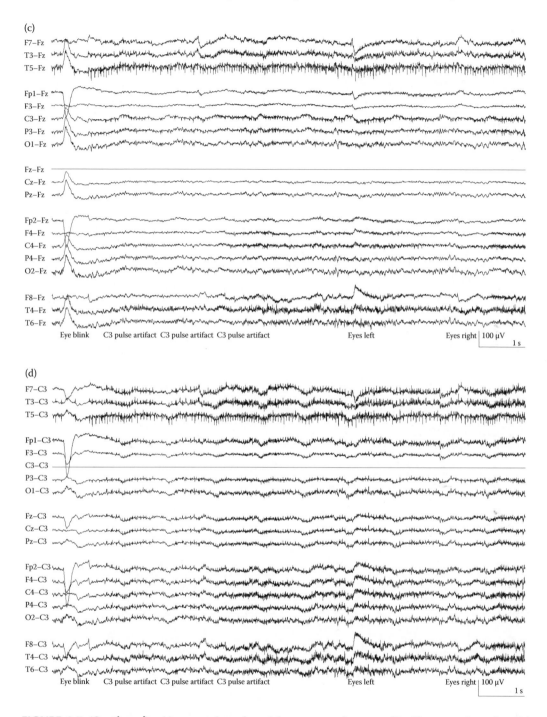

FIGURE 2.7 (Continued)   (c) monopolar referential montage, reference = Fz; (d) monopolar referential montage, reference = C3.

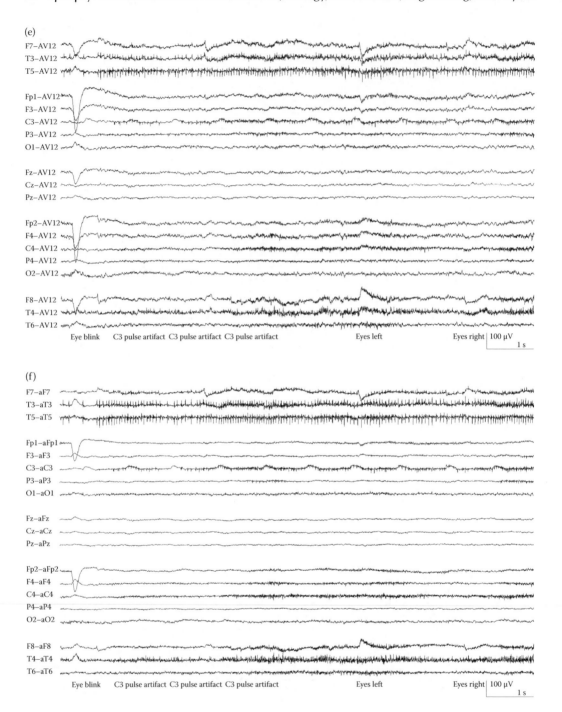

FIGURE 2.7 (Continued) (e) common average referential montage, reference = common average (of electrodes F3, F4, T3, C3, C4, T4, T5, P3, P4, T6, O1, and O2); and (f) Laplacian referential montage.

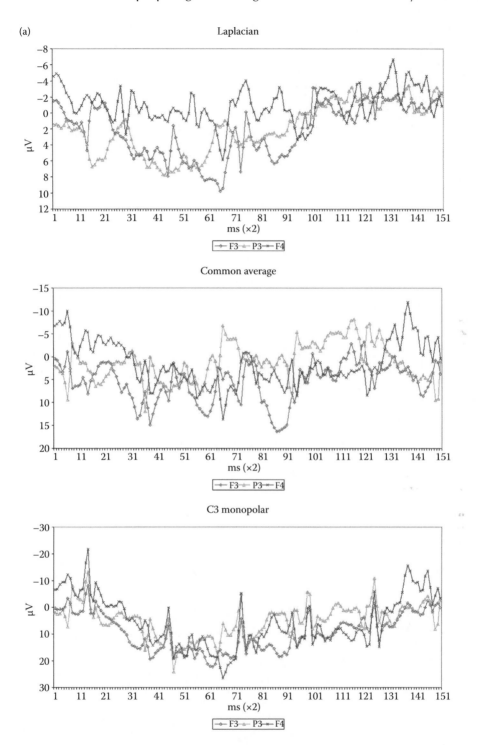

FIGURE 2.8 **(See color insert.)** EEG time series data points from a 300-millisecond epoch at the time of the first labelled "C3 pulse artifact" segment of EEG depicted in Figure 2.7. Shown are the data points for electrode positions F3, P3, and F4 (as first inputs into the referential or bipolar channels). Examination of the fluctuations in differences between, for example, F3 and P3 or F3 and F4, with respect to the choice of reference electrode, highlights the potential for difficulties in definitively elucidating coordinated activity between different brain regions using scalp EEG signal.

(b)

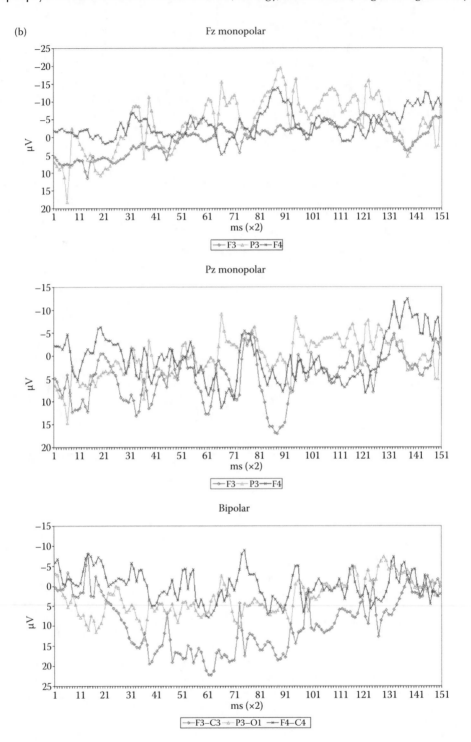

FIGURE 2.8 (Continued)

sampled at 500 Hz—are presented for three electrode positions (F3, P3, and F4). An investigator might be interested in the degree of synchronization evident between the left frontal and parietal regions, or perhaps between the left and right frontal regions, and choose to study this EEG time series data looking for evidence of correlation or coherency or phase locking between F3 and P3 or F3 and F4. As can be seen by comparing the different graphs, it is not immediately obvious that the same measure of synchronization between these pairs of electrodes would be found for each of the six different recording arrays.

2.7   LIMITATIONS OF SCALP EEG RECORDING

Noninvasive EEG recording from the scalp has a number of limitations compared with intracranial EEG recording, including the abundance of extracranial (especially muscle) artifacts, the inability to accurately record lower amplitude faster frequencies in the beta and gamma bands, and, especially for the study of epilepsy, the fact that EEG cannot "see" deep into the brain. Spontaneous activity in, for example, the mesial temporal regions, interhemispheric frontal lobe structures, or thalamus is *not* apparent on scalp EEG. This should be of no surprise given the accepted superficial cortical source of the potentials detected by EEG, described in more detail above. However, somewhat surprisingly, this is a point that continues to provoke debate among some clinical electroencephalographers. A number of papers have shown that only a small minority of the total interictal and

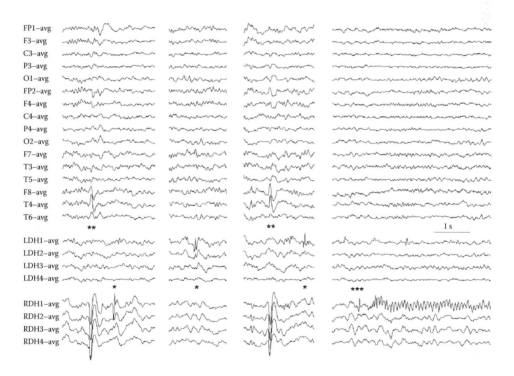

FIGURE 2.9 Comparison of intracranial interictal and ictal epileptiform activity recorded during sleep with simultaneous scalp EEG. Focal spikes in left and right hippocampus (*), electrode contacts LHD1 and RHD1, show no scalp EEG correlates; more diffuse right temporal spike and wave complexes (**) apparent at multiple contacts of right temporal depth electrode (RHD 1–4) are associated with visible epileptiform potentials on scalp EEG (F8, T4). Focal electrographic seizure in right hippocampus (***) has no scalp EEG correlate. Referential montage; reference = common average 10–20 electrodes. Top 16 channels = scalp EEG. Channels 17–20 and 21–24 = left and right, respectively, temporal-depth electrode recordings; four-contact orthogonally implanted depth electrodes, deepest contact of each electrode in the hippocampus, most superficial contact in the lateral temporal neocortex. Sensitivity = 15 μV/mm for scalp EEG, 50 μV/mm for intracranial recordings.

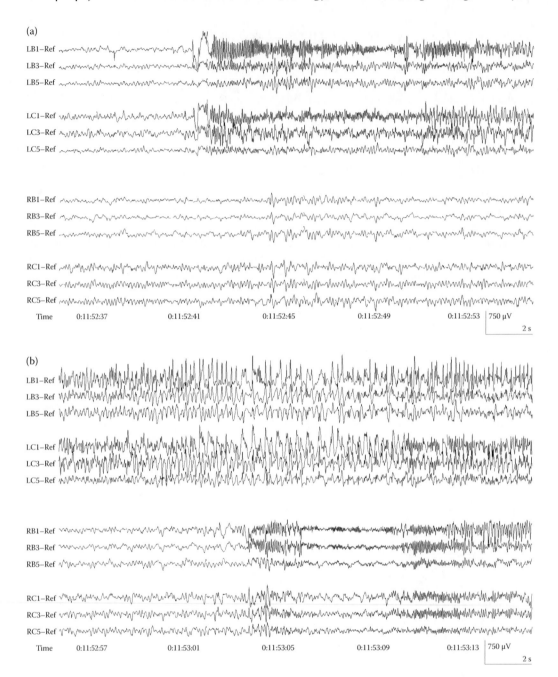

FIGURE 2.10 Intracranial EEG recording obtained from bilateral orthogonally implanted eight-contact depth electrodes; only the signal from the first (deepest), third, and fifth contacts of each depth electrode are shown, with LB1/RB1 situated in the hippocampi and LC1/RC1 in the parahippocampal gyri. Left mesial temporal regional seizure onset at 11:52:41 (a) in hippocampus and parahippocampal gyrus (LB1 > 3 > 5 and LC1 > 3 > 5), with subsequent propagation ("spread") to contralateral right mesial temporal region at 11:53:05 (b). Note that the contralateral "propagation" is actually the abrupt onset of an independent seizure in the right temporal lobe—the ictal activity is not synchronous between the two hemispheres. Seizure prediction signal analysis techniques need to uncover predictive features from the type of signal seen in the left-sided electrodes before 11:52:41 and in the right-sided electrodes before 11:53:05.

ictal epileptiform activity that can be detected with intracranial depth electrodes in patients with temporal lobe epilepsy is detectable with scalp EEG, and that focal mesial temporal epileptiform activity is never detected by scalp EEG (Alarcon et al. 1994; Merlet and Gotman, 1999; Wennberg and Lozano 2003; Gavaret et al. 2004; Nayak et al. 2004).

Figure 2.9 shows examples of interictal and ictal epileptiform activity recorded using simultaneous scalp EEG and intracranial depth electrodes in a patient with bitemporal epilepsy.

2.8 INTRACRANIAL EEG

The inability of scalp EEG to see deep into the brain is the reason one needs to occasionally perform, for clinical purposes, intracranial EEG. Indeed, intracranial EEG recordings acquired during the course of investigations for possible epilepsy surgery compose most of the time series provided for nonclinicians to work with in the hunt for features predictive of seizures. Benefits of the intracranial EEG signal include (usually) far less contamination with artifact, greater proximity to sites of seizure generation, and lesser problems with reference electrode issues—provided an

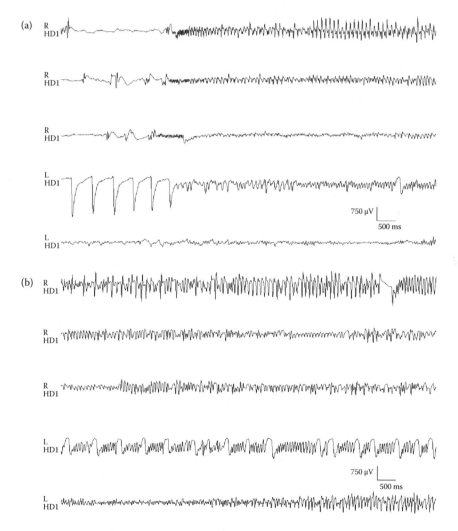

FIGURE 2.11 Five different seizure onsets recorded from intracranial-depth electrodes in one patient over 24 h. (a) Focal seizure onsets either in right hippocampus (RHDI) or left hippocampus (LHDI). (b) Continuation of same seizures.

extracranial reference is used. An extracranial reference greatly minimizes the reference contribution to the signal because of an approximately tenfold difference in amplitude between intracranial and extracranial potentials. If one or more of the intracranial electrodes are used as reference, as has often been done in published reports, the situation is then identical to that described above for noninvasive scalp EEG recordings.

Figure 2.10 shows an intracranial EEG recording of a relatively typical left mesial temporal seizure onset, with subsequent propagation or "spread" to the contralateral right temporal lobe. As can easily be seen in the figure, the rhythmic ictal activity does not really "spread" (the term used by clinical electroencephalographers), but instead a second, independent seizure arises on the right side. This is a consistent, repeated occurrence following unilateral onsets in this (and other) patients. This patient became seizure-free after a left anterior temporal surgical resection. It is from the type of signal seen before the seizure onsets on either side that information will need to be derived to predict seizures from EEG data.

In the hunt for features predictive of seizures, and in the analysis of EEG time series in the hope of understanding seizure dynamics, it is important to note a few further potential complicating factors. First, not all seizures look alike in terms of morphology, frequency, or even site of onset—even in a given patient and even when the outward clinical manifestations may be identical. This suggests that seizure foci may contain a multiplicity of potential ictal generators (Wennberg et al. 2002). Figure 2.11 shows five different seizure onsets that occurred in one patient during a 24-h period; a patient ultimately rendered seizure-free after a right anterior temporal resection. In addition, the state of a patient at the time of EEG recording may significantly affect the background activity present in the signal, and fluctuations in state may need to be considered in the signal analysis, whether it be sleep–wake states, hormonal fluctuations, changing levels of antiepileptic medications, or the postictal effects of previous seizures.

REFERENCES

Alarcon G., C. N. Guy, C. D. Binnie, S. R. Walker, R. D. C. Elwes, and C. E. Polkey. 1994. Intracerebral propagation of interictal activity in partial epilepsy: Implications for source localization. *J. Neurol. Neurosurg. Psychiatry* 57:435–449.

Cooper R., A. L. Winter, H. J. Crow, and W. G. Walter. 1965. Comparison of subcortical, cortical and scalp activity using chronically indwelling electrodes in man. *Electroencephalogr. Clin. Neurophysiol.* 18:217–228.

Daly, D. D., and T. A. Pedley, eds. 1990. *Current practice of clinical electroencephalography*, 2nd ed. New York: Raven Press.

Gavaret, M., J.-M. Badier, P. Marquis, F. Bartolomei, and P. Chauvel. 2004. Electric source imaging in temporal lobe epilepsy. *J. Clin. Neurophysiol.* 21:267–282.

Gloor, P. 1985. Neuronal generators and the problem of localization in electroencephalography: Application of volume conductor theory to electroencephalography. *J. Clin. Neurophysiol.* 2:327–354.

Guevara, R., J. L. Pérez Velazquez, V. Nenadovic, R. Wennberg, G. Senjanovič, and L. García Dominguez. 2005. Phase synchronization measurements using electroencephalographic recordings: What can we really say about neuronal synchrony? *Neuroinformatics* 3:301–314.

Lopes da Silva, F., and A. Van Rotterdam. 1993. Biophysical aspects of EEG and magnetoencephalogram generation. In *Electroencephalography: Basic principles, clinical applications, and related fields*, 3rd ed. E. Niedermeyer and F. Lopes da Silva, eds., 78–91. Baltimore: Williams & Wilkins.

Merlet, I., and J. Gotman. 1999. Reliability of dipole models of epileptic spikes. *Clin. Neurophysiol.* 110: 1013–1028.

Nayak, D., A. Valentín, G. Alarcón, J. J. G. Seoane, F. Brunnhuber, J. Juler, C. E. Polkey, and C. D. Binnie. 2004. Characteristics of scalp electrical fields associated with deep medial temporal epileptiform discharges. *Clin. Neurophysiol.* 115:1423–1435.

Niedermeyer, E., and F. Lopes da Silva, eds. 1993. *Electroencephalography: Basic principles, clinical applications, and related fields*, 3rd ed. Baltimore: Williams & Wilkins.

Nunez, P. L., R. Srinivasan, A. F. Westdorp, R. S. Wijesinghe, D. M. Tucker, R. B. Silberstein, and P. J. Cadush. 1997. EEG coherency: I. Statistics, reference electrode, volume conduction, Laplacians, cortical imaging, and interpretation at multiple scales. *Electroencephalogr. Clin. Neurophysiol.* 103:516–527.

Pedley, T. A., and R. D. Traub. 1990. Physiological basis of the EEG. In *Current practice of clinical electro-encephalography,* 2nd ed. D. D. Daly and T. A. Pedley, eds., 107–137. New York: Raven Press.

Speckmann, E.-J., and C. E. Elger. 1993. Introduction to the neurophysiological basis of the EEG and DC potentials. In *Electroencephalography: Basic principles, clinical applications, and related fields*, 3rd ed. E. Niedermeyer and F. Lopes da Silva, eds., 15–26. Baltimore: Williams & Wilkins.

Steriade, M. 1993. Cellular substrates of brain rhythms. In *Electroencephalography: Basic principles, clinical applications, and related fields*, 3rd ed. E. Niedermeyer and F. Lopes da Silva, eds., 27–62. Baltimore: Williams & Wilkins.

Tao, J. X., M. Baldwin, S. Hawes-Ebersole, and J. S. Ebersole. 2007. Cortical substrates of scalp EEG epileptiform discharges. *J. Clin. Neurophysiol.* 24:96–100.

Wennberg R., F. Arruda, L. F. Quesney, and A. Olivier. 2002. Preeminence of extrahippocampal structures in the generation of mesial temporal seizures: Evidence from human depth electrode recordings. *Epilepsia* 43:716–726.

Wennberg, R. A., and A. M. Lozano. 2003. Intracranial volume conduction of cortical spikes and sleep potentials recorded with deep brain stimulating electrodes. *Clin. Neurophysiol.* 114:1403–1418.

Whitham, E. M., K. J. Pope, S. P. Fitzgibbon, T. Lewis, C. R. Clark, S. Loveless, M. Broberg, A. Wallace, D. DeLosAngeles, P. Lillie, A. Hardy, R. Fronsko, A. Pulbrook, and J. O. Willoughby. 2007. Scalp electrical recording during paralysis: Quantitative evidence that EEG frequencies above 20 Hz are contaminated by EMG. *Clin. Neurophysiol.* 118:1877–1888.

Whitham, E. M., T. Lewis, K. J. Pope, S. P. Fitzgibbon, C. R. Clark, S. Loveless, D. DeLosAngeles, A. K. Wallace, M. Broberg, and J. O. Willoughby. 2008. Thinking activates EMG in scalp electrical recordings. *Clin. Neurophysiol.* 119:1166–1175.

Zaveri, H. P., R. B. Duckrow, and S. S. Spencer. 2000. The effect of a scalp reference signal on coherence measurements of intracranial electroencephalograms. *Clin. Neurophysiol.* 111:1293–1299.

3 Basic Mechanisms of Seizure Generation

John G. R. Jefferys and Premysl Jiruska

CONTENTS

3.1 INTRODUCTION

Seizures are due to excessively synchronous and/or excessively intense activity of neuronal circuits in the brain, particularly in the cerebral cortex. In epilepsy, seizures occur spontaneously, repeatedly, and usually suddenly. The unpredictability of seizures represents one of the main disabling features of epilepsy. This chapter will address mechanisms of focal (partial) epilepsy and seizures, i.e., cases where clinical evidence can identify some degree of localized seizure onset. The clinical consequences—the ictal semiology—will depend critically on where the focus is located and to some extent on the nature of the underlying molecular, structural, or functional pathology. The electrographic properties of seizures are a bit more consistent between cortical regions than are the associated semiologies, and, apart from distinguishing between hippocampus and neocortex, this review will not delve into regional differences.

We can all experience "symptomatic" seizures where some identifiable trigger such as acute trauma, ingesting convulsant substances, or severe sleep deprivation can alter the functioning of normal brain circuitry in ways that trigger symptomatic seizures. However, patients with epilepsy have brains with alterations of their circuitry that cause them to experience seizures with no identifiable trigger. Where an epileptic focus can be identified clearly, patients who fail to respond to anticonvulsant drugs can benefit from the surgical removal of the focus, if it is in a region that can be removed without causing significant disability (i.e., avoiding "eloquent" cortex). Experience gained during presurgical examination has led to the classification of the cortical zones in the epileptic brain (Rosenow and Luders 2001). The main zone is the *epileptogenic zone*, which is the region that needs to be totally removed or disconnected to achieve complete seizure-freedom. Usually the epileptogenic zone includes the *seizure onset zone*, which is the tissue where seizures start. The *ictal symptomatogenic zone* is the area of cortex that generates the initial clinical (behavioral) seizure symptoms. The *irritative* zone is the cortical tissue that generates interictal discharges, i.e., brief (dozens to hundreds of milliseconds), spatially restricted, synchronous discharges characteristic of focal epilepsies and distinct from seizures, which last longer (tens of seconds to a few minutes). *Epileptogenic lesions* are structural lesions "causally related to epilepsy," although the direction of causality often is far from clear. Finally, there is the *functional deficit zone*, which is the region of the brain that is functionally abnormal between seizures, a reminder that seizures are not the only problem in epilepsies. In this chapter, we will concentrate on description of cellular and network phenomena that can be observed in the irritative and seizure onset zones, in particular in the context of temporal lobe epilepsy.

3.2   EXPERIMENTAL MODELS

Experimental models play a key role in discovering mechanisms of seizure generation. These models need to exhibit the features of seizures we wish to investigate. The requirements are very different from those needed for screening potential anticonvulsant drugs, where the key issue is to identify a measurable phenomenon that predicts anticonvulsant activity. Models used to identify seizure mechanisms need to have stronger parallels in their electrographic and physiological properties to clinical seizures. Space prohibits an extensive review of experimental models, but a monograph on the topic has been published recently (Pitkänen et al. 2005). Experimental models that have contributed to our understanding of seizures can be divided into acute and chronic.

3.2.1   ACUTE MODELS

Acute models use some convulsant treatment to trigger epileptic activity in normal brain tissue. Convulsant drugs and electrical stimulation can be used both in vivo and in vitro. Several of the most common convulsants block $GABA_A$ receptor-mediated inhibition; penicillin, bicuculline, and picrotoxin are examples. However, other convulsants do not (e.g., potassium channel blockers such as 4-aminopyridine and mast cell degranulating peptide). Changes in extracellular ion composition, including increased K^+, decreased Mg^{2+}, or decreased Ca^{2+}, are much easier to implement in vitro in the brain slice preparation, where ~0.4mm sections are cut from the brains of freshly (and humanely) killed experimental animals, usually rodents (Jefferys 2007). However, disrupted ion homeostasis, for instance due to dietary or metabolic problems, can cause symptomatic seizures (Morris 1992).

3.2.2   CHRONIC MODELS

Acute models have produced substantial insights into epileptic synchronization, but they essentially tell us how *normal* brains can generate epileptiform phenomena (both interictal discharges and seizures); chronic models are more relevant for understanding seizures in the *epileptic* brain. These models are a closer fit to clinical epilepsy in that the functional organization of the brain is changed to become epileptic, ideally generating spontaneous seizures. In the case of temporal lobe epilepsy, chronic models can be divided into two broad groups: those that use a period of status epilepticus as a precipitating factor, and those that do not.

The status models use kainic acid (intrahippocampal or systemic), pilocarpine (usually systemic), or prolonged repetitive electrical stimulation to trigger status epilepticus, which is allowed to continue many tens of minutes up to several hours. This leads to significant neuronal loss and morphological reorganization of the hippocampus, which shares some features with hippocampal sclerosis (an epileptogenic lesion often found in patients with mesial temporal lobe epilepsy). In these models, the initial status epilepticus is followed by a latent period of a week or two before the onset of spontaneous seizures, which then recur intermittently for the rest of the life of the animal. During the latent period, the normal brain transforms into an epileptic brain through the still poorly understood process of epileptogenesis.

Chronic foci that do not depend on an initial status epilepticus are not associated with early structural lesions. One of the most extensively used is kindling, in which initially subconvulsive stimulations are repeated, often daily, and evoke progressively stronger responses culminating in generalized epileptic seizures. This reduction in seizure threshold persists for most of the lifetime of the animal, but in most cases the seizures do not occur spontaneously. The need to evoke kindled seizures by stimulation is a shortcoming as a model of epilepsy (Stables et al. 2002; Stables et al. 2003), although it can be convenient for some types of experiments.

The model we have used most involves injecting a very small dose of tetanus toxin into the hippocampus (Jefferys and Walker 2006). This method does not trigger initial status epilepticus, nor does it create early lesions. As in the other chronic models, there is a latent period of 1–2 weeks between the initial insult (the tetanus toxin injection) and the onset of spontaneous seizures. The majority of animals gain remission from seizures after 6–8 weeks, a process that seems to be active because several of the cell-level pathophysiologies of the active seizure phase are preserved. In this model, seizure remission is not a return to normal function as there are permanent impairments of normal brain function. A minority of animals may develop lesions of dorsal CA1, which resemble hippocampal sclerosis, some time after spontaneous seizures have started. There is a substantial overlap between rats with hippocampal sclerosis and seizure persistence.

3.3 CELLULAR PATHOPHYSIOLOGY OF CHRONIC EPILEPTIC FOCI

There is no simple universal pathophysiology for epileptic foci. However, there are some recurring themes that we will outline here. While it is tempting to think that all functional and structural changes identified in chronic epileptic foci might contribute to the generation of epileptic activity, some of these changes may prove to be potentially antiepileptic responses to repeated seizures, while others could just be epiphenomena with no functional impact on epileptic activity.

It is increasingly clear that epileptic foci are associated with abnormal functioning of voltage-gated ion channels. Several rare genetic epilepsies have turned out to be due to mutations in voltage-gated ion channels (Baulac and Baulac 2009; Catterall et al. 2008), known as channelopathies, for instance, some that cause increases in persistent Na^+ currents. Channelopathies do not have to be inherited, but can also be acquired. Good examples include increased persistent forms of voltage-gated Na^+ currents (in clinical as well as experimental conditions) (Vreugdenhil et al. 2004), changes in Ca^{2+} channel expression (Becker et al. 2008), decreased I_A (a K^+ current) (Bernard et al. 2004), and decreased hyperpolarization-activated cation current (I_h) (Huang et al. 2009; Jung et al. 2007), all of which have been demonstrated in chronic models of temporal lobe epilepsy.

Synaptic properties also change in epileptic foci. Again, there are several examples of inherited channelopathies of ligand-gated ion channels (synaptic receptor subunits) (Baulac and Baulac 2009; Catterall et al. 2008). In our own work on the tetanus toxin model, we have found impaired recruitment of inhibitory neurons in active epileptic foci (Whittington and Jefferys 1994), and weakened excitatory transmission in foci following remission from seizures (Vreugdenhil et al. 2002). One subtle change in $GABA_A$ receptor-mediated inhibition is a shift of the reversal potential to be more depolarizing, because of changes in chloride homeostasis, perhaps due to a reversion of the Cl^- transporter to an early developmental type (Huberfeld et al. 2007).

Selective loss of inhibitory interneurons has been proposed as a mechanism for epileptic foci over several decades. Evidence for a general selective loss of inhibitory neurons is sparse, but there are accounts of losses of some classes of inhibitory interneuron, such as the chandelier cell in human foci (DeFelipe 1999) and O-LM cells in rodent models of TLE (Cossart et al. 2005).

Also, changes in the network connectivity occur in epileptic foci. Perhaps the best documented is mossy fiber sprouting where the axons of hippocampal dentate granule cells project into layers they do not normally occupy (Sutula and Dudek 2007). This sprouting is easily detected because these axons contain a high concentration of zinc, and because they project to very specific layers, but the phenomenon of sprouting occurs in other pathways too (Prince et al. 2009; Vreugdenhil 2002).

3.4 THE IRRITATIVE ZONE: GENERATION OF INTERICTAL DISCHARGES

Interictal discharges are clinically and experimentally used as a reliable marker of epilepsy. They are population phenomena, which are in EEG described as high-amplitude electrographic transients of brief duration (50–200 ms), often followed by slow waves. Cellular and network mechanisms of interictal discharges have been extensively studied over the past decades. Intracellularly, they are associated with the *paroxysmal depolarizing shift*, which is an abrupt depolarization of a few tens of mV that lasts throughout the interictal discharge. The replication of this phenomenon in brain slices in vitro led to a rather complete account of the synaptic and intrinsic neuronal mechanisms that are responsible. In short, paroxysmal depolarization shifts depend on the mutual, glutamatergic, and excitatory connections within cortical regions. The central idea is that if these connections are sufficiently effective, and if the connections are divergent, and if the connected population is large enough, then a chain reaction of excitation can propagate throughout the population, resulting in the excessive synchronization of the interictal discharge (Jefferys 2007; Traub et al. 1999). The analogy of the chain reaction extends to the existence of a critical mass—or minimum aggregate— for epileptic discharges, which can be 1000–2000 pyramidal cells in the case of the hippocampal slice in vitro.

The effectiveness of the excitatory connections depends both on the strength of the synapses, which in the hippocampal CA3 region can reach 1 mV per synapse, and on the ability of certain neurons to generate "intrinsic bursts," which are bursts of action potentials driven by slower voltage-dependent inward currents, mediated by Ca^{2+} channels in CA3 and by persistent Na^+ in CA1. In a normal brain, the majority of neurons are regular-firing. Intrinsic bursting neurons are more common in certain brain areas, including pyramidal cells in area CA3 in the hippocampus or in layer V of the neocortex. However, regular-firing neurons can be converted to bursting by simply changing the ionic composition of extracellular fluid, in acute in vitro models of epileptic seizures, for example (Jensen and Yaari 1997). The incidence of intrinsic bursting neurons can be greater than normal in slices from chronic experimental epileptic foci (Wellmer et al. 2002). Changes of this kind are probably due to molecular changes in voltage-gated channels in chronic epilepsy, such as those mentioned in the previous section. Intrinsic bursts in presynaptic neurons will cause postsynaptic summation of EPSPs, which can increase the probability of the target neuron firing by a factor of ten or so, contributing to further recruitment of neurons into the interictal discharge.

We should not forget that neurons are not the only cell type in the brain, and in recent years astrocytes (a type of glial cell) have proved to be rather more active participants in brain function than was previously thought. In the case of chronic experimental models of epilepsy, there is evidence of changes in astrocytes that result in impaired clearance of extracellular K^+, which is predicted to increase neuronal excitability and synchronization (David et al. 2009).

The interconnections between excitatory neurons are not present to make us susceptible to seizures, but rather are a necessary component of the cortical circuitry required for normal function. This feedback excitatory circuitry is normally controlled by the minority of cortical neurons that are inhibitory (less than 20% of the total), which explains why depressing $GABA_A$ receptor-mediated inhibition is such a potent convulsant, both in experiments and in some kinds of clinical symptomatic seizures. The inhibitory neurons come in a variety of types, classified in part by their synaptic targets. Interneurons that mediate feedback inhibition play a key role in preventing the excessive build-up of excitatory activity because each is excited by many excitatory neurons in its vicinity, and each inhibits a substantial proportion of the population of excitatory neurons.

One of the features of interictal discharges is that they are brief. A variety of mechanisms have been implicated in terminating them, including depletion of excitatory synaptic vesicles from the "readily releasable pool," synaptic inhibition, accumulation of neuromodulators such as adenosine, afterhyperpolarizations due to activation of K^+ channels, and electrogenic ion pumps. It is likely that there are multiple mechanisms of termination, with variable contributions in each particular condition.

Interictal discharges are a good indicator of epilepsy, but similar activity can be seen under physiological conditions. In the rat at least, these "sharp waves" are associated with specific behavioral states and can therefore be discriminated from morphologically similar interictal discharges because the latter are not restricted to those behavioral states and often are more frequent. Interictal discharges typically lack any obvious behavioral correlate, but detailed studies show that they can cause detectable impairments of cognition, both transient (Aldenkamp and Arends 2004) and longer-term (Holmes and Lenck-Santim 2006).

What is the role of interictal discharges in epilepsy and seizure generation? They provide markers for the presence of epileptogenic cortex, and are often used to localize epileptic foci. In some cases, the location of interictal discharges corresponds to the epileptogenic zone. However, in some cases, the irritative zone can extend far from the primary focus and can be distributed over wide areas of cortex, or even in the contralateral hemisphere. One complication is that interictal discharges can spread via normal anatomical connections between cortical areas, through the commissural fibers or via subcortical structures, which may lead to widespread or bilateral occurrence of interictal discharges. In this case, distinguishing primary irritative areas from secondary can be difficult and may require advanced signal processing techniques. One approach to distinguish primary (red) spikes from secondary (propagated green) spikes is by the presence of superimposed pathological high-frequency oscillations (Engel et al. 2009).

Several observations suggest interictal discharges may have beneficial roles in preventing seizure initiation. It has been suggested that the inhibition that follows the interictal spike is also associated with a decreased probability of seizure occurrence. It has been shown that interictal discharges, or stimulation-mimicking interictal discharges, may delay seizure onset in both hippocampal slices and entorhino-hippocampal slices (Barbarosie and Avoli 1997; D'Antuono et al. 2002). However, clinical and some experimental investigations suggest that the relationship between interictal discharges and seizures is more complex, both in terms of their underlying mechanisms and their functional interactions (De Curtis and Avanzini 2001).

3.5   SEIZURE GENERATION

As discussed above, interictal discharges (irritative zone) and seizures (seizure onset zone) may coexist or arise from completely distinct areas of brain. In hippocampal slices in vitro, the irritative zone is usually in the CA3 region, but the epileptogenic (seizure-generating) zone is more variable. It can be in CA3 too, in some cases (Borck and Jefferys 1999; Dzhala and Staley 2003), or it may be in either CA1 (Jensen and Yaari 1988; Traynelis and Dingledine 1988) or the entorhinal cortex (Barbarosie et al. 2002; Weissinger et al. 2000). We will return to the epileptogenic zone in vivo after a brief discussion of what prolongs the hypersynchronous epileptic discharge beyond the first few seconds. As mentioned before, interictal discharges normally are terminated within a few hundred milliseconds by several different mechanisms. A more prolonged depolarizing mechanism, outlasting the termination mechanisms of the interictal discharge, is required to produce seizures lasting tens of seconds to minutes.

One of the more robust of these prolonging mechanisms is the accumulation of extracellular K^+, resulting from several sources including membrane repolarization after action potentials, and perhaps quantitatively more importantly, the various transporters that use the K^+ gradient across neuronal and glial membranes to remove neurotransmitters and to restore ion gradients. These mechanisms can produce prolonged seizure-like events in brain slices in vitro. Barbarosie and Avoli (2002) found two distinct classes of interictal discharge in the acute 4-aminopyridine model, one predominantly glutamatergic (as outlined previously), and one with a substantial GABAergic component that was associated with a much more substantial K^+ transient (presumably due to the combination of reuptake of GABA and the removal of Cl^-). These GABAergic events tended to trigger seizure discharges. In an acute disinhibition model in vitro, we also found that reaching a critical K^+ level could trigger the transition to seizure-like events (Borck & Jefferys 1999; Traub et al. 1996).

Elevations of K^+ also play a central role in the nonsynaptic seizure-like events found in the low-Ca^{2+} model in vitro (Bikson et al. 2003b; Jefferys 1995; Yaari et al. 1983). Increasing extracellular K^+ leads to a slow negative potential (also known as a DC shift), which is related to the uptake of K^+ by glia as part of the buffering of the ion (Lux et al. 1986). Such negative shifts are seen during seizures, both experimental (see above) and clinical (Vanhatalo et al. 2003).

The close association of increased extracellular K^+ with seizures may be functionally important. Elevated K^+ increases neuronal excitability by depolarizing the resting membrane potential, which is mainly determined by the K^+ gradient across the neuronal membrane, and indirectly by mechanisms such as reducing the driving force for various transporters, including the Cl^- transporter KCC2, and GABA and glutamate transporters. So extracellular K^+ increases as a result of neuronal activity, and, in turn, the increased K^+ increases excitability. The time course of these changes in extracellular K^+ is of the order of several seconds (Bikson et al. 2003b; Jefferys 1995; Yaari et al. 1983), which is long enough to outlast most of the terminating mechanisms of interictal discharges. As noted before, the clearance of extracellular K^+ in epileptic foci may be compromised by dysfunction of astrocytes (David et al. 2009).

While local factors such as increases in extracellular K^+ undoubtedly play a role in seizures, it is important to remember that seizures in vivo generally involve multiple neuronal populations. In general, they consist of rhythmic bursts of activity, from a few Hz to 20–30 Hz, separated by intervals of little obvious activity. This led to the concept of reentrant, or reverberatory, loops. In the case of the mesial temporal lobe, there was the idea that waves of excitation would spread around a loop including the hippocampus, parahippocampal areas (subicular complex and entorhinal cortex), and dentate gyrus, with further loops from the entorhinal cortex to other parts of the cortex and limbic system (Lothman 1994; Paré et al. 1992). However, other experimental investigations on the relative timing of activity in structures that become engaged in limbic seizures show that phase lags between structures are close to zero, more like coupled oscillators (perhaps better called coupled epileptic generators) than a reentrant loop (Bragin et al. 1997; Finnerty and Jefferys 2002). These longer range connections provide a network of structures that may play a role in sustaining seizures beyond a few seconds, with one or more sites generating excitatory input to re-excite areas as the refractory period following each cycle of activity ends. Essentially, the difference between the reentrant loop idea and the coupled epileptic generators is whether that re-excitation follows an orderly sequence or whether the structure of discharges lacks such order.

Focal seizures can become secondarily generalized and spread much more widely through circuits that will include the thalamus, which in turn provides for widespread propagation through the brain (Bertram et al. 1998).

Seizures are, perhaps inevitably, a good deal more complex than interictal discharges, depending on combinations of intrinsic and synaptic mechanisms in multiple classes of inhibitory and excitatory neurons, nonsynaptic interactions between neurons, regulation of the size and composition of the extracellular space, and network dynamics on both local and long-range scales. The diversity of these mechanisms may suggest that targeting seizure initiation may stand a better chance of success than trying to stop seizures once they are underway.

3.6 HIGH-FREQUENCY ACTIVITY

Until the early 1990s, most reports on the electrophysiology of clinical seizures had remarkably narrow bandwidths, generally with high-frequency cut-offs of 60 or 80 Hz. Two clinical publications in 1992 raised the prospect that much higher frequencies might play important roles in focal epilepsies, including in the initiation of seizures (Allen et al. 1992; Fisher et al. 1992).

High-frequency activity, in the band from 80 to 500 Hz, can be further divided into ripples and fast ripples (Bragin et al. 1999; Staba et al. 2002). The cut-off between the two varies between 170 and 300 Hz in different studies, and we do not propose to try to pin it down here. The key point is that ripples occur in a normal brain, during "sharp-wave ripples" in the hippocampus, for example, where the ripples are attributed to fast inhibitory inputs and are superimposed on the sharp wave, which is due to a slower excitatory input (Ylinen et al. 1995). In contrast, fast ripples appear to be more specific for epileptic foci (Staba et al. 2002; Jiruska et al. 2010b), and appear to be generated by the activity of pyramidal cells. Our work on in vitro models supports the idea that high-frequency activity is produced by pyramidal cells, each firing on a small proportion of cycles, and that each cycle depends on multiple pyramidal cells synchronized by nonsynaptic mechanisms (Bikson et al. 2003a; Jiruska et al. 2010a). Indeed, it is hard to see what could synchronize activity on a millisecond to submillisecond time scale, apart from electrical interactions, either ephaptic or gap junction electrotonic coupling (Dudek et al. 1998; Jefferys 1995). Chemical synapses generally work on time-scales of ~2 ms upward, and extracellular K^+ operates on a time-scale of seconds.

One idea for why fast ripples should be good markers for epileptic foci is that they depend on neuronal loss, and particularly on hippocampal sclerosis: the relative power of fast ripples over ripples correlates with neuronal loss in clinical temporal lobe cases (Staba et al. 2007). A similar correlation was found in chronic experimental epileptic foci in vitro, where the hypothesis was that the loss of neurons fragments electrical or ephaptic coupling of neurons into multiple small groups (Foffani et al. 2007). However, we have recently found that fast ripples are generated by epileptic foci in the absence of hippocampal sclerosis (Jiruska et al. 2010b), which suggests that neuronal loss is not a necessary condition for their generation and also raises the prospect of their use for focus localization in nonlesional epilepsies.

The epilepsy-related high-frequency activity occurs during periods of excitation, during interictal discharges, at seizure onset, and during seizures themselves. Much of the analysis of the association between fast ripples and epileptic foci rests on evidence from interictal activity. What may be more interesting from the point of view of this conference, and for our understanding of the transition to seizure, is its presence at or even before seizure onset. The original studies in 1992 focused on high-frequency activity during the onset or early parts of seizures (Allen et al. 1992; Fisher et al. 1992), although their recordings were limited to below 150 Hz. More recent studies have also shown fast activity (mainly in the ripple and not the fast ripple band) preceding seizure onset (Jiruska et al. 2008; Ochi et al. 2007; Traub et al. 2001; Worrell et al. 2004), which may suggest that the preictal buildup of HFA may be used as a marker of an approaching seizure. The presence of relatively prolonged (seconds to tens of seconds) ripple-band activity is not typical of physiological ripple activity, which makes it potentially useful for short-range seizure prediction of the sort that can be useful for rapid therapeutic intervention, closed-loop brain stimulation, for example. From the point of view of the mechanisms of transition to seizure, the synchronized high-frequency activity of pyramidal cells will promote spatial and temporal summation, and various kinds of synaptic potentiation, at their target structures, which could tip the balance from normal activity to propagating seizure activity.

3.7 CONCLUDING REMARKS

Epileptic foci are associated with abnormal synchronization of neuronal activity as a result of multiple mechanisms, including abnormal activity of voltage-gated ion channels, abnormal synaptic activity, altered synaptic connectivity, and so on. The brief interictal discharges are probably due to a chain reaction of excitatory connections. The more prolonged hypersynchronous hyperactivity of the seizure depends on slower mechanisms that can sustain excitation beyond the refractory period that typically follows a synchronous discharge. These slower mechanisms can be local—the extracellular accumulation of K^+ ions, for example—or more widespread—mutual excitation between multiple epileptogenic sites that can be distributed in a particular structure or that can be distributed across multiple brain regions. Full-blown seizures involve cascades of interconnected and mutually enhancing mechanisms, which suggest the phenomena that precede seizure onset may prove to be promising targets for the development of more effective therapy.

REFERENCES

Aldenkamp, A. P., and J. Arends. 2004. Effects of epileptiform EEG discharges on cognitive function: Is the concept of "transient cognitive impairment" still valid? *Epilepsy & Behavior* 5:S25–S34.

Allen, P. J., D. R. Fish, and S. J. M. Smith. 1992. Very high-frequency rhythmic activity during SEEG suppression in frontal lobe epilepsy. *Electroencephalography and Clinical Neurophysiology* 82:155–159.

Barbarosie, M., and M. Avoli. 1997. CA3-driven hippocampal-entorhinal loop controls rather than sustains *in vitro* limbic seizures. *Journal of Neuroscience* 17:9308–9314.

Barbarosie, M., J. Louvel, M. D'Antuono, I. Kurcewicz, and M. Avoli. 2002. Masking synchronous GABA-mediated potentials controls limbic seizures. *Epilepsia* 43:1469–1479.

Baulac, S., and M. Baulac. 2009. Advances on the genetics of Mendelian idiopathic epilepsies. *Neurologic Clinics* 27:1041–1061.

Becker, A. J., J. Pitsch, D. Sochivko, T. Opitz, M. Staniek, C. C. Chen, K. P. Campbell, S. Schoch, Y. Yaari, and H. Beck. 2008. Transcriptional upregulation of Ca(v)3.2 mediates epileptogenesis in the pilocarpine model of epilepsy. *Journal of Neuroscience* 28:13341–13353.

Bernard, C., A. Anderson, A. Becker, N. P. Poolos, H. Beck, and D. Johnston. 2004. Acquired dendritic channelopathy in temporal lobe epilepsy. *Science* 305:532–535.

Bertram, E. H., D. X. Zhang, P. Mangan, N. Fountain, and D. Rempe. 1998. Functional anatomy of limbic epilepsy: A proposal for central synchronization of a diffusely hyperexcitable network. *Epilepsy Research* 32:194–205.

Bikson, M., J. E. Fox, and J. G. R. Jefferys. 2003a. Neuronal aggregate formation underlies spatiotemporal dynamics of nonsynaptic seizure initiation. *Journal of Neurophysiology* 89:2330–2333.

Bikson, M., P. J. Hahn, J. E. Fox, and J. G. R. Jefferys. 2003b. Depolarization block of neurons during maintenance of electrographic seizures. *Journal of Neurophysiology* 90:2402–2408.

Borck, C., and J. G. R. Jefferys. 1999. Seizure-like events in disinhibited ventral slices of adult rat hippocampus. *Journal of Neurophysiology* 82:2130–2142.

Bragin, A., J. Csisvari, M. Penttonen, and G. Buzsáki. 1997. Epileptic afterdischarge in the hippocampal-entorhinal system: Current source density and unit studies. *Neuroscience* 76:1187–1203.

Bragin, A., J. J. Engel, C. L. Wilson, I. Fried, and G. W. Mathern. 1999. Hippocampal and entorhinal cortex high-frequency oscillations (100–500 Hz) in human epileptic brain and in kainic acid-treated rats with chronic seizures. *Epilepsia* 40:127–137.

Catterall, W. A., S. Dib-Hajj, M. H. Meisler, and D. Pietrobon. 2008. Inherited neuronal ion channelopathies: New windows on complex neurological diseases. *Journal of Neuroscience* 28:11768–11777.

Cossart, R., C. Bernard, and Y. Ben Ari. 2005. Multiple facets of GABAergic neurons and synapses: Multiple fates of GABA signalling in epilepsies. *Trends in Neurosciences* 28:108–115.

D'Antuono, M., R. Benini, G. Biagini, G. D'Arcangelo, M. Barbarosie, V. Tancredi, and M. Avoli. 2002. Limbic network interactions leading to hyperexcitability in a model of temporal lobe epilepsy. *Journal of Neurophysiology* 87:634–639.

David, Y., L. P. Cacheaux, S. Ivens, E. Lapilover, U. Heinemann, D. Kaufer, and A. Friedman. 2009. Astrocytic dysfunction in epileptogenesis: Consequence of altered potassium and glutamate homeostasis? *Journal of Neuroscience* 29:10588–10599.

De Curtis, M., and Avanzini, G. 2001. Interictal spikes in focal epileptogenesis. *Progress in Neurobiology* 63:541–567.

DeFelipe, J. 1999. Chandelier cells and epilepsy. *Brain* 122:1807–1822.

Dudek, F. E., T. Yasumura, and J. E. Rash. 1998. 'Non-synaptic' mechanisms in seizures and epileptogenesis. *Cell Biology International* 22:793–805.

Dzhala, V. I., and K. J. Staley. 2003. Transition from interictal to ictal activity in limbic networks in vitro. *Journal of Neuroscience* 23:7873–7880.

Engel, J., A. Bragin, R. Staba, and I. Mody. 2009. High-frequency oscillations: What is normal and what is not? *Epilepsia* 50:598–604 .

Finnerty, G. T., and J. G. R. Jefferys. 2002. Investigation of the neuronal aggregate generating seizures in the rat tetanus toxin model of epilepsy. *Journal of Neurophysiology* 88:2919–2927.

Fisher, R. S., W. R. S. Webber, R. P. Lesser, S. Arroyo, and S. Uematsu. 1992. High-frequency EEG activity at the start of seizures. *Journal of Clinical Neurophysiology* 9:441–448.

Foffani, G., Y. G. Uzcategui, B. Gal, and L. M. de la Prida. 2007. Reduced spike-timing reliability correlates with the emergence of fast ripples in the rat epileptic hippocampus. *Neuron* 55:930–941.

Holmes, G. L., and P. P. Lenck-Santim. 2006. Role of interictal epileptiform abnormalities in cognitive impairment. *Epilepsy & Behavior* 8:504–515.

Huang, Z., M. C. Walker, and M. M. Shah. 2009. Loss of dendritic HCN1 subunits enhances cortical excitability and epileptogenesis. *Journal of Neuroscience* 29:10979–10988.

Huberfeld, G., L. Wittner, S. Clemenceau, M. Baulac, K. Kaila, R. Miles, and C. Rivera. 2007. Perturbed chloride homeostasis and GABAergic signaling in human temporal lobe epilepsy. *Journal of Neuroscience* 27:9866–9873.

Jefferys, J. G. R. 1995. Non-synaptic modulation of neuronal activity in the brain: Electric currents and extracellular ions. *Physiological Reviews* 75:689–723.

Jefferys, J. G. R. 2007. Epilepsy in Vitro: Electrophysiology and Computer Modeling. In *Epilepsy: A Comprehensive Texbook*, 2nd ed. J. Engel, Jr. et al., eds. Philadelphia: Lippincott, Willams & Wilkins.

Jefferys, J. G. R., and M. C. Walker. 2006. Tetanus toxin model of focal epilepsy. In *Models of Seizures and Epilepsy*, A. Pitkänen, P. A. Schwartzkroin, and S. L. Moshé, eds., 407–414. Amsterdam: Elsevier Academic Press.

Jensen, M. S., and Y. Yaari. 1988. The relationship between interictal and ictal paroxysms in an in vitro model of focal hippocampal epilepsy. *Annals of Neurology* 24:591–598.

Jensen, M. S., and Y. Yaari. 1997. Role of intrinsic burst firing, potassium accumulation, and electrical coupling in the elevated potassium model of hippocampal epilepsy. *Journal of Neurophysiology* 77:1224–1233.

Jiruska P., J. Csicsvari, A. D. Powell, J. E. Fox, W. C. Chang, M. Vreugdenhil, X. Li, M. Palus, A. F. Bujan, R. W. Dearden, and J. G. R. Jefferys. 2010a. High-frequency network activity, global increase in neuronal activity and synchrony expansion precede epileptic seizures in vitro. *Journal of Neuroscience* 30:5690–5701.

Jiruska, P., G. T. Finnerty, A. D. Powell, N. Lofti, R. Cmejla, and J. G. R. Jefferys. 2010b. High-frequency network activity in a model of non-lesional temporal lobe epilepsy. *Brain* 133:1380–1390.

Jiruska, P., M. Tomasek, D. Netuka, J. Otahal, J. G. R. Jefferys, X. Li, and P. Marusic. 2008. Clinical impact of high-frequency seizure onset zone in a case of bi-temporal epilepsy. *Epileptic Disorders* 10:231–238.

Jung, S., T. D. Jones, J. N. Lugo, A. H. Sheerin, J. W. Miller, R. D'Ambrosio, A. E. Anderson, and N. P. Poolos. 2007. Progressive dendritic HCN channelopathy during epileptogenesis in the rat pilocarpine model of epilepsy. *Journal of Neuroscience* 27:13012–13021.

Lothman, E. W. 1994. Seizure circuits in the hippocampus and associated structures. *Hippocampus* 4:286–290.

Lux, H. D., U. Heinemann, and I. Dietzel. 1986. Ionic changes and alterations in the size of the extracellular space during epileptic activity. *Advances in Neurology* 44:619–639.

Morris, M. E. 1992. Brain and CSF magnesium concentrations during magnesium deficit in animals and humans: Neurological symptoms. *Magnesium Research* 5:303–313.

Ochi, A., H. Otsubo, E. J. Donner, I. Elliott, R. Iwata, T. Funaki, Y. Akizuki, T. Akiyama, K. Imai, J. T. Rutka, and O. C. Snead. 2007. Dynamic changes of ictal high-frequency oscillations in neocortical epilepsy: Using multiple band frequency analysis. *Epilepsia* 48:286–296.

Paré, D., M. DeCurtis, and R. Llinás. 1992. Role of the hippocampal-entorhinal loop in temporal lobe epilepsy: Extra- and intracellular study in the isolated guinea pig brain *in vitro. Journal of Neuroscience* 12:1867–1881.

Pitkänen, A., P. A. Schwartzkroin, and S. L. Moshé. 2005. *Models of Seizures and Epilepsy.* Amsterdam: Elsevier.

Prince, D. A., I. Parada, K. Scalise, K. Graber, X. M. Jin, and F. Shen. 2009. Epilepsy following cortical injury: Cellular and molecular mechanisms as targets for potential prophylaxis. *Epilepsia* 50:30–40.

Rosenow, F., and H. Luders. 2001. Presurgical evaluation of epilepsy. *Brain* 124:1683–1700.

Staba, R. J., L. Frighetto, E. J. Behnke, G. Mathern, T. Fields, A. Bragin, J. Ogren, I. Fried, C. L. Wilson, and J. Engel. 2007. Increased fast ripple to ripple ratios correlate with reduced hippocampal volumes and neuron loss in temporal lobe epilepsy patients. *Epilepsia* 48:2130–2138.

Staba, R. J., C. L. Wilson, A. Bragin, I. Fried, and J. Engel, Jr. 2002. Quantitative analysis of high-frequency oscillations (80-500 Hz) recorded in human epileptic hippocampus and entorhinal cortex. *Journal of Neurophysiology* 88:1743–1752.

Stables, J. P., E. Bertram, F. Dudek, G. Holmes, G. Mathern, A. Pitkanen, and H. White. 2003. Therapy discovery for pharmacoresistant epilepsy and for disease-modifying therapeutics: Summary of the NIH/NINDS/AES Models II Workshop. *Epilepsia* 44:1472–1478.

Stables, J. P., E. H. Bertram, H. White, D. A. Coulter, M. A. Dichter, M. P. Jacobs, W. Loscher, D. H. Lowenstein, S. L. Moshe, J. L. Noebels, and M. Davis. 2002. Models for epilepsy and epileptogenesis: Report from the NIH workshop, Bethesda, Maryland. *Epilepsia* 43(11):1410–1420.

Sutula, T. P., and F. E. Dudek. 2007. Unmasking recurrent excitation generated by mossy fiber sprouting in the epileptic dentate gyrus: An emergent property of a complex system. *Progress in Brain Research* 163:541–563.

Traub, R. D., C. Borck, S. B. Colling, and J. G. R. Jefferys. 1996. On the structure of ictal events *in vitro*. *Epilepsia* 37:879–891.

Traub, R. D., J. G. R. Jefferys, and M. A. Whittington. 1999. *Fast Oscillations in Cortical Circuits*. Cambridge: The MIT Press.

Traub, R. D., M. A. Whittington, E. H. Buhl, F. E. LeBeau, A. Bibbig, S. Boyd, H. Cross, and T. Baldeweg. 2001. A possible role for gap junctions in generation of very fast EEG oscillations preceding the onset of, and perhaps initiating, seizures. *Epilepsia* 42:153–170.

Traynelis, S. F., and R. Dingledine. 1988. Potassium-induced spontaneous electrographic seizures in the rat hippocampal slice. *Journal of Neurophysiology* 59:259–276.

Vanhatalo, S., M. D. Holmes, P. Tallgren, J. Voipio, K. Kaila, and J. W. Miller. 2003. Very slow EEG responses lateralize temporal lobe seizures: An evaluation of non-invasive DGEEG. *Neurology (NY)* 60:1098–1104.

Vreugdenhil, M., S. P. Hack, A. Draguhn, and J. G. R. Jefferys. 2002. Tetanus toxin induces long-term changes in excitation and inhibition in the rat hippocampal CA1 area. *Neuroscience* 114:983–994.

Vreugdenhil, M., G. Hoogland, C. W. van Veelen, and W. J. Wadman. 2004. Persistent sodium current in subicular neurons isolated from patients with temporal lobe epilepsy. *European Journal of Neuroscience* 19:2769–2778.

Weissinger, F., K. Buchheim, H. Siegmund, U. Heinemann, and H. Meierkord. 2000. Optical imaging reveals characteristic seizure onsets, spread patterns, and propagation velocities in hippocampal-entorhinal cortex slices of juvenile rats. *Neurobiology of Disease* 7:286–298.

Wellmer, J., H. Su, H. Beck, and Y. Yaari. 2002. Long-lasting modification of intrinsic discharge properties in subicular neurons following status epilepticus. *European Journal of Neuroscience* 16:259–266.

Whittington, M. A., and J. G. R. Jefferys. 1994. Epileptic activity outlasts disinhibition after intrahippocampal tetanus toxin in the rat. *Journal of Physiology* 481:593–604.

Worrell, G. A., L. Parish, S. D. Cranstoun, R. Jonas, G. Baltuch, and B. Litt. 2004. High-frequency oscillations and seizure generation in neocortical epilepsy. *Brain* 127:1496–1506.

Yaari, Y., A. Konnerth, and U. Heinemann. 1983. Spontaneous epileptiform activity of CA1 hippocampal neurons in low extracellular calcium solutions. *Experimental Brain Research* 51:153–156.

Ylinen, A., A. Bragin, Z. Nádasdy, G. Jandó, I. Szabo, A. Sik, and G. Buzsáki. 1995. Sharp wave-associated high-frequency oscillation (200 Hz) in the intact hippocampus: Network and intracellular mechanisms. *Journal of Neuroscience* 15:30–46.

4 An Introduction to Epileptiform Activities and Seizure Patterns Obtained by Scalp and Invasive EEG Recordings

Andreas Schulze-Bonhage

CONTENTS

Brains of patients with epileptic seizures generate characteristic field potentials that can be recorded as epileptiform discharges in the electroencephalogram (EEG). Such epileptiform EEG activity is an important contributor to the electroclinical definition of epilepsies. Importantly, epileptiform discharges can be recorded not only during behavioral seizures, but also when patients appear asymptomatic. Depending on the duration, time of day, and inclusion of sleep recordings, such interictal EEG patterns are visible in surface EEG recordings in the vast majority of patients. Even though the absence of interictal epileptiform activity does not exclude the presence of an epileptic disorder, the sensitivity of the surface EEG is surprisingly high when considering the fact that only field potentials originating from the dorsolateral convexities of the brain are accessible to standard scalp recordings. Patterns of epileptiform activity historically made important contributions to the classification of diseases as epilepsies, and presently contribute to their syndromatic classification as focal or generalized epilepsies and particular syndromes within these groups, e.g., rolandic epilepsy or absence epilepsy.

4.1   INTERICTAL EPILEPTIFORM DISCHARGES IN SURFACE AND INTRACRANIAL EEG

4.1.1   OCCURRENCE OF INTERICTAL DISCHARGES

By definition, epilepsy is a disease characterized by paroxysmally occurring symptoms. Usually, the treating physician has access to ictal phenomena only by history. Thus, the existence of interictal abnormalities in the EEG of patients with epilepsy was a major step forward in confirming the diagnosis of suspected epilepsy (Gibbs et al. 1935). Depending on the duration of EEG recordings and the inclusion of different states of vigilance, interictal epileptiform discharges can be shown in up to 90% of patients in the surface EEG using standard electrodes placed according to the international 10-20 system (Jasper 1958). This is remarkable since surface EEG electrodes cover only the dorsolateral convexity of the brain and are incapable of accessing activity in basal, mesial, or deep areas, in structures with a closed electric field like the hippocampus, and in structures without cortical layering (e.g., basal ganglia or nuclei of the amgydala). Interictal epileptiform discharges, however, often extend over a large irritative zone that allows them to be assessed with high sensitivity. Simultaneous intracranial and surface recordings show that the majority of epileptiform discharges recorded directly at the cortical surface are not visible on surface EEG or at least do not show their epileptic nature with sufficiently clarity (Figure 4.1a and b) (Tao et al. 2007). There is a wide variability in the frequency of occurrence of interictal discharges in the surface EEG ranging from a few per day to a continuous presence, and they represent only the tip of an iceberg of ongoing epileptic discharges in the brain.

4.1.2   MORPHOLOGY AND FIELD OF INTERICTAL SPIKES AND SHARP WAVES

Interictal epileptiform discharges are defined as highly stereotyped potentials that clearly differ from background activity, are typically surface-negative, and ideally consist of a steep transient (spike or sharp wave, used synonymously here) and a subsequent slow wave. Not uncommonly, and depending on the distance and orientation of generators and recording sites, only sharp waves or only slow waves may be visible (Figure 4.2).

Patients with idiopathic generalized epilepsy show spike and wave discharges that commonly involve extended parts of the brain almost simultaneously; these discharges may, however, show an occipital or frontal predominance or may appear only (multi)regional. Their appearance and extent depends critically on the presence or absence of antiepileptic drugs (AEDs) (Rocamora et al. 2006; Oehl et al. 2010). There is some experimental evidence that generalized spikes may be triggered locally; standard EEG recordings are not able to show such a spread with brief latencies (Meeren et al. 2005).

In patients with focal epilepsy, interictal sharp waves appear mostly regional, encompassing only one lobe in the posterior or anterior brain regions. Depending on the state of vigilance and other factors influencing the spread of epileptic activity (including AEDs), the field involved in the irritative zone varies; there may be propagation to the homologic contralateral brain regions (secondary bilateral synchrony) or to areas to which major pathways project the activity, e.g., to the multimodal temporal or frontal association cortex. Spikes may be superimposed with high-frequency oscillations that are detectable using intracranial recordings (Figure 4.3). Less frequently, paroxysmal beta and gamma activity may be recorded using scalp electrodes over the focus area, often indicating local generation of ictal activity in the neocortex.

In focal epilepsy, interictal epileptiform discharges are considered to have a different mechanism of generation from ictal activity (Gotman and Marciani 1985) and may even serve to control epilepsy (de Curtis and Avanzini 2001). At least in certain pathologies, steepness, frequencies, and amplitudes of interictal spiking recorded by intracranial electrodes correlate to the seizure onset zone (Asano et al. 2003).

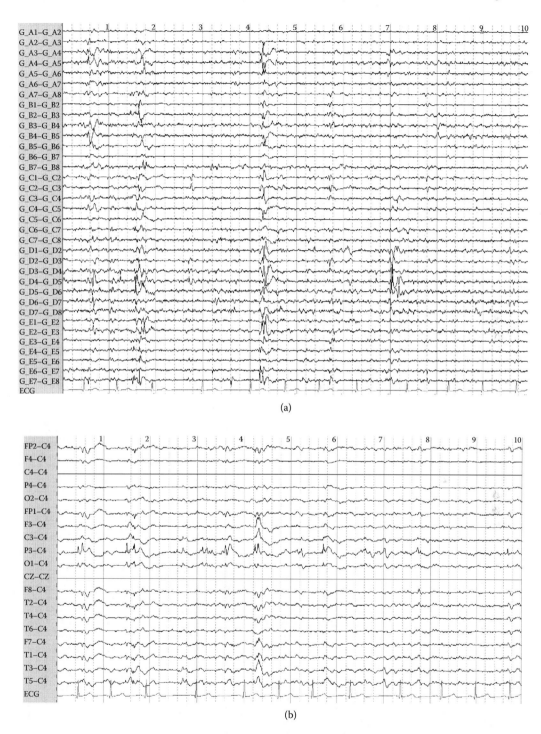

(a)

(b)

FIGURE 4.1 Simultaneous subdural grid recording (a) and surface EEG (b). Only spikes with extended fields are visible on the surface. (If not indicated otherwise, 10 s of EEG are displayed.)

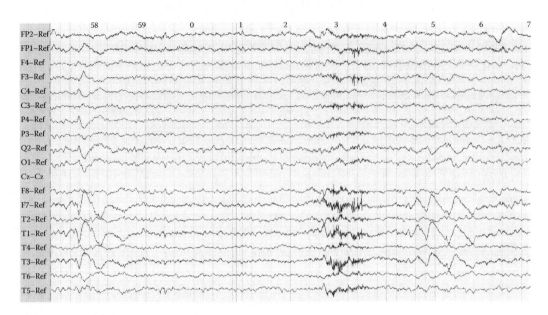

FIGURE 4.2 Sharp wave (left) and epileptic temporal intermittent rhythmic delta activity (TIRDA) (right) in a patient with left temporal lobe epilepsy (surface EEG).

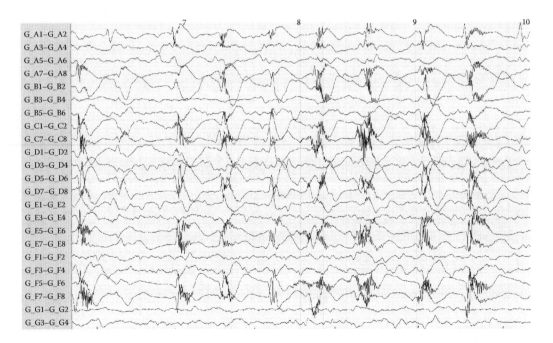

FIGURE 4.3 High-frequency oscillations partially superimposed on spikes in a patient with cortical dysplasia (intracranial grid recording).

4.1.3   OTHER FORMS OF INTERICTAL EPILEPTIC ACTIVITY

In addition to spikes, temporal intermittent rhythmic delta waves with asymmetrical rhythmic waves are a marker of subclinical epileptic activity, most typically found in the temporal regions during drowsiness in patients with temporal lobe epilepsy (Figure 4.2) (Geyer et al. 1999; Reiher et al. 1989). This stereotyped and intermittent rhythmic pattern differs from unspecific polymorphic slowing, which is unspecifically associated with various types of brain lesions.

In recent years, high-frequency oscillations (HFOs) in the frequency range of 100–500 Hz have been increasingly recognized to occur in epileptogenic brain areas (Bragin et al. 2002, 2004). HFOs have been shown to have localizing value not only in hippocampal seizure origin (Jirsch et al. 2006) but in a variety of pathologies as an epileptic phenomenon that can serve as a marker of epileptogenicity (Bragin et al. 2004; Jacobs et al. 2009). HFOs occur both in isolation and associated with interictal spikes (Figure 4.3). Further investigation will be necessary to compare their yield as markers of epileptogenicity in comparison to spikes and ictal discharges, to distinguish them from physiologically occurring oscillations (e.g., during memory consolidation), and to better understand the synaptic and extrasynaptic mechanisms underlying their generation.

4.1.4   POTENTIALS TO BE DIFFERENTIATED FROM INTERICTAL EPILEPTIC ACTIVITY

A spectrum of potentials exist that fulfil some criteria for spikes or sharp waves but are not associated with epilepsy. Such potentials include wicket spikes, sharply contoured theta or alpha activity over temporal regions, 6-s or 14-s groups of positive spikes in the form of arcades, positive occipital transients of sleep, small sharp spikes, and lambda waves occurring during visual exploration over the occipital cortex. Many of these potentials are more common during drowsiness, and these spiky potentials typically lack the association with a slow wave. Potentials with spiky appearance may also result from biological (ECG, EMG) or technical (electrode pops) artifacts.

4.2   ICTAL EPILEPTIFORM DISCHARGES IN THE SURFACE AND INTRACRANIAL EEG

During seizures, patients with idiopathic generalized epilepsies display similar, often longer generalized spike wave patterns at certain frequencies of repetition (e.g., 3 s in absence epilepsy or 4 s in juvenile myoclonic epilepsy) and additional generalized patterns consisting of polyspikes, spike runs, or rhythmic beta activity (Figure 4.4). Whereas polyspikes associated with myoclonic jerks may be quite brief, rhythmic spike and wave patterns in absence epilepsy must persist for at least 3 to 5 s to have obvious behavioral correlates (prolonged reaction times or loss of consciousness).

In contrast, focal epileptic seizures may be accompanied by a variety of patterns in surface EEG. Simple partial seizures with preserved consciousness often have limited spread of epileptic activity that often is not detectable on surface EEG at all (Verma and Radke 2006). More extended spread of epileptic discharges leading to complex partial seizures with impairment of consciousness are accompanied by ictal surface EEG correlates in more than 80% of cases and secondarily generalized seizures always have surface EEG patterns that may, however, be widely obscured by muscle artifacts.

4.2.1   MORPHOLOGY OF ICTAL PATTERNS

EEG patterns occurring during focal seizures are polymorphic. Typical patterns are listed in Table 4.1. These patterns typically show an evolution of frequency, the field involved, and amplitude. Patterns often show transitions within a given seizure.

Similar patterns can also be recorded using intracranial electrodes, which are sometimes needed to determine the epileptogenic area for planning of epilepsy surgery. Most commonly, subdural strip or grid electrodes and/or multicontact depth electrodes are used for determination of the irritative

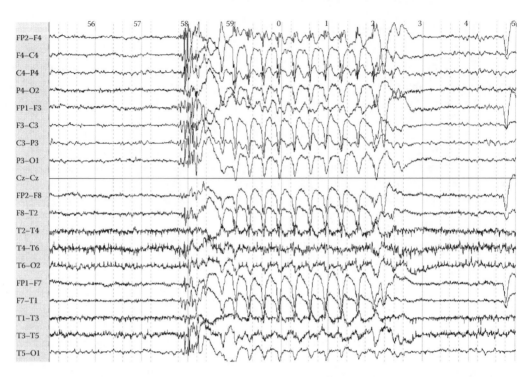

FIGURE 4.4   Generalized polyspikes and rhythmic 3-s spike wave complexes in a patient with idiopathic generalized epilepsy (juvenile myoclonic epilepsy).

zone, seizure onset zone, and of eloquent brain regions (Figure 4.5). Electrode placement in the epileptogenic region is believed to be mostly accompanied by patterns I through III (Figures 4.6 and 4.7) (Bancaud et al. 1973; Alarcon et al. 1995; Toczek et al. 1997) as the initial sign, whereas sinusoidal waves—particularly of low frequency—indicate that some propagation and recruitment of more extended networks has already occurred. The electrodecremental pattern (type I) sometimes poses problems in reliability of identification as amplitude modulation often occurs physiologically or secondary to clinical seizure manifestations. This is particularly the case in changes in vigilance during epileptic arousal or secondary to eye opening or closure secondary to an ictal behavioral change.

Characteristically, ictal activity shows an evolution between and within patterns. Seizures may thus be divided into stages with a predominant pattern. Transitions between patterns may be I>III> IV, II>V>IV, etc. (Figure 4.8). Factors influencing the spread of epileptic activity include the localization and connectivity of the onset area (Goetz-Trabert et al. 2008) and the underlying neuropathology (Spencer et al. 1992). Not uncommonly, seizures with extensive spread of ictal activity may by asynchronous at different cortical areas, thus a pattern type III may be prominent in one region

TABLE 4.1
Commonly Observed Patterns in Surface EEG during Focal Epileptic Seizures

I. Attenuation or amplitude depression of ongoing EEG activity

II. Low-amplitude, high-frequency discharges

III. Repetitive spiking or sharp waves

IV. Sinusoidal waves at various frequencies

V. Sharply contoured theta waves

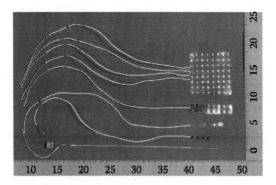

FIGURE 4.5   (See color insert.) Commonly used intracranial electrode types—subdural grids (top), strips, and depth electrodes (bottom)—for invasive presurgical evaluation.

when there is a pattern type IV in contralateral cortical areas. Rapid spread may indicate more extended epileptogenicity (Kutsy et al. 1999).

Within a pattern, there is a typical evolution with an increase in amplitude, a decrease in frequency of type IV patterns representing increasing synchronization of neuronal pools involved in the generation of ictal discharges, and regional spread indicated by an increasing extension of electrodes capturing the ictal pattern. Within a given seizure, ictal patterns may differ between brain regions, reflecting differences in the physiology of networks involved.

The area of cortical tissue that should be involved in order to show epileptic activity in the surface EEG is considered to be 6 to 15 cm^2 (Figure 4.1a and b) (Pacia and Eversole 1997; Tao et al. 2007; Hashiguchi et al. 2007; Ramantani et al. 2010). Thus, considerable spread of epileptic activity has to occur before an interictal or ictal epileptiform pattern becomes clearly recognizable. A given pattern represents the dynamics of extended networks rather than of only the epileptic focus. For example, rhythmic theta patterns frequently observed in patients with mesial temporal lobe epilepsy due to hippocampal sclerosis evolves when hippocampal-neocortical networks are involved during a seizure (Figure 4.8) (Risinger et al. 1989). Possible exceptions to this are high-frequency spikes and gamma activity recorded locally in patients with neocortical seizure generators with cortical dysplasia.

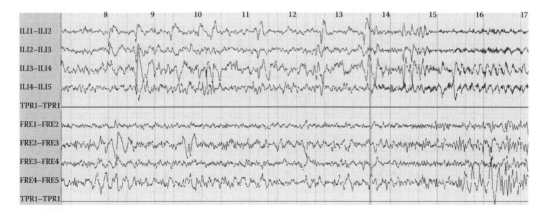

FIGURE 4.6   Ictal neocortical onset as recorded with subdural electrodes: rhythmic (preictal) spiking gives way to low-amplitude fast activity in the upper contacts.

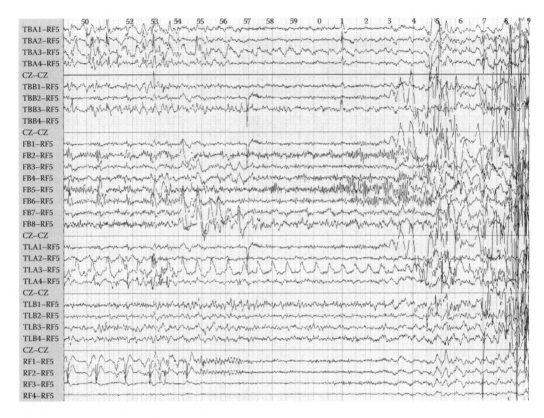

FIGURE 4.7 Ictal neocortical onset as recorded with subdural electrodes: rhythmic beta activity over electrode contact FB5 with rapid spread.

4.2.2 TIMING OF SEIZURE ONSET IN SURFACE EEG

Depending on the time to involvement of symptomatogenic areas of the brain, surface EEG patterns may become visible before or after the onset of clinical symptomatology, i.e., subjective seizure symptoms (auras) or objective semiological signs. The localizing value of surface EEG patterns thus depends considerably on the area and extension of ictal activity at seizure onset and on the speed and extent of spread. Whereas neocortical seizures with lateral onset may be visible early on surface EEG, those with mesiotemporal or temporobasal onset may have considerable latencies or not be visible at all (Tao et al. 2007). Some patients have a remarkably long period during which ictal activity remains in one cortical region (e.g., the temporal lobe), and even patterns occurring 10–20 s after the onset of clinical symptomatology may have a high localizing value (Risinger et al. 1989). Connectivity of the focus and recruitability of projection areas modulate propagation, and the initial surface EEG patterns may represent not only spread to the cortical surface of this region but to other lobes (e.g., in parietal or occipital foci) or even to the contralateral hemisphere (e.g., in mesially located seizure onset zones).

4.2.3 SUBCLINICAL ELECTROGRAPHIC SEIZURE PATTERNS

Ictal patterns without clinical correlates are often regarded as subclinical electrographic seizure patterns. There is, however, a considerable dependence on the exactness of clinical observation that may influence this distinction; often seizures without clear visible symptoms may be accompanied by vegetative signs (like changes in heart rate, flush, piloerection, sweating, etc.) or with subtle cognitive signs that become overt only when particular tests are performed (e.g., continuous monitoring of reaction times, tests on word retrieval, etc.). Figure 4.9 illustrates an example of a hippocampal ictal pattern associated with anomia. In focal epilepsy, subclinical patterns generally tend to be shorter in time and more limited compared to ictal patterns accompanied by clinical signs.

4.2.4 INTERICTAL-ICTAL TRANSITIONS

In most cases, the new onset of an ictal pattern can be clearly distinguished from the normal background EEG activity. Thus, the build-up of rhythmic discharges can be determined with a high interrater reliability within a range of a few seconds or less. Not uncommonly, the earliest visible EEG change is a sharp wave or a slow negative direct current (DC) shift (sometimes called an early

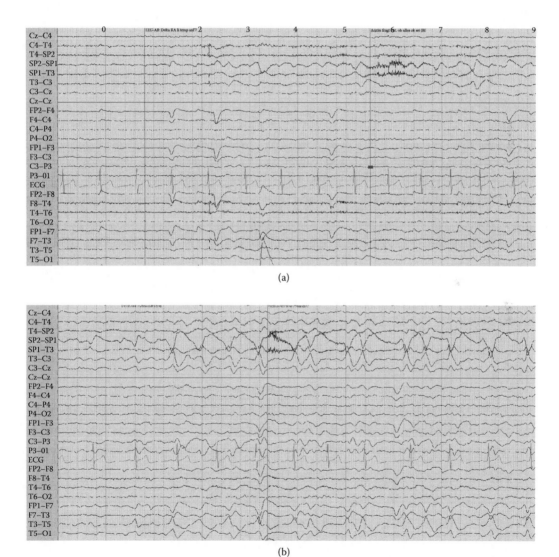

(a)

(b)

FIGURE 4.8 Transition of ictal patterns in a continuous surface recording (a through c): Initial rhythmic sinusoidal delta activity changes into rhythmic sharp waves and then to rhythmic 6-s activity over the left temporal region. Such a pattern is often observed in patients with temporal neocortical onset and spreads to the hippocampus.

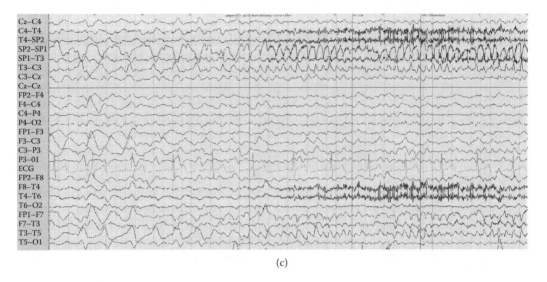

(c)

FIGURE 4.8 (Continued)

ictal event) (Alarcon et al. 1995) followed by an electrodecremental pattern with desynchronized low-amplitude fast activity that gradually synchronizes to build up a rhythmic pattern in the beta band and then slower rhythms of increasing amplitude (Figure 4.10) (Alarcon et al. 1995; Velasco et al. 2000, Kim et al. 2009).

Less well-defined, rhythmic spiking at a frequency of 1–2 s in an area that also shows abundant interictal spikes may occur as the first EEG chance indicative of an eminent seizure, particularly in a hippocampus with severe neuronal loss in the Ammon's horn (Velasco et al. 2000). As spiking

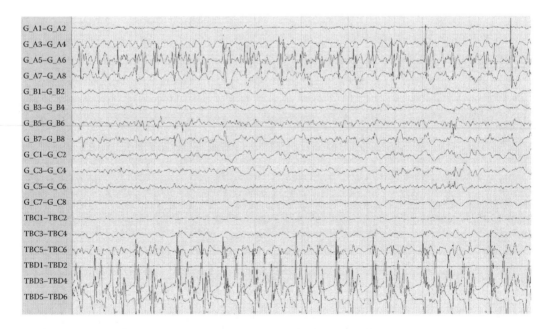

FIGURE 4.9 Continuous repetitive spiking in temporobasal and posterior channels that was associated with word-finding difficulties evident by appropriate testing.

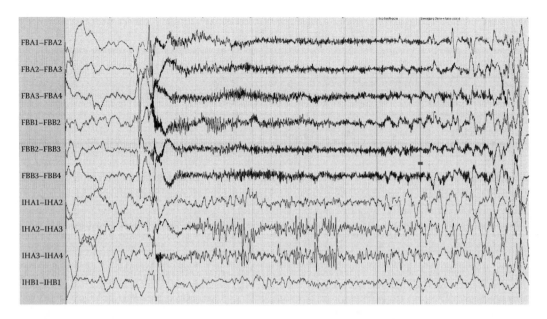

FIGURE 4.10 Seizure initiation by a sharp wave and slow depolarization followed by low-amplitude fast activity that rapidly spreads. Shown here are segments of subdural recordings with frontobasal seizure onset (upper channels) and spread to frontomesial regions (lower channels).

may occur at high frequency over long periods, there is presently no consensus as to what degree of rhythmicity of such spiking indicates that it will be followed by a seizure and if this repetitive spiking is to be considered part of the seizure or as an exceptional visible marker of a preictal phase (Spencer et al. 1992). Seizures with rhythmic spiking tend to spread more slowly and less often than those with initial low-amplitude fast activity; they typically evolve into a regional low-amplitude fast activity (Spencer 1999, Velasco et al. 2000).

In neocortical seizures, initial rhythmic delta activity, often with wide extension that may involve multiple lobes of the brain, is often found in the surface EEG (Tao et al. 2007; Blumenfeld et al. 2004), suggesting the spread of activity in the majority of cases. Further studies are needed to identify which network properties contribute to this pattern observed with subdural and surface recordings. Sampling with depth electrodes sometimes shows that local (e.g., sulcal) high-frequency discharges precede extended oscillatory activity recorded from gyral crowns.

Pragmatically, electrodes from which the first ictal patterns are recorded are regarded as the seizure onset zone, independent of the specific pattern. It is often assumed, however, that rhythmic oscillatory patterns indicate a certain distance to the initial seizure generator and that, due to restrictions in sampling by intracranial electrodes, these are not placed in the exact area where interictal-ictal transitions occur (Gotman et al. 1995). Although this may be correct, the recording of an initial low-amplitude high-frequency pattern does not guarantee that the electrodes capturing this are indicative of close proximity to a very local seizure generator as these patterns may also be recorded from relatively extended areas. Limited spatial sampling is a problem for the planning of epilepsy surgery, even if somewhat extended resections are performed, and poses a principal problem for definition of epileptogenic zones from a scientific view.

4.2.5 Practical Implications of Ictal Onset Zone and Interictal Spiking

In presurgical evaluation, the so-called epileptogenic zone, which needs to be resected to render a patient seizure free following epilepsy surgery, is an ill-defined concept. It is generally assumed to encompass seizure-onset zones, regions of consistent early ictal spread, regions with frequent independent interictal spiking, and structurally abnormal brain regions (Lüders et al. 1993). As seizure outcome is a post hoc parameter, there is uncertainty as to how extended such resections have to be, and most correlations to any of the parameters used for defining the resection zone are valid only at a group level and do not predict individual outcome.

4.2.6 Seizure Termination

Termination of seizures is characterized by resolution of rhythmic activity, often after a phase of sharp slow waves, and the appearance of slow or flat background activity that may show interictal discharges. At variable latency and often last in the region where ictal activity was particularly prominent, the usual interictal activity reappears (Figure 4.11). Prolonged persistence of ictal activity in regions different from the seizure onset area that outlast ictal discharges in the seizure onset zone may point to an additional epileptogenic potential of these regions (Spencer and Spencer 1996).

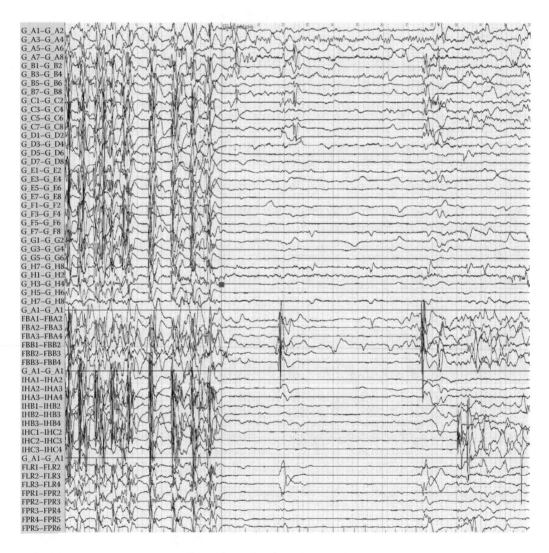

FIGURE 4.11 Brief postictal amplitude depression in an intracranial subdural recording following widespread ictal activity (left) with regionally different reappearance of interictal activity (20 s displayed).

4.2.7 EFFECTS OF ANTIEPILEPTIC DRUGS ON EEG PATTERNS

The effects of AEDs on EEG patterns have not been studied sufficiently. In idiopathic generalized epilepsy, AEDs reduce the duration and frequency of spike wave complexes, and abortive complexes with less well-defined rhythmicity and variable amplitudes of spikes and waves appear (Rocamora et al. 2006; Oehl et al. 2010). Focal epileptic discharges, in contrast, are influenced only by some AEDs, whereas others may accentuate interictal spiking even if they effectively control seizures (e.g., carbamazepine). They may, however, be partially or completely suppressed and less extended in their field and distribution, depending on the type of medication given. AEDs may also affect the spread of ictal activity (Tilz et al. 2006) and the duration of postictal EEG changes. Changes in medication during periods of monitoring are considered a common cause of nonstationarity in the majority of available long-term recordings in epilepsy patients.

4.2.8 PATTERNS TO BE DISTINGUISHED FROM ICTAL ACTIVITY

Variants of normal EEG activity that are not related to ictal activity include rhythmic midtemporal theta of drowsiness (RMTD) and subclinical rhythmic epileptiform discharges of adulthood (SREDA). RMTD consists of bursts of 4- to 7-Hz theta activity frequently notched that occur unilaterally but, at times, with changing lateralization over temporal regions. SREDA is a rhythmic pattern with sudden onset and offset occurring in the parietal and temporal regions, often at 4–5 Hz. The pattern remains highly stereotyped without any evolution and is not accompanied by any clinical symptomatology.

Not infrequently, artifacts occur during the long-term EEG recordings related to rhythmic movements or to electrode dysfunction. Such patterns often involve only one electrode or electrodes that do not cover a physiologically plausible field. Moreover, rhythmic activity generally lacks the evolution of frequency, amplitude, and field described earlier. Artifacts have a high variability of visual appearance, however, and often need detailed analysis for proper recognition.

REFERENCES

Alarcon, G., C. D. Binnie, R. D. C. Elwes, and C. E. Polkey. 1995. Power spectrum and intracranial EEG patterns at seizure onset in partial epilepsy. *Electroenceph. Clin. Neurophysiol.* 94:326–337.

Asano E., O. Muzik, A. Shah, C. Juhász, J. Shah, D. C. Chugani, O. Muzik, S. Sood, and H. T. Chugani. 2003. Quantitative interictal subdural EEG analyses in children with neocortical epilepsy. *Epilepsia* 44:425–434.

Bancaud, J., J. Talairach, S. Geier, and J. M. Scarabin. 1973. *EEG et SEEG dans les tumours cérébrales et l'épilepsie*. Paris: Ed. Edifor.

Blume, W. T., G. B. Young, and J. F. Lemieux. 1984. EEG morphology of partial epileptic seizures. *Electroenceph. Clin. Neurophysiol.* 57:295–302.

Blumenfeld, H., M. River, K. A. McNally, K. Davis, D. D. Spencer, and S. S. Spencer. 2004. Ictal neocortical slowing in temporal lobe epilepsy. *Neurology* 63:1015–1021.

Bragin, A., C. L. Wilson, J. Almajano, I. Mody, and J. Engel, Jr. 2004. High-frequency oscillations after status epilepticus: Epileptogenesis and seizure genesis. *Epilepsia* 45:1017–1023.

Bragin, A., C. L. Wilson, R. J. Staba, M. Reddick, I. Fried, and J. Engel, Jr. 2002. Interictal high-frequency oscillations (80–500 Hz) in the human epileptic brain: Entorhinal cortex. *Ann. Neurol.* 52:407–415.

De Curtis, M., and G. Avanzini. 2001. Interictal spikes in focal epileptogenesis. *Prog. Neurobiol.* 6:541–567.

Geyer, J. D., E. Bilir, R. E. Faught, R. Kuzniecky, and F. Gilliam. 1999. Significance of interictal temporal lobe ictal delta activity for localization of the epileptogenic region. *Neurology* 52:202–205.

Gibbs, F. A., H. Davies, and W. G. Lennox. 1935. The electroencephalogram in epilepsy and in conditions of impaired consciousness. *Arch. Neurol. Psychiatry* 34:1133–1148.

Goetz-Trabert K., C. Hauck, K. Wagner, S. Fauser, and A. Schulze-Bonhage. 2008. Spread of ictal activity in focal epilepsy. *Epilepsia* 49:1594–1601.

Gotman J., V. Levtova, and A. Olivier. 1995. Frequency of the electroencephalographic discharge in seizures of focal and widespread onset in intracerebral recordings. *Epilepsia* 36:697–703.

Gotman, J., and M. G. Marciani. 1985. Electroencephalographic spiking activity, drug levels, and seizure occurrence in epileptic patients. *Ann. Neurol.* 17:597–603.

Hashiguchi, K., T. Morioka, and F. Yoshida. 2007. Correlation between scalp-recorded electro-encephalographic and electrocorticographic activities during ictal period. *Seizure* 16:238–247.

Jacobs, J., P. Levan, C. E. Châtillon, A. Olivier, F. Dubeau, and J. Gotman. 2009. High frequency oscillations in intracranial EEGs mark epileptogenicity rather than lesion type. *Brain* 132:1022–1037.

Jasper, H. H. 1958. The ten-twenty electrode system of the International Federation. *Electroenceph. Clin. Neurophysiol.* 10:371–375.

Jirsch, J. D., E. Urrestarazu, P. LeVan, A, Olivier, F. Dubeau, and J. Gotman. 2006. High frequency oscillations during human focal seizures. *Brain* 129:1593–1608.

Kim, W., J. W. Miller, J. G. Ojemann, and K. J. Miller. 2009. Ictal localization by invasive recordings of infraslow activity with DC-coupled amplifiers. *J. Clin. Neurophysiol.* 26:135–144.

Kutsy, R. L., D. F. Farrell, and G. A. Ojemann. 1999. Ictal patterns of neocortical seizures monitored with intracranial electrodes: Correlation with surgical outcome. *Epilepsia* 40:257–266.

Lieb, J. P., G. O. Walsh, T. L. Babb, R. D. Walter, and P. H. Crandall. 1976. A comparison of EEG seizure patterns recorded with surface and depth electrodes in patients with temporal lobe epilepsy. *Epilepsia* 17:137–160.

Lüders, H. O., J. Engel, Jr., and C. Munari. 1993. General principles. In *Surgical treatment of epilepsies,* 2nd ed. J. Engel, Jr., ed., 137–153. New York: Raven Press.

Meeren, H., G. van Luijtelaar, F. Lopes da Silva, and A. Coenen. 2005. Evolving concepts on the pathophysiology of absence seizures. The cortical focus theory. *Arch. Neurol.* 62:371–376.

Oehl, B., K. Götz-Trabert, A. Brandt, C. Lehmann, and A. Schulze-Bonhage. 2010. Latencies to first typical generalized spike-wave discharge in idiopathic generalized epilepsies during video-EEG monitoring. *J. Clin. Neurophysiol.* 27:1–6.

Ojeman, G. A., and J. Engel, Jr. 1987. Acute and chronic intracranial recording and stimulation. In *Surgical treatment of epilepsies,* 2nd ed. J. Engel, Jr., ed., 263–288. New York: Raven Press.

Pacia, S. V., and J. S. Ebersole. 1997. Intracranial EEG substrates of scalp ictal patterns from temporal lobe foci. *Epilepsia* 38:642–654.

Ramantani, G., D. M. Altenmüller, T. Ball, A. Schulze-Bonhage, and M. Dümpelmann. 2010. Subdural EEG substrates of scalp EEG interictal spikes in frontal lobe epilepsy. *Epilepsia* 51: 140.

Reiher, J., M. Beaudry, and C. P. Leduc. 1989. Temporal intermittent rhythmic delta activity (TIRDA) in the diagnosis of complex partial epilepsy: Sensitivity, specificity and predictive value. *Can. J. Neurol. Sci.* 16:398–401.

Risinger, M. W. J. Engel, P. C. van Ness, T. R. Henry, and P. H. Crandall. 1989. Ictal localization of temporal lobe seizures with scalp/phenoidal recordings. *Neurology* 39:1288–1293.

Rocamora, R., K. Wagner, and A. Schulze-Bonhage. 2006. Levetiracetam reduces frequency and duration of epileptic activity in patients with refractory primary generalized epilepsy. *Seizure* 15:428–433.

Spencer, S. S., D. Marks, A. Katz, J. Kim, and D. D. Spencer. 1992. Anatomic correlates of interhippocampal seizure propagation time. *Epilepsia* 33:862–873.

Spencer, S. S., P. Guimaraes, A. Katz, J. Kim, and D. Spencer. 1992. Morphological patterns of seizures recorded intracranially. *Epilepsia* 33:537–545.

Spencer, S. S., and D. D. Spencer. 1996. Implications of seizure termination location in temporal lobe epilepsa. *Epilepsia* 37:455–458.

Tao, J. X., M. Baldwin, A. Ray, S. Hawer-Ebersole, and J. S. Ebersole. 2007. The impact of cerebral source area and synchrony on recording scalp electroencephalography ictal patterns. *Epilepsia* 48:2167–2176.

Tilz, C., H. Stefan, R. Hopfengärtner, F. Kerling, A. Genow, and Y. Wang-Tilz. 2006. Influence of levetiracetam on ictal and postictal EEG in patients with partial seizures. *Eur. J. Neurol.* 13:1352–1358.

Tokczek, M. T., M. J. Morrell, M. W. Risinger, and L. Shuer. 1997. Intracranial ictal recordings in mesial frontal lobe epilepsy. *J. Clin. Neurophysiol.* 14:499–506.

Velasco, A. L., C. L. Wilson, T. L. Babb, and J. Engel, Jr. 2000. Functional and anatomic correlates of two frequently observed temporal lobe seizure-onset patterns. *Neural Plast.* 7:49–63.

Verma, A., and R. Radke. 2006. EEG of partial seizures. *J. Clin. Neurophysiol.* 23:333–339.

5 Seizures and Epilepsy: An Overview

Steven Weinstein

CONTENTS

The study of seizures and epilepsy requires a common language to describe what happens to the patient. Agreed upon definitions are necessary. Observations detailing seizure behavior remain the standard of care since EEG recordings are generally not available for a sufficiently long duration to capture an event in the great majority of patients. The patients undergoing the most intense study are usually the most difficult to control, often with significant comorbidity. Nonetheless, even if recordings were available, their interpretation would need to be replicable. This chapter will present some of the explanations as to why studies have significant variability.

5.1 EPILEPSY CATEGORIZATION

The word *epilepsy* is derived from a Greek word meaning "take hold of" or "seize," as a consequence of the belief that an individual having seizures was being possessed by the gods or evil spirits (Reynolds 2000). Today epilepsy is conceptualized as a manifestation of an underlying chronic neurologic condition characterized by an "enduring predisposition to generate epileptic seizures and their associated consequences." A seizure presents as a paroxysmal clinical event occurring in conjunction with an abnormal cortical electrographic discharge: "a transient occurrence of signs and/or symptoms due to abnormal excessive or synchronous neuronal activity in the brain" (Fisher 2005). There is a stereotypic electrographic and/or behavioral onset, evolution, and conclusion. Epilepsy is not defined by the severity of the events nor its prognosis.

The incidence of epilepsy varies greatly in differing populations, in part due to the inadequacy of clinically reliable observations and accurate reporting, as well as the need to compare frequencies at differing ages. Further, if the epilepsy is a consequence of an acquired brain abnormality, it is important to know when the disruption of normal function occurred, its site, severity, adequacy of therapy, as well as duration of follow-up. In assessing whether the disorder is active (intractable), the availability of antiepileptic drugs, their appropriate use, and the patient's socioeconomic status need to be considered. Genetic susceptibility to epilepsy and if, secondary to injury, knowledge of triggered acute injurious cascades and repair processes could alter the development of the disorder. The impact of these factors is evident in Banerjee's review of the worldwide literature finding significant variability in incident rates of 16/100,000 to 111/100,000 and age-adjusted prevalence rates of 2.2/1000 to 41/1000 in door-to-door surveys (Banerjee 2009).

These conditions are sufficiently nebulous to prompt the International League against Epilepsy (ILAE) to establish criteria for classifying both seizures and epilepsies that would be of practical use by clinicians to "facilitate communication and clinical and epidemiological investigation. Indeed it may encourage further research . . . [but] do not, in the least, [to] consider it fixed for all time . . . as our knowledge increases, it will undoubtedly require modification . . . . " (Merlis 1970). Clinical and EEG criteria are considered in defining the epilepsy (syndrome): 1. Ictal phenomenology; 2. Seizure-type pathology; 3. Epilepsy syndrome; 4. Etiology; and 5. Other impairments (Engel 2001). Debate regarding the fundamental basis of this classification proposal has revolved primarily around the need for detailed seizure descriptions with the requirement for recording the seizure as well as the use of the construct of the epilepsy syndrome (Engel 2005). Multiple iterations of categorization have subsequently been presented with the most recent, changing some terminology and concepts, but "not having a tangible impact on how clinicians use electroclinical syndromes" (Berg 2010). Epilepsy syndromes are a construct to describe these multiple features and allow for concise and simple communication of the patient's disorder. The transient nature of categorization by assigning seizure type and epilepsy syndrome is demonstrated in the pediatric entities with age-dependent evolution (see below).

5.2   SEIZURES AND EEG

Seizures are categorized as having either an apparently generalized onset (simultaneously involving widespread cortical structures over both brain hemispheres and subcortical regions) or a partial onset (arising in a limited brain region, cortical or subcortical, with variable degrees of spread). This distinction is important for guiding the diagnostic evaluation and prescribing appropriate therapy. Correct identification of the seizure onset characteristics can be difficult because the patient often has a poor memory of the initial features and accurate observations of others are lacking. These historical details are problematic and are further complicated by the onset not being witnessed or recognized, poor environmental conditions for detailed observations that are further clouded by the caretaker's fears, lack of specific testing for altered consciousness by a bystander, and a tendency to only remember the more dramatic features of the seizure. At times, the epileptic event cannot be distinguished from nonepileptic events that appear similar (Benbadis 2009). There is an assumption that physicians or other trained personnel who witness the event without simultaneous EEG will improve seizure description and categorization. This is correct for the most part; however, a significant number of events remain of uncertain origin.

Interictal EEG has served as the cornerstone of epilepsy evaluation. The interictal EEG does allow for reasonable classification of epilepsy syndrome, however, the routine recording does not always accurately identify the nature of a seizure; normal individuals may have epileptiform activity (isolated or multifocal) and patients with seizures can be normal (Eeg Olofsson 1971; Cavazzuti 1980; Asano 2005). The requirement for actually capturing the seizure is governed by whether there is an absolute need to know if it is generalized or partial, a nonepileptic event, or for planning epilepsy surgery.

Clinical interictal EEG can nonetheless be meaningful, but there should be agreement in its interpretation among multiple readers. There have been several attempts to demonstrate replicability of the EEG interpretations; however, since neurologists without formal training read most recordings, these investigations may not be relevant for the greater community. The quality of the tracing is important, with identification of the epileptic zone dependent upon the presence of specific interictal waveforms that require artifact-free traces of sufficient duration to include both wakefulness and sleep (Asano 2005). The introduction of digital EEG has improved interrater reliability of EEG interpretation but it is not perfect, even using the simple classification of normal, equivocal, or abnormal (Levy 1998). However, weighted kappa statistics among four observers was far better for the digital recordings as compared to paper traces in categorizing the EEG as normal. Agreement fell to the fair to good range across the readers using only the digital tracing, with or without reformatting, to describe whether it was focal. Agreement on the presence of epileptiform activity as well as lateralization and localization of the focal features was not reported.

Performing a similar assessment of interrater reliability in classifying seizures is dependent upon capturing an event necessitating that the patient have sufficiently frequent seizures to prompt continuous monitoring. These recordings are most often done to assess paroxysmal behaviors following acute brain injury, presurgical evaluations during antiepileptic drug (AED) weaning, or in individuals with epileptic encephalopathies or frequent seizures (often of a mixed variety). Outcome measures should include a comparison of both lateralization and localization of the seizure onset and, if necessary, comparison of extracranial and intracranial electrodes.

Although in many circumstances extracranial recordings may be adequate for identification of the required seizure features, intracranial electrodes may be necessary to assist in defining its site of onset and evolution. Any improved sensitivity in identifying seizures may, however, be at the expense of lack of specificity. This is frequently seen in recordings obtained in the ICU (Carrera 2008; Chong 2005). There can be slow rhythmic activity and runs of epileptiform discharges, including stimulus-induced epileptiform discharges—neither of which are associated with behavioral changes. These electrographic events, however, may just represent epiphenomena reflecting underlying episodic alterations in brain metabolism, blood flow, or electrolyte/water homeostasis and reflect acute brain dysfunction that might precede the onset of the seizure. Similar concerns are discussed below in children with Lennox Gastaut syndrome or electrographic status epilepticus in sleep.

Agreement of the zone of seizure onset among three interpreters was studied in 35 patients with extracranial recordings. Walczak calculated kappa statistics that were fair to good for temporal lobe, but poor for extratemporal lobe seizures, in part explicable by muscle artifact in many of the traces (Walczak 1992). Comparing extracranial to intracranial recordings by three independent reviewers of 54 patients, Spencer found 58%–60% agreement for lobe of seizure onset and 64%–74% for lateralization (Spencer 1985). Overall accuracy for scalp recordings compared to depth recordings was 21%–38% for lobe and 46%–49% lateralization. About one third of the extracranial record seizure onsets were not correctly identified. However, surgical outcomes to validate the depth recordings were not reported. Haut reported a multicenter study of nonrandomized patients with intractable seizures undergoing evaluation for surgery (Haut 2002). Temporal lobe onsets were found in 87.6% and extratemporal in 12.4%. Interrater reliability among the six centers demonstrated that extracranial EEG lateralization correlation was excellent, but only good for localization, whereas the intracranial recording agreement was excellent.

The attempt to capture seizures for categorization is dependent upon their overall frequency and whether they cluster and also upon understanding that patients undergoing monitoring may not reflect what occurs in the majority of outpatients (Haut 2006; Asano 2005). This has relevance for studies investigating the preseizure or seizure-permissive state. In a community-based study, Hart distributed questionnaires through practitioners to 2528 patients receiving AEDs (Hart 1995). Seizure frequency ranged from none in the previous year (46%) to one or two (33%) and more than 50 (8%). It would appear that during a brief period of monitoring it would be unlikely that the preictal state could be distinguished from the interictal state and, if captured, may reflect proximity to the previous seizure. Furthermore, if seizures were truly random then studying prediction on most patients would be meaningless. However, patients frequently argue that their events are not random and, although rarely identifying a specific precipitant, they believe that stress, sleep deprivation, menses, infection, or fever predispose them to their seizures (Petitmengin 2006; Haut 2007; Fang 2008; Sperling 2008). These states last hours and/or days and, therefore, should not be viewed as the trigger; however, performing prolonged continuous EEGs during these states and comparing them to those recorded in times of wellness might prove informative.

The inability to obtain electrographic confirmation should not deter treatment or further studies. Seizure description may be all that is available. An attempt to prove reliability of the clinical description of the event was performed in semistructured telephone interviews (Ottman 1993). Trained lay personnel contacted 1957 probands, as well as first-degree relatives who may have witnessed the event. A subset received calls by a study neurologist. There was "substantial to almost perfect" agreement in the overall group of partial onset seizures (although poor for simple partial), and only "moderate to substantial" for the group of generalized seizures (with a substantial difference between neurologist and lay reviewers for myoclonic and atonic seizures). Direct observation of 66 patients with nocturnal frontal lobe seizures during polysomnography by physicians demonstrated only fair to substantial agreement for seizure versus no seizure. Paroxysmal arousals accounted for most of the disagreement but polysomnogram data was not reported (Vignatelli 2007). Pediatric seizures schema classifications fare no better. Kappa statistics were calculated for a senior neurologist compared to residents on seizure classification based upon clinical description with relatively poor agreement noted (Bodensteiner 1988). The addition of video EEG in infants less than 26 months of age still provided poor agreement of seizure classification based on clinical observations, alone with the EEG somewhat improving focal and generalized onset identification (Nordli 1997).

Although there is fair agreement data scoring observational studies, the very importance of the observations has been questioned (Hoppe 2007). Hoppe evaluated the accuracy of self-identification of seizure occurrence in 91 patients in a monitoring unit. Only 45% of 582 partial seizures were self identified, including only 58.3% of those that secondarily generalized. Sleep accounted for most of the underreporting, although 33% of seizures during wakefulness went unreported. The investigators did not utilize a family member, as would happen at home, to simultaneously report the seizure occurrence. This discrepancy in EEG data and observations raises questions regarding the validity of seizure diaries, both in seizure prediction and demonstrating efficacy of therapies, especially adults living by themselves.

5.3 NATURAL HISTORY

Short-term variability in seizure frequency may increase with behavioral and chemical alterations, often in association with new significant illnesses or drug exposure (Brust 2007; Hesdorffer 2009). The seizure and epilepsy presentation can be impacted by maturation, especially in infancy given the rapidity of functional and anatomic development.

The presence of seizures and/or epilepsy does not necessarily predict a life-long need for medication. Epilepsy may be a self-limited condition that is expressed during a specific developmental phase and may resolve over several years. This process may reflect changing gene product expression, molecular maturation of channels, anatomic developmental changes, or altered exposure to environmental triggers (Hauser 1992; Lu 2004; Berkovic 2006; Giedd 2009; Liao 2010). Distinguishing pharmacoresistant epilepsy at onset from a transient disorder may be difficult. Epilepsies that are an expression of a structural or degenerative process are more likely to persist, whereas some of the childhood epilepsy syndromes will resolve, as do individuals responding early to medication (Spooner 2006; Berg 2009; Sillanpää 2009). However, the natural history of epilepsy is highly variable among individuals, even with the same seizure semiology and epilepsy syndrome.

Kwan and Brodie followed a referral population of 525 patients for a median of 5 years (range 2–16 years) with only 63% seizure-free at the last follow-up (the frequency of continuing seizures was unreported) (Kwan 2000). At the last visit, the most frequently used AEDs were carbamazepine (30%), valproate (24%), and lamotrigine (19%). Patients were considered to be free of seizures if they had not had an event for a minimum of 1 year, either off AEDs or on AEDs but without a change in dosage during this interval. Compliance was monitored, but at the time of a breakthrough seizure drug levels are not mentioned. The refractory group represents the minority of patients, but probably constitutes those who are best known to neurologists and are suffering the greatest morbidity (Birbeck 2002). Although debatable whether the seizures do biologic harm, there is no doubt they produce significant psychosocial and economic consequences with an impaired quality of life.

Pharmacoresistant epilepsy is probably the term to describe the latter intractable group. There is, however, no general agreement when epilepsy should be deemed pharmacoresistant and surgery should be considered (Choi 2008; Berg 2009). The generally accepted criteria include failure of at least two drugs (if not more) within a relatively short period of time, such as 18 months to 2 years. However, some patients with devastating epilepsy will eventually come under control. Huttenlocher reviewed his clinical practice of children having at least 2 years of continuing seizures and with a 5- to 20-year follow-up (Huttenlocher 1990). Children with borderline to normal intelligence (≥70) became seizure free or had at most one seizure per year at a rate of 4% per year compared to a rate of 1.5% for those with mental retardation; this perhaps reflecting more diffuse and profound brain dysfunction. Approximately 25% continued to have >1 seizure per year. In a broader population followed for a longer period, Sillanpää tracked 102 children in Finland for a mean of 40 years (range 11–42 years) (Sillanpää 2006). Terminal remission at last follow-up was found in 67%, with or without medication, and there was pharmacoresistance in 19% (2006). Some had a remitting-relapsing course with a small subgroup (5.6%) having remission followed by recurrence to intractability. The latter finding raises the possibility that patients never truly "outgrow" their epilepsy, and possibly should not be weaned off their medication. There is no evidence that the relapse could have been prevented with continued therapy (Sillanpää 2006).

Several factors modulate the predisposition for seizures during development and the threshold at which they occur. An acute brain insult has the potential to trigger a seizure in any person, but not every person will have a seizure when presented with the same insult. This differing susceptibility reflects genetic and acquired factors effecting structural and chemical brain interactions (Frucht 2000; Briellmann 2001; Sanchez 2001; Baulac 2009; Andrade 2009). The age of the child, reflecting the brain development stage, also influences the seizure threshold, perhaps explaining why some children "grow into" and "outgrow" epilepsy and it might also explain the age dependency of seizure semiology at differing developmental states (e.g., neonatal period or adolescence). There are maturational morphologic and chemical changes in networks (Kriegstein 1999). Alterations in typical brain development could affect the subsequent cascade of processes leading to brain dysfunction and predispose the child to epilepsy, either from the primary insult or as a consequence of a reparative process.

There is much debate about the potential adverse effects of clinical seizures on brain structure and function (Hermann 2008). Motor and certain seizures frequently produce cyanosis, but it could be cellular energy failure, initiation of excessive apoptosis/necrosis, network/cellular reorganization, or transformation of genetic expression that leads to worsening seizures and other morbidity. An example is the controversy regarding the consequences of febrile seizures. It could be argued that morbidity is produced by any seizure (even the first) or is due only to status epilepticus, a specific virus producing an encephalitis, the height of fever, the nature of the host's immunologic response, or a preexisting abnormality (Millichap 1959; Marusic 2007; Theodore 2008; Shinnar 2008). Some of these seizures in a small group of children seem to be associated with mesial temporal sclerosis with network reorganization that can lead to memory difficulties and new learning deficits. Whereas AEDs can inhibit febrile seizures, they do not protect the brain from pathological reorganization or from epileptogenesis (Tempkin 2001; Rakhade 2009). These processes in febrile seizures are not necessarily generalizable to other seizures and epilepsies.

Further complicating understanding of the natural history of epilepsy is the instability of the diagnosis of seizure and epilepsy type as children mature. For example, the majority of infants with infantile spasms, 50%–60%, transition to syndromes of partial seizures (cryptogenic or symptomatic) or generalized syndromes (Dulac 2002). Shinnar followed a cohort of 182 children aged 1 month to 19 years for at least 2 years from the first seizure (Shinnar 1999). The epilepsy syndrome classification was changed in 33 children (18%), especially in those with frequent seizures. The EEG also has a similar age-dependent evolution. Hughes reviewed 1645 interictal EEGs in 224 patients with intractable epilepsy (Hughes 1985). The first recording was obtained at a mean age of 15.8 ± 12.5 years, and patients were followed for a minimum of 15 years. In 6–8 years, the spike focus appeared to change in 89.3% of patients. Temporal lobe foci tended to become bilateral and there was an equal incidence between anterior to posterior foci migration. The relationship of these changes to seizure semiology is not well defined given the unavailability of continuous EEG recordings. This migratory behavior of the presumptive focus gave pause to doing resective surgery early in life.

5.4 CLASSIFICATION OF EPILEPSY AND EPILEPSY SYNDROMES

5.4.1 PARTIAL SEIZURES

Partial seizures are the most common type of seizure disorder in childhood, accounting for almost 60% of all cases (Shinnar 1999). These seizures often start with an aura and/or an abrupt and unprovoked alteration in behavior. A simple partial seizure may be limited to a small region of one hemisphere with maintenance of normal alertness. The event becomes complex partial as it spreads and alters consciousness. This type of seizure may occur with or without subsequent secondary generalization into a generalized motor seizure. The distinction between partial seizure groups, however, has been recently questioned by the ILAE (Berg 2010).

The signs and symptoms of complex partial seizures are the result of the brain's impaired ability to rapidly plan motor activities and process environmental stimuli. Typically, the movements become purposeless and slowed, the earliest signs providing a clue to the region of onset (Rossetti 2009). Frequently temporal lobe onset is accompanied by motor automatisms including eye blinking, lip smacking, facial grimacing, groaning, chewing, and unbuttoning and buttoning of clothing. During other events, especially with frontal onset, the individual may appear agitated.

Surround inhibition may allow the brain to confine the epileptic event to a limited area and its capacity to assist in extinguishing the seizure may explain the limited spread and duration of most seizures (Van Paesschen 2007). When these mechanisms fail, seizures may become generalized or prolonged.

5.4.2 PRIMARY GENERALIZED SEIZURES

Primary generalized seizures occur less often than partial seizures during childhood (Shinnar 1999). These seizures do not have a recognizable focus of onset; instead, large areas of each cortex (or subcortical structures) appear to be simultaneously affected. These seizures may be hypomotor with decreased motor activity (e.g., absence seizures) or occur with vigorous abnormal motor behaviors (e.g., myoclonic, tonic, and tonic-clonic seizures), or loss of postural tone only (atonic).

5.4.3 ABSENCE SEIZURES

Absence seizures—especially childhood onset syndromes—occur frequently throughout the day; nonetheless, they are among the most benign seizure type (Loiseau 1992; Hughes 2009). They appear to arise simultaneously in broad regions of the brain, mediated in part through the thalamus. The onset is usually between 3 and 12 years old and characterized by a brief (usually <30 s) behavioral arrest with impaired consciousness. During this time, a characteristic 3-Hz spike and wave discharge occurs on EEG with no or minimal dysfunction postictally. During the seizure, the child may continue to perform a simple motor activity, walking or looking at something, but is unable to rapidly respond to a novel task. Rarely is the behavioral arrest just a motionless blank stare; it typically occurs with a glazed eye appearance, associated eye blinks, and/or changes in head and extremity tone. Unlike simple daydreaming, absence seizures are difficult to interrupt by verbal or tactile stimulation (Loiseau 1992). It may not be the biologic consequences of this seizure but its frequency that interferes with learning, the susceptibility to harm from physical hazards such as continuing to walk while being unaware of the environment, or the psychosocial impact when witnessed by family and friends.

Longer absence seizures and those associated with prominent motor activity may be a manifestation of a more complex absence epilepsy syndrome such as juvenile myoclonic epilepsy (JME), atypical absence epilepsy seen with Lennox Gastaut syndrome (LGS), or eyelid myoclonus (Jeavon's syndrome). Absence epilepsy may evolve into an epilepsy that includes generalized tonic-clonic seizures that initially appear in adolescence or adulthood (Mayville 2000).

5.4.4 MYOCLONIC AND ATONIC SEIZURES

These seizures may have positive or negative motor components. Myoclonic seizures are lightning fast motor events frequently involving neck, trunk, or extremities. Some are subtle—with just head nodding—whereas others appear to abruptly pull the child over. Myoclonic, brief tonic, atonic, and any other type of seizure that leads to an abrupt loss of posture constitute a drop seizure, often occurring during a polyspike discharge or period of abrupt attenuation of the EEG. Consciousness is usually impaired and there is no self-protective behavior during the fall, commonly resulting in head trauma. These seizures, like the absence, occur frequently, often in clusters, and are exceedingly difficult to control. Individuals with these epilepsies typically have significant cognitive impairment, wear a protective helmet, and have numerous physical signs of trauma (see *Infantile Spasms and Myoclonic Syndromes*). The public often perceives this child to be the prototype "epileptic."

5.4.5 TONIC-CLONIC AND TONIC SEIZURES

Tonic-clonic and tonic seizures can arise focally and then secondarily generalize or appear to simultaneously involve both hemispheres due to rapid electrographic propagation. Clonic motor activity often occurs within the same seizure, sometimes starting with a Jacksonian march as the electrographic discharge spreads in an orderly sequence from one region of the motor cortex (e.g., that controlling the thumb) to contiguous regions such as those controlling the hand, forearm, arm, and face. An ensuing period of transient weakness (Todd's paralysis) after the end of the seizure may be misinterpreted as a stroke or other acute brain injury (Szabo 2000). During the seizure, observers frequently hear raspy breathing or see no breathing effort at all, and mistakenly assume the tongue has been swallowed. Tonic-clonic seizures are often associated with an unusual cry, cyanosis, and incontinence. A period of sleep is required for recovery, and the individual usually has no memory of the event.

5.5 EPILEPSY SYNDROMES

5.5.1 BENIGN PARTIAL SEIZURES

The ILAE has questioned the terminology of catastrophic (too many emotional overtones) and benign (minimizes concerns regarding associated cognitive and behavioral difficulties, although not necessarily caused by the seizures) epilepsy (Berg 2010). Regardless, there are increasing numbers of these syndromes being described with the distinctions becoming increasingly small.

Typically benign partial seizures appear around the age of school entry and remit by adolescence, although many of these syndromes can be recognized in infancy (Chahine 2006a, 2006b). The children have previously been developing normally but underlying metabolic and structural etiologies need to be excluded. Their benign natures can only be suspected, but not proven, until seizures are easily controlled and eventually disappear as the child matures. There is frequently a characteristic EEG that can be mistaken for a more serious disorder (epileptic encephalopathy) given the abundant spikes seen during sleep. A strong genetic history of benign seizures in near relatives is reassuring and these syndromes are unassociated with major long-term sequelae.

Rolandic epilepsy is the most common benign epilepsy syndrome of childhood. The seizures typically start in the central-temporal motor cortex that controls tongue, face, and hand movements. The presence of the interictal discharges is of great interest to investigators because they are abundant, especially in sleep, and there appears to be a horizontal dipole. The seizures occur most commonly at sleep onset or at arousal and often become rapidly generalized. When seizures start in sleep, the partial component is unobserved and children may be interpreted as having a primary generalized tonic-clonic seizure (Lerman 1992). This epilepsy is associated with numerous childhood static encephalopathies such as attention deficit or learning disabilities.

Other benign partial seizures may arise from the occipital lobe. Seizure semiology during wakefulness is frequently associated with visual disturbances (Gastaut type) and may be triggered by photic stimuli. At night, the occipital seizures can simulate a migraine-like event, with vomiting and severe headache (Panayiotopoulos type). Another expression of occipital epilepsy is infrequent episodes of status epilepticus expressed primarily with autonomic features (Chahine 2006).

5.5.2 INFANTILE SPASMS AND MYOCLONIC EPILEPSY SYNDROMES

Myoclonic epilepsy is typically a nonspecific manifestation of a serious underlying condition; however, both benign nonepileptic and epileptic myoclonus can be present in infancy (Dara 2006; Marx 2008; Caraballo 2009). Cryptogenic and idiopathic myoclonic epilepsy is strongly associated with a family history of epilepsy and an EEG showing spike and polyspike and wave discharges.

Infantile spasms are a separate syndrome with myoclonus occurring in flurries of brief flexor or extensor seizures, a unique high-voltage chaotic EEG background, and an ictal signature of slow waves followed by diffuse attenuation that typically cluster upon arousal from sleep but also sporadically throughout the day. The spasm occurrence may be precipitated by a concomitant complex partial seizure (Carrazana 1993). Onset is typically between 4 and 8 months of age and is difficult to control with standard AEDs (Hancock 2008). More than three-quarters of infants have a defined underlying disorder such as tuberous sclerosis or other cerebral dysplasias, genetic or metabolic abnormalities, or acquired brain disorders from hypoxia-ischemia or early CNS infection. A small percentage of the children have cryptogenic spasms and rapidly respond to therapy. Both groups have the same initial seizure semiology, EEG, and interictal behavioral phenotype of increasing lethargy/irritability, poor visual fixation, and plateau/loss of milestones, but they differ in outcome. The symptomatic group develops cognitive impairment and frequently evolves into the LGS, whereas the cryptogenic group can frequently have normal development and no subsequent seizures.

LGS is also associated with many of the underlying disorders seen in spasms. The EEG demonstrates a characteristic pattern of generalized slow spike and wave and multiple seizures types including nocturnal tonic as well as awake absence/atypical absence, drop (including myoclonic), tonic-clonic, and complex partial seizures. There are associated intellectual and behavioral disturbances. Phenotypically, the entity is similar to multifocal encephalopathy that does not prominently express the generalized discharges (Blume 1978; Yamatogi 2003). Both entities need to be distinguished from the more benign myoclonic-astatic of epilepsy of Doose, which may have similar seizures but enter remission between the ages of 3 and 8 years (Stephani 2006). This latter disorder is associated with the SCN1A gene, as are a wide range of other epilepsy syndromes.

JME also demonstrates myoclonic seizures. It commonly begins during adolescence as morning myoclonus and medical attention is frequently not sought until the individual experiences a generalized tonic-clonic seizure. JME is a benign genetic disorder and is not associated with intellectual disability.

5.5.3 Epileptic Encephalopathy

Landau-Kleffner syndrome, or acquired epileptic aphasia, is characterized by a progressive encephalopathy, the hallmark of which is loss of language skills. The disorder is manifested by an auditory agnosia, language regression (or the inability to attain language skills), and behavioral disturbances including inattention. Occasionally, the clinical picture resembles an autism spectrum disorder that has prompted many to perform prolonged EEGs in that population to exclude nocturnal discharges. Clinically evident seizures may be infrequent or absent but, when the child is asleep, the EEG pattern shows continuous abnormal epileptiform activity that obscures the normal sleep pattern. A similar electrographic pattern, electrographic status epileptics during sleep (ESES), can be seen in normal children, those with significant seizures, or in individuals with cognitive impairment (Tassinari 2009). Treatment of the epileptic encephalopathies—hypsarrhythmia, LGS, Landau-Kleffner, and ESES—is commonly not only focused on seizure control, but also on the normalization of the EEG.

5.6 TREATMENT

The decision to treat patients focuses on seizure prophylaxis and not suppressing the interictal discharges, except in certain epilepsies (epileptic encephalopathy). These interictal electrographic markers—spikes, polyspikes, and sharp waves—are statistically associated with seizures, but do not necessarily transform into them. These waveforms are assumed to represent underlying hypersynchronous network activity, excitatory or inhibitory, with that network having the capacity to generate a seizure under the appropriate clinical circumstance. What triggers the network to produce the seizure is unknown, even with study of the reflex epilepsies and the use of intracranial electrodes.

If, however, treatment is deemed necessary then which AED is most effective in preventing a seizure, much less suppressing interictal discharges, is uncertain. Due to methodological differences and the diverse populations involved, it is difficult to conclusively determine the superiority of one AED over the other. Most double-blind controlled studies, particularly for newer AEDs have seizures (clinical and or electrographic) as the primary efficacy measure and do not address the issue of clearing up the EEG. The potential toxicities of each agent have been ignored for the same reasons.

5.7 SUMMARY

Epilepsy is not a disease, but an expression of abnormal cerebral network activity. The seizure semiologies vary among individuals due to biologic disparities in genes, intrauterine and extrauterine environments, and maturational (or degenerative) processes. Animal models are important, and although one can control for litter and environment, including how the seizures are produced, they remain only a construct to study seizures in a species at a given age. Computer modeling is necessary, but will not replicate the complexity of a brain. The study of epilepsy will ultimately need to be performed in humans. This writing has tried to demonstrate that not only biology is important, but also the conditions in which observations are made, with the hope that everyone agrees to what is seen.

REFERENCES

Andrade, D. 2009. Genetic basis in epilepsies caused by malformations of cortical development and in those with structurally normal brain. *Hum. Genet.* 126(1):173–193.

Asano, E., C. Pawlak, A. Shah, J. Shah, A. F. Luat, J. Ahn-Ewing, and H. T. Chugani. 2005. The diagnostic value of initial video-EEG monitoring in children—Review of 1000 cases. *Epilepsy Res.* 66(1–3):129–135.

Banerjee, P. N., D. Fillippi, and W. A. Hauser. 2009. The descriptive epidemiology of epilepsy—A review. *Epilepsy Res.* 85(1):31–45.

Baulac, S., and M. Baulac. 2009. Advances on the genetics of mendelian idiopathic epilepsies. *Neurol. Clin.* 27(4):1041–1061.

Benbadis, S. R., W C. LaFrance, Jr., G. D. Papandonatos, K. Korabathina, K. Lin, and H. C. Kraemer. 2009. Interrater reliability of EEG-video monitoring. *Neurology* 73(11):843–846.

Berg, A. T., S. R. Levy, F. M. Testa, and S. Shinnar. 1999. Classification of childhood epilepsy syndromes in newly diagnosed epilepsy: Interrater agreement and reasons for disagreement. *Epilepsia* 40(4):439–444.

Berg, A. T. 2009. Identification of pharmacoresistant epilepsy. *Neurol. Clin.* 27(4):1003–1013.

Berg, A. T., G. W. Mathern, R. A. Bronen, R. K. Fulbright, F. DiMario, F. M. Testa, and S. R. Levy. 2009. Frequency, prognosis and surgical treatment of structural abnormalities seen with magnetic resonance imaging in childhood epilepsy. *Brain* 132(Pt 10):2785–2797.

Berg, A. T., S. F. Berkovic, M. J. Brodie, J. Buchhalter, J. H. Cross, B. W. van Emde, J. Engel, J. French, T. A. Glauser, G. W. Mathern, S. L. Moshé, et al. 2010. Revised terminology and concepts for organization of seizures and epilepsies: Report of the ILAE Commission on Classification and Terminology, 2005–2009. *Epilepsia* 51(4):676–685.

Brust, J. C. 2008. Seizures, illicit drugs, and ethanol. *Curr. Neuro. Neurosci. Rep.* 8(4):333–338.

Berkovic, S. F., L. Harkin, J. M. McMahon, J. T. Pelekanos, S. M. Zuberi, E. C. Wirrell, D. S. Gill, X. Iona, J. C. Mulley, and I. E. Scheffer. 2006. De-novo mutations of the sodium channel gene SCN1A in alleged vaccine encephalopathy: A retrospective study. *Lancet Neurol.* 5(6):488–492.

Birbeck, G. L., R. D. Hays, X. Cui, and B. G. Vickrey. 2002. Seizure reduction and quality of life improvements in people with epilepsy. *Epilepsia* 43(5):535–538.

Blume, W. T. 1978. Clinical and electroencephalographic correlates of the multiple independent spike foci pattern in children. *Ann. Neurol.* 4(6):541–547.

Bodensteiner, J. B., R. D. Brownsworth, J. R. Knapik, M. C. Kanter, L. D. Cowan, and A. Leviton. 1988. Interobserver variability in the ILAE classification of seizures in childhood. *Epilepsia* 29(2):123–128.

Briellmann, R. S., G. D. Jackson, Y. Torn-Broers, and S. F. Berkovic. 2001. Causes of epilepsies: Insights from discordant monozygous twins. *Ann. Neurol.* 49(1):45–52.

Caraballo, R. H., G. Capovilla, F. Vivevano, F. Beccaria, N. Speccho, and N. Fejerman. 2009. The spectrum of benign myoclonus of early infancy: Clinical and neurophysiologic features in 102 patients. *Epilepsia* 50(5):1176–1183.

Carrazana, E. J., C. T. Lombroso, M. Mikati, S. Helmers, and G. L. Holmes. 1993. Facilitation of infantile spasms by partial seizures. *Epilepsia* 34(1):97–109.

Carrera, E., J. Claassen, M. Oddo, R. G. Emerson, S. A. Mayer, and L. J. Hirsch. 2008. Continuous electroencephalographic monitoring in critically ill patients with central nervous system infections. *Arch. Neurol.* 65(12):1612–1618.

Cavazzuti, G. B., L. Cappella, and A. Nalin. 1980. Longitudinal study of epileptiform EEG patterns in normal children. *Epilepsia* 21:43–55.

Chahine, L. M., and M. A. Mikati. 2006a. Benign pediatric localization related epilepsies. Part I. Syndromes in infancy. *Epileptic Disord.* 8(3):169–183.

Chahine, L. M., and M. A. Mikati. 2006b. Benign pediatric localization related epilepsies. Part II. Syndromes in childhood. *Epileptic Disord.* 8(4):243–258.

Choi, H., R. L. Sell, L. Lenert, P. Muennig, R. R. Goodman, F. G. Gilliam, and J. B. Wong. 2008. Epilepsy surgery for pharmacoresistant temporal lobe epilepsy: A decision analysis. *JAMA* 300(21):2497–2505.

Chong, D. J., and L. J. Hirsch. 2005. Which EEG patterns warrant treatment in the critically ill? Reviewing the evidence for treatment of periodic epileptiform discharges and related patterns. *J. Clin. Neurophysiol.* 22(2):79–91.

Darra, F., E. Fiorini, L. Zoccante, L. Mastella, C. Torniero, S. Cortese, L. Meneghello, E. Fontana, and B. D. Bernardina. 2006. Benign myoclonic epilepsy in infancy (BMEI): A longitudinal electroclinical study of 22 cases. *Epilepsia* 47(Suppl 5):31–35.

Duluc, O., and I. Tuxhrn. 2002. Infantile spasms and West syndrome. In *Epileptic Syndromes in Infancy, Childhood and Adolescence,* 3rd ed. J. Roger, M. Bueau, C. H. Dravet, P. Genton, C. A. Tassinari, and P. Wolf, eds., 47–63. Eastleigh: John Libby & Co. Ltd.

Eeg-Olofsson, O., I. Petersen, and U. Sellden. 1971. The development of the electroencephalogram in normal children from the age of 1 through 15 years. Paroxysmal activity. *Neuropadiatrie* 2:375–404.

Engel, J., and International League Against Epilepsy (ILAE). 2001. A proposed diagnostic scheme for people with epileptic seizures and with epilepsy: Report of the ILAE Task Force on Classification and Terminology. *Epilepsia* 42(6):796–803.

Engel, J. 2005. Classification is not EZ (Invited Editorial Comment). *Epileptic Disord.* 7(4):317–320.

Fang, P. C., Y. J. Chen, and I. C. Lee. 2008. Seizure precipitants in children with intractable epilepsy. *Brain Dev.* 30(8):527–532.

Fisher, R. S., W. van Emde Boas, W. Blume, C. Elger, P. Genton, P. Lee, and J. Engel, Jr. 2005. Epileptic seizures and epilepsy: Definitions proposed by the International League against Epilepsy. *Epilepsia* 46(4):470–472.

Frucht, M. M., M. Quigg, C. Schwaner, and N. B. Fountain. 2000. Distribution of seizure precipitants among epilepsy syndromes. *Epilepsia* 41(12):1534–1539.

Giedd, J. N., F. M. Lalonde, M. J. Celano, S. L. White, G. L. Wallace, N. R. Lee, and R. K. Lenroot. 2009. Anatomical brain magnetic resonance imaging of typically developing children and adolescents. *J. Am. Acad. Child Adolesc. Psychiatry* 48(5):465–470.

Hancock, E. C., J. P. Osborne, and S. W. Edwards. 2008. Treatment of infantile spasms. *Cochrane Database Syst. Rev.* 4:CD001770.

Hart, Y. M., and S. D. Shorvon. 1995. The nature of epilepsy in the general population. I. Characteristics of patients receiving medication for epilepsy. *Epilepsy Res.* 21(1):43–49.

Hauser, W. A. 1992. Seizure disorders: The changes with age. *Epilepsia* 33(Suppl 4):S6–S14.

Haut, S. R., A. T. Berg, S. Shinnar, H. W. Cohen, C. W. Bazil, M. R. Sperling, J. T. Langfitt, S. V. Pacia, T. S. Walczak, and S. S. Spencer. 2002. Interrater reliability among epilepsy centers: Multicenter study of epilepsy surgery. *Epilepsia* 43(11):1396–1401.

Haut, S. R. Seizure clustering. 2006. *Epilepsy Behav.* 8(1):50–55.

Haut, S. R., C. B. Hall, J. Masur, and R. B. Lipton. 2007. Seizure occurrence: Precipitants and prediction. *Neurology* 69(20):1905–1910.

Hermann, B., M. Seidenberg, and J. Jones. 2008. The neurobehavioural comorbidities of epilepsy: Can a natural history be developed? *Lancet Neurol.* 7(2):151–160.

Hesdorffer, D., E. K. Benn, G. Cascino, and W. A. Hauser. 2009. Is a first acute symptomatic seizure epilepsy? Mortality and risk for recurrent seizure. *Epilepsia* 50(5):1102–1108.

Hitiris, N., R. Mohanraj, J. Norrie, G. J. Sills, and M. J. Bodie. 2007. Predictors of pharmacoresistant epilepsy. *Epilepsy Res.* 75(2–3):192–196.

Hoppe, C., A. Poepel, and C. E. Elger. 2007. Epilepsy: Accuracy of patient seizure counts. *Arch. Neurol.* 64(11):1595–1599.

Hughes, J. R. 1985. Long-term clinical and EEG changes in patients with epilepsy. *Arch. Neurol.* 42(3):213–223.

Hughes, J. R. 2009. Absence seizures: A review of recent reports with new concepts. *Epilepsy Behav.* 15(4):404–412.

Huttenlocher, P. R., and R. J. Hapke. 1990. A follow-up study of intractable seizures in childhood. *Ann. Neurol.* 28(5):699–705.

Kotagal, P. 2000. Tonic-clonic seizures. In *Epileptic seizures: Pathophysiology and clinical semiology.* H. O. Luders and S. Noachtar, 652–657. New York: Churchill Livingstone.

Kriegstein, A. R., D. F. Owens, and M. Avoli. 1999. Ontogeny of channels, transmitters and epileptogenesis. *Adv. Neurol.* 79:145–159.

Kwan, P., and M. J. Brodie. 2000. Early identification of refractory epilepsy. *N. Engl. J. Med.* 342(5):314–319.

Levy, S. R., A. T. Berg, F. M. Testa, E. J. Novotny, and K. H. Chiappa. 1998. Comparison of digital and conventional EEG interpretation. *J. Clin. Neurophysiol.* 15(6):476–480.

Liao, Y., L. Deprez, S. Maljevic, J. Pitsch, L. Claes, D. Hristova, A. Jordanova, S. Ala-Mello, A. Bellan-Koch, D. Blazevic, S. Schubert, et al. 2010. Molecular correlates of age-dependent seizures in an inherited neonatal-infantile epilepsy. *Brain* 133(Pt 5):1403–1414.

Loiseau, P. 1992. Childhood absence epilepsy. In *Epileptic Syndromes in Infancy, Childhood and Adolescence,* 3rd ed. J. Roger, M. Bueau, C. H. Dravet, P. Genton, C. A. Tassinari, and P. Wolf, eds., 285–303. Eastleigh: John Libby & Co. Ltd.

Lu, T., Y. Pan, S. Y. Kao, C. Li, I. Kohane, J. Chan, and B. A. Yankner. 2004. Gene regulation and DNA damage in the ageing human brain. *Nature* 429(6994):883–891.

Marusic, P., M. Tomásek, P. Krsek, H. Krijtová, J. Zárubová, J. Zámecník, M. Mohapl, V. Benes, M. Tichý, and V. Komárek. 2007. Clinical characteristics in patients with hippocampal sclerosis with or without cortical dysplasia. *Epileptic Disord.* 9 (Suppl. 1):S75–S82.

Marx, C., M. R. Masruha, E. Garzon, and L. C. Vilanova. 2008. Benign neonatal sleep myoclonus. *Epileptic Disord.* 10(2):177–180.

Mayville, C., T. Fakhoury, and B. Abou-Khalil. 2000. Absence seizures with evolution into generalized tonic-clonic activity: Clinical and EEG features. *Epilepsia* 41(4):391–394.

Merlis, J. K. 1970. Proposal for an international classification of the epilepsies. *Epilepsia* 11(1):114–119.

Millichap, J. G. 1959. Studies in febrile seizures. I. Height of body temperature as a measure of the febrile-seizure threshold. *Pediatrics* 23(Pt. 1):76–85.

Millichap, J. G., and J. J. Millichap. 2006. Role of viral infections in the etiology of febrile seizures. *Pediatr. Neurol.* 35(3):165–172.

Neuschlová, L., K. Šterbová, J. Žácková, and V. Komárek. 2007. Epileptiform activity in children with developmental dysphasia: Quantification of discharges in overnight sleep video-EEG. *Epileptic Disord.* 9(Suppl. 1):S28–S35.

Nordli, D. R., C. W. Bazil, M. L. Scheuner, and T. A. Pedley. 1997. Recognition and classification of seizures in infants. *Epilepsia* 38(5):553–560.

Ottman, R., J. H. Lee, W. A. Hauser, S. Hong, D. Hesdorffer, N. Schupf, T. A. Pedley, and M. L. Scheuer. 1993. Reliability of seizure classification using a semistructured interview. *Neurology* 43(12):2526–2530.

Petitmengin, C., M. Baulac, and V. Navarro. 2006. Seizure anticipation: Are neurophenomenological approaches able to detect preictal symptoms? *Epilepsy Behav.* 9(2):298–306.

Rakhade, S. N, and F. Jensen. 2009. Epileptogenesis in the immature brain: Emerging mechanisms. *Nature Rev. Neurol.* 5(7):380–391.

Reynolds, E. H. 2000. The ILAE/IBE/WHO global campaign against epilepsy: Bringing epilepsy "out of the shadows." *Epilepsy Behav.* 1(4):S3–S8.

Rossetti, A. O., and P. W. Kaplan. 2009. Seizure semiology: An overview of the 'inverse problem.' *Eur. Neurol.* 63(1):3–10.

Sanchez, R. M., and F. E. Jensen. 2001. Maturational aspects of epilepsy mechanisms and consequences for the immature brain. *Epilepsia* 42(5):577–585.

Shinnar S., C. O'Dell, and A. T. Berg. 1999. Distribution of epilepsy syndromes in a cohort of children prospectively monitored from the time of their first unprovoked seizure. *Epilepsia* 40(10):1378–1383.

Shinnar, S., D. C. Hesdorffer, D. R. Nordli, Jr., J. M. Pellock, C. O'Dell, D. V. Lewis, L. M. Frank, S. L. Moshé, L. G. Epstein, A. Marmarou, and E. Bagiella. 2008. Phenomenology of prolonged febrile seizures: Results of the FEBSTAT study. *Neurology* 71(3):70–76.

Sillanpää, M., and D. Schmidt. 2006. Natural history of treated childhood-onset epilepsy: Prospective, long-term population-based study. *Brain* 129(Pt 3):617–624.

Sillanpää, M., and D. Schmidt. 2006. Prognosis of seizure recurrence after stopping antiepileptic drugs in seizure-free patients: A long-term population-based study of childhood-onset epilepsy. *Epilepsy Behav.* 8(4):713–719.

Sillanpää, M., and D. Schmidt. 2009. Early seizure frequency and aetiology predict long-term medical outcome in childhood-onset epilepsy. *Brain* 132(Pt 4):989–998.

Spencer, S. S., P. D. Williamson, S. L. Bridgers, R. H. Mattson, D. V. Cicchetti, and D. D. Spencer. 1985. Reliability and accuracy of localization by scalp ictal EEG. *Neurology* 35(11):1567–1575.

Sperling, M. R., C. A. Schilling, D. Glosser, J. I. Tracy, and A. A. Asadi-Pooya. 2008. Self-perception of seizure precipitants and their relation to anxiety level, depression, and health locus of control in epilepsy. *Seizure* 17(4):302–307.

Spooner, C. G., S. F. Berovic, L. A. Mitchell, J. A. Wrennall, and A. S. Harvey. 2006. New-onset temporal lobe epilepsy in children: Lesion on MRI predicts poor seizure outcome. *Neurology* 67(12):2147–2153.

Stephani, U. 2006. The natural history of myoclonic astatic epilepsy (Doose syndrome) and Lennox-Gastaut syndrome. *Epilepsia* 47(Suppl. 2):53–55.

Szabo, C. A., and H. O. Luders. 2000. Todd's paralysis and postictal aphasia. In *Epileptic Seizures: Pathophysiology and Clinical Semiology.* H. O. Luders and S. Noachtar, eds., 652–657. New York: Churchill Livingstone.

Tassinari, C. A., G. Cantalupo, L. Rios-Pohl, E. D. Giustina, and G. Rubboli. 2009. Encephalopathy with status epilepticus during slow sleep: "The Penelope syndrome." *Epilepsia* 50(Suppl. 7):4–8.

Temkin, N. 2001. Antiepileptogenesis and seizure prevention trials with antiepileptic drugs: Meta-analysis of controlled trials. *Epilepsia* 42(4):515–524.

Theodore, W., L. Epstein, W. Gaillard, S. Shinnar, M. S. Wainwright, and S. Jacobson. 2008. Human herpes virus 6B: A possible role in epilepsy? *Epilepsia* 49(11):1828–1837.

Van Paesschen, W., P. Dupont, S. Sunaert, K. Goffon, and K. Van Laere. 2007. The use of SPECT and PET in routine clinical practice in epilepsy. *Curr. Opin. Neurol.* 20(2):194–202.

Vignatelli, L., F. Bisulli, F. Provini, I. Naldi, F. Pittau, A. Zaniboni, P. Montagna, and P. Tinuper. 2007. Interobserver reliability of video recording in the diagnosis of nocturnal frontal lobe seizures. *Epilepsia* 48(8):1506–1511.

Walczak, T. S., R. A. Radtke, and D. V. Lewis. 1992. Accuracy and interobserver reliability of scalp ictal EEG. *Neurology* 42(12):2279–2285.

Yamatogi, Y., and S. Ohtahara. 2003. Severe epilepsy with multiple independent spike foci. *J. Clin. Neurophysiol.* 20(6):442–448.

Section II

Foundations of Engineering,
Math, and Physics

6 Intracranial EEG
Electrodes, Filtering, Amplification, Digitization, Storage, and Display

Hitten P. Zaveri and Mark G. Frei

CONTENTS

6.1 INTRODUCTION

Intracranial EEG (icEEG) acquisition has evolved considerably over the past five decades. This chapter presents an exposition of current icEEG acquisition methodology. The chapter's focus is on icEEG acquisition rather than EEG acquisition as the former presents a significant challenge and allows more in-depth exposition of methods, and the latter is covered in standard EEG textbooks. We proceed by considering each of the components of a contemporary icEEG acquisition system from intracranial electrodes, through signal conditioning, sampling, storage, and display. This is followed by a brief discussion of some other aspects of an icEEG acquisition system that are separate from this data path and conclusions.

6.2 SIGNAL ACQUISITION

A modern EEG acquisition system can be considered, principally, to be composed of the components connected in series in Figure 6.1. In this section, we shall discuss each component individually—its role in icEEG acquisition and the evolution of salient aspects of the component through time. Before we proceed, we will highlight a few general aspects of the signal path indicated in Figure 6.1. In a modern acquisition system, the signals are low amplitude until they reach stage 3 (preamplification) if a preamplifier is present or stage 4 (amplification) if stage 3 is not present. The signals are analog up until stage 5, where the amplified signals are converted from analog to digital form. Prior

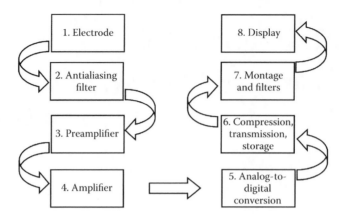

FIGURE 6.1 The signal path for icEEG acquisition. The primary components of the data acquisition system consist of the 8 serially connected stages shown here. The signals from the electrode contact are filtered, amplified, and digitized prior to transmission to a data collection computer. Here, the digital data stream is reconstituted from the data packets received, stored, and displayed.

to this conversion, the signals are filtered, typically with a band-pass filter. The signals are open band until they undergo filtering in stage 2. The acquisition system performs in real-time in stages 1 through 5. Once the data have been digitized in stage 5, the system is not strictly real-time, though high sampling rates and extremely high processor speeds make the delays associated with digital data transmission and processing virtually negligible. For practical purposes, the system appears to perform in real-time. Stages 7 and 8 are typically performed offline.

6.2.1 Intracranial Electrodes

Intracranial EEG electrodes have undergone a considerable change in the past five decades. At the start of this period, icEEG electrodes were essentially bundles of twisted wires where the insulation of the individual wires was removed at specific depths to expose the metal wire and thus create an electrode contact. The electrode contacts were the sections of the wire where the insulation had been removed. These wire bundle depth electrodes were constructed in laboratories at surgical centers. The wire bundle depth electrodes were replaced in the 1970s and 1980s with the first commercially available intracranial depth electrodes. The first commercially available intracranial depth electrodes, just as the wire bundle depth electrodes that preceded them, were rigid in construction. Since these rigid depth electrodes were placed in a manner where they were fixed with respect to the skull, any change in the position of the brain with respect to the skull could result in injury to the brain. The rigid depth electrodes were replaced in the 1970s and 1980s with flexible depth electrodes for this reason.

Modern icEEG electrodes are typically available as depth electrodes or subdural strip or grid electrodes (Figure 6.2). The electrodes consist of a small exposed metal surface constituting the electrode contact (Plonsey 1969) embedded in silastic with wires connected to each electrode contact that exits the electrode as one or more wire bundles. The electrode contacts are typically placed in a simple geometric arrangement in linear (subdural strip and depth electrodes) or array like (subdural grid electrodes) forms. A standard subdural strip or grid electrode is constructed with disk electrode contacts placed with a center-to-center spacing of 1 cm. A depth electrode is constructed with metal contacts placed on the outer surface of a silastic cylinder with the same intercontact distance. Depth electrodes are stereotactically implanted using MRI guidance and a computerized planning system that allows interactive previewing of electrode trajectories while subdural grids are placed after a craniotomy to expose the target location. The flexible depth electrodes are

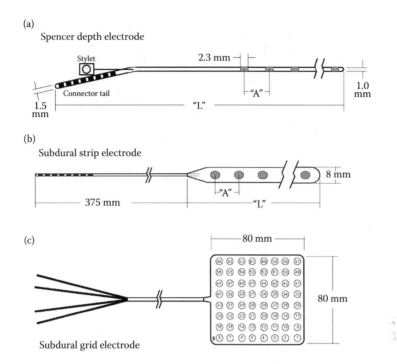

FIGURE 6.2 Schematics of example electrodes from AdTech Medical Instrument Corporation: (a) depth; (b) subdural strip; and (c) subdural grid electrode. The depth and subdural strip and grid electrodes consist of electrode contacts (metal contacts) placed at regular intervals (typically 10-mm center-to-center spacing) on a planar surface (subdural strip and grid electrodes) or on the outer surface of a flexible cylinder (depth electrodes). The flexible depth electrode is introduced with a rigid stylet that is removed subsequent to electrode placement. Wires emanate from these electrodes as wire bundles, and these are connected to the front end of the acquisition instrumentation. In (a) and (b) A represents the intercontact distance and L represents the length of the electrode.

introduced with a rigid stylet that is removed after the depth electrode has been placed. Subdural strip electrodes are placed freehand by the neurosurgeon using fluid to create and irrigate a channel in subdural space and slide the subdural strip to its target location. An example of the placement of subdural strip and grid electrodes is shown in Figure 6.3. As indicated above, wires emanate from the electrodes, one for each electrode contact and are combined together as a wire bundle. Typically, 1–4 wire bundles exit from each intracranial electrode. The wire bundles are anchored for strain relief both inside and outside the skull, and exit the skull through a water tight seal constructed by the neurosurgeon during surgery to place electrodes.

6.2.2 FILTERS

A filter alters the spectral content of a signal by emphasizing or deemphasizing specific frequency bands. There are a number of types of filters. A low-pass filter, for example, passes all signals below a certain frequency, and rejects all signals at higher frequencies. A high-pass filter passes signals above a specified frequency and rejects all other signals at lower frequencies. A low-pass filter in series with a high-pass filter can create a band-pass filter, which is a filter that rejects signals lower than the cutoff frequency specified for the high-pass filter and higher than the cutoff frequency specified for the low-pass filter. Other kinds of filters are band-reject and notch filters, the latter of which can be constructed to reject signals in a very narrow frequency band such as line noise.

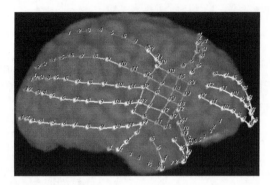

FIGURE 6.3 An example of subdural grid and strip electrode placement on the right hemisphere. Here, a craniotomy was performed to place a 5 × 4 contact subdural grid. The neurosurgeon has introduced subdural strips in a starburst pattern from the edges of the craniotomy to sample a larger extent of the hemisphere. The electrode contacts are displayed on the preop MRI of the patient's brain.

A useful analogy is the operation of a light dimming circuit, which consists of a variable resistor in series with a lamp. As the dimmer switch in the circuit is turned to increase the resistance of the variable resistor, the illumination offered by the lamp decreases. This is because the increased resistance of the variable resistor causes an increasingly larger voltage drop across the variable resistor and a correspondingly smaller voltage drop across the resistance of the light bulb (which causes a smaller amount of illumination). A filter is akin to this, but with the difference that the reduction of voltage across an output element occurs in a manner that is a function of the frequency of the signal. Suppose, for example, that a circuit element's resistance decreases as the frequency of a signal increases (i.e., the circuit element provides increasingly lower resistance to higher frequencies in comparison to lower frequencies). Now suppose that this circuit element is in series with a fixed resistor and the output voltage is measured across the circuit element. The output voltage would then decrease as the frequency of the input signal increases. This is because, as the resistance offered by the circuit element decreases with frequency, a smaller amount of the voltage will be measured across it and a larger amount of the voltage will be measured across the resistor. That is, the circuit will behave like a low-pass filter by allowing voltages at low frequencies and not allowing voltages at high frequencies. Alternatively, if the output of this circuit were measured across the resistor instead of the circuit element, the output voltage would increase as the signal frequency increases. That is, the circuit would function like a high-pass filter. The operation of simple resistor-capacitor (RC) circuits, which function as low-pass and high-pass filters, are illustrated in Figure 6.4.

A filter is typically applied to the EEG at two points, prior to digitization and during display. Given the mixed analog and digital nature of the icEEG acquisition system, the first filter is analog (implemented in hardware such as the filters shown in Figure 6.4), while the second filter is digital (implemented in software). Here, we consider the analog filter that is in the path of the signal prior to digitization. There is a need to filter the EEG prior to its amplification and digitization. The role of this filter is to condition the input signal so that the signal fits the design specification of the acquisition system. This is done for two purposes. The first is so that the signal does not contain components at frequencies higher than the design specification of the data acquisition system. If it did, this would cause aliasing in the subsequent digitization. Aliasing is the erroneous mapping of signal power to a frequency different from the true frequency. Because of its role, this first filter is known as an anti-aliasing filter. The filter also ensures that the signal does not contain power at lower frequencies than those specified for the instrumentation. This may be important if the instrumentation has not been designed for very low frequencies (e.g., from 0.5 Hz down to 0 Hz or "DC"). DC and very low-frequency signals can be large in amplitude and have persistent drift or bias. If these large signals are admitted to the instrumentation without first applying a high-pass filter to remove them, they can saturate the amplifiers. The filter placed in the path of the analog input signal

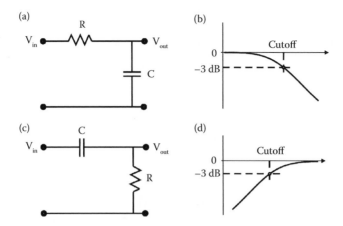

FIGURE 6.4 An example of resistor-capacitor (RC) circuits: (a) low-pass filter; (b) filter characteristics of the low-pass filter in the frequency domain; (c) high-pass filter; (d) filter characteristics of the high-pass filter in the frequency domain. In (b) and (d), the horizontal (x) axis is frequency and the vertical (y) axis is power.

before it is digitized is thus either a low-pass filter (anti-aliasing filter) or a band-pass filter designed both to serve as an anti-aliasing filter and also to reject low-frequency signal components. A notch filter may also be included at this stage to remove line noise.

6.2.3 AMPLIFICATION

The icEEG field potentials of interest are very small (μVs) and may ride on top of a larger (mVs) common field potential. EEG acquisition systems amplify the observed signal with respect to a reference signal to remove the common signal. A bioinstrumentation or differential amplifier is built using operational amplifiers (op-amps) (Webster 1978). An op-amp is a circuit element idealized in the manner depicted in Figure 6.5 (Jung 1986). The op-amp has two inputs, termed the negative and positive inputs, and a single output. The op-amp subtracts the negative input from the positive input and amplifies the resulting difference signal. The performance of a differential amplifier is characterized by the differential gain (DG), common mode gain (CMG), and the common mode rejection ratio (CMRR). The DG represents the differential amplification of the input signals, the CMG represents the common mode amplification of the input signals, and the CMRR (defined as DG / CMG) characterizes the ability of the amplifier to reject common mode signals. The CMRR of conventional equipment is typically >100 dB, which indicates that the ratio of DG to CMG is greater than 100,000, indicating a considerable ability to reject a common mode signal. Amplifiers are further characterized in a number of manners, for example by their passband and input impedance (that

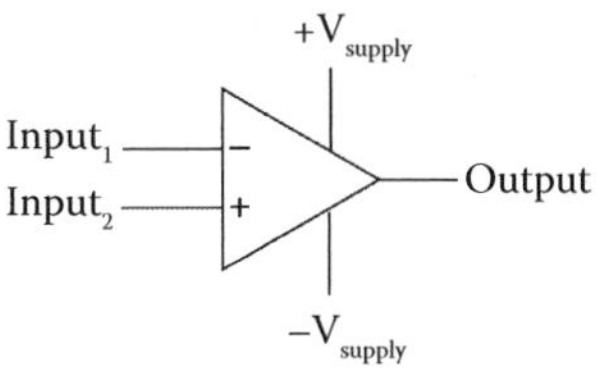

FIGURE 6.5 An operational amplifier (op-amp). This circuit element has two input terminals (inverting and noninverting) and one output terminal. The output is proportional to the difference between the two input voltages. An ideal op-amp has infinite voltage gain, infinite bandwidth, infinite input impedance, no input current, and no input offset voltage.

is, the resistance offered to current entering the input terminal of the amplifier). Bioinstrumentation amplifiers typically pass signals in a wide band ranging from less than 1 Hz to several hundred hetz or kilohertz and are designed to meet the specifications of the acquisition system. The input impedance should be as high as possible, since the amplifier is placed in a circuit that includes the electrode. It is important that the amplifier does not present a path from the electrode for electrical current to pass through; otherwise, it would alter the phenomenon being studied. A high input impedance prevents current from flowing into the amplifier.

In Figure 6.1, signal amplification is depicted in two positions—preamplification and amplification. The purpose of a preamplifier is to amplify the signal acquired as close to the source as possible to minimize the contamination of the low level icEEG signals by noise. Preamplifiers may be worn by the patient or placed in the head-dressing. The large number of channels used has precluded the widespread use of preamplifiers and, indeed, preamplifiers are seldom used. This may change with the greater interest in high-frequency activity, which tends to be very low in amplitude and thus susceptible to contamination by artifact, and the miniaturization of preamplifiers.

The amplification provided by the bank of amplifiers used for icEEG acquisition must be matched across all channels; otherwise, a differential gain between channels could be introduced by the equipment. Amplifiers are built from discrete electronic components. It is possible that these components may not be matched across the channels of a multichannel icEEG acquisition system. It is also possible that the characteristics of electronic components could change fractionally over time. There is a need, thus, to normalize the performance of the different amplifiers in an icEEG acquisition system. This is accomplished by calibration. Modern equipment typically has a built in calibration circuit that injects a known current and measures the response to this known input. The different channels are calibrated by the use of this measured output to a known signal so that all channels amplify a given signal to the same level. Calibration is typically performed at the start of a recording session.

6.2.4 SAMPLING AND ANALOG-TO-DIGITAL CONVERSION

The next stage in the acquisition of brain activity is digitization by an analog-to-digital converter (ADC). There are two important aspects to this stage. First, the amplitude of the signal is converted from an infinite range of continuous values in analog form to one of a sequence of discrete values ranging between a minimum and a maximum. This represents the discretization of the signal amplitude. The number of distinct values that can be represented depends on the number of bits or resolution of the ADC. The greater the number of bits, the greater the number of discrete values possible. If the ADC has a resolution of k bits then 2^k discrete levels are possible (see Table 6.1). We describe the parameters of this digitization process in terms of the dynamic range (maximum value − minimum value), the resolution (the value k) and the resulting accuracy of a noise-free measurement, which is typically given as (maximum − minimum)/2^k with units of millivolts per bit. An

---

TABLE 6.1
Resolution of an ADC (Number of Bits) and Corresponding Number of Discrete Levels

ADC Bits	Levels
8	256
10	1024
12	4096
16	65,536
22	4,194,304
24	16,777,216

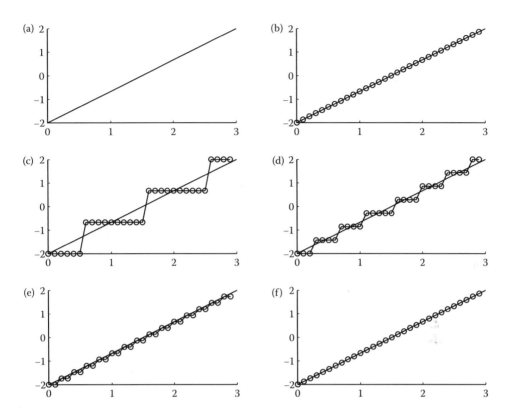

FIGURE 6.6 An illustration of the discretization of an input analog signal by ADCs with increasing resolution (number of bits). In these panels, the horizontal (x) axis represents time in seconds, and the vertical axis (y) represents voltage in mV. (a) An input analog signal (a ramp waveform); (b) the input analog signal sampled at $f_s = 10$ Hz (10 samples per second), with amplitude represented with infinite resolution; (c–f) the analog signal digitized with (c) a 2-bit resolution over 4 mV (4 discrete levels); (d) a 3-bit resolution over 4 mV (8 discrete levels); (e) a 4-bit resolution over 4 mV (16 discrete levels); and (f) an 8-bit resolution over 4 mV (256 discrete levels). With increasing ADC resolution, the digitized values increasingly approximate the sampled values shown in (b).

example illustrating the effect of increasing the number of bits in an ADC is displayed in Figure 6.6. A second part of this stage is the sampling, in time, of the signals. While the first part of this stage can be considered to be a discretization of analog values in the y (amplitude) axis, the second can be considered the discretization along the x (time) axis.

The sampling of the data is governed by the sampling or Nyquist-Shannon theorem (Nyquist 1928; Shanon 1949). This theorem holds that a signal can be reconstructed from a sampled signal if the sampled signal was obtained with a sampling frequency (f_s) that is at least twice the highest frequency (f_{max}) of the input signal. The sampling frequency defines a number of aspects of the instrumentation system. It places an upper bound on the highest signal frequencies that can be acquired by the instrumentation. Indeed, the acquisition system is designed with this band limitation in mind. It also affects design decisions on the anti-aliasing filter, the amplifier passband, the bandwidth of the data path, and the storage capacity required. While the sampling theorem states that f_s should be at least twice f_{max}, a common rule of thumb is that the sampling frequency should be at least 5 times f_{max}. We illustrate the sampling of an analog signal with a simple example in Figures 6.7 and 6.8, and in Figures 6.9 and 6.10 we show the joint effect of varying k, the number of bits in an ADC, and f_s, the sampling frequency.

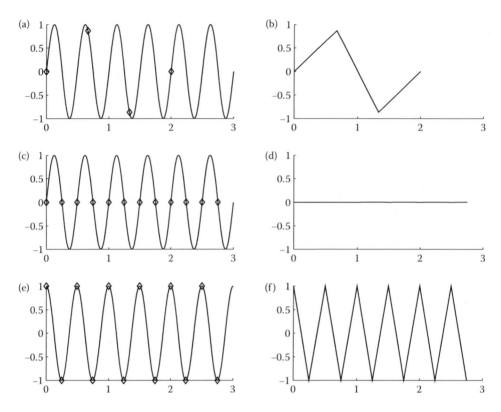

FIGURE 6.7 An illustration of the sampling of an input analog signal with increasing sampling frequency. In these panels, the horizontal (x) axis represents time in seconds, and the vertical axis (y) represents voltage in mV. In (a), (c), and (e), we display the input analog signal (2-Hz sine wave) and samples of this signal obtained with (a) $f_s = 1.5$ Hz, (c) $f_s = 4$ Hz, and (e) $f_s = 4$ Hz. In (b), (d), and (f), we display the corresponding signal recreated from the digital samples when the samples were obtained with (b) $f_s = 1.5$ Hz, (d) $f_s = 4$ Hz, and (f) $f_s = 4$ Hz. Note that sampling at less than twice the maximum frequency—as in (a) and (b)—results in an aliased version of the input signal. When the input signal is sampled at exactly twice its frequency (as stipulated by the sampling theorem), we observe either a flat line (d) or a triangular wave (f). In the latter instance, we captured the minima and maxima of the sine wave while, in the former, we sampled its zero crossings. Through this example we wish to note that to adequately capture a signal, it is recommended that the sampling frequency be much greater than that stipulated by the sampling theorem.

6.2.5 COMPRESSION, TRANSMISSION, AND STORAGE

The output of the previous stage of the system is a steady stream of numbers, most likely in integer form. The scaling factors obtained during calibration are stored and transmitted separately. The digital data stream is typically transmitted out of the ADC stage using a standard internet protocol suite—transmission control protocol (TCP)/internet protocol (IP)—or over a universal serial bus (USB) to a computer system (the collector) where the data are received and saved to files, either on a local or networked storage site. Digital EEG data are either transmitted in a compressed manner or in an uncompressed manner. If compressed, the formats in use are typically lossless, that is the original digital EEG signal is fully recovered from the compressed data. At this stage, the data collection is not truly real-time because the data are broken into packets of information and are transmitted as such and reconstituted by the collector as a stream of data samples; however, for all practical purposes, the system works at near real-time. The amount of data which is stored varies

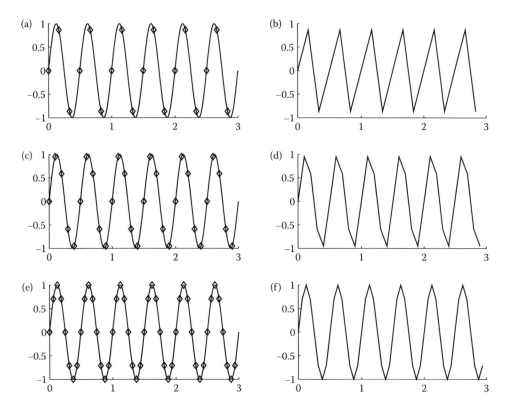

FIGURE 6.8 A further illustration of the sampling of an input analog signal with increasing sampling frequency. As in Figure 6.7, the horizontal (x) axis represents time in seconds, the vertical axis (y) represents voltage in mV. In (a), (c), and (e), we display the input analog signal and samples of this signal obtained with (a) $f_s = 6$ Hz, (c) $f_s = 10$ Hz, (e) $f_s = 16$ Hz. In (b), (d), and (f), we display the corresponding signal recreated from the digital samples when the samples were obtained with (b) $f_s = 6$ Hz, (d) $f_s = 10$ Hz, (f) $f_s = 16$ Hz. It is clear that with increasing sampling frequency, the signal reconstructed from the sampled values increasingly approximates the input signal. Note that the values of f_s in (d) and (f) are 5 and 8 times f_{max}, respectively.

widely depending on the sampling frequency used. While the largest storage component is used for video, the EEG storage requirement can be substantial depending on the sampling frequency used, the number of channels recorded, and the ADC resolution (number of bits per sample).

6.2.6 MONTAGE AND FILTERS

The data stored on the collection or server system is typically open-band (to the limit of the data acquisition system) and recorded with respect to the specified reference electrode. The data can now be flexibly remontaged, filtered, and displayed. While the data were recorded with respect to a specified reference electrode, these data can now be rereferenced to another electrode or to a computed reference. Modern digital systems perform all display conversions after the data have been collected. The montage that is applied can be a general montage or it can be specific to the patient. As indicated above, a second set of filters can be applied at this stage to facilitate signal display. These filters are digital, and can be designed with greater flexibility than the analog filters incorporated at the front-end of the signal path.

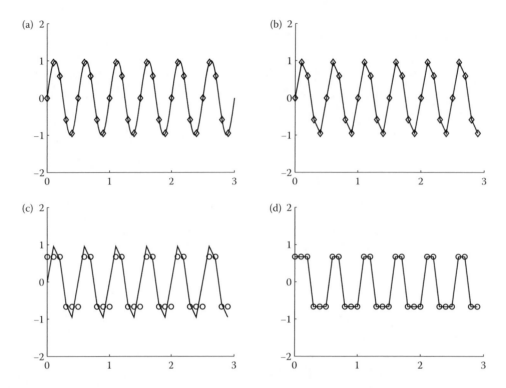

FIGURE 6.9 An illustration of the effect of sampling and digitization. In these panels the horizontal (x) axis represents time in seconds, and the vertical axis (y) represents voltage in mV. (a) An input analog signal (a 2-Hz sine wave that was displayed in Figures 6.7 and 6.8), and samples of this signal obtained with $f_s = 10$ Hz; (b) the signal reconstructed from the samples displayed in (a); (c) the signal displayed in (b) and its digitization with an ADC with 2 bits of resolution over 4 mV (4 discrete levels); and (d) the signal recreated from the sampled and digitized values displayed in (c).

6.2.7 DISPLAY

The montage and filters are applied to the recorded EEG and the resultant display is created. While the display of EEG is nominally simple, there are a few subtle considerations. We have thus far not discussed the video data stream, which is acquired along its own data path. The display must now be created in a manner that allows the synchronization of the video and EEG streams. Since the two data streams are acquired separately (though typically by the same collection system) and have different sampling frequencies and compression and storage requirements, the display software must bring the two data streams synchronously together when presenting them. A lack of synchronization is possible between video and EEG, and typically the two data streams must continuously be checked for synchronization. Importantly, most modern systems execute on computer platforms where the operating system (OS) is not real-time. The fact that the OS is not real-time means that it may take a variable amount of time to complete the same task. For example, the OS may take a few milliseconds or it may take a few seconds to close a data file. The program that is requesting the closure of the file may stipulate the time required to close the file, but it cannot completely control the OS and force timeliness. The lack of a real-time OS can impact tasks such as keeping the video and EEG data streams synchronized and requires attention on the part of system designers.

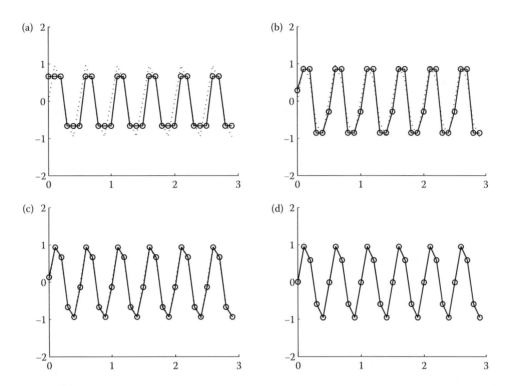

FIGURE 6.10 A further illustration of the effect of sampling and digitization. In these panels, the horizontal (*x*) axis represents time in seconds, and the vertical axis (*y*) represents voltage in mV. In each of the panels here, the sampled signal shown in Figure 6.9b is reproduced (with a dashed line). In (a)–(d), the signal is digitized with (a) a 2-bit resolution over 4 mV (4 discrete levels), (b) a 3-bit resolution over 4 mV (8 discrete levels), (c) a 4-bit resolution over 4 mV (16 discrete levels), and (d) an 8-bit resolution over 4 mV (256 discrete levels). With increasing ADC resolution, the digitized values increasingly approximate the sampled values.

An important aspect of display is the extraordinary amount of data that can be collected. If, for example, 250 channels are being sampled at 10 kHz, this results in $250 \times 10,000 = 2,500,000$ data samples per second, and 100,000 data samples per channel for a 10-s interval. Current large-format monitors have a maximum resolution of approximately 2000 pixels in the horizontal dimension. If we intend to place at least one sample per pixel (though it would be better to have more than one pixel assigned to a sample), we would be able to display only 200 ms of data on the monitor for each channel. The extremely high number of data samples also makes it difficult to scroll or page through icEEG. It would then be necessary to have a systematic procedure where the data are first reviewed, possibly at a lower sampling frequency (e.g., at 512 Hz, 1 kHz, or 2 kHz) to detect seizures and segments and channels of interest. Following this, select high-frequency data segments can be reviewed to answer hypotheses regarding the selected segments and channels.

We note that if the data are downsampled, the possibility exists that there could be aliasing. This can happen in two ways. First, it can occur when downsampling the data from, for example, 10 to 1 kHz. Second, it can happen if the data are mapped on too few pixels (which also represents a downsampling). The data must therefore be passed through an anti-aliasing filter constructed for the new sampling frequency before they are downsampled, and the display routines should not map data onto the monitor in a manner where a data sample is represented by less than 1 pixel.

6.3 VIDEO, DIGITAL PROCESSING, AND FUNCTIONAL MAPPING

The components described above are part of a serial sequence of stages that are deployed to condition and acquire icEEGs. There are a few components that are outside this serial sequence of stages. These are discussed here.

Since the advent of standard protocols for compressing and playing back digital video, the decrease in cost of the hardware and software to support digital video and the increasingly powerful acquisition workstations, collection systems have switched from analog to digital video. The cameras typically are color and have an IR illuminator for use when lights have been turned off. The frame rate employed varies, but a rate of 30 frames per second (fps) is typical. Greater resolution, more cameras, and faster frame rates are all beneficial. While earlier there were concerns with the use of mainstream video recording and playback solutions for very exacting clinical purposes, improvements in video compression protocols have eased these concerns.

It should be increasingly clear that the modern acquisition system is a hybrid system containing both instrumentation and computer-based aspects. The acquisition system needs to be both a sophisticated instrumentation solution and a sophisticated IT solution. The IT aspects of an icEEG acquisition system touch upon many critical aspects of system function, ranging from data storage, review, and archival. We note that a number of reviewers evaluate the data, and that these reviewers range in technical expertise from monitor watchers, EEG technicians, residents, fellows, and attending physicians. The data acquisition and display system must allow different roles with rights and restrictions to different users and must allow simultaneous access by different users, who are possibly in different geographic locations.

A component that has historically not received enough attention but may be getting more attention recently is the methodology employed for functional mapping. Functional mapping has historically been conducted by stimulating pairs of electrodes to interrupt function. This is a laborious, time-consuming effort that has not been aided by equipment manufacturers other than to allow the connection of a stimulator to the electrode headbox. More recently, we have seen the advent of integrated stimulators that allow a user to deliver a sequence of stimuli between electrode contacts, for example from the bedside using a handheld device, in a manner where the stimulus time, strength, and contacts stimulated are documented by the data collection system. This is a considerable improvement from previous solutions and should lead to decreased effort and greater accuracy through more complete documentation and the minimization of errors.

6.4 CONCLUSION

Intracranial EEG acquisition has evolved steadily over the past several decades, reflecting both the need for more complete monitoring of brain activity and the facilitation of solutions by the advance of technology. Based on these advances, it is not difficult to envision further considerable future improvements in icEEG acquisition with the advance of technology. Advances can indubitably be expected in types of sensors, electrode contact sizes, materials for electrodes and contacts, coverage of the brain, signal passband, amplification, sampling frequency, storage, and display.

What remains a challenge is the interpretation of the icEEG signal, which—like the EEG—remains poorly understood 100 years after the discovery of the EEG. Further, it is not clear what limitations may exist for the accuracy of visual interpretation of thousands of samples of icEEG per second from tens and hundreds of channels collected over long periods of time. It is clear that strategies can be evolved so that there is a focused, programmatic evaluation of the collected data and so that each step of the review answers defined questions, with detailed review being conducted to answer very specific hypotheses. However, it is also clear that, with the advent of fully digital systems with high sampling frequencies, and high numbers of channels, there is both a considerable need and place for computer-assisted analysis of the collected data.

REFERENCES

Jung, W. G. 1986. *IC Op-Amp Cookbook*. Upper Saddle River, NJ: Prentice Hall.

Nyquist, H. 1928. Certain topics in telegraph transmission theory. *Trans. AIEE* 47:617–644.

Plonsey, D., and G. Fleming. 1969. *Bioelectric phenomena*. McGraw-Hill Series in Bioengineering. New York: McGraw-Hill.

Shannon, C. E. 1949. Communication in the presence of noise. *Proc. Institute of Radio Engineers* 37(1):10–21.

Webster, J. G. 2009. *Medical instrumentation: Application and design*. New York: Wiley.

7 Time-Frequency Energy Analysis

Piotr J. Franaszczuk

CONTENTS

This chapter presents basic methods of analysis of signals in the time-frequency domain. Since it is primarily directed to researchers interested in analysis of electrical activity of the human brain, we use a sample recording of nonstationary epileptiform activity as a real world example of the application of time-frequency methods.

This chapter starts with an introduction to spectral analysis in the frequency domain and then introduces the common time-frequency methods of short-time Fourier transform (STFT) and continuous wavelet transform (CWT). The last section provides a short description of the matching pursuit decomposition method (MP), one of the adaptive methods of time-frequency analysis. We do not discuss quadratic time-frequency energy methods beyond the Wigner-Ville distribution used as part of MP analysis. A more detailed theoretical basis for time-frequency methods is presented by Williams (1998), Carmona et al. (1998), and Mallat (1999).

7.1 FOURIER TRANSFORM

One of the common methods of investigation of natural processes is recording specific measurements over time. These measurements, represented as a function of time, are called signals or time series. Many natural phenomena (e.g., light and sound waves) exhibit periodicity. Recordings of electrical activity of the brain—electroencephalogram (EEG)—also show some degree of rhythmic activity. The analysis and interpretation of the spectral properties of these rhythms, known as alpha, beta, gamma, delta, and theta, have proved useful for diagnostic purposes. Also, frequency content of less rhythmic components can provide important information about a signal. Spectral analysis can provide this information. Almost all spectral analysis methods are related to Fourier series or Fourier transforms.

In the early nineteenth century, Joseph Fourier showed that any periodic function can be represented by the sum of sines and cosines representing elementary oscillations.

$$x(t) = \frac{a_0}{2} + a_1\cos(2\pi f_T t) + b_1\sin(2\pi f_T t) + a_2\cos(4\pi f_T t) + b_2\sin(4\pi f_T t) + \ldots$$

In more compact form

$$x(t) = \frac{a_0}{2} + \sum_{n=1}^{\infty} [a_n\cos(2\pi nf_T t) + b_n\sin(2\pi nf_T t)], \qquad (7.1)$$

where $x(t)$ is a periodic function with period $T = 1/f_T$. The oscillations with frequencies equal to multiples of the base frequency f are called harmonics. Coefficients $\{a_n\}$ and $\{b_n\}$ can be computed from Fourier integrals on one period of $x(t)$. Sometimes it is more convenient to use one cosine or sine function for each oscillation frequency (i.e., $A_n\cos(2\pi nf_T t + \varphi_n)$ where A_n is the amplitude and φ_n is the phase). The energy of one period of the signal associated with oscillation of frequency nf_T is represented by the sum of squares of a_n and b_n (or, in an alternative representation, the square of the absolute value of the amplitude, $|A_n|^2$). Energy values divided by the length of the time period represent the power of the signal. The power spectrum is the distribution of power of the signal as a function of frequency. For strictly periodic signals, power is nonzero only for discrete values of frequency resulting in a so-called line spectrum. However, continuous periodic functions may have an infinite number of spectral components. Figure 7.1 shows a simple periodic signal and the first four components of its power spectrum. The Fourier transform analysis is an extension of a harmonic

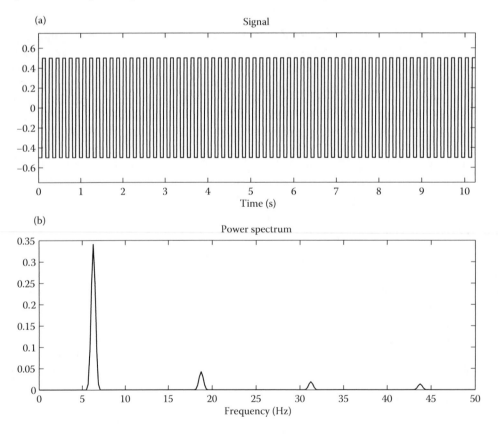

FIGURE 7.1 A simple periodic signal (a) and its power spectrum showing peaks at base frequency as well as 3 harmonics (b).

analysis of Fourier series for nonperiodic functions. In this case, there is no base frequency of the signal, and components of any frequency can be present in the decomposition. To represent this situation, we need to replace the sum of frequency components with an integral over all frequencies. To take advantage of easier calculations of complex exponential functions, as compared to trigonometric functions, we will use the relationship $X(f)\exp(i2\pi ft) = X(f)(\cos(2\pi ft) + i\sin(2\pi ft))$.

$$x(t) = \int\limits_{-\infty}^{\infty} df X(f)\exp(i2\pi ft). \tag{7.2}$$

The term $dfX(f)$ is analogous to coefficients a and b in Equation 7.1. $X(f)$ is a Fourier transform of $x(t)$ and can be calculated from

$$X(f) = \frac{1}{2\pi} \int\limits_{-\infty}^{\infty} dt\, x(t)\exp(-i2\pi ft). \tag{7.3}$$

$|X(f)|^2$ represents power density (power per frequency unit). In this case, the power spectrum is not restricted to specific frequencies but can be a continuous function of f.

In Equation 7.3, $X(f)$ is calculated from the values of $x(t)$ for all times t. This implies that $x(t)$ has the same spectral representation for all times. As long as $x(t)$ is a stationary and ergodic process, this describes the spectral properties of the signal well. In real applications, we have a limited time interval for observation. Simply cutting the integration interval in Equation 7.3 to the available signal epoch (e.g., $[-T/2,T/2]$) will produce spurious high-frequency components due to the abrupt changes of the signal at the beginning and end of the interval. The procedure of abruptly cutting the signal can be interpreted as multiplying it by the rectangular window function $h(t) = 1$ in the interval $[-T/2,T/2]$ and $h(t) = 0$ outside of it. To smooth the transition at the edges of the interval, we can apply different window functions—also called tapers—defined on a finite interval. Commonly used window functions are based on cosine (Hamming, Blackman), square of cosine (Hanning), or Gaussian defined on the interval $[-T/2,T/2]$. Formulas and properties of these window functions can be found in basic signal processing textbooks (Mallat 1999). The resulting Fourier transform can be represented as

$$X(f) = \frac{1}{2\pi} \int\limits_{-\infty}^{\infty} dt\, h(t)x(t)\exp(-i2\pi ft) = \frac{1}{T} \int\limits_{-T/2}^{T/2} dt\, h(t)x(t)\exp(-i2\pi ft), \tag{7.4}$$

where $h(t)$ is the smoothing window.

For purely deterministic signals, estimation of the Fourier transform and power spectrum from the whole window may provide satisfactory results. However, with stochastic signals and signals with significant noise components, the resulting estimate is characterized by large variability. To diminish this variability, repeated realizations of the same stochastic process can be averaged. If only one realization of the process is available, the common approach is to divide the original interval to smaller, possibly overlapping, epochs, calculate the power spectrum estimate for each epoch using a suitable window function and the Fourier Transform, and then average these results. Welch's method (Welch 1967) is the most popular method to estimate the power spectral density using this approach. It is implemented in the MATLAB® software package as the function *pwelch*. This method introduces a significant bias if the epochs are relatively short. To diminish this bias, a multitaper method (Thompson 1982) can be used, which, instead of dividing the whole interval into small epochs, uses an orthogonal set of window (taper) functions on the whole available signal length. Estimates obtained with each taper are then averaged. Both methods rely on the stationarity of the signal in the whole analysis period, and both methods reduce variance (noise) in the power spectrum estimate in exchange for reducing spectral resolution. In multitaper methods, this loss of resolution is due to the shape of the tapers and this will be easier to understand after we introduce the uncertainty principle later in this chapter. However, loss of resolution in methods using averaging of short segments can be explained by reexamining the Fourier series (Equation 7.1). If we treat each segment of length T as one period of a periodic function, we can represent it using the Fourier series from Equation 7.1. The minimal separation between each frequency component (i.e., frequency resolution) is thus equal to $1/T$. If we divide the interval T into two intervals of length $T/2$, the frequency resolution will be twice the previous value. Additional multiplication by a window function decreases this further, depending on the shape of the window.

The above approach works well if the assumption of stationarity of the signal is fulfilled. However, epileptiform activity is nonstationary and transient events characteristic for this activity need to be analyzed. Figure 7.2 shows an example of intracranial EEG recording at a seizure onset and its power spectrum calculated using Welch's method. Clearly, the signal is not stationary. The power spectrum shows the predominant frequency at the seizure onset but does not provide any information about whether this frequency component was present during the whole period of analysis, or if it started or stopped during this interval. To better describe spectral content of dynamically changing signals, we need to apply time-frequency methods of analysis.

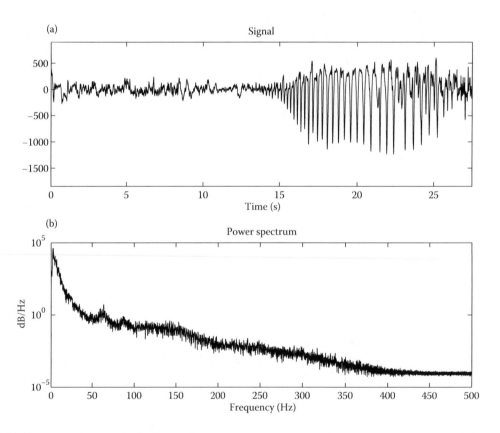

FIGURE 7.2 An intracranial EEG recording at a seizure onset (a), and its power spectrum calculated using Welch's method (b). The power spectrum is calculated assuming stationarity of the signal through the whole period of analysis. For nonstationary signals, only the most prominent frequency components are visible in the power spectrum.

7.2   WINDOWED FOURIER TRANSFORM

Welch's method of power spectrum estimation described in the previous section divides the epoch of analysis into smaller intervals and averages the results of each window, with the assumption that the signal is stationary. For nonstationary signals, we may employ a similar approach by using the Fourier transform defined in Equation 7.4. However, instead of dividing the interval into sub-intervals, we will apply a moving window and obtain a windowed Fourier transform in the time-frequency plane (u, f):

$$Sx(u, f) = \frac{1}{2\pi} \int\limits_{-\infty}^{\infty} dt \, h(t - u) \, x(t) \exp(-i2\pi ft). \tag{7.5}$$

This transform is also known as the Gabor transform, or STFT if the window $h(t)$ has nonzero values on small intervals as compared to the length of the analyzed epoch. The value of $E(u, f) = |Sx(u, f)|^2$ represents the distribution of energy of signal x on the time-frequency plane. For a dynamically

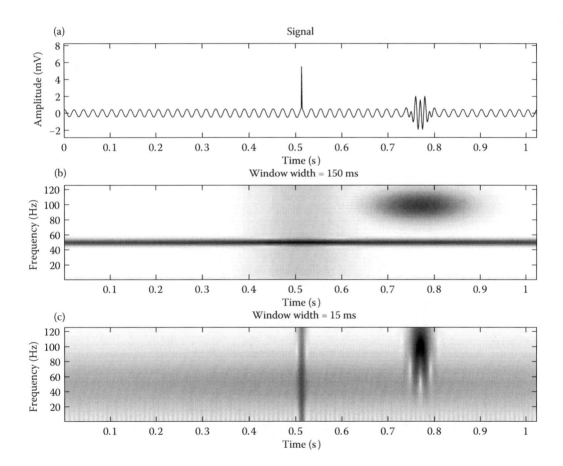

FIGURE 7.3 Effects of choosing different widths of the Gaussian window. An artificial signal composed of one sine function, one Gabor function, and one delta function (a). Energy distribution computed with a wider window ($w = 150$ ms) results in a better representation of the periodic sine function, but time localization of the delta function is very poor (b). A narrow window ($w = 15$ ms) is better suited for representing the energy of the delta function, which is well localized in time, but the energy of the sine function is spread over a broader frequency band (c).

changing nonstationary signal, we choose a window $h(t)$ with a maximum value of $t = 0$ and diminishing fast for both positive and negative values of t (e.g., a Gaussian function). The significant contribution from $x(t)$ to Sx for time u occurs only when $h(t - u)x(t)$ is significantly larger than zero. This means that values of $E(u,f)$ represent spectral content of signal $x(t)$ in the neighborhood of time $t = u$. Choosing a narrow window quickly diminishing to zero will improve the time resolution of $E(u,f)$. However, this will diminish the frequency resolution since the Fourier transform is computed from an effectively shorter interval. Figure 7.3 illustrates the effects of choosing different widths for the Gaussian window on an artificial signal composed of one sine, one Gabor, and one delta function. The energy distribution computed with a wider window better represents the periodic sine function, but time localization of the delta function is poor. A narrow window is better suited for representing the energy of the delta function well localized in time, but the energy of the sine function is spread over a broad frequency band. It is clear that the choice of window influences the resulting estimator of the energy distribution. Results of time-frequency analysis of intracranial recording are shown in Figure 7.4. Figure 7.4b shows the several rhythmic components with relatively well-defined low frequencies. Figure 7.4c, obtained with a much narrower window, shows the

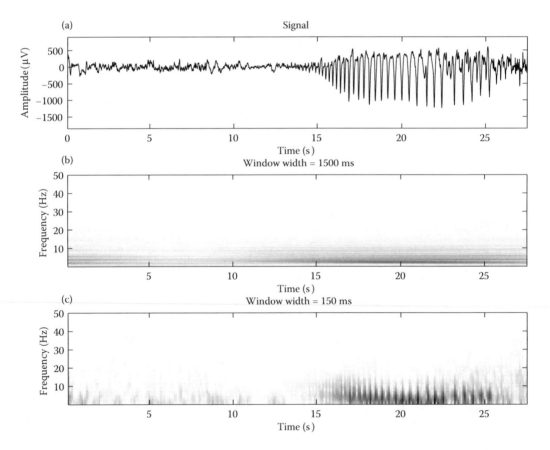

FIGURE 7.4 An example of intracranial EEG recording (a) showing time-frequency energy distribution of the signal using a wide window ($w = 1500$ ms) (b). After seizure onset, a horizontal energy concentration representing rhythmic components is visible. Time-frequency energy distribution calculated with a narrower window ($w = 150$ ms) (c). It shows a series of transient broadband components representing local maxima in the signal. The grey scale reflects the amount of energy associated with specific time-frequency components; here, black is maximum and white is minimum.

signal as a series of transient broadband components. It is clear that increasing the time resolution decreases the frequency resolution and vice-versa. We will discuss this property and its importance in interpreting time-frequency energy distributions in Section 7.4.

7.3   SIGNAL DECOMPOSITION INTO TIME-FREQUENCY ATOMS

In Section 7.1, we introduced the Fourier series (Equation 7.1), which is a representation of a signal in time as a sum of elementary oscillations. The inverse Fourier transform (Equation 7.2) represents an integral version of this sum and can be interpreted as the decomposition of $x(t)$ using oscillatory functions $g_f(t) = \exp(i2\pi ft)$. The notation $g_f(.)$ indicates that there is a separate function of time for each frequency f.

Similarly (see Mallat (1999) for necessary assumptions), we can use a reconstruction formula to represent signal $x(t)$ in terms of a windowed Fourier transform:

$$x(t) = \int_{-\infty}^{\infty} df \int_{-\infty}^{\infty} du\ Sx(u,f)h(t-u)\exp(i2\pi ft). \tag{7.6}$$

In this formula, $x(t)$ is decomposed into time-frequency components $g_{u,f}(t) = h(t-u)\exp(i2\pi ft)$ called time-frequency atoms or waveforms. We can interpret function $g_{u,f}(t)$ as a translation by u in time and f in frequency of basic window $h(t)$.

Substituting $g_{uf}(t)$ in Equation 7.5 we obtain

$$Sx(u,f) = \frac{1}{2\pi} \int_{-\infty}^{\infty} dt x(t) g_{u,f}^*(t) = \left\langle x, g_{u,f} \right\rangle \tag{7.7}$$

where * indicates complex conjugation.

This formula represents the correlation of functions $x(t)$ and $g_{u,f}(t)$ and can also be interpreted as their inner product $<x,g_{u,f}>$ in Hilbert space. Using this notation in Equation 7.6, we obtain

$$x(t) = \int_{-\infty}^{\infty} df \int_{-\infty}^{\infty} du \left\langle x, g_{u,f} \right\rangle g_{u,f}(t). \tag{7.8}$$

Thus signal $x(t)$ is represented by its inner products with functions $g_{u,f}$.

This representation is also useful in application to discrete finite signals. A discrete signal $x[t]$, $t = 1..N$, can be treated as a vector in N-dimensional Euclidian space. The discrete functions $g_{u,f}[t]$, $t = 1..N$ are also N-dimensional vectors, and inner product $<x,g_{u,f}>$ is defined as a dot product of these vectors:

$$\left\langle x, g_{u,f} \right\rangle = \sum_{t=1}^{N} x[t] g_{u,f}[t]. \tag{7.9}$$

For functions $g_{u,f}[t] = h[t-u]\exp[i2\pi ft]$, Equation 7.9 represents a discrete Fourier transform of $x(t)h(t-u)$ and can be efficiently computed using the fast Fourier transform algorithm (FFT).

We can choose a family of discrete functions $\{g_n\}_{n\in\Gamma}$ and represent signal $x[t]$ as a linear combination of these functions with inner products as coefficients:

$$x[t] = \sum_{n \in \Gamma} \langle x, g_n \rangle g_n[t]. \tag{7.10}$$

Dependent on the choice of this family, the representation may be approximate, complete, or redundant. If the functions are orthonormal (i.e., the inner product of every pair is equal to zero and the norm of each function is one), this family forms a basis of an N-dimensional space and the representation of $x[t]$ in Equation 7.10 is complete and unique in this base. The orthonormality requirement is not necessary for a family of functions to define a complete and stable representation of the signal. These families of functions are called frames, and frame theory provides the formalism for analyzing their properties (Mallat 1999).

We are interested here in the representation of the energy of the signal on the time-frequency plane. In this case, we need to choose functions g_n with good localization in time and frequency, so that their inner products with signal will represent spectral content well for a given time. It means we need functions that are well localized in time and whose power is concentrated in a narrow frequency band. Figure 7.5 shows some examples of Gabor functions,

$$g_{u,f}(t) = A \exp\left(-\frac{(t-u)^2}{2s^2}\right)\cos(2\pi f t + \varphi), \tag{7.11}$$

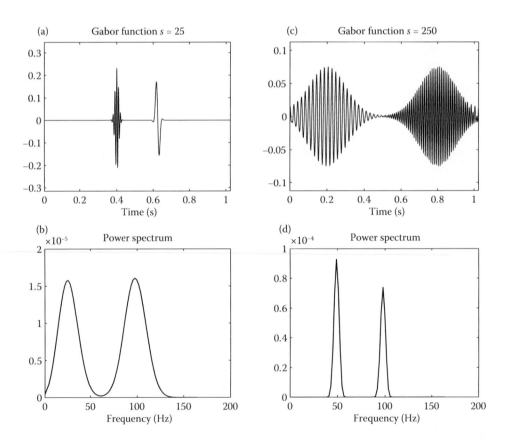

FIGURE 7.5 An example of Gabor atoms with different scales and frequencies and their power spectra: (a) and (b) with a Gaussian of $s = 25$, and (c) and (d) with a Gaussian of $s = 250$. The width of peak in power spectra is inversely proportional to the width of the Gabor function in the time domain.

and their power spectra. It is clear that a narrower function in time has a broader peak in power spectrum. This is in agreement with results of STFT analysis illustrated in Figures 7.3 and 7.4, where the time and frequency spread of components was determined by the window used.

7.4 UNCERTAINTY PRINCIPLE

In previous sections we illustrated with examples that improving localization of energy of the signal in time is accompanied by a decrease in frequency resolution. It suggests that we cannot characterize signal energy simultaneously in time and frequency with arbitrary precision. This restriction in precision of observation was first described by Werner Heisenberg in relation to uncertainty of the position and momentum of a free particle. In the context of distribution of the energy of a signal, we can formalize this uncertainty by defining the temporal spread σ_t and the frequency spread σ_ω of a function $g(t)$ as

$$\sigma_t = \left[\frac{1}{2\pi\|g\|^2} \int_{-\infty}^{\infty} dt\, (t - t_0)^2 |g(t)|^2 \right]^{1/2} \tag{7.12}$$

and

$$\sigma_\omega = \left[\frac{1}{\|G\|^2} \int_{-\infty}^{\infty} d\omega\, (\omega - \omega_0)^2 |G(\omega)|^2 \right]^{1/2} \tag{7.13}$$

where, $G(\omega)$ denotes the Fourier transform of $g(t)$, $\|g\|$ denotes the norm of the function $g(t)$ and $\|G\|$ denotes the norm of $G(\omega)$; both are equal to the total energy of the signal, $\omega = 2\pi f$, and t_0 and ω_0 denote the center of energy concentration on the time-frequency plane, defined as

$$t_0 = \frac{1}{2\pi\|g\|^2} \int_{-\infty}^{\infty} dt\, t\, |g(t)|^2 \tag{7.14}$$

and

$$\omega_0 = \frac{1}{\|G\|^2} \int_{-\infty}^{\infty} d\omega\, \omega\, |G(\omega)|^2. \tag{7.15}$$

Note that Equations 7.12 through 7.15 can be interpreted as standard deviations and means of the true location of the energy concentration if we treat $|g(t)|^2/\|g\|^2$ and $|G(\omega)|^2/\|G\|^2$ as probability density functions in time and frequency space, respectively.

The uncertainty principle (Heisenberg uncertainty) in time-frequency space states that

$$\sigma_t \sigma_\omega \geq \frac{1}{2}. \tag{7.16}$$

The inequality in Equation 7.16 is an equality if function $g(t)$ is a Gabor function, i.e., modulated Gaussian (see Equation 7.11). The time-frequency boxes (Heisenberg boxes), are rectangles of size

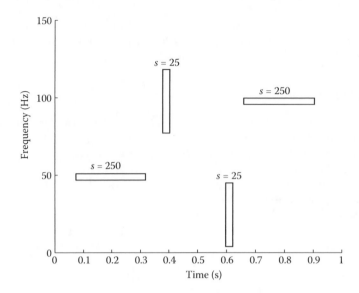

FIGURE 7.6 Time-frequency boxes representing the energy spread of Gabor atoms shown in Figure 7.5. The size of rectangles is the same for Gabor atoms with the same s, independent of translations in time and frequency.

σ_t by σ_ω, and can be used to illustrate the time-frequency resolution of signal components. Figure 7.6 shows Heisenberg boxes for the Gabor functions shown in Figure 7.5. The sizes of these boxes for any functions used in windowed Fourier transforms are independent of location on the time-frequency plane. They depend only on the shape of window function $h(t)$. This means that the time-frequency energy distribution obtained with STFT has a constant time-frequency resolution determined by choice of window (Figures 7.3 and 7.4). In the next sections, we will show that this is not always the case.

7.5 WAVELETS AND WAVELET TRANSFORMS

The short-time Fourier transform provides decomposition of signals into time-frequency components from the family of functions $g_{u,f} = h(t - u)\exp(i2\pi ft)$ (Equations 7.6 and 7.8). Functions in this family are generated by translating function $h(t)$ in time and frequency by u and f, respectively. Wavelets are another set of functions well suited to represent nonstationary signals. A wavelet is a function with zero average over the whole time scale, which is well localized in time (i.e., it has large values only in a relatively small interval). In this respect, it is similar to functions discussed in the previous section. Different families of wavelets are generated by scaling a basic function known as the mother wavelet $w(t)$ by factor s, and translating it in time by u:

$$w_{u,s}(t) = \frac{1}{\sqrt{s}} w\left(\frac{t-u}{s}\right). \tag{7.17}$$

There are many functions suitable for generating wavelet families. Figure 7.7 shows an example of Morlet wavelets with different scales and their power spectra.

We can define the continuous wavelet transform (CWT) of $x(t)$ similarly to the windowed Fourier transform in Equation 7.7 as

$$Vx(u,s) = \frac{1}{2\pi} \int\limits_{-\infty}^{\infty} dt\, x(t) w_{u,s}(t) = \left\langle x, w_{u,s} \right\rangle. \tag{7.18}$$

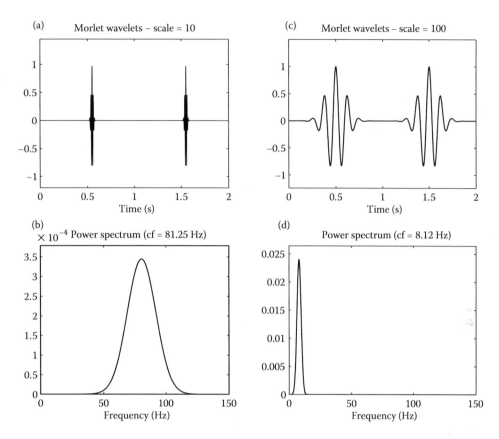

FIGURE 7.7 An example of Morlet wavelets with different scales and their power spectra: (a) and (b) with scale $s = 10$, and (c) and (d) with scale $s = 100$. The central frequency of each wavelet is determined by the wavelet type and its scale: 81.25 Hz for a Morlet wavelet with scale $s = 10$, 8.125 Hz for a scale that is 10 times larger, $s = 100$.

The wavelet transform is a function of time and scale. Time-frequency resolution defined by spread in time and frequency can be computed for wavelets using Equations 7.12 through 7.15 by substituting wavelet $w_{u,s}(t)$ and its Fourier transform for $g(t)$ and $G(\omega)$, respectively. The translation parameter u defines the center of energy concentration of $w_{u,s}(t)$ similarly to functions $g_{u,f}(t)$ in previous sections. Spread in time is proportional to scale and equals $s\sigma_t$, where σ_t is the time spread of the mother wavelet. Its Fourier transform $W_{u,s}(\omega)$ can be calculated from

$$W_{u,s}(\omega) = \exp(-iu\omega)\sqrt{s}\,W(s\omega),\tag{7.19}$$

where $W(\omega)$ is the Fourier transform of mother wavelet $w(t)$. We are interested only in positive frequencies and use the analytic wavelet for which $W(\omega) = 0$ for $\omega < 0$. From Equation 7.19, we can deduce that if the energy of $W(\omega)$ is concentrated at positive frequency ν with spread σ_ω, the energy of $W_{u,s}(\omega)$ is concentrated at frequency ν/s with spread σ_ω/s. Figure 7.8 illustrates how the time-frequency resolution of wavelets from the same family changes with scale. For smaller scales, the time spread is smaller but the frequency spread is larger. Since the energy of wavelets of smaller scales is centered in higher frequencies, the time-frequency resolution of energy distribution in wavelet representation is different for different frequencies. It is well suited for many "real world" signals, including EEG, where lower frequency rhythmic activity occurs in longer intervals than high frequency components representing short transients.

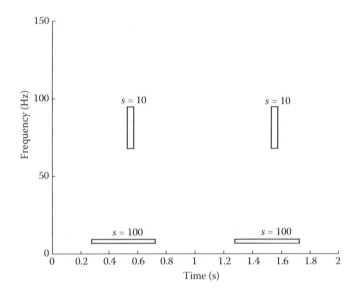

FIGURE 7.8 Time-frequency boxes representing the energy spread of wavelets shown in Figure 7.7. For smaller scales ($s = 10$), the boxes are longer in frequency than in time; for larger scales ($s = 100$), the boxes are longer in time than in frequency. The location on the time-frequency plane is dependent on scale. The boxes for smaller scales are centered in higher frequencies. This indicates that wavelet decompositions have higher time resolution and lower frequency resolution for higher frequencies than for lower frequencies.

The stable and complete signal representation can be achieved by appropriate selection of discrete families of wavelet or windowed Fourier functions. The time-frequency boxes of functions from these families provide a tiling of the time-frequency plane with full, sometimes redundant coverage. The windowed Fourier function family $\{g_{u,f}\}$ generated by uniform translations in time and frequency provide coverage with boxes of constant size on the whole plane. The wavelet functions with discrete scale provide coverage of the time-frequency plane with boxes of the same size for a given scale, but changing with frequency. If the scale is made discrete along a dyadic sequence (i.e., $s_k = 2^k$), the numerical calculations can be performed with a fast computational algorithm.

Figure 7.9 shows the time-frequency energy distribution of the intracranial EEG example shown in Figures 7.2a and 7.4a, calculated with the CWT (using MATLAB function *cwt*). Figure 7.9b shows the energy distribution in time-scale plane. The frequency scale in Figure 7.9c is in fact a pseudo-frequency calculated from the scales using formula $f = f_0/s$, where f_0 is the center frequency of the mother wavelet (we used the MATLAB function *scal2frq*).

There are other families of functions (e.g., wavelet packet bases, or local cosine bases) providing different coverage of time-frequency space (Mallat 1999) that might be suitable for specific purposes.

7.6 MATCHING PURSUIT TIME-FREQUENCY DECOMPOSITION

A signal can be represented completely in Fourier or wavelet basis. In real world applications, the signal contains noise that is not relevant to the phenomenon under investigation. We are interested in a time-frequency decomposition of the relevant part of the signal and not the noise. In this case, in Equation 7.10, we may use only a subset of all functions needed for full representation. If the subset is chosen a priori, the approximation is linear. However, if the subset of functions approximating the signal is chosen in a manner where it depends on the signal, the approximation is nonlinear. The functions best describing the features of the signal are selected based on predetermined criterion.

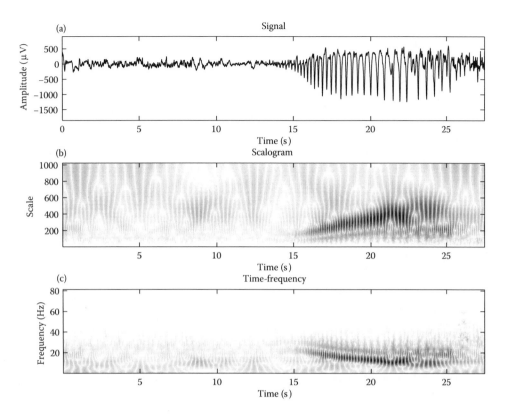

FIGURE 7.9 The energy distribution of an intracranial EEG recording calculated using continuous wavelet transform with Morlet wavelets. An example of intracranial EEG recording (the same as that shown in Figures 7.2a and 7.4a (a). A scalogram representing the proportion of energy of each wavelet component on the time-scale plane (b). A time-frequency energy representation of CWT coefficients (c). The frequency scale is representing the central frequency of wavelets for each scale. The grey scale reflects the amount of energy associated with specific wavelets normalized to total energy; here, black is maximum and white is minimum.

One such adaptive method of signal approximation with time-frequency atoms, called matching pursuit, was introduced by Mallat and Zhang (1993).

A matching pursuit algorithm starts by choosing from the large dictionary of functions $\{g_n\}_{n\in\Gamma}$ function g_{n0} such that absolute value of its inner product with signal $x(t)$ $|\langle x, g_{n0}\rangle|$ is maximum. Then, the signal is represented by

$$x(t) = \left\langle x, g_{n0}\right\rangle g_{n0}(t) + R_x^1(t) \tag{7.20}$$

where $R_x^1(t)$ is a residue. The next step finds the function g_{n1}, which maximizes $|\langle R_{x1}, g_{n1}\rangle|$. After m steps, the mth residue $R_x{}^m(t)$ is

$$R_x^m(t) = \left\langle R_x^m, g_{nm}\right\rangle g_{nm}(t) + R_x^{m+1}(t). \tag{7.21}$$

Substituting $R_x^m(t)$ for $m = 1$ to $M - 1$, and $x = R_0^m$ in Equation 7.20 results in the representation of signal $x(t)$ as:

$$x(t) = \sum_{m=0}^{M-1} \left\langle R_x^m, g_{nm}\right\rangle g_{nm}(t) + R_x^M(t), \tag{7.22}$$

and its energy as

$$\|x\|^2 = \sum_{m=0}^{M-1} \left| \left\langle R_x^m, g_{nm} \right\rangle \right|^2 + \left\| R_x^M \right\|^2. \tag{7.23}$$

The last equation assumes that the functions g_{nm} have norms equal to 1. The pursuit is stopped after a preset number of iterations, after a certain proportion of the total energy of the signal is accounted for, or when the energy of the function g_{nm} at the last step falls below a certain threshold.

If the dictionary functions are well localized in time and frequency, matching pursuit decomposition provides a good representation of the signal in time-frequency space. A common choice for the dictionary of functions is a family of scaled Gabor functions:

$$g_{u,f,s}(t) = A_s \exp\left(-\frac{(t-u)^2}{2s^2} \right) \cos(2\pi ft + \varphi), \tag{7.24}$$

where A_s is a coefficient normalizing function to unit norm. These functions are generated from the Gaussian function by translation in time and frequency as well as scaling, thus combining properties of windowed Fourier functions and wavelets. This dictionary is usually supplemented by sine functions and Dirac delta functions characterized by perfect localization in frequency and time, respectively.

To represent the energy of the signal as a sum of energy of dictionary functions in the time-frequency plane, we can use the Wigner-Ville distribution for each component, which is defined as

$$Pf(u,\omega) = \int_{-\infty}^{\infty} f\left(u + \frac{t}{2} \right) f\left(u - \frac{t}{2} \right) \exp(-it\omega)d\omega. \tag{7.25}$$

This distribution belongs to a larger family of quadratic time-frequency transformations (Williams 1998). They can be used to analyze time-frequency structures of the signal itself. Here, we use them to represent time-frequency energy concentrations of each component of the signal decomposition in Equation 7.22.

The Wigner-Ville distribution preserves the center of energy concentration of the function, and its time-frequency spread is equal to the time-frequency spread of the function. Gabor functions are represented as 2D Gaussians, sine functions as lines parallel to the time axis, and delta functions as lines parallel to the frequency axis.

The application of the matching pursuit method to analysis of intracranial EEG provides useful information about dynamical changes in spectral content before, during, and after seizures (Franaszczuk et al. 1998). Figure 7.10 shows a time-frequency energy distribution of the same intracranial EEG shown in Figures 7.2a, 7.4a, and 7.9a, calculated using matching pursuit decomposition with the Gabor function dictionary. This time-frequency representation describes well the dynamic character of the signal.

The application of stopping criterion in matching pursuit algorithm, not based on number of iterations, results in a different number of components in the decomposition of different signals. The number of components needed to explain the same proportion of energy for a simple periodic signal is smaller than for a highly irregular, chaotic signal. This measure was successfully applied to intracranial recordings of seizures (Bergey and Franaszczuk, 2001). The Gabor atom density (GAD) (Jouny et al. 2003) is a normalized measure of complexity based on matching pursuit, and proved useful in characterizing the onset of seizures (Jouny et al. 2010).

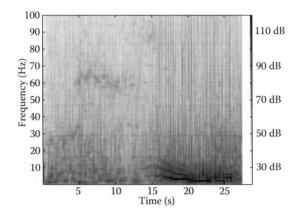

FIGURE 7.10   (See color insert.) A time-frequency distribution of energy of the same intracranial EEG signal shown in Figures 7.2a, 7.4a, and 7.9a. The matching pursuit analysis was performed with the $n = 1000$ iterations. The dictionary functions included Gabor functions, Dirac delta functions, and sine functions. The color scale represents a logarithm of energy associated with each point on the time-frequency plane calculated using the Wigner-Ville distribution for all atoms in the decomposition. Here, black represents the maximum and white the minimum. In color plate red represents maximum and blue the minimum.

7.7   SUMMARY

Time-frequency methods are powerful tools for the analysis of nonstationary signals. They can provide detailed information about dynamical changes in EEG recordings from epileptic patients. However, there is no unique representation of the signal in time-frequency space. The choice of a particular method and its parameters should be done carefully, depending on the particular application. The STFT method has the advantages of a straightforward interpretation of results in time-frequency space, and the existence of a fast computational algorithm. The choice of window in this method has a great influence on the results. The methods based on wavelets also have the advantage of fast computational algorithms, and the coverage of the time-frequency plane is more suited to the analysis of typical signals, including EEG. However, these are essentially time-scale methods and interpretation in time-frequency space may not be straightforward. The methods best suited for the investigation of time-frequency properties of noisy signals are adaptive methods like matching pursuit. The decompositions into a variable number of time-frequency atoms provide a measure of signal complexity in addition to time-frequency distributions. However, they are much more computationally intensive and not yet suitable for applications requiring fast analysis in real time.

REFERENCES

Bergey, G. K., and P. J. Franaszczuk. 2001. Epileptic seizures are characterized by changing signal complexity. *Clin. Nerophysiol.* 112:241–249.

Carmona, R., W.-L. Hwang, and B. Torresani. 1998. *Gabor and wavelet transforms with an implementation in S.* Vol. 9 of *Practical time-frequency analysis.* San Diego: Academic Press.

Franaszczuk, P. J., G. K. Bergey, P. J. Durka, and H. M. Eisenberg. 1998. Time-frequency analysis using the matching pursuit algorithm applied to seizures originating from the mesial temporal lobe. *Electroencephalogr. Clin. Neurophysiol.* 106:513–521.

Jouny, C. C., P. J. Franaszczuk, and G. K. Bergey. 2003. Characterization of epileptic seizure dynamics using Gabor atom density. *Clin. Neurophysiol.* 114:426–437.

Jouny, C. C., G. K. Bergey, and P. J. Franaszczuk. 2010. Partial seizures are associated with early increases in signal complexity. *Clin. Neurophysiol.* 121:7–13.

Mallat, S., and Z. Zhang. 1993. Matching pursuits with time-frequency dictionaries. *IEEE Trans. Signal. Proc.* 41:3397–3415.

Mallat, S. 1999. *A wavelet tour of signal processing,* 2nd ed. San Diego: Academic Press.

Thomson, D. J. 1982. Spectrum estimation and harmonic analysis. In *Proceedings of the IEEE* 70:1055–1096.

Welch, P. D. 1967. The use of fast Fourier transform for the estimation of power spectra: A method based on time averaging over short, modified periodograms. *IEEE Transactions on Audio Electroacoustics* AU-15:70–73.

Williams, J. W. 1998. Recent advances in time-frequency representations: Some theoretical foundation. In *Time frequency and wavelets in biomedical signal processing,* ed. M. Akay, 3–43. New York: IEEE Press.

8 Neurodynamics and Ion Channels
A Tutorial

John G. Milton

CONTENTS

8.1 INTRODUCTION

Autosomal dominant nocturnal frontal lobe epilepsy (NFLE) is associated with a defect in the alpha 4 subunit of the nicotinic acetylcholine-gated ion channel (Mann and Mody 2008). Since the engine for neuronal excitability is the membrane ion channel, it is not surprising that there would be a relationship between epilepsy and ion channel abnormalities. The puzzle arises because the ion channel abnormality is a fixed deficit, yet seizures are paroxysmal events.

At the most fundamental level, a seizure represents a change in the activity of neurons. Physiological (Abeles et al. 1990) and metabolic (Lennie 2003) considerations indicate that the cortical interictal state is primarily characterized by low-frequency periodic neuronal spiking. The initial event for seizures is likely associated with a change in the spiking pattern of a subpopulation of neurons (Babb et al. 1989; Prince 1969; Shusterman and Troy 2008). The spiking frequency becomes higher and the pattern more periodic. Although computational neuroscientists have identified the mechanisms by which changes in neural activity occur, the fact that these concepts are expressed using mathematical terminologies creates a formidable barrier for clinical epileptologists. Thus, discussions between those who have access to clinical data and those who develop models are inhibited and scientific progress impeded.

The purpose of this chapter is to introduce to clinicians the way that scientists who develop mathematical models think about epilepsy, neurons, and ion channels. The discussion focuses on the success story of computational neuroscience, namely, the development of the Hodgkin–Huxley (HH) model that describes the spiking activities of neurons. Although these equations can be easily derived using the basic principles of electricity taught in introductory physics courses required for admission into medical school, it is surprising that few clinicians appreciate the rich range of dynamical behaviors that single neurons can generate. A fundamental problem is that the mathematical models are sufficiently complex that solutions cannot be obtained analytically, i.e., using paper and pencil. If this point is kept in mind, it become easier to appreciate that much of the effort in modeling is directed toward identifying ways to test predictions even though the solutions can only be obtained using computer simulations. This observation motivates a consideration of qualitative techniques that can be used even by nonmathematically oriented epileptologists to propose key experiments that can be tested at the bench top and bedside. This procedure is illustrated in Chapter 38 (Milton et al. 2010).

8.2 DYNAMICAL SYSTEMS

Dynamics is concerned with the description of how variables change as a function of time. A variable is anything that can be measured. Examples include the number of neurons, the membrane voltage potential, the number and types of membrane receptors, and so on. All variables in the nervous system change as a function of time. Figure 8.1 shows a time series for a hypothetical variable. In the physical sciences, the hypothesis that the change in a variable, rather than the magnitude of the variable itself, is important for the development of mathematical models has enjoyed great success. Consequently, the starting point for models of the nervous system is based on the same premise.

The change in the variable x between times t_1 and t_2, denoted $\Delta x/\Delta t$, as

$$\frac{\Delta x}{\Delta t} = \frac{x(t_2) - x(t_1)}{t_2 - t_1} = \frac{x(t+h) - x(t)}{h}, \tag{8.1}$$

$$= \frac{\text{rise}}{\text{run}}, \tag{8.2}$$

where h is the time step, i.e., $h = t_2 - t_1$. It is natural to ask what happens as the time interval becomes as small as possible, i.e., as $h \to 0$. This is the question that Newton posed and leads to the definition of the derivative

$$\frac{dx}{dt} = \lim_{h \to 0} \frac{x(t+h) - x(t)}{h}. \tag{8.3}$$

The change in the variable as a function of time, i.e., the predicted time series, is the solution of the differential equation

$$\dot{x} \equiv \frac{dx}{dt} = f(x), \tag{8.4}$$

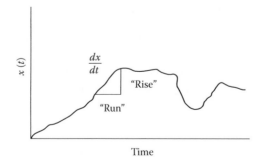

FIGURE 8.1 Plot of a variable, x, as a function of time.

where we have introduced the dot notation for the derivative. The left-hand side of Equation 8.4 reiterates the concept that the change in the variable per unit of time is the important variable. The right-hand side states the hypothesis proposed to govern the changes in the variable.

A key point is that the time scale, i.e., the time taken for a significant part of the change to occur, differs markedly for different variables. For example, the binding of a drug to a membrane receptor takes $\leq 10^{-9}$ s, whereas the number of receptors changes on time scales of hours, days, and even weeks. Thus what we observe at any instant in time is the consequence of many processes evolving on many time scales. One way to handle this complexity is to introduce the concept of a parameter. A parameter is a variable that changes so slowly, or so quickly, that in comparison to the time scale of variables of interest that it can be regarded as constant. Thus, for example, if we are interested in the dynamics of drug binding to a receptor, the number of receptors can be regarded to be a parameter.

The mantra of dynamical systems approaches to the nervous system is that dynamics can qualitatively change as parameters change (Guckenheimer and Holmes 1990). This observation is closely tied to the concept of stability, i.e., the resistance to change. Historically, many experiments were performed on systems that were either at equilibrium or steady state. Mathematically, this corresponds to setting $\dot{x} = 0$ in Equation 8.4: The values of x for which $\dot{x} = 0$ are called the fixed-points, $\hat{x}$, since if we choose the initial value $x = \hat{x}$, the system remains there indefinitely. Now suppose we gently perturb a system at steady state and define $u = x - \hat{x}$. If u is small enough, we can anticipate that the effects of this perturbation can be descried as

$$\dot{u} = ku, \tag{8.5}$$

where k is a parameter. The solution is

$$u(t) = U_0 e^{kt}, \tag{8.6}$$

where U_0 represents the initial deviation from the fixed-point due to a perturbation.

Figure 8.2 shows that the qualitative behavior of this solution depends on the value of k. If $k < 0$, then the system returns eventually to its fixed-point value after the perturbation. Thus we say that the fixed-point is stable to the perturbation. On the other hand, if $k > 0$, then the system diverges away from the fixed-point and, hence, the fixed-point is said to be unstable to the perturbation. The value $k = 0$ is called a bifurcation point, or stability boundary, since as k crosses this boundary there is an abrupt, qualitative change in behavior, i.e., following the perturbation the deviation away

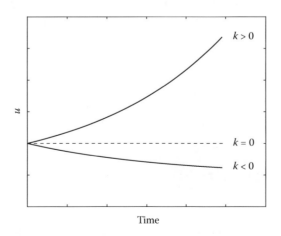

FIGURE 8.2 Comparison of the dynamics of Equation 8.6 for different choices of the parameter k.

from the fixed-point either increases or decreases. The advent of the computer led to the remarkable discovery that persistent periodic and aperiodic (quasiperiodic, chaotic) solutions could arise in situations in which fixed-points were unstable. This observation serves to remind us that discussions of the stability of a fixed-point are local, i.e., the perturbed system must be close enough to the fixed-point so that the approximation given by Equation 8.5 holds. Neural systems contain important nonlinearities that operate far from equilibrium that result in stable nonstationary behaviors. This added complexity of nonlinear systems motivated the use of the term attractor for behaviors that are stable to perturbations and repellor for those behaviors that are not. Thus we can have a fixed-point attractor, a limit cycle attractor, a limit cycle repellor, and so on.

8.3 NEURODYNAMICS

The very same principles used to understand an RC circuit in a physics lab can be used to understand how an action potential is generated by a neuron (Guevara 2003; Hille 2001; Hodgkin and Huxley 1954). The key experimental observation was that the electrical resistance of a neuron is variable (~20–10^6 Ωcm) and has values that lie intermediate between those obtained for ionic solutions (~20 Ωcm) and those obtained for pure lipid bilayers (~10^{15} Ωcm) (Hille 2001). This led to the concept that membrane resistance was determined by the opening and closing of holes in the membranes, i.e., the ion channels (Guevara 2003; Hodgkin and Huxley 1954): Resistance falls when the channels are open and increases when channels close (Figure 8.3a). In other words, compared to the RC circuit (Figure 8.3b), the resistance of a neuronal membrane, R, is not a parameter but a variable.

Suppose initially that all of the ion channels are closed. Then, the capacitance, C, across the neuron membrane is

$$Q = CV, \tag{8.7}$$

where Q is the charge stored across the membrane and V is the potential, or voltage, difference. When an ion channel opens, current, I, flows, and C changes according to

$$C\frac{dV}{dt} = \frac{dQ}{dt} \equiv I. \tag{8.8}$$

Hodgkin and Huxley (1954) hypothesized that the generation of an action potential is determined by the opening and closing of three types of ion channels: sodium (Na^+), potassium (K^+), and a leak (L) ion channel. Since ion channels function independently, Equation 8.8 becomes

$$C\dot{V} = -\left[I_{Na} + I_K + I_L \right], \tag{8.9}$$

(a)   (b)

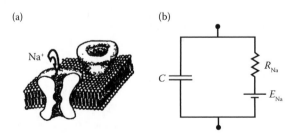

FIGURE 8.3 (a) Schematic representation of ion channel inserted into lipid bilayer. (b) Electrical equivalent RC circuit.

where the negative sign reflects the convention that the inside of the neuron is negative with respect to the outside. Next, they assumed that for each channel (linear membrane hypothesis) (Hille 2001)

$$I_X = g_X(V - V_X),$$ (8.10)

where $g_x = R^{-1}$ is the conductance of the ion channel, and hence Equation 8.9 becomes

$$C\dot{V} = -\left[g_{Na}(V - E_{Na}) + g_K(V - E_K) + g_L(V - E_L) \right],$$ (8.11)

where the terms E_X are the equilibrium potentials determined for each ion from the Nernst equation. However, experimental measurements indicate that the conductances are not constant and, hence, the g_X are not parameters, as assumed in a RC circuit in physics, but are variables. What this means is that we must supplement Equation 8.11 with differential equations that describe the changes in the conductances as a function of time.

The differential equations that describe the ion conductances were determined by using the patch clamp technique to isolate single ion channels and then measuring the openings of ion channels when the membrane potential was clamped at values from -100 mV to $+50$ mV. It is observed that each ion channel (even those of the same type) opens after a variable delay and remains open until the voltage clamp is turned off. An ensemble average is obtained by averaging many such trials for each type of ion channel. Hodgkin and Huxley assumed that the average of multiple trials recorded from the same ion channel is the same as the average for one response measured at the same time for a large number of different channels. In other words, the time course of activation of the ensemble average that appears in the mathematical model is connected with the variable latencies to first opening individual ion channels. Thus we see that a differential equation, such as Equation 8.4, describes the average behavior of a stochastic dynamical system.

Since changes in membrane resistance are due to the opening and closing of ion channels it was assumed that

$$g_X(t) = \overline{g}_K n(t),$$

where n is a gating variable that controls the opening and closing of the channel. Curve-fitting techniques combined with simple models describing ion channel dynamics were used to conclude that for the K$^+$channel

$$g_K(t) = \overline{g}_K n^4(t)(V - E_K)$$
$$\dot{n} = \alpha_n(1 - n) - \beta_n n$$

where α_n and β_n are the rate constants for, respectively, the opening and closing of the ion channel.

The same approach was applied to determining the contribution of the Na-current. This current is more complex than the K-current since two gating variables are required to describe its dynamics: an activation gating variable, m, and an inactivation gating variable, h. The equations that describe the Na-current are

$$g_{Na} = g_{Na} m^3 h(V - E_{Na})$$
$$\dot{m} = \alpha_m(1 - m) - \beta_m m.$$
$$\dot{h} = \alpha_h(1 - h) - \beta_h h$$

The general form of the HH equation is a system of equations: one is of the form of Equation 8.11 and the others are differential equations that take into account the openings and closings of the various types of ion channels. The specific equations obtained by Hodgkin and Huxley are

$$C\dot{V} = -\left[g_{Na}m^3 h(V - E_{Na}) + g_k n^4 (V - E_K) + g_L (V - E_L) + I_{ext} \right]$$

$$\dot{m} = \alpha_m(V)(1 - m) - \beta_m(V)m \qquad\qquad (8.12)$$

$$\dot{h} = \alpha_h(1 - h) - \beta_m(V)h$$

$$\dot{n} = \alpha_n(1 - n) - \beta_n(V)n.$$

In the context of this model, the generation of the action potential arises because the time scales for the sodium and potassium conductances are different; in particular, the sodium conductance changes on time scales shorter than the potassium conductance.

The framework developed by Hodgkin and Huxley has proven to be very robust for the description of the spiking behaviors of neurons and even other excitable cells including cardiac, pancreatic β–cells, and muscle cells. As new types of ion currents have been discovered, these have been incorporated into the framework by using the patch clamping and model-based curve-fitting techniques to model the gating of these channels. The resulting equations rapidly become exceedingly complex. For example, the HH-type model that describes the leech heartbeat neuronal network includes two different sodium currents, three different potassium currents, and two calcium currents resulting in fourteen differential equations that operate on multiple time scales that vary from a few ms to s (Hill et al. 2001). In parallel, the possible neuronal spiking behaviors become more complex including, for example, bursting and chattering behaviors (Izhikevich 2004; Wilson 1999). However, it has become clear that all experimentally observed cortical neuron spiking patterns can be described by simplified mathematical models based on the HH framework (Izhikevich 2004; Wilson 1999). Thus, mathematically guided computational studies have become a necessary tool for the investigation of neurodynamics.

8.4 SINGLE NEURON DYNAMICS

It is likely that every student who has taken an introductory biology course is familiar with the statement that the action potential of a neuron is described by the HH equation. However, few realize that this equation successfully predicted that a wide range of dynamical behaviors can be generated by neurons, all of which have been observed experimentally.

8.4.1 EXCITABILITY

Spiking neurons generate action potentials when the membrane potential exceeds a spiking threshold. Figure 8.4a shows the effect of single excitatory inputs of differing magnitude on the membrane potential of a HH neuron. When the stimulus intensity is less than threshold, the membrane potential following the stimulus (dotted line) decays approximately exponentially back toward its resting membrane potential. However, when the stimulus intensity exceeds the threshold, the generation of an action potential produces a very different time course for the return of the membrane potential to its resting value (solid line). The term excitability refers to the behavior of threshold–type systems in which small perturbations result in a short-path return to baseline, whereas a sufficiently large stimuli results in a disproportionately longer path return to baseline.

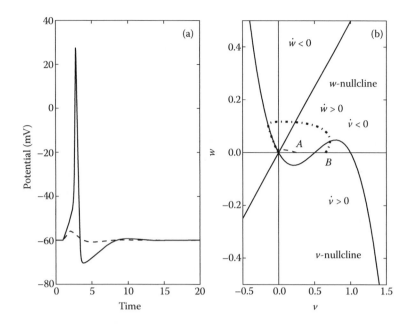

FIGURE 8.4   (a) Simulation of the response of the membrane potential of a Hodgkin–Huxley neuron in the excitable regime to current pulses of different intensity: 4 mA (dotted line) and 10 mA (solid line). Simulation was done using XPPAUT (Ermentrout 2002) following an exercise described by Guevara (2003). (b) Phase plane for a Fitzhugh–Nagumo model of a neuron whose parameters have been tuned to the excitable regime.

Excitability is one of the properties of a neuron that can be attributed to the presence of the so-called cubic nonlinearity (van der Pol and van der Mark 1928). This terminology is most easily understood by reducing the HH equations to a simpler form. Since the time scale for the relaxation of m is much faster than n and h, we can assume that $\dot{m} = 0$. If, in addition, we assume that $h = h_0$ is a constant, then the resulting reduced system of two differential equation retains many of the features observed experimentally [for an alternate derivation based on a simplified model of the cell membrane see (Keener and Sneyd, 1998)]. This system of equations is called the Fitzhugh–Nagumo equation (FitzHugh 1961; Nagumo et al. 1962)

$$\dot{v} = f(v) - w + I_{\text{ext}}, \tag{8.13}$$

$$\dot{w} = bv - \gamma w,$$

where

$$f(v) = v(a - v)(v - 1),$$

where $0 < a < 1$, $b > 0$ and $\gamma > 0$. In the Fitzhugh–Nagumo equation, v plays the role of the membrane potential and w plays the role of a recovery variable.

Figure 8.4b shows the phase plane of the Fitzhugh–Nagumo equation, when the parameters are tuned to the excitable regime. The phase plane is a plot of $v(t)$ versus $w(t)$, i.e., time does not appear explicitly. The basic principle of phase plane analysis is to use the conditions that $\dot{v} = 0$ and $\dot{w} = 0$ to divide the phase plane into distinct regions and to identify the fixed-points. Following this procedure, when $I_{\text{ext}} = 0$, we have for $\dot{v} = 0$

$$w = v(a - v)(v - 1).$$

This equation is called the v-nullcline and describes the cubic nonlinearity shown in Figure 8.4b. Since $\dot{v} = f(v) - w$, we see that above the v-nullcline we have $\dot{v} < 0$ and below the v-nullcline we have $\dot{v} > 0$. Similarly, the w-nullcline is obtained by setting $\dot{w} = 0$ and is

$$w = \frac{b}{\gamma} v.$$

Thus, it is a straight line through the origin of the phase plane with slope b/γ. From Equation 8.13, we have, above the w-nullcline, $\dot{w} < 0$ and, below, $\dot{w} > 0$. In other words, the v and w nullclines divide the phase plane into four regions in terms of the relative signs of $\dot{v}$ and $\dot{w}$. The points at which the v and w nullcline intersect on the phase plane are the fixed-points. In this case, there is one fixed-point that corresponds to the resting membrane potential of the neuron. Since the time constant for the changes in v is larger than that for the changes in w by a factor of about 12.5 (Haken 2002), the stability of the fixed-point is determined by slope of the v-nullcline. In particular the fixed-point is stable if the v-nullcline slopes downward at the fixed-point, and unstable if it slopes upward. Thus in the excitable regime, the resting membrane potential is stable; hence, it is a fixed-point attractor.

We can see the relationship between excitability and a cubic nonlinearity by picking different starting points, i.e., different initial values of $v(t)$ and $w(t)$, and seeing what happens. Suppose we applied to the neuron in its resting an excitatory input of magnitude A to obtain the initial condition $v(0) = A$ and $w = 0$. Since $\dot{w} > 0$ and $\dot{v} < 0$, we know that v must decrease and w must increase in such a way that the effects on v outweigh the effects on w. The next effect is that v (and, hence, w) decays monotonically toward the fixed-point (Figure 8.4b). However, a completely different scenario occurs if we apply an excitatory input large enough to cross the v-nullcline, e.g., point B in Figure 8.4b. Since both $\dot{v} > 0$ and $\dot{w} > 0$, both v and w must increase. However, at some point, the v-nullcline will be crossed and, consequently, $\dot{v} < 0$. Consequently, v, w are drawn back to the fixed-point. These are the typical time courses that define an excitable system.

The spiking threshold is measured experimentally as the amplitude of a brief current input that produces an action potential 50% of the time (Verveen and DeFelice 1974). Surprisingly, it has only been recently recognized that the spiking threshold of a HH–type neuron has a very complex structure, typical of that seen in systems that exhibit chaotic dynamics, namely, a fractal basin boundary (Guckenheimer and Oliva 2002). Although this observation raises the possibility of chaos in the HH-neuron, it must be kept in mind that these equations are an approximation of a process that should, strictly speaking, be described by a stochastic differential equation. As we have discussed, this stochasticity arises because the mathematical expressions for ionic conductances are averages obtained by measuring the variable openings of ion channels. Thus it is possible that this mathematical observation points to a limitation of the ensemble averaging procedure used by Hodgkin and Huxley for describing neural excitability. In any case, these observations remind us that there is a degree of unpredictability about the responses of neurons to stimulation.

8.4.2 Periodic Spiking

The question of how neurons change their spiking behavior is of critical importance for understanding the occurrence of an epileptic seizure. Indeed, current mechanisms for neural synchronization of large populations of neurons require that individual neurons exhibit some form of periodic spiking behavior (Strogatz 2003). We can use our phase plane approach to understand why a Fitzhugh–Nagumo neuron generates periodic spiking when an external current is injected (Figure 8.5). The effect of $I_{ext} > 0$ is to simply move the position of the v-nullcline upward. Thus, there is no longer a fixed-point at (0,0). The v-nullcline at the new fixed-point is positive; however, it is unstable. This instability must only be true close to the fixed-point. Why? If, for example, v becomes a very large positive, $\dot{v} < 0$, and hence both v and w must contract toward the fixed-point. The same would be

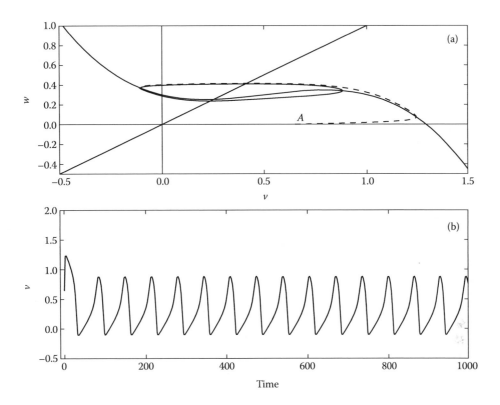

FIGURE 8.5 (a) Phase plane and (b) spiking pattern for the Fitzhugh–Nagumo equation tuned to the periodic regime.

true if we imagined that v became a very large negative. In other words, the trajectory (v, w) can neither come too close or too far away from the fixed-point, but must both be confined to a region, shaped roughly like a doughnut, in which the fixed-point lies somewhere inside the hole. Since every solution of a differential equation is determined uniquely by its initial conditions, the trajectory that describes (v, w) can never cross itself. The only possible trajectory that can remain confined within a two-dimensional region and never cross itself is a closed curve such as a circle or ellipse. This means that the solution must be periodic; this periodic solution is called a limit cycle, i.e., a limit cycle attractor. It should be noted that on the limit cycle for the Fitzhugh–Nagumo neuron, trajectory movements in the horizontal direction are typically much faster than those in the vertical direction. This is the property of relaxation-type oscillators.

A change in neuronal activity from a resting, but excitable, state to a periodic spiking behavior is a bifurcation. A great deal of mathematical effort has been expended to show that there are only a few types of bifurcations that produce limit cycles (Guckenheimer and Holmes 1990). This fact greatly simplifies the task for experimental investigations since it becomes possible to systematically examine each mechanism and decide which are the most likely candidates by directly comparing observation to prediction.

The most common bifurcation that produces periodic activity in single neurons (Guevara 2003) and neural populations (Wilson and Cowan 1972) is the subcritical Hopf bifurcation (Figure 8.6a). A bifurcation diagram is constructed by determining the long time or steady state behavior for different choices of the initial conditions as the bifurcation parameter is changed (in this case, μ). Our convention is that solid lines correspond to fixed-point attractors (stable), dotted lines to fixed-point repellors (instability), black dots to a stable limit cycle, and so on. Imagine that we slowly increase μ from -0.4. Two interesting phenomena occur. First, oscillation onset occurs abruptly (there is a bifurcation) and, once the oscillation occurs, its amplitude does not significantly change

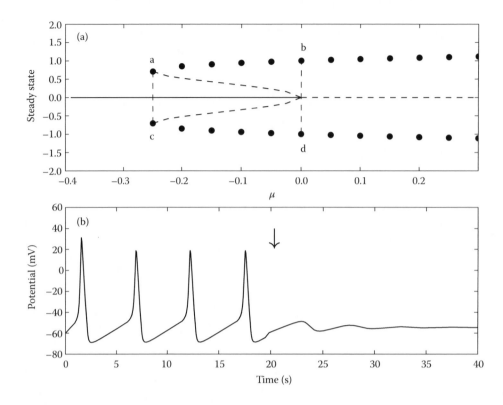

FIGURE 8.6 (a) Bifurcation diagram for a generic subcritical Hopf bifurcation. The region whose vertices are a,b,c,d denotes the parameter range for which bistability occurs. (b) Simulation using XPPAUT (Ermentrout 2002) following an exercise described by Guevara (2003) demonstrating that a regular spiking Hodgkin–Huxley neuron can be stopped by applying a single 0.4-s, 10-mA pulse (↓).

as μ increases. Neurons whose oscillation onset behaves in this manner are said to possess type II excitable membranes. Second, in the region outlined by abcd in Figure 8.6a, two stable behaviors coexist: a fixed-point attractor and a limit cycle attractor. The coexistence of more than one attractor is referred to as multistability; bistability refers to the special case when two attractors coexist. In Chapter 38 (Milton et al. 2010), we show that the occurrence of bistability in a neuron is intimately associated with the presence of a cubic nonlinearity.

The existence of multistability in a neuron (or neural population) suggests a simple mechanism for starting and stopping neural rhythms. Suppose that μ is in the range between c and d in Figure 8.6a. Then a single, carefully honed perturbation would either annihilate an existing oscillation or initiate one. As is shown in Figure 8.6b, this prediction can be verified using a simulation of the HH equation. In support of this hypothesis, qualitative changes in the dynamics of single neurons (Guttman et al. 1980; Tasaki 1959) and neural circuits (Foss and Milton 2000; Kleinfeld et al. 1990) can be induced by applying appropriately timed, brief electrical stimuli. Moreover, single DC pulses terminate seizures in rats (Osorio and Frei 2009), brief electrical stimuli block after-discharges (Motamedi et al. 2002), and sensory stimuli can abort seizures in humans with epilepsy (Milton 2000). These observations suggest that, at least for short intervals of time, living neurons and neural populations behave as would be expected for a multistable dynamical system.

8.5 DISCUSSION

A fundamental postulate of dynamic systems theory is that the dynamics of large populations of neurons can be captured by the dynamics of much simpler systems (Guckenheimer and Holmes

1990; Lytton 2008). This is because complex systems with many components exhibit a tendency to self-organize at, or near, stability boundaries in parameter space. The closer the dynamical system is to the stability boundary, the better the dynamics can be approximated by a simpler dimensional system. In other words, phenomena we have described for single neurons, such as excitability, periodic spiking, multistability, and even chaos, all have their counterparts in the dynamics of large neural populations. Of course, neural populations exhibit additional properties that arise because they are large (Milton 2010), including the effects of energy constraints, the finiteness of dendritic/axonal conduction velocities, anatomical pathways for propagation, and the tendency of a periodically spiking neuron to synchronize with other similarly spiking neurons. However, if we believe that the ultimate triggering event for a seizure is a qualitative change in the activity of individual neurons, then these larger scale phenomena shape the resultant seizure, but do not initiate it. An implication of this observation is that little seizures involving just a few neurons may be as important as large, clinically significant seizures for understanding seizure initiation and the development of methods for seizure prediction. This conclusion is supported by the following observations.

Measurements of the time, Δt, between successive seizure recurrences indicate the presence of a power law of the form

$$P(\Delta t) \sim \frac{1}{t^{\alpha}}, \tag{8.14}$$

where $P(\Delta t)$ is the probability that the time between successive seizures is Δt and α is a constant equal to the power law exponent (Osorio et al. 2009, 2010). This equation indicates that a log–log plot of $P(\Delta t)$ versus Δt is linear with slope α. Theoretical studies indicate that power laws generically arise in systems of relaxation oscillators, such as neurons, which are tuned very close to stability boundaries, i.e., the edge of stability (Sornette 2004). The fact that a power law is observed is quite disturbing since it implies that there is no characteristic time scale. This means that we would obtain the same power law exponent whether we measured events at an accuracy of ms or years and, hence, distinctions between parameters and variables become increasingly blurred. An important observation is that some systems that exhibit power laws may also exhibit long-range correlations (Milton 2010; Sornette 2004); the existence of these correlations raises the possibility of some degree of predictability. Finally, since the size of seizure events also can exhibit a power law (Osorio et al. 2009, 2010), there is no characteristic scale for epileptic seizures. In other words, even a single spike slow wave discharge may be as important for understanding the dynamics of seizure recurrence in patients with epilepsy as the occurrence of clinically significant seizures.

Current mathematical models typically express neural activities in terms of action potentials, either timing or rate, but the clinical methods available to evaluate these models do not directly measure action potentials. There are two types of methods available to test predictions of mathematical models for epilepsy: 1) those that indirectly monitor neuronal metabolism (fMRI, PET); and 2) the electroencephalogram (EEG), a method that measures the extracellular currents generated by excitatory and inhibitory postsynaptic potentials. Although optical methods have the potential of measuring action potentials, these are likely to be of limited clinical usefulness (Haglund et al. 1992). Here lies the barrier that inhibits progress into the application of computational methods to study the level of precision required to validate computational models of seizures (Lytton 2008; Soltesz and Staley 2008). Thus it is absolutely essential that computational neuroscientists begin to develop models that offer predictions related to the clinically accessible variables. Current attempts to interpret the epileptic spike measured by the surface EEG in terms of depth EEG signals are notable steps forward (Cosandier-Rimélé et al. 2007).

The spread of neural activity during a seizure is highly dependent on the topology of the neuronal connectivity. In contrast to the obvious complexity of neural pathways in the brain, mathematical models of populations of neurons consider topologies (random, exponentially distributed, small world) that are obviously overly naive. Consequently, issues related to neuroanatomy and neurophysiology will always remain of fundamental importance to the study of epilepsy. At the very least, this observation emphasizes that progress requires the efforts of interdisciplinary teams involving clinicians and modelers so that critically important experiments can be designed and conducted.

ACKNOWLEDGMENTS

The author acknowledges discussions with U. an der Heiden, F. Andermann, J. D. Cowan, B. G. Ermentrout, L. Glass, P. Gloor, J. Gotman, M. R. Guevara, M. C. Mackey, I. Osorio, J. Rinzel and D. Sornette who over the years shaped my views about epilepsy. The author acknowledges support from the William R. Kenan, Jr. Foundation and the National Science Foundation (NSF–0617072 and NSF-1028970).

REFERENCES

Abeles, M., E. Vaadia, and H. Bergman. 1990. Firing patterns of single units in the prefrontal cortex and neural network models. *Network* 1(1):13–25.

Babb, T. L., C. L. Wilson, and M. Isokawaakesson. 1989. Firing patterns of human limbic neurons during stereoencephalography (SEEG) and clinical temporal-lobe seizures. *Electroenceph. Clin. Neurophysiol.* 66:467–482.

Cosandier-Rimélé, D., J.-M. Badier, P. Chauvel, and F. Wendling. 2007. A physiologically plausible spatio–temporal model for EEG signals recorded with intracerebral electrodes in human partial epilepsy. *IEEE Trans. Biomed. Eng.* 54:380–388.

Ermentrout, B. 2002. *Simulating, analyzing and animating dynamical systems: A guide to XPPAUT for researchers and students.* Philadelphia: SIAM.

FitzHugh, R. 1961. Impulses and physiological states in theoretical models of nerve membranes. *Biophys. J.* 1:445–466.

Foss, J., and J. Milton. 2000. Multistability in recurrent neural loops arising from delay. *J. Neurophysiol.* 84:975–985.

Guckenheimer, J., and P. Holmes. 1990. *Nonlinear oscillations, dynamical systems, and bifurcations of vector fields.* New York: Springer–Verlag.

Guckenheimer, J., and R. A. Oliva. 2002. Chaos in the Hodgkin–Huxley model. *SIAM J. Appl. Dyn. Sys.* 1:105–114.

Guevara, M. R. 2003. Dynamics of excitable cells. In *Nonlinear dynamics in physiology and medicine*, eds. A. Beuter, L. Glass, M. C. Mackey, and M. S. Titcombe, 87–121. New York: Springer.

Guttman, R., S. Lewis, and J. Rinzel. 1980. Control of repetitive firing in squid axon membrane as a model for a neuron oscillator. *J. Physiol. (London)* 305:377–395.

Haglund, M. M., G. A. Ojemann, and D. W. Hochman. 1992. Optical imaging of epileptiform activity from human cortex. *Nature (London)* 358:668–671.

Haken, H. 2002. *Brain dynamics: Synchronization and activity patterns in pulse-coupled neural nets with delays and noise.* New York: Springer.

Hill, A., J. Lu, M. Masino, O. Olsen, and R. L. Calabrese. 2001. A model of a segmental oscillator in the leech heart beat neuronal network. *J. Comp. Neurosci.* 10:281–302.

Hille, B. 2001. *Ion channels of excitable membranes,* 3rd ed. Sunderland, MA: Sinnauer Associated.

Hodgkin, A. L., and A. F. Huxley. 1954. A quantitative description of membrane current and its application to conduction and excitation in nerve. *J. Physiol. (London)* 117:500–544.

Izhikevich, E. M. 2004. Which model to use for cortical spiking neurons? *IEEE Trans. Neural Networks* 15:1063–1070.

Keener, J., and J. Sneyd. 1998. *Mathematical Physiology.* New York: Springer.

Kleinfeld, D., F. Raccuia-Behling, and H. J. Chiel. 1990. Circuits constructed from identified *Aplysia* neurons exhibit multiple patterns of persistent activity. *Biophys. J.,* 57:697–715.

Lennie, P. 2003. The cost of cortical computing. *Curr. Biol.* 13:493–497.

Lytton, W. W. 2008. Computer modeling in epilepsy. *Nature Rev. Neurosci.* 9:626–637.

Mann, E. O. and I. Mody. 2008. The multifaceted role of inhibition in epilepsy: Seizure-genesis through excessive GABAergic inhibition in autosomal dominant nocturnal frontal lobe epilepsy. *Curr. Opin. Neurol.* 21:155–160.

Milton, J. G. 2000. Epilepsy: Multistability in a dynamical disease. In *Self-organized biological dynamics & nonlinear control*, ed. J. Walleczek, 374–386. New York: Cambridge University Press.

Milton, J. G. 2010. Epilepsy as a dynamic disease: A tutorial of the past with an eye to the future. *Epil. Behav* 18(1–2):33–44.

Milton, J. G., A. Quan, and I. Osorio. 2010. Nocturnal frontal lobe epilepsy: Metastability in a dynamic disease? In *Epilepsy: The intersection of neurosciences, biology, mathematics, engineering and physics*, eds. I. Osorio, H. P. Zanvari, M. G. Frei, and S. Arthurs, 501–510. Boca Raton: CRC Press.

Motamedi, G. K., R. P. Lesser, D. L. Miglioretti, Y. Mizuno-Matsumoto, B. Gordon, W. R. S. Webber, D. C. Jackson, J. P. Sepkuty, and N. C. Crone. 2002. Optimizing parameters for terminating cortical after-discharges with pulse stimulation. *Epilepsia* 43:836–846.

Nagumo, J. S., S. Arimoto, and S. Yoshizawa. 1962. An active pulse transmission line simulating a nerve axon. *Proc. IRE*, 50:2061–2070.

Osorio, I., and M. G. Frei. 2009. Seizure abatement with single DC pulses: Is phase resetting at play? *Int. J. Neural Sys.* 19:1–8.

Osorio, I., M. G. Frei, D. Sornette, and J. Milton. 2009. Pharamco–resistant seizures: Self-triggering capacity, scale-free properties and predictability? *Eur. J. Neurosci.* 30:1554–1558.

Osorio, I., M. G. Frei, D. Sornette, J. Milton, and Y.-C. Lai. 2010. Epileptic seizures: Quakes of the brain? *Phys. Rev.* 82: 021919.

Prince, D. A. 1969. Microelectrode studies of penicillin foci. In *Basic mechanisms of the epilepsies,* eds. H. H. Jasper, A. A. Ward, and A. Pope, 320–328. Boston: Little, Brown.

Shusterman, V., and W. C. Troy. 2008. From baseline to epileptiform activity: A path to synchronized rhythmicity in large-scale neural networks. *Phys. Rev. E* 77:061911.

Soltesz, I., and K. Staley. 2008. *Computational neuroscience in epilepsy.* San Diego: Academic Press.

Sornette, D. 2004. *Critical phenomena in natural sciences: Chaos, fractals, self-organization and disorder: Concepts and tools.* New York: Springer.

Strogatz, S. H. 2003. *SYNC: The emerging science of spontaneous order.* New York: Hyperion.

Tasaki, I. 1959. Determination of two stable states of the nerve membrane in potassium-rich media. *J. Physiol. (London)* 148:306–331.

Van der Pol, B., and J. van der Mark. 1928. The heartbeat considered as a relaxation oscillator and an electrical model of the heart. *Phil. Mag.* 6:763–775.

Verveen, A. A., and L. J. DeFelice. 1974. Membrane noise. *Prog. Biophys. Mol. Biol.* 28:189–265.

Wilson, H. R. 1999. Simplified dynamics of human and mammalian neocortical neurons. *J. Theoret. Biol.* 200:375–388.

Wilson, H. R., and J. D. Cowan. 1972. Excitatory and inhibitory interactions in localized populations of model neurons. *Biophys. J.,* 12:1–23.

9 Nonlinear Time Series Analysis in a Nutshell

Ralph G. Andrzejak

CONTENTS

9.1 INTRODUCTION

Nonlinear time series analysis is a practical spinoff from complex dynamical systems theory and chaos theory. It allows one to characterize dynamical systems in which nonlinearities give rise to a complex temporal evolution. Importantly, this concept allows extracting information that cannot be resolved using classical linear techniques such as the power spectrum or spectral coherence. Applications of nonlinear time series analysis to signals measured from the brain contribute to our understanding of brain functions and malfunctions and thereby help to advance cognitive neuroscience and neurology. In this chapter, we show how a combination of a nonlinear prediction error and the Monte Carlo concept of surrogate time series can be used to attempt to distinguish between purely stochastic, purely deterministic, and deterministic dynamics superimposed with noise.

The framework of nonlinear time series analysis comprises a wide variety of measures that allow one to extract different characteristic features of a dynamical system underlying some measured signal (Kantz and Schreiber 1997). These include the correlation dimension as an estimate of the number of independent degrees of freedom, the Lyapunov exponent as a measure for the divergence of similar system states in time, prediction errors as detectors for characteristic traits of deterministic dynamics, or different information theory measures. The aforementioned nonlinear time series measures are univariate, i.e., they are applied to single signals measured from individual dynamics. In contrast, bivariate measures are used to analyze pairs of signals measured simultaneously from two dynamics. Such bivariate time series analysis measures aim to distinguish whether the two dynamics are independent or interacting through some coupling. Some of these bivariate measures aim to extract not only the strength, but also the direction of these couplings. The Monte Carlo concept of surrogates allows one to test the results of the different nonlinear measures against well-specified null hypotheses.

We should briefly define the different types of dynamics. Let $\mathbf{x}(t)$ denote a q-dimensional vector that fully defines the state of the dynamics at time t. For purely deterministic dynamics, the future temporal evolution of the state is unambiguously defined as a function f of the present state:

$$\dot{\mathbf{x}}(t) = \mathbf{f}(\mathbf{x}(t)). \tag{9.1}$$

Here $\dot{\mathbf{x}}(t)$ denotes the time derivative of the state $\mathbf{x}(t)$. We further suppose that a univariate time series s_i is measured at integer multiples of a sampling time Δt from the dynamics using some measurement function h:

$$s_i = h(\mathbf{x}(i\Delta t)). \tag{9.2}$$

The combination of Equation 9.1 and Equation 9.2 results in the purely deterministic case. When some noise $\psi(t)$ enters the measurement:

$$s_i = h(\mathbf{x}(i\Delta t),\psi(i\Delta t)), \tag{9.3}$$

we obtain the case of deterministic dynamics superimposed with noise. Here it is important to note that the noise only enters at the measurement. The deterministic evolution of the system state $\mathbf{x}(t)$ via Equation 9.1 is not distorted. In contrast to such measurement noise stands dynamical noise $\xi(t)$ for which we have:

$$\dot{\mathbf{x}}(t) = \mathbf{f}(\mathbf{x}(t),\xi(t)). \tag{9.4}$$

Due to the noise term, the future temporal evolution of $\mathbf{x}(t)$ is not unambiguously determined by the present state. Accordingly, Equation 9.4 represents stochastic dynamics. We again suppose that a measurement function, with or without measurement noise (Equations 9.3 and 9.2, respectively), is used to derive a univariate time series s_i from this stochastic dynamics.

As stated above, we aim to distinguish between purely stochastic, purely deterministic, and deterministic dynamics superimposed with noise. Accordingly, we want to design an algorithm A to be applied to a time series s_i that fulfills the following criteria:

    i. $A(s_i)$ should attain its highest values for time series s_i measured from any purely stochastic process.
    ii. $A(s_i)$ should attain its lowest values for time series s_i measured from a noise-free deterministic dynamics.[*,†]
    iii. $A(s_i)$ should attain some value in the middle of its scale for time series s_i measured from noisy deterministic dynamics.

In particular, the ranges obtained for ii. and iii. should not overlap with the one of i. For iii., $A(s_i)$ should ideally vary with the relative noise amplitude.

In order to study our problem under controlled conditions, we use a time series of mathematical model systems that represent these different types of dynamics. While this chapter is restricted to these model systems, the underlying lecture included results for exemplary intracranial electroencephalographic (EEG) recordings of epilepsy patients (Andrzejak et al. 2001). These EEG times

---

[*] The following text contains a number of footnotes. These are meant to provide background information. The reader might safely ignore them without risk of losing the essentials of the main text.
[†] The polarity of A does not matter. It could likewise attain lowest values for i. and highest values for ii.

series as well as the source codes to generate the different time series studied here, the generation of surrogates, and the nonlinear prediction error are freely available.[*]

9.2 DYNAMICS AND TIME SERIES

9.2.1 NONLINEAR DETERMINISTIC DYNAMICS—LORENZ DYNAMICS

As nonlinear deterministic dynamics we use the following Lorenz dynamics.[†]

$$\dot{x}(t) = 10(y(t) - x(t)) \tag{9.5}$$

$$\dot{y}(t) = 39x(t) - y(t) - x(t)z(t) \tag{9.6}$$

$$\dot{z}(t) = x(t)y(t) - (8/3)z(t). \tag{9.7}$$

This set of first-order differential equations can be integrated using, for example, a fourth order Runge–Kutta algorithm.[‡] To obtain time series x_i, y_i, z_i for $i = 1,\ldots,N = 2048$, we sample the output of the numerical integration at an interval of $\Delta t = 0.03$ time units.[§] We denote the resulting temporal sequence of three-dimensional vectors as $L_i = (x_i, y_i, z_i)$. In our analysis, we restrict ourselves to the first component x_i (see Table 9.1 and Figure 9.1). Hence, we have an example of purely deterministic dynamics (Equation 9.1). The discretization in time is obtained from the numerical integration of the dynamics. This integration can be considered as part of the measurement function (Equation 9.2), which furthermore consists of the projection to the first component.

9.2.2 NOISY DETERMINISTIC DYNAMICS—NOISY LORENZ DYNAMICS

By adding white noise to the first component of our Lorenz time series x_i, we obtain noisy deterministic dynamics:

$$n_i = x_i + 18\psi_i, \tag{9.8}$$

where ψ_i denotes uncorrelated noise with uniform amplitude distribution between 0 and 1 (see Table 9.1 and Figure 9.1). Our time series n_i is an example of a signal generated by deterministic dynamics (Equation 9.1) where noise is superimposed on the measurement function Equation 9.3. To roughly assess the strength of the noise, we have to know that the standard deviation of the noise term ψ_i and the Lorenz time series x_i is approximately 0.29 and 9.6, respectively. With the prefactor of 18, the standard deviation of the noise[¶] amounts to approximately 5.2.

---

[*] www.meb.uni-bonn.de/epileptologie/science/physik/eegdata.html; www.cns.upf.edu/ralph.

[†] Equations 9.5 through 9.7 result in chaotic dynamics. However, this is not of our concern here. Rather, we use the Lorenz dynamics as a representative of some aperiodic deterministic dynamics.

[‡] Here we use an algorithm with fixed step size of 0.005 time units. Starting at random initial condition we use some arbitrary but high number (e.g., 10^6) of preiterations to diminish transients.

[§] In the signal analysis literature, one often finds 1024, 2048, 4096, etc. for the number of samples. The reason is that these numbers are integer powers of two: $N^* = 2^n$ with n being an integer. For these numbers of samples, one can use the very efficient fast Fourier transform algorithm. For other numbers of samples, one has to use algorithms that divide the number of samples in prime factors and that are slower.

[¶] In assessing the relative strength of noise superimposed to the Lorenz dynamics, it is important to note that a major contribution to the overall standard deviation originates from the switching behavior of this dynamics. The standard deviation of the actual aperiodic oscillations in the two wings is smaller (see Figures 9.1–9.2).

TABLE 9.1
Overview of Different Time Series Studied

Type		Origin and Parameter
a_i	Stochastic	AR process, Equation 9.9 with $r = 0$
b_i	Stochastic	AR process, Equation 9.9 with $r = 0.95$
c_i	Stochastic	AR process, Equation 9.9 with $r = 0.98$
x_i	Deterministic	x-component of Lorenz dynamics
n_i	Noisy deterministic	x_i superimposed with white noise

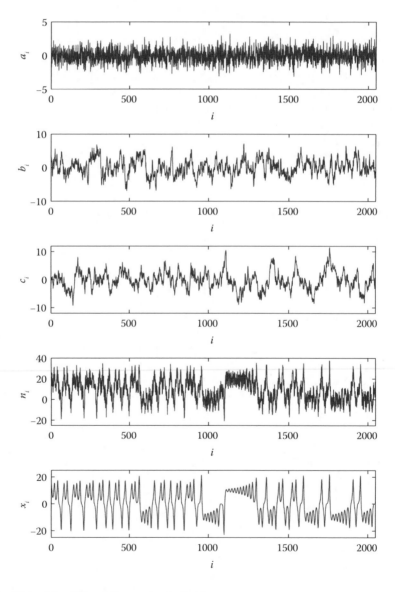

FIGURE 9.1 Plots of the different time series studied here.

9.2.3 Stochastic Processes—Linear Autoregressive Models

As a stochastic process, we use a simple but still very instructive example—a first-order linear autoregressive process with varying degrees of autocorrelation:

$$\zeta_{i+1} = r\zeta_i + \xi_i, \tag{9.9}$$

for $i = 1, \ldots, N = 2048$, where ξ_i denotes uncorrelated Gaussian noise with 0 mean and unit variance.[*] We can readily see that for $r = 0$, Equation 9.9 results in white, i.e., uncorrelated noise. Upon the increase of r, within the interval $0 < r < 1$, the strength of the autocorrelation increases. For this chapter, we study three different time series derived from Equation 9.9 using different values of r (see Table 9.1 and Figure 9.1).

In contrast to the time-continuous stochastic dynamics expressed in Equation 9.4, the autoregressive process of Equation 9.9 is defined for discrete time. What is common to both formulations, and what is the important point here, is that the noise is intrinsic to the dynamics; no additional noise enters through the measurement. We should note that the fact that we can write down a rule to generate an autoregressive process (as in Equation 9.9) does not conflict with its stochastic nature in any way. The point is that due to the noise term, ξ_i, an initial condition does not unambiguously determine the future temporal evolution of the dynamics. Furthermore, without this noise term, the autoregressive process would not result in sustained dynamics but rather decay to 0.

9.3 DETERMINISTIC VERSUS STOCHASTIC DYNAMICS

Plotting the three components of the Lorenz dynamics L_i in a three-dimensional state space, we obtain the trajectory of the dynamics that form the famous Lorenz attractor (Figure 9.2, left). Once we specify some state of the Lorenz dynamics $x(0)$, $y(0)$, $z(0)$ at $t = 0$ as an initial condition the future temporal evolution of the dynamics is unambiguously determined by Equations 9.5–9.7. As a result of this deterministic nature of the Lorenz dynamics, its trajectory cannot intersect with itself.[†] A trajectory crossing would imply that the state at the crossing would have two distinct future evolutions, which would contradict with the uniqueness of the solution of Equations 9.5–9.7. Inspecting the trajectory in the left panel of Figure 9.2, we can appreciate that neighboring trajectory segments are aligned. That means that similar present states of the deterministic Lorenz dynamics result in similar states in the near future.

In opposition to deterministic dynamical systems are stochastic dynamics. When we start Equation 9.9 twice for an identical initial condition, $\zeta 0$, and the same r value, we get two completely different solutions. Moreover, similar present states of stochastic dynamics will typically not result in similar future states. Accordingly, trajectories of stochastic dynamics can intersect, and neighboring segments are typically not aligned. This is illustrated in the right panel of Figure 9.2, where we plot the trajectory of simple stochastic dynamics derived from Equation 9.9 in a three dimensional state space.

9.4 DELAY COORDINATES

In the previous section, we saw that the alignment of neighboring trajectory segments can be used as a criterion to try to distinguish deterministic from stochastic dynamics (see the two panels of Figure 9.2). Before we can quantify this criterion we have to realize, however, that in experimental measurements from real-world dynamics we typically cannot access all variables of a dynamical

---

[*] To avoid transients, we again run some arbitrary but high number of preiterations (e.g., 10^4). That is, we start Equation 9.9 at $\zeta_{-10000} = 0$ before recording the samples of the process starting from $\zeta 1$.

[†] The edginess at the outer corners of the wings results from the fixed nonzero step size of the numerical integration.

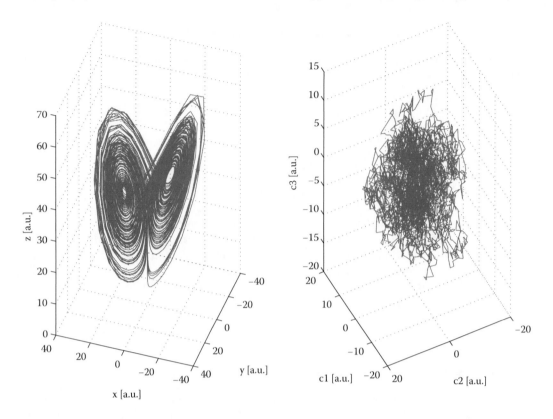

FIGURE 9.2 (left) The three components of the Lorenz dynamics x_i, y_i, and z_i plotted in a three-dimensional state space. (right) Three independent realizations of an autoregressive process (Equation 9.9 for $r = 0.98$) plotted in a three-dimensional state space.

system. Suppose for example that we study a real-world dynamical system that is fully described by the Lorenz differential Equations 9.5–9.7. Suppose, however, that we cannot measure all three components of L_i, but rather only $s_i = x_i$ is assessed by our measurement function. In this case, we can use the method of delay coordinates to obtain an estimate of the underlying dynamics (Takens 1981):

$$v_i = (s_i, \ldots, s_{i-(m-1)\tau}),\tag{9.10}$$

with embedding dimension m and delay τ for $i = 1 + (m - 1)\tau, \ldots, N = \eta, \ldots, N$. Where we define $\eta = 1 + (m - 1)\tau$ to simplify the notation below. An introduction to the embedding theorem underlying the reconstruction of dynamics using such delay coordinates can be found in Kantz and Schreiber (1997). We restrict ourselves here to a hand-waving argument that Equation 9.10 allows us to obtain a structure that is topologically similar to the original dynamics. In Figure 9.3, we contrast the full Lorenz dynamics L_i with delay coordinate reconstructions obtained from its first component x_i using $m = 3$ and different values of τ. Comparing these structures, we find a striking similarity between the original dynamics and its reconstruction for not too high values of τ. For low values of τ, the dynamics is somewhat stretched out along the main diagonal and is not properly unfolded. At an intermediate value of τ, the two-wing structure of the Lorenz dynamics is clearly discernible from the reconstruction. For too high a value of τ, the dynamics is overfolded.

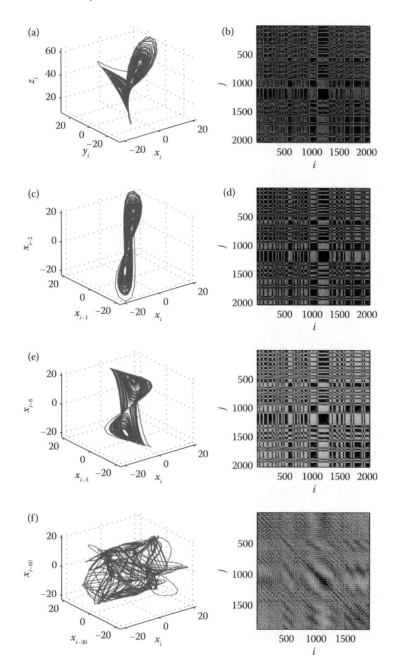

FIGURE 9.3 (a) The three components of the Lorenz dynamics $L_i = (x_i, y_i, z_i)$ plotted in a three-dimensional state space. Hence, this panel is the same as the left panel of Figure 9.2, but viewed from a different angle. (b) Euclidean distance matrix calculated for the Lorenz dynamics L_i. (c) Delay reconstruction of the Lorenz dynamics based on the first component, x_i. Here we use a time delay of $\tau = 1$. For obvious reasons, the maximal embedding dimension that we can use for this illustration is $m = 3$. (d) Euclidean distance matrix calculated from such an embedding, but for $m = 6$. (e and f) Same as second row, but for $\tau = 6$ and $\tau = 30$, respectively.

We can appreciate this similarity further by inspecting Euclidean distance matrices derived from the full Lorenz dynamics and its delay coordinate reconstruction.[*] For a set of q-component vectors w_i for $i = \eta, \ldots, N$, such a matrix contains in its element Dij the Euclidean distance between the vectors w_i and w_j. The Euclidean distance between two q-component vectors w_1 and w_2 is defined as

$$d(w_1, w_2) = \sqrt{\sum_{l=1}^{q} (w_{1,l} - w_{2,l})^2}. \tag{9.11}$$

Figure 9.3 shows the distance matrix for the full Lorenz dynamics L_i, which can also be seen as a set of N vectors with $q = 3$ components. This is contrasted to the distance matrix derived from delay coordinate reconstructions v_i, which is a set of N vectors with $q = m$ components. While the overall scales of the different matrices are different, we can observe an evident similarity between the structures found in these matrices for the original and its delay coordinate reconstruction for not too high values of τ. Importantly, this similarity implies that close and distant states in the full Lorenz dynamics are mapped to close and distant states in the reconstructed dynamics, respectively. This topological similarity implies that our criterion of distinction between deterministic and stochastic dynamics, namely the alignment of neighboring trajectory segments, carries over from the original dynamics to the reconstructed dynamics obtained by means of delay coordinates. The different distance matrices derived from the delay reconstructions are, however, not identical to the original distance matrix and the degree of their similarity depends on the embedding parameters m and τ. These aspects will be discussed in more detail below. To get a thorough understanding of the quality of the embedding requires playing with different parameters and viewing the result from different angles. This can be done using the source code referred to in the introduction.

9.5 NONLINEAR PREDICTION ERROR

Above we have identified the alignment of neighboring trajectory segments as a signature of deterministic dynamics. We will now introduce the nonlinear prediction error as a straightforward way to quantify the degree of this alignment (see Kantz and Schreiber (1997) and references therein).

Before we begin, we first normalize our time series s_i to have zero mean and unit standard deviation and variance.[†] Subsequently, we form a delay reconstruction v_i from our time series for $i = \eta, \ldots, N$. To calculate the nonlinear prediction error with horizon h we carry out the following steps for $i_0 = \eta, \ldots, N - h$. We take v_i as reference point and look up the time indices of the k nearest neighbors of the reference point: $\{i_g\}_{(g=1, \ldots, k)}$. These k nearest neighbors are simply those points that have the k smallest distances to our reference point in the reconstructed state space. In other words, $\{i_g\}_{(g=1, \ldots, k)}$ are the indices of the k smallest entries in row i_0 of the distance matrix derived from v_i.[‡] Now we use the future states of these nearest neighbors to predict the future states of our reference point and quantify the error that we make in doing so by

$$\varepsilon_{i_0} = s_{i_0} + h - \frac{1}{k} \sum_{g=1}^{k} s_{i_g} + h. \tag{9.12}$$

---

[*] The choice of the Euclidean distance is arbitrary. The same arguments hold when other definitions of distances are used.

[†] In this way we do not have to normalize the prediction error itself and the subsequent notations become simpler.

[‡] Since distance matrices are symmetric, these indices coincide with the k smallest entries in the column i_0 of the matrix. Note however that if v_{i1} is among the k nearest neighbors of v_{i2}, this does not imply that v_{i2} is among the k nearest neighbors of v_{i1}.

Finally, we take the root-mean-square over all reference points:

$$E = \sqrt{\frac{1}{N-h-\eta+1} \sum_{i_0=\eta}^{i_0=N-h} \varepsilon_{i_0}^2}.$$ (9.13)

Note that while the nearest neighbors are determined according to their distance in the reconstructed state space, the actual prediction in Equation 9.12 is made using the scalar time series values. Regarding Equation 9.12, it also becomes clear that we must exclude nearest neighbors with indices $i_r > N - h$. Furthermore, we should aim to base our prediction on neighboring trajectory segments rather than on the preceding and subsequent piece of the trajectory on which our reference point v_{i_0} is located. For this purpose we apply the so-called Theiler correction with window length W (Theiler 1986), by including only points to the set of k nearest neighbors that fulfill $|i_g - i_0| > W$.

For the parameters, we use the following values. We fix the embedding dimension used for the delay coordinates to $m = 6$, the number of nearest neighbors to $k = 5$, the prediction horizon to $h = 5$ samples, and the length of the Theiler correction to $W = 30$. The delay used for the delay coordinates is varied according to $\tau = 1, 2, \ldots, 29$.

Figure 9.4 shows results obtained for our set of exemplary time series. The white noise time series a_i results in E values around 1.1. For this stochastic process, points close to some reference points do not carry any predictive information about the future evolution of this reference point. Neighboring state space trajectory segments of this stochastic process are not aligned. Hence, for white noise the predictions s_{i_r+h} used in Equation 9.12 are no better than guessing a set of k points randomly drawn from s_i.

For the noise-free deterministic Lorenz time series x_i, neighboring trajectory segments are well aligned. Points close to some reference points have a similar future evolution as this reference point. The predictions s_{i_r+h} used in Equation 9.12 are good predictions for s_{i_0+h}. Therefore, we get small values of E. For increasing τ the prediction error increases, reflecting the overfolding of the reconstructed dynamics (see Figure 9.3). Nonetheless, as of now, E seems to fulfill the criteria i and ii.

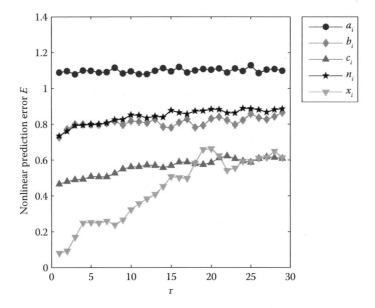

FIGURE 9.4 Dependence of the nonlinear prediction error, E, on the time delay, τ, used for the state space reconstruction. Here, results are shown only for the original time series.

formulated in the Introduction. It allows us to clearly tell apart our deterministic Lorenz time series from white noise. It can be objected that this distinction is not terribly difficult, and could readily be achieved by visual inspection of our data.

We should therefore proceed to our noisy deterministic time series n_i and time series of autoregressive processes with stronger autocorrelation (b_i, c_i). When we look at their results in Figure 9.4, we must realize that E is in fact not a good candidate for Figure 9.4A. For the time series n_i it is clear that the noise must degrade the quality of the predictions of Equation 9.12 as compared to the noise-free case of x_i. This increase in E by itself would not disqualify E, as it is compatible with criterion iii. The problem is that E values for b_i overlap with those for n_i. Results for c_i are even lower than those for n_i and for higher τ, even similar to those of x_i. The reason why these autocorrelated time series exhibit a higher predictability as opposed to white noise is the following: v_{i_1} being a close neighbor of v_{i_2} implies that the scalar s_{i_1} is typically similar to the scalar s_{i_2}. Furthermore, s_{i_1} being similar to s_{i_2} implies, to a certain degree, that s_{i_1+h} is similar to s_{i_2+h}. It is instructive to convince oneself that it is just the strength of the autocorrelation that determines the strength of the "to a certain degree" in the previous statement.

As a consequence, stochastic time series with stronger autocorrelation and noisy deterministic dynamics cannot be told apart by our nonlinear time series algorithm E. We are still missing one very substantial ingredient to reach our aim, which will we introduce in Section 9.6.

9.6 THE CONCEPT OF SURROGATE TIME SERIES

Nonlinear time series analysis was initially developed studying complex low-dimensional deterministic model systems such as the Lorenz dynamics. The brain is certainly not a low-dimensional deterministic dynamics and will typically not behave as one. Rather, it might be regarded as a particularly challenging high-dimensional and complicated dynamical system that might have both deterministic and stochastic features. Problems that arise in applications to signals measured from the nervous system continue to trigger the refinement of existing algorithms and development of novel approaches and, thereby, help to advance nonlinear time series analysis. One important concept in nonlinear time series analysis, which has some of its roots in EEG analysis (Pijn et al. 1991), is the Monte Carlo framework of surrogate time series (Theiler et al. 1992; Schreiber and Schmitz 2000).

Suppose that we have calculated the nonlinear prediction error for some experimental times series and consider the following scenarios. First, we obtain a relatively low E value but have doubts that this is due to some deterministic structure in our time series. Second, we obtain a relatively high E value but we suspect that our time series should have some deterministic structure that is possibly distorted by some measurement noise. In any case, we want to know what a relatively high or a relatively low value of the nonlinear prediction error means. A way to address this problem is to scrutinize our time series further by testing different null hypotheses about it by means of surrogate time series. Such surrogate time series are generated from a constrained randomization of an original time series.[*] The surrogates are designed to share specific properties with the original time series, but are otherwise random. In particular, the properties that the surrogates share with the original time series are selected such that the surrogates represent signals that are consistent with a well-specified null hypothesis.

Here, we use the simple but illustrative example of phase-randomized surrogates (Theiler et al. 1992). To construct these surrogates, first calculate the complex-valued Fourier transform of the original time series. In the next step, randomize all phases of the complex-valued Fourier coefficients between 0 and 2π.[†] Finally, the inverse Fourier transform of these modified Fourier

---

[*] See Schreiber and Schmitz (2000) for the alternative to using a typical randomization to generate surrogates.

[†] In this randomization it is very important to preserve the antisymmetry between Fourier coefficients representing positive and negative frequencies. This antisymmetry holds for real-valued signals. If it is not preserved the inverse Fourier transform will not result in a real-valued time series.

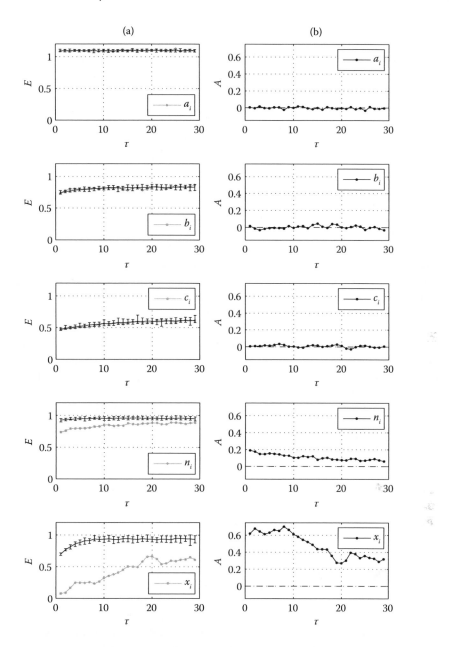

FIGURE 9.5 (a) Dependence of the nonlinear prediction error, E, on the time delay, τ, used for the state space reconstruction. Here, results are shown for the original time series (grey lines with symbols) and 19 surrogate time series (black lines with error bars). The error-bar denotes the range obtained from the surrogates. (b) Dependence of the surrogate-corrected nonlinear prediction error, A, on the time delay, τ. Each row corresponds to results of one specific time series (see legends).

coefficients results in a surrogate time series. An ensemble of surrogates is generated using different randomization of the phases. Randomizing the phase of a complex number does not affect its absolute value, that is, the length of the corresponding vector in the complex plane is preserved. Therefore, the power of the resulting signal and its distribution over all frequencies, as measured via the periodogram, is identical to the one of the original time series. Consequently, the surrogates also have the same autocorrelation function as the original time series. However, regardless of the nature of the original time series, the surrogate time series is a stationary linear stochastic Gaussian process. Whether the original was deterministic or stochastic, stationary or nonstationary, Gaussian or nonGaussian, the surrogate is stochastic, stationary, and Gaussian. In that sense, the surrogate algorithm projects from the space of all possible time series with a certain periodogram to the space of all stationary linear stochastic time Gaussian time series with this same periodogram. To see this point, it is instructive to convince oneself that a phase-randomized surrogate of a phase-randomized surrogate cannot be distinguished from a phase-randomized surrogate of the original time series.

In Figure 9.5, we again show values of the nonlinear prediction error for our exemplary time series (see again Figure 9.4). Here, however, we not only show results for the original time series but also contrast these to results obtained for an ensemble of 19 phase-randomized surrogate time series generated for each original time series. For all time series from autoregressive processes a_i, b_i, and c_i, the results for the surrogates match those of the original time series. The autoregressive processes generated by Equation 9.9 are stationary linear stochastic Gaussian processes. The r value determines the strength of the autocorrelation of this process and thereby the shape of the periodogram. A phase-randomized surrogate of a stationary linear stochastic Gaussian process with any periodogram is a stationary linear stochastic Gaussian process with this very periodogram. Accordingly, phase-randomized surrogates constructed from a_i, b_i, and c_i cannot be distinguished from their corresponding original time series. The null hypothesis is correctly accepted for all three cases.[*]

For both the noise-free Lorenz time series x_i and the noisy Lorenz time series n_i, we find that the original time series results in lower E values as compared to their corresponding ensemble of surrogates. Here, we deal with a purely deterministic time series and a noisy deterministic one. When we construct surrogates from them, their deterministic structures are destroyed. In particular, the randomization destroys the local alignment of neighboring trajectory segments. Regarding the significant difference between the E values for the original time series and the mean values obtained for the surrogates, we can conclude that the null hypothesis is correctly rejected for both x_i and n_i.

From these results, it is only a small step for us to arrive at our aim by defining

$$A = \langle E \rangle \text{surr} - E_{\text{orig}}, \tag{9.14}$$

where the brackets denote the average taken across the ensemble of 19 surrogates. The resulting surrogate-corrected nonlinear prediction error values A are shown in Figure 9.5, and we can see that the definition of Equation 9.14 fulfills criteria i., ii., and iii. that we defined as the aim of this chapter.

9.7 INFLUENCING PARAMETERS

In this chapter, we have plotted the nonlinear prediction error only in dependence on the time delay τ used for the delay coordinates. To fully understand how nonlinear techniques and surrogates work, it is always important to study the influence of all parameters in detail. In our case, these parameters are the delay τ and embedding dimension m used for the delay coordinates, the number of nearest neighbors k and the prediction horizon h, and, finally, the length of the Theiler-correction W. In general, the influence of these different parameters will not be independent from another. For example, it is often the so-called embedding window $\tau(m - 1)$ that matters, rather than the two parameters τ and m independently. There exist very useful guidelines and recipes for an adequate choice of these different parameters (see Kantz and Schreiber (1997) and references therein). However, whenever one is analyzing some type of experimental data, one should play with these parameters and see

---

[*] A good check of whether one is still on track in understanding this chapter is to find an explanation of why our different autoregressive processes result in different E values, and why the surrogates match closely with their corresponding original time series in these E values, rather than just undiscriminatingly covering the entire range of E values obtained for a_i, b_i, and c_i.

how they influence results. Certainly, if one finds some effect in experimental data using nonlinear time series analysis techniques, the significance of this effect should not critically depend on the choice of the different parameters.

We have studied our problem based on only single realizations of our time series. However, whenever we have controlled conditions, we should make use of them and generate many independent realizations of our time series. In this way, we can get more reliable results and also learn more about our measures by regarding their variability across realizations. Another important issue that we have neglected here is that it can be very informative to study the influence of noise for a whole range of noise amplitudes and for different types of noise.

9.8 WHAT CAN AND CANNOT BE CONCLUDED?

First, we can draw a very positive conclusion: We reached the aim that we stated in the Introduction. Using a combination of a nonlinear prediction error and surrogates, we were able to distinguish between purely stochastic, purely deterministic, and deterministic dynamics superimposed with noise. We demonstrated this distinction under controlled conditions using mathematical model systems with well-defined properties. However, how do these results carry over once we leave the secure environment of controlled conditions? Under controlled conditions, we know what dynamics underlie our time series. Here, we can draw conclusions such as the null hypothesis is correctly accepted or rejected, but what conclusions can be drawn when facing real-world experimental time series from some unknown dynamical system?

Let us suppose we want to test the following working hypothesis consisting of two parts. First, the dynamics exhibited by an epileptic focus is nonlinear deterministic. Second, the dynamics exhibited by healthy brain tissue is linear stochastic. Suppose that to test this hypothesis we compare EEG time series measured from within an epileptic focus against EEG time series measured outside the focal area. We calculate the nonlinear prediction error for these EEG time series and phase-randomized surrogates constructed from them. Suppose further that for the focal EEG time series we find significant differences between the results for original EEG time series and those of the surrogates, whereas for the nonfocal EEG we find that the results for the original are not significantly different from the surrogates. (In the lecture underlying this chapter we looked at such results for real EEG time series.) Have we proven our working hypothesis? The answer is no, we have not—neither its first nor its second part.

First, we need to recall that the rejection of a null hypothesis at say $p < 0.05$ in no way proves that the null hypothesis is wrong. It only implies that if the null hypothesis was true, the probability to get our result (or a more extreme result) is less than 5%. Putting that aside, let us assume that the difference between the focal EEG and surrogates is so pronounced that we deem the underlying null hypothesis as indeed very implausible. Still, this does not prove that the dynamics are nonlinear deterministic. Recall that we used phase-randomized surrogates representing the null hypothesis of a linear stochastic Gaussian stationary process. We have to keep in mind that the complement to our null hypothesis is the world of all dynamics that do not comprise all the properties: linear, stochastic, Gaussian, and stationary. Hence, if we deal, for example, with a stationary linear stochastic nonGaussian dynamics, our null hypothesis is wrong. Our test should reject it. Likewise, we might have nonstationary linear stochastic Gaussian dynamics or nonlinear deterministic process. We might also deal with stationary linear stochastic Gaussian dynamics measured by some nonlinear measurement function. We have to realize that the dynamics exhibited by an epileptic focus is nonlinear deterministic and is among these alternatives. However, it is only one of many explanations of our findings. Given only the results of our test, our working hypothesis is not more likely than any of the other explanations of the rejection of our null hypothesis.

We should furthermore note that our test cannot prove the second part of our null hypothesis either. The fact that we could not reject the null hypothesis for our nonfocal EEG does not prove the correctness of this null hypothesis. We cannot rule out that our nonfocal EEG originates from some nonlinear deterministic dynamics, but our test is not sufficiently sensitive to detect this feature. Our time series might be too short or too noisy for the prediction error to detect the alignment of neighboring trajectories, or we might have used inadequate values of our parameters m, τ, and k.

It is important to keep these limitations in mind when interpreting results derived from nonlinear measures in combination with surrogates.[*] Nonetheless, nonlinear measures in combination with surrogates can still be very useful in characterizing signals measured from the brain. Andrzejak et al. studied the discriminative power of different time series analysis measures to lateralize the seizure-generating hemisphere in patients with medically intractable mesial temporal lobe epilepsy (Andrzejak et al. 2006). The measures that were tested comprised different linear time series analysis measures, different nonlinear time series analysis measures, and a combination of these nonlinear time series analysis measures with surrogates. Subject to the analysis were intracranial electroencephalographic recordings from the seizure-free interval of 29 patients. The performance of both linear and nonlinear measures was weak, if not insignificant. A very high performance in correctly lateralizing the seizure-generating hemisphere was, however, obtained by the combination of nonlinear measures with surrogates. Hence, the very strategy that brought us closest to the aim formulated in the introduction of this chapter—to reliably distinguish between stochastic and deterministic dynamics in mathematical model systems—also seems key to a successful characterization of the spatial distribution of the epileptic process. The degree to which such findings carry over to the study of the predictability of seizures was among the topics discussed at the meeting in Kansas.

REFERENCES

Andrzejak, R. G., K. Lehnertz, F. Mormann, C. Rieke, P. David, and C. E. Elger. 2001. Indications of nonlinear deterministic and finite-dimensional structures in time series of brain electrical activity: Dependence on recording region and brain state. *Phys. Rev. E* 64:061907.

Andrzejak, R. G., G. Widman, F. Mormann, T. Kreuz, C. E. Elger, and K. Lehnertz. 2006. Improved spatial characterization of the epileptic brain by focusing on nonlinearity. *Epilepsy Res.* 69:30–44.

Kantz, H., and T. Schreiber. 1997. *Nonlinear time series analysis.* Cambridge: Cambridge Univ. Press.

Pijn, J. P., J. van Neerven, A. Noes, and F. Lopes da Silva. 1991. Chaos or noise in EEG signals; dependence on state and brain site. *Clin. Neurophysiol.* 79:371–381.

Schreiber, T., and A. Schmitz. 2000. Surrogate time series. *Physica D* 142:346–382.

Takens, F. 1981. Detecting strange attractors in turbulence. In *Dynamical systems and turbulence*, Vol. 898 of *Lecture notes in mathematics*, eds. D. A. Rand and L.-S. Young, 366–381. Berlin: Springer-Verlag.

Theiler, J. 1986. Spurious dimensions from correlation algorithms applied to limited time-series data. *Phys. Rev. A* 34:2427–2432.

Theiler, J., S. Eubank, A. Longtin, B. Galdrikian, and J. D. Farmer. 1992. Testing for nonlinearity in time series: The method of surrogate data. *Physica D* 58:77–94.

---

[*] Consider as a further principle limitation, a periodic process with a very long period that exhibits no regularities within each period. We would need to observe at least two periods to appreciate the periodicity of the process. In principle, we need an infinite amount of data to rule out that we misinterpret periodic, and hence deterministic, processes as stochastic processes.

10 How to Detect and Quantify Epileptic Seizures

Sridhar Sunderam

CONTENTS

10.1 INTRODUCTION

The seizure detection algorithm (SDA) has become an increasingly common tool in epileptology since the advent of automated medical signal analysis. SDAs find utility beyond clinical epilepsy management in basic science as an analysis tool, and more recently in the development of devices for automated seizure warning and treatment. This chapter provides an introduction to the formulation, evaluation, and optimization of SDAs, with examples chosen from the literature. It is directed more at clinicians than at those whose primary expertise is mathematics, engineering, or physics—but will hopefully also motivate practicing engineers and mathematicians to revisit the ways in which they design and implement these algorithms. It is emphasized that seizure detection is important not only for timely intervention but for understanding seizure dynamics and evaluating treatment as well. To this end, approaches for seizure quantification are discussed. The goal is to provide readers with the tools for understanding the workings of any SDA within a generic framework, and to pinpoint critical considerations for correct SDA use and performance assessment.

10.2 WHAT TO DETECT AND WHY

10.2.1 Targets of Epilepsy Therapy

The goal of epilepsy therapy is to alleviate spontaneously recurring seizures, the hallmark of the disease. But for the noticeable impairment of consciousness that accompanies clinical seizures, epilepsy may even go undetected (e.g., Scher et al. 2003). A closer look at epileptic brain activity reveals other electrophysiological markers of epilepsy than clinical seizures alone, such as interictal spikes (Spencer et al. 2008), afterdischarges elicited by electrical stimulation as an index of seizure threshold (Lesser et al. 1999), high-frequency oscillations (Zelmann et al. 2009), and subclinical (i.e., without an identified behavioral correlate) seizures (Scher et al. 2003). Any of these markers may have a degenerative effect on the brain and the patient's quality of life; in fact, some are being investigated as surrogate markers for the diagnosis of seizure etiology (Marsh et al. 2010) or prognosis of treatment outcome (Jacobs et al. 2009). But we are mostly interested in identifying and treating clinical seizures, because of the immediate consequences of their occurrence.

Before we can treat seizures, we must detect them. What exactly causes or constitutes a seizure is a subject of much inquiry; but for our purpose, it is enough to know what characteristics define a seizure in continuous measurements of brain activity and behavior. And since electrophysiological measurements are the de facto standard for characterizing brain dynamics, we focus here on the automated detection of unequivocal epileptic seizures from measurements such as surface (scalp) or intracranial EEG. The principles discussed, however, are equally relevant to other markers of epilepsy and other measurements that reflect activity in the brain.

In Figure 10.1, we see the EEG onset of a clinical seizure in the mesial temporal lobe of a human patient. In general, a seizure may last anywhere from a few seconds to several minutes. Electrographic seizure onset (EO), the time at which the earliest signs of seizure-like activity are visible on the EEG, may be followed by the onset of behavioral manifestations recognizable by an expert or other observer familiar with the patient's seizures, i.e., clinical onset (CO), as the activity spreads across the cortex. This spread may appear to be instantaneous in generalized epilepsies, or more gradual in focal epilepsies (e.g., temporal lobe seizures). The goal of seizure detection is to use information about the morphology and dynamics of EEG to accurately identify seizure onset in monitored EEG signals and quantify the severity and dynamical progression of seizure activity.

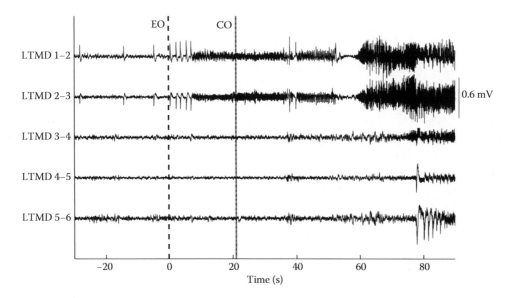

FIGURE 10.1 Intracranial EEG of a secondarily generalized complex partial seizure. The EEG was recorded from the left mesial temporal lobe of a human patient (FHS009; data courtesy of Flint Hills Scientific, LLC) using a left temporal mesial depth (LTMD) electrode with 6 contacts (1 to 6). Earliest electrographic onset (EO) is evident on LTMD2 at $t = 0$ s, followed by clinical onset (CO) and spread of activity to other locations of the brain.

10.2.2 UTILITY OF SEIZURE DETECTION

Why detect seizures? First, to pinpoint their source and region of influence, which helps determine an appropriate course of treatment. This usually involves analysis of a chronic EEG recording, possibly during evaluation for surgical treatment, or an experimental trial for a new drug or other therapy. Automated seizure detection further enables a quantitative comparison of the frequency and nature of ictal activity with and without treatment. Such an evaluation may even be feasible for outpatients with chronically implanted monitoring devices (Skarpaas and Morrell 2009). Finally, in emerging therapies such as closed-loop electrical stimulation, seizure detection may itself trigger therapy delivery, and must therefore be timely and highly specific to the activity being targeted. Seizure detection thus plays a central role in most aspects of epilepsy care. In addition, there is growing interest in detecting hypothetical states that may have predictive value with respect to seizure onset (Mormann et al. 2007). This is beyond the scope of this chapter and will be the subject of the chapter by Mormann et al.

10.2.3 PERFORMANCE GOALS AND CRITERIA

Before designing and deploying a seizure detection algorithm (SDA), we must specify measures of performance that define successful detection. Performance criteria are tied to the needs of the application for which seizures are to be detected. Whether the goal is seizure diagnosis or responsive intervention, we would like to detect seizures early, and with high sensitivity and specificity; high sensitivity means rare misses (i.e., most true seizures are detected) and high specificity means rare false alarms. A low false alarm rate minimizes patient anxiety if a warning is issued in response to seizure detection, and avoids treatment overdose if electrical stimulation or drug infusion is triggered. For implanted monitoring and control devices, a low computational cost/power requirement for the detection algorithm is also desirable to prolong battery life and minimize heat generation (Sunderam et al. 2010). Conventionally, detection refers to accurate determination of seizure onset, but for objective treatment evaluation it may be important to have meaningful quantitative descriptors of seizure intensity, duration, spread, and other dynamical characteristics as well.

10.3 SEIZURE DETECTION PRIMER

10.3.1 EVERYONE'S FAVORITE SEIZURE DETECTION ALGORITHM

Suppose that we are in a situation where automatic seizure detection would be advantageous. We search the literature for a suitable algorithm and find methods that make use of wavelets (Khan and Gotman 2003), radial basis function networks (Schuyler et al. 2007), support vector machines (Gardner et al. 2006), Gaussian process models (Faul et al. 2007), genetic programming (Firpi et al. 2007), and mixed-band wavelet-chaos-neural networks (Ghosh-Dastidar et al. 2007) among other computational devices, each more daunting than the last. Perhaps we care less about the algorithm employed in each study than the circumstances in which it was tested: intracranial (Khan and Gotman 2003) versus scalp EEG (van Putten et al. 2005); adult (Slooter et al. 2006) versus pediatric (Lee et al. 2007) or neonatal patients (Faul et al. 2007); focal (Grewal and Gotman 2005) versus generalized (Adeli et al. 2003) epilepsy; or real-time (Chan et al. 2008) versus retrospective (Schiff et al. 2000) analysis. Also, we are reluctant to learn and implement a particular algorithm that just may not work on our data; perhaps another algorithm would do just as well or better.

What are we to do? Rather than panic, we should recognize that despite the variety of methods proposed, most SDAs share the general computational framework presented in Figure 10.2. EEG signals are acquired from an appropriate source (e.g., scalp or cortex); the signals are conditioned (e.g., antialiasing, line noise reduction) to satisfy certain quality criteria; and then filtered to extract one or more relevant features. The features are chosen to capture the contrast between the dynamics of the EEG during seizure (ictal) and normal (interictal) behavior. Features may be further transformed numerically or pruned—a process known as feature reduction—to yield a subset that best conveys the instantaneous seizure content of the signal. A postprocessing step that we shall refer to here as *augmentation* may be applied (e.g., median filtering, background normalization) to further amplify ictal–interictal contrast, i.e., the difference between typical ictal and interictal feature values that enables us to distinguish the two states, or reduce unwanted variability. In the final step, i.e., event detection, a classifier compares the filtered output against reference values or an interictal

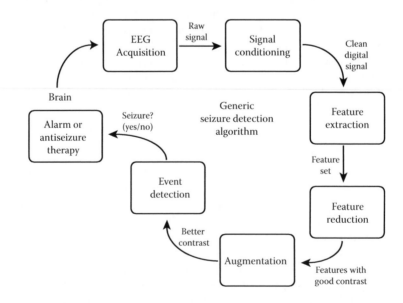

FIGURE 10.2 Generic computational framework of a seizure detection algorithm (SDA).

baseline to determine whether a seizure has started or occurred. This usually involves comparison of the SDA output against preset threshold and duration constraints, parameters that may be optimized individually or for a cohort. The end goal of seizure detection may be to trigger an alarm or responsive treatment (e.g., drugs, stimulation, etc.), or to mark the time of an event for subsequent evaluation by a human expert.

The operation of almost any SDA can be described by this sequence. Note that this description applies mainly to SDAs intended for early onset detection; seizure quantification involves additional considerations that will be addressed later. Moreover, though we emphasize seizure detection from EEG signals, the use of video (Karayiannis et al. 2006), motion (Nijsen et al. 2007), heart rate variability (Malarvili and Mesbah 2009), or neurochemical (Hillered et al. 2005) measurements may prove useful as an adjunct to EEG or exclusively in situations where EEG is impractical. Now that we have an overview of a typical SDA, we will take a closer look at each stage.

10.3.2 SIGNAL ACQUISITION

First, we must acquire one or more EEG signals from an appropriate scalp or intracranial source, including locations suspected or known to cover the onset and spread of seizure activity, and taking care to choose a quiet reference (Schiff 2005). If a common reference is used for many electrodes, it is a good idea to eliminate common mode signals and localize recorded activity by taking the difference of potentials measured in neighboring contacts before analysis; this and other alternatives are discussed by Nunez and Srinivasan (2006). The signals should capture the range of ictal activity (seizures), interictal activity (spikes, brief interictal discharges, etc.), and normal activity (sleep–wake rhythms) as well as possible. This means that the signals have adequate temporal and amplitude resolution—measured in samples per second (hertz) and bits, respectively—with good dynamic range. Clinical EEG systems typically amplify and record data at a sampling rate f_s ranging anywhere between 100 and 2000 Hz, with a digital resolution b of 10 to 16 bits (2^b possible values). Of course, adequate spatial resolution is critical as well. Good resolution ensures that we can reconstruct the signal well and clearly identify the signatures of the events of interest. The use of a suitable low-pass analog filter before digitization is essential to avoid aliasing, i.e., false superposition of signal power above the Nyquist frequency (one-half of f_s) at lower frequencies; this further constrains the frequency bandwidth of the signal. A high cutoff of 35–40 Hz imposed by the antialiasing filter is the bare minimum required for preserving the most commonly inspected brain rhythms (delta, theta, etc.) in the EEG. In addition, a low cutoff of 0.1–0.5 Hz is usually applied (AC coupling) to prevent signals from drifting and saturating the recording amplifiers. For instance, the recorded EEG may have frequency content restricted to 0.1–70 Hz. A good introduction to signal sampling issues can be found in Smith (1999).

10.3.3 SIGNAL CONDITIONING

Any analysis is only as good as the data; therefore, clean signals are highly desirable. Artifacts are common in EEG and it is often necessary to use filters to reject or compensate for ocular, muscle, or motion artifact (specifically in scalp recordings), saturation, sensor failure, electrical stimulation artifact, or excessive power line noise prior to analysis so that the results are not contaminated. Saab and Gotman (2005) present several illustrative examples of such artifacts.

Line noise can be minimized in real time using a simple notch filter (Gotman 1982), which only attenuates signal components close to or at the frequency of the power supply (either 50 or 60 Hz, depending on the prevailing standard). For offline analysis, multitaper spectral methods have been proposed for estimating line components in a data window using regression and subtracting them from the raw signal (Thomson 1982); this could be adapted for real-time use through Kalman estimation.

Contraction of muscles on the face or scalp produces a distinctive artifact on EEG that is within the spectral range of normal brain rhythms; movement of the eyeball generates a prominent potential visible mainly on frontal and other proximal electrodes. Electrodes placed at suitable locations on the face or neck and beside the eyes to record an electromyogram (EMG) and electrooculogram (EOG), respectively, can pick up activity from these spurious sources; this is standard protocol in sleep studies for distinguishing REM sleep from the waking state. But actual removal of ocular and muscle artifacts from the EEG leaving only neural activity behind is notoriously difficult. Novel methods for separating these sources from EEG such as regression analysis (Schlogl et al. 2007) and independent component analysis (Hoffmann and Falkenstein 2008; Shackman et al. 2009) continue to be proposed, but are hard to validate since the neural signal underlying the artifact is unknown.

Removal of electrical stimulation artifacts is also problematic because it depends on the waveform applied, though promising adaptive cancellation techniques have been developed (Aksenova et al. 2009; Rolston et al. 2009). In common practice, however, the recording amplifier is blanked (i.e., shut off) during stimulation to prevent the artifact from contaminating the EEG (Alvarez et al. 2010), losing both signal and artifact, and possibly introducing a reconnection artifact in the process. For low-frequency continuous waveform (e.g., sinusoidal) stimulation the techniques suggested previously for line noise reduction may prove beneficial.

To conclude, artifacts in EEG come from various sources, each with its idiosyncrasies. But regardless of the source, artifactual data must be decontaminated or flagged before analysis to avoid confounding the results.

10.3.4 Feature Selection and Extraction

10.3.4.1 What Is a Feature?

A feature is a characteristic of the EEG signal(s) from which seizures are to be detected. One or more features must be selected and extracted that have distinctive values during seizure as compared to the interictal state. In other words, the distributions of ictal and interictal feature values must have relatively low overlap; only then is the feature likely to be useful for seizure detection. One important caveat: Two or more poor features, when considered together, may provide good ictal–interictal separation in a multidimensional feature space. The most relevant EEG characteristics employed to date may be loosely categorized in terms of spectral properties (e.g., band power, edge frequency), signal morphology (e.g., wave amplitude or sharpness), and statistical measures (e.g., entropy, correlation time, line length). This scheme is adopted here merely for convenience; the boundaries are ill-defined and many features rightfully belong to more than one class. Members of each class may be univariate or multivariate, where the latter pertains to properties derived using two or more signals at the same time (e.g., coherence, mutual information). A distinction can also be made between linear and nonlinear measures, or time domain and frequency domain measures.

Any dynamical property of the signal may be considered a candidate feature for seizure detection. The key is to identify those features that show distinctive and appreciable changes during seizure, and do so reliably as well. The next step is to perform feature extraction, or in general terms, filtering. In the following sections, we describe each category of features.

10.3.4.2 Spectral Features

Visual examination of the EEG reveals oscillations at different timescales or frequencies that reflect fluctuations in potential associated with neural activity. It seems natural, therefore, to think of the signal as a mixture of periodic—even sinusoidal—waveform components. This is the basic assumption of Fourier spectral analysis: that an arbitrarily complex time domain signal $x(t)$ can be expressed as the linear summation of pure sinusoids over a range of different frequencies f, with each component weighted in proportion to its relative contribution to the signal. Expressed mathematically,

$$x(t) = \int_{-\infty}^{\infty} X(f)e^{i2\pi ft}\, df, \tag{10.1}$$

where $i = \sqrt{-1}$. The complex exponential terms $e^{i2\pi ft} = \cos 2\pi ft + i \sin 2\pi ft$ constitute a basis of unit vectors (or phasors) rotating in the complex plane with angular frequency $2\pi f$ as t changes. The Fourier coefficients $X(f)$ give the amplitude and phase of each basis function, and are computed using the so-called Fourier transform F:

$$X(f) = F[x(t)] = \int_{-\infty}^{\infty} x(t)e^{-i2\pi ft}\,\mathrm{d}t, \qquad (10.2)$$

$X(f)$ and its inverse, $x(t)$, are called a Fourier transform pair.

The energy content of $x(t)$ can be computed in the time or frequency domain as follows:

$$P = \int_{-\infty}^{\infty} |x(t)|^2\,\mathrm{d}t = \int_{-\infty}^{\infty} |X(f)|^2\,\mathrm{d}f, \qquad (10.3)$$

which is known as Parseval's identity. The sequence of coefficients $P(f) = |X(f)|^2$, appropriately scaled by the signal duration, is known as the power spectral density (PSD) or power spectrum. Equation 10.3 illustrates how the PSD embodies the relative contributions of different frequency components to the time domain signal $x(t)$. This is fortunate since the band power BP, i.e., the power contained in a particular range of frequency, say $[f_1, f_2]$, can be determined by integrating the PSD over that interval:

$$BP = P(f_1, f_2) = \int_{f_1}^{f_2} |X(f)|^2\,\mathrm{d}f. \qquad (10.4)$$

BP is an extremely useful measure for signal state detection. Note that the foregoing equations are defined for continuous-valued time and frequency. In practice, we work with finite duration signals of some length T, sampled at discrete intervals with frequency f_s from the analog EEG measurements. Equivalent forms of the Fourier transform have been formulated for dealing with discrete time samples of analog signals estimated at discrete values of frequency. The resolution of the frequency grid is fixed by the number of time samples used—this can be increased by padding with zeros (which does not actually produce more information but only interpolated values)—and the frequencies range from zero up to $f_s/2$, the Nyquist frequency. This limitation makes intuitive sense, since we cannot reconstruct oscillations of some period T unless at least two samples are available for every period.

A word of caution is necessary regarding spectral estimation: In spectral analysis, individual waves are irrelevant and it is implicitly assumed that the frequency content of the signal is stationary (i.e., its properties do not vary significantly) over the course of the data window. Properties may be computed in windows as brief as 1 s in duration, or in larger windows of a few s duration, with overlap in consecutive windows to produce an effective resolution of 1-s or lower. As a rule of thumb, it is meaningless to estimate the contribution of a particular frequency to the spectrum in a segment that will not accommodate at least 5–10 wavelengths; therefore, a 1-s window will not provide much useful information below about 5 Hz.

The above is a sketchy overview of Fourier analysis and overlooks many practical considerations in spectral estimation, such as the need to minimize spectral leakage by using appropriate data tapers when making short-time estimates of the Fourier transform, but hopefully will provide enough insight into the practical relevance of spectral analysis.

The power spectrum represents the distribution of power as a function of frequency. A pure sinusoid has a line spectrum since, by definition, it is zero everywhere except for an impulse function (a line) at a unique frequency, whereas the spectrum of white noise is flat since it contains all frequencies in equal measure. The EEG spectrum is, of course, somewhere in the middle: It is an eclectic mixture of components that hint at distinct cortical generators of different behaviors, along with noise from diffuse neural sources and measurement noise. EEG is thus a window into the brain. Spectral features of the EEG help us characterize the instantaneous brain state from a finite window of data.

Judging from the dramatic increase in EEG amplitude and frequency at the onset of the seizure in Figure 10.1, it seems reasonable to suppose that the raw signal power P would be useful as an SDA feature. Let us examine the data more closely. In Figure 10.3, we take the differential signal left temporal mesial depth (LTMD)1–2 (Figure 10.3a), and compute the PSD of two segments, each 5 s long: one (labeled *Int*) from the baseline well before electrographic seizure onset, and the other (*Sz*) from the body of the seizure but before clinical onset at $t \sim 20$ s (Figure 10.3a). The PSD (Figure 10.3c) appears much greater for *Sz* than for *Int* in the 10- to 20-Hz interval; hence, this frequency band is possibly more distinctive of seizure activity than the broadband power P (Figure 10.3b) across the entire spectrum. Power in the 10- to 20-Hz band can be estimated by applying a time domain linear filter to $x(t)$ that is selective to EEG activity in this frequency range to produce an output $x_f(t)$, and then computing the instantaneous band power BP as $|x_f(t)|^2$. This is a common approach for real-time band power estimation in SDAs (e.g., Osorio et al. 1998). Here, we use a 10- to 20-Hz Butterworth filter and compute the mean of BP in a moving 0.5-s window for the filtered output $x_f(t)$ (Figure 10.3e). The frequency response of the filter is shown overlaying the power spectra of *Int* and *Sz* in Figure 10.3c.

Both P and BP are elevated after seizure onset, but with one critical difference: The preseizure values of P contain isolated spikes that are of comparable amplitude to the seizure period, whereas that of BP appears almost flat. An objective way to evaluate the ictal–interictal contrast afforded by each feature is to measure the standard distance (Flury 1997) between the ictal and interictal

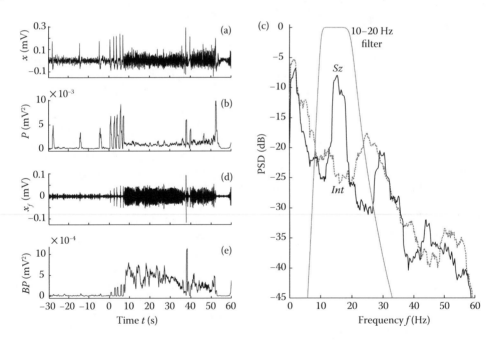

FIGURE 10.3 Spectral band power as a detection feature. (a) Differential EEG of LTMD1–2 in the vicinity of a seizure (onset at $t = 0$ s). (b) Raw signal power P in a moving 0.5-s window. (c) Comparison of PSD for 5-s interictal (*Int*, dotted line) and ictal (*Sz*, solid line) signal samples. Amplitude response of a 10–20 Hz Butterworth filter (solid gray) highlights frequency band in which *Sz* is relatively high compared to *Int*. (d) EEG after application of a 10-20 Hz (–3-dB cutoff) Butterworth filter. (e) Mean band power BP in a moving 0.5-s window of the filtered output in (d) produces greater ictal–interictal contrast D_{12} of 1.3 compared to 0.9 for P.

samples of that feature, i.e., the absolute difference between their sample means m_1 and m_2, normalized by s, the standard deviation of both samples pooled together:

$$D_{12} = \frac{m_1 - m_2}{s}. \tag{10.5}$$

Samples 1 and 2 refer to sets of feature values computed from any two epochs; for instance, in 0.5-s windows of the 60-s periods (see Figure 10.3a) immediately before (not all shown) and after seizure onset, respectively. For the feature P (Figure 10.3b), the standard distance $D_{12} = 0.9$, whereas for BP it is about 50% greater at 1.3. Clearly, the narrowband power BP improves the contrast between baseline and seizure activity.

Other interesting and potentially useful spectral properties (i.e., candidate features) are the center frequency (the centroid), spectral edge frequency (i.e., the frequency below which some percentage, say 90%, of the total signal power is concentrated), and spectral entropy (flatness of the spectrum), all of which help describe how power is distributed. But unlike BP, computation of these features usually requires that the entire PSD be estimated without the option of simpler time domain filtering.

10.3.4.3 Morphology

Fourier analysis is an extremely powerful technique, but does have limitations. First, reconstructing an arbitrary waveform, especially a sharp spike, requires more and more Fourier coefficients (i.e., summed sinusoids) as the waveform gets more complex in shape and deviates from a sinusoid. Second, epileptic spike-waveform units seen during seizures can retain their general shape but evolve in duration and amplitude as they progress in a sequence; this dilutes their frequency content further—making them difficult to represent using a small set of frequencies/sinusoids—and has the effect of spreading the power spectrum, thereby increasing the overlap between interictal and ictal activity. The reduction in available contrast makes band power less specific to seizures. Both limitations are readily confirmed by looking closely at the sequence of spikes at seizure onset in Figure 10.3a. It is therefore natural to look for features that describe the morphology of individual EEG waveforms in the time domain.

The wavelet transform offers the capability to look at signals over different scales, and wavelet-based features have found widespread use in seizure detection (Osorio et al. 1998; Shoeb et al. 2004; Saab and Gotman 2005; Ghosh-Dastidar et al. 2007; Talathi et al. 2008). Unlike the Fourier transform, which is localized only in the frequency domain, wavelets are localized in frequency and time. Starting with a function known as the mother wavelet, this analysis generates a family of orthogonal daughter wavelets by applying a scaling function that morphs the mother wavelet over a range of time and frequency scales. The coverage of time-frequency information provided by the wavelet transform enables us to detect finite (i.e., time-limited) waveforms that are modulated over time as we see during seizures.

Of course, there are many other ways to analyze signal morphology. Nonparametric alternatives may be as simple as the peak amplitude, slope, curvature, or sharpness. An early instance of this is Gotman (1982), in which dominant maxima and minima in the EEG signal are identified, and each half-wave is characterized by its amplitude and duration to make a judgment about whether or not it is typical of seizure activity. Since then, more sophisticated methods of decomposing a signal into its natural scales have been proposed and applied toward seizure detection (Frei and Osorio 2007; Orosco et al. 2009).

10.3.4.4 Statistical Features

Although the Fourier power spectrum is essentially a statistical property of the signal—it represents the distribution of signal variance by frequency—we set it apart because Fourier analysis is an extensive field in itself. It is harder to generalize the properties of other statistical features that may be useful for seizure detection. Obvious choices are the central moments of the signal, which are defined for a finite sample of some random variable X as

$$m_k = \frac{1}{n} \sum_{i=1}^{n} (x(i) - \mu_1)^k; \mu_k = E[m_k], \quad k > 1, \tag{10.6}$$

where $\mu_1 = E(X)$ denotes the expected value of X for the sample (the underlying mean of the distribution is typically unknown). The central moments μ_k for $k = 2, 3,$ and 4 represent the variance (spread), skewness (asymmetry), and kurtosis (sharpness), respectively, of the distribution of X; note that variance is another way of expressing the signal power that remains after the mean is removed. Other statistics of interest are the minimum, median, and maximum of the sample (so-called rank filters).

If we extend these moments to include temporal correlations between measurements separated in time by a delay τ, we get the autocorrelation function (ACF) $R(\tau)$:

$$r(\tau) = \frac{1}{n - \tau + 1} \sum_{k=1}^{n-\tau} x(k)x(\tau + k); \quad R(\tau) = E[r(\tau)], \quad \tau \geq 0. \tag{10.7}$$

In Equation 10.7, it is assumed that X has zero mean, which can be achieved by subtracting out the first moment μ_1. The ACF is directly related to the power spectrum through the Weiner–Khinchin theorem, which states that the PSD of a random signal (that satisfies conditions of stationarity) is merely the Fourier transform of its ACF. A signal that has strong serial correlation varies slowly and therefore has relatively low frequency content; conversely, a broad spectrum signal tends to have weak serial correlation. Hence, it is a matter of convenience whether to use the ACF or PSD as the basis for feature extraction.

Other statistical features that have been proposed for use in SDAs include measures of complexity, information, or entropy (Hu et al. 2006); measures of instantaneous signal energy or activity such as the line length feature (Esteller et al. 2001)

$$LL = \frac{1}{n} \sum_{i=1}^{n} |x(i) - x(i-1)|, \tag{10.8}$$

and the Teager energy (Kaiser 1990; Zaveri et al. 2010)

$$TE = \frac{1}{n-2} \sum_{i=2}^{n-1} x(i)^2 - x(i-1)x(i+1), \tag{10.9}$$

estimated for a sample of size n; and measures of rhythmicity (Harrison et al. 2008).

In Figure 10.4a, we examine three statistical features: the sample variance SV, line length LL, and Teager energy TE, each computed in a moving 0.5-s window for the EEG signal in Figure 10.3. Each feature is sensitive to different aspects of the signal, but all appear to possess the requisite contrast for distinguishing the seizure from its preictal baseline.

10.3.4.5 Multivariate Features

Multivariate features capture interactions or correlations between two or more EEG channels; for instance, measures of synchronization, crosscorrelation, or mutual information (Pereda et al. 2005).

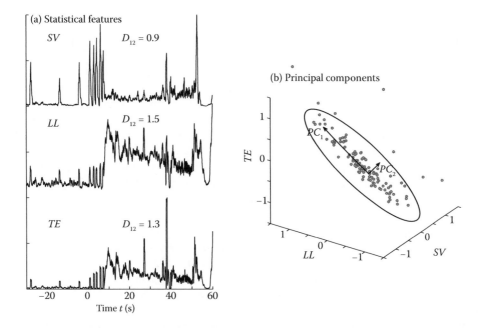

FIGURE 10.4 (a) Statistical features for seizure detection. Comparison of sample variance (SV), line length (LL), and Teager energy (TE) computed in a 0.5-s moving window for LTMD1–2. D_{12} gives the available ictal–interictal contrast. (b) Feature reduction using principal component analysis (PCA). Scatter plot of SV, LL, and TE in a 60-s interictal period (not in picture) shows significant intercorrelation. Directed line segments give the orientations of the first (PC_1) and second (PC_2) principal components of the sample, which capture 76% and 15% of the total variance, respectively.

Although multivariate features are likely to capture distinctive dynamics at seizure onset, they have not been applied to seizure detection as much as to investigations of seizure predictability (Mormann et al. 2007) or the analysis of seizure dynamics beyond onset (Schiff et al. 2005). We will not discuss them further here.

10.3.4.6 Feature Extraction in Real Time

SDA features are commonly computed from successive short windows or data epochs, 1–2 s long, for example, sometimes with overlap from one window to the next. We have seen that a relatively large window is needed to estimate power accurately at low frequencies. As we pointed out earlier in Section 10.3.4.2, there is no need to estimate the entire PSD; a digital filter can be designed with an appropriate pass band (or other differential frequency weighting) and used to process the data in the time domain in small chunks, perhaps even one point at a time. The output of the filter can be squared and averaged over an epoch to estimate band power.

A digital filter is any operation that takes a finite sequence of data points and manipulates them to produce an output with specific properties. Filtering may be linear (e.g., finite or infinite impulse response) or nonlinear (e.g., sample median, max or min). A linear filter obeys the principle of superposition; a weighted average of several inputs will produce a similarly weighted average of their individual filter outputs. A major advantage of linear filters is that if the filter's response to a single pulse input is known, the response to any arbitrary signal can be determined by the super-position principle using a convolution integral. It is important to design stable filters, meaning in general that their response is predictable and bounded for normal input. For processing sequential data, the final state of the filter is retained for use as the initial condition in the next window to ensure continuous real-time operation. Techniques for designing linear digital filters can be found in standard signal processing textbooks.

10.3.5 Feature Reduction

The advantage of identifying several (reasonably independent) features with potential for use in SDAs is that, when used in combination, it is increasingly easy to distinguish ictal and interictal waveforms in a higher dimensional feature space. The use of a larger feature set may therefore improve our chances of discriminating true epileptiform from normal EEG. But this presumes that each feature captures some unique property of the EEG that the others do not. It should be obvious by now that many different features tend to reflect similar signal properties, which means they are at least moderately correlated with each other and, therefore, somewhat redundant. It is worthwhile to try to pare down the feature space to a minimal set of features that represent unique signal characteristics, or to find a subset of combinations of the initial features that serve the same purpose. Sometimes, several features may be condensed into one, two, or a few that presumably capture the salient characteristics of seizures and, therefore, track the seizure content of the EEG. This is the goal of feature reduction.

Principal component analysis (PCA) is a popular statistical method of feature reduction often used for this purpose (Flury 1997). Assume that we have sample measurements of n normally distributed features from an EEG signal obtained at multiple time points. If the features have no functional relationship with one another, the samples will form a cloud of points with spherical symmetry in the feature space. But if a subset of the features tends to covary, the sphere will be stretched into an ellipsoid along the axes of significant correlation. PCA computes linear combinations of the original features so that the resulting features are uncorrelated and oriented along the directions of greatest variation. By choosing only the principal components with the largest eigenvalues (i.e., the feature combinations with the greatest variation) as a newly defined feature set, we can reduce the dimensionality of the feature space and focus on the reduced number of combination features with the greatest dynamic range in the feature space. For example, SV, LL, and TE in a finite time window all respond to the level of activity in a signal and should be positively correlated with each other. When subjected to PCA, the feature set can be collapsed by linear combination into one or two new features that explain a greater amount of the variability in the signal than any one of the original features alone, which is advantageous for use in an SDA. This is confirmed by Figure 10.4, in which we see that LL and TE are strongly correlated. As expected, PCA of a 60-s interictal baseline gives a first principal component PC_1 that accounts for about 76% of the variance of the signal, while the second, third, and fourth components account for 15%, 8%, and 1%, respectively. Note that this does not tell us anything about potential contrast between the baseline EEG and seizure onset, but is a useful way to reduce the dimensionality of the feature space and make it more tractable. Fortunately, it also turns out that the standard distance D_{12} for PC_1 (evaluated for two 60-s windows, one interictal and one ictal) is about 1.4, higher than for any of the other features studied in Figure 10.4 for the same pair of windows with the exception of LL ($D_{12} = 1.5$), though the difference is probably not significant.

10.3.6 Augmentation

Features computed from a finite EEG window are susceptible to variations in noise, background power, and spectral changes with the patient's circadian and homeostatic drives (Borbely 2005), sleep-wake state (Broughton 1999), and drug taper (Zaveri et al. 2010). The frequency of interictal spiking itself can change with sleep stage in epilepsy patients (Malow et al. 1998).

Moreover, features with good ictal–interictal contrast will probably fluctuate in response to any instantaneous epileptiform activity that is of comparable duration to the processing window. While we wish to detect only true seizures, the chosen features may be sensitive to relatively brief epileptiform events such as isolated spikes or interictal rhythmic discharges as well. It may not always be possible to successfully design a filter or select a feature that can resolve the difference between desirable and undesirable events and still work on a timescale small enough to be relevant for application in real time.

For the above reasons, an augmentation step is almost always necessary to reduce noise or unwanted variability; i.e., to clean up SDA output. Correcting for changes in the background may be done by normalizing feature values with reference to baseline values updated on a slower timescale. For example, Gotman (1982) used the average amplitude of half-waves in a 16-s background segment ending 12 s before the current window to normalize feature values. Osorio et al. (1998) devised a computationally efficient exponential updating technique to compute an instantaneous background that is heavily biased in favor of historical values on a long timescale for the same purpose. Meng et al. (2004) used a Gaussian mixture model to describe the multimodal feature distribution of the EEG background for normalizing instantaneous feature values. Methods used for feature reduction may simultaneously increase contrast as well.

Isolated noise can be removed by decimation and median filtering techniques (e.g., Osorio et al. 1998) that are relatively insensitive to outliers. The median is the fiftieth percentile point of a probability distribution and is estimated by the middle value in a rank-ordered sample. In Figure 10.5, we see that the feature PC_1, derived in the previous section through PCA, is sensitive to interictal spikes in the preictal period. Nothing prevents the brief surges in PC_1 due to interictal spikes from being mistaken by the classifier for seizure onset. A 2-s median filter applied to the output successfully exterminates the spikes without adversely affecting the seizure period. All of the above processing steps perform the vital function of refining the output of the SDA so as to enhance its specificity to true seizure activity.

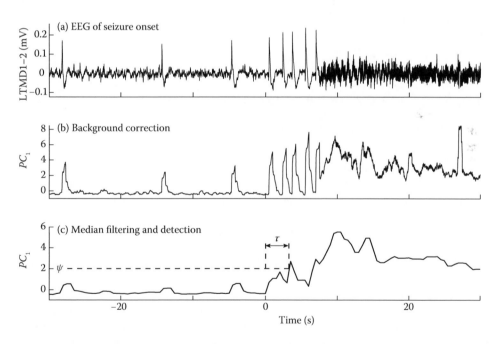

FIGURE 10.5 Feature augmentation and seizure detection. (a) Differential EEG of onset channel LTMD1–2 in the vicinity of a seizure. (b) Feature PC_1 generated as the first principal component of *BP*, *SV*, *LL*, and *TE* for the signal in (a), but normalized to correct for slow background variation by computing the standard distance of the instantaneous PC_1 in a 0.5-s moving window with respect to the mean of a 60-s interictal baseline sample (similar to a Z score). (c) Output of a 2-s moving median filter applied to the feature trace in (b) extinguishes interictal spikes prior to seizure onset. Application of a threshold ψ enables seizure detection with a short delay τ.

10.3.7 Event Detection

At this stage, suppose that a promising set of features has been selected and extracted from the EEG and distilled by the processes of feature reduction and augmentation into one or more features that promise to track the instantaneous seizure content of the signal. The final task to be accomplished by the SDA is to discriminate feature values during true seizures from interictal values and artifacts. This is achieved by drawing a boundary between the two classes in the feature space: Points on one side of this boundary are classified as interictal EEG, and points on the other side are classified as seizures. Successive windows of EEG are mapped onto the feature space and compared against the detection boundaries to detect potential seizures. The boundaries are not necessarily linear; their only qualification is that they enclose points typical of ictal EEG and exclude interictal pretenders as accurately as possible.

If we are using a single feature, the boundary is typically a scalar threshold value or a bounded interval. When the instantaneous feature value crosses the threshold, as is shown in Figure 10.5 for the feature PC_1, a seizure is detected. A multidimensional feature space, on the other hand, requires boundaries for regions in the higher dimensional space. In addition to an amplitude threshold, detections may have to satisfy a minimum duration constraint to increase the specificity of the SDA to true seizures and avoid brief seizure-like discharges that are not of interest.

The boundary may be selected by visual inspection of the feature space, but is better done by numerically maximizing ictal–interictal contrast and minimizing error rates using training samples selected from seizure and nonseizure segments (see Section 10.4 on SDA optimization). This is known as supervised classification. Unsupervised approaches exist and may generate valuable insights into seizure patterns, but must ultimately withstand visual expert validation.

Detection of a seizure may trigger an alarm or responsive treatment (drug infusion, electrical stimulation, etc.). Detection parameters may be optimized individually or for a cohort. The SDA may be designed for early detection or for quantification, which has different requirements.

10.4 ASSESSMENT OF SDA PERFORMANCE

10.4.1 The Detection Problem: Events and Outcomes

In essence, the problem of seizure detection is one of distinguishing between signals from two states, corrupted by noise and variability across the sleep-wake cycle, and corresponding to normal (or interictal) behavior and seizures, respectively. Depending on the level of noise and available ictal–interictal contrast, existing techniques are not always able to correctly determine brain state from the signal without error.

In the language of statistics, the null hypothesis is that the brain is in the interictal state. The state is determined from features of successive short data windows of the signal. If feature values are atypical of their interictal distribution, the null hypothesis may be rejected and the brain deemed to be in the seizure state. Depending on the actual state (interictal or seizure), one of four outcomes is possible: a true seizure is (a) correctly detected (true positive, *TP*) or (b) missed altogether (false negative, *FN*); an interictal epoch is (c) wrongly flagged as a seizure (false positive, *FP*) or (d) rejected as deserved (true negative, *TN*). Examples of different outcomes as a consequence of applying a scalar threshold c to a simulated SDA feature are demonstrated in Figure 10.6. Furthermore, when the outcome is *TP*, it may be important to know whether the SDA response delay τ (see Figure 10.5) satisfied the demands of the situation. Accordingly, detection outcome may be classified as early or late.

For convenient assessment of SDA performance, it is common practice to assign a unique label to a signal epoch that is of comparable duration to a typical seizure, or some fixed duration, e.g., 10s, which defines the temporal resolution of the classifier: if the epoch is interictal, it could be labeled *FP* or *TN*; if a seizure, it is called *TP* or *FN*. These epochs must not overlap with each other and are usually sequential, but not necessarily so.

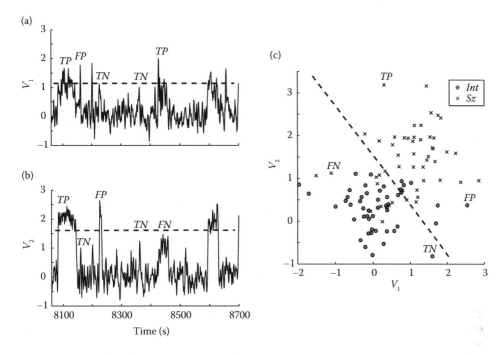

FIGURE 10.6 Outcomes of seizure detection. (a) Univariate seizure detection with a simulated SDA feature V_1 with relatively poor ictal–interictal contrast. Comparison with a threshold (dashed line) spawns detections that are categorized as *TP*, *FP*, *TN*, or *FN* depending on the truth label for that epoch. (b) Detection outcomes using a feature V_2 with better contrast. (c) Bivariate seizure detection with features V_1 versus V_2 computed for ictal (×) and interictal (•) EEG samples. Imposition of a detection boundary (dashed line) segregates sample points into different labeled outcomes.

Sometimes, the output of an SDA may fluctuate during a seizure so that it crosses the detection threshold more than once between onset and termination. These are typically clustered so that each event in an epoch has a unique outcome label. In the case of a seizure, this would be *TP* or *FN*; otherwise, the label would be *FP* or *TN*. Hence, multiple detections beyond onset are ignored if they are deemed close enough to each other (within approximately 30–60 s), and for practical purposes, part of the same event.

10.4.2 MEASURES OF PERFORMANCE

Let us examine some indices by which the performance of an SDA is commonly judged.

10.4.2.1 Detection Delay or Latency, τ

Depending on how much of the signal we need to see before we make a determination, there may be an error of delay because we are sacrificing detection speed for accuracy. The detection latency τ is the time interval from electrographic seizure onset to actual detection (see Figure 10.5). The choice of analysis window size or the imposition of a minimum duration constraint influences detection latency as well as outcome. A larger analysis window tends to have a smoothing effect on feature values if operations like the sample mean or median are applied; extreme values typical of ictal activity are less likely to break through unless they persist for most of the duration of the window. This may help avoid brief interictal events, but delays the moment of detection.

10.4.2.2 Sensitivity Φ

Detection latency only applies to detected seizures, i.e., *TPs*. For a true seizure that goes undetected altogether or is detected too late, there is a risk of an *FN* outcome. Sensitivity expresses the likelihood that a true seizure is detected by the SDA (*TP*). The number of *TPs* as a fraction f_{TP} of all true seizure epochs estimates the true sensitivity Φ of the SDA from a finite sample of epochs. In the large sample limit, assuming that feature distributions are stationary, f_{TP} should approach Φ:

$$f_{TP} = \frac{TP}{TP + FN}; \Phi = E[f_{TP}]. \tag{10.10}$$

10.4.2.3 Specificity κ

A detector may be extremely sensitive, perhaps never missing a single seizure, but its value is diminished if it frequently generates *FPs*. Specificity describes how selective the SDA is to seizure patterns; i.e., how unlikely it is to mistake interictal activity for seizures. Similar to f_{TP}, the *TN* fraction f_{TN}, estimates the true specificity κ of a detector and approaches κ in the large sample limit.

$$f_{TN} = \frac{TN}{FP + TN}; \kappa = E[f_{TN}]. \tag{10.11}$$

10.4.2.4 False Positive Rate *FPR*

The number of false positives occurring per unit time (e.g., per hour), denoted by *FPR*, is a critical measure of performance. It is conceivable that, when used in a seizure warning device, each patient will have a subjective threshold *FPR* that she or he is willing to tolerate as a trade-off for timely seizure warning. However, *FPR* by itself means little without information about the true seizure rate, the size of epoch used in the classifier, and the sensitivity and specificity of the SDA. For instance, an *FPR* of 0.1/h may not seem excessive, unless seizures occur only once per week! Conversely, an *FPR* of 0.3/h may be acceptable if a patient's seizures occur almost every hour. The perception of what an acceptable *FPR* is may be still more stringent if the SDA has doubtful sensitivity; when an increasing number of seizures go undetected (*FN*), the burden associated with each *FP* grows, even if the *FPR* stays the same. Similar considerations apply for the use of SDAs in automated treatment devices.

Several detections may be part of the same event. How we deal with them affects the perceived false positive rate. If a small epoch size is chosen, clustering detections that are close together may reduce *FPs*, but increase *TNs* without increasing either *TPs* or *FNs*. Therefore, clustering may exaggerate specificity unless the epoch size is fixed appropriately. Concerns about overall treatment dose, and the fact that an *FP* can cause anxiety for some period of time after it is issued may be addressed by measuring the fraction of interictal time spent under false positive.

10.4.2.5 Positive Prediction Value *PPV*

Specificity is defined with respect to the number of interictal events or epochs, i.e., *TNs*. However, the estimate is dependent on the choice of epoch size, which may range from a fraction of a second to the duration of a typical seizure. Furthermore, *TNs* are difficult to assess because, in practice, EEG does not monitor neural activity in all possible locations; also behavioral impairment associated with subclinical seizures may go unrecognized. The need to define a *TN* may be avoided by using measures such as the positive prediction value (*PPV*), which is the fraction of detections that are true seizures (Adjouadi et al. 2004):

$$PPV = \frac{TP}{TP + FP}. \tag{10.12}$$

PPV may be viewed as a measure akin to specificity, since a low *FP* would give a high value; however, it does not tell us how many seizures are likely to be missed (*FNs*). Hence, although *PPV* is a useful determinant of SDA performance, it must be employed with caution. Some alternative measures have been proposed in the literature (Brown et al. 2007).

10.4.3   WHAT IS A TRUE NEGATIVE?

Sensitivity is quantified as the fraction of all seizures or epochs that were correctly flagged by the SDA. Specificity is the fraction of all nonseizure events (negatives) that were correctly ignored. But the label negative event may seem like an oxymoron. Should it refer to the entire signal record after all seizure epochs are excluded? Or to the entire interval between two seizures—in which case it would have a variable duration? Neither alternative seems reasonable. One way out is to pick one or more seizure-free intervals for every seizure interval of equal or comparable duration, and use this set of epochs for performance assessment. If any detection is made in the seizure-free interval, it would be an *FP*; else it is a *TN*. Another option is to divide the entire signal record into segments of equal duration—such that an epoch cannot contain more than one seizure—and label them as positives and negatives for SDA assessment. Some investigators circumvent the need to define *TNs* by using *PPV* or *FPR* as indirect measures of specificity. Another approach is to obtain a superset of detections by relaxing the constraints on the SDA to make it extremely sensitive. This will likely approach or attain 100% *TPs*, but also generate a large set of *FPs* that can serve as *TNs* in our evaluation of SDA performance. This approach benchmarks specificity against *potential FPs* and therefore produces a more stringent index of performance.

These indices capture the performance of the SDA. Since most or all of them are dependent on SDA parameters that can be manipulated over a range of values, it follows that there must be a unique value or a subset of values that is in some sense optimal for the chosen application. SDA optimization is discussed in Section 10.5.

10.4.4   GROUND REALITY

Classifying detection outcome and assessing SDA performance implies that we know the true state of the EEG signal (i.e., the ground truth label) at any given time. SDA accuracy must be evaluated against ground truth, which is most commonly a human expert score, but could conceivably be the output (seizure onset time, duration, or zone of influence) of another previously validated SDA for the same set of epochs. If any other standard were available or if there was reasonable consensus among experts about what a seizure should look like, the collective heuristics could simply be coded to produce the world's best SDA. Sadly, this is not the case; interrater agreement is not perfect (Wilson et al. 2003), nor can it be, given our limited understanding of the mechanisms of seizure generation and the measurements available for studying them. For these reasons, visual scoring of the EEG remains largely a subjective art; but for want of a better standard, it is the only benchmark available for grading SDA outcome.

It must be emphasized that in SDA performance evaluation, artifacts should not be considered events. Sources of artifact (motion, line noise, clipping, signal dropout, crosstalk, etc.) can mask genuine epileptiform activity or activate a false response if it falls within the pass band of any filter used for feature extraction. It is important to have a good prefilter to check signal quality and flag potential artifacts. The detector must then ignore these data segments while, at the same time, making sure that events are not missed because of the interrupted information flow. Artifactual detections cannot be used to systematically assess SDA performance unless the artifact is both stereotypical and unavoidable in the available system configuration. Time rates of true or false detections must be normalized by the duration of available good quality signal.

10.5 SDA TRAINING AND OPTIMIZATION

10.5.1 WHY OPTIMIZE?

The reasons for wanting to optimize SDA performance are self-evident: 1. To minimize *FP* and *FN* error rates; 2. To achieve a compromise between conflicting constraints (e.g., high sensitivity with low *FPR*); 3. To make *TP* detections early (i.e., close to electrographic onset); and 4. To tailor the performance of a generic SDA to the peculiarities of a signal source, individual, or cohort. Other objectives include attempting to obtain a better description of seizure dynamics, increase diagnostic yield of monitoring, or assess/improve treatment efficacy.

10.5.2 OVERVIEW OF ROC ANALYSIS

A chosen performance index (e.g., sensitivity Φ) may respond to changes in one or more detection parameters; this provides an opportunity for optimization. However, maximization of Φ comes at the price of specificity κ. For instance, lowering the detection threshold makes the SDA more sensitive to seizures, but may cause it to pick up brief interictal discharges as well (see Figure 10.6). On the other hand, increasing the threshold may improve κ, but not without lowering Φ and increasing detection latency τ. We therefore need to consider how two or more conflicting measures respond to parametric changes at the same time. One way to do this is through receiver operating characteristic (ROC) analysis.

ROC analysis originated in communication theory in the mid-twentieth century. It is a graphical method of examining the trade-off between the sensitivity and specificity of a detector, achieved by varying some detection parameter (see Figure 10.7). Let us illustrate this for a single feature that has

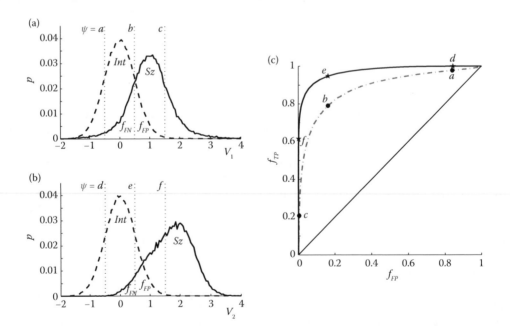

FIGURE 10.7 Receiver operating characteristic (ROC) analysis. (a) Interictal, *Int*, and ictal, *Sz*, probability distributions, *p*, for an SDA feature V_1. Error rates f_{FN} (or $1 - f_{TP}$) and f_{FP} are measured by the areas to the left and right of a detection threshold ψ under $p(V_2)$ for *Sz* and *Int*, respectively. (b) Similar distributions to those in (a), but for a feature V_2 with better ictal–interictal contrast than V_1. (c) ROCs constructed for features V_1 and V_2 by varying ψ from low to high values (e.g., *a*, *b*, and *c* for V_1; *d*, *e*, and *f* for V_2) and plotting the outcome in terms of f_{TP} against f_{FP}. Greater ictal–interictal contrast for V_2 translates into a more optimal ROC.

high values during seizure but is relatively low otherwise. The shape and location of the ROC curve indicates what need or scope there is for improvement.

In an ROC, the sensitivity (estimated as f_{TP}) is plotted against the reverse specificity or non-specificity (estimated as $f_{FP} = 1 - f_{TN}$) as a function of a detection parameter, for instance, a feature threshold ψ. FPR may be used instead of specificity; but as ψ is lowered, in the limit only one enormous (clustered) detection will remain (i.e., a constant state of active detection), so the ROC will not monotonically increase but reach a plateau and then fall off as the total number of detections decreases. Other possible forms for the ROC could plot f_{TP} against PPV or f_{TP} against τ, with possibly unique functionality as well.

Assume that we have a training data set consisting of many epochs, with truth labels for seizure (Sz) and interictal (Int) samples and feature values V_1 generated by an SDA. V_1 is relatively high during seizure and low otherwise, as shown by its probability distribution p in Figure 10.7a. A threshold ψ applied to V_1 will generate a set of detections, with an outcome for each labeled epoch. Comparing the outcome with the event tells us whether an epoch is a TP, FN, FP, or TN, from which the sensitivity f_{TP} and nonspecificity f_{FP} achieved for that threshold can be computed and plotted against each other as an ROC (Figure 10.7c). For a sufficiently low ψ, all epochs will be TP or FP, which means perfect sensitivity but zero specificity; the operating point of the SDA is at the top right of the ROC. As ψ is gradually increased, we start to lose some FPs but also some TPs, and we move toward the bottom left of the curve. For an arbitrarily high threshold ψ, we get no FPs, but no TPs either—perfect specificity with zero sensitivity, the bottom left of the ROC. By repeating this exercise for various intermediate thresholds (e.g., $\psi = a$, b, or c in Figure 10.7), we can construct an ROC that describes the effect of ψ on SDA performance. Note that the area under the probability distributions of Int and Sz to the right and left of ψ, respectively, are identical to the error rates f_{FP} and f_{FN} that go into the ROC. It is worthwhile to examine the effect of varying ψ on these regions of the feature distributions.

10.5.3 Optimization Strategies

Our goal is to find a set of detection parameter values that would correspond to an optimal operating point for the SDA. To reach this goal, we must formulate an objective function, which tells us how performance is going to vary when we tune the parameters. We can then optimize the objective function mathematically or numerically depending on which approach is feasible.

The ROC is a graphical objective function. The theoretical optimum—perfect sensitivity and specificity—is at the top left corner; this occurs only if the ictal and interictal feature distributions in Figure 10.7 have no overlap. The objective function can be optimized either by changing the threshold ψ to move the ROC operating point to the location on the knee closest to the theoretical optimum (top left corner), or by selecting or transforming the features (and their probability distributions thereby) to amplify contrast so that the ROC itself is stretched closer to the theoretical optimum. Since the area under the ROC is increased by the latter, it is obvious that ROC area is a useful index of SDA performance, although it does not quantify the speed of detection.

As we pointed out above, a static ROC can be optimized by varying the detection threshold. The advantage to be gained by this is limited by the ROC area. If the ROC area is inadequate, it could possibly be increased by selecting a different detection feature or redefining it to improve feature contrast. This is illustrated in Figure 10.7; if we had initiated our analysis using a feature V_1 (e.g., total signal power P), we might improve the ROC by using a different feature V_2 (e.g., 10–20-Hz band power BP) that has greater ictal–interictal contrast, or by designing a filter specifically attuned to activity in the first few seconds of seizure onset. Discussions of SDA optimization strategies can be found in Qu and Gotman (1997), Meng et al. (2004), Haas et al. (2007), and Talathi et al. (2008).

Imposing a minimum duration constraint for which the feature needs to cross the threshold to trigger the SDA will also affect the ROC. But it is not a manipulation of the feature threshold ψ and does not necessarily have a similar effect to shifting the threshold, nor is it a modification of the feature itself. The trade-off is a delay in *TP* detection.

When the feature space has more than one dimension (a vector of features), the detection boundary is typically a hyperplane (or a curve/surface that cordons off one or more regions in multidimensional space) rather than a scalar threshold (see Figure 10.6c), and its location with respect to ictal and interictal training samples needs to be optimized. The boundary is not necessarily linear and may even be designed to circumscribe the samples of one of the classes. Again, an ROC can be constructed as a function of the location/orientation of the boundary but this is not usually very practical or useful given the larger number of degrees of freedom. Instead, a mathematical statement of the objective function is typically proposed and optimized using a numerical procedure such as maximum likelihood estimation (MLE), with the help of labeled training samples. Unlike the ROC, which weighs them equally, transformations can be applied to the sensitivity and specificity to control their weights and make the optimum more relevant to a particular application. Statistical classification techniques often go by the name machine learning, based on the notion that a computational machine inspects training samples to learn the decision boundary in the feature space that is expected to minimize classification error when applied to new, similarly distributed random samples. Examples include linear discriminant analysis (LDA), artificial neural networks (ANNs), and support vector machines (SVMs), all of which have been applied to the problem of seizure detection.

In LDA (Flury 1997), a linear boundary is identified that separates samples of the two classes in the feature space by maximizing the ratio of the variance between classes to the average variance within each class, similar to the *F* statistic in standard analysis of variance. In Fisher's discriminant analysis, feature reduction is simultaneously accomplished by identifying linear combinations of features that maximize the objective function. Jerger et al. (2005) describe seizure detection using LDA from intracranial EEG recordings.

ANNs are nonlinear computational devices with a structure reminiscent of biological neuronal networks. In their most basic form, ANNs have an input layer (a set of nodes for receiving inputs), an output layer, and several intermediate hidden layers with weighted nodes. Each node has a threshold for activation by its summed inputs. Given samples with feature values and truth labels, the weights of the nodes in each layer are optimized to reproduce the input–output behavior. Examples of the use of ANNs in SDAs are found in Ghosh-Dastidar et al. (2007), Tzallas et al. (2007), and Patnaik and Manyam (2008). The decision boundary generated by an ANN is not linear and great flexibility can be obtained by increasing the number of layers and/or nodes in the network. However, there is a risk of overtraining with small training sets; data boosting procedures may be useful in dealing with sparse training data.

SVMs are a more recent development in machine learning. In this method, a subset of training samples or support vectors (hence the name) closest to the decision boundary in the feature space are retained for classifying future data. The decision boundary is found by maximizing the distance or margin between support vectors of the two classes. Since only points bordering the two classes are of interest, SVMs are found to be suitable for problems such as seizure detection in which one of the classes (seizures) is sparsely sampled and a reasonable probability model cannot be formulated. As in other classification domains, SVMs are quickly becoming popular for seizure detection (Esteller et al. 2001; Chan et al. 2008).

In general, these methods try to implement a classification rule that labels any sample it is presented with as ictal or interictal. The parameters of this rule are determined by minimizing the total classification error rate, which is a function of the detection parameter(s). Practical constraints may necessitate the imposition of weights or penalties for misclassification of one or the other class; for instance, *FN*s may be considered more detrimental than *FP*s. None of the methods are restricted to binary detection and each can be designed to work for multiple classes. This may be useful if we wish to distinguish between two or more patterns of seizure onset, or correct for seizure onset during different sleep–wake states.

10.5.4   POWER TRAINING

In our discussion so far, it has been assumed that the SDA will be optimized on one training data set based on its performance on the same data set. In reality, signal and seizure characteristics change over time, and so will SDA performance. The robustness of the SDA to such variation must be assessed in training using out-of-sample measures such as the leave-one-out error, twofold cross-validation error, and so on, in which several trials are performed using a subset of the data to train the SDA and the remainder to test it (Flury 1997). The distribution of out-of-sample errors tells us what performance we expect to see in real-time operation.

The first requirement for optimizing an SDA is representative training data; but the need for adequate training data must be stressed as well. An ROC only approximates SDA behavior and a small sample leads to a poor estimate of its properties. Good performance on a small training sample is not a guarantee of future performance. If possible, statistical power analysis (Faul et al. 1997) should be performed to determine the minimum sample size required for testing the ability of a classifier to distinguish between ictal and interictal epochs. Statistical power determines our ability to (1) assess whether performance is significant, (2) train the SDA to be robust, (3) estimate the ROC curve accurately and optimize parameters, and (4) evaluate the treatment effect.

10.6   HOW TO QUANTIFY SEIZURES

Why do we need to quantify seizures?

1. To attempt to assess their severity and thereby evaluate more subtle effects of stimuli and treatment than a simple change in seizure frequency
2. To understand why some discharges become seizures and propose strategies to prevent them from doing so
3. To provide dynamical descriptions of seizure types associated with a specific etiology and semiology, e.g., complex partial seizures of temporal origin with secondary generalization
4. To identify measurable dynamical parameters that have analogues in mathematical models of neuronal networks and use this to test hypotheses regarding seizure generation and control

Seizure quantification poses different challenges from real-time (or even retrospective) seizure detection. What do we wish to quantify? We would like to have objective and meaningful measures of seizure intensity, duration, and spread, and descriptors of seizure evolution and progression.

Measures of intensity, duration, and spread are relatively easy to formulate if we already have a feature(s) that reliably detects seizure activity from onset to termination and provides a reasonable measure of the instantaneous seizure content of the EEG. If this is the case, we can simply use the integral of this feature (or its peak) over the duration of the seizure as a measure of seizure intensity. Seizure duration is then the duration of the detection and seizure spread is measured by the number of recording contacts that detected the same event (Osorio et al., 1998). In Figure 10.8a, we see an intensity map over all LTMD contacts for the seizure in Figure 10.1 that conveys the progression and severity of the seizure. In this way, we can generate a coarse description of seizure dynamics.

The above approach does not convey spectral changes or dynamical interactions between brain regions that occur during seizure. More sophisticated representations of seizure dynamics have been proposed using time-frequency analysis.

We have seen how the Fourier power spectral density (PSD) gives the distribution of power across frequency in a single data window. A spectrogram is a visual representation in which PSDs are computed in successive data windows and plotted as columns of an array in which each row is indexed to frequency and each column to time. PSD values are color-coded by their relative

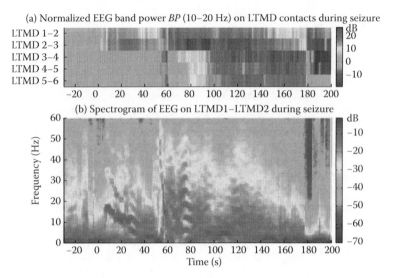

FIGURE 10.8 (See color insert.) Quantification of seizure dynamics. (a) Seizure intensity as measured by 10–20 Hz band power, *BP*, mapped against time for differential recordings from all contacts on a depth electrode LTMD. (b) Spectrogram for the EEG signal on seizure onset location LTMD1–2 shows time-frequency distribution of signal power. Each column is a PSD computed for a moving 4-s window of data with a 3-s overlap, using the multitaper spectral analysis method.

amplitude. The seizure EEG visualized as a spectrogram (Figure 10.7b) provides an interesting time-frequency representation of its dynamics with a highly distinctive signature.

Schiff et al. (2000) have used the term brain "chirp" to describe the spectrographic signature of a seizure. In that work, they use a matched filter to compare a moving snapshot of the spectrogram against the template of a previously computed signature of a typical seizure to detect stereotypical seizures in chronic intracranial recordings from epilepsy patients being evaluated for surgery. By combining information from the spectrogram with an instantaneous spatial map of when the chirp was detected on different contacts on an electrode grid, the authors were able to study both seizure propagation and localization. More recently, Schiff et al. (2005) applied Fisher's multivariate discriminant analysis to confirm that human seizure EEG could be divided into dynamically distinct stages—a beginning, middle, and end—using both linear and nonlinear measures of synchronization, correlation, and phase.

The short time Fourier transform is limited in its ability to resolve nonstationary dynamical activity that is changing relatively rapidly, as happens during a seizure. Bergey and Franaszczuk (2001) proposed the use of a matching pursuit algorithm to provide continuous decompositions of dynamical activity during seizures. In this algorithm, the EEG signal at a given time is expressed in terms of a set of elementary functions called Gabor atoms that best match its time-frequency structure using an iterative procedure. The number of time-frequency atoms required to reconstruct the signal is known as the Gabor atom density (GAD), which is a measure of the complexity of the signal. When the matching pursuit algorithm was applied to epileptic EEG (Jouny et al. 2004), it was found that GAD increased appreciably at the onset of complex partial seizures, and the earliest increase occurred near the focus. GAD patterns were distinctive of the seizure type studied, and were not merely correlated with signal power. It is therefore felt to be a promising technique for analyzing seizure dynamics.

These examples are representative of tools being developed to help quantify epileptic seizures and understand their dynamics. It is hoped that the emerging capability for seizure quantification will help us understand the causes and mechanisms of seizure generation and provide clues for preventing or interrupting them using drugs or other therapies.

10.7 THE COST OF POOR DETECTION

The benefits of accurate seizure detection are self-evident. But what are the costs associated with using a poor SDA? Detection errors can lead to patient anxiety, unwarranted or missed treatment (e.g., closed-loop stimulation or other treatment of *FP*s or nonstimulation of *FN*s), or flawed assessment of treatment outcome (Sunderam et al. 2010). In the first case, apart from safety concerns, excessive responsive stimulation of *FP*s could blur the distinction between closed-loop and open-loop treatment protocols. A hypothetical instance of the latter case would be empirical tuning of closed-loop stimulation parameters to optimize treatment. If the design involves stimulation of alternate detections until a two sample *t* test can be used to compare the duration of stimulated versus control detections before changing parameters, and it turns out that some of those detections were *FP*s, the post hoc statistical power of the test—i.e., the probability that an effect will be detected when there really is one—is lowered. Likewise, if there are *FN*s in this sample, and they are converted to *TP*s in a retrospective analysis by decreasing the detection threshold, they change the true effect size of the test since they were not detected by the initial design and therefore not treated. These are complications that can arise from erroneous detection.

10.7.1 SO WHICH SDA IS THE BEST—OR AT LEAST GOOD ENOUGH?

It may be argued that any SDA that precedes clinical onset with adequate lead time and good specificity (good sensitivity is not as elusive) is a good seizure detector, because: 1. We are most interested in identifying and/or eliminating the cognitive impairment associated with seizures; and 2. Given that electrographic onset may itself be preceded by other dynamical changes, we do not really know when seizures begin. On the other hand, there may be a benefit to treating subclinical events (e.g., interictal spikes, HFOs) or using them for treatment evaluation if correlated with seizure recurrence (Sunderam et al. 2010). Also, degenerative changes (network, behavior, cognition) may be associated with subclinical events that warrant further investigation.

In conclusion, the onus is on practicing clinicians and engineers to address the questions of what to detect, why, and how, in a manner that benefits individuals with epilepsy in their care.

ACKNOWLEDGMENTS

The author is grateful for support from the National Institute of Neurological Disorders and Stroke (NIH Grant No. 5R03NS065451-03) and Epilepsy Foundation during the writing of this manuscript. EEG data used in the examples were drawn from the database developed by Flint Hills Scientific, LLC, with the support of NIH/NINDS Grant No. 3R01NS046602-03S1.

REFERENCES

Adeli, H., Z. Zhou, and N. Dadmehr. 2003. Analysis of EEG records in an epileptic patient using wavelet transform. *J. Neurosci. Methods* 123(1):69–87.

Adjouadi, M., D. Sanchez, M. Cabrerizo, M. Ayala, P. Jayakar, I. Yaylali, and A. Barreto. 2004. Interictal spike detection using the Walsh transform. *IEEE Trans. BME* 51(5):868–872.

Aksenova, T. I., D. V. Nowicki, and A. L. Benabid. 2009. Filtering out deep brain stimulation artifacts using a nonlinear oscillatory model. *Neural Comput.* 21(9):2648–2666.

Alvarez, I., A. de la Torre, M. Sainz, and C. Roldan. 2010. Reducing blanking artifact in electrically evoked compound action potentials. *Comput. Methods Programs Biomed.* 97(3):257–263.

Bergey, G. K., and P. J. Franaszczuk. 2001. Epileptic seizures are characterized by changing signal complexity. *Clin. Neurophysiol.* 112(2):241–249.

Borbély, A. A., and P. Achermann. 2005. Sleep homeostasis and models of sleep regulation. In *Principles and practice of sleep medicine*, eds. M. H. Kryger, T. Roth, and W. C. Dement, 405–417. Philadelphia: Elsevier Saunders.

Broughton, R. J. 1999. Polysomnography: Principles and applications in sleep and arousal disorders. In *Electroencephalography basic principles, clinical applications and related fields*, 4th ed., eds. E. Niedermeyer and F. Lopes da Silva, 858–895. Baltimore: Lippincott, Williams, and Wilkins.

Brown III, M. W., B. E. Porter, D. J. Dlugos, J. Keating, A. B. Gardner, P. B. Storm Jr., and E. D. Marsh. 2007. Comparison of novel computer detectors and human performance for spike detection in intracranial EEG. *Clin. Neurophysiol.* 118:1744–1752.

Chan, A. M., F. T. Sun, E. H. Boto, and B. M. Wingeier. 2008. Automated seizure onset detection for accurate onset time determination in intracranial EEG. *Clin. Neurophysiol.* 119:2687–2696.

Esteller, R., J. Echauz, T. Tcheng, B. Litt, and B. Pless. 2001 Line length: An efficient feature for seizure onset detection. *Proc. IEEE Conf. Eng. Med. Biol. Soc.* 2001:1707–1710.

Faul, S., G. Gregorcic, G. Boylan, W. Marnane, G. Lightbody, and S. Connolly. 2007. Gaussian process modeling of EEG for the detection of neonatal seizures. *IEEE Trans. BME* 54(12):2151–2162.

Firpi, H., E. D. Goodman, and J. Echauz. 2007. Epileptic seizure detection using genetically programmed artificial features. *IEEE Trans. BME* 54(2):212–224.

Flury, B. 1997. *A first course in multivariate statistics.* New York: Springer-Verlag, Inc.

Frei, M. G., and I. Osorio. 2007. Intrinsic time-scale decomposition: Time-frequency-energy analysis and real-time filtering of non-stationary signals. *Proc. R. Soc. A* 463(2078):321–342.

Gardner, A. B., A. M. Krieger, G. Vachtsevanos, and B. Litt. 2006. One-class novelty detection for seizure analysis from intracranial EEG. *J. Machine Learning Res.* 7:1025–1044.

Ghosh-Dastidar, S., H. Adeli, and N. Dadmehr. 2007. Mixed-band wavelet-chaos-neural network methodology for epilepsy and epileptic seizure detection. *IEEE Trans. BME* 54(9):1545–1551.

Gotman, J. 1982. Automatic recognition of epileptic seizures in the EEG. *Electroencephalogr. Clin. Neurophysiol.* 54:530–545.

Grewal, S., and J. Gotman. 2005. An automatic warning system for epileptic seizures recorded on intracerebral EEGs. *Clin. Neurophysiol.* 116:2460–2472.

Haas, S. M., M. G. Frei, and I. Osorio. 2007. Strategies for adapting automated seizure detection algorithms. *Med. Eng. Phys.* 29:895–909.

Harrison, M. A., M. G. Frei, and I. Osorio. 2008. Detection of seizure rhythmicity by recurrences. *Chaos* 18(3):033124.

Hoffmann, S., and M. Falkenstein. 2008. The correction of eye blink artifacts in the EEG: A comparison of two prominent methods. *PLoS One* 3(8):e3004.

Hu, J., J. Gao, and J. C. Principe. 2006. Analysis of biomedical signals by the Lempel-Ziv complexity: The effect of finite data size. *IEEE Trans. Biomed. Eng.* 53(12 Pt 2):2606–2609.

Hillered, L., P. M. Vespa, and D. A. Hovda. 2005. Translational neurochemical research in acute human brain injury: The current status and potential future for cerebral microdialysis. *J. Neurotrauma.* 22(1):3–41.

Jacobs, J., M. Zijlmans, R. Zelmann, C. E. Chatillon, J. Hall, A. Olivier, F. Dubeau, and J. Gotman. 2009. High-frequency electroencephalographic oscillations correlate with outcome of epilepsy surgery. *Ann. Neurol.* 67(2):209–220.

Jerger, K. K., S. L. Weinstein, T. Sauer, and S. J. Schiff. 2005. Multivariate linear discrimination of seizures. *Clin. Neurophysiol.* 116:545–551.

Jouny, C. C., P. J. Franaszczuk, B. Adamolekun, and G. K. Bergey. 2004. Gabor atom density as a measure of seizure complexity. *Conf. Proc IEEE Eng. Med. Biol. Soc.* 1:310–312.

Kaiser, J. F. 1990. On a simple algorithm to calculate the 'energy' of a signal. *ICASSP* 1990:381–384.

Karayiannis, N. B., Y. Xiong, J. D. Frost Jr., M. S. Wise, R. A. Hrachovy, and E. M. Mizrahi. 2006. Automated detection of videotaped neonatal seizures based on motion tracking methods. *J. Clin. Neurophysiol.* 23(6):521–531.

Khan, Y. U., and J. Gotman. 2003. Wavelet based automatic seizure detection in intracerebral electroencephalogram. *Clin. Neurophysiol.* 114:898–908.

Lee, H. C., W. van Drongelen, A. B. McGee, D. M. Frim, and M. H. Kohrman. 2007. Comparison of seizure detection algorithms in continuously monitored pediatric patients. *J. Clin. Neurophysiol.* 24(2):137–146.

Lesser, R. P., S. H. Kim, L. Beyderman, D. L. Miglioretti, W. R. Webber, M. Bare, B. Cysyk, G. Krauss, and B. Gordon. 1999. Brief bursts of pulse stimulation terminate afterdischarges caused by cortical stimulation. *Neurology* 53(9):2073–2081.

Malarvili, M. B., and Mesbah, M. 2009. Newborn seizure detection based on heart rate variability. *IEEE Trans. BME* 56(11):2594–2603.

Malow, B. A., X. Lin, R. Kushwaha, and M. Aldrich. 1998. Interictal spiking increases with sleep depth in temporal lobe epilepsy. *Epilepsia* 39(12):1309–1316.

Marsh, E. D., B. Peltzer, M. W. Brown III, C. Wusthoff, P. B. Storm Jr., B. Litt, and B. E. Porter. 2010. Interictal EEG spikes identify the region of electrographic seizure onset in some, but not all, pediatric epilepsy patients. *Epilepsia* 51(4):592–601.

Meng, L., M. G. Frei, I. Osorio, G. Strang, and T. Q. Nguyen. 2004. Gaussian mixture models of ECoG signal features for improved detection of epileptic seizures. *Med. Eng. Phys.* 26:379–393.

Mormann, F., R. G. Andrzejak, C. E. Elger, and K. Lehnertz. 2007. Seizure prediction: The long and winding road. *Brain* 130(Pt 2):314–333.

Nijsen, T. M., P. J. Cluitmans, J. B. Arends, and P. A. Griep. 2007. Detection of subtle nocturnal motor activity from 3-D accelerometry recordings in epilepsy patients. *IEEE Trans. BME* 54(11):2073–2081.

Nunez, P. L., and R. Srinivasan. 2006. *Electric fields of the brain. The neurophysics of EEG*, 2nd ed. New York: Oxford University Press.

Orosco, L., E. Laciar, A. G. Correa, A. Torres, and J. P. Graffigna. 2009. An epileptic seizures detection algorithm based on the empirical mode decomposition of EEG. *Conf. Proc. IEEE Eng. Med. Biol. Soc.* 2009:2651–2654.

Osorio, I., M. G. Frei, and S. B. Wilkinson. 1998. Real-time automated detection and quantitative analysis of epileptic seizures and short-term prediction of clinical onset. *Epilepsia* 39(6):615–627.

Patnaik, L. M., and O. K. Manyam. 2008. Epileptic EEG detection using neural networks and post-classification. *Comput. Methods Program Biomed.* 91(2):100–109.

Pereda, E., R. Q. Quiroga, and J. Bhattacharya. 2005. Nonlinear multivariate analysis of neurophysiological signals. *Prog. Neurobiol.* 77:1–37.

Qu, H., and J. Gotman. 1997. A patient-specific algorithm for the detection of seizure onset in long-term EEG monitoring: Possible use as a warning device. *IEEE Trans BME* 44(2):115–122.

Rolston, J. D., R. E. Gross, and S. M. Potter. 2009. A low-cost multielectrode system for data acquisition enabling real-time closed-loop processing with rapid recovery from stimulation artifacts. *Front. Neuroengineering* 2:12.

Saab, M. E., and J. Gotman. 2005. A system to detect the onset of epileptic seizures in scalp EEG. *Clin. Neurophysiol.* 116:427–442.

Scher, M. S., J. Alvin, L. Gaus, B. Minnigh, and M. J. Painter. 2003. Uncoupling of EEG-clinical neonatal seizures after antiepileptic drug use. *Pediatric Neurology* 28(4):277–280.

Schiff, S. J. 2005. Dangerous phase. *Neurinformatics* 3:315–317.

Schiff, S. J., D. Colella, G. M. Jacyna, E. Hughes, J. W. Creekmore, A. Marshall, M. Bozek-Kuzmicki, G. Benke, W. D. Gaillard, J. Conry, and S. Weinstein. 2000. Brain chirps: Spectrographic signatures of epileptic seizures. *Clin. Neurophysiol.* 111:953–958.

Schiff, S. J., T. Sauer, R. Kumar, and S. L. Weinstein. 2005. Neuronal spatiotemporal pattern discrimination: The dynamical evolution of seizures. *Neuroimage* 28(4):1043–1055.

Schlogl, A., C. Keinrath, D. Zimmermann, R. Scherer, R. Leeb, and G. Pfurtscheller. 2007. A fully automated correction method of EOG artifacts in EEG recordings. *Clin. Neurophysiol.* 118(1):98–104.

Schuyler, R., A. White, K. Staley, and K. J. Cios. 2007. Identification of ictal and pre-ictal states using RBF networks with wavelet-decomposed EEG data. *IEEE EMBS Mag.* 74–81.

Shackman, A. J., B. W. McMenamin, H. A. Slagter, J. S. Maxwell, L. L. Greischar, and R. J. Davidson. 2009. Electromyogenic artifacts and electroencephalographic inferences. *Brain Topogr.* 22(1):7–12.

Shoeb, A., H. Edwards, J. Connolly, B. Bourgeois, S. T. Treves, and J. Guttag. 2004. Patient-specific seizure onset detection. *Epilepsy Behav.* 5(4):483–498.

Skarpaas, T. L., and M. J. Morrell. 2009. Intracranial stimulation therapy for epilepsy. *Neurotherapeutics* 6(2):238–243.

Slooter, A. J. C., E. M. Vriens, F. S. S. Leijten, J. J. Spijkstra, A. R. J. Girbes, A. C. van Huffelen, and C. J. Stam. 2006. Seizure detection in adult ICU patients based on changes in EEG synchronization likelihood. *Neurocrit. Care* 5:186–192.

Smith, S. W. 1999. *The scientist's and engineer's guide to digital signal processing*, 2nd ed. San Diego: California Technical Publishing.

Spencer, S. S., I. I. Goncharova, R. B. Duckrow, E. J. Novotny, and H. P. Zaveri. 2008. Interictal spikes on intracranial recording: Behavior, physiology, and implications. *Epilepsia* 49(11):1881–1892.

Sunderam, S., Gluckman, B. J., Reato, D., and Bikson, M. 2010. Toward rational design of electrical stimulation strategies for epilepsy control. *Epilepsy Behav.* 17(1):6–22.

Talathi, S. S., D. U. Hwang, M. L. Spano, J. Simonotto, M. D. Furman, S. M. Myers, J. T. Winters, W. L. Ditto, and P. R. Carney. 2008. Non-parametric early seizure detection in an animal model of temporal lobe epilepsy. *J. Neural. Eng.* 5(1):85–98.

Thomson, D. J. 1982. Spectrum estimation and harmonic analysis. *Proc. IEEE* 70:1055–1096.

Tzallas, A. T., M. G. Tsipouras, and D. I. Fotiadis. 2007. Automatic seizure detection based on time-frequency analysis and artificial neural networks. *Comput. Intell. Neurosci.* 2007:80510.

Van Putten, M. J. A. M., T. Kind, F. Visser, and V. Lagerburg. 2005. Detecting temporal lobe seizures from scalp EEG recordings: a comparison of various features. *Clin. Neurophysiol.* 166: 2480–2489.

Wilson, S. B., M. L. Scheuer, C. Plummer, B. Young, and S. Pacia. 2003. Seizure detection: correlation of human experts. *Clin. Neurophysiol.* 114: 2156–2164.

Zaveri, H. P., S. M. Pincus, I. I. Goncharova, E. J. Novotny, R. B. Duckrow, D. D. Spencer, H. Blumenfeld, and S. S. Spencer. 2010. Background intracranial EEG spectral changes with anti-epileptic drug taper. *Clin. Neurophysiol.* 121(3):311–317.

Zelmann, R., M. Zijlmans, J. Jacobs, C. E. Châtillon, and J. Gotman. 2009. Improving the identification of high frequency oscillations. *Clin. Neurophysiol.* 120(8):1457–1464.

11 Automated Prediction and Assessment of Seizure Prediction Algorithms

Florian Mormann, Ralph G. Andrzejak, and Klaus Lehnertz

CONTENTS

The sudden and apparently unpredictable nature of seizures is one of the most disabling aspects of the disease epilepsy. A method of predicting the occurrence of seizures from the electroencephalogram (EEG) of epilepsy patients would open new therapeutic possibilities. Since the 1970s, investigations on the predictability of seizures have advanced from preliminary descriptions of seizure precursors to controlled studies applying prediction algorithms to continuous multiday EEG recordings. While most of the studies published in the 1990s and around the turn of the millennium yielded rather promising results, more recent evaluations could not reproduce these optimistic findings. The current literature is at best inconclusive as to whether seizures are predictable by prospective algorithms.

11.1 PART I: CONCEPTUAL ISSUES

The first part of this chapter addresses issues related to the design of seizure prediction studies.

11.1.1 Prediction, Forecasting, or Anticipation?

In the strict sense of the words, predicting or forecasting an event means the ability to determine in advance the time of its occurrence with a certain precision. The term *anticipation* implies more of an uncertainty as to when exactly an event will occur. This latter concept better fits the design of seizure prediction algorithms, which usually assume a seizure to occur within a certain time period after an alarm is issued, without knowing its exact onset time. As in the majority of publications in this field, we will use the three different terms interchangeably.

11.1.2 The Events to Be Predicted: Clinical or Electrographic Seizures?

An important issue is the selection of the ictal events that are to be anticipated by an algorithm (Lehnertz and Litt 2005). While the benchmark for clinical application would clearly be the forecasting of clinical seizure events, subclinical seizures today are mostly regarded as a milder variant of the same dynamical event that constitutes a clinical seizure. It is therefore arguable whether it is reasonable to exclude subclinical ictal events in a prediction algorithm. Nevertheless, most studies so far have restricted themselves to the analysis of clinical seizures.

Similarly, the onset time of a seizure can be determined either from the first clinical signs or from the first visible EEG changes. Since there is often some uncertainty in the assessment of clinical symptoms, particularly in complex partial and absence seizures, it is reasonable to determine the seizure onset electrographically, especially if intracranial recordings from the seizure onset zone are available. In cases where the seizure onset zone is not located within eloquent cortex, seizure activity may need to propagate into eloquent cortex before clinical symptoms can be observed.

11.1.3 Seizure Prediction versus Early Seizure Detection

Algorithms that aim at an early detection of the electrographic seizure onset, which may occur several seconds before the first clinical symptoms, should not be regarded as seizure prediction algorithms, but rather as early seizure detection algorithms (Osorio et al. 1998). In contrast to seizure prediction, which aims at the identification of a sufficiently long preictal state before the electrographic seizure onset, early seizure detection does not provide extensive time for intervention. If implemented within a closed-loop intervention system endowed with sufficient seizure abatement strategies, early detection algorithms may prove useful as a basis for responsive intervention, provided the epileptic brain is not yet beyond a point of no return from which it will inevitably evolve into a clinical seizure. The clinical utility of both prediction-triggered therapy and early-detection-triggered therapy remains to be conclusively demonstrated (Osorio et al. 2005).

11.1.4 The Type of EEG: Intracranial or Surface Recordings?

While the majority of seizure prediction studies to date have been carried out on intracranial recordings, there are some studies that analyzed surface recordings. Intracranial recordings bear the advantage of a higher signal-to-noise ratio and a better spatial resolution and are typically far less contaminated by artifacts and more representative of brain activity than data obtained from the scalp. They also bear the potential advantage of allowing one to record directly from the seizure-generating region. On the other hand, surface recordings are less invasive and could, in principle, be used in an ambulatory setting to monitor a patient's seizure situation in his or her usual environment. This would require a high degree of compliance on the part of a patient owing to the inconvenience of constantly wearing an EEG cap. Potential problems associated with chronic surface recordings are skin erosion and discomfort as well as maintaining good skin-electrode interface.

If seizure anticipation algorithms proved to be successful, they would most likely be implemented in an implantable, closed-loop warning or intervention system. Since interventional therapies are mostly based on electrical stimulation, they usually require implantation of a stimulation electrode, which can then also be used to acquire the signal for seizure prediction or detection (Osorio et al. 2005). Intracranial recordings may also allow earlier prediction or detection of seizures than scalp recordings. The technical feasibility of intracranial intervention systems has already been proven by responsive brain-stimulation devices that are currently being tested in clinical trials for their ability to reduce seizure frequency (Morrell 2006; Sun et al. 2008). Many groups in the field therefore regard the usefulness of scalp EEG recordings for studies on seizure prediction as rather limited compared with intracranial recordings.

11.1.5 DATA REQUIREMENTS

While it is desirable for analysis to use data sets that contain a large number of seizures, it is also desirable to have a sufficient time interval between consecutive seizures, so that they can be regarded as independent events. If seizures are too closely spaced (clustered seizures) it becomes difficult to separate the postictal period from a presumed preictal state (Jouny et al. 2005) as the exact duration of either is unknown. It may be noted in this context that the average seizure frequency in a monitoring unit of up to three events per day (Haut et al. 2002) is approximately 30 times higher than the mean seizure frequency of three per month under normal circumstances (Bauer and Burr 2001). If a certain false prediction rate in the epilepsy monitoring unit corresponds to a situation where every other alarm is a false alarm (positive predictive value of 50%), then the same false prediction rate under normal circumstances would mean that only 1 out of 60 alarms is a correct warning (positive predictive value of 1.7%) (Winterhalder et al. 2003).

EEG recordings used for studies on seizure prediction should ideally comprise EEG data recorded continuously over several days. Recording gaps due to diagnostic procedures during the presurgical work-up (e.g., structural MRI to verify electrode placement) are usually unavoidable and are not considered a major drawback. Since during the presurgical monitoring, patients are constantly undergoing changes that could have a confounding influence on characterizing measures of the EEG (e.g., tapering of medication), it is advisable to use all interictal control data available since a restriction (e.g., the first 24 h of an EEG) could introduce a confounding bias.

11.2   PART II: ASSESSING THE PERFORMANCE OF A PREDICTION ALGORITHM

In order to compare the relative merit of the different studies on seizure prediction published to date, it is necessary to realize how the performance of a seizure prediction technique is assessed. In this part we will discuss some of the problems and pitfalls involved in the evaluation of an algorithm for seizure prediction.

11.2.1 MOVING WINDOW ANALYSIS

Most EEG-based prediction techniques use a moving-window analysis in which some (linear or nonlinear) characterizing measure (Kantz and Schreiber 2003) is calculated from a window of EEG data with a predefined length, and then the subsequent window of EEG is analyzed. The duration of these analysis windows typically ranges between 10 and 40 s. Depending on whether the measure is used to characterize a single EEG channel or relations between two or more channels, it is referred to as a univariate, bivariate, or multivariate measure, respectively. The moving-window analysis thus renders time profiles of a characterizing measure for different channels or channel combinations (Figure 11.1).

11.2.2 STATISTICAL VERSUS ALGORITHMIC APPROACHES

The analysis design used to evaluate these time profiles in the following step can be either statistical or algorithmic. A statistical design is retrospective by nature and compares the amplitude

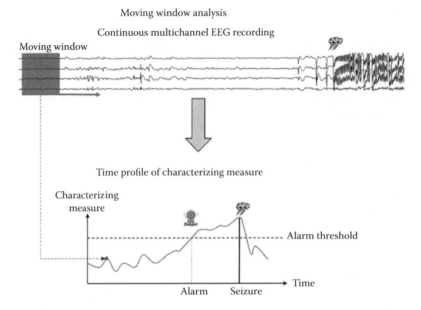

FIGURE 11.1 Seizure prediction from EEG time series. Continuous multichannel EEG recordings are analyzed by means of a moving-window analysis. The data covered by the gray window is transformed into a single value in the time profile of a multivariate characterizing measure. When this time profile crosses a certain predefined threshold, an alarm is issued. (From Mormann, F., *Scholarpedia*, 3(10), 5770, 2008.)

distributions of the characterizing measures from the interictal with those from the assumed preictal period in one way or another. The temporal structure of the time profiles is usually not preserved in this type of analysis. Such a design can be useful for investigating and comparing the potential predictive performance of different characterizing measures under different conditions.

On the other hand, an algorithmic analysis uses a design that produces a time-resolved output (i.e., an output for every point of a time profile). With respect to practical application, the algorithm should ideally be prospective (i.e., its output at a given time should be a function of the information available at this time). Prediction algorithms usually employ certain thresholds. If the time profile of a characterizing measure crosses the threshold, an alarm is issued (Figure 11.1). This alarm can be either true or false, depending on whether it is actually followed by a seizure or not. For this distinction, it is necessary to define a prediction horizon or warning time, i.e., the period after an alarm within which a seizure is expected to occur. If an alarm is followed by a seizure within the prediction horizon, it is classified as a true alarm (true positive), otherwise it is regarded as a false alarm (false positive) (Figure 11.2). Prediction horizons reported in the literature range from several minutes to a few hours.

In addition, it may be useful to require a minimum time interval between an alarm and a seizure occurrence in order to count this alarm as a successful prediction if the algorithm is to be used for seizure prevention. This minimum intervention time can be introduced as an additional constraint. [It may be noted that in the literature, different definitions are sometimes used for these quantities, e.g., one group has used the term seizure occurrence period instead of prediction horizon and seizure prediction horizon instead of minimum intervention time (Aschenbrenner-Scheibe et al. 2003; Winterhalder et al. 2003; Maiwald et al. 2004)]. In studies that employ a statistical instead of an algorithmic design, the prediction horizon corresponds to the assumed preictal period.

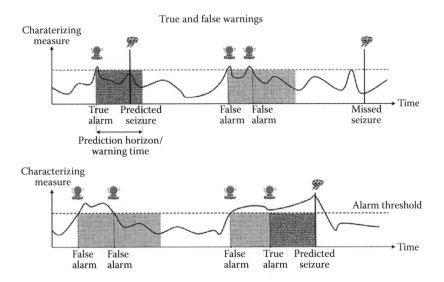

FIGURE 11.2 Seizure prediction from EEG time series. When the time profile of some characterizing measure rises above a predefined threshold, the prediction algorithm issues an alarm that is classified as true or false depending on whether it is followed by a seizure within a specified warning time—termed the prediction horizon. The area shaded in light gray represents time during which the patient is under false warning, i.e., awaiting a nonoccurring seizure. The area shaded in dark gray represents time under a correct warning. In principle, alarms are issued continuously for as long as the measure profile remains above the threshold. For better legibility, not all of these continuous alarms are shown. For the last seizure, alarms begin too early with respect to the prediction horizon so they are classified as false warnings until the seizure falls within the prediction horizon. The time period shown on the *x*-axis could typically represent 1 day. (From Mormann, F., *Scholarpedia*, 3(10), 5770, 2008.)

11.2.3 SENSITIVITY AND SPECIFICITY

If a seizure is not preceded by an alarm within the prediction horizon, this will be counted as a false negative. A less trivial question is how to quantify true negatives. In principle, every single window of the moving-window analysis that is outside the duration of the assumed preictal period (i.e., one prediction horizon before a seizure) and does not produce an alarm could be counted as a true negative. However, since sensitivity is usually quantified as the number of seizures with at least one alarm within the preceding prediction horizon divided by the total number of seizures, it is reasonable to define specificity based on the prediction horizon. If the prediction horizon is 3 h, the sensitivity quantifies the fraction of correctly classified preictal 3-h segments, while the specificity measures the fraction of correctly classified (consecutive) interictal 3-h segments.

In order to avoid any ambiguity in statistically quantifying the specificity of a prediction algorithm, most groups have instead reported specificity rates measured as false predictions per hour. Unfortunately, even for false prediction rates, different definitions are found in the literature. Several groups have determined false prediction rates by counting all false positives and dividing this number by the total duration of the analyzed recording (Iasemidis et al. 2003; Niederhauser et al. 2003; Chaovalitwongse et al. 2005; Esteller et al. 2005; Iasemidis et al. 2005). This definition ignores the fact that for each seizure contained in the recording, there is a preictal period (i.e., the prediction horizon) during which every alarm is counted as a true prediction, and false predictions cannot occur by definition. Therefore, other groups have used corrected false prediction rates that were calculated only for the interictal period (Aschenbrenner-Scheibe et al. 2003; Mormann et al. 2003a, b; Winterhalder et al. 2003; Maiwald et al. 2004).

In this context it is important to realize that a reported false prediction rate cannot be judged independent from the prediction horizon, since in a prospective prediction algorithm a false alarm will leave the patient mistakenly awaiting a seizure for the duration of the prediction horizon. It is only after this duration that the patient will know if the alarm was a false warning.

As an example from the literature, consider an algorithm with a 2-h prediction horizon that yields a sensitivity of $9/11 = 82\%$ and an uncorrected false prediction rate of $6/41$ h $= 0.15/h$ (Iasemidis et al. 2005). If we take into account that the uncorrected false prediction rate includes the preictal periods during which no false prediction can occur by definition, the corrected false prediction rate (assuming that the preictal periods of the different seizures are nonoverlapping) is $6/19$ h $= 0.32/h$, thus, more than twice as high. Furthermore, if we consider that after each false prediction, the patient needs to wait 2 h before knowing if it was a false prediction, the algorithm of our example may (assuming that false predictions are not spaced closer than the prediction horizon) leave a patient spending $6 \cdot (2/19) = 63\%$ of the interictal period waiting for a seizure that will not occur while still failing to anticipate every fifth seizure. An algorithm yielding the same results for a prediction horizon of 10 min would instead leave the patient in futile expectation of a seizure only in 3% of his or her seizure-free time. This example shows that a prediction rate should be judged in view of the prediction horizon used by the algorithm and that it is the product of these two quantities that should be compared across studies.

A better way to assess the specificity of a prediction algorithm would therefore be to report the portion of time from the interictal period (i.e., the interseizure interval without the preictal period) during which the patient is not in a state of falsely awaiting a seizure.

In general, any algorithm can be tuned (e.g., by varying the alarm threshold) to yield a higher sensitivity at the cost of a lower specificity and vice versa. For a closed-loop intervention system, the desired relation between these two quantities will depend on the invasiveness of the intervention technique under consideration. If the intervention does not impair the patient, a higher false prediction rate will be tolerated up to a point where even a constant intervention such as a chronic or scheduled stimulation from implantable brain-stimulation devices (Theodore and Fisher 2004; Stacey and Litt 2008) is possible and could be performed without a prediction algorithm.

11.2.4 THE PROBLEM OF IN-SAMPLE OPTIMIZATION

Another important issue in the evaluation of a prediction algorithm is the use of a posteriori information. For a prospective prediction algorithm, this type of information is not available. Two typical cases of using a posteriori information are found in the literature: (i) in-sample optimization of parameters of the algorithm, and (ii) retrospective selection of one or more channels with particularly good performance.

In-sample optimization or training of parameters is present whenever parameters used for the calculation of the characterizing measure of the EEG or of the prediction algorithm itself are adjusted to produce optimal performance of the algorithm for a given set of data. Such an optimization is likely to result in an overestimated performance, which will not be reproducible when applying the algorithm to other, out-of-sample testing data that were not used in the optimization process. In order to assess the true performance of a prediction or detection algorithm, it is mandatory to test it on out-of-sample data.

Another way of using a posteriori information relates to the selection of channels that are able to discriminate an interictal from a preictal state. The great majority of studies have shown that out of the available number of recording channels, only a limited number carry information that can actually be used for the detection of a preseizure state, while the remaining channels are likely to increase the number of false detections without contributing to the detection sensitivity of an algorithm. The task at hand is to decide in advance which channels are best suited for the purpose. Several studies have attempted to tackle the problem of channel selection by using the first few seizures to select the appropriate channels and/or parameters for the algorithm before trying to detect precursors of the seizures that follow (D'Alessandro et al. 2003, 2005; Esteller et al. 2005; Le Van Quyen et al. 2005). Such a procedure implies that the spatiotemporal dynamics preceding a seizure do not change from seizure to seizure. Iasemidis et al. (2003, 2005) designed an algorithm using a selection of channels that is readjusted after every seizure so that it would have been optimal for the seizure that has just occurred. Such a procedure is based on the implicit assumption that preictal dynamics change to a certain degree from seizure to seizure, but the preictal dynamics of a seizure still depend on the dynamics of the previous one. If these algorithms reliably prove to be better than a random prediction, they could, in addition to being beneficial for patients, provide valuable clues for new theories on the mechanisms involved in ictogenesis.

11.2.5 The Need for Statistical Validation

If an algorithm is designed to run prospectively, its quasiprospective out-of-sample performance can be tested retrospectively on continuous long-term recordings that were not previously used for parameter optimization or channel selection. Once this quasiprospective performance (in terms of correct alarms and false alarms with respect to the given prediction horizon) has been assessed, it remains to be tested whether it is indeed superior to naive prediction schemes such as periodic or random predictors. For this, researchers designed a framework to assess the performance of such a random predictor (Winterhalder et al. 2003; Schelter et al. 2006; Wong et al. 2007; Snyder et al. 2008).

In retrospective statistical studies on predictability, however, it may be desirable to investigate and compare the potential predictive performance of different characterizing measures for various thresholds and parameters (Mormann et al. 2005). In this case, the use of a random predictor for statistical validation would require corrections for multiple testing that can be difficult to perform since the data used for the different tests are usually not independent. Here, the concept of seizure time surrogates as introduced by Andrzejak et al. (2003) can provide a means for statistical validation. In this process, artificial seizure onset times are generated by randomly shuffling the original interseizure intervals. Using these surrogate seizure onset times instead of the original onset times, the EEG data are then subjected to the same algorithms or prediction statistics that were used for the original onset times. Only if the performance of the algorithm for the original seizure times is significantly better than the performance for a number of independent realizations of the surrogate seizure times, can the null hypothesis, namely, that a given algorithm cannot detect a preseizure state with a performance above chance level, be rejected. The advantage of this type of statistical validation is that it can be applied to any type of analysis, algorithmic or statistical. A modification of this surrogate test has recently been proposed on the basis of a constrained randomization of the time profile of the characterizing measure (Kreuz et al. 2004). Another surrogate test, based on alarm time surrogates, has recently been introduced along with a comprehensive evaluation and comparison of existing bootstrapping and analytical approaches for statistical validation (Andrzejak et al. 2009).

The simplest way to assess the performance of a prospective prediction algorithm is to compare the time under warning with the number of seizures that were successfully predicted. For instance, if a warning light is on 60% of the time, we would expect 60% of the seizures to be predicted by chance. Only if the percentage of predicted seizures (i.e., the sensitivity) significantly exceeds the percentage of time under warning can a given algorithm be assumed to bear predictive power above chance level (Snyder et al. 2008). The statistical significance of this performance, i.e., the probability of predicting S out of N seizures by chance for a given portion of time under warning of q is simply given by the binomial probability

$$P(N,S,q) = \sum_{K \geq S} \binom{N}{K} \cdot q^K \cdot (1-q)^{N-K}.$$

11.3 SUMMARY

The more rigorous methodological design in many recent seizure prediction studies has shown that many of the measures previously considered suitable for prediction perform no better than a random predictor. On the other hand, evidence has accumulated that certain measures, particularly measures quantifying relations between recording sites to characterize interaction between different brain regions, show a promising performance that exceeds the chance level as evidenced by statistical validation. The few studies that have used prediction algorithms in a quasiprospective manner (i.e., without the use of a posteriori information) either did not include a statistical validation or did not apply it correctly.

The design and evaluation of prospective seizure prediction algorithms involve numerous caveats that need to be considered. The current literature allows no definite conclusion as to whether seizures are predictable by prospective algorithms. To answer this question, future studies need to rely on sound and strict methodology and include a rigorous statistical validation.

In order to assure the methodological quality of future studies on seizure prediction, we propose the following guidelines (Mormann et al. 2007):

- Prediction algorithms should be tested on continuous long-term recordings covering several days of EEG in order to comprise the full spectrum of physiological and pathophysiological states for an individual patient. Studies should assess both sensitivity and specificity and should report these quantities with respect to the applied prediction horizon. Rather than false prediction rates, the portion of time under false warning should be reported. If false prediction rates are reported, they should be reported only for the seizure-free interval.
- Results should be tested using statistical validation methods based on Monte Carlo simulations or naive prediction schemes to prove that a given prediction algorithm performs indeed above chance level. This is particularly important for studies that contain in-sample optimization such as retrospective adjustment of parameters or selection of EEG channels.
- If prediction algorithms are optimized using training data (in-sample), they should be tested on independent testing data (out-of-sample). If part of the data from an individual patient are used for patient-specific parameter adjustment or EEG channel selection, these data must be excluded when evaluating the performance out-of-sample. Performance of an algorithm should always be reported separately for the testing data.

The next logical step in the field of seizure prediction will be to test on long-term recordings whether any of the prediction algorithms devised to date are able to perform better than a random prediction in a quasiprospective setting on out-of-sample data. This step is an indispensable prerequisite for justifying prospective clinical trials involving invasive seizure-intervention techniques such as electrical brain stimulation in patients based on seizure prediction.

REFERENCES

Andrzejak, R. G., F. Mormann, T. Kreuz, C. Rieke, A. Kraskov, C. E. Elger, and K. Lehnertz. 2003. Testing the null hypothesis of the nonexistence of a preseizure state. *Phys. Rev. E Stat. Nonlin. Soft Matter Phys.* 67:010901.

Andrzejak, R. G., D. Chicharro, C. E. Elger, and F. Mormann. 2009. Seizure prediction: Any better than chance? *Clin. Neurophysiol.* 120:1465–1478.

Aschenbrenner-Scheibe, R., T. Maiwald, M. Winterhalder, H. U. Voss, J. Timmer, and A. Schulze-Bonhage. 2003. How well can epileptic seizures be predicted? An evaluation of a nonlinear method. *Brain* 126:2616–2626.

Bauer, J., and W. Burr. 2001. Course of chronic focal epilepsy resistant to anticonvulsant treatment. *Seizure* 10:239–246.

Chaovalitwongse, W., L. D. Iasemidis, P. M. Pardalos, P. R. Carney, D. Shiau, and J. C. Sackellares. 2005. Performance of a seizure warning algorithm based on the dynamics of intracranial EEG. *Epilepsy Res.* 64:93–113.

D'Alessandro, M., R. Esteller, G. Vachtsevanos, A. Hinson, J. Echauz, and B. Litt. 2003. Epileptic seizure prediction using hybrid feature selection over multiple intracranial EEG electrode contacts: A report of four patients. *IEEE Trans. Biomed. Eng.* 50:603–615.

D'Alessandro, M., G. Vachtsevanos, R. Esteller, J. Echauz, S. Cranstoun, G. Worrell, L. Parish, and B. Litt. 2005. A multi-feature and multi-channel univariate selection process for seizure prediction. *Clin. Neurophysiol.* 16:506–516.

Esteller, R., J. Echauz, M. D'Alessandro, G. Worrell, S. Cranstoun, G. Vachtsevanos, and B. Litt. 2005. Continuous energy variation during the seizure cycle: Towards an on-line accumulated energy. *Clin. Neurophysiol.* 116:517–526.

Haut, S. R., C. Swick, K. Freeman, and S. Spencer. 2002. Seizure clustering during epilepsy monitoring. *Epilepsia* 43:711–715.

Iasemidis, L. D., D. Shiau, W. Chaovalitwongse, J. C. Sackellares, P. M. Pardalos, J. C. Principe, P. R. Carney, A. Prasad, B. Veeramani, and K. Tsakalis. 2003. Adaptive epileptic seizure prediction system. *IEEE Trans. Biomed. Eng.* 50:616–627.

Iasemidis, L. D., D. Shiau, P. M. Pardalos, W. Chaovalitwongse, K. Narayanan, A. Prasad, K. Tsakalis, P. R. Carney, and J. C. Sackellares. 2005. Long-term prospective on-line real-time seizure prediction. *Clin. Neurophysiol.* 116:532–544.

Jouny, C. C., P. J. Franaszczuk, and G. K. Bergey. 2005. Signal complexity and synchrony of epileptic seizures: Is there an identifiable preictal period? *Clin. Neurophysiol.* 116:552–558.

Kantz, H., and T. Schreiber. 2004. *Nonlinear time series analysis*, 2nd ed. Cambridge: Cambridge University Press.

Kreuz, T., R. G. Andrzejak, F. Mormann, A. Kraskov, H. Stögbauer, C. E. Elger, K. Lehnertz, and P. Grassberger. 2004. Measure profile surrogates: A method to validate the performance of epileptic seizure prediction algorithms. *Phys. Rev. E Stat. Nonlin. Soft Matter Phys.* 69:061915.

Le Van Quyen, M., J. Soss, V. Navarro, R. Robertson, M. Chavez, M. Baulac, and J. Martinerie. 2005. Preictal state identification by synchronization changes in long-term intracranial EEG recordings. *Clin. Neurophysiol.* 116:559–568.

Lehnertz, K., and B. Litt. 2005. The first international collaborative workshop on seizure prediction: Summary and data descriptions. *Clin. Neurophysiol.* 116:493–505.

Maiwald, T., M. Winterhalder, R. Aschenbrenner-Scheibe, H. U. Voss, A. Schulze-Bonhage, and J. Timmer. 2004. Nonlinear phenomena: Comparison of three nonlinear seizure prediction methods by means of the seizure prediction characteristic. *Physica D* 194:357–368.

Mormann F., R. G. Andrzejak, T. Kreuz, C. Rieke, P. David, C. E. Elger and K. Lehnertz. 2003a. Automated detection of a pre-seizure state based on a decrease in synchronization in intracranial EEG recordings from epilepsy patients. *Phys. Rev. E Stat. Nonlin. Soft Matters Phys.* 67:021912.

Mormann, F., T. Kreuz, R. G. Andrzejak, P. David, K. Lehnertz, and C. E. Elger. 2003b. Epileptic seizures are preceded by a decrease in synchronization. *Epilepsy Res.* 53:173–185.

Mormann, F., T. Kreuz, C. Rieke, R. G. Andrzejak, A. Kraskov, P. David, C. E. Elger, and K. Lehnertz. 2005. On the predictability of epileptic seizures. *Clin. Neurophysiol.* 116:569–587.

Mormann, F., R. Andrzejak, C. E. Elger, and K. Lehnertz. 2007. Seizure prediction: The long and winding road. *Brain* 130:314–333.

Mormann, F. 2008. Seizure prediction. *Scholarpedia* 3(10):5770.

Morrell, M. 2006. Brain stimulation for epilepsy: Can scheduled or responsive neurostimulation stop seizures? *Curr. Opin. Neurol.* 19:164–168.

Niederhauser, J. J., R. Esteller, J. Echauz, G. Vachtsevanos, and B. Litt. 2003. Detection of seizure precursors from depth-EEG using a sign periodogram transform. *IEEE Trans. Biomed. Eng.* 50:449–458.

Osorio, I., M. Frei, and S. Wilkinson. 1998. Real-time automated detection and quantitative analysis of seizures and short-term prediction of clinical onset. *Epilepsia* 39:615–627.

Osorio, I., M. G. Frei, S. Sunderam, J. Giftakis, N. C. Bhavaraju, S. F. Schaffner, and S. B. Wilkinson. 2005. Automated seizure abatement in humans using electrical stimulation. *Ann. Neurol.* 57:258–268.

Pikovsky, A. S., M. Rosenblum, and J. Kurths. 2001. *Synchronization: A universal concept in nonlinear sciences.* Cambridge: Cambridge University Press.

Schelter, B., M. Winterhalder, T. Maiwald, A. Brandt, A. Schad, A. Schulze-Bonhage, and J. Timmer. 2006. Testing statistical significance of multivariate time series analysis techniques for epileptic seizure prediction. *Chaos* 16:013108.

Snyder, D. E., J. Echauz, D. B. Grimes, and B. Litt. 2008. The statistics of a practical seizure warning system. *J. Neural. Eng.* 5:392–401.

Stacey, W. C., and B. Litt. 2008. Technology insight: Neuroengineering and epilepsy-designing devices for seizure control. *Nat. Clin. Pract. Neurol.* 4:190–201.

Sun, F. T., M. J. Morrell, and R. E. Wharen. 2008. Responsive cortical stimulation for the treatment of epilepsy. *Neurotherapeutics* 5:68–74.

Theodore, W. H., and R. S., Fisher. 2004. Brain stimulation for epilepsy. *Lancet Neurol.* 3:111–118.

Winterhalder, M., T. Maiwald, H. U. Voss, R. Aschenbrenner-Scheibe, J. Timmer, and A. Schulze-Bonhage. 2003. The seizure prediction characteristic: A general framework to assess and compare seizure prediction methods. *Epilepsy Behav.* 4:318–325.

Wong, S., A. B. Gardner, A. M. Krieger, and B. Litt. 2007. A stochastic framework for evaluating seizure prediction algorithms using hidden Markov models. *J. Neurophysiol.* 97:2525–2532.

12 Autonomous State Transitions in the Epileptic Brain
Anticipation and Control

Stiliyan N. Kalitzin, Demetrios N. Velis,
and Fernando Lopes da Silva

CONTENTS

12.1 INTRODUCTION

Epilepsy is a neurological condition characterized by intermittent periods of abnormal states or seizures that in most cases strike abruptly and without clear warning. In some cases, it is possible to localize the source of the abnormal behavior in the brain (localization-related epilepsy); in others, this does not appear to be feasible. The possibility to predict these dynamic transitions both in time and location in the brain may lead to a new realm of epilepsy treatment, hopefully including the treatment of those patients where medication has failed. For a review on the possible impact of seizure prediction we refer to Litt and Lehnertz (2002). Here, we note that one of the most realistic future therapies, namely the control of epileptic seizures using electrical stimulation, is still in the early stages of development. Existing clinical trials utilize intracranial electrical stimulation consisting of periodic high-frequency (100- to 200-Hz) bursts that are administered at regular intervals (Velasco et al. 2007). Such schemes are referred to as open-loop paradigms as they do not take into account the state of the neuronal system since they simply interfere with the ongoing activity of the tissue that is being stimulated. More advanced, state-dependent schemes called closed-loop paradigms are in development (Osorio et al. 2005). Their success relies on the possibility to either foresee an impending seizure or to detect its occurrence very early and only then apply electrical stimulation. A special class of these closed-loop approaches uses properties of the measured EEG signal to determine the exact timing of the stimulation (Osorio and Frei 2009) or even to generate an appropriate seizure suppression stimulation waveform (Ullah and Schiff 2009). We refer to the latter as state-reactive control paradigms. For a recent survey on the current approaches for seizure control by electrical stimulation and the ongoing clinical trials see Jobst (2009).

The success of the closed-loop seizure control approach is dependent on our ability to robustly identify states preceding epileptic seizures, the so-called preictal states. This challenge, however, has been more elusive than originally expected. Some of the early enthusiastic promises appear not to be widely reproducible and in some cases the success of prediction could not be validated as significantly better than a random guess. Accordingly, the seizure prediction community has diverted its attention from sophisticated signal analytic methods—for a review, see Mormann et al. (2007)—to the science of statistical validation of the prediction schemes (Andrzejak et al. 2003; Snyder et al. 2008; Sackellares et al. 2006). The very fundamental assumption about predictability of seizures, taken naively for granted before, has now been questioned. The predictive power of an observable feature statistically unexplainable as a lucky guess is assumed to constitute precious evidence that seizures can, in principle, be predicted.

In this contribution, we take a step outside pure phenomenological descriptions and consider two fundamental questions. First, what sort of dynamics governs the transitions from normal neurological state or states to epileptic ones? Second, to what extent can we predict or anticipate these state transitions? Different dynamic scenarios give different prospects of finding a prediction paradigm. In the following section, we present some basically different possibilities, including those that may arise in stochastic and intermittent systems. We distinguish between parameter-driven transitions where the transitions are due to shifts in one or more parameters and multistable dynamic systems where input fluctuation or noise alone is sufficient to trigger the transition. Statistics of interictal interval durations indicates that models based on fluctuation (noise)-driven transitions cannot be ruled out. In order to illustrate this issue, we briefly present simulation examples using analytical and realistic models with bistability. We present data from both in vivo models and from clinical cases. In the next section, we address our second question, the quest of defining predictability of dynamic transitions in general. Our measure of predictive power is defined as the nonlinear association between the value of a given feature and the interval between the time of the measurement of this value and onset of the next state transition. If such an association is proven as statistically significant, then a proper practical warning paradigm can be established that will certainly outperform a chance-based paradigm. According to such a broader definition, predictability allows us to estimate the acute risk of epileptic transitions. Even if the transitions occur due to random (or unobservable) fluctuations in multistable systems, the existence and the closeness of an alternative state can be detected. A general analytical approach to this topic has been explored by Scheffer et al. (2009). We briefly describe our method of sensing the emergence of new dynamical states by using a stimulation paradigm and deriving from the system's response the relative phase clustering index (rPCI).

Our last topic concerns the primary reason that seizure prediction is a relevant issue: the challenge of how to stop or prevent a transition to the epileptic state by using either continuous or pulsed electrical stimulation. We demonstrate that in the context of multistable systems, the transitions from a normal to epileptic attractor can be reversed by an appropriate, state-dependent counter-stimulus in the case of multistable system dynamics. Computer simulation results, and some preliminary trials with electrically induced afterdischarges are presented as well as supporting results from animal experiments done by others (Osorio and Frei 2009).

12.2 COMPUTER MODELS AND PHENOMENOLOGICAL VALIDATION FOR THE GENERIC SCENARIOS OF TRANSITION FROM "NORMAL" TO EPILEPTIC STATES

12.2.1 General Classification

The knowledge about precipitating factors for epileptic seizures is growing rapidly, still no single theory or model can account for this neurological condition in general. Not surprisingly, the realm of seizure prediction has been predominantly focused on both phenomenology and on elegant mathematical concepts often borrowed from other fields such as weather forecasting and prediction of earthquakes. Here we attempt to take into account different dynamic concepts or scenarios that may result in different prediction concepts and strategies. We distinguish two major classes of systems capable of generating epileptic transitions. The first one is characterized by one or more asymptotically stable manifolds in the phase space of the system, called attractors. Epileptic seizures can be due to either a pathological deformation of the attractor or to transitions between different attractors. The illustration in Figure 12.1, upper plot, shows a parameter-driven deformation of steady-state dynamics, interpreted as normal behavior, resulting in a limit cycle type of dynamics that represents a model epileptic seizure. While this model of transitions might be plausible, it is not autonomous because it leaves open the question of how the control parameter changes. The latter is, by definition, outside the dynamics of the system and, therefore, such a scenario teaches us little about seizure predictability. There is, however, another possibility as illustrated in the middle plot of Figure 12.1, where more than one attractor coexists in the phase space of the system. Transitions in these systems can occur due to external perturbations—reflex epilepsies may be an example of this type of transition—or internal noisy fluctuations in the system's dynamics. No parameter changes need to be involved in the process. We will refer to this transition class as a bifurcation scenario (Lopes da Silva et al. 2003a, 2003b). It may be argued that those transitions are not exactly autonomous because of the involvement of external factors such as noise or input. Strictly speaking this is true, but remembering that (1) the brain is always an open system and (2) fluctuating parameters affect the system on all levels of organization, we have little difficulty accepting this scenario as realistic.

The more general, combined scenario is illustrated in the bottom plot of Figure 12.1. Parameter changes can create and annihilate attractors in the phase space of the system. A transition can occur either due to the disappearance of the attractor currently occupied by the system, or due to stochastic leaps from one attractor to another. While in the first case, some anticipation of an impending transition can still be possible (Scheffer et al. 2009), in the latter case, transitions will be totally unpredictable.

It has seldom been assumed explicitly, but nonlinear methods (Elger and Lehnertz 1998) such as stability analysis based on Lyapunov exponents (Lehnertz 1999; Iasemidis et al. 1990), correlation dimension estimates (Lehnertz and Elger 1997), etc. should be most appropriate in a second generic class of dynamical systems exhibiting autonomous transitions to epileptic-like behavior, namely the systems with intermittent dynamics. Those systems, although governed by deterministic laws, contain instability regions in their phase space, causing them to exhibit apparent changes of dynamic patterns when they enter those regions. We can illustrate this with a simple five-unit cellular automaton schematically presented in Figure 12.2 (top). It is driven by first-order discrete dynamics (Ohayon et al. 2004b) that, for certain parameter ranges, behaves intermittently. As we can see from Figure 12.2 (bottom), the transition from laminar to chaotic phase is always preceded by an increase of the number of positive Lyapunov exponents. From Figure 12.2 (bottom), we see that the maximal exponent significantly increases before the transition as well. The attractive feature in this scenario is that transitions will occur without any external influence or noisy parameters, which makes the predictions reliable. When more than one Lyapunov exponent becomes positive, a point of no return has been crossed. That said, there are also some limitations associated with these models. Dynamic intermittency is a relatively rare phenomenon. Models showing intermittent dynamics are fine-tuned mathematical constructs. In a realistic system of neurons with a general connection topology, the probability of intermittent patterns is infinitesimal, as some studies have shown (Ohayon et al. 2004a). We are aware that intermittent behavior can still explain paroxysmal events in some simple biological models and even seizures in human recordings (Velazquez et al. 1999), but as yet no biologically realistic computer models have been known to show this feature. Perhaps the coupled oscillator models (Zalay and Bardakjian 2008) comes the closest to this goal. While we cannot completely reject intermittency as a relevant theory of epileptic transitions, the rest of this chapter will concentrate on systems with multistability, which are more commonly encountered.

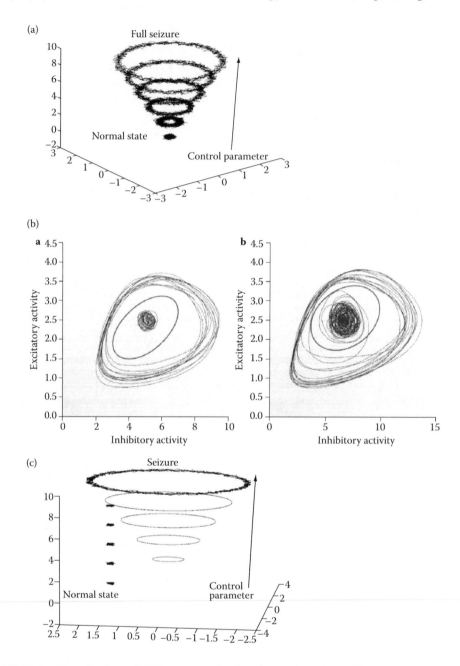

FIGURE 12.1 **(See color insert.)** Different scenarios for seizure generation. (a) Deformation of the system's attractor from a steady-state normal state to a limit cycle (seizure). The vertical axis is the control parameter and the two horizontal axes represent the two-dimensional phase space of the system. (b) Situation where two attractors are present for the same value of the control parameters (bifurcation scenario). Subframe a illustrates a healthy system where the two attractors are well separated by the red closed curve (the separatrix). On subframe b, transitions take place resulting in occasional seizure-type behavior. (c) Combined scenario for seizure generation where, for different critical values of the control parameter, attractors are created or destroyed. (From Kalitzan et al., *Epilepsy and Behavior*, 17(3), 310–323, 2010.)

(a)

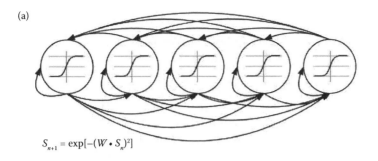

$$S_{n+1} = \exp[-(W \cdot S_n)^2]$$

(b)

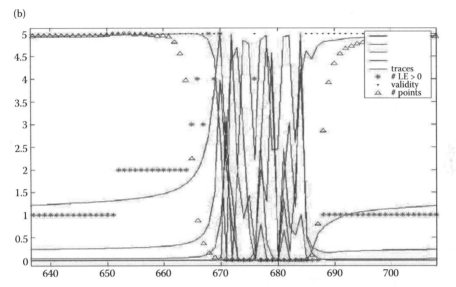

FIGURE 12.2 **(See color insert.)** (a) Schematic presentation of a discrete-step cellular automaton model. The state-update function is given by the formula in the inset. The five-dimensional state of the system is denoted as $S_n \in [01]$, $n = 1...5$, and the weight coefficients are contained in the 5×5 matrix W. (b) Transition from laminar to turbulent pattern of the automaton. The five unit traces are given with green lines. Stationary (nonwindow-based) Lyapunov exponent analysis was performed for the local neighborhood of each stimulated point in the five-dimensional phase space. The red stars represent the number of positive exponents (unstable directions) in the local neighborhood, the blue triangles give the number of points detected in the corresponding neighborhood, and the black dots give the points where the Lyapunov exponent spectrum was possible to compute from the simulated data. We see that in the turbulent phase, the analysis is unreliable but at the beginning there is a clear increase of instability. (From Kalitzan et al., *Epilepsy and Behavior*, 17(3), 310–323, 2010.)

12.2.2 METAPHORIC AND REALISTIC MODELS

Modeling a system with a deformable attractor or with multiple dynamic attractors is relatively simple. A theoretical example may be used to exemplify such a dynamical behavior. Consider the second-order (first-order complex) nonlinear system described by the following ODE:

$$\frac{d}{dt}Z = -F(Z,\bar{Z}); F = 3aZ^4Z + 2bZ^2Z + cZ + i\omega Z + \mu Z^2 + \eta(t). \tag{12.1}$$

Here a,b,c,μ,ω are real parameters and $\eta(t)$ is an additive noise (random Gaussian process); $\bar{Z}$ is the complex conjugated of Z, and $|Z| \equiv \sqrt{Z\bar{Z}}$.

Straightforward analysis of the dynamic properties of Equation 2.1 reveals that the system can have different attractor properties. For some parameter settings, we can have a fixed-point steady-state attractor, for others a limit cycle, and for a third parameter choice, both limit cycle and fixed-point patterns are possible. Additionally, the integrity of the limit cycle can be broken for some values of the parameter μ and reduced to a fixed point. Three different transitions are illustrated in Figure 12.3. The first two are attractor deformations due to parameter changes. The upper plot shows a signal that resemblances very much EEG signals recorded in the human brain at seizure onset in temporal lobe epilepsy (TLE); the signal in the middle plot resembles field potentials that are recorded in brain slices in vitro that exhibit seizure-like activity when exposed to a medium with low Mg^{2+}. The lowest plot in Figure 12.3 represents the sudden change in behavior due to a single perturbation of the system in its bistable regime, resembling EEG signals recorded during idiopathic absence seizures.

The above analytical model is perhaps the simplest model that incorporates transitions between limit cycle and steady-state behavior. It is two-dimensional, which implies that the limit cycle lies in the same plane as the steady-state point. This may have consequences for designing a seizure control strategy, as we shall present in Section 12.4. We now present a slightly extended three-dimensional model, including DC shifts.

$$\frac{d}{dt}Z = -Z\left(|Z|^2 - u\right) + i\omega Z$$

$$\frac{d}{dt}u = -\lambda u(u^2 - 1)$$

(12.2)

Here, Z is a complex variable and u is a real one. The parameter ω defines the angular speed in the Z plane where the λ parameter introduces an additional time constant governing the speed of the interstate transitions. The phase plot of the model is illustrated in Figure 12.4a. It can be seen that this system has two stable attractors, one at $u = -1; Z = 0$ (steady-state; red dot) and one at $u = 1; |Z| = 1$ (limit cycle; red circle). The stationary state $u = 0; Z = 0$ is unstable. The surface $u = 0$ is the separatrix between the two attractors. Examples of state transitions are given in the top plot of Figure 12.4b. In general, we can assume that our measurements are taken neither along the u-axis nor in the Z plane. A mixture of the Z and u signals is depicted in the bottom plot. We compare this bottom plot with EEG recordings from a patient with implanted electrodes in the right hippocampus as presented in the bottom plot of Figure 12.4b. Here we can see the large DC shift preceding the fast oscillatory behavior in the EEG corresponding to a seizure, that afterwards returns to baseline level.

The simple models of Equations 12.1 and 12.2 are analytical exercises that, though they may reveal some generic properties of multistable systems and the transitions that can occur there, have no direct biologically relevant interpretation. Next, we consider more realistic neuronal models used to explore seizure generation in corticothalamic circuits (Suffczynski et al. 2001) or in the hippocampus (Wendling et al. 2001). We do not intend to describe any of these models in detail here. These models were constructed at the lumped scale of detail, i.e., instead of modeling every individual neuron, a whole ensemble of neurons having similar properties are represented by a single lumped unit. The models can describe the bistable type of dynamics (Suffczynski et al. 2004) that might be responsible for the autonomous, fluctuation-driven seizure generation in biological neuronal circuits such as the corticothalamic and those present in the hippocampus structures.

Our model assumptions for epileptic transitions have been based on the multistability scenario where stochastic fluctuations cause the transitions. A simple test of the validity of our model's assumptions can be performed by comparing the distributions of interseizure intervals as well as seizure durations predicted according to the model to those measured in various experimental and clinical conditions (Suffczynski et al. 2006). To quantify the comparison between the distributions, we have fitted a gamma distribution to the collected data:

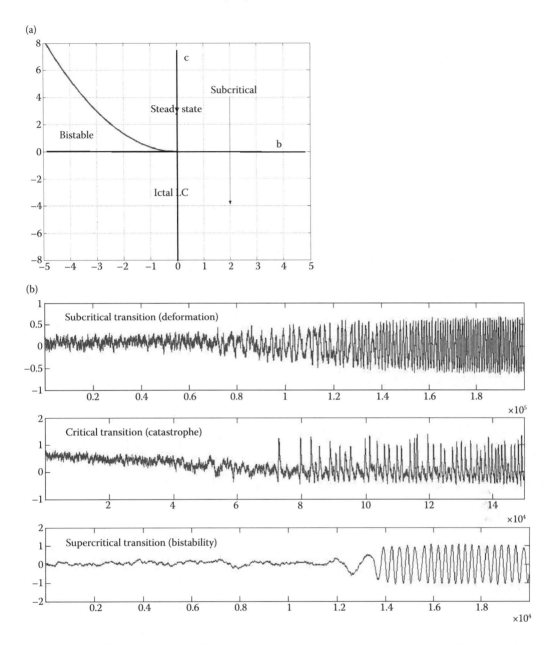

FIGURE 12.3   (See color insert.) (a) Parameter plot of the complex z^6 model described in the text. The vertical axis is the c parameter (Equation 12.1) controlling the transition from a normal to epileptic state. The horizontal axis is the b parameter that determines the critical nature of the transition. The model has different dynamic properties (attractor topologies) depending on the parameter configuration. In the $c < 0$ half-plane, the model has one limit cycle attractor (rotator); for $c > 0$ and $b > 0$, the model has a single steady-state point (nonlinear oscillator). In the area $c > 0$ and $b < 0$, under the blue curve, the model has two attractors (bistability), one steady-state point, and one limit cycle. Different paths of parameter changes will result in various types of dynamic transitions. (b) Traces illustrating the different transitions in the z^6 model. Top trace: Changing the c parameter deforms the steady-state attractor into a limit cycle. Middle trace: By changing the μ parameter fluctuations-induced paroxysmal spikes that join into a limit cycle. Bottom trace: In the bistable case phase fluctuations alone can trigger sudden transitions from steady-state to limit cycle behavior. (From Kalitzan et al., *Epilepsy and Behavior*, 17(3), 310–323, 2010.)

(a)

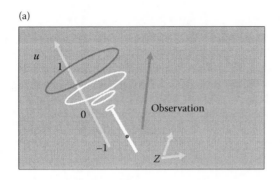

(b)

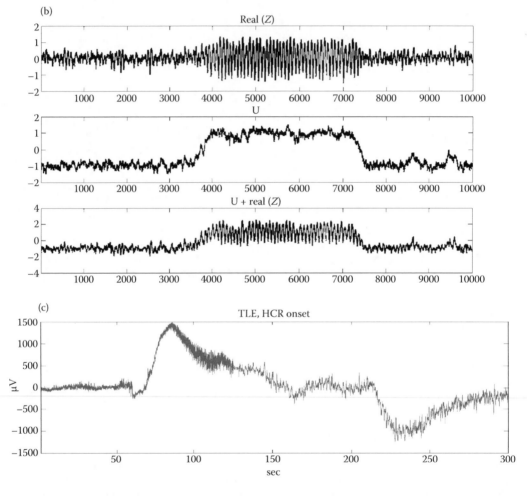

FIGURE 12.4 (See color insert.) (a) Illustration of the phase space topology of the model described by Equation 12.2. The red dot ($u = 0$, $Z = 0$) represents a stable steady state. The red circle ($u = 1$, $|Z| = 1$) is a stable limit cycle. The pink arrow represents an observation projection in the phase space. (b) Traces generated by the model when performing fluctuation-induced transitions between point attractor and limit cycle. The upper plot is the real part of the Z-complex variable, the middle plot is the u component representing the direct current shift, and the bottom plot is a linear combination of both, illustrating a realistic measurement situation. (c) Seizure recorded intracranially from the right hippocampus of a human subject with temporal lobe epilepsy; we point to its similarity with (b), bottom trace. (From Kalitzan et al., *Epilepsy and Behavior*, 17(3), 310–323, 2010.)

$$y = Cx^{\alpha-1}e^{-x/\beta}. \tag{12.3}$$

In the above formula, y is the fraction of observations that does not contain seizure transitions within time interval x. The two distribution parameters, α and β, are the shape and the inverse rate parameters, correspondingly. One example of a statistical match between our computer model simulations and data from animal models is shown in Figure 12.5a. Equation 2.3 generalizes the Poisson exponential distribution (the nuclear decay law), which corresponds to $\alpha = 1$. In the case of $\alpha = 1$, the random transitions between the states are instantaneous and the system has no memory. When $\alpha < 1$, the probability of transition is greater immediately after the previous transition, which can be interpreted as indicative of closeness to the separator (threshold) between the two attractors in a multistable system. The opposite $\alpha > 1$ cases may indicate processes that, in time, repulse the system from its current state. We can safely assume that for distributions obeying Equation 12.3 with $\alpha \leq 1$, the hypothesis of transitions driven by pure stochastic random walk cannot be rejected. Our results summarized in Figure 12.5b show that this might be the predominant case, at least for the interictal intervals. For the seizure durations, however, the distributions with $\alpha > 1$ are more frequent. This suggests the presence of some state-dependent mechanism attempting to terminate the seizure.

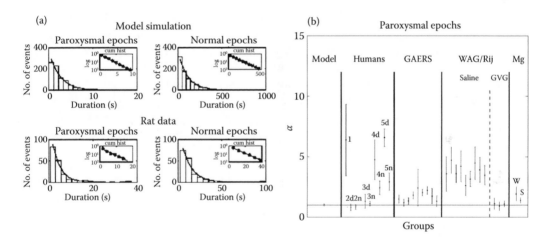

FIGURE 12.5 (a) Comparison between model simulation and measurement from a particular rat model of epilepsy. Distributions of lengths of normal and paroxysmal epochs of the simulated signal (upper panel) and of real EEG signal recorded from the rat (lower panel). On each plot, an exponential function fitted to the histogram is shown. Model histograms were obtained by simulating 24 h of activity using the reference parameters set. Histograms of rat data were obtained from 30-min recording of EEGs of WAG/Rij rat after administration of a high dose of vigabatrin. Insets in each window depict cumulative histograms shown on logarithmic scale and straight lines fitted to histogram points. In all four plots, the lines fit well with the histograms' points, further confirming exponential distributions. (b) Graphical presentation of the values of shape parameter α in Equation 12.3, with 95% confidence intervals of gamma distribution of normal epoch for the model and different experimental conditions. The horizontal line denotes the value of $\alpha = 1$. In cases where 95% confidence intervals include $\alpha = 1$, the initiation of ictal epochs is consistent with a Poisson process. Finding $\alpha < 1$ at 95% confidence intervals suggests that seizure initiation occurs according to a random-walk process. Numbers on the graph denote numbers of the patients. d, daytime recording; n, nighttimes recording; GAERS, genetic absence epilepsy rats from Strasbourg; WAG/Rij, Wistar albino Glaxo from Rijswijk; GVG, vigabatrin; Mg, low-magnesium hippocampal model; W, whole hippocampal recording; S, hippocampal slice recording. (From (a) Suffczynski et al., *Neuroscience*, 126(2), 467–484, 2004; (b) Suffczynski et al., *Journal of Clinical Neurophysiology*, 22(5), 288–299, 2005.)

Recent analysis performed independently by other authors (Osorio et al. 2009) has explored the distribution of interictal intervals in patients with seizures originating from the temporal and frontal lobe. Their results suggest that in 51 of 60 patients, the probability density function (the x-derivative of Equation 12.3) can be approximated by a power law. This corroborates our findings as the exponential factor in Equation 12.3 modulates the power law factor predominantly for large interictal intervals that fall in the tails of the statistical distribution. It has to be noted, however, that in that study and in our own report, in a fraction of patients the power was violated at certain time scales, indicating in our view the existence of transient processes with predetermined or at least statistically preferred time durations.

12.3 PREDICTION AND PREDICTABILITY

12.3.1 DEFINITION

If the stochastic scenario means fluctuation-induced transitions between coexisting attractors, does it also mean that any attempt to predict these state transitions is doomed? Stochastic dynamics has been studied in the context of epileptic networks (Prusseit and Lehnertz 2007), but not in the context of phase transitions. As mentioned earlier, recent analysis has shown that precursory features indicating transitions in multistable systems might be found within specific models (Scheffer et al. 2009) but those studies have focused on parametric-driven transitions and not on fluctuation-driven ones.

We argue that stochasticity and predictability do not necessary exclude each other. It depends rather on how we define predictability. As we mentioned in the Introduction, a great deal of effort has been devoted to demonstrating that some seizure prediction techniques are performing better than a random guess. While in most of those studies, prediction is associated with the exact timing of the transition (the prediction horizon), here we take a broader view, abolish the concept of a prediction horizon, and define predictability as "the existence of a set of measurements that can be performed during an arbitrary finite interictal time interval, such that a statistically significant association between the outcomes of these measurements and the time interval to the first seizure following these measurements can be demonstrated." The only thing that remains to be added to this definition is an appropriate measure to estimate the required association. One can use different approaches; we have selected the nonlinear association index h^2 (Kalitzin et al. 2007). This index relates the total variation of one sequence to the conditional variance of the same sequence with respect to another one. The definition in our case can be summarized as follows:

$$h^2 = 1 - \frac{\sum_f \mathrm{var}^2(TS \mid F = f)}{\mathrm{var}^2(TS)}. \tag{12.4}$$

In Equation 12.4, F is a set of measurements (features) grouped in bins with label f, and TS is a set of times to the next ictal transition for the same time points where the Fs are measured.

Although this is not the only choice, association measures such as mutual information can be used in the same way; for example, we have selected the h^2 index because it detects nonlinear, unidirectional, and, most important, single-mode functional relations between data sets. This last property, which is not a virtue of the mutual information analysis (Sabesan et al. 2009), which will prevent ambiguous predictions to be qualified as successful. Depending on the application, one can develop and utilize an appropriate association quantifier. The significance level of the association in Equation 12.4 can be tested by sufficient number of surrogate data sets obtained by random permutations of the measurement time points while keeping the TS data fixed.

In the context of a multistability, fluctuation-driven scenario of epileptic transitions, one can go one step further in extending the definition of predictability. If we can identify a property of the system that is associated with any measure of a system's susceptibility to ictal transition (and not only to the time of next seizure), we can also call this property a predictor or, more appropriately, a risk estimator. Other researchers have arrived at the same conclusion and introduced the term epileptogenicity index (Bartolomei et al. 2008). The possibility of estimating the vulnerability of the system for seizing (or conversely, its resilience against seizing) extends the scope of the concept of seizure prediction. Indeed, if we deal with open systems that are affected by exogenous and endogenous perturbations, any prediction can only be a statistical one as we can never predict the random processes affecting the system. In Section 12.3.2, we introduce one possible observable that, in some cases, has been found to fulfill the role of risk estimator and statistical predictor.

12.3.2 Relative Phase Clustering Index

Predicting the exact time of an event triggered by a random fluctuation is a futile task. Worse than that, if seizure activity is due to transitions to salient states in a multistable system, those states will generally be unobservable until the actual transition takes place. From data recorded from the system, we can only infer properties from its current state or attractor. How can we estimate then the risk for transitions to nonobservable states? In this respect, we propose a route based on the concept of perturbing the brain dynamics through external stimulation. This may give us a measure of the presence or absence of a possible epileptic seizure state and of the risk of transition to such a state (functional closeness), hopefully without triggering an actual transition by itself.

We reintroduce here the phase clustering index (PCI) (Kalitzin et al., 2002) and the associated rPCI as a measure of a time-variant propensity of the system to phase-lock its response to an external stimulus. In a typical evoked-response type of analysis, the system is exposed to the same stimulus at some discrete time points $t_s \in \left\{ t_1, \ldots t_{N_s} \right\}$ where N_s is the number of stimuli. Typically, we use trains of biphasic pulses as shown in Figure 12.6a. If we denote the system's response (this can be any measured multidimensional physiological signal such as EEG, MEG, etc.) as $S^c(t)$, $c = 1 \ldots N_c$, where N_c denotes the number of channels, then the stimulus triggered response (STR) function is defined as:

$$R_s^c(t) \equiv S^c(t + t_s). \tag{12.5}$$

In this definition, the time argument t can be restricted, but it can also be taken along the whole time axis, thus creating a redundant representation of the original signal. In the particular case of periodic stimulations where all consecutive stimuli are separated by a constant period $T \equiv t_{s+1} - t_s$, $s = 1 \ldots N_s$, it is natural to select the response interval $t \in (0,T)$ in Equation 12.5. Our further definitions are not restricted to this choice but the previous applications of PCI and the examples given below assume this simple nonredundant representation. The general theory is also perfectly applicable to the general nonperiodic stimuli. Below we also omit the channel index c for notation simplicity.

Next we Fourier-transform (12.5) along the t-axis:

$$Z_f^s \equiv \frac{1}{\sqrt{N_t}} \sum_t e^{2\pi i f t} R_s(t)), \tag{12.6}$$

where f denotes the discrete frequency associated with the complex coefficients Z and N_t is the number of time samples for each response epoch. In our applications, we assume $f \in \{1, 2, \ldots N_f\} f_0$ where f_0 is the fundamental frequency, typically the frequency of the stimulation in case of periodic stimuli and N_f is the highest harmonic frequency considered. Please note that we will omit the f_0 notation,

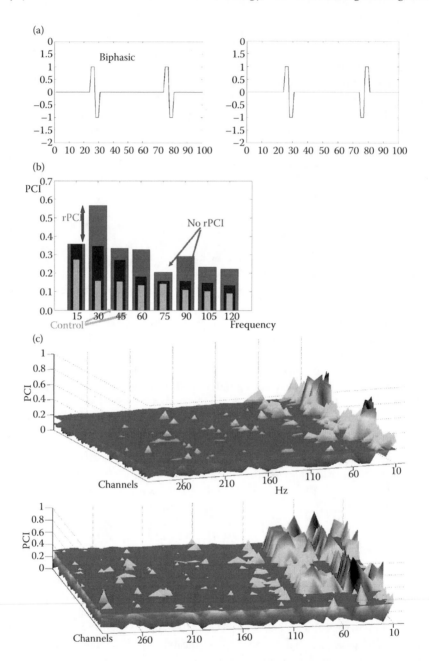

FIGURE 12.6 **(See color insert.)** (a) Different stimulation patterns. The left plot shows trains of biphasic pulses and the right one illustrates the idea of cyclic polarity alternation. (b) Histogram of PCI values measured from the MEG of a photosensitive epileptic patient during stimulation with intermittent light at 15 Hz; the relative PCI (rPCI) is the difference between the PCI value at any frequency minus that at the fundamental frequency; red bars correspond to paroxysmal response, blue bars correspond to stimulation of photosensitive patient in whom an epileptic seizure was not elicited during the session, and green bars present data from the control subject. The horizontal axis represents the frequency of the Fourier component in Hz, the stimulation frequency was 15 Hz, and the data are from one MEG trace from the occipital part of the head. (c) The full PCI spectra of all 152 MEG channels compared between nonprovocative (upper plot) and provocative (bottom plot) trials recorded from a patient with known visual sensitivity. The stimulation frequency was 10 Hz. The blue sea surface represents the statistical significance level of PCI ($p < 0.05$) obtained from random phase surrogates. The islands above are the statistically significant PCI values, noticeably higher for the provocative trial.

unless otherwise stated, and we shall refer to the frequency as an integer number. However, the true frequency scale is presented in the figures.

Averaging either Equation 12.5 or Equation 12.6 over the stimuli s yields the classic averaged evoked response (or evoked potential if the signal represents local field potentials). We introduce here the definition of the PCI:

$$\left| \left\langle Z_f^s \right\rangle_s \right| = PCI_f \left\langle \left| Z_f^s \right| \right\rangle_s \equiv PCI_f A_f. \tag{12.7}$$

Here and later, $\langle \rangle_s$ means averaging over all periods (stimuli). In essence, the factorization in Equation 12.7 represents the spectral evoked response as a product of an averaged amplitude spectrum and PCI. In this way, one can obtain one more degree of information related to the evoked responses. Explicitly, the factors are:

$$PCI_f = \frac{\left| \left\langle Z_f^s \right\rangle_s \right|}{\left\langle \left| Z_f^s \right| \right\rangle_s}. \tag{12.8}$$

$$A_f = \left\langle \left| Z_f^s \right| \right\rangle_s$$

From Equations 12.7 and 12.8, it is clear that in Fourier representation the STR is a product of two factors. One of the factors represents the averaged spectral power of the response and the second, the PCI, accounts for the phase of the response relative to the stimulus. To compare the quantities in Equations 12.7 and 12.8 in case of recorded data, we refer to the upper plot of Figure 12.10.

The definition in Equation 12.8 is not unique. We introduce a family of generalized PCI, some of which are related to similar techniques used by other authors.

$$PCI_f^q = \frac{\left| \left\langle Z_f^s \right\rangle_s \right|}{\left(\left\langle \left| Z_f^s \right|^q \right\rangle_s \right)^{1/q}} \equiv \frac{\left| \left\langle Z_f^s \right\rangle_s \right|}{A_f^q}, \quad A_f^q = \left(\left\langle \left| Z_f^s \right|^q \right\rangle_s \right)^{1/q}. \tag{12.9}$$

In the above formula q can be any positive real number. The case where $q = 2$ corresponds to the magnitude-square coherence (MSC) introduced by Dobie and Wilson (1989). In general, the quantities in Equation 12.9 reflect the phase consistency of the sequences of complex amplitudes (in our context, derived as Fourier coefficients Z).

Furthermore, we note that those quantities are also sensitive to the distribution of the magnitudes of the complex Fourier coefficients. To illustrate the point, assume that a sequence of complex numbers has the same complex phase but different magnitudes. Then we have from Equation 12.9:

$$Z^s = A^s e^{i\phi} \Rightarrow PCI^q = \frac{\left\langle A^s \right\rangle_s}{\left(\left\langle \left(A^s \right)^q \right\rangle_s \right)^{1/q}} \begin{cases} \leq 1, q > 1 \\ = 1, q = 1 \\ \geq 1, q < 1 \end{cases}. \tag{12.10}$$

It follows that only the case of $q = 1$, $PCI = 1$ will unambiguously reflect a perfect phase alignment, independent of the magnitudes. For other choices of the parameter, q, the index will be influenced by the distribution of magnitudes. In a sense, the higher the value $q > 1$ is, the more it penalizes unequal magnitudes and vice versa. For small $q < 1$, the index of Equation 12.10 encourages larger magnitude spread. Obviously, when all responses have equal magnitudes A, the PCI features produce adequate measures of phase alignment for all values of q.

One question that arises with these and similar analytical techniques proposed for (preictal) state identification is whether there is any dynamical model that would validate the use of the calculated feature beyond pure phenomenology. We present in the next paragraph results from model studies and discuss in a more general frame the meaning of the PCI. In the context of a generic linear system, the response can be represented by

$$Z_f^s = T_f^s x_f^s + Q_f^s. \tag{12.11}$$

Here T is the input transfer function, x is the input, and Q is the background activity. For the case when $q = 2$, one can derive from Equations 12.9 and 12.11:

$$PCI_f = \left| \frac{\langle Z_f^s \rangle_s}{\sqrt{\langle |Z_f^s|^2 \rangle_s}} \right| = \left(1 + \frac{S_f^2}{D_f^2} \right)^{(-1/2)}. \tag{12.12}$$

In Equation 12.12, $D_f^2 \equiv \left| \langle Z_f^k \rangle_k \right|^2 = |T_f|^2 |x_f|^2$ is the driving of the system and $S_f^2 \equiv \langle |Q_f^s|^2 \rangle_s$ is the variance of the background activity. Therefore, our quantity is little more than a functional transformation of the ratio between the induced and the background activity for linear systems described in Equation 12.11. In realistic systems with nonlinear response features, the induced activity will include the neuronal activation response and volume conductance effects that contain no information about the biological system. To filter out that linear contamination and keep only the nonlinear response effects, we have applied in some cases the cyclic (polarity) alternating patterns (CAP) paradigm illustrated in Figure 12.6a, right plot. If every second pulse is of inverted polarity, or $x_f^s = -x_f^{s+1}$, in formula (12.11), then the average linear response will cancel, $D_f \equiv 0$ and, therefore, $PCI_f \equiv 0$. Any measured result greater than zero will be due to deviation from the linearity assumption in Equation 12.11. Assuming the modified response function

$$Z_f^s = T_f x_f + Q_f^s + R_f(x) \tag{12.13}$$

with a nonlinear response spectral amplitude is denoted as $R_f(x)$ and modifying appropriately the definition of Equation 12.8, we can derive the following relation

$$PCI_f^{alt} \equiv \left| \frac{\langle Z_f^{2s-1} + Z_f^{2s} \rangle_s}{\sqrt{\langle |(Z_f^{2s-1} + Z_f^{2s})|^2 \rangle_s}} \right| = \left(1 + \frac{S_f^2}{R_f^2(x)} \right)^{(-1/2)}. \tag{12.14}$$

In Equation 12.14, the background activity is compared to the nonlinear physiological response only.

We have shown previously (Parra et al. 2003) that a particular combination of the PCI spectrum, called rPCI, emerged as a good identifier of an impending transition to an epileptic seizure in patients suffering from epilepsy with visual sensitivity submitted to periodic visual stimulation. We define the rPCI as

$$rPCI = \max_f(PCI_f) - PCI_1. \qquad (12.15)$$

In this definition, the spectral maximum of PCI at frequency f is compared to the phase consistency of the first Fourier component or the base driving, as we later refer to it. An illustration of Equation 12.15 and three cases of stimulation are shown on the bottom plot of Figure 6b.

Our final remark in this section concerns the statistical significance test of the measured PCI values. Using the same magnitudes A_f as defined in Equations 12.7 and 12.8, but generating a large number of surrogate phases for each frequency component of the response, we can compute a set of PCI values in the absence of phase locking to the stimulus. Then the 95 percentile PCI value defines the threshold for significant PCI ($p < 0.05$). An illustration of a 152-channel MEG analysis of the statistically significant PCI is shown in Figure 12.6b. Here, we have compared two measurements from a photosensitive patient, one that resulted in a photo paroxysmal discharge and one that did not. The blue background represents the $p < 0.05$ significance level and we see in the paroxysmal case the existence of high-frequency islands of statistically significant PCI values, predominantly between 40–60 Hz. As all data are collected prior to the clinical reaction to the stimulation, we conclude that the rPCI quantity measures the susceptibility of the system to paroxysmal reaction.

Applying the same technique responses to direct intracranial stimulation, we have observed (Kalitzin et al. 2005) that rPCI can estimate both the spatial location and the time period of a high-risk of impending seizure in patients with TLE. In Figure 12.7, we have our published results.

12.3.3 COMPUTER MODEL EXPLANATION OF rPCI

The quantity of Equation 12.15 will be our candidate feature for seizure-risk estimation. In the case of the computer model introduced earlier, we can explore its exact predictive powers (Suffczynski et al. 2009). The model has many parameters and some of these can control transitions to high-amplitude oscillatory states that model epileptic seizures. For detail we refer the reader to the referenced publication and here we present the analysis of only two parameters. It is commonly accepted that epileptic seizures are generally caused by unbalanced excitation and inhibition. We can verify that this is indeed the case in our model, both increasing excitation and/or decreasing inhibition can deform the system's attractor to a limit cycle (Wendling et al. 2003). Therefore, much as in the example presented in Figure 12.9, we have two-parameter transition dynamics. The seizure threshold in the model was quantified by measuring the value of the average rate of afferent action potentials (formally the DC of the noise input signal) that has to be reached in order to switch the network's behavior from normal activity of low amplitude to large-amplitude limit cycle oscillations corresponding to seizure activity.

The results of our simulations are presented in Figure 12.8a and b. The color code on Figure 12.8a shows the threshold and, therefore, the risk of transition (red is high, blue is low) due to external or internal perturbations. Along the vertical axis, several potential measurable estimators of that risk are depicted, such as the variance of the spontaneous activity (S), the response amplitude (D) and the above-introduced rPCI quantity. A good risk quantifier is a measurable feature that can be unambiguously associated with transition threshold. We see from Figure 12.8 that the spontaneous signal variance can reflect the seizure risk only along the direction of excitation. The signal response amplitude (the driving), on the other hand, detects variations of the inhibition of the system. When both system parameters are involved, none of the above two features is an adequate seizure predictor. In contrast to these features, the rPCI quantity is associated with the risk of transition in a ubiquitous way, so at least for this model, it possibly has a better predictive

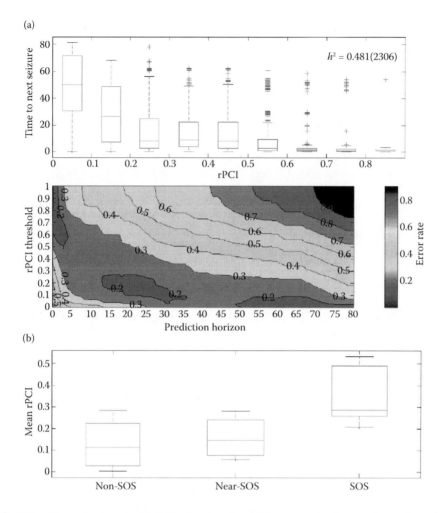

FIGURE 12.7 **(See color insert.)** (a) Collective data from five contacts in three patients where significant association between rPCI and time to seizure has been established. The upper frame gives the distributions of times to next seizure as a function of the measured rPCI, sorted in 10 equidistant bins. The horizontal red bars represent the mean and the boxes are the 10–90 percentile data intervals. The bottom frame gives the prediction error rate (color scale) as a function of a selected rPCI threshold and a selected prediction horizon. The prediction is successful if, when the measured rPCI is higher than the selected threshold, then the next seizure is recorded to happen earlier than the selected prediction horizon and vice-versa, if rPCI is lower than the threshold, the next seizure has come later than the prediction horizon. (b) Combined statistics of the interictal (more than 24 h away from seizure) rPCI values derived from 18 traces in 6 patients presented in the study of Parra et al. (2005). Median (upper plot) and mean (lower plot) interictal rPCI for each trace are computed and grouped according to SOS contacts, near contacts in the SOS-ipsilateral hemisphere, and SOS-contralateral contacts (denoted as non-SOS), as found from ictal EEG. A median rPCI threshold of 0.25 gives the optimal discrimination between SOS and non-SOS. (From Kalitzin et al., *Clinical Neurophysiology*, 116(3), 718–728, 2005.)

power. As described in the beginning of this section, the predictive power can be quantified by the nonlinear association index between the studied feature and the threshold to transition. The numbers are given in the corresponding scatterplots in Figure 12.8b. In the bottom-right plot of the figure, we present for comparison the scattered plot of rPCI versus the time to next seizure from a TLE patient.

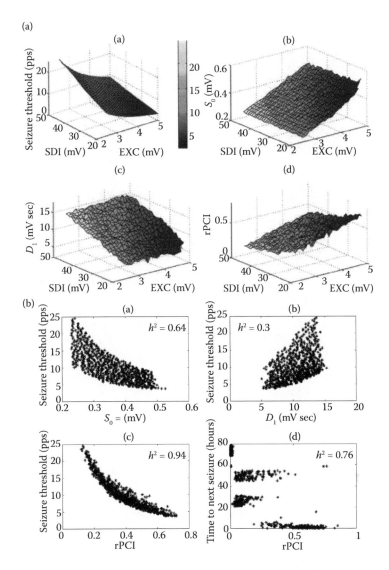

FIGURE 12.8 (**See color insert.**) (a) Dependence of the seizure threshold (subframe a) and of three observable signal features (subframes b–d) on the hippocampus model parameters. The z-axis represents the corresponding feature values, except in plot (a) where the threshold is considered. The x-axis is the inhibition parameter (SDI) and the y-axis is the excitation control parameter (EXC). In all four plots, color scale corresponds to the value of the seizure threshold. Subframe a shows the dependence of the seizure threshold (in pulses per second, pps) and a color bar showing color scale used. Subframe b shows the dependence of the standard deviation of spontaneous activity S_0. One can see that this measure is sensitive only to change of the EXC but not to change of the SDI parameter. Subframe c shows the dependence of the system's triggered response D_1. One can see that this measure is sensitive only to change of the SDI but not to change of the EXC parameter. Subframe d shows the dependence of the rPCI showing that this measure is sensitive to change of both the EXC and the SDI parameters. (b) Subframe a–c shows the scatterplots of the seizure threshold (pulses per second, pps) and measurable quantities (S_0, D_1 and rPCI, respectively) obtained for all pairs of the EXC and SDI parameters in the model. In each graph, the value of nonlinear association measure h^2 is used to quantify the correlation between the respective variables. The h^2 values show that the rPCI measure is best correlated with the seizure threshold, while its scatterplot hyperbolic shape resembles that of the real data. Subframe d shows the scatterplot of the time to next seizure and rPCI value in a TLE patient [Patient 3 as described in Parra et al. (2005)]. (From Suffczynski et al., *Phys. Rev. E Stat. Nonlin. Soft Matter Phys.*, 78(5 Pt 1), 051917, 2008. With permission.)

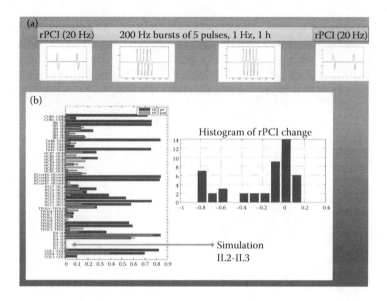

FIGURE 12.9 **(See color insert.)** (a) Schematic presentation of our low-frequency repetitive electrical stimulation protocol. Before and after 1 h of stimulation with 200-Hz bursts of 5 biphasic bursts, 1 burst every second, the rPCI was measured using 20-Hz biphasic stimulation through the same set of electrode contacts. The contacts were selected close to the confirmed epileptic onset area. The amplitude for both rPCI measurement and burst stimulation was 1 mA. (b) On the left plot, the rPCI values for the responses per contact are represented as horizontal bars, blue for the rPCI before the LFrEC session and red for the rPCI after. The right plot gives the number of channels (vertical axis) for a given rPCI change (horizontal axis). The rPCI changes are represented in bins from −1 to +1 of width 0.1. Note that the cluster of electrode sites that show the largest difference after the LFrES stimulation are those of the left hippocampus HCL1 to HCL5, but the same pattern can be seen at other sites (TPOL 3-5, HCcor R2 - R4, and IR 5-7).

We conclude this chapter by presenting another possible application of the rPCI quantity as a possible detector of proictality or seizure vulnerability of the system. We have used rPCI to measure the effect of a prospective therapeutic procedure, in this case, a low-frequency (one burst per second) repetitive electrical stimulation (LFrES) consisting of short (five pulses per burst) bipolar bursts at 200 Hz. The result, presented in Figure 12.9, shows decreased rPCI values after the LFrES session that can be interpreted as a decreased seizure risk. We have not followed the long-term carry-over effects of the LFrES, so our conclusions are only indicative although encouraging at this moment.

12.4 STATE-REACTIVE CONTROL OF MODEL SEIZURES IN MULTISTABLE SYSTEMS

One of the most attractive points of the multistable scenario of seizure generation is that it allows for an acute intervention. We have shown in realistic corticothalamic models that, if detected in time, an epileptic seizure can be aborted with even a single stimulation pulse (Suffczynski et al. 2004). Here we expand the concept by including continuously operating additional modules designed to constantly monitor and react to epileptic transitions. We use the same analytic models as described in Section 12.3.

Our first test is presented in Figure 12.10, where a Simulink® model is used to simulate the neural network (2.1) with a feedback extension (the module in red ellipse). The output signal is first

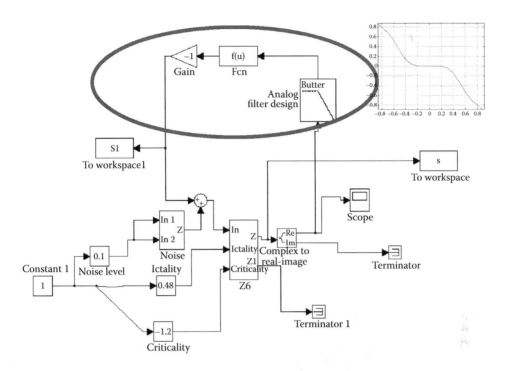

FIGURE 12.10 **(See color insert.)** A Simulink model of the analytical complex model defined by Equation 12.1 enhanced with a continuous feedback module (in the red ellipse). The nonlinear function (the upper-right inset) provides seizure rejection while keeping the interictal dynamics practically unaltered.

filtered (to remove slow fluctuations) and then transformed point-wise with a nonlinear transformation shown in the inset. The feedback coupling (the gain block) is the parameter that controls the level of the interference with the spontaneous dynamics of the system. The idea behind this model concept is to feed the inverted output of the system back as an input, as in some vibration cancellation devices. But in our case, we only want to do this when the output signal exceeds certain thresholds, as during transition to a limit cycle of behavior. This necessitates the use of a nonlinear function that tolerates small values and gives progressive rates of cancellation of larger amplitudes.

The results are displayed in Figure 12.11, where the traces (a) and the auto-correlation function (b) for different levels and types of feedback are shown. We can see that the nonlinear feedback prevents transitions to the limit cycle (model ictal activity) but leaves the point attractor (interpreted in the model as interictal dynamics) almost intact.

For intended application in clinical trials, we simulated a simplified feedback paradigm where instead of using a continuous signal, individual pulses of both polarities are fed into the system. The setup is similar to that in Figure 12.10. After the linear filtering as before, the signal is simply compared to a set of thresholds, one positive and the other negative. When the signal crosses any of the thresholds, a stimulation pulse of opposite sign is administered to the system. In Figure 12.12a, we see the overall performance, the detailed pulse action, and the two-dimensional phase diagram with the stimulated (red) and original (blue) trajectories. We see that the stimulation paradigm manages to contain the system in the center of the Z plane, thus keeping it far from the limit cycle.

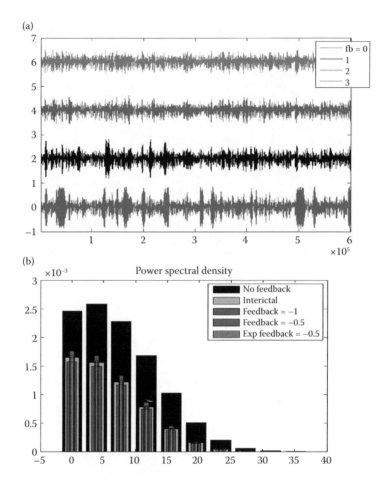

FIGURE 12.11 **(See color insert.)** (a) Output from the model on Figure 12.10 for various levels of feedback coupling. (b) The power spectral density of the ongoing signal (the autocorrelation function) for different feedback couplings and nonlinear functions. The black bars include both model ictal and interictal epochs as measured without the feedback module engaged. The green bars represent only the interictal epochs segmented out from the signal without feedback engaged. The blue, red, and purple bars give the power spectrum with activated feedback module with various feedback coupling strengths and nonlinear functions. We observe that the feedback function produced seizure suppression with minimal alteration of the power spectrum in comparison with that of the interictal state. The vertical axes of both plots represent arbitrary units.

Similar results were obtained for the system described by Equation 12.2. Here we used monopolar pulses opposite to the DC shift. We assumed that the measured and stimulated degrees of freedom are obtained from the same mixture of the Z and u variables that serves as observation variable. An illustration of the simulated DC shift counter-stimulation is shown in Figure 12.12b.

Finally, these model paradigms have been tested on patients with implanted electrodes undergoing routine diagnostic stimulation to determine the area of seizure onset. Those stimulations (unipolar trains of 40 Hz, increasing pulse amplitude up to 10 mA) are provocative; they cause afterdischarges (AD) and may result in an electrographic, or even clinical, seizure. We have applied some of the acute seizure reactive control methods as prescribed from the models in attempts to stop or limit the duration and amplitude of ADs. One successful trial is shown in Figure 12.13, together with the overall procedure.

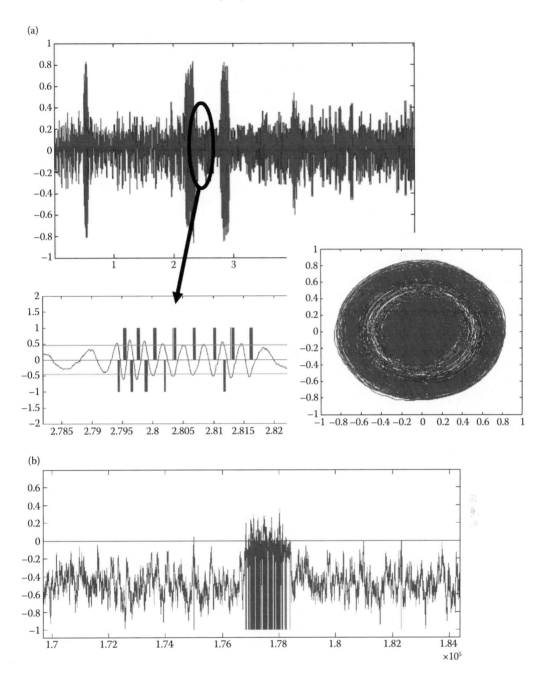

FIGURE 12.12   (See color insert.) (a) Results from the model presented in Figure 12.11 but with discrete pulse counter-stimulation instead of continuous feedback. The upper plot represents two traces, the blue one is without activation of feedback and the red one is with feedback activation. Both traces are taken as the real component of the complex model. The two-dimensional plot at the lower right shows the same traces on the complex plane. The blue (no feedback) trace intermittently enters the limit cycle, whereas the red one (active feedback) stays clear of the limit cycle. The lower left-hand plot is an enlargement of the trace with feedback (blue line) and the sequence of stimuli (red bars) that act in counterphase with the signal. (b) Result of a simulation similar to that of (a), but with the model described by Equation 12.2. The reactive control here is directed against the DC shift.

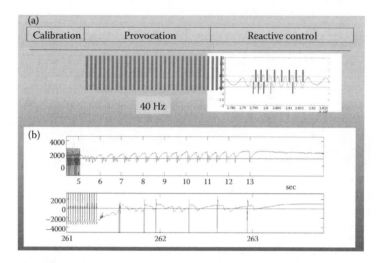

FIGURE 12.13 **(See color insert.)** (a) Schematic representation of the counter-stimulation paradigm applied to suppress afterdischarges (AD). AD are provoked with stimulation pulses of 40 Hz as part of a standard diagnostic routine in our facility in SEIN aimed to outline the brain area that might be responsible for igniting the epileptic seizures. The closed-loop software subsequently attempts to suppress the ADs, thus making the diagnostic tool safer for the patient. (b) The upper trace represents an AD lasting 8 s after provocation without suppression while reactive control was not engaged. Lower plot shows a successfully aborted AD by the reactive control within 2 s after provoking it.

12.5 CONCLUSIONS AND DISCUSSION

In this chapter, we have presented a particularly biased approach to seizure prediction. This may be in contrast to most of the earlier approaches where the focus was on phenomenological statistics without a clear dynamic model in sight. We have based our analysis on the assumption that epileptic seizures are due to interstate transitions in dynamic systems having multiple attractors. We fully realize though that this is not the only possible scenario; parameter changes and chaotic transitions (intermittency) might play a role in certain cases of epilepsy. To validate our a priori assumptions, we explored the statistics of ictal and interictal durations and we found that in a variety of animal models and clinical cases, the multistability model could not be ruled out, at least not for the transitions to an epileptic state. This finding does not represent, of course, proof for our paradigm. We derived further prescriptions for seizure control that, if confirmed experimentally, will also provide a posterior validation of the concept. Indeed, the work of Osorio and Frei (2009) provides possible operational evidence of the mechanism described, for example, by Equation 12.2. The possibility of reversing an impending transition to a seizure state attractor is a topological problem. If we assume that we can measure and affect only one particular degree of freedom of the system, the relative position of the nonictal and the epileptic attractors projected onto the available dimension is crucial. More specifically, we can reverse the transition to an epileptic attractor only if the stimulation direction crosses the separatrix, i.e., the hypersurface that, when crossed, brings the system from one attractor to the other. If, for example, in Figure 12.4a, the measurement and stimulation dimension is in the Z plane, no stimulation algorithm would ever succeed to move the system back to the steady state. Indeed, one can easily derive from Equation 12.2 that the separatrix is the $u = 0$ plane, which does not intersect with the $u = 1$ plane, where the limit cycle resides and therefore cannot be reached by the stimulation operating within this plane. On the contrary, in Equation 12.1, the separatrix is a circle in-between the point attractor and the limit cycle and can be penetrated by appropriate stimulation. Therefore, although the seizure depicted in the bottom plot in Figure 12.3b (Equation 12.1) and the one shown in the top plot in Figure 12.4b (Equation 12.2), are similar, a phase-dependent counter-stimulation will abort the seizure in Figure 12.3, but not the one in Figure 12.4.

In our scenario, external influences, including thermal noise, govern the transition between normal and paroxysmal states. This circumstance does not forbid seizure predictability in its broader context. If the stochastic process of transition depends on the values of parameters that we can measure, then we can estimate the state-dependent probability of an instantaneous seizure. If we base a prediction scheme on this probability, then a better than chance warning scheme will be possible. In reality, finding the true critical parameters and thresholds is not straightforward. We have based our attempts on active paradigms involving stimulation of the neural tissue. We found that a particular measure of phase locking of the system response to the stimulus, the rPCI, can be a suitable feature associated with the risk of seizure. Clinical studies, as well as computer model simulations, have so far confirmed our hypothesis. Replacing the concept of preictal state—defined as a state that deterministically precedes an epileptic seizure—with a broader concept that we call proictal state—or state with an increased risk of developing into a seizure—poses rightfully the question of the clinical relevance of identifying such state whenever possible. As we have shown in the case study illustrated in Figure 12.14, the ability to modify the susceptibility of the system to undergo a dynamic transition might be helpful in developing new treatment techniques other than acute contrastimulation at the time of seizure onset. In this context, identifying proictal state features may be instrumental for an adequate strategy to interfere whenever necessary.

What about the possibility of a true deterministic prediction of epileptic transitions? In terms of our definition in Section 12.3, this would mean a nearly exact functional relation (corresponding to $h^2 \sim 1$) between the time interval to next seizure event and the outcome of the proposed measurement. Theoretically, this cannot be ruled out if a system is governed by intermittent types of state changes as we have shown in Section 12.2. Certain measurable features, the number of positive Lyapunov exponents in our model, may indeed serve as a countdown identifier of the next transition. Real systems are always subject to external perturbations, noise, and fluctuating parameters. Nonlinear systems behaving intermittently are nice mathematical constructs, but normally they loose some of their properties under the above conditions inasmuch as the exact predictability.

Our final remark concerns the use of computer models in order to understand the dynamics of epileptic seizures and their impact on the efforts at seizure prediction (Wendling 2008). If we have a good realistic model of epilepsy in silico, we would be able to test the predictive powers of various observables as we have shown in Section 12.4. But there is a second and perhaps more important application of such computer modeling. It is the reverse engineering that allows us, from known phenomenology, to infer the mechanisms of seizure generation. If, for example, the Lyapunov exponents can predict seizure onsets in an almost deterministic way, we consider intermittency, as briefly explained in Section 12.2, as the leading mechanism for seizure generation. If, on the other hand, the signal's variation, or the rate of accumulated energy (Litt et al. 2001), is a predictive feature then according to our realistic model, we would expect increasing excitation to be the main factor. Therefore, simulation models can help both in finding the optimal feature to predict seizures and reconstructing the mechanism leading to those seizures wherever such a feature has been successfully validated.

REFERENCES

Andrzejak, R. G., F. Mormann, T. Kreuz, C. Rieke, A. Kraskov, C. E. Elger, and K. Lehnertz. 2003. Testing the null hypothesis of the nonexistence of a preseizure state. *Phys. Rev. E Stat. Nonlin. Soft Matter Phys.* 67:010901.

Bartolomei, F., P. Chauvel, and F. Wendling. 2008. Epileptogenicity of brain structures in human temporal lobe epilepsy: A quantified study from intracerebral EEG. *Brain* 131:1818–1830.

Dobie, R. A., and M. J. Wilson. 1989. Analysis of auditory evoked potentials by magnitude-squared coherence. *Ear Hear* 10:2–13.

Elger, C. E., and K. Lehnertz. 1998. Seizure prediction by non-linear time series analysis of brain electrical activity. *Eur. J. Neurosci.* 10:786–789.

Iasemidis, L. D., J. C. Sackellares, H. P. Zaveri, and W. J. Williams. 1990. Phase space topography and the Lyapunov exponent of electrocorticograms in partial seizures. *Brain Topogr.* 2:187–201.

Jobst, B. 2009. Brain stimulation for surgical epilepsy. *Epilepsy Res.* 89(1):154–161.

Kalitzin, S., J. Parra, D. N. Velis, and F. H. L. d. Silva. 2002. Enhancement of phase clustering in the EEG/EMG gamma frequency band anticipates transitions to paroxysmal epileptiform activity in epileptic patients with known visual sensitivity. *IEEE Trans. Biomed. Eng.* 49:1279–1286.

Kalitzin, S., D. Velis, P. Suffczynski, J. Parra, and F. L. da Silva. 2005. Electrical brain-stimulation paradigm for estimating the seizure onset site and the time to ictal transition in temporal lobe epilepsy. *Clin. Neurophysiol.* 116:718–728.

Kalitzin, S. N., J. Parra, D. N. Velis, and F. H. L. d. Silva. 2007. Quantification of unidirectional nonlinear associations between multidimensional signals. *IEEE Trans. Biomed. Eng.* 54:454–461.

Kalitzin, S. N., D. N. Velis, and F. H. da Silva. 2010. Stimulation-based anticipation and control of state transitions in the epileptic brain. *Epilepsy Behav.* 17(3):310–323.

Lehnertz, K. 1999. Non-linear time series analysis of intracranial EEG recordings in patients with epilepsy—an overview. *Int. J. Psychophysiol.* 34:45–52.

Lehnertz, K., and C. E. Elger. 1997. Neuronal complexity loss in temporal lobe epilepsy: Effects of carbamazepine on the dynamics of the epileptogenic focus. *Electroencephalogr. Clin. Neurophysiol.* 103:376–380.

Litt, B., R. Esteller, J. Echauz, M. D'Alessandro, R. Shor, T. Henry, P. Pennell, C. Epstein, R. Bakay, M. Dichter, and G. Vachtsevanos. 2001. Epileptic seizures may begin hours in advance of clinical onset: A report of five patients. *Neuron.* 30:51–64.

Litt, B., and K. Lehnertz. 2002. Seizure prediction and the preseizure period. *Curr. Opin. Neurol.* 15:173–177.

Lopes da Silva, F., W. Blanes, S. N. Kalitzin, J. Parra, P. Suffczynski, and D. N. Velis. 2003a. Epilepsies as dynamical diseases of brain systems: Basic models of the transition between normal and epileptic activity. *Epilepsia* 44:(Suppl 12):72–83.

Lopes da Silva, F. H., W. Blanes, S. N. Kalitzin, J. Parra, P. Suffczynski, and D. N. Velis. 2003b. Dynamical diseases of brain systems: Different routes to epileptic seizures. *IEEE Trans. Biomed. Eng.* 50:540–548.

Mormann, F., R. G. Andrzejak, C. E. Elger, and K. Lehnertz. 2007. Seizure prediction: The long and winding road. *Brain* 130:314–333.

Ohayon, E. L., S. Kalitzin, P. Suffczynski, F. Y. Jin, P. W. Tsang, D. S. Borrett, W. M. Burnham, and H. C. Kwan. 2004a. Charting epilepsy by searching for intelligence in network space with the help of evolving autonomous agents. *J. Physiol. Paris* 98:507–529.

Ohayon, E. L., H. C. Kwan, W. M. Burnham, P. Suffczynski, and S. Kalitzin. 2004b. Emergent complex patterns in autonomous distributed systems: Mechanisms for attention recovery and relation to models of clinical epilepsy. *IEEE International Conference on Systems, Man and Cybernetics 2004* 2:2066–2072.

Osorio I., and M. G. Frei. 2009. Seizure abatement with single DC pulses: Is phase resetting at play? *Int. J. Neural. Syst.* 19:149–156.

Osorio, I., M. G. Frei, D. Sornette, and J. Milton. 2009. Pharmaco-resistant seizures: Self-triggering capacity, scale-free properties and predictability? *Eur. J. Neurosci.* 30:1554–1558.

Osorio, I., M. G. Frei, S. Sunderam, J. Giftakis, N. C. Bhavaraju, S. F. Schaffner, and S. B. Wilkinson. 2005. Automated seizure abatement in humans using electrical stimulation. *Ann. Neurol.* 57:258–268.

Parra, J., S. N. Kalitzin, J. Iriarte, W. Blanes, D. N. Velis, and F. H. Lopes da Silva. 2003. Gamma-band phase clustering and photosensitivity: Is there an underlying mechanism common to photosensitive epilepsy and visual perception? *Brain* 126:1164–1172.

Prusseit, J., and K. Lehnertz. 2007. Stochastic qualifiers of epileptic brain dynamics. *Phys. Rev. Lett.* 98:138103.

Sabesan, S., L. B. Good, K. S. Tsakalis, A. Spanias, D. M. Treiman, and L. D. Iasemidis. 2009. Information flow and application to epileptogenic focus localization from intracranial EEG. *IEEE Trans. Neural Syst. Rehabil. Eng.* 17:244–253.

Sackellares, J. C., D. S. Shiau, J. C. Principe, M. C. Yang, L. K. Dance, W. Suharitdamrong, W. Chaovalitwongse, P. M. Pardalos, and L. D. Iasemidis. 2006. Predictability analysis for an automated seizure prediction algorithm. *J. Clin. Neurophysiol.* 23:509–520.

Scheffer, M., J. Bascompte, W. A. Brock, V. Brovkin, S. R. Carpenter, V. Dakos, H. Held, E. H. van Nes, M. Rietkerk, and G. Sugihara. 2009. Early-warning signals for critical transitions. *Nature* 461:53–59.

Snyder, D. E., J. Echauz, D. B. Grimes, and B. Litt. 2008. The statistics of a practical seizure warning system. *J. Neural. Eng.* 5:392–401.

Suffczynski, P., S. Kalitzin, F. Lopes da Silva, J. Parra, D. Velis, and F. Wendling. 2008. Active paradigms of seizure anticipation—a computer model evidence for necessity of stimulation. *Phys. Rev. E Stat. Nonlin. Soft Matter Phys.* 78(5 Pt. 1):051917.

Suffczynski, P., S. Kalitzin, and F. H. Lopes Da Silva. 2004. Dynamics of non-convulsive epileptic phenomena modeled by a bistable neuronal network. *Neuroscience* 126:467–484.

Suffczynski, P., S. Kalitzin, G. Pfurtscheller, and F. H. Lopes da Silva. 2001. Computational model of thalamocortical networks: Dynamical control of alpha rhythms in relation to focal attention. *Int. J. Psychophysiol.* 43:25–40.

Suffczynski, P., F. H. Lopes da Silva, J. Parra, D. N. Velis, B. M. Bouwman, C. M. van Rijn, P. van Hese, P. Boon, H. Khosravani, M. Derchansky, P. Carlen, and S. Kalitzin. 2006. Dynamics of epileptic phenomena determined from statistics of ictal transitions. *IEEE Trans. Biomed. Eng.* 53:524–532.

Ullah, G., and S. J. Schiff. 2009. Tracking and control of neuronal Hodgkin-Huxley dynamics. *Phys. Rev. E Stat. Nonlin. Soft Matter Phys.* 79:040901.

Velasco, A. L., F. Velasco, M. Velasco, D. Trejo, G. Castro, and J. D. Carrillo-Ruiz. 2007. Electrical stimulation of the hippocampal epileptic foci for seizure control: A double-blind, long-term follow-up study. *Epilepsia* 48:1895–1903.

Velazquez, J. L., H. Khosravani, A. Lozano, B. L. Bardakjian, P. L. Carlen, and R. Wennberg. 1999. Type III intermittency in human partial epilepsy. *Eur. J. Neurosci.* 11:2571–2576.

Wendling, F. 2008. Computational models of epileptic activity: A bridge between observation and pathophysiological interpretation. *Expert Rev. Neurother.* 8:889–896.

Wendling, F., F. Bartolomei, J. J. Bellanger, J. Bourien, and P. Chauvel. 2003. Epileptic fast intracerebral EEG activity: Evidence for spatial decorrelation at seizure onset. *Brain* 126:1449–1459.

Wendling, F., F. Bartolomei, J. J. Bellanger, and P. Chauvel. 2001. Interpretation of interdependencies in epileptic signals using a macroscopic physiological model of the EEG. *Clin. Neurophysiol.* 112:1201–1218.

Zalay, O. C., and B. L. Bardakjian. 2008. Mapped clock oscillators as ring devices and their application to neuronal electrical rhythms. *IEEE Trans. Neural Syst. Rehabil. Eng.* 16:233–244.

Section III

The Challenge of Prediction

13 Prediction

Didier Sornette and Ivan Osorio

CONTENTS

This chapter first presents a rather personal view of some different aspects of predictability, going in crescendo from simple linear systems to high-dimensional nonlinear systems with stochastic forcing, which exhibit emergent properties such as phase transitions and regime shifts. Then, a detailed correspondence between the phenomenology of earthquakes, financial crashes, and epileptic seizures is offered. The presented statistical evidence provides the substance of a general phase diagram for understanding the many facets of the spatiotemporal organization of these systems. A key insight is to organize the evidence and mechanisms in terms of two summarizing measures: (i) amplitude of disorder or heterogeneity in the system and (ii) level of coupling or interaction strength among the system's components. On the basis of the recently identified remarkable correspondence between earthquakes and seizures, we present detailed information on a class of stochastic point processes that has been found to be particularly powerful in describing earthquake phenomenology and which, we think, has a promising future in epileptology. The so-called self-exciting Hawkes point processes capture parsimoniously the idea that events can trigger other events, and their cascades of interactions and mutual influence are essential to understand the behavior of these systems.

13.1 BRIEF CLASSIFICATION OF PREDICTABILITY

Characterizations of the predictability (or unpredictability) of a system provide useful theoretical and practical measure of its complexity (Boffetta et al. 1997; Kantz and Schreiber 2004). It is also a grail in epileptology, as advanced warnings by a few minutes may drastically improve the quality of life of these patients.

13.1.1 PREDICTABILITY OF LINEAR STOCHASTIC SYSTEMS

Consider a simple dynamical system with the following linear autoregressive dynamics:

$$r(t) = \beta r(t-1) + \varepsilon(t),
\tag{13.1}$$

where $0 < \beta < 1$ is a constant and $\varepsilon(t)$ is an independently identically distributed (i.i.d.) random variable (i.e., a noise) with variance σ_ε^2. The dependence structure between successive values of $r(t)$ is entirely captured by the correlation function. The correlation coefficient between the random variable at some time and its realization at the following time step is nothing but β. Correspondingly, the covariance of $r(t-1)$ and $r(t)$ is $\beta \times \sigma_r^2$, where $\sigma_r^2 = \sigma_\varepsilon^2/(1-\beta^2)$ is the variance of $r(t)$. More generally, consider an extension of expression (13.1) into a linear autoregressive process of larger order, so that we can consider an arbitrary covariance matrix $C(t, t')$ between $r(t)$ and $r(t')$ for all possible instant pairs t and t'. A simple mathematical calculation shows that the best linear predictor m_t for $r(t)$ at time t, knowing the past history $r_{t-1}, r_{t-2}, \ldots, r_i, \ldots$ is given by

$$m_t \equiv \frac{1}{B(t,t)} \sum_{i<t} B(i,t) r_i,
\tag{13.2}$$

where $B(i,t)$ is the coefficient (i,t) of the inverse matrix of the covariance matrix $C(t, t')$. This formula (13.2) expresses that each past value r_i impacts on the future r_t in proportion to its value with a coefficient $B(i,t)/B(t,t)$ which is nonzero only if there is nonzero correlation between the realization of the variable at time i and time t. This formula (13.2) provides the best linear predictor in that it minimizes the errors in a variance sense. Armed with this prediction, useful operational strategies can be developed, depending on the context. For instance, if the set $\{r(t)\}$ denotes the returns of a financial asset, then one could use prediction (13.2) to invest: buy if $m_t > 0$ (expected future price increase) and sell if $m_t < 0$ (expected future price decrease).

Such a predictor can be applied to general moving average and autoregressive processes with long memory whose general expression reads

$$\left(1 - \sum_{i=1}^{p} \phi_i L^i\right)\left(1 - L\right)^d r(t) = \left(1 + \sum_{j=1}^{q} \theta_j L^j\right)\varepsilon(t), \tag{13.3}$$

where L is the lag operator defined by $Lr(t) = r(t - 1)$ and p, q, and d can be arbitrary integers (Hamilton 1994). Such predictors are optimal or close to optimal as long as there is no change of regime, that is, if the process is stationary and the coefficients $\{\phi_i\}$ and $\{\theta_j\}$ and the orders of moving average q, autoregression p, and fractional derivation d do not change during the course of the dynamics. Otherwise, other methods, including Monte Carlo Markov chains, are needed (Hamilton 1994). In a case where the initial conditions or observations during the course of the dynamics are obtained with noise or uncertainty, Kalman filtering and, more generally, data assimilation methods provide significant improvements in predicting the dynamics of the system (Ide et al. 1997).

13.1.2 PREDICTABILITY OF LOW-DIMENSIONAL DETERMINISTIC CHAOTIC SYSTEMS

There is an enormous amount of literature on this subject since the 1970s (Belacicco et al. 1995; Kravstov and Kadtke 1996; Kantz and Schreiber 2004; Orrell 2007; Smith 2007). The idea of how to develop predictors for low-dimensional deterministic chaotic systems is very natural: Because of determinism, and provided that the dynamics is in some sense sufficiently regular, the short-time evolution remembers the initial conditions, so that two trajectories that are found in a neighborhood of each other remain close to each other for a time t_f roughly given by the inverse of the largest Lyapunov exponent. Thus, if one monitors past evolution, however complicated, a future path that comes in the vicinity of a previously visited point will then evolve along a trajectory shadowing the previous one over a time of the order of t_f (Sugihara and May 1990; Sornette et al. 1991; Robinson and Thiel 2009). The previously recorded dynamical evolution of a domain over some short-time horizon can thus provide, in principle, a short-term prediction through the knowledge of the transformation of this domain.

However, in practice, there are many caveats to this idealized situation. Model errors and noise, additive and/or multiplicative (also called parametric), complicate and limit predictability. Model errors refer to the generic problem that the used model is at best only an approximation of the true dynamics, and more generally neglects some possibly important ingredients, making prediction questionable.

In the simplest case of additive noise decorating deterministic chaotic dynamics, it turns out that the standard statistic methods for the estimation of the parameters of the model break down. For instance, the application of the maximum likelihood method to unstable nonlinear systems distorted by noise has no mathematical ground so far (Pisarenko and Sornette 2004). There are inherent difficulties in the statistical analysis of deterministically chaotic time series due to the tradeoff between the need of using a large number of data points in the maximum likelihood analysis to decrease the bias and to guarantee consistency of the estimation, on the one hand, and the unstable nature of dynamical trajectories with exponentially fast loss of memory of the initial condition, on the other hand. The method of statistical moments for the estimation of the parameter seems to be the unique method whose consistency for deterministically chaotic time series is proved so far theoretically (and not just numerically), but the method of moments is known to be relatively inefficient.

13.1.3 PREDICTABILITY OF SYSTEMS WITH MULTIPLICATIVE NOISE

The presence of multiplicative (or parametric) noise makes the dynamics much richer . . . and complex. New phenomena appear, such as stochastic resonance (Gammaitoni et al. 1998), coherence resonance (Pikovsky and Kurths 1997), noise-induced phase transitions (Horsthemke and Lefever 1983; Sancho and García-Ojalvo 2000), noise-induced transport (Hänggi and Marchesoni 2009), and its game-theoretical version, the Parrondo's Paradox (Abbott 2010). The predictability is then a nonmonotonous function of the noise level. Even the simplest possible combination of nonlinearity and noise can utterly transform the nature of predictability. Consider, for instance, the bilinear stochastic dynamical system, arguably the simplest incarnation of nonlinearity (via bilinear dependence on the noise) and stochasticity:

$$r(t) = e(t) + b\, e(t - 1)e(t - 2), \tag{13.4}$$

where $e(t)$ is an i.i.d. noise. The dynamics (13.4) is the simplest implementation of the general Volterra discrete series of the type

$$r(t) = H_1\, [e(t)] + H_2\, [e(t)] + H_3\, [e(t)] + \& + H_n\, [e(t)] + \cdots, \tag{13.5}$$

where

$$H_n\big[\varepsilon(t)\big] = \sum\nolimits_{j_1 = 0}^{+\infty} \ldots \sum\nolimits_{j_n = 0}^{+\infty} h_n(j_1,\ldots,j_n)\varepsilon(t - j_1)\ldots\varepsilon(t - j_n). \tag{13.6}$$

By construction, the time series $\{r(t)\}$ generated by expression (13.4) has no linear predictability (zero two-point correlation) but a certain nonlinear predictability (nonzero three-point correlation) (Pisarenko and Sornette 2008). It can thus be considered as a paradigm for testing the existence of a possible nonlinear predictability in a given time series. Notwithstanding its remarkable simplicity, the bilinear stochastic process (13.4) exhibits remarkably rich and complex behavior. In particular, the inversion of the key nonlinear parameter b and of the two initial conditions necessary for the implementation of a prediction scheme exhibits a quite anomalous instability: in the presence of a random large impulse of the exogenous noise $e(t)$, the ensuing dynamics exhibits superexponential sensitivity for the inversion of the innovations (Pisarenko and Sornette 2008).

13.1.4 Higher Dimensions, Coherent Flows, and Predictability

Going bottom-up in the complexity hierarchy, we have low-dimensional chaos → spatiotemporal chaos (Cross and Hohenberg 1993) → turbulence (Frisch 1996). It turns out that, contrary to naive expectation, increasing dimensionality and introducing spatial interactions does not necessarily destroy predictability. This is due to the organization of the spatiotemporal dynamics in so-called coherent structures, corresponding to coherent vortices in hydrodynamic flows (Boffetta et al. 1997). It has been shown that the full nonlinearity acting on a large number of degrees of freedom can, paradoxically, improve the predictability of the large scale motion, giving a picture opposite to the one largely popularized by Lorenz for low-dimensional chaos. The mechanism for improved predictability is that small local perturbations can progressively grow to larger and larger scales by nonlinear interaction and finally cause macroscopic organized persistent structures (Robert and Rosier 2001).

13.1.5 Fundamental Limits of Predictability and the Virtue of Coarse-Graining

Algorithmic information theory combines information theory, computer science, and meta-mathematic logic (Li and Vitanyi 1997). In the context of system predictability, it has profound implications. Indeed, a central result of algorithmic information theory obtained as a synthesis of the efforts of Solomonoff (2009), Kolmogorov and Chaitin (1987), Martin-Löf and Burgin, and others states roughly that most dynamical systems evolve according to and/or produce outputs that are utterly unpredictable. In this sense, "most dynamical systems" indicates that this property holds with probability 1 when choosing, at random, a dynamical system from the space of all possible dynamical systems. Specifically, the data series produced by most dynamical systems have been proved to be computationally irreducible, i.e., the only way to decide about their evolution is to actually let them evolve in time. There is no way you can compress their dynamics and the resulting information into generation rules or algorithms that are shorter than the output itself. Then, the only strategy is to let the system evolve and reveal its complexity, without any hope of predicting or characterizing in advance its properties.

The future time evolution of most complex systems thus appears inherently unpredictable. This is the foundation for the approach pioneered by S. Wolfram (2002) to basically replace the search for mathematical laws and predictability with a search for cellular automata that have universal computational abilities (like the so-called Turing machines) and that can reproduce any desired pattern.

Such views are almost shocking to most scientists, whose job is to find patterns that can be captured in coherent models that provide a reduced encoding of the observed complexity, in direct apparent contradiction with the central result of algorithmic information theory. Israeli and Goldenfeld have provided an insightful and elegant procedure, based on renormalization group theory, to reconcile the two view points (Israeli and Goldenfeld 2004, 2006). The key idea is to ask only for approximate answers, which, for instance, allows physics to work unhampered by computational irreducibility. By adopting the appropriate coarse-grained perspective of how to study the system, Israeli and Goldenfeld found that even the known computational irreducible cellular automaton—Rule 110 in Wolfram's (1983) classification—becomes relatively simple and predictable. In physics, this comes as no surprise. Each trajectory of the approximately 10^{25} molecules in an office room follows an utterly chaotic trajectory that loses predictability after a few intermolecular collisions. But the coarse-grained large-scale properties of the gas are well-captured by the law of ideal gas $pV = nRT$, or van der Waals' equation if one wants a bit more precision, where p is the pressure in the enclosure of volume V at temperature T, and n is the number of moles of gas, while R is a constant. By asking questions involving different scales, computationally irreducible systems can become predictable at some level of description. The challenge is to find how to coarse-grain: What is the optimal level of description, and what effective macroscopic interactions and patterns emerge from this procedure? There are promising developments in this direction to elaborate a general theory of hierarchical dynamics (Nilsson Jacobi and Görnerup 2009; Görnerup and Nilsson Jacobi 2010; Edlund and Nilsson Jacobi 2010) using the renormalization group as a constructive metatheory of model building (Wilson 1979).

13.1.6 DRAGON-KINGS

Predictability may come from another source, that is, directly from specific transient structures developing in the system that we refer to as "dragon-kings" (Sornette 2009; Satinover and Sornette 2010).

The concept of dragon-kings has been introduced as a frontal refutal to the claim that black swans characterize the dynamics of most systems (Taleb 2007). According to the black swan hypothesis, highly improbable events with extreme sizes or impacts are thought to occur randomly, without any precursory signatures. Black swan events are thought to be events of large sizes associated with the tail of distributions such as power laws. Because the same power law distribution is thought to describe the whole population of event sizes, including the black swans of great impact, the argument is that there are not distinguishing features for these black swans except their great sizes and, therefore, no way to diagnose their occurrence in advance. In this story, for instance, a great earthquake is just an event that started as a small earthquake and did not stop growing. Its occurrence is argued to be inherently unpredictable because there is no way to distinguish the nucleation of the myriads of small events from the rare ones that will grow to great sizes by chance (Geller et al. 1997).

In contrast, the dragon-king hypothesis proposes that extreme events from many seemingly unrelated domains may be plausibly understood as part of a different population than that composing the large majority of events. This difference may result from amplifying mechanisms such as positive feedbacks, which are active only transiently, leading to the emergence of nonstationarity structures. The term dragon refers to the mythical animal that belongs to a different animal kingdom beyond the normal, with extraordinary characteristics. The term king had been introduced earlier (Laherrere and Sornette 1998) to emphasize the importance of those events, which are beyond the extrapolation of the fat tail distribution of the rest of the population. This is in analogy with the sometimes special position of the fortune of kings, which appear to exist beyond the Pareto law distribution of wealth of their subjects (Sornette 2009). The concept of dragon-kings has been argued to be relevant under a broad range of conditions in a large variety of systems, including the distribution of city sizes in certain countries such as France and the United Kingdom, the distribution of acoustic emissions associated with material failure, the distribution of velocity increments in hydrodynamic turbulence, the distribution of financial drawdowns, the distribution of the energies of epileptic seizures in humans and in model animals, the distribution of the earthquake energies, and the distribution of avalanches in slowly driven systems with frozen heterogeneities. A detailed presentation of these various examples and the related bibliographies may be found in works by Sornette (2009) and Satinover and Sornette (2010).

13.1.7 LANDAU-GINZBURG MODEL OF SELF-ORGANIZED CRITICAL AVALANCHES COEXISTING WITH DRAGON-KINGS

The following model provides a quite generic set-up for the emergence of dragon-kings under wide parameter conditions, coexisting with a self-organized critical regime under different parameter conditions (Gil and Sornette 1996). This model is relevant to a large number of systems, including systems of coupled neurons. Consider an extended system, whose local state at position $\vec{r}$ and time t is characterized by the local order parameters $S(\vec{r},t)$. The order parameter S is zero in absence of activity and nonzero otherwise; its amplitude $S(\vec{r},t)$ quantifies the level of activity at $\vec{r}$ and time t.

The simple and general dynamical equation that captures the process of jumps from a zero to a nonzero activity state consists of the normal form of the subcritical pitchfork bifurcation of codimension 1:

$$\frac{\partial S}{\partial t} = \chi \left(\mu S + 2\beta S^3 - S^5 \right). \tag{13.7}$$

The parameter χ sets the characteristic time scale $1/\chi$ of the dynamics of S. The parameter β is taken to be positive, corresponding to the subcritical pitchfork bifurcation regime. In absence of stabilizing the $-S^5$ term, the nonzero-fixed points (for the relevant regime $\mu < 0$) given by $S_{\pm}^* = \pm\sqrt{-\mu / 2\beta}$ are unstable, while the fixed point $S^0 = 0$ is locally stable. These two unstable fixed points correspond to the dashed line in Figure 13.1. The term locally reflects the fact that a sufficiently large perturbation that pushes S above S_{+}^* or below S_{-}^* will be amplified leading to a diverging amplitude $|S|$ at long times. In the presence of the $-S^5$ term, two new fixed points exist, which are stable. They correspond to the upper solid line in Figure 13.1. The bifurcation diagram of these fixed points as a function of μ shown in Figure 13.1 is similar to the bifurcation diagram of the Hodgkin-Huxley model (describing how action potentials in neurons are initiated and propagated), for which the transmembrane voltage is the order parameter S and the external potassium concentration is the control parameter μ.

Now, imagine that the normal form (13.7) describes the local state $S(\vec{r},t)$ at $\vec{r}$ and time t, which may be different from point to point because the control parameter μ is actually dependent on position $\vec{r}$ and time t. We thus have as many dynamical equations of the form (13.7) as there are points $\vec{r}$ in the system. For each point $\vec{r}$, the local control parameter $\mu(\vec{r},t)$ is assumed to be an affine function of the gradient of a local concentration h:

$$\mu(r,t) = g_c - \frac{\partial h}{\partial r}. \tag{13.8}$$

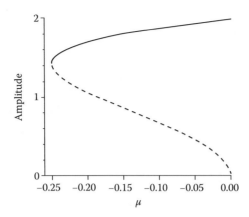

FIGURE 13.1 Bifurcation diagram in the positive domain $S > 0$ of the normal form (13.7) plotting the amplitude $|S|$ as a function of the control parameter μ.

We consider a cylindrical (or one-dimensional) geometry so that a single spatial coordinate r is sufficient (and we can drop the arrow on $\vec{r}$). Here, g_c is the critical value of the gradient at which the zero-fixed point $S^0 = 0$ becomes linearly unstable. The model in the work by Gil and Sornette (1996) assumed a slightly different technical form $\left(\mu(r,t) = g_c - \left(\frac{\partial h}{\partial r}\right)^2\right)$, which does not change the main regimes and results described below.

Because we think of $h(r,t)$ as a diffusing field, its equation of evolution is generically

$$\frac{\partial h}{\partial t} = -\frac{\partial F\left(S, \frac{\partial h}{r}\right)}{\partial r} + n(r,t). \tag{13.9}$$

This equation expresses that the rate of change of h is equal to the gradient of a flux that ensures the conservation of the concentration, up to an external fluctuation noise $n(r,t)$ acting on the system. The last ingredient of the model consists of writing that the flux is proportional to the gradient of the field:

$$F\left(S, \frac{\partial h}{r}\right) = -\alpha\, S^2\, \frac{\partial h}{\partial r}, \tag{13.10}$$

where α is another inverse time scale controlling the diffusion rate of the field within the system. The proportionality between the flux F and the gradient $-\frac{\partial h}{\partial r}$ of the field is simply Fick's law. The nonstandard ingredient stems from the fact that the coefficient of proportionality, usually defining the diffusion coefficient, is controlled by the amplitude S^2 of the order parameter. In absence of activity $S = 0$, the local flux F is here zero and the field does not change, up to noise perturbations. This corresponds to a strong feedback of the order parameter onto the control parameter, which has been shown to be one of the possible mechanisms for the emergence of self-organized criticality (Sornette 1992; Fraysse et al. 1993; Gil and Sornette 1996). Recall that standard formulations of the dynamics and bifurcation patterns of evolving systems in terms of normal forms assume the existence of control parameters that are exogenously determined. Here, the order parameter of the dynamics has an essential role in determining the value of the control parameter, which becomes itself an endogenous variable.

The analysis of the dynamics described by expressions (13.7), (13.8), (13.9), and (13.10) presented by Gil and Sornette (1996) shows that a self-organized critical (SOC) regime (Bak 1996) appears under the condition of small driving noise and when the diffusive relaxation is faster than

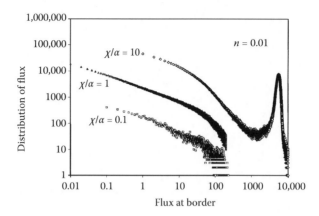

FIGURE 13.2 Distribution of flux amplitudes at the open border of the one-dimensional system obeying the dynamics described by expressions (13.7), (13.8), (13.9), and (13.10). The standard deviation n of the noise term $n(r,t)$ is equal to 0.01 (small driving noise regime). (Reproduced from Gil, L., and Sornette, D., *Phys. Rev. Lett.*, 76, 3991–3994, 1996.)

the instability growth rate: $\alpha > \chi$. The SOC dynamics can be shown to be associated with a renormalized diffusion equation at large scale with an effective negative diffusion coefficient (Gil and Sornette 1996), expressing that small scale fluctuations are the most unstable and cascade intermittently to large scale avalanches. This SOC regime is exemplified by the power law distributions of avalanche sizes shown in Figure 13.2 for $\chi/\alpha = 0.1$ and 1. More interesting for our purpose is the fact that, when $\alpha < \chi$, characteristic large scale events appear that coexist with a crowd of smaller events, themselves approximately distributed according to a power law with an exponent larger than in the SOC regime. The dragon-kings correspond to the peak on the right of Figure 13.2, associated with the run-away avalanches of size comparable to the size of the system.

This constitutes an example of what we believe to be a generic behavior found in systems made of heterogeneous coupled threshold oscillators, such as sandpile models, Burridge-Knopoff block-spring models (Schmittbuhl et al. 1993), and earthquake-fault models (Sornette et al. 1994b, 1995b; Dahmen et al. 1998): a power law regime (self-organized critical) (Figure 13.11, right lower half) is coextensive with one of synchronization (Strogatz 2004) with characteristic size events (Figure 13.11, upper left half). We will discuss this generic phase diagram in our attempt to compare the dynamics and resulting statistical regularities observed in earthquakes, financial fluctuations, and epileptic seizures.

13.1.8 BIFURCATIONS, DRAGON-KINGS, AND PREDICTABILITY

The existence of dragon-kings punctuating the dynamics of a given system suggests mechanisms of self-organization otherwise not apparent in the distribution of their smaller siblings. Therefore, this opens the potential for predictability, based on the hypothesis that these specific mechanisms that are at the origin of the dragon-kings could leave precursory fingerprints usable for forecasts.

The dynamical system (13.7, 13.8, 13.9, and 13.10) presented in the previous section shows an example in which the dragon-kings appear in a large range of parameters in the presence of small scale subcritical bifurcation dynamics that are renormalized at large scales into a change of regime, a bifurcation of behavior, more generally a transition of phase. In other words, dragon-kings are commonly associated with a phase transition. If a phase transition can be detected before it occurs, it may be understood as an abrupt increase in the probability, or risk, of an extreme event. Practical examples include ruptures in materials and the bursting of financial bubbles.

Mathematicians have proved (Thom 1972; Arnol'd 1988) that, under fairly general conditions, the local study of bifurcations of almost arbitrarily complex dynamical systems can be reduced to a few archetypes. More precisely, it is proved that there exist reduction processes, series expansions, and changes of variables of the many complex microscopic equations such that, near the fixed point (i.e., for small values of the order parameter S), the behavior is described by a small number of ordinary differential equations depending only a few control parameters, like μ in expression (13.7) for a subcritical pitchfork bifurcation. The result is nontrivial since a few effective numbers such as μ represent the values of the various relevant control variables and a single (or just a few)-order parameter(s) is(are) sufficient to analyze the bifurcation instability. The remarkable consequence is that the dynamics of the system in the vicinity of the bifurcation is reducible and thus predictable to some degree. This situation can be described as a reduction of dimensionality or of complexity that occurs in the vicinity of the bifurcation. Such reduction of complexity may occur dynamically and intermittently in large dimensional out-of-equilibrium systems, such as in hierarchically coupled Lorenz systems (Lorenz 1991) or in agent-based models of financial markets (Andersen and Sornette 2005).

As an illustration, consider expression (13.7) where β is now assumed to be negative. Since the cubic term $2\beta S^3$ is now stabilizing, the quintic term $-S^5$ can be dropped. An interesting so-called supercritical bifurcation occurs at $\mu = 0$, separating the regime for $\mu < 0$ where the zero-fixed point $S^0 = 0$ is unique and stable, from the regime $\mu > 0$, where two symmetric stable fixed points appear at $S_\pm^* = \pm\sqrt{\mu/2|\beta|}$, and the zero-fixed point $S^0 = 0$ becomes unstable. Consider the dynamics of such a system slightly perturbed by an external noise $n(t)$ with zero mean and variable σ^2, so that its dynamics reads

$$\frac{dS}{dt} = \mu S - 2|\beta|S^3 + n(t). \tag{13.11}$$

For $\mu < 0$, the average value $\langle S(t) \rangle$ vanishes, but its variance can be calculated explicitly from the solution of (13.11). Indeed, to a very good approximation, we can drop also the $2|\beta|S^3$ term since S is exhibiting only small fluctuating excursions around 0, for $\mu < 0$, leading to

$$S(t) = \int_{-\infty}^{t} e^{-|\mu|(t-\tau)}\, n(\tau)\, d\tau. \tag{13.12}$$

Its variance $\langle [S(t)]^2 \rangle$ is then given by

$$\left\langle [S(t)]^2 \right\rangle = \sigma^2 \int_{-\infty}^{t} e^{-2|\mu|(t-\tau)}\, d\tau = \frac{\sigma^2}{2|\mu|}. \tag{13.13}$$

This result (13.13) shows that the variance $\langle [S(t)]^2 \rangle$ of the fluctuations of the order parameter diverges as the critical bifurcation point is approached from below: $\mu \to 0^-$. $\langle [S(t)]^2 \rangle$ plays the role of a susceptibility, whose divergence on the approach to the critical point suggests a general predictability, for instance, obtained by monitoring the growth and correlation properties of the system fluctuations. This method has been used in particular for material failure (e.g., recording of microdamage via acoustic emissions) (Anifrani et al. 1995; Garcimartin et al. 1997; Johansen and Sornette 2000), human parturition (proposed recording of the mother-fetus maturation process via Braxton–Hicks contractions of the uterus) (Sornette et al. 1994a, 1995a), financial crashes (monitoring of bursts of price acceleration and various risk measures via options) (Sornette et al. 1996; Johansen and Sornette 1999; Johansen et al. 1999; Sornette 2003) and earthquakes (monitoring of precursory seismic, electromagnetic, and chemical activity) (Sornette and Sammis 1995; Johansen et al. 1996; Bowman et al. 1998). We believe that this phase transition approach bears great potential to predict catastrophic events, recognizing precursors in time series associated with finite-time singularities (Johansen and Sornette 2001; Sammis and Sornette 2002; Ide and Sornette 2002; Bowman et al. 1998), hierarchical power law precursors (Sornette 2002), critical slowing down (Dakos et al. 2008), and other types of precursors (Sornette 2006; Scheffer et al. 2009).

13.2 PARALLELS BETWEEN EARTHQUAKES, FINANCIAL CRASHES, AND EPILEPTIC SEIZURES

How can the concepts described in the previous section be applied to real systems, and in particular to the prediction of epileptic seizures? To put this question in a broader perspective, we present in this section an original attempt (Osorio et al. 2009, 2010) to draw parallels between seemingly drastically different systems and phenomena, based on both qualitative and quantitative evidence.

13.2.1 INTRODUCTION TO EARTHQUAKES, FINANCIAL CRASHES, AND EPILEPTIC SEIZURES

Earthquakes occur mostly in the fragile thin upper layer of the Earth, called the upper crust. A complex system of slowly moving tectonic plate boundaries delineate their most probable location, as shown in Figure 13.3. Recent syntheses of compendium of geological and seismic data (Bird 2003) suggest that the system of tectonic plates covering the earth's surface is fractal (Sornette and Pisarenko 2003), i.e., composed of a broad (power law) distribution of plates. Even more interesting is the fact that, in broad region around the tectonic plate boundaries, earthquakes are clustered on networks of faults forming rich hierarchical structures from the thousand kilometer scale to the meter scale and below (Ouillon et al. 1996), as shown in Figure 13.4. At a qualitative level (and supported quantitatively by some models (Sornette et al. 1994b; Lyakhovsky et al. 2001)), it is thought that the fault networks are self-organized by the repetitive action of earthquakes.

Financial crashes occur in organized markets trading assets, such as equities of firms, commodities such as oil or gold, and bonds (debts of firms or of countries). By their varying and heterogeneous demand and supply, investors are responsible for the observed price variations. Investors come in a very broad distribution of sizes (and therefore market impacts), from the individual private household to the largest pension and mutual funds commanding up to hundreds of billions of dollars. These investors are interacting with other investors as well as with market makers, with commercial and investment banks, and more recently with sovereign funds. This variety of sizes, needs, and goals provides a fertile ground for rich behaviors, including systemic instabilities and crippling crashes.

Epileptic seizures occur within what many exaggeratedly refer to as the most complex system of this universe, the human brain. The human brain is organized in an exceedingly rich set of

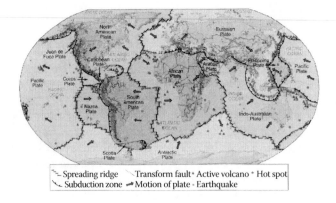

FIGURE 13.3   (See color insert.) Standard model of 12 major tectonic plates showing their relative motions (thick red arrows), and the plate boundaries that concentrate a large fraction of seismic and volcanic activity. The three types of place boundaries are indicated in the legend.

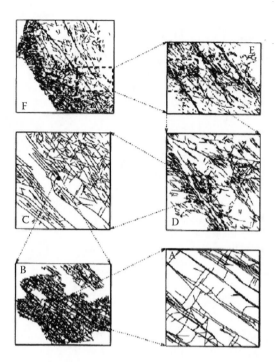

FIGURE 13.4   Example illustrating the hierarchical organization of faults from a 400-km scale (upper left panel) down to a 1-m scale (lower right panel). (Reproduced from Ouillon, G. et al., *J. Geophys. Res.*, 101(B3), 5477–5487, 1996.)

topographic and functional divisions at many scales, from the lobes and complex folded structures down to columns and to neurons (see Figure 13.5 for a partial insight into this rich organization). The networking and function of these units reflect both encoded development programs as well as the impact of learning and experience that provide feedback on the development processes.

13.2.2   Common Properties between Earthquakes, Financial Crashes, and Epileptic Seizures

Earthquakes, financial crashes, and epileptic seizures are characterized by several strikingly similar mechanisms and properties: they occur on hierarchically organized structures, with many interconnected scales; their distribution in sizes are heavy-tailed; and extreme events are typical. There is a strong entanglement between the growth and properties of the supporting structures and the

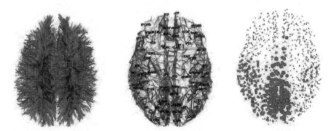

FIGURE 13.5   (See color insert.) Illustration of the complex hierarchical network structure of the brain. Left: fiber pathways of the human cerebral cortex; middle: network of connections in the human cortex, with lines between brain regions indicating the strengths of the connections; right: location of highly connected hub nodes forming the structural core. (Reproduced from Hagmann, P. et al., *PLoS Biol.*, 6(7), e159, 2008.)

(a)
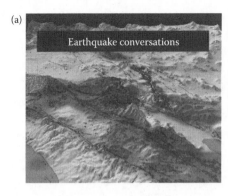

(b)
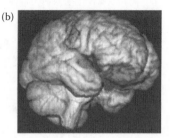

Summary of neuroanatomical structures involved in social cognition.

FIGURE 13.6 **(See color insert.)** (a) Cover picture of the *Scientific American* journal in which R. S. Stein reviews the evidence for earthquake conversations, that is, the predominant hypothesis that earthquakes trigger earthquakes. (b) Several important structures involved in social cognition and interactions—ventromedial prefrontal cortex (green), amygdala (red), right somatosensory cortex (blue), and insula (purple). (Reproduced from Hagmann, P. et al., *PLoS Biol.*, 6(7), e159, 2008.)

spatiotemporal organization of the events themselves. The supporting structures and the events interorganize, as in a chicken-and-egg problem: earthquakes occur on faults and faults grow and form networks shaped by the repetition of earthquakes; financial crashes occur on financial markets acted on by investors, whose actions and impacts result from the cumulative growth of their fortunes shaped by past financial performance that feedbacks on future performance; and young brains grow with epileptic regimes (e.g., absence seizures) and there are many feedbacks between structures and functions. This suggests that a genuine understanding of the generating processes and of the properties of earthquakes, financial crashes, and epileptic seizures can only be obtained by studying the joint organization of these events and their evolving self-organized carrying structures. The bases for this important statement are not well developed for seizures.

Past earthquakes trigger future earthquakes: it is estimated that between 50% and nearly 100% of earthquakes are triggered by past earthquakes (and not just the aftershocks). This is illustrated by the concept that earthquakes have conversations, similar to the exchanges between different areas of the brain when developing cognitive tasks (see Figure 13.6). Most of the volatility of financial markets is probably the result of endogenous amplification of past returns on future returns rather than the direct exogenous effect of external news as, for instance, exemplified by the so-called "excess volatility" effect. The concept that seizures beget seizures has a long history and new recent empirical evidence supports revisiting this hypothesis (Osorio et al. 2009, 2010).

Within a coarse-grained approach to the modeling of these systems, they can be represented as made of coupled threshold oscillators of relaxation (faults going to rupture, investors going to investment decisions, neurons going to a firing state).

There is some evidence that these three systems are characterized by the coexistence of scaling (power law distribution of event sizes) and regimes with large characteristic events (Sornette 2009).

Finally, there is a lot of interest in our modern societies in diagnosing and predicting large catastrophic events in order to alleviate the damage associated with earthquakes, the losses of financial crashes, and also to help patients recover normal lives in the presence of intermittent seizures.

Figure 13.7 summarizes the main statistical laws that have been documented in seismology (see the work by Sornette (1991) and references therein). The Gutenberg–Richter law describes the probability density function (pdf) of earthquakes of a given energy E, as being a power law with a small exponent $\beta \approx 2/3$.

The Omori law states that the rate of aftershocks following an earthquake (usually improperly referred to as a main shock) exhibits a burst immediately after the main shock and decays slowly in time afterward as the inverse of time raised to an exponent p, which is close to 1 for large earthquakes.

1 Gutenberg–Richter law: $\sim 1/E^{1+\beta}$ (with $\beta \approx 2/3$)

2 Omori law $\sim 1/t^p$ (with $p \approx 1$ for large earthquakes)

3 Productivity law $\sim E^a$ (with $a \approx 2/3$)

4 PDF of fault lengths $\sim 1/L^2$

5 Fractal/multifractal structure of fault networks $\zeta(q), f(\alpha)$

6 PDF of seismic stress output $\sim 1/s^{2+\delta}$ (with $\delta \geq 0$)

7 Distribution of interearthquake times

8 Distribution of seismic rates

FIGURE 13.7 Survey of the major statistical laws in seismicity. The line numbers provide the correspondence for the mapping between the statistical laws in the three systems shown in Figures 13.7–13.9.

The productivity law describes how the average number of triggered earthquakes depends on the energy E of the triggering earthquake: The larger an earthquake, the more earthquakes it triggers, according to a power law with an exponent a probably slightly smaller than β (Helmstetter 2003).

Because earthquakes occur on faults, and faults grow by earthquakes, it is important also to characterize the properties of fault networks. It is well-documented that the probability density distribution $P(L)$ of fault lengths in a given area is described by a power law $P(L) \simeq 1/L^f$ with exponent f not far from 2. Several studies have documented that fault networks exhibit fractal, multifractal, or better multiscale hierarchical properties (Ouillon et al. 1996).

Earthquakes result from deformations that produce complex stress fields, which are one of the important fields at the origin of the nucleation of earthquakes. The distribution of water (brine) in the crust is also thought to play a crucial role, albeit we have only indirect and incomplete information (Sornette 1999, 2000). The distribution of stress amplitudes have been documented from the focal source mechanism of earthquakes to be close to a Cauchy distribution, i.e., with a power law tail $\simeq 1/s^{2+\delta}$ and δ small (Kagan 1994a).

The distribution of waiting times between earthquakes in a given region is also characterized by a fat tail, approximately quantified by a power law, indicative of a broad range of interevent intervals. However, recent studies suggest that the pdf of interearthquake intervals has several regimes and may not be describable by a simple power law (Saichev and Sornette 2006c, 2007b; Sornette et al. 2008). The distribution of seismic rates (number of earthquakes per unit time) in fixed regions is also well described by a power law function (Saichev and Sornette 2006a).

Figure 13.8 presents the most important statistical laws that characterize the regularities found in financial time series of returns.

The distribution of financial returns (or relative price variations) is fat tailed with a tail approximately described by a power law, but the exponent is in the range of 2–4 and thus much larger than for earthquake energies (whose exponent is $\simeq 2/3$). Hence, the returns have a well-defined variance.

The relaxation of the level of activity of price fluctuations (called financial volatility) after a burst is also found to decay approximately as a power law (Lillo and Mantegna 2003; Sornette et al. 2003), similar to the Omori law of earthquake aftershocks.

The analog of the productivity law of earthquakes is the price impact function, which relates the price change to the volume of stocks of a given transaction: The larger the demand for a stock, the more the price is pushed up. Prices fluctuate because investors place orders. The size of the orders play an important role, as just said. The sizes of orders are obviously related to the sizes of the investors: a large mutual fund managing 100 billion dollars has much more impact on the market than an individual managing a modest portfolio. The size distribution of individuals' wealth, of firm sizes, of mutual fund portfolios, or university endowments are all found to be power laws. Characterizing the distributions by the pdf, it is found to be of the form $\simeq 1/W^f$ with the exponent f close to 2, which

1 Heavy-tail pdf of returns

2 Omori law and long-memory of volatility

3 Price impact function   Price $\sim V^{\beta}$ with $\beta = 0.2–0.6$

4 Pareto distribution of wealth

5 Multifractal structure of returns

6 PDF of news' sizes?

7 Distribution of intershock times

8 Distribution of limit order sizes

9 "Leverage" effect

FIGURE 13.8   Survey of the major statistical laws in financial markets. The line numbers provide the correspondence for the mapping between the statistical laws in the three systems shown in Figures 13.7–13.9.

corresponds to Zipf's law (Saichev et al. 2009). For such exponents, the mean is either not defined or converges poorly in typical statistical estimations.

The size distribution of portfolios plays a role similar to the fault distribution in earthquakes: Portfolio sizes impact the size and nature of orders that move prices; reciprocally, the cumulative effect of price moves controls the performance of investment portfolios and, thus, whether the size increases or decreases. Again we encounter again the chicken-and-egg structure. There is also ample evidence that financial time series of returns are characterized by multifractal scaling. The analogy of stress would be news, but we are only starting to understand what is a news size and how to quantify it via the response function of social networks (Roehner et al. 2004; Deschatres and Sornette 2005; Sornette 2005; Crane and Sornette 2008).

The distribution of time intervals between high levels of volatility has a similar structure as the interearthquake time distribution. The distribution of limit-order sizes, analogous to the distribution of seismicity rates, is also a power law (Gopikrishnan et al. 2000; Gabaix et al. 2006). However, the so-called leverage effect in which past losses (large negative returns) tend to increase future volatility (and not reciprocally) (Perello and Masoliver 2003), does not seem to have any counterpart in seismicity.

Figure 13.9 reviews a number of statistical laws that have been found to characterize focal seizures in humans and generalized seizures in animals (Osorio et al. 2009, 2010).

The analogy with earthquakes is particularly striking for the Gutenberg–Richter distribution of event sizes, the Omori and inverse Omori laws, and the distribution of interevent intervals, as shown in Figure 13.10.

1 Heavy-tail pdf of seizure energies

2 Omori law and long-memory of seizure activity

3 Productivity law?

4 Hierarchy of brain structure

5 Multifractal structure?

6 PDF of seizures source sizes or amplitudes?

7 Distribution of interseizure times

8 Distribution of seizure rates?

FIGURE 13.9   Survey of the major statistical laws known in epileptology. The line numbers provide the correspondence for the mapping between the statistical laws in the three systems shown in Figures 13.7–13.9.

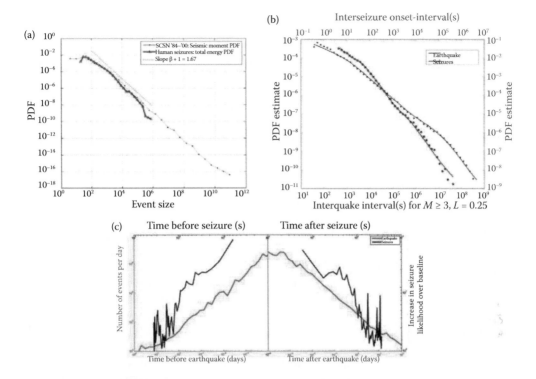

FIGURE 13.10 (See color insert.) (a) Empirical probability density function (pdf) estimates of seismic moments (SCSN catalog 1984–2000) (blue curve) and of seizure energies of 60 human subjects (red curve) originating from different epileptogenic regions. (b) Empirical probability density function estimates of the interevent times between earthquakes (blue circles; blue lower left scales) and seizures in humans (red circles; red upper right scales). (c) Superimposed epoch analysis of seizures (red line) and earthquakes (blue line) to test for the existence in seizures of aftershocks (Omori-like behavior) and foreshocks (inverse Omori-like behavior). (Reproduced from Osorio, I. et al., *Phys. Rev. E.* 82, 021919, 2010.)

While these events occur in drastically different systems, they may nevertheless be described at a coarse-grained level by similar models of coupled heterogeneous threshold oscillators of relaxation. This provides inspiration to investigate the possible existence of other statistical laws, such as productivity. One can suspect that the triggering ability of a seizure to promote future seizures (Osorio et al. 2009, 2010) might depend on its duration, amplitude, and/or energy. This remains to be tested.

We have already mentioned the hierarchical structure of the brain as the structure supporting the spatiotemporal organization of brain activities and the seizures. But it is not known whether it can be characterized with multifractal properties. The analog to stress sources in earthquakes would be the electric current field within the brain or GABA or other chemical compound concentration fields. It remains to be quantified whether these fields present interesting statistical properties, that may be used to better constrain modeling and perhaps be used for diagnostic purposes.

The distribution of seizure rates has not yet been quantified in a systematic manner. And there is no obvious analogy with the leverage effect in finance. It is possible that similar asymmetric dynamical effects reveal at a collective level the asymmetry between excitatory and inhibitory processes in the brain.

13.2.3 Rationales for the Analogy between Earthquakes, Financial Crashes, and Epileptic Seizures

The previous section has documented (and also extended conjectures on) a number of quantitative and qualitative correspondences between earthquakes, financial crashes, and seizures. It is perhaps a priori counterintuitive to compare earthquakes, financial fluctuations and seizures (the events), or fault networks, financial markets and neuron assemblies (the events' supporting structures), due to the systems' large differences in scales and in their constituent matter. However, the proposed correspondence may be motivated and at least partially explained on the grounds that these phenomena occur in systems composed of interacting heterogeneous threshold oscillators.

Consider first that the textbook model of an earthquake represents a single fault slowly loaded by cm/year tectonic deformations until a threshold is reached at which meter-scale displacements occur in seconds. This textbook model ignores the recent realization that earthquakes do not occur in isolation but are part of a complex multiscale organization in which earthquakes occur continuously at all spatiotemporal scales according to a highly intermittent, frequent energy release process (Kagan 1994b; Ouillon and Sornette 2005). Indeed, the earth's crust is in continuous jerky motion almost everywhere but due to the relative scarcity of recording devices, only the few sufficiently large earthquakes are detected, appearing as isolated events. In this sense, the dynamics of earthquakes is similar to the persistent barrages of subthreshold oscillations and of action potentials in neurons, which sometimes coalesce into seizures.

Market investors continuously place limit and market orders, with buyers tending to push prices up and sellers tending to push prices down. Early on, Takayasu et al. (1992) noticed that trading strategies lead to dynamics belonging to the larger class of threshold dynamics with mean-reversal behavior, akin to the outcome of coupled threshold oscillators of relaxation. Traders and investors enter and exit financial markets at many different time scales, from milliseconds for the most modern electronic automatic platforms to years for investors with long horizons. The evolution of their impact is on the order of years, which is the time scale for growth or decay of fortunes. Furthermore, market rules and regulations, such as the Glass–Steagall act of 1932–33 or the Sarbanes–Oxley act of 2002, appear as reactions to extreme market regimes such as financial crashes (the 1929 crash and ensuing depression for the former and the accounting scandals revealed by the collapse of market capitalization of new technology firms in 2000), illustrating another process for the evolution of supporting structures coevolving with the dynamics of events.

The separation of time scales in epileptogenic neuronal assemblies is similar (milliseconds to years) to financial markets (milliseconds to years), but smaller than in-fault networks (fraction of seconds to millennia), but the organization of coupled threshold oscillators is not very sensitive to the magnitude of the separation of time scales—as long as there is one—a property that characterizes relaxational processes.

The term relaxational process is applied here to phenomena with a disproportionately long (hours to years) charging/loading process vis-a-vis the very short (seconds to minutes) discharge of the accumulated seismic energy, money/assets, or neuronal membrane potentials. For instance, in the case of earthquakes, the slow motion of tectonic plates at typical velocities of a few cm/year accumulates stress in the cores of locked faults over hundreds to thousands of years, which are suddenly relaxed by the meter-size slips occurring in seconds to minutes that define large earthquakes. Thus, one fault taken in isolation is genuinely a single relaxation threshold oscillator, alternating long phases of loading and short slip relaxations (the earthquakes). While less well studied than earthquakes, the long (hours to years) interval between seizures and their short duration (rarely over 2 min) interpreted in light of the fact that the brain is composed of relaxational threshold oscillators (neurons) supports the notion that seizures are also relaxational phenomena. The relaxation nature of investment dynamics can be seen as the result of the competition between different strategies available to each investor and their collective output. This is particularly evident for first-entry games (Rapoport et al. 1998) and minority games (Challet et al. 2005; Coolen 2005), in which agents with bounded rationality are continuously oscillating between different strategies, creating collectively large market price fluctuations and crashes (Harras and Sornette 2010).

13.2.4 GENERIC PHASE DIAGRAM OF COUPLED THRESHOLD OSCILLATORS OF RELAXATION

It is well known in statistical physics and in dynamical systems theory that ensembles of interacting heterogeneous threshold oscillators of relaxation generically exhibit self-organized behavior with

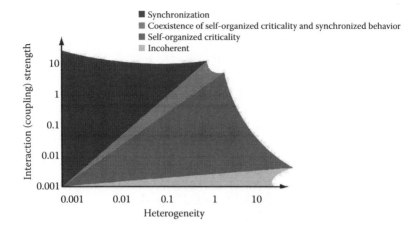

FIGURE 13.11 (See color insert.) Qualitative phase diagram illustrating the effect of changes in coupling strength (y-axis) and heterogeneity (x-axis) on the behavior of systems (such as the brain) composed of interacting threshold oscillators. Marked increases in excitatory coupling drives the system toward the synchronized regime. Slight increases in coupling drive the system toward the power law regime, indicative of self-organized criticality. (Reproduced from Osorio, I. et al., *Phys. Rev. E.* 82, 021919, 2010b.)

nonGaussian statistics (Rundle et al. 1995; Zhao and Chen 2002; Kapiris et al. 2005). The cumulative evidence presented in Figures 13.7, 13.8, and 13.9 provides a strong case for the dynamical analogy between earthquakes, financial fluctuations, and seizures (i.e., the existence of an underlying universal organization principle captured by the sand pile avalanche paradigm and the concept of self-organized criticality) (Bak 1996).

A generic qualitative phase diagram (Figure 13.11) depicts the different main regimes found in systems made of heterogeneous coupled threshold oscillators, such as sandpile models, Burridge-Knopoff block-spring models (Schmittbuhl et al. 1993), and earthquake-fault models (Sornette et al. 1994b, 1995b; Dahmen et al. 1998). A power law regime (probably self-organized critical) (Figure 13.11, right lower half) is coextensive with one of synchronization (Strogatz 2004) with characteristic size events (Figure 13.11, upper left half). This phase diagram embodies the principal qualitative modes that result from the competition between strong coupling leading to coherence and weak coupling manifesting as incoherence. Coupling (or interaction strength) is dependent upon features such as the distance between constituent elements (synaptic gap size in the case of neurons), their type (excitatory or inhibitory) and the extent of contact (number of synapses and their density), and the existence and size of delays in the transmission of signals as well as their density and flux rate between constituent elements. Heterogeneity, the other determinant of the systems organization, may be present in the natural frequencies of the oscillators (when taken in isolation), in the distribution of the coupling strengths between pairs of oscillators, in the composition and structure of the substrate (earth or neuropil), and in their topology among others. As shown in Figure 13.11 for very weak coupling and large heterogeneity, the dynamics are incoherent. Increasing the coupling strength (and/or decreasing the heterogeneity) leads to the emergence of intermediate coherence and of a power law regime (self-organized criticality (SOC)); further increases in coupling strength (and/or decreases in heterogeneity) force the system toward strong coherence/synchronization and periodic behavior.

The specific boundaries between these different regimes depend on the system under study and on the details of the constituting elements and their interactions. In addition, these boundaries may have multiple bifurcations across a hierarchy of partially synchronized regimes within the system. The diagram of Figure 13.11 is adapted from the study of a system of coupled fault elements subjected to a slow tectonic loading with quenched disorder in the rupture thresholds (Sornette et al. 1994b). In the SOC regime, the extreme events are not different from smaller ones, making the former practically unpredictable or at best very weakly predictable (Taleb 2007). In contrast, in the synchronized regime, the extreme events are different, i.e., they are outliers or dragon-kings (Laherrere and Sornette 1998; Sornette 2009), occurring as a result of some additional amplifying mechanism. These outliers, unlike those in the SOC regime, have a degree of predictability (Sornette 2002). The model described in Section 13.1.7 constitutes a nice example of a system that can be described by the phase diagram shown in Figure 13.11. The correspondence works as follows:

- The heterogeneity dimension corresponds to the amplitude of the noise n defined in Equation (13.9).
- The coupling strength is quantified by the ratio $\frac{\chi}{\alpha}$ of the instability growth rate divided by the diffusive relaxation rate.

A large ratio $\frac{\chi}{\alpha}$ corresponds to a large coupling strength because the local order parameters $S(r,t)$ then exhibits large fluctuations because the full amplitude between the two branches of the subcritical pitchfork bifurcation can be sampled, and these large fluctuations have a proportionally strong influence on neighboring locations. This rationalized the results that dragon-kings emerge only for relatively small noise levels n and large ratios $\frac{\chi}{\alpha}$.

13.3 SELF-EXCITED HAWKES PROCESS FOR EPILEPTIC SEIZURES

The analogy with earthquakes and financial fluctuations, and in particular the evidence that seizures may trigger other seizures (see inverse and direct Omori laws shown in Figure 13.10), motivates the presentation of a class of stochastic processes that is specifically formulated to account for triggering, also called self-excitation. But, before diving into the formalism, some caveats and definitions must be presented.

13.3.1 PARTICLES VERSUS WAVES

While clinical seizures are rather unambiguous objects on the basis of the often dramatic observable symptoms, continuous voltage recordings directly from the brains of human subjects via electrocorticogram (ECoG) show the existence of many so-called subclinical seizures (Osorio et al. 1998, 2002). Subclinical seizures have ECoG patterns that are undistinguishable from their clinical siblings (except perhaps for their durations and extent of spread), but they are without obvious manifestations. In textbooks, ictal events are classified as having clinical manifestations and interictal events as lacking visible behavioral changes in the usual sense of clinical manifestations. But the definition and characterization of relevant patterns that can be used for diagnosing incoming clinical seizures remains elusive. For instance, the textbook concepts of ictal and interictal events turn out to be quite fuzzy, given the demonstration that their durations do not form two well-separated classes (long durations for ictal events and short durations for interictal events), but a continuum better characterized by scale-free power law statistics (Osorio et al. 2009, 2010). In addition, interictal events also comprise what have been coined as spikes and bursts of spikes. Figure 13.12 shows a trace of a continuous recording from the brain of a rat that received injections of a convulsant. One can observe at the top a pattern that qualifies as an epileptic seizure, followed by bursts of spikes or by single spikes. In some cases, interictal spikes appear to arise from a different location (in a given brain) from the site of seizure initiation, which has led some to propose that they are quite distinct mechanistically. As better recording methods are available and longer time series of ECoG provide data for more sophisticated statistical analyses, understanding the relationship between spikes, bursts, and seizures is highly relevant, given the growing realization of the fuzziness of past classifications based mainly on clinical criteria. Moreover, one should not exclude the possibility that spikes and bursts could be relevant diagnostics or even precursory signals announcing clinical seizures, since they also constitute signatures of the excitatory activity of the brain.

In the following, we formulate a model of self-excitation that remains as general as possible, keeping open the possibility for interactions between spikes, bursts, and seizures. Similarly to

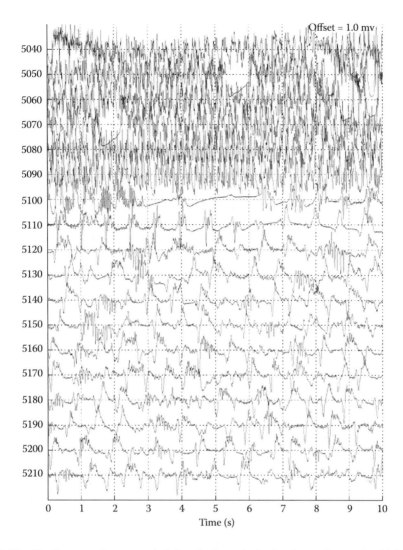

FIGURE 13.12 Continuous voltage recorded directly from the brain of a rat that received injections of an epileptogenetic substance.

earthquakes or financial crashes, the key idea is to view the activity of a brain, as measured by ECoG, as a wave-like background on which particle-like structures appear and possibly interact. We refer to this view as the particle approach, as opposed to the wave approach. The wave approach consists of viewing the ECoG as a continuous signal and then applying various signal analysis techniques, for instance, those derived from the theory of dynamical systems and chaos (Mormann et al. 2007). In contrast, the particle approach assumes that coherent structures or patterns exist on the noisy wave background, allowing us to treat them as individuals or events. The formalism is then constructed to describe the relationships between these discrete events.

13.3.2 BRIEF CLASSIFICATION OF POINT PROCESSES

When using the particle point of view, the relevant mathematical language is that of so-called point processes (also known as shot noise in physics or jump processes in finance). Daley and Vere-Jones (2003) provide a rigorous development of the theory of point processes.

The (conditional) rate $\lambda(t \mid H_t)$ (also called conditional intensity) of a point process is defined by

$$\lambda(t \mid H_t) = \lim_{\Delta \to 0} \frac{1}{\Delta} \Pr(\text{event occurs in } [t, t+\Delta] \mid H_t), \tag{13.14}$$

where $\Pr(X)$ means probability that event X occurs. The symbol H_t represents the entire history up to time t, which includes all previous events. This definition is straightforward to generalize for space-dependent intensities $\lambda(t, \vec{i} \mid H_t)$ and to include marks such as amplitudes or magnitudes. The Poisson process is the special case such that $\lambda(t \mid H_t)$ is constant. Recall that the simplest point process is the memoryless Poisson process, in which events occur continuously and independently of one another. The term conditional refers to the fact that, in general, the rate $\lambda(t \mid H_t)$ is not constant, but may depend on past history, i.e., on the specific realization of past events.

Let us define $f(t \mid H_t)$ as the pdf of the time until the next event (possibly dependent on more than just the last event, when the process is nonMarkovian) and $F(t \mid H_t)$ as the corresponding survivor function (or complementary cumulative distribution function). The relationship between the conditional intensity and these two quantities is given by

$$\lambda(t \mid H_t) = \frac{f(t \mid H_t)}{F(t \mid H_t)}. \tag{13.15}$$

The probability of an event in the time interval $[t_c, t_c + s]$ is given by

$$P(t_c; s \mid H_{t_c}) = 1 - \exp\left(-\int_{t_c}^{t_c+s} \lambda(u \mid H_{t_c}) \, du \right). \tag{13.16}$$

When an event occurs, the history H_t changes and therefore $\lambda(t \mid H_t)$ may change abruptly, as it is defined as a piecewise continuous function between events. Another useful relationship relates the pdf $f(t_i \mid H_t)$ for the i-th event to the conditional density, by differentiation of Equation (13.16):

$$f(t_i \mid H_{t_i}) = \lambda(t_i \mid H_{t_i}) \exp\left(-\int_{t_{i-1}}^{t_i} \lambda(u \mid H_u) \, du \right) I(t_i - t_{i-1}), \tag{13.17}$$

where $I(\cdot)$ is the Heaviside step function.

13.3.2.1 Renewal Processes

Renewal processes constitute the simplest class of point processes. A renewal process is a particular class of temporal point process in which the probability of occurrence of the next event depends only on the time since the last event. The pdf of the waiting time from the $(i-1)$-th event to the i-th one is defined by

$$f(t_i \mid H_{t_i}) = f(t_i \mid t_{i-1}) = f(t_i - t_{i-1}) \, I(t_i - t_{i-1}), \tag{13.18}$$

expressing the fact that the history H_{t_i} is reduced to the knowledge of t_{i-1}. Renewal processes are to point processes what Markov processes are to general stochastic processes.

One can equivalently define renewal processes by the fact that their conditional intensity at time $t > t_i$, where i is the index of the last event, depends only on the occurrence time of the last event t_i:

$$\lambda(t_i \mid H_{t_i}) = \lambda(t - t_i). \tag{13.19}$$

The Poisson process is the simplest renewal process, and corresponds to the specification

$$f_{\text{Poisson}}(\tau;\lambda) = \lambda \exp(-\lambda\tau), \tag{13.20}$$

where we note $\tau = t - t_i$, the running time since the last event i. This exponential form of the waiting time distribution of the Poisson process is uniquely associated with its memoryless property, which can be quantified by asking for instance "What is the average remaining waiting time at present time t, given that a time $t - t_i$ has passed since the last event?" It turns out that the Poisson process is the only process such that the average remaining time remains equal to $1/\lambda$ at all times t. The conditional distribution of the remaining time, conditional on the waiting time $t - t_i$ since the last event, also remains unchanged in the form (13.20). Sornette and Knopoff have offered a systematic classification of renewal processes into three classes (Sornette and Knopoff 1997).

When the pdf $f(\tau)$ has a tail decaying faster than exponential, the longer the time since the last event, the shorter the average remaining waiting time till the next event.

When the pdf $f(\tau)$ has an exponential tail, the average remaining waiting time till the next event is independent of the time that has elapsed since the last event (this is the Poisson process).

When the pdf $f(\tau)$ has a tail decaying slower than exponential, the longer the time since the last event, the longer the average remaining waiting time till the next event.

These statements can be made more precise by calculating explicitly the full shape of the conditional distribution of waiting time till the next event, conditional on the waiting time $t - t_i$ since the last event that occurred at t_i. See Sornette and Knopoff (1997) for detailed information. Osorio et al. have used this statistic as one of the diagnostics to characterize the sequence of epileptic seizures and to compare them with earthquake sequences (Osorio et al. 2009, 2010).

13.3.2.2 Clustering Models

These models rely on the general observation that earthquakes, financial volatility, and seizures occur in bursts, that is, according to patterns exhibiting much more clustering or grouping than predicted by renewal processes. Clustering models are usually constructed from two processes: a cluster center process, which is often a renewal process, and a cluster member process. In simple terms, the center events are main events or sources, from which the member events derive. The cluster member process consists of events that are triggered by the cluster centers via a triggering function $h(t - t_i, \xi)$, which usually depends only on the time $t - t_i$ since the occurrence time t_i of the cluster center, and on a stochastic amplitude ξ drawn from a distribution usually chosen to be invariant in time. In other words, cluster centers are parents and the cluster members are their corresponding offspring; a given parent triggers only his cohort of offsprings and has no influence on the offsprings of other parents (center sources).

An example is given by the simple aftershock model, which considers that there are main shocks distinctly different from their aftershocks. The former are the cause of the latter, which cluster strongly after them. Such aftershock model is the standard textbook model for main earthquakes and their aftershocks. It consists in writing the conditional intensity as

$$\lambda(t \mid H_t^c, \Theta) = \lambda_c + \sum_{i_c \mid t_{i_c} < t} h(t - t_i, \xi), \tag{13.21}$$

where H_t^c is the history up to time t that needs only include information about the cluster centers $\{t_{i_c}, \xi_{i_c}\}_{1 \le i_c \le N}$, as cluster members do not trigger their own events and do not influence the future. In the specification (13.21), we have assumed for simplicity that the cluster center process is a Poisson process with constant rate λ_c. The triggering process from centers to members is described by the set of parameters Θ characterizing the kernel $h(t - t_i, \xi)$ quantifying the ability of centers to trigger their offsprings.

13.3.2.3 Self-Excited Models

These models were first introduced by Hawkes (1971a, b) and Hawkes and Oakes (1974). They generalize the cluster models by allowing each event, including cluster members, i.e., aftershocks, to trigger their own events according to some memory kernel $h(t - t_i, \xi)$.

$$\lambda(t \mid H_t, \Theta) = \lambda_c + \sum_{i|t_i < t} h(t - t_i, \xi), \tag{13.22}$$

where the history $H_t = \{t_i\}_{1 \le i \le N}$ now includes all events and the sum in expression (13.22) runs over all triggered events. The term λ_c means that there are still external background sources occurring according to a Poisson process with constant intensity λ_c but all other events can be both triggered by previous events and can themselves trigger their offspring. This gives rise to the existence of many generations of events.

13.3.2.4 Marked Self-Excited Point Processes

This class is a multidimensional extension of the former self-excited process. The generalization consists in associating with each event some marks (possible multiple traits), drawn from some distribution $p(m)$, usually a chosen invariant as a function of time:

$$\lambda(t, m \mid H_t, \Theta) = p(m) \left(\lambda_c + \sum_{i|t_i < t} h(t - t_i, \xi, m_i) \right), \tag{13.23}$$

where the mark m_i of a given previous event now controls the shape and properties of the triggering kernel describing the future offsprings of that event i. The history now consists in the set of occurrence times of each triggered event and their marks: $H_t = \{t_i, m_i\}_{1 \le i \le N}$. The first factor $p(m)$ in the right-hand side of expression (13.23) writes that the marks of triggered events are drawn from the distribution $p(m)$, independently of their generation and waiting times. This is a simplifying specification that can be relaxed. The inclusion of a spatial kernel to describe how distance impacts triggering efficiency is straightforward.

A particularly well-studied specification of this class of marked self-excited point processes is the so-called epidemic-type aftershock sequence (ETAS) model (Kagan and Knopoff 1981; Ogata 1988):

$$\lambda(t, m \mid H_t, \Theta) = p(m) \left(\lambda_c + \sum_{i|t_i < t} \frac{k e^{a(m_i - m_0)}}{(t - t_i + c)^{1 + \theta}} \right), \tag{13.24}$$

where $p(m)$ is given by the Gutenberg–Richter law with exponent β (discussed in Section 13.2.2). The memory kernel is chosen as the power law (called the Omori law) with exponent $1 + \theta$. The lower magnitude cut-off m_0 is such that events with marks smaller than m_0 do not generate offspring. This is necessary to make the theory convergent and well-defined, otherwise the crowd of small events may actually dominate the generating process. The time constant c ensures normalization and finiteness of the triggering rate immediately following any event. Each event (of magnitude m) triggers other events with a rate $\sim e^{am}$, which defines the so-called fertility or productivity law. The set of parameters is $\Theta = \{\beta, \lambda_c, k, a, m_0, c, \theta\}$. Figure 13.13 shows a typical realization of a sequence of events generated with the ETAS model.

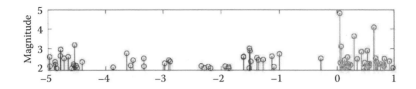

FIGURE 13.13 **(See color insert.)** Typical realization of a sequence of events using the ETAS model. The horizontal axis shows time and the vertical axis shows the magnitude m (or mark) of each event.

An observed aftershock sequence in the ETAS model is the sum of a cascade of events in which each event can trigger more events. The triggering process may be caused by various mechanisms that either compete with each other or combine. The ETAS model is parsimonious as it lumps all the complications of physical and biological properties as well as geometric structural geometry in a few key parameters quantifying the Omori law, the Gutenberg–Richter law, and the productivity law. This is particularly important as seismic and seizure data are relatively sparse and have limited precision accuracy. Further, the characterization of the properties of these dynamical processes is bound to be full of misleading paths if solid theoretical and analytical guidelines do not constrain the research on empirical data.

This class of marked self-excited point processes is now considered as the benchmark that best describes the statistical properties of spatiotemporal earthquake catalogs. In particular, the textbook classification of foreshocks, mainshocks, and aftershocks is now considered obsolete by many seismologists due to the cumulative evidence that any earthquake may trigger other earthquakes through a variety of physical mechanisms but that this does not allow one to put a tag on them (Felzer et al. 2002). The textbook classification of foreshocks, mainshocks, and aftershocks is essentially a human-made construction that is open to revision as a function of the development of the sequences of earthquake magnitudes. For instance, if an aftershock happens to have a larger magnitude that the earthquake that was qualified previously as a mainshock, it is reclassified as the new mainshock of the unfolding sequence, and the previous mainshock becomes one of its foreshocks. The fact that many small earthquakes occur after large mainshocks, and are thus classified as aftershocks, is simply due to the fact that large earthquakes trigger many earthquakes and most earthquakes are small. Thus, it is improbable (but not impossible) that a large earthquake is followed in close succession by still larger earthquakes.

Rather than keeping the textbook classification that foreshocks are precursors of mainshocks and that mainshocks trigger aftershocks, the self-excited class of models starts from the hypothesis that a parsimonious description of seismicity does not require the division between foreshocks, mainshocks, and aftershocks that are indistinguishable from the point of view of their physical processes (Felzer et al. 2002, 2004). It is, however, sometimes convenient to use the time-honored foreshocks-mainshocks-aftershocks terminology, as long as it is understood that the model refers only to events which may trigger other earthquakes. But the story is not written as, recently, some evidence of a difference between spontaneous and triggered earthquakes was obtained (Zhuang et al. 2008).

We propose that a similar approach may be a useful starting point in epileptology. Single epileptiform discharges (spikes), bursts of spikes, and seizures may not be, as often claimed, distinct phenomena but simply reflect the heterogeneous manifestations of processes governed by the same mechanisms or laws, while having different self-triggering capacity or degree of fertility. This seemingly radical shift in conceptualization may provide a deeper and more fruitful insight into the dynamics of ictiogenesis.

The ETAS model and other related models developed on similar principles are popular with statisticians interested in the characterizations of complex spatiotemporal patterns (particularly with applications to seismicity) (Kagan 1991; Musmeci and Vere-Jones 1992; Rathbun 1993; Ogata 1998; Console and Murru 2001; Zhuang et al. 2002; Console et al. 2003; Ogata et al. 2003; Ogata 2004; Zhuang et al. 2004) using maximum likelihood methods for parameter estimations and residual analysis (Ogata 2004; Ogata et al. 2003) for the detection of deviations from normal seismicity. We believe that these statistical techniques could be usefully applied to seizure time series. A detailed understanding of the observable properties of the marked self-excited point processes has been developed in the past decade that we briefly summarize in the next section.

13.3.3 MAIN PROPERTIES OF THE ETAS MODEL

We stress that the advantage of the ETAS model is to offer a very parsimonious description of the complex spatiotemporal organization of systems characterized by self-excitation of bursty events, without the need to invoke ingredients other than the well-documented stylized facts reported in the previous section: distribution of event sizes, Omori law, and productivity law. An important insight is that the Omori law may come in different forms, which can be derived from the same model only via the change of a crucial parameter, the branching ratio n. The parameter n is defined as the mean number of events of first generation triggered per event. Using the notation of expression (13.24), the branching ratio is given by

$$n = K \frac{\beta}{\beta - a},$$
(13.25)

where

$$K := \frac{k}{\theta c^{\theta}}.$$

The variability of the apparent Omori's exponent p is then obtained as a result of the relative importance of cascades of aftershocks, of aftershocks of aftershocks, and so on, possibly over many generations (Marsan and Lengline 2008). The branching ratio n can vary with time and from location to location. In the context of epileptic seizures, it can be used as a diagnostic of the susceptibility of the brain to trigger epileptic seizures.

While the results summarized in the next section pertain to earthquakes, the method used to obtain them can be applied to seizure time series as well as financial fluctuations. For an early attempt at the latter, see the work by Chavez-Demoulin et al. (2005).

13.3.3.1 Subcritical, Critical, and Supercritical Regimes

Precise analytical results and numerical simulations show the existence of three time-dependent regimes, depending on the branching ratio n and on the sign of θ. This classification is valid for the range of parameters $a < \beta$. When the productivity exponent a is larger than the exponent β of the Gutenberg–Richter law, an explosive regime occurs leading to stochastic finite time singularities (Sornette and Helmstetter 2002)—a regime that we do not consider further, but that is relevant to describe the accelerated damage processes leading to global systemic failures in possibly many different types of systems (Sornette 2002).

For $n < 1$ (subcritical regime), the rate of events triggered by a given shock decays according to an effective Omori power law, characterized by a crossover from an Omori exponent $p = 1 - \theta$ for $t < t^*$ to a larger exponent $p = 1 + \theta$ for $t > t^*$ (Sornette and Sornette 1999; Helmstetter and Sornette 2002b), where t^* is a characteristic time $t^* \sim c/(1 - n)^{1/\theta}$, which is controlled by the distance from n to 1.

For $n > 1$ and $\theta > 0$ (super-critical regime), one finds a transition from an Omori decay law with exponent $p = 1 - \theta$ at early times after the mainshock to an explosive exponential increase of the seismicity rate (Sornette and Sornette 1999; Helmstetter and Sornette 2002b; Saichev and Sornette 2009). In the case $\theta < 0$, there is a transition from an Omori law with exponent $1 - |\theta|$, similar to the local law, to an exponential increase at large times, with a crossover time τ different from the characteristic time t^* found in the case $\theta > 0$.

These results may open the road for the discovery of new types of seizure precursors. These could include (i) variable p-values, in particular the suggestion that a small p-value may be a precursor of a large event, (ii) relative seizure quiescence in some spatial domain preceding the occurrence of large seizures, and (iii) an exponential increase in seizure activity in some other spatial domains preceding large events.

13.3.3.2 Importance of Small Events for Triggering Other Events of Any Size

In the context of earthquakes for which the productivity exponent a is estimated smaller than, but close to, the Gutenberg–Richter exponent β, small events have been found to provide a dominating contribution to the overall activity, as their number more than compensates their relatively smaller individual impact (Helmstetter 2003; Helmstetter et al. 2005). This is due to the structure of the model in which all events can trigger other events. This realization came as a big surprise to experts, who have been accustomed to the concept that only large and great earthquakes needed to be considered since they overwhelmingly dominate the overall release of energy in the earth's crust—not so for the triggering ability, as is now understood. Could there be a similar situation for epileptic seizures, for whom the myriad single spikes, bursts of spikes, and subclinical seizures play an important role in the triggering of clinical seizures?

13.3.3.3 Effects of Undetected Seismicity: Constraints on the Size of
the Smallest Triggering Event from the ETAS Model

The mechanism of event triggering together with simple assumptions of self-similarity, as captured in the simple ETAS specification, make obligatory the existence of a minimum magnitude m_0 below which events do not or only weakly trigger other events (Sornette and Werner 2005b). It is possible to estimate an order of magnitude of m_0 by noting that the magnitude m_d of completeness of empirical catalogs has no reason to be the same as m_0, and by using diverse empirical data based on maximum likelihood inversions of observed aftershock sequences of real catalogs with the ETAS model. The obtained constraint $m_0 \simeq -1 \pm 2$ is loose and reflects the many uncertainties in the model calibrations and model errors.

13.3.3.4 Apparent Earthquake Sources and Clustering Biased by Undetected Seismicity

In models of triggered seismicity, the detection threshold m_d is commonly equated to the magnitude m_0 of the smallest triggering earthquake. This unjustified assumption neglects the possibility that shocks below the detection threshold may trigger observable events. Distinguishing between the detection threshold m_d and the minimum triggering earthquake $m_0 \leq m_d$, and considering the branching structure of one complete cascade of triggered events, an apparent branching ratio n_a and an apparent background source S_a can be determined from the exact calculation of the sequence of observed triggered events with marks above the detection threshold m_d (Sornette and Werner 2005a; Saichev and Sornette 2006b). The presence of smaller undetected events that are capable of triggering larger events is the cause for the renormalization. One could imagine that triggering between seizures could be similarly renormalized when not taking into account structures such as spikes and bursts of spikes if the latter have some triggering effects on seizures.

13.3.3.5 Cascades of Triggered Events

By comparison between synthetic catalogs generated with the ETAS model and real seismicity, it is now understood that a surprisingly large fraction of earthquakes in real seismicity are probably triggered by previous events. Recent conservative lower bounds suggest that at least 60% and perhaps up to 99% of earthquakes are triggered by previous earthquakes (Helmstetter and Sornette 2003b; Sornette and Werner 2005a,b; Marsan and Lengline 2008). This fraction is nothing but the so-called average branching ratio n or mean number of triggered event per earthquake, averaged over all magnitudes defined by expression (13.25) (Helmstetter and Sornette 2003b). In addition, within the picture that earthquakes can trigger events that themselves trigger new events and so on, according to the same basic physics, then, most triggered events within a sequence should be triggered indirectly through cascades (Helmstetter and Sornette 2003b). Therefore, previous observations that a significant fraction of earthquakes are triggered earthquakes imply that most aftershocks are indirectly triggered by the mainshocks. In the class of ETAS models, this has the implication that the observed Omori law is obtained from a renormalization of the direct Omori law (describing the direct interactions between triggering and triggered earthquakes) to the global law with different exponent p (Sornette and Sornette 1999; Helmstetter and Sornette 2002b). The cascades of secondary triggering provide a mechanism for slow aftershock subdiffusion (Helmstetter and Sornette 2002a; Helmstetter et al. 2003a) and slow foreshock migration (Helmstetter et al. 2003b; Helmstetter and Sornette 2003a).

13.3.3.6 Other Results Available for Marked Self-Excited Point Processes

A number of other interesting mathematical and statistical results have been derived for the ETAS model, which show that the model has nonstandard properties resulting from the interplay between the triggered cascades and the two power laws characterizing the distribution of sizes and the productivity process. These results have been obtained by rigorous mathematical derivations using probability generating functions:

- Non-mean field anomalous exponents for the distribution of cluster sizes due to the interplay between cascades of generation and the power laws of productivity and of marks (magnitudes) (Saichev et al. 2005);
- Non-mean field distributions of lifetimes and total number of generations before extinctions of aftershock sequences emanating from isolated main shocks (Saichev and Sornette 2004);
- Distribution of waiting times between events in a given region characterized by an approximate power law (Saichev and Sornette 2006c; 2007b; Sornette et al. 2008); and
- Stochastic reconstruction of the genealogy of the cascades of triggered events (Zhuang et al. 2002, 2004; Marsan and Lengline 2008; Sornette and Utkin 2009).

13.3.4 Forecasts Using Self-Excited Marked Point Processes

The understanding of the importance of cascades of triggered seismicity has led to important improvements of existing methods of earthquake forecasts (Kagan and Jackson 2000), based on variations of the ETAS model, by taking into account the cascades of secondary triggering (Helmstetter and Sornette 2003c; Helmstetter et al. 2006; Werner et al. 2009).

As a quantitative theoretical check, the number r of earthquakes in finite space-time windows is often taken as the target for forecasts; for instance, within the regional earthquake likelihood models (RELM) project in Southern California (www.relm.org), a forecast is expressed as a vector of earthquake rates specified for each multidimensional bin (Schorlemmer et al. 2007), where a bin is defined by an interval of location, time, magnitude, and focal mechanism, and the resolution of a model corresponds to the bin sizes. The full theory of this observable within the ETAS model has been developed using the formalism of generating probability functions (GPF) describing the space-time organization of earthquake sequences (Saichev and Sornette 2006a, 2007a). The calibration of the theory to the empirical observations for the California catalog shows that it is essential to augment the ETAS model by taking account of the preexisting frozen heterogeneity of spontaneous earthquake sources. This seems natural in view of the complex multiscale nature of fault networks on which earthquakes nucleate. The extended theory is able to account for the empirical observation satisfactorily. In particular, the pdf $P_{data}(r)$ of the number r of earthquakes in finite space-time windows for the California catalog over fixed spatial boxes 5×5 km^2, 20×20 km^2, and 50×50 km^2, and time intervals $d = 1$, 10, 100, and 1000 days, have been determined. One finds a stable power law tail compatible with $P_{data}(r) \sim 1/r^{1+(\beta/a)}$ (Saichev and Sornette 2006a, 2007a). This result recovers previous estimates with different statistical methods and for large space and time windows (Corral 2003; Mega et al. 2003; Scaffetta and West 2004), while proposing a simple and generic explanation in terms of cascades of triggering of earthquakes. This example and others (Helmstetter and Sornette 2004) show the power of the simple concept of triggered seismicity to account for many (most?) empirical observations.

The Working Group on Regional Earthquake Likelihood Models (RELM) has invited long-term (5-year) forecasts for California in a specific format to facilitate comparative testing (Field 2007a,b; Schorlemmer and Gerstenberger 2007; Schorlemmer et al. 2007, 2009). Building on RELM's success, the Collaboratory for the Study of Earthquake Predictability (CSEP, www.cseptesting.org) inherited and expanded RELM's mission to regionally and globally test prospective forecasts (Schorlemmer et al. 2009; Zechar et al. 2009; Werner et al. 2010b). Many of the competing models are based on concepts of earthquake triggering embodied in the marked self-excited conditional point processes described above.

New developments for point processes include the adaptation of data assimilation methods (Werner et al. 2010a). Recall that, in meteorology, engineering, and computer sciences, data assimilation is routinely employed as the optimal way to combine noisy observations with prior model information for obtaining better estimates of a state and, thus, better forecasts than can be achieved by ignoring data uncertainties. Earthquake forecasting as well as seizure prediction suffer from measurement errors and from model information that is limited, and may thus gain significantly from data assimilation. Werner et al. have presented perhaps the first fully implementable data assimilation method for forecasts generated by a point process model (Werner et al. 2010a). The method has been tested on a synthetic and pedagogical example of a renewal process observed in noise, which is relevant to the seismic gap hypothesis, models of characteristic earthquakes, and to recurrence statistics of large quakes inferred from paleoseismic data records. In order to address the nonGaussian statistics of earthquakes, it was necessary to use sequential Monte Carlo methods, which provide a set of flexible simulation-based methods for recursively estimating arbitrary posterior distributions. Extensive numerical simulations have demonstrated the feasibility and benefits of forecasting earthquakes based on data assimilation. The forecasts based on the sequential importance resampling particle filter are found to be significantly better than those of a benchmark forecast that ignores uncertainties in the observed event times. We predict that data assimilation will also become an important tool for seizure predictions in the future.

13.3.5 PRELIMINARY ATTEMPT TO GENERATE SYNTHETIC ECoG WITH THE ETAS MODEL

The following is a modest example of how to generate synthetic time series that look like electrocorticogram (ECoG) using the ETAS model defined by the conditional intensity given by expression (13.24). We imagine that the elementary events are spikes and that spikes can excite other spikes following the ETAS specification. Sequences of closely occurring spikes may then define bursts, and seizures can perhaps be observed when bursts are sufficiently clustered.

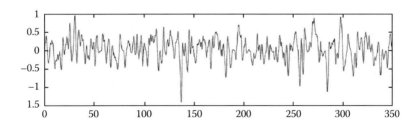

FIGURE 13.14 Synthetic electrocorticogram constructed with the ETAS model and using the spike pattern given by expression (13.26) with (13.27) for the following parameters: $\{\lambda_c = 1, \beta = 2/3, n = 0.5, a = 0.2, m_0, c = 60, \theta = 0.5, a = 0.001, d = 0.001\}$.

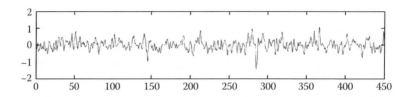

FIGURE 13.15 Synthetic electrocorticogram constructed with the ETAS model and using the spike pattern given by expression (13.26) with (13.27) for the following parameters: $\{\lambda_c = 1, \beta = 2/3, n = 0.5, a = 0.2, m_0, c = 60, \theta = 0.05, a = 0.001, d = 0.001\}$.

The synthetic ECoG are generated as follows. For a given choice of the parameter set $\Theta = \{\lambda_c, \beta, n, a, m_0, c, \theta\}$, we generate a time series of events $\{t_i, m_i\}$ in which each event i is characterized by its occurrence time t_i and its mark m_i. Note that we use n instead of k, but the two are related directly through expression (13.25).

Then, we assume that each event i is associated with a spike pattern in a virtual ECoG recording given by

$$F(t - t_i) = \text{sign}_i \cdot f_i \cdot \frac{1}{\sqrt{2\pi\tau_i^3}}(t - t_i)\exp\left(-\frac{(t - t_i)^2}{2\tau_i^2}\right),\tag{13.26}$$

where

$$f_i = f_0 \cdot 10^{\alpha m_i}, \quad \tau_i = \tau_0 \cdot 10^{dm_i}.\tag{13.27}$$

The signal $F(t - t_i)$ is thus a derivative of a Gaussian function and shows a typical dipole structure with a positive arch followed by a negative arch or vice-versa, depending on the sign term sign_i that is chosen here at random and independently for each event. The mark m_i of event i is assumed to control the amplitude f_i of the spike and its duration τ_i according to the expressions (13.27).

Figures 13.14 and 13.15 show two realizations with the same parameters, except for the memory exponent $\theta = 0.5$ in the former, and $\theta = 0.05$ in the latter. The comparison between the two figures illustrates the impact of the memory in the triggering of spikes by previous spikes. Figure 13.14 corresponds to a shorter lived memory and a more spiky regime, compared with Figure 13.15.

Figure 13.16 shows a synthetic ECoG obtained by changing the branching ratio from a low value $n = 0.1$ to a large value $n = 0.9$ abruptly in the middle of the graph. For $n = 0.1$, the cumulative effect of ten spikes is needed on average to directly trigger an additional spike. For $n = 0.9$, each spike, almost by itself, directly triggers an additional spike. This corresponds to a much more intense

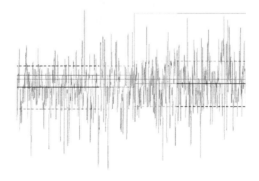

FIGURE 13.16 Synthetic electrocorticogram constructed with the ETAS model and using the spike pattern given by expression (13.26) with (13.27) for the following parameters: $\{\lambda_c = 1, \beta = 2/3, n = 0.1 \rightarrow 0.9, a = 0.2, m_0, c = 60, \theta = 0.05, a = 0.001, d = 0.001\}$.

activity, with more correlations and burstiness. Seizure-like patterns can be obtained by decreasing λ_c and increasing further the value of n toward the critical value 1.

13.4 CONCLUDING REMARKS

The inherent value of predicting seizures has led to many efforts to fulfill this aim, efforts that have been unsuccessful to date. In particular, from our point of view, an acute limitation is the absence of understanding of the spatiotemporal behavior of seizures and the absence of corresponding models. While the reasons behind this state of affairs are multiple, there has been progress, if only in recognizing the challenges (after a first wave of overoptimism) as well as the need for rigorous testing procedures and for new approaches.

The task of predicting the occurrence of recurrent aperiodic events such as seizures would benefit from development of models that recognize the value of multiscale approaches aimed at excluding features that unnecessarily increase algorithmic complexity and the danger of computational irreducibility and its associated unpredictability. This chapter's central predicates are that seizures may be statistically predictable and that application of tools from statistical physics such as renormalization group theory may facilitate identification of the scale of observation (likely to be coarse-grained) that is the most informative. Through systematic quantitative comparisons with earthquakes, conceptual groundwork (neurons and faults are treated as coupled threshold oscillators of relaxation) is laid out that allows epileptology to adopt approaches with potential usefulness from the more mature fields of seismology and finance. Among those approaches briefly presented in this chapter, self-excited marked point processes are, in our opinion, worthy of investigation.

The perspectives we provide in this chapter and the approaches we propose are intended to stimulate new research directions to increase the knowledge of epilepsy dynamics and, with it, the likelihood of predicting seizures in a manner that improves the quality of life of those afflicted.

GLOSSARY

Bifurcation: The phenomenon in which a change in a so-called order parameter causes a qualitative change (from one regime to another) in the system's dynamical behaviors. The theory of bifurcations has led to a classification of regime changes, which turn out to be reducible to a limited number of cases. The mathematical description of a bifurcation is called a normal form, which is a differential equation representing the time evolution of the order parameter, given the value(s) of the control parameter(s).

Coarse-graining: The procedure that removes uninformative (for the task at hand) degrees of freedom to obtain a description of a system at a more integrated and computationally manageable level. Coarse-graining provides a range of techniques to bridge the gap between the microscopic and macroscopic levels.

Point process: In the field of stochastic processes, one must distinguish between two broad classes: (i) Continuous or discrete time processes and (ii) point processes. An example of the former class is the so-called random walk (or Wiener process in mathematical parlance). Point processes generate events that are distinct from background activity. In other words, the value of a point process is zero, except when the event occurs. In contrast, in the first class of processes, the activity is present at all times or occurs in continuing steps. Point processes, also called shot noise in physics and jump processes in finance, are thus particularly suitable to describe and model system dynamics characterized by the occurrence of events, such as earthquakes, financial crashes, and epileptic seizures.

Renormalization group: The metatheory developed by L. Kadanoff, M. Fisher, K. Wilson, and many others, which allows the construction of macroscopic theories of critical phenomena at the macro-level, from the knowledge of constituents and interactions at the microscopic level.

Self-organized criticality: A concept introduced in 1987 by P. Bak, C. Tang, and K. Wiesenfeld, according to which many out-of-equilibrium dynamical spatiotemporal systems, which are slowly driven and which exhibit threshold-like responses, tend to self-organize to a dynamical state characterized by a broad range of avalanche sizes quantified by a power law distribution. A subclass of self-organized critical systems can be shown to be made of systems functioning at or close to a standard critical point (in the sense of phase transitions in statistical physics). It is the nonstandard type of slow driving of the order parameters that leads to the attraction of the dynamics to the usually unstable critical point.

REFERENCES

Abbott, D. Asymmetry and disorder. 2010. A decade of parrondo's paradox. *Fluct. Noise Lett.* 9:129–156.

Andersen, J. V., and D. Sornette. 2005. A mechanism for pockets of predictability in complex adaptive systems. *Europhys. Lett.* 70(5):697–703.

Anifrani, J.-C., C. Le Floc'h, D. Sornette, and B. Souillard. 1995. Universal logperiodic correction to renormalization group scaling for rupture stress prediction from acoustic emissions. *J. Phys. I France* 5:631–638.

Arnol'd, V. 1988. *Geometrical methods in the theory of ordinary differential equations*, 2nd ed. New York: Springer.

Bak, P. 1996. *How nature works: The science of self-organized criticality*. New York: Copernicus.

Bellacicco, A., G. Koch, and A. Vulpiani, eds. 1995. *Forecasting and modelling for chaotic and stochastic systems*. Collana scientifica 61. Collana scientifica (Franco Angeli editore).

Bird, P. 2003. An updated digital model of plate boundaries. *Geochem. Geophys. Geosyst.* 4(3):1–52.

Boffetta, G., A. Celani, A. Crisanti, and A. Vulpiani. 1997. Predictability in two-dimensional decaying turbulence. *Phys. Fluids* 9(3):724–734.

Boffetta, G., M. Cencini, M. Falcioni, and A. Vulpiani. 2002. Predictability: A way to characterize complexity. *Physics Rep.* 356:367–474.

Bowman, D. D., G. Ouillon, C. G. Sammis, A. Sornette, and D. Sornette. 1998. An observational test of the critical earthquake concept. *J. Geophys. Res.* 103:24359–24372.

Chaitin, G. J. 1987. *Algorithmic information theory*. Cambridge: Cambridge University Press.

Challet, D., M. Marsili, and Y.-C. Zhang. 2005. Minority games: Interacting agents in financial markets. New York: Oxford University Press USA.

Chavez-Demoulin, V., A. C. Davison, and A. C. McNeil. 2005. Estimating value-at-risk: A point process approach. *Quant. Fin.* 5(2):227–234.

Console, R., and M. Murru. 2001. A simple and testable model for earthquake clustering. *J. Geophys. Res.* 166:8699–8711.

Console, R., M. Murru, and A. M. Lombardi. 2003. Refining earthquake clustering models. *J. Geophys. Res.* 108(B10):2468.

Coolen, A. C. C. 2005. The mathematical theory of minority games: Statistical mechanics of interacting agents. New York: Oxford University Press USA.

Corral, A. 2003. Local distributions and rate fluctuations in a unified scaling law for earthquakes. *Phys. Rev. E* 6803:5102.

Crane, R., and D. Sornette. 2008. Robust dynamic classes revealed by measuring the response function of a social system. *Proc. Nat. Acad. Sci. USA* 105(41):15649–15653.

Cross, M. C., and P. C. 1993. Hohenberg. Pattern formation outside of equilibrium. *Rev. Mod. Phys.* 65:851–1112.

Dahmen, K., D. Ertas, and Y. Ben-Zion. 1998. Gutenberg-richter and characteristic earthquake behavior in simple mean-field models of heterogeneous faults. *Phys. Rev. E.* 58:1494–1501.

Dakos, V., M. Scheffer, E. H. van Nes, V. Brovkin, V. Petoukhov, and H. Held. 2008. Slowing down as an early warning signal for abrupt climate change. *Proc. Natl. Acad. Sci. USA* 105(38):14308–14312.

Daley, D. J., and D. Vere-Jones. 2003. An introduction to the theory of point processes, Vol. I. New York: Springer, USA.

Deschatres, F., and D. Sornette. 2005. The dynamics of book sales: Endogenous versus exogenous shocks in complex networks. *Phys. Rev. E* 72:016112.

Edlund, E., and M. Nilsson Jacobi. 2010. Renormalization of cellular automata and self-similarity. *J. Stat. Phys.* 139(6):972–984.

Felzer, K. R., R. E. Abercrombie, and G. Ekström. 2004. A common origin for aftershocks, foreshocks, and multiplets. *Bull. Seismol. Soc. Am.* 94:88–99.

Felzer, K. R., T. W. Becker, R. E. Abercrombie, G. Ekström, and J. R. Rice. (2002). Triggering of the 1999 mw 7.1 hector mine earthquake by aftershocks of the 1992 mw 7.3 landers earthquake. *J. Geophys. Res.* 107 (B09):doi:10.1029/2001JB000911.

Field, E. H. 2007a. Overview of the working group for the development of regional earthquake likelihood models (RELM). *Seismol. Res. Letts.* 78(1):7–16.

Field, E. H. 2007b. A summary of previous working groups on California earthquake probabilities. *Bull. Seismol. Soc. Am.* 97(4):1033–1053.

Fraysse, N., A. Sornette, and D. Sornette. 1993. Critical transitions made self-organized: Proposed experiments. *J. Phys. I France* 3:1377–1386.

Frisch, U. 1996. Turbulence: The legacy of A. N. Kolmogorov. Cambridge: Cambridge University Press.

Sugihara, G., and R. May. 1990. Nonlinear forecasting as a way of distinguishing chaos from measurement error in time series. *Nature* 344:734–741.

Gabaix, X., P. Gopikrishnan, V. Plerou, and H. E. 2006. Stanley. Institutional investors and stock market volatility. *Q. J. Econ.* 121:461–504.

Gammaitoni, L., P. Hänggi, P. Jung, and F. Marchesoni. 1998. Stochastic resonance. *Rev. Mod. Phys.* 70(1):223–287, Jan.

Garcimartin, A., A. Guarino, L. Bellon, and S. Ciliberto. 1997. Statistical properties of fracture precursors. *Phys. Rev. Lett.* 79:3202–3205.

Geller, R. J., D. D. Jackson, Y. Y. Kagan, and F. Mulargia. 1997. Earthquakes cannot be predicted. *Science* 275(5306):1616–1617.

Gil, L., and D. Sornette. 1996. Landau-Ginzburg theory of self-organized criticality. *Phys. Rev. Lett.* 76:3991–3994.

Gluzman, S., and D. Sornette. 2002. Classification of possible finite-time singularities by functional renormalization. *Phys. Rev. E* 6601:016134.

Gopikrishnan, P., X. Gabaix, H. E. Plerou, and V. Stanley. 2000. Statistical properties of share volume traded in financial markets. *Phys. Rev. E.* 62:R4493–R4496.

Görnerup, O., and M. Nilsson Jacobi. 2010. An algorithm for aggregating variables in linear dynamical systems. *Adv. Compl. Syst.* 13(2):199–215.

Hagmann, P., L. Cammoun, X. Gigandet, R. Meuli, C. J. Honey, V. J. Wedeen, and O. Sporns. 2008. Mapping the structural core of human cerebral cortex. *PLoS Biol.* 6(7):e159.

Hamilton, J. D. 1994. *Time series analysis.* Princeton: Princeton University Press.

Hänggi, P., and F. Marchesoni. 2009. Artificial brownian motors: Controlling transport on the nanoscale. *Rev. Mod. Phys.* 81(1):387–442.

Harras, G., and D. Sornette. 2010. How to grow a bubble: A model of myopic adapting agents. *J. Economic Behavior and Organization* in press. http://arXiv.org/abs/0806.2989.

Hawkes, A. G. 1971a. Point spectra of some mutually exciting point processes. *Royal Stat. Soc. Series B (Meth.)* 33(3):438–443.

Hawkes, A. G. 1971b. Spectra of some self-exciting and mutually exciting point processes. *Biometrika* 58(1):83–90.

Hawkes, A. G., and D. Oakes. 1974. A cluster process representation of a self-exciting process. *J. Appl. Prob.* 11(3):493–503.

Helmstetter, A. 2003. Is earthquake triggering driven by small earthquakes? *Phys. Res. Lett.* 91:058501.

Helmstetter, A., Y. Y. Kagan, and D. Jackson. 2005. Importance of small earthquakes for stress transfers and earthquake triggering. *J. Geophys. Res.* 110(B05S08):10.1029/2004JB003286.

Helmstetter, A., Y. Y. Kagan, and D. D. Jackson. 2006. Comparison of short-term and long-term earthquake forecast models for southern California. *Bull. Seism. Soc. Am.*, 96:90–106.

Helmstetter, A., G. Ouillon, and D. Sornette. 2003a. Are aftershocks of large Californian earthquakes diffusing? *J. Geophys. Res.*, 108(2483):10.1029/2003JB002503.

Helmstetter, A., and D. Sornette. 2002a. Diffusion of epicenters of earthquake aftershocks, Omori's law, and generalized continuous-time random walk models. *Phys. Rev. E* 6606:061104.

Helmstetter, A., and D. Sornette. 2002b. Sub-critical and supercritical regimes in epidemic models of earthquake aftershocks. *J. Geophys. Res.* 107(B10):2237.

Helmstetter, A., and D. Sornette. 2003a. Foreshocks explained by cascades of triggered seismicity. *J. Geophys. Res. (Solid Earth)* 108(B10):10.1029/2003JB002409 01a.

Helmstetter, A., and D. Sornette. 2003b. Importance of direct and indirect triggered seismicity in the ETAS model of seismicity. *Geophys. Res. Lett.*, 30(11):doi:10.1029/2003GL017670.

Helmstetter, A., and D. Sornette. 2003c. Predictability in the ETAS model of interacting triggered seismicity. *J. Geophys. Res. (Solid Earth)*, 108(2482):10.1029/2003JB002485.

Helmstetter, A., and D. Sornette. 2004. Comment on "power-law time distribution of large earthquakes," *Phys. Rev. Lett.*, 92:129801.

Helmstetter, A., D. Sornette, and J.-R. Grasso. 2003b. Mainshocks are aftershocks of conditional foreshocks: How do foreshock statistical properties emerge from aftershock laws? *J. Geophys. Res. (Solid Earth)*, 108(B10, 2046):doi:10.1029/2002JB001991.

Horsthemke, W., and R. Lefever. 1983. Noise-induced transitions: Theory and applications in physics, chemistry, and biology. Springer Series in Synergetics. New York: Springer.

Ide, K., P. Courtier, M. Ghil, and A. Lorenc. 1997. Unified notation for data assimilation: Operational, sequential and variational. *J. Meteor. Soc. Japan*, 75:181–189.

Ide, K., and D. Sornette. 2002. Oscillatory finite-time singularities in finance, population and rupture. *Physica A* 307(1–2):63–106.

Israeli, N., and N. Goldenfeld. 2004. Computational irreducibility and the predictability of complex physical systems. *Phys. Rev. Letts.* 92(7):074105.

Israeli, N., and N. Goldenfeld. 2006. Coarse-graining of cellular automata, emergence, and the predictability of complex systems. *Phys. Rev. E* 73:026203.

Johansen, A., and D. Sornette. 1999. Critical crashes. *Risk* 12:91–94.

Johansen, A., and D. Sornette. 2000. Critical ruptures. *Eur. Phys. J. B*, 18:163–181.

Johansen, A., and D. Sornette. 2001. Finite-time singularity in the dynamics of the world population and economic indices. *Physica A* 294(3–4):465–502.

Johansen, A., D. Sornette, and O. Ledoit. 1999. Predicting financial crashes using discrete scale invariance. *J. Risk* 1:5–32.

Johansen, A., D. Sornette, G. Wakita, U. Tsunogai, W. I. Newman, and H. Saleur. 1996. Discrete scaling in earthquake precursory phenomena: Evidence in the kobe earthquake, Japan. *J. Phys. I France* 6:1391–1402.

Kagan, Y. Y., and L. Knopoff. 1981. Stochastic synthesis of earthquake catalogs. *J. Geophys. Res.* 86:2853–2862.

Kagan, Y. Y. 1991. Likelihood analysis of earthquake catalogues. *Geophys. J. Int.* 106:135–148.

Kagan, Y. Y. 1994a. Distribution of incremental static stress caused by earthquakes. *Nonlinear Processes Geophys* 1:172–181.

Kagan, Y. Y. 1994b. Observational evidence for earthquakes as a nonlinear dynamic process. *Physica D* 77:160–192.

Kagan, Y. Y., and D. D. Jackson. 2000. Probabilistic forecasting of earthquakes. *Geophys. J. Int.* 143:438–453.

Kantz, H., and T. Schreiber. 2004. *Nonlinear time series analysis*, 2nd ed. Cambridge: Cambridge University Press.

Kapiris, P. G., J. Polygiannakis, X. Li, X. Yao, and K. A. Eftaxias. 2005. Similarities in precursory features in seismic shocks and epileptic seizures. *Europhys. Lett.* 69:657–663.

Kravtsov, Y. A., and J. B. Kadtke. 1996. *Predictability of complex dynamical systems*. Springer Series in Synergetics. New York: Springer-Verlag.

Laherrere, J., and D. Sornette. 1998. Stretched exponential distributions in nature and economy: "Fat tails" with characteristic scales. *Eur. Phys. J. B*, 2:525–539.

Li, M., and P. Vitanyi. 1997. *An introduction to Kolmogorov complexity and its applications*. New York: Springer-Verlag.

Lillo, F., and R. N. Mantegna. 2003. Power-law relaxation in a complex system: Omori law after a financial market crash. *Phys. Rev. E* 68:016119.

Lorenz, E. N. 1991. Dimension of weather and climate attractors. *Nature* 353:241–244.

Lyakhovsky, V., Y. Ben-Zion, and A. Agnon. 2001. Earthquake cycle, fault zones and seismicity patterns in a rheologically layered lithosphere. *J. Geophys. Res.* 106:4103–4120.

Marsan, D., and O. Lengline. 2008. Extending earthquakes' reach through cascading. *Science* 319(5866):1076–1079.

Mega, I. M., P. Allegrini, P. Grigolini, L. Latora, L. Palatella, A. Rapisarda, and S. Vinciguerra. 2003. Power-law time distribution of large earthquakes. *Phys. Rev. Lett.* 90:188501.

Mormann, F., R. G. Andrzejak, C. E. Elger, and K. Lehnertz. 2007. Seizure prediction: The long and winding road. *Brain* 130:314–333.

Musmeci, F., and D. Vere-Jones. 1992. A space-time clustering model for historical earthquakes. *Ann. Inst. Stat. Math.* 44:1–11.

Nilsson Jacobi, M., and O. Görnerup. 2009. A spectral method for aggregating variables in linear dynamical systems with application to cellular automata renormalization. *Adv. Compl. Syst.* 12(2):1–25.

Ogata, Y. 1988. Statistical models for earthquake occurrence and residual analysis for point processes. *J. Am. Stat. Assoc.* 83:9–27.

Ogata, Y. 1998. Space-time point process models for earthquake occurrences. *Ann. Inst. Stat. Mech.* 50:379–402.

Ogata, Y. 2004. Space-time model for regional seismicity and detection of crustal stress changes. *J. Geophys. Res.* 109:B03308.

Ogata, Y., K. Katsura, and M. Tanemura. 2003. Modelling heterogeneous space-time occurrences of earthquake and its residual analysis. *J. Roy. Statist. Soc. Ser. C* 52:499–509.

Orrell, D. 2007. *The future of everything: The science of prediction.* 1st Thunder's Mouth Press Edition. New York: Basic Books.

Osorio, I., M. G. Frei, J. Giftakis, T. Peters, J. Ingram, M. Turnbull, M. Herzog, M. Rise, S. Schaffner, R. Wennberg, T. Walczak, M. Risinger, and C. Ajomone-Marsan. 2002. Performance re-assessment of a real-time seizure detection algorithm on long ecog series. *Epilepsia* 43(12):1522–1535.

Osorio, I., M. G. Frei, D. Sornette, and J. Milton. 2009. Pharmaco-resistant seizures: Self-triggering capacity, scale-free properties and predictability? *Eur. J. Neurosci.* 30:1554–1558.

Osorio, I., M. G. Frei, D. Sornette, J. Milton, and Y.-C. Lai. 2010. Epileptic seizures, quakes of the brain. *Phys. Rev. E* 82(2 Pt 1):021919.

Osorio, I., M. G. Frei, and S. B. Wilkinson. 1998. Real time automated detection and quantitative analysis of seizures and short term predictions of clinical onset. *Epilepsia* 39(S16):615–627.

Ouillon, G., C. Castaing, and D. Sornette. 1996. Hierarchical scaling of faulting. *J. Geophys. Res.* 101(B3):5477–5487.

Ouillon, G., and D. Sornette. 2005. Magnitude-dependent Omori law: Theory and empirical study. *J. Geophys. Res.*, 110:B04306.

Perello, J., and J. Masoliver. 2003. Random diffusion and leverage effect in financial markets. *Phys. Rev. E.*, 67:037102.

Pikovsky, A., and J. Kurths. 1997. Coherence resonance in a noise-driven excitable system. *Phys. Rev. Lett.* 78:775–778.

Pisarenko, V. F., and D. Sornette. 2004. On statistical methods of parameter estimation for deterministically chaotic time-series. *Phys. Rev. E* 69:036122.

Pisarenko, V. F., and D. Sornette. 2008. Properties of a simple bilinear stochastic model: Estimation and predictability. *Physica D* 237(4):429–445.

Rapoport, A., D. A. Seale, I. Erev, and J. A. Sundali. 1998. Equilibrium play in large group market entry games. *Manag. Sci.* 44(1):119–141.

Rathbun, S. L. 1993. Modelling marked spatio-temporal point patterns. *Bull. Int. Stat. Inst.* 55(Book 2):379–396.

Robert, R., and C. Rosier. 2001. Long range predictability of atmospheric flows. *Nonlinear Processes Geophys.* 8:55–67.

Robinson, G., and M. Thiel. 2009. Recurrences determine the dynamics. *Chaos* 19:023104.

Roehner, B. M., D. Sornette, and J. V. Andersen. 2004. Response functions to critical shocks in social sciences: An empirical and numerical study. *Int. J. Mod. Phys. C* 15(6):809–834.

Rundle, J., W. Klein, S. Gross, and D. L. Turcotte. 1995. Boltzmann fluctuations in numerical simulations of nonequilibrium lattice threshold systems. *Phys. Rev. Lett.* 75:1658–1661.

Saichev, A., A. Helmstetter, and D. Sornette. 2005. Anomalous scaling of offspring and generation numbers in branching processes. *Pure Appl. Geophys.* 162:1113–1134.

Saichev, A., Y. Malevergne, and D. Sornette. 2009. Theory of Zipf's law and beyond. In *Lecture notes in economics and mathematical systems*, Vol. 632. New York: Springer.

Saichev, A., and D. Sornette. 2004. Anomalous power law distribution of total lifetimes of aftershock sequences. *Phys. Rev. E* 70:046123.

Saichev, A., and D. Sornette. 2006a. Power law distribution of seismic rates: Theory and data. *Eur. Phys. J. B* 49:377–401.

Saichev, A., and D. Sornette. 2006b. Renormalization of the ETAS branching model of triggered seismicity from total to observable seismicity. *Eur. Phys. J. B*, 51(3):443–459.

Saichev, A., and D. Sornette. 2006c. "Universal" distribution of inter-earthquake times explained. *Phys. Rev. E*, 97:078501.

Saichev, A., and D. Sornette. 2007a. Power law distribution of seismic rates. *Tectonophysics* 431:7–13.

Saichev, A., and D. Sornette. 2007b. Theory of earthquake recurrence times. *J. Geophys. Res.* 112:B04313.

Saichev, A., and D. Sornette. 2009. Generation-by-generation dissection of the response function in long memory epidemic processes. *Eur. Phys. J. B* 75:343–355.

Sammis, S. G., and D. Sornette. 2002. Positive feedback, memory and the predictability of earthquakes. *Proc. Natl. Acad. Sci. USA*, 99(SUPP1):2501–2508.

Sancho, J. M., and J. García-Ojalvo. 2000. Noise-induced order in extended systems: A tutorial. *Lecture Notes in Physics*, eds. J. A. Freund and T. Pöschel, 557:235–246. Berlin: Springer-Verlag.

Satinover, J., and D. Sornette. 2010. Taming manias: On the origins, inevitability, prediction and regulation of bubbles and crashes. In *Governance and control of financial systems: A resilience engineering perspective*, part of the *Resilience engineering perspectives* series, eds. Gunilla Sundström and Erik Hollnagel. Farnham, Surrey, UK: Ashgate Publishing Group.

Scafetta, N., and B. J. West. 2004. Multiscaling comparative analysis of time series and a discussion on "earthquake conversations" in California. *Phys. Rev. Lett.* 92:138501.

Scheffer, M., J. Bascompte, W. A. Brock, V. Brovkin, S. R. Carpenter, V. Dakos, H. Held, E. H. van Nes, M. Rietkerk, and G. Sugihara. 2009. Early-warning signals for critical transitions. *Nature* 461:53–59.

Schmittbuhl, J., J.-P. Vilotte, and S. Roux. 1993. Propagative macrodislocation modes in an earthquake fault model. *Europhys. Lett.* 21:375–380.

Schorlemmer, D., and M. Gerstenberger. 2007. RELM testing center. *Seismol. Res. Lett.* 78(1):30.

Schorlemmer, D., M. Gerstenberger, S.Wiemer, D. D. Jackson, and D. A. Rhoades. 2007. Earthquake likelihood model testing. *Seismol. Res. Lett.* 78(1):17–29.

Schorlemmer, D., J. D. Zechar, M. J. Werner, E. Field, D. D. Jackson, T. H. Jordan, and the RELM Working Group. 2009. First results of the regional earthquake likelihood models experiment. *Pure Appl. Geophys.* 167(8):859–876.

Smith, L. 2007. *Chaos: A very short introduction*. New York: Oxford University Press, USA.

Solomonoff, R. J. 2009. Algorithmic probability: Theory and applications, information theory and statistical learning, eds. F. Emmert-Streib and M. Dehmer. New York: Springer.

Sornette, A., J. Dubois, J. L. Cheminée, and D. Sornette. 1991. Are sequences of volcanic eruptions deterministically chaotic? *J. Geophys. Res.* 96(B7):11931–11945.

Sornette, A., and D. Sornette. 1999. Renormalization of earthquake aftershocks. *Geophys. Res. Lett.*, 26:1981–1984.

Sornette, D. 1991. Self-organized criticality in plate tectonics. In *Proceedings of NATO ASI Spontaneous formation of space-time structures and criticality-Geilo, Norway April 2–12, 1991*, eds. T. Riste and D. Sherrington, 57–106, Dordrecht, Boston: Kluwer Academic Press.

Sornette, D. 1992. How to make critical phase transitions self-organized: A dynamical system feedback mechanism for self-organized criticality. *J. Phys. I France* 2:2065–2073.

Sornette, D. 1999. Earthquakes: From chemical alteration to mechanical rupture. *Phys. Rep.* 313(5):238–292.

Sornette, D. 2000. Mechanochemistry: An hypothesis for shallow earthquakes. In *Earthquake thermodynamics and phase transformations in the Earth's interior*, Vol. 76 of International Geophysics Series, eds. R. Teisseyre and E. Majewski, 329–366. Cambridge: Cambridge University Press.

Sornette, D. 2002. Predictability of catastrophic events: Material rupture, earthquakes, turbulence, financial crashes and human birth. *P. Natl. Acad. Sci. USA* 99(SUPP1):2522–2529.

Sornette, D. 2003. *Why stock markets crash (Critical events in complex financial systems)*. Princeton, NJ: Princeton University Press.

Sornette, D. 2005. Endogenous versus exogenous origins of crises. In *Extreme events in nature and society*, eds. Albeverio, S., Jentsch, V., and Kantz, H., 329–366. Berlin: Springer.

Sornette, D. 2006. *Critical phenomena in natural sciences*, 2nd ed. Springer Series in Synergetics. Heidelberg: Springer.

Sornette, D. 2009. Dragon-kings, black swans and the prediction of crises. *Int. J. Terraspace Sci. Eng.* 2(1):1–18.

Sornette, D., D. Carbone, F. Ferré, C. Vauge, and E. Papiernik. 1995a. Modèle mathématique de la parturition humaine: implications pour le diagnostic prenatal. *Médecine/Science* 11:1150–1153.

Sornette, D., F. Ferré, and E. Papiernik. 1994a. Mathematical model of human gestation and parturition: implications for early diagnostic of prematurity and post-maturity. *Int. J. Bifurcation and Chaos* 4:693–699.

Sornette, D., and A. Helmstetter. 2002. Occurrence of finite-time-singularity in epidemic models of rupture, earthquakes and starquakes. *Phys. Rev. Lett.* 89(15):158501.

Sornette, D., A. Johansen, and J.-P. Bouchaud. 1996. Stock market crashes, precursors and replicas. *J. Phys. I France* 6:167–175.

Sornette, D., and L. Knopoff. 1997. The paradox of the expected time until the next earthquake. *Bull. Seism. Soc. Am.* 87(4):789–798.

Sornette, D., Y. Malevergne, and J.-F. Muzy. 2003. What causes crashes? *Risk* 16(2):67–71.

Sornette, D., P. Miltenberger, and C. Vanneste. 1994b. Statistical physics of fault patterns self-organized by repeated earthquakes. *Pure Appl. Geophys.* 142(3/4):491–527.

Sornette, D., P. Miltenberger, and C. Vanneste. 1995b. Statistical physics of fault patterns self-organized by repeated earthquakes: synchronization versus self-organized criticality. In *Recent progresses in statistical mechanics and quantum field theory*, 329–366 and Proceedings of the Conference Statistical Mechanics and Quantum Field Theory, USC, Los Angeles, May 16–21, 1994, eds. P. Bouwknegt, P. Fendley, J. Minahan, D. Nemeschansky, K. Pilch, H. Saleur and N. Warner, 313–332. Singapore: World Scientific, Singapore.

Sornette, D., and V. F. Pisarenko. 2003. Fractal plate tectonics. *Geophys. Res. Lett.*, 30(3):1105.

Sornette, D., and C. G. Sammis. 1995. Complex critical exponents from renormalization group theory of earthquakes: Implications for earthquake predictions. *J. Phys. I France* 5:607–619.

Sornette, D., and S. Utkin. 2009. Limits of declustering methods for disentangling exogenous from endogenous events in time series with foreshocks, main shocks and aftershocks. *Phys. Rev. E* 79:061110.

Sornette, D., S. Utkin, and A. Saichev. 2008. Solution of the nonlinear theory and tests of earthquake recurrence times. *Phys. Rev. E* 77:066109.

Sornette, D., and M. J. Werner. 2005a. Apparent clustering and apparent background earthquakes biased by undetected seismicity. *J. Geophys. Res.* 110:B09303.

Sornette, D., and M. J. Werner. 2005b. Constraints on the size of the smallest triggering earthquake from the ETAS model, Baath's law, and observed aftershock sequences. *J. Geophys. Res.* 110:B08304.

Stein, R. S. 2003. Earthquake conversations. *Sci. Am.* 288:72–79.

Strogatz, S. H. 2004. *Sync: How order emerges from chaos in the universe. Nature, and daily life.* New York: Hyperion.

Sugihara, G., and R. M. May. 1990. Nonlinear forecasting as a way of distinguishing chaos from measurement error in time series. *Nature* 344:734–741.

Takayasu, H., H. Miura, T. Hirabayashi, and K. Hamada. 1992. Statistical properties of deterministic threshold elements: the case of market price. *Physica A* 184(1–2):127–134.

Taleb, N. N. 2007. *The black swan: The impact of the highly improbable.* Random House, New York.

Thom, R. 1972. *Stabilité structurelle et morphogénèse.* New York: Benjamin.

Werner, M. J., A. Helmstetter, D. D. Jackson, and Y. Y. Kagan. 2009. High resolution long- and short-term earthquake forecasts for California. *Bull. Seismol. Soc. Am.* November 20.

Werner, M. J., K. Ide, and D. Sornette. 2010a. Earthquake forecasting based on data assimilation: Sequential Monte Carlo methods for renewal processes. Nonlinear Processes in Geophysics, http://arxiv.org/abs/0908.1516

Werner, M. J., J.D. Zechar, W. Marzocchi, S. Wiemer, and the CSEP-Italy Working Group. 2010b. Retrospective evaluation of the five-year and ten-year CSEP-Italy earthquake forecasts. 2010b. Submitted 3 March 2010 to the CSEP-Italy special issue of the Annals of Geophysics, arXiv:1003.1092.

Wilson, K. G. 1979. Problems in physics with many scales of length. *Sci. Am.* 241(2):140–157.

Wolfram, S. 1983. Statistical mechanics of cellular automata. *Rev. Mod. Phys.* 55:601–644.

Wolfram, S. 2002. *A new kind of science.* Champaign, IL: Wolfram Media.

Zechar, J. D., D. Schorlemmer, M. Liukis, J. Yu, F. Euchner, P. J. Maechling, and T. H. Jordan. 2009. The collaboratory for the study of earthquake predictability: perspective on computational earthquake science. *Concurr. Comp.-Pract. E.* 22(12):1836–1847.

Zhao, X., and T. Chen. 2002. Type of self-organized criticality model based on neural networks. *Phys. Rev. E* 65:026114.

Zhuang, J., J. A. Christophersen, M. K. Savage, D. Vere-Jones, Y. Ogata, and D. D. Jackson. 2008. Differences between spontaneous and triggered earthquakes: Their influences on foreshock probabilities. *J. Geophys. Res.* 113:B11302.

Zhuang, J., Y. Ogata, and D. Vere-Jones. 2002. Stochastic declustering of space-time earthquake occurrences. *J. Am. Stat. Assoc.* 97:369–380.

Zhuang, J., Y. Ogata, and D. Vere-Jones. 2004. Analyzing earthquake clustering features by using stochastic reconstruction. *J. Geophys. Res.* 109(B5):B05301.

Section IV

The State of Seizure Prediction:
Seizure Prediction and Detection

14 Impact of Biases in the False-Positive Rate on Null Hypothesis Testing

Ralph G. Andrzejak, Daniel Chicharro, and Florian Mormann

CONTENTS

14.1 NULL HYPOTHESIS TESTING FOR SEIZURE PREDICTION ALGORITHMS

To test whether a prediction algorithm has any true predictive power, it is necessary to compare its performance with that expected under various well-defined null hypotheses. These null hypotheses must include the assumption that the prediction algorithm lacks any true predictive power. In the context of studies investigating the predictability of epileptic seizures, two approaches have been introduced for this purpose: analytical performance estimates (Schelter et al. 2006a; Snyder et al. 2008; Winterhalder et al. 2003; Wong et al. 2007) and seizure predictor surrogates based on constrained randomizations of the original seizure predictor (Andrzejak et al. 2003; Kreuz et al. 2004).

We recently extended the Monte Carlo framework of seizure predictor surrogates by introducing the concept of alarm times surrogates (Andrzejak et al. 2009). This new type of seizure predictor surrogates is as straightforward to implement as seizure time surrogates (Andrzejak et al. 2003), while offering the same flexibility as measure profile surrogates (Kreuz et al. 2004). Therefore, alarm times surrogates combine the advantages of existing seizure predictor surrogates and remedy some of their shortcomings. The resulting flexibility in formulating distinct, testable null hypotheses is the key advantage of alarm times surrogates over analytical performance estimates. In our previous work (Andrzejak et al. 2009), we illustrated this by using artificial seizure time sequences and artificial original seizure predictors specifically constructed to be consistent or inconsistent with various null hypotheses. This allowed us to determine the frequency of null hypothesis rejections obtained from the analytical performance estimate and different types of alarm times surrogates under controlled conditions. The flexibility of alarm times surrogates was illustrated by including postseizure nonstationarities in the null hypothesis, which is not possible for analytical performance estimates. Furthermore, we showed that for Poisson predictors, which fulfill the null hypothesis of analytical performance estimates, the frequency of false-positive null hypothesis rejections substantially exceeds the significance level for long mean interalarm intervals, revealing an intrinsic bias of these estimates. Finally, we showed that alarm times surrogates and analytical performance estimates have a similar high statistical power to recognize seizure predictors with true predictive power. We extend this previous work here by illustrating the robustness of alarm times surrogates against a biased estimation of the specificity of the original predictor. Furthermore, we show that for analytical performance estimates, in contrast, such biases can lead to false-positive null hypothesis rejections.

14.2 UNBIASED AND BIASED FALSE-POSITIVE RATES

Suppose that a multichannel EEG recorded from an epilepsy patient has a total duration of d hours and includes Q seizures. Let us further assume that some measure was applied in a moving window to produce a temporal profile with values corresponding to each respective window. The temporal profile of this measure is evaluated for signatures that are assumed to be predictive for impending seizures, resulting in a univariate temporal sequence of alarm times. To analyze whether these alarms indeed carry any true predictive power, one first has to quantify their sensitivity and specificity.

For this purpose, we define the periods directly preceding the seizures as prediction horizons. The length of these prediction horizons in hours is denoted by h and is the same for all seizures. If, however, the interval between two particular consecutive seizures is shorter than h, then the length of the prediction horizon assigned to the second seizure is reduced to the interval enclosed by the two seizures. Cases where at least one alarm is raised within the prediction horizon of a given seizure are counted as true positive predictions. The sensitivity is given by the ratio of the total number of true positive predictions (P^+) to the total number of seizures:

$$ S = \frac{P^+}{Q} . \tag{14.1} $$

Seizures for which no alarm is raised during the prediction horizon count as false-negative predictions. All alarms outside of any prediction horizon are counted as false-positive predictions. In our simulations, subsequent false-positive predictions are not grouped. They are counted separately, regardless of how close they are in time. In particular, it makes no difference whether the time between two subsequent false-positive alarms is shorter or longer than the length of the prediction horizon h. In both cases, both false-positive alarms are counted. In real applications, sequences of alarms promptly following each other can for example be obtained when a measure profile frequently crosses a threshold and this crossing is used to trigger the alarms. Several ways exist to derive a specificity value from false-positive alarms. To relate to other studies (Aschenbrenner-Scheibe et al. 2003; Chaovalitwongse et al. 2005; Iasemidis et al. 2003, 2005; Maiwald et al. 2004; Mormann et al. 2003; Sackellares et al. 2006; Schad et al. 2008; Schelter et al. 2006a, 2006b, 2007; Winterhalder et al. 2003, 2006) and to connect to our previous work (Andrzejak et al. 2009), we use the false-positive rate here. This should be determined by dividing the total number of false-positive alarms (P^-) by the total time in hours during which such false alarms can occur. Evidently, this time is given by the total duration of the recording minus the total time covered by the prediction horizons, since by construction no false alarms can occur during the prediction horizons:

$$ F = \frac{P^-}{d - Qh} . \tag{14.2} $$

Nonetheless, in a number of studies (Chaovalitwongse et al. 2005; Iasemidis et al. 2003, 2005), the false-positive rate was calculated by dividing the total number of false positives by the total duration of the recording, including the time covered by the prediction horizons:

$$F^* = \frac{P^-}{d}.$$ (14.3)

Obviously, Equation 14.3 must result in an underestimation of the actual false-positive rate.

Sensitivity and specificity can be combined into a single measure of performance. As definition of the performance, we use (Chaovalitwongse et al. 2005):

$$P(S,F) = 1 - \sqrt{(1-S)^2 + \frac{F^2}{F_0^2}}$$ (14.4)

for the unbiased false-positive rate, or

$$P^*(S,F^*) = 1 - \sqrt{(1-S)^2 + \frac{F^{*2}}{F_0^2}}$$ (14.5)

for the biased false-positive rate. Here we use $F_0 = 1h^{-1}$ in order to turn the second summand in the square-root into a dimensionless quantity and, since $1h^{-1}$ is commonly used as a unit for false-positive rates (Aschenbrenner-Scheibe et al. 2003; Chaovalitwongse et al. 2005; Iasemidis et al. 2003, 2005; Maiwald et al. 2004; Mormann et al. 2003; Sackellares et al. 2006; Schad et al. 2008; Schelter et al. 2006a, 2006b, 2007; Winterhalder et al. 2003, 2006). Note that F_0 also determines the relative weight assigned to the sensitivity and the specificity in the performance estimate. For $S = 1$ and $F = 0h^{-1}$, we get $P = 1$, while lower values are obtained for deviations from this ideal predictor. However, the performance is not normalized since, for poor predictors, negative values can be obtained.

We now illustrate the impact of the bias in the false-positive rate on alarm times surrogates and analytical performance estimates. As in our previous work (Andrzejak et al. 2009), we generate artificial seizure time sequences and artificial measure profiles. This allows us to generate large ensembles of data from which we can estimate the frequency of null hypothesis rejections under well-defined conditions. While the entire analysis presented here is based on artificial data, we will use terms such as patients, EEG recording, and seizure times to make the analogy clear.

We generate artificial seizure time sequences to represent the seizure times included in a continuous EEG recording from an individual patient. In total, we generate artificial data for an arbitrary but high number of $K = 100{,}000$ patients. For each patient, we draw $Q = 15$ random interseizure intervals, the lengths of which are uniformly distributed between 2 and 14 h. The concatenation of these intervals results in a random sequence of 15 seizure times. Accordingly, across patients, these recordings have an average duration of 5 days = 120 h. We use a homogeneous Poisson process to generate artificial alarm times. Such a Poisson process has an exponential interalarm interval distribution, and its only parameter is the mean alarm rate, F^P, which we specify in units of h^{-1}. The parameters of the artificial seizure time sequences are taken from our previous work (Andrzejak et al. 2009). The choice of a uniform interseizure interval distribution is arbitrary. We expect qualitatively similar results for other artificial distributions or real interseizure interval distributions from epilepsy patients.

An analytical approach to test null hypotheses for epileptic seizure predictors was proposed by Winterhalder et al. (2003), extended by Schelter et al. (2006a) and Wong et al. (2007), and applied by Aschenbrenner-Scheibe et al. (2003), Maiwald et al. (2004), Schad et al. (2008), Schelter et al. (2006b, 2007), and Winterhalder et al. (2006). Winterhalder et al. (2003) derived an analytical sensitivity estimate for the null hypothesis* H_0^V that the alarms arise from a homogenous Poisson process. This corresponds to assuming alarms arise from an uncorrelated random process with a time-independent mean alarm rate and an exponential interalarm interval distribution. The analytical sensitivity estimate corresponds to the maximal sensitivity expected under this H_0^V for a given false-positive rate of the original seizure predictor. Any higher value of the sensitivity would be expected under H_0^V only with the probability specified by the significance level of the test. Here we use $\alpha = 5\%$ as significance level. For any performance measure defined as a function of the sensitivity and specificity, the analytical sensitivity estimate of Winterhalder et al. (2003) can readily be transformed into an analytical performance estimate (see Equations 5–8 in our previous work). Here, we here use $P(S, F)$ (Equation 14.4) as a performance measure based on the unbiased false-positive rate F (Equation 14.2). Beyond that, we also use $P^*(S, F^*)$ (Equation 14.5), which is based on the biased false-positive rate F^* (Equation 14.3). Note that the false-positive rate enters into the analytical performance estimate in two ways. It is used to determine the probability that the predictor raises at least one alarm in a prediction horizon, and it is used to determine the performance (see Equation 5 in our previous work and Equation 14.5 here).

In our previous work, we devised a group of algorithms to generate alarm times surrogates that allow one to test a variety of different null hypotheses. These included surrogates to test the null hypothesis H_0^{III} of a stationary, unspecific, naive predictor, the state of which is not influenced by the event of an actual seizure. No assumptions are made about the distribution of the interalarm intervals generated by this predictor. Note that the null hypothesis H_0^V is a special case of H_0^{III}, in which the interalarm interval distribution is exponential. In consequence, a test-based H_0^{III} alarm time surrogate should not be rejected if H_0^V is valid. Here, we use H_0^{III} alarm time surrogates.

14.3 IMPACT OF THE BIAS IN THE FALSE-POSITIVE RATE

It is evident a priori that the bias in the false-positive rate must lead to problems in the application of the analytical performance estimate. Using the analytical performance estimate derived from the biased instead of the correct false-positive rate will result in an underestimation of the original predictor performance under H_0^V and, therefore, to false-positive rejections of this null hypothesis when compared to the actual higher predictor performance. This becomes particularly relevant for long prediction horizons. To illustrate this point, we compare results for $d = 2$h and $d = 3$h. Recall that the interseizure intervals used here are uniformly distributed between 2h and 14h. Accordingly, some of the interseizure intervals are shorter than the prediction horizon of 3h. (Note that this is also the case in the works by Chaovalitwongse et al. (2005) and Iasemidis et al. (2003), where a prediction horizon of 3h was used as well). This leads to a further problem with the application of the analytical performance estimate. For its derivation, the length of the prediction horizon is assumed to be constant for all seizures. However, this assumption is violated when some interseizure intervals are shorter than the prediction horizon. This bias will in turn contribute to an overestimation of the expected performance. These considerations show that the analytical performance estimate in this case cannot be reliable due to biases and flaws in the evaluation procedure. Nonetheless, it is instructive to consider this case to illustrate the use of the analytical performance estimate and alarm times surrogates under such conditions, since one might not be aware of such biases when defining the evaluation procedure.

Results for a Poisson predictor with $F^P = 0.25$h^{-1} and 0.36h^{-1} are shown in Figure 14.1 for $d = 2$h. When extracted patient-wise and averaged across the entire patient ensemble, the unbiased false-positive rate F matches the actual values of the predictor F^P in both cases. Indeed, the asymptotic value of the average F for $K \to \infty$ must be F^P. In contrast, for the biased false-positive rates, we obtain $F^* = 0.16$h^{-1} and $F^* = 0.23$h^{-1} instead of $F^P = 0.25$h^{-1} and $F^P = 0.36$h^{-1}, respectively. This illustrates

---

* We here use the same notation for the different null hypothesis as in Andrzejak et al. 2009.

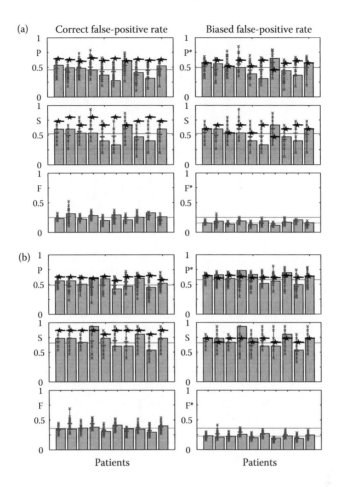

FIGURE 14.1 Results obtained for 10 exemplary patients. (a) Left panels: Results for a Poisson predictor with $F^P = 0.25\text{h}^{-1}$ using a prediction horizon of $d = 2\text{h}$ and for the unbiased false-positive rate. Top: performance P; Middle: sensitivity S; Bottom: false-positive rate F. Bars represent results obtained for individual patients from the original predictor. Grey horizontal lines depict the corresponding mean value taken across the entire group of $K = 100,000$ patients. Small grey crosses represent values obtained for 19 H_0^{III} surrogates, the corresponding mean value for individual patients is indicated by small horizontal grey bars. Black bars with asterisks indicate analytical sensitivity and performance estimates for individual patients. Right panels: Same as left panels, but for the biased false-positive rate F^* and resulting biased performance P^*. In the plot for F^*, the upper horizontal line corresponds to actual mean alarm rate of the Poisson predictor. The lower horizontal line corresponds to the mean value of F^* across all patients. (b) Same as (a), but for a Poisson predictor with $F^P = 0.36\text{h}^{-1}$.

how substantial the bias of the false-positive rate can become. While the sensitivity values are unaffected by the definition of the false-positive rate, the performance values are higher when calculated from the biased false-positive rate. This holds in particular for higher F^P (see Figure 14.1a versus 14.1b). In consequence, for a number of individual patients, the performance P is smaller than the analytical performance estimate. However, for these same patients, the performance P^* is higher than the analytical performance estimate when the biased false-positive rate is used. Due to this spurious performance increase, H_0^{V} is rejected for these patients.

For the results for the entire patient ensemble we restrict ourselves to $F^P = 0.25\text{h}^{-1}$ but use for the prediction horizon apart from $h = 2\text{h}$ also $h = 3\text{h}$ (Figure 14.2). Based on F^* and P^* and using $h = 3\text{h}$, the frequency of H_0^{V} rejections is substantially above the applied significance level of 5%, thereby clearly rejecting this null hypothesis on the patient ensemble level. However, in the setting

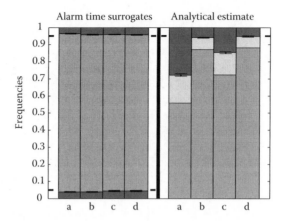

FIGURE 14.2 Results obtained for the entire patient group for H_0^{III} alarm times surrogates as well as for the H_0^V analytical performance estimate. For the alarm times surrogates, the fractions represent the percentages of the 100,000 patients for which the performance of the original predictor was lower than the minimal value for all 19 surrogates (lowest fraction, dark grey), within the surrogate distribution (middle fraction, mid grey), and higher than the maximal value for all 19 surrogates (upper fraction, dark grey). For the analytical performance estimate, the fractions correspond to the percentage for which the performance of the original predictor was lower (lower fraction, mid grey), equal (middle fraction, light grey), and higher (upper fraction, dark grey) than the analytical performance estimate. Note that for both the alarm times surrogates and the analytical performance estimate, the dark grey fractions correspond to rejections of the respective null hypotheses. To derive confidence intervals all values were determined from 10 nonoverlapping subdivisions of 10,000 patients each. All fractions correspond to the mean values across these subdivisions; error bars depict the corresponding ranges; dashed horizontal lines indicate the 5% significance levels. All results are obtained from a Poisson predictor with $F^P = 0.25h^{-1}$. (a) Biased false-positive rate F^* with prediction horizon of $h = 3h$; (b) unbiased false-positive rate F with $h = 3h$; (c) biased false-positive rate F^* with prediction horizon of $h = 2h$; (d) unbiased false-positive rate F with $h = 3h$.

used here, this null hypothesis is correct and it is falsely rejected due to the bias in the false-positive rate. This effect is reduced for a shorter prediction horizon of $h = 2h$ because the overall magnitude of the bias is determined by the total time covered by all prediction horizons. In general, only if the unbiased definitions F and P are used is the rejection frequency of the analytical performance estimate close to its significance level, and H_0^V is correctly accepted.

Both on the level of individual patients and for the patient ensemble, results for H_0^{III} surrogates match the original predictor within the expected level of accuracy (Figures 14.1 and 14.2). The crucial difference to results for the analytical estimate is that this correct null hypothesis acceptance is not affected by the bias in the false-positive rate. The biased false-positive rate is used for both the original predictor and the predictor surrogates. Hence, the bias affects the original and the surrogate predictor in the same way and it does not impair the proper hypothesis testing. In contrast to analytical performance estimates, alarm times surrogates are robust against potential biases in the sensitivity, specificity, and performance.

Alarm times surrogates offer a straightforward way to test well-defined null hypotheses about seizure predictors derived from the EEG of epilepsy patients. At the same time, alarm times surrogates are in no way restricted to this particular application. An interesting application of alarm times surrogates would be to validate prospective seizure predictions given by the patients themselves (Haut et al. 2007; Schulze-Bonhage et al. 2006). Furthermore, analogies exist between the problem of seizure prediction and the prediction of earthquakes or other extreme events. In particular, these analogies exist with regard to the assessment and validation of prediction algorithms (Console 2001; Kagan 1997; Stark 1997; Zechar and Jordan 2008). Future efforts in these fields should therefore be aimed at a transfer of knowledge between different prediction disciplines. Apart from probing a transfer of the fundamental ideas underlying the different algorithms used for prediction (Eftaxias et al. 2006), this interdisciplinary work could further promote and improve the null hypothesis testing concepts used in these fields.

REFERENCES

Andrzejak, R. G., F. Mormann, T. Kreuz, C. Rieke, A. Kraskov, C. E. Elger, and K. Lehnertz. 2003. Testing the null hypothesis of the nonexistence of a preseizure state. *Phys. Rev. E* 67:010901.

Andrzejak, R. G., D. Chicharro, C. E. Elger, and F. Mormann 2009. Seizure prediction: Any better than chance? *Clin. Neurophysiol.* 120:1465–1478.

Aschenbrenner-Scheibe, R., T. Maiwald, M. Winterhalder, H. U. Voss, J. Timmer, and A. Schulze-Bonhage. 2003. How well can epileptic seizures be predicted? An evaluation of a nonlinear method. *Brain* 126:2616–2626.

Chaovalitwongse, W. A., L. D. Iasemidis, P. A. Pardalos, P. R. Carney, D. S. Shiau, and J. C. Sackellares. 2005. Performance of a seizure warning algorithm based on the dynamics of intracranial EEG. *Epilepsy Res.* 64:93–113.

Console, R. 2001. Testing earthquake forecast hypotheses. *Tectonophysics* 338:261–268.

Eftaxias, K. A., P. G. Kapiris, G. T. Balasis, A. Peratzakis, K. Karamanos, J. Kopanas, G. Antonopoulos, and K. D. Nomicos. 2006. Unified approach to catastrophic events: From the normal state to geological or biological shock in terms of spectral fractal and nonlinear analysis. *Nat. Hazards Earth Syst. Sci.* 6:205–228.

Haut, S. R., C, B. Hall. A. J. LeValley, and R. B. Lipton. 2007. Can patients with epilepsy predict their seizures? *Neurology* 68:262–266.

Iasemidis, L. D., D. S. Shiau, W. Chaovalitwongse, J. C. Sackellares, P. A. Pardalos, J. C. Principe, P. R. Carney, A. Prasad, B. Veeramani, and K. Tsakalis. 2003. Adaptive epileptic seizure prediction system. *IEEE Trans. Biomed. Eng.* 50:616–627.

Iasemidis, L. D., D. S. Shiau, P. M. Pardalos, W. Chaovalitwongse, K. Narayanan, A. Prasad, K. Tsakalis, P. R. Carney, and J. C. Sackellares. 2005. Long-term prospective on-line real-time seizure prediction. *Clin. Neurophysiol.* 116:532–544.

Kagan, Y. Y. 1997. Are earthquakes predictable. *Geophys. J. Int.* 131:505–525.

Kreuz, T., R. G. Andrzejak, F. Mormann, A. Kraskov, H. Stogbauer, C. E. Elger, K. Lehnertz, and P. Grassberger. 2004. Measure profile surrogates: A method to validate the performance of epileptic seizure prediction algorithms. *Phys. Rev. E* 69:061915.

Maiwald, T., M. Winterhalder, R. Aschenbrenner-Scheibe, H. U. Voss, A. Schulze-Bonhage, and J. Timmer. 2004. Comparison of three nonlinear seizure prediction methods by means of the seizure prediction characteristic. *Physica D* 194:357–368.

Mormann, F., R. G. Andrzejak, T. Kreuz, C. Rieke, P. David, C. E. Elger, and K. Lehnertz. 2003. Automated detection of a preseizure state based on a decrease in synchronization in intracranial electroencephalogram recordings from epilepsy patients. *Phys. Rev. E* 67:021912.

Sackellares, J. C., D. S. Shiau, J. C. Principe, M. C. K. Yang, L. K. Dance, W. Suharitdamrong, W. Chaovalitwongse, P. M. Pardalos, and L. D. Iasemidis. 2006. Predictability analysis for an automated seizure prediction algorithm. *J. Clin. Neurophysiol.* 23:509–520.

Schad, A., K. Schindler, B. Schelter, T. Maiwald, A. Brandt, J. Timmer, and A. Schulze-Bonhage. 2008. Application of a multivariate seizure detection and prediction method to non-invasive and intracranial long-term EEG recordings. *Clin. Neurophysiol.* 119:197–211.

Schelter, B., M. Winterhalder, T. Maiwald, A. Brandt, A. Schad, A. Schulze-Bonhage, and J. Timmer. 2006a. Testing statistical significance of multivariate time series analysis techniques for epileptic seizure prediction. *Chaos* 16:013108.

Schelter, B., M. Winterhalder, T. Maiwald, A. Brandt, A. Schad, J. Timmer, and A. Schulze-Bonhage. 2006b. Do false predictions of seizures depend on the state of vigilance? A report from two seizure-prediction methods and proposed remedies. *Epilepsia* 47:2058–2070.

Schelter, B., M. Winterhalder, H. F. G. Drentrup, J. Wohlmuth, J. Nawrath, A. Brandt, A. Schulze-Bonhage, and J. Timmer. 2007. Seizure prediction: The impact of long prediction horizons. *Epilepsy Res.* 73:213–217.

Schulze-Bonhage, A., C. Kurth, A. Carius, B. J. Steinhoff, and T. Mayer. 2006. Seizure anticipation by patients with focal and generalized epilepsy: A multicentre assessment of premonitory symptoms. *Epilepsy Res.* 70:83–88.

Snyder, D. E., J. Echauz, D. B. Grimes, and B. Litt. 2008. The statistics of a practical seizure warning system. *J. Neural. Eng.* 5:392–401.

Stark, P. B. 1997. Earthquake prediction: The null hypothesis. *Geophys. J. Int.* 131:495–499.

Winterhalder, M., T. Maiwald, H. U. Voss, R. Aschenbrenner-Scheibe, J. Timmer, and A. Schulze-Bonhage. 2003. The seizure prediction characteristic: A general framework to assess and compare seizure prediction methods. *Epilepsy Behav.* 4:318–325.

Winterhalder, M., B. Schelter, T. Maiwald, A. Brandt, A. Schad, A. Schulze-Bonhage, and J. Timmer. 2006. Spatio-temporal patient-individual assessment of synchronization changes for epileptic seizure prediction. *Clin. Neurophysiol.* 117:2399–2413.

Wong, S., A. B. Gardner, A. M. Krieger, and B. Litt. 2007. A stochastic framework for evaluating seizure prediction algorithms using hidden markov models. *J. Neurophysiol.* 97:2525–2532.

Zechar, J. D., and T. H. Jordan. 2008. Testing alarm-based earthquake predictions. *Geophys. J. Int.* 172:715–724.

15 Seizure Prediction
An Approach Using Probabilistic Forecasting

*Bjoern Schelter, Hinnerk Feldwisch-Drentrup,
Andreas Schulze-Bonhage, and Jens Timmer*

CONTENTS

15.1 INTRODUCTION

Epilepsy is one of the most common chronic disorders of the central nervous system, leading to severe shortcomings in quality of life in about 1% of the world's population. Abnormal synchronization of discharges of epileptogenic tissue may involve extended brain areas and provoke seizures with subjective as well as objective symptoms like the loss of motor control and consciousness. About 30% of all epilepsy patients cannot be treated adequately by continuous antiepileptic medication. Surgical resection of the epileptogenic brain tissue as a means for epilepsy treatment is only applicable in a subgroup of patients. Therefore, a reliable prediction of epileptic seizures could enable novel therapeutic strategies. If the time of seizure onset could be predicted in a timely manner, the patient could be alerted or a short-term intervention would become possible.

Various kinds of prediction methods for epileptic seizures have been developed, either based on linear or nonlinear time series analysis techniques (for a review, see Mormann et al., 2007). These prediction methods detect changes in the dynamics of electroencephalography (EEG) recordings and raise alarms that are supposed to be predictive of the upcoming seizure. This approach realizes a yes/no decision about possibly emerging seizures. With the availability of continuous long-term EEG data, statistical evaluation of proposed prediction methods has evolved and several approaches have been introduced. These validation approaches separate into analytic approaches such as the seizure prediction characteristic (Winterhalder et al. 2003; Maiwald et al. 2004; Schelter et al. 2006a; Snyder et al. 2008) and validation methods based on surrogates such as seizure times surrogates (Andrzejak et al. 2003), measure profile surrogates (Kreuz et al. 2004), or alarm times surrogates (Andrzejak et al. 2009). Studies based on these evaluation strategies have so far demonstrated that seizure prediction techniques have not yet achieved a clinically relevant performance. Speculating that this is partly due to the binary yes/no decisions, we suggest a probabilistic approach here. As in weather forecasting where typically probabilities for certain weather conditions are provided, in the field of seizure prediction in epilepsy, probabilistic forecasting will provide the probability for upcoming seizures. This approach has presumably at least two advantages. First, the patients can decide based on their actual situation whether they are taking actions against a possibly occurring, predicted seizure depending on the prediction probability. In a business meeting, for instance, a probability for an upcoming seizure of 50% might already be too high to take the risk, while when in bed, a probability of 50% might be acceptable without taking any actions against the seizure. Second, yes/no decisions suffer from the disadvantage that a yes decision predicts a seizure in a certain interval. Imagine now that the situation, such as the state of vigilance in which the patient is, changes; this might prevent a seizure from occurring. In this case, the yes decision has to be judged as a false prediction. A probabilistic approach giving a seizure prediction probability of below 100% might be superior compared to the yes/no decision here. In the suggested probabilistic approach, the probability can either increase or decrease dependent on new information arising from the continuous input of EEG signals and prediction measure profiles. So, if the state of vigilance changes, the probability that might have been rather high before, might drop to a very few percent taking into account the change in the patient's situation (Schelter 2006b).

The challenges in probabilistic seizure prediction are to derive a probabilistic forecasting and to statistically evaluate seizure prediction performance for probabilistic seizure prediction. To this end, we present a method to combine numerous measure profile time series into a single measure profile, which can be interpreted as the probability of seizure occurrence. We suggest an approach to assess prediction performance of this novel strategy by using the so-called Brier score, a mean-square error measure for probabilistic predictors together with the method of seizure times surrogates. Based on a set of five patients, we demonstrate the applicability of the approach for two measures—the mean phase coherence and the dynamical similarity index.

15.2 METHODS

In this section, we introduce the data used, the measure profiles utilized for seizure prediction, the probabilistic forecasting approach, and the statistical evaluation thereof using the Brier score together with the method of seizure times surrogates.

15.2.1 Data Acquisition and Patient Details

The analyzed data consists of recordings of five patients suffering from medically intractable focal epilepsy. The retrospective evaluation of the data received prior approval by the Ethics Committee, Medical Faculty, University of Freiburg, Germany. Informed consent was obtained from each patient. The EEG data were acquired using a Neurofile NT digital video EEG system with 128 channels, a sampling rate of 256 or 512 Hz, and a 16-bit A/D converter. To eliminate possible line noise and low-frequency components, the EEG data sets were preprocessed by a 50-Hz notch filter and a high-pass filter at 0.5 Hz. A subset of six electrode contacts was selected prior to the analysis by visual inspection by an experienced epileptologist (ASB). Three focal electrode contacts, i.e., three recording sites initially involved in ictal activity, and three extra-focal electrode contacts, i.e., recording sites not involved in seizure or in most cases involved late during spread of ictal activity, were selected for analysis.

On average, 188 h of continuous recordings were analyzed in this study, with a range between 158.5 and 228.5 h per patient (Table 15.1). During these recordings, on average, 24 seizures occurred, varying between 7 and 40 per patient. A first part of the data of each patient, including at least 40% of the seizures and at least 40 h at the beginning of the recordings, was used for training of parameters (see below). The analyzed seizures occurred spontaneously, but antiepileptic medication

TABLE 15.1
Patient and EEG Characteristics

Pat	Age	Sex	Seizure Types	Focus	# Seizures	Length of Recording (h)	Length of Training (h)
1	29	M	SP, CP	NC	28	228.5	104
2	19	M	CP, GTC	NC, HC	7	208.0	180
3	18	M	SP, CP, GTC	NC	28	158.5	40
4	11	M	SP, CP, GTC	HC	40	186.5	132
5	42	F	SP	NC	18	158.5	72
mean	24				24	188.0	106

Note: Seizure types: SP, simple partial; CP, complex partial; GTC, generalized tonic–clonic. Seizure focus: NC, neocortex; HC, hippocampus.

was reduced in the majority of patients during the recording period to facilitate the occurrence of seizures.

15.2.2 Measure Profiles

We evaluated two seizure prediction measures, i.e., the mean phase coherence (Mormann et al. 2003a) as well as the dynamical similarity index (Le van Quyen et al. 1999) on long-term continuous intra-cranial EEG data. The mean phase coherence measures the interaction between two channels while the dynamical similarity index compares the actual segment of EEG data to a reference snippet.

For the mean phase, coherence preictal changes have been reported using a sliding window technique (Mormann et al. 2000, 2003a, 2003b). Here, a sliding window of 32-s duration was used, shifted by 5 s for each estimation. Subsequently, a median filter of window length 240 s was applied to remove outliers and to smooth the measure profiles. The mean phase coherence was estimated for all 15 channel combinations that are possible for the six preselected EEG channels. For the dynami-cal similarity index, the length of the reference window was chosen as 300 s and the sliding window length was 25 s. The dynamical similarity index was calculated for the EEG data of three electrode channels inside and three channels distant to the epileptic focal area. This provides six measure profiles for the dynamical similarity index.

15.2.3 Measure Profile Combination by Logistic Regression

To obtain a weighted combination of prediction measures, we made use of logistic regression. The method of logistic regression can be used to determine the probability of occurrence of an event given several measure profiles. The combination will result in a probabilistic decision, which can be interpreted as the probability of seizure occurrence for a certain time window of length T_{pred}. The seizure warning is provided with probabilistic information between 0% and 100% for a seizure in the following time interval of duration T_{pred}.

For practical applications, the running prospective time interval $T_+ = [t, t + T_{pred}]$, for which the prediction is made, is in the range of minutes to hours, thus it is much longer than a typical sampling interval T_f of a measure profile $f(n)$. The raw measure profiles $f(n)$ are averaged block-wise to obtain the re-sampled measure profiles

$$\overline{f}(T_-) = \frac{T_f}{T_{pred}} \sum_{n=1}^{T_{pred}/T_f} f\left(\frac{T_{pred}}{T_f} t - n\right).$$

By this averaging the information contained in the time interval $T_- = [t - T_{pred}, t]$ is used to generate a probabilistic forecast for the subsequent time interval T_+. Given a D-dimensional measure profile vector $f(T_-)$ as input to the logistic regression model, the probability of seizure occurrence in the time span T_+ is defined as

$$P_f(t) = P\left[\text{seizure within } T_+\right] = \left[1 + e^{-\left(\delta + \boldsymbol{\beta} \, f(T_-)\right)}\right]^{-1}.$$

The weights δ and the vector $\boldsymbol{\beta}$ are estimated in a training phase. The schematic drawing in Figure 15.1 illustrates this procedure. We emphasize that irrespective of the number of measure profiles used, $P_f(T_-)$ is one-dimensional.

As an alternative to the combination of all measure profiles into the probabilistic measure profile, a method that allows us to incorporate only such measure profiles that exhibit major contribution to the prediction performance is accomplished by using a penalized estimation algorithm for the logistic regression (Goeman 2008, Technical Report). Thus, measure profiles with a very small component β_i of the vector $\boldsymbol{\beta}$ are eliminated.

15.2.4 THE BRIER SCORE

The Brier score (Brier 1950) measures the accuracy of probabilistic forecasting methods, especially for rare events such as earthquakes, storms, or epileptic seizures. It has widely been applied in analyzing weather forecasts (Murphy 1991, 1993). It is defined as the average quadratic deviation between the predicted probabilities for an event and its outcomes, so a lower score represents higher accuracy. We review the computations only for a dichotomic event, such as the absence or presence of a seizure in the time interval T_+. The Brier score is defined as

$$B_f = \frac{1}{T} \sum_{t=1}^{T} \left[P_f(t) - I(T_+)\right]^2,$$

where $I(T_+)$ is the indicator, whether at least one seizure occurred in the interval T_+.

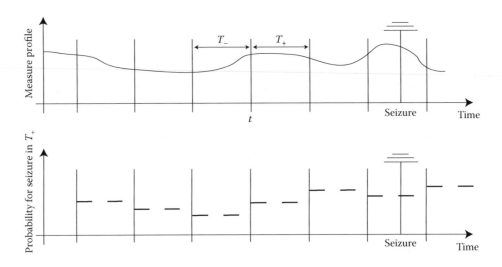

FIGURE 15.1 Schematic functioning of logistic regression. A measure profile is transformed into the probability for an upcoming seizure.

The constantly indecisive predictor, which yields a forecast of 50% and thus is practically not applicable, yields a Brier score of 0.25. Thus, a practically useful forecasting system must deliver a Brier score below 0.25 in order to show any predictive power. But the constant predictor, which yields a constant forecast probability following the natural seizure frequency of the patient, achieves a rather small Brier score in the case of rare events. The constant-null predictor is also a valid forecasting scheme for rare occurrences. Its Brier score is rather small because a contribution to it is made only occasionally in the rare case of a seizure occurrence. These characteristics of the Brier score call for setting up a statistical validation. We used the method of seizure times surrogates here (Andrzejak 2003).

15.3 APPLICATION

From the patient collection defined in Table 15.1, we computed the measure profiles of the mean phase coherence and the dynamical similarity index. For both, we trained the logistic regression with the data from the predefined training phase. For the data of the test phase, we evaluated the Brier scores of the two resulting seizure prediction methods and we tested for statistical significance by use of seizure times surrogates with 1000 trials.

In Table 15.2, the results for the five patients are shown for prediction windows T_{pred} = 5, 10, 15, 20, and 30 min. As significance level, we have chosen α = 1%, such that we correct for multiple testing for the five T_{pred}-values. For three out of five patients, the Brier score was significantly lower as with the surrogates for several T_{pred}, when the MPC was used as prediction measure. For the DSI, two out of the five patients showed significant performance.

15.4 DISCUSSION AND CONCLUSION

In this chapter, we proposed a new approach toward epileptic seizure prediction. In contrast to yes/no alarm-based systems, we present a probabilistic prediction method that provides a probability of seizure occurrence by incorporating all available measure profiles. The amount of measure profiles depends on the type of the underlying prediction measure. Univariate measures, such as the dynamical similarity index (DSI), have the same number of measure profile channels as EEG channels, while bivariate measures like the mean phase coherence (MPC) deliver a measure profile

---

TABLE 15.2
Brier Scores Obtained for Five Patients for Several Values of T_{pred}

pat	Measure	5 min	10 min	15 min	20 min	30 min
1	MPC	0.013	0.017*	0.034	0.063*	0.074
	DSI	0.008*	0.016*	0.033*	0.042*	0.054
2	MPC	0.012	0.045	0.051	0.069	0.101
	DSI	0.064	0.029	0.045	0.063	0.162
3	MPC	0.019*	0.019*	0.020	0.023*	0.430
	DSI	0.006	0.140	0.036	0.035	0.069
4	MPC	0.025	0.138	0.091	0.098	0.163
	DSI	0.028	0.055	0.084	0.098	0.140
5	MPC	0.014	0.022	0.034	0.041	0.038*
	DSI	0.010	0.019*	0.025*	0.033	0.039*

Note: Significant values are marked by an asterisk (α = 1%). Mean phase coherence (MPC), dynamical similarity index (DSI).

for each possible channel combination. Thus, as we used 6 preselected intracranial EEG channels, 6 time series for the DSI and 15 profiles of the MPC have been obtained. For yes/no alarm-based systems, the high amount of available measure profile channels leads to the problem of multiple testing, and thus to an increased sensitivity of the random predictor, which can hardly be beaten by an actual predictor (Schelter et al. 2006a). Our alternative approach, however, combines all available measure profiles into a single measure profile, which can be interpreted as the probability of seizure occurrence. Thus, the issue of multiple testing is circumvented. The measure profiles are combined into this probabilistic quantity by means of a logistic regression, whose weights are determined in a training phase. After this training, the proposed approach runs continuously on EEG signals prospectively. The probabilistic forecaster does not provide an alarm prior to expected seizures, but it delivers continually the information of how likely a seizure occurrence will be for a specified time interval $T_+ = [t, t + T_{pred}]$, where t is the actual time. This probability is computed from the obtained measure profile/profiles in a time span $T_- = [t - T_{pred}, t]$ preceding the actual time point. A means to assess and evaluate the seizure prediction performance is provided by the Brier score together with the method of seizure times surrogates. We have demonstrated the applicability of our novel approach based on long-term recordings of five epilepsy patients. We like to emphasize that the results obtained are gained after splitting the data into training and testing data. In other words, we demonstrated that the proposed approach operates quasi prospectively as parameters are not optimized in the testing data. In two or three out of five patients, we obtained a seizure prediction performance above chance level. These first results are promising and call for an extended study.

ACKNOWLEDGMENT

We acknowledge the contribution of and discussions with M. Jachan. We would like to thank R. Schoop and H. Binder for stimulating discussions on statistical evaluation. We would like to thank Carolin Gierschner and the team of EEG technicians at the Epilepsy Center Freiburg for their technical support, especially concerning the extensive data acquisition and handling. B.S. is indebted to the Landesstiftung Baden-Wuerttemberg for the financial support by the Eliteprogramme for Postdocs. This work was supported by the EU (EPILEPSIAE 211713).

REFERENCES

Andrzejak, R. G., F. Mormann, T. Kreuz, C. Rieke, C. E. Elger, and K. Lehnertz. 2003. Testing the null hypothesis of the nonexistence of a preseizure state. *Phys. Rev. E* 67:010901.

Andrzejak, R. G., D. Chicharro, C. E. Elger, and F. Mormann. 2009. Seizure prediction: Any better than chance? *Clin. Neurophysiol.* 120:1465–1478.

Brier, G. W. 1950. Verification of forecasts expressed in terms of probability. *Month. Weath. Rev.* 78:1–3.

Goeman, J. J. 2008. An efficient algorithm for L1 penalized estimation. Technical report, Department of Medical Statistics and Bioinformatics. Leiden: University of Leiden.

Kreuz, T., R. G. Andrzejak, F. Mormann, A. Kraskov, H. Stoegbauer, C. E. Elger, K. Lehnertz, and P. Grassberger. 2004. Measure profile surrogates: A method to validate the performance of epileptic seizure prediction algorithms. *Phys. Rev. E* 69:061915.

Le Van Quyen, M., J. Martinerie, M. Baulac, and F. Varela. 1999. Anticipating epileptic seizures in real time by a non-linear analysis of similarity between EEG recordings. *Neuroreport* 10:2149–2155.

Maiwald, T., M. Winterhalder, R. Aschenbrenner-Scheibe, H. U. Voss, A. Schulze-Bonhage, and J. Timmer. 2004. Comparison of three nonlinear seizure prediction methods by means of the seizure prediction characteristic. *Physica D* 194:357–368.

Mormann, F., K. Lehnertz, P. David, and C. E. Elger. 2000. Mean phase coherence as a measure for phase synchronization and its application to the EEG of epilepsy patients. *Physica D* 144:358–369.

Mormann, F., T. Kreuz, R. G. Andrzejak, P. David, K. Lehnertz, and C. E. Elger. 2003a. Epileptic seizures are preceded by a decrease in synchronization. *Epilepsy Res.* 53:173–185.

Mormann, F., R. G. Andrzejak, T. Kreuz, C. Rieke, P. David, C. E. Elger, and K. Lehnertz. 2003b. Automated detection of a preseizure state based on a decrease in synchronization in intracranial electroencephalogram recordings from epilepsy patients. *Phys. Rev. E* 67:021912.

Mormann, F., R. G. Andrzejak, C. E. Elger, and K. Lehnertz. 2007. Seizure prediction: The long and winding road. *Brain* 130:314–333.

Murphy, A. H. 1991. Probabilities, odds, and forecasts of rare events. *Weath. Forecast.* 6:302–307.

Murphy, A. H. 1993. What is a good forecast? An essay on the nature of goodness in weather forecasting. *Weath. Forecast.* 8:281–293.

Schelter, B., M. Winterhalder, T. Maiwald, A. Brandt, A. Schad, A. Schulze-Bonhage, and J. Timmer. 2006a. Testing statistical significance of multivariate time series analysis techniques for epileptic seizure prediction. *Chaos* 16:013108.

Schelter, B., M. Winterhalder, T. Maiwald, A. Brandt, A. Schad, J. Timmer, and A. Schulze-Bonhage. 2006b. Do false predictions of seizures depend on the state of vigilance? A report from two seizure-prediction methods and proposed remedies. *Epilepsia* 47:2058–2070.

Snyder, D. E., J. Echauz, D. B. Grimes, and B. Litt. 2008. The statistics of a practical seizure warning system. *J. Neu. Eng* 5:392–402.

Winterhalder, M., T. Maiwald, H. U. Voss, R. Aschenbrenner-Scheibe, J. Timmer, and A. Schulze-Bonhage. 2003. The seizure prediction characteristic: A general framework to assess and compare seizure prediction methods. *Epilepsy Behav.* 4:318–325.

16 Seizure Prediction and Detection Research at Optima Neuroscience

Deng-Shan Shiau, Jui-Hong Chien, Ryan T. Kern, Panos M. Pardalos, and J. Chris Sackellares

CONTENTS

16.1 INTRODUCTION AND BACKGROUND

Optima Neuroscience is a biomedical technology company that develops a line of revolutionary brain monitoring products designed for use in clinical settings in which brain function monitoring is necessary such as emergency rooms, operating rooms, and intensive care units. In U.S. emergency rooms, more than 10 million patients are admitted every year due to acute brain injuries including head trauma, transient ischemic attack (TIA), and stroke, and no reliable technologies exist to automatically predict and detect seizures in these patients. As a result, up to 27% of these patients are misdiagnosed annually, resulting in preventable brain damage and loss of neurological function due to ongoing but untreated seizures. Optima's innovative solutions will allow nonspecialist hospital staff to monitor seizures and brain function, helping to protect the most important organ: the brain. We believe that our automated systems will quickly become the standard of care for all brain injury patients.

16.2 SEIZURE PREDICTION AND MONITORING TECHNOLOGY

The essential question as to what causes seizures to come and go remains unanswered. In 1988, Sackellares and Iasemidis initiated research to test the hypothesis that the transition from an interictal state to a seizure (ictal state) is similar to the state transitions that had been observed in chaotic systems (Iasemidis et al. 1988a, 1988b, 1990; Iasemidis and Sackellares 1990). In the course of these investigations, they found that the spatiotemporal patterns of the ictal states were consistently more ordered than that of the interictal and postictal states. Furthermore, there was a significant change in measures of spatial order among EEG signals that precede seizures by periods on the order of an hour. Thus, by combining measures of temporal order, spatial order, and signal energy and frequency, it is possible to develop devices that can predict as well as detect seizures.

These scientific discoveries led to the development and testing of patented methods for automated seizure detection and prediction algorithms in long-term scalp EEG recordings. The methods utilize spatiotemporal patterns derived from EEG dynamics, including linear and nonlinear, univariate and bivariate EEG descriptors. Important findings are: (1) an apparent significant increase of signal regularity during the ictal state, and (2) the apparent transition before a seizure occurrence (i.e., the so-called preictal transition), which can be characterized by progressive convergence of the mean signal regularity among specific anatomical areas (mean value entrainment).

16.2.1 SEIZURE DETECTION

It has long been recognized that designing a computer algorithm to detect seizures from scalp EEG is much more challenging than that from intracranial/depth recordings (Harding 1993; Osorio et al. 2002; Khan and Gotman 2003; Gardner et al. 2006). Scalp EEGs are more sensitive to the recording environment (e.g., electrode failure, electrical artifacts), muscle and movement artifacts (e.g., chewing), as well as to the normal physiological state changes (e.g., sleep-awake cycle). The EEG patterns of these artifacts and activities share a certain degree of signal characteristics similar to those of ictal EEGs, such as signal amplitude and frequency, and thus could generate numerous false detections. Therefore, a successful scalp EEG seizure detection algorithm must include a robust artifact rejection module that is able to distinguish, spatially or temporally, the signal characteristics between artifact and true ictal EEG patterns.

Optima has developed a novel automated seizure detection algorithm for accurate and rapid analysis of long-term scalp EEG recordings to identify ictal EEG segments (Shiau et al. 2010; Kelly et al. 2010; Halford et al. 2010). The algorithm analyzes recorded scalp EEG records sequentially for each nonoverlapping 5.12 second EEG epoch and reports a seizure detection when sufficient criteria are met. For each computation epoch, the analysis involves translating each of the EEG channels into six EEG descriptors: (1) pattern-match regularity statistic (PMRS); (2) local maximal frequency (LFmax), estimated by the number of occurrences of positive zero crossings (after mean removal) in a one-second epoch within the calculation window; (3) amplitude variation (AV); (4–5) local minimal and maximal amplitude variations (LAVmin and LAVmax); and (6) maximal amplitude in a high-frequency band (AHFmax). The spatiotemporal patterns of these descriptors are used for detecting ictal EEG epochs as well as rejecting the epochs with significant artifacts or unwanted normal patterns. PMRS is used as the primary detector, whereas the other descriptors are used for artifact rejections. Figure 16.1 demonstrates the behavior of PMRS values from one EEG channel (T3 in the 10-20 labeling system) that was involved in an ictal discharge. Because of the more rhythmic signals during the seizure period, PMRS values drop significantly when compared with the periods before and after the seizure. Therefore, with a proper threshold, the seizure can be easily detected with the change of PMRS values.

In order to recognize as many spatiotemporal dynamic patterns as possible, the detection algorithm generates over 800 descriptor traces (both discrete/binary and continuous variables) across 16 EEG channels. The parameters of the algorithm were trained and optimized based on the detection performance, in terms of detection sensitivity and false detection rate, in 47 long-term scalp EEG recordings (47 subjects, 141 seizures in a total of 3652.5 h of recording). Its detection performance was then validated in a separate test EEG dataset (55 subjects, 146 seizures in a total of 1208 h of recording) in which each of the EEG segments was independently reviewed by three epileptologists. The overall results showed that the detection algorithm was able to detect 80% of the test set electrographic seizure activities agreed upon by the majority of epileptologists with a mean false detection rate of just 2 per day (0.086/h). In fact, more than half of the test subjects had a false detection rate of less than 1 per day.

This algorithm processes scalp EEG recordings at a speed of approximately 50 times faster than real time on a standard dual-core desktop computer. In other words, it takes less than 30 min to process and report detections for a 24-h scalp EEG recording. Computer software based on such an

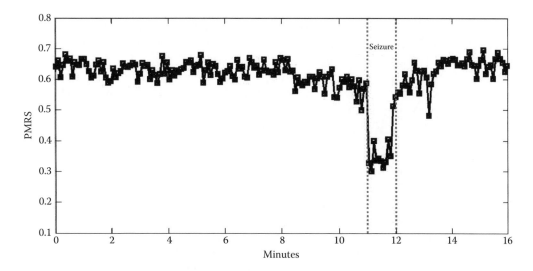

FIGURE 16.1 An example of a PMRS curve before, during, and after a seizure (between the two vertical dotted lines). PMRS values drop significantly during the ictal period compared to the other periods.

algorithm would greatly enhance the efficiency of long-term EEG monitoring in epilepsy monitoring units (EMUs) and intensive care units (ICUs).

16.2.2 Seizure Prediction

Results from several studies based on the analysis of intracranial EEG (Iasemidis et al. 1997; Lehnertz and Elger 1998; Elger and Lehnertz 1998; Le Van Quyen et al. 1999, 2000; Sackellares et al. 2000, 2006; Litt et al. 2001; Iasemidis et al. 2001, 2002, 2003a, 2003b, Mormann et al. 2003) and fMRI (Federico et al. 2005) data suggest the existence of a preictal transition between an interictal and ictal state. However, detecting a preictal transition using EEG signals from scalp electrodes may be more difficult due to the attenuating, spatial distortion, and filtering effects of the skull and soft tissues. Nevertheless, if a seizure warning system can be developed for scalp EEG, it could have a wide range of clinical diagnostic and therapeutic applications due to its portability, relatively low cost, and safety.

By applying a nonlinear similarity index, Le Van Quyen, Martinerie and colleagues studied preictal EEG dynamics in 26 scalp EEG recordings obtained from 23 patients with temporal lobe epilepsy. In five patients with simultaneous scalp and intracranial recordings, changes of similarity index values were observed during a preictal period in both types of EEG recordings, and 25 out of the 26 EEG segments showed changes prior to the occurrence of seizures (mean 7 min). Although this study did not evaluate the specificity of the similarity index change in long-term EEG recordings (the test recordings contained only 50 min before seizure onsets), it rendered an encouraging result using only scalp EEG to achieve seizure prediction (Le Van Quten et al. 2001). Another study by Hively and Protopopescu used L_1-distance and χ^2 statistic to estimate the dissimilarity in density functions between the base-windows and the test-windows in 20 scalp EEG recordings (Hively et al. 2000). The result showed preseizure changes in all datasets with the forewarning times ranged from 10 to 13,660 s (mean = 2961.5, standard deviation = 4265.6). However, this study only examined one selected channel in each data set, and specificity of the method was not reported. In two follow-up studies by the same group, one used all available recording channels and the other used a fixed channel, showed similar results (Protopopescu et al. 2001; Hively and Protopopescu 2003). While the results seemed promising, no prospective validation study has been reported for this method. In the study reported by Corsini et al. (2006), 20 sets of simultaneous scalp and intracranial

EEG recordings were analyzed using blind source separation and short-term Lyapunov exponent (STLmax) (Corsini et al. 2006). They observed changes of STLmax values before seizure onsets and that the scalp EEG may give better predictive power over intracranial EEG when the intracranial electrodes did not record the electrical activity in the epileptic focus. However, the practicality of this method on long-term scalp EEG recordings may be limited due to the lack of an automatic procedure to select the most relevant source component. Schad et al. (2008) investigated seizure detection and prediction in 423 h of long-term simultaneously scalp and intracranial EEG recordings from six epileptic patients. The method used techniques based on simulated leaky integrate-and-fire neurons. The study reported that 59%/50% of the 22 seizures were predicted using scalp/invasive EEGs given a maximum number of 0.15 false predictions per hour. In the study by Bruzzo et al. (2008), a small sample of scalp EEG recordings (115 h from THREE epileptic patients) was analyzed using permutation entropy (PE). By examining the area under the receiver operating characteristic (ROC) curve, they reported that the decrease of PE values was correlated with the occurrence of seizures. However, the authors also concluded that the dependency of PE changes on the vigilance state may restrict its possible application for seizure prediction. More recently, Zandi et al. (2009) reported a prediction method based on the positive zero-crossing interval series. The method was applied to a 21.5-h scalp EEG dataset (16 bipolar channels) recorded from 4 patients with temporal lobe epilepsy. They reported a training result of 87.5% sensitivity (16 seizures) with a false prediction rate of 0.28/h, and an average prediction time of approximately 25 min. James and Gupta (2009) analyzed long-term continuous scalp EEG recordings from nine patients (five in the training set and four in the test dataset). The data were processed by a sequence of techniques consisting of independent component analysis, phase locking value, neuroscale, and Gaussian mixture model. The prediction performance of this method achieved a sensitivity of 65%–100% and specificity of 65%–80% as the prediction horizon ranged from 35–65 min in the test dataset. These encouraging results, though not being prospectively and statistically validated, have prompted the researchers in the community to devote efforts in developing scalp EEG based seizure warning algorithms that can be applied clinically for seizure monitoring systems or closed-loop seizure control devices.

Optima Neuroscience is currently conducting clinical research aimed to develop a clinically useful seizure warning system for brain condition monitoring application. Our current algorithm identifies preictal transitions by detecting dynamic entrainment based on the convergence of PMRS time series among multiple electrode sites. By further applying paired t-statistic (denoted as "T-index" hereafter) that quantifies the convergence of two PMRS time series over time, we constructed an automated seizure prediction algorithm that monitors the change of T-index and issues a warning of an impending seizure when the T-index curve exhibits the pattern defined by the algorithm. Figure 16.2 shows an example of a seizure warning triggered by an occurrence of PMRS convergence. In this demonstration, a PMRS time series of the two frontal-temporal channels (F7 and F8; black line and gray line in Figure 16.2, respectively) were monitored for detecting preictal transitions of a seizure with a right-temporal onset zone. Over the period of 3 h before the seizure (between the two vertical dotted red lines), two PMRS time series were apart most of the time. They became convergent about 30 min before the seizure onset, both dropped during the seizure, and immediately became divergent after the seizure ended. This behavior was quantified by a single T-index curve (bottom panel), which exhibited gradual decrease before the seizure and sharp rise after the seizure. This pattern of gradual decrease in T-index curve was used for detection of a preictal transition in our seizure warning algorithm. To increase the detection sensitivity, the algorithm was further designed to monitor multiple T-index curves simultaneously.

This seizure warning algorithm has been trained in a dataset consisting of 21 long-term scalp EEG recordings. The primary training was to optimize two parameters that define the pattern of "gradual decrease" in a T-index curve: duration and degree of the convergence. After these two parameters were optimized such that the algorithm achieved a sensitivity of at least 70% with the lowest possible false positive rate, its performance was evaluated in a separate test dataset consisting of long-term recordings from 31 different patients with a total of 60 seizures. With the parameters

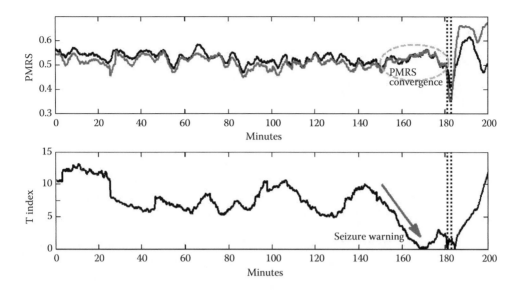

FIGURE 16.2   An example of PMRS convergence between two EEG channels (F7 and F8) over a period of 200 min, approximately 3 h before the seizure and 20 min after the seizure (between the two vertical dotted lines). PMRS values of the two EEG channels started getting convergent to each other about 30 min before the seizure onset, and immediately became divergent after the seizure ended. This behavior can be quantified by a single T-index curve that showed gradual decrease before the seizure and sharp rise after the seizure.

fixed from the training study for all test data sets, the overall sensitivity was 68.3% with an overall false positive rate of 0.235/h.

This performance was compared with a random warning scheme designed to issue random warnings (uniformly within the recording segment) with the same number as that issued by the test algorithm for each segment in the test dataset. Each random trial ran through all of the patients in the test dataset, and an overall sensitivity was calculated. The random scheme repeated 1000 times

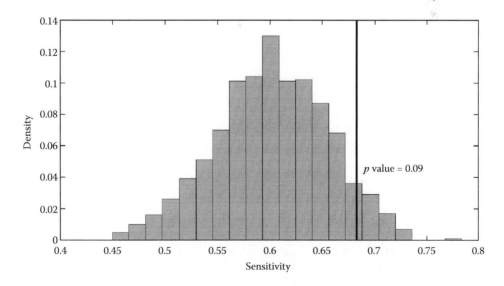

FIGURE 16.3   The distribution of the overall sensitivities achieved by the random warning scheme (generated from 1000 trials). The overall sensitivity of the test algorithm is denoted by the vertical line, which was better than 91% of the trials by the random scheme.

on the test dataset and thus generated 1000 overall sensitivity estimates. Its empirical distribution is shown in Figure 16.3. The sensitivity of the test algorithm exceeded the sensitivity of 91% of the random warning trials (i.e., p-value = 0.09).

16.3 CLINICAL APPLICATIONS

Based on these technologies, Optima Neuroscience is producing a line of brain monitoring products designed for use in hospital emergency rooms and intensive care units. In U.S. emergency rooms, over 10 million patients are admitted every year due to acute brain injuries including head trauma, TIAs, and stroke. Due to the limited availability of EEG diagnostics, an alarming percentage of these patients develop unrecognized seizure activity causing further neuronal damage and loss of function. Optima's monitors will help automate the process of EEG interpretation, reducing the current time delay to treatment and greatly improving the identification of subclinical seizures.

Optima's seizure detection and prediction technology will also be used to improve the efficiency of reviewing and identifying seizures in long-term EEG recordings. This translates to a dramatic time savings for any neurologist or technician tasked with manually screening multiple days of EEG recordings. When used online in an epilepsy monitoring unit, our software can help the attending physician discharge the patient sooner by notifying the staff as soon as the requisite number of events has been recorded.

Similar algorithms are being developed by our research team to detect a wide range of normal and abnormal brain wave patterns. These algorithms are applied to the detection of EEG patterns that occur during stroke, impending stroke, hypoxia, hypoglycemia, and other metabolic disorders that alter brain function. These algorithms will be incorporated into the same brain monitoring systems used to detect and predict seizures. The systems can be used in a variety of settings including special diagnostic and treatment units, intensive care units, emergency departments, postoperative recovery rooms, general care hospital beds, emergency vehicles, and even in the home.

ACKNOWLEDGMENT

This work was supported by the grants 5R01NS050582 (JCS) and 1R43NS064647 (DSS) from NIH-NINDS.

REFERENCES

Bruzzo, A. A., B. Gesierich, M. Santi, C. A. Tassinari, N. Birbaumer, and G. Rubboli. 2008. Permutation entropy to detect vigilance changes and preictal states from scalp EEG in epileptic patients. A preliminary study. *Neurol. Sci.* 29:3–9.

Corsini, J., L. Shoker, S. Sanei, and G. Alarcon. 2006. Epileptic seizure predictability from scalp EEG incorporating constrained blind source separation. *IEEE Trans. Biomed. Eng.* 53:790–799.

Elger, C. E., and K. Lehnertz. 1998. Seizure prediction by non-linear time series analysis of brain electrical activity. *Eur. J. Neurosci.* 10:786–789.

Federico, P., D. F. Abbott, R. S. Briellmann, A. S. Harvey, and G. D. Jackson. 2005. Functional MRI of the preictal state. *Brain* 128:1811–1817.

Gardner, A. B., A. M. Krieger, G. Vachtsevanos, and B. Lih. 2006. One-class novelty detection for seizure analysis from intracranial EEG. *J. Mach. Learn Res.* 7:1025–1044.

Halford, J. J., D.-S. Shiau, R. T. Kern, et al. 2010. Seizure detection software used to complement the visual screening process for long-term EEG monitoring. *Am. J. Electroneurodiag. Tech. (AJET)*, in press.

Harding, G. W. 1993. An automated seizure monitoring system for patients with indwelling recording electrodes. *Electroencephalogr. Clin. Neurophysiol.* 86:428–437.

Hively, L. M., and V. A. Protopopescu. 2003. Channel-consistent forewarning of epileptic events from scalp EEG. *IEEE Trans. Biomed. Eng.* 50:584–593.

Hively, L. M., V. A. Protopopescu, and P. C. Gailey. 2000. Timely detection of dynamical change in scalp EEG signals. *Chaos* 10:864–875.

Iasemidis, L. D., and J. C. Sackellares. 1990. Long time scale temporo-spatial patterns of entrainment of preictal electrocorticographic data in human temporal lobe epilepsy. *Epilepsia* 31(5):621.

Iasemidis, L. D., P. M. Pardalos, J. C. Sackellares, and D.-S. Shiau. 2001. Quadratic binary programming and dynamical system approach to determine the predictability of epileptic seizures. *J. Comb. Optim.* 5:9–26.

Iasemidis, L. D., P. M. Pardalos, D.-S. Shiau, W. Chaowolitwongse, K. Narayanan, S. Kumar, P. R. Carney, and J. C. Sackellares. 2003a. Prediction of human epileptic seizures based on optimization and phase changes of brain electrical activity. *Optim. Methods Softw.* 18(1):81–104.

Iasemidis, L. D., J. C. Principe, J. M. Czaplewski, et al. 1997. Spatiotemporal transition to epileptic seizures: A nonlinear dynamical analysis of scalp and intracranial EEG recordings. In *Spatiotemporal models in biological and artificial systems*, eds. F. H. Lopes de Silva, J. C. Principe and L. B. Almeida, 81–88. Amsterdam: IOS Press.

Iasemidis, L. D., J. C. Sackellares, H. P. Zaveri, and W. J. Williams. 1990. Phase space topography of the electrocorticogram and the Lyapunov exponent in partial seizures. *Brain Topogr.* 2:187–201.

Iasemidis, L. D., D.-S. Shiau, W. Chaowolitwongse, J. C Sackellares, P. M. Pardalos, J. C. Principe, P. R. Carney, A. Prasad, B. Veeramani, and K. Tsakalis. 2003b. Adaptive epileptic seizure prediction system. *IEEE Trans. Bio. Eng.* 15(5):616–627.

Iasemidis, L. D., D.-S. Shiau, P. M. Pardalos, and J. C. Sackellares. 2002. Phase entrainment and predictability of epileptic seizures. In *Biocomputing*, eds. P. M. Pardalos and J. C. Principe, 59–84. Dordrecht: Kluwer Academic Publishers.

Iasemidis, L. D., H. P. Zaveri, J. C. Sackellares, and W. J. Williams. 1988a. Phase space analysis of EEG data in temporal lobe epilepsy. Proceedings of IEEE Engineering and Medicine & Biology Society 10th Annual International Conference. New Orleans, LA.

Iasemidis, L. D., H. P. Zaveri, J. C. Sackellares, and W. J. Williams. 1988b. Linear and nonlinear modeling of ECoG in temporal lobe epilepsy. *25th Annual Rocky Mountain Bioengineering Symposium* 24:187–193.

James, C. J., and D. Gupta. 2009. Seizure prediction for epilepsy using a multi-stage phase synchrony based system. *Conf. Proc. IEEE Eng. Med. Biol. Soc.* 2009:25–28.

Kelly, K. M., D.-S. Shiau, R. T. Kern, J. H. Chen, M. C. Yang, K. A Yandora, J. P. Valeriano, J. J. Halford, and J. C. Sackellares. 2010. Assessment of a Scalp EEG-based automated seizure detection system. *Clin. Neurophysiol.* 121(11):1832–1843.

Khan, Y. U., and J. Gotman. 2003. Wavelet-based automatic seizure detection in intracerebral electroencephalogram. *Clin. Neurophysiol.* 114(5):898–908.

Le Van Quyen, M., C. Adam, J. Martinerie, M. Baulac, S. Clémenceau, and F. Varela. 2000. Spatio-temporal characterizations of non-linear changes in intracranial activities prior to human temporal lobe seizures. *Eur. J. Neuroscience* 12:2124–2134.

Le Van Quyen, M., J. Martinerie, M. Baulac, F. Varela. 1999. Anticipating epileptic seizures in real time by non-linear analysis of similarity between EEG recordings. *NeuroReport* 10: 2149–2155.

Le Van Quyen, M., J. Martinerie, V. Navarro, et al. 2001. Anticipation of epileptic seizures from standard EEG recordings. *The Lancet* 357:183–188.

Lehnertz, K., and C. E. Elger. 1998. Can epileptic seizures be predicted? Evidence from nonlinear time series analysis of brain electrical activity. *Phys. Rev. Lett.* 80(2):5019–5022.

Litt, B., R. Esteller, J. Echauz, M. D'Alessandro, R. Shor, T. Henry, P. Pennell, C. Epstein, R. Bakay, M. Dichter, and G. Vachtsevanos. 2001. Epileptic seizures may begin hours in advance of clinical onset: A report of five patients. *Neuron* 30:51–64.

Mormann, F., R. G. Andrzejak, T. Kreuz, C. Rieke, P. David, C. E. Elger, and K. Lehnertz. 2003. Automated detection of a preseizure state based on a decrease in synchronization in intracranial electroencephalogram recordings from epilepsy patients. *Phys. Rev. E* 67:021912-1-10.

Osorio, I., M. G. Frei, J. Giftakis, T. Peters, J. Ingram, M. Tumball, M. Herzog, M. T. Rise, S. Schaffner, R. A. Wennberg, T. S. Walczak, M. W. Risinger, and C. Ajmone-Marsan. 2002. Performance reassessment of a real-time seizure detection on long ECoG series. *Epilepsia* 43:1522–1535.

Protopopescu, V. A., L. M. Hively, and P. C. Gailey, 2001. Epileptic event forewarning from scalp EEG. *J. Clin. Neurophysiol.* 18:223–245.

Sackellares, J. C., L. D. Iasemidis, D.-S. Shiau, R. L. Gilmore, and S. N. Roper. 2000. Epilepsy—When chaos fails. In *Chaos in the brain?* eds. K. Lehnertz, J. Arnhold, P. Grassberger, and C. E. Elger, 112–133. Singapore: World Scientific.

Sackellares, J. C., D.-S. Shiau, J. C. Principe, M. C. K. Yang, L. K. Dance, W. Suharitdamrong, W. Chaowolitwongse, P. M. Pardalos, and L. D. Lasemidis. 2006. Predictability analysis for an automated seizure prediction algorithm. *J. Clin. Neurophysiol.* 29(6):509–520.

Schad, A., K. Schindler, B. Schelter, et al. 2008. Application of a multivariate seizure detection and prediction method to non-invasive and intracranial long-term EEG recordings. *Clin. Neurophysiol.* 119:197–211.

Shiau, D.-S., J. J. Halford, K. M. Kelly, et al. 2010. Signal regularity-based automated seizure detection system for scalp EEG monitoring. *Cybern. Syst. Anal.*, in press.

Zandi, A. S., G. A. Dumont, M. Javidan, and R. Tafreshi. 2009. An entropy-based approach to predict seizures in temporal lobe epilepsy using scalp EEG. *Conf. Proc. IEEE Eng. Med. Biol. Soc.* 2009:228–231.

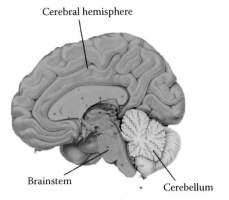

Cerebral hemisphere

Brainstem

Cerebellum

FIGURE 1.1   (See caption in text.)

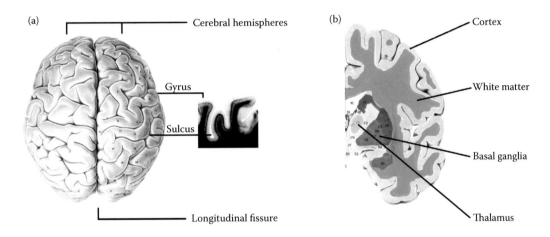

(a)

Cerebral hemispheres

Gyrus

Sulcus

Longitudinal fissure

(b)

Cortex

White matter

Basal ganglia

Thalamus

FIGURE 1.2   (See caption in text.)

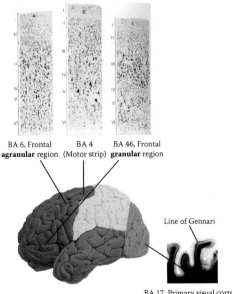

BA 6, Frontal **agranular** region

BA 4 (Motor strip)

BA 46, Frontal **granular** region

Line of Gennari

BA 17, Primary visual cortex

FIGURE 1.4   (See caption in text.)

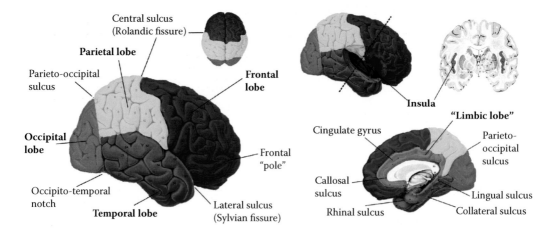

FIGURE 1.5 (See caption in text.)

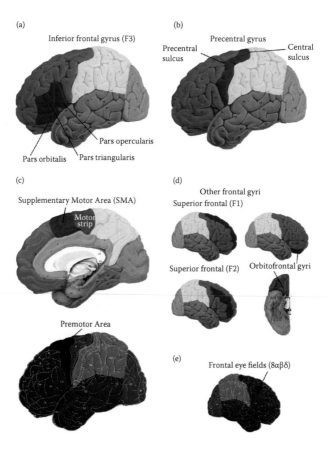

FIGURE 1.6 (See caption in text.)

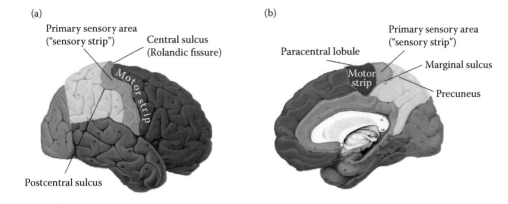

(a)

Primary sensory area
("sensory strip")

Central sulcus
(Rolandic fissure)

Motor strip

Postcentral sulcus

(b)

Paracentral lobule

Primary sensory area
("sensory strip")

Marginal sulcus

Motor strip

Precuneus

FIGURE 1.7 (See caption in text.)

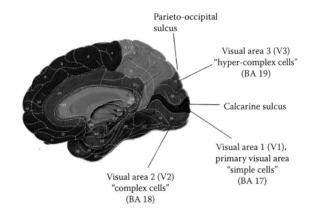

Parieto-occipital
sulcus

Visual area 3 (V3)
"hyper-complex cells"
(BA 19)

Calcarine sulcus

Visual area 1 (V1),
primary visual area
"simple cells"
(BA 17)

Visual area 2 (V2)
"complex cells"
(BA 18)

FIGURE 1.8 (See caption in text.)

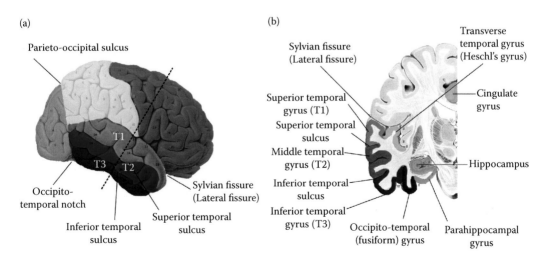

(a)

Parieto-occipital sulcus

T1

T3   T2

Occipito-
temporal notch

Sylvian fissure
(Lateral fissure)

Superior temporal
sulcus

Inferior temporal
sulcus

(b)

Sylvian fissure
(Lateral fissure)

Transverse
temporal gyrus
(Heschl's gyrus)

Superior temporal
gyrus (T1)

Cingulate
gyrus

Superior temporal
sulcus

Middle temporal
gyrus (T2)

Inferior temporal
sulcus

Hippocampus

Inferior temporal
gyrus (T3)

Occipito-temporal
(fusiform) gyrus

Parahippocampal
gyrus

FIGURE 1.9 (See caption in text.)

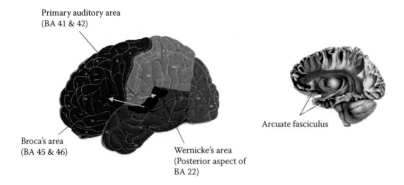

FIGURE 1.10 (See caption in text.)

(a)

(b)

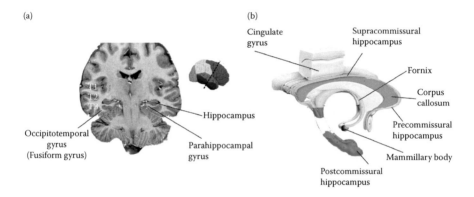

FIGURE 1.11 (See caption in text.)

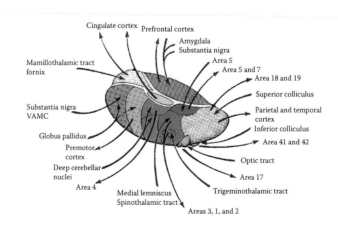

FIGURE 1.12 (See caption in text.)

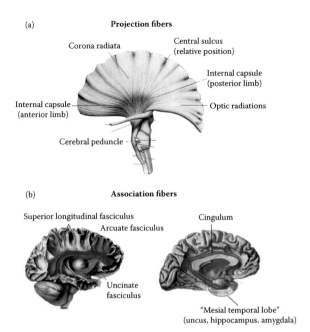

(a) **Projection fibers**

Corona radiata

Central sulcus
(relative position)

Internal capsule
(posterior limb)

Internal capsule
(anterior limb)

Optic radiations

Cerebral peduncle

(b) **Association fibers**

Superior longitudinal fasciculus

Arcuate fasciculus

Cingulum

Uncinate
fasciculus

"Mesial temporal lobe"
(uncus, hippocampus, amygdala)

FIGURE 1.13   (See caption in text.)

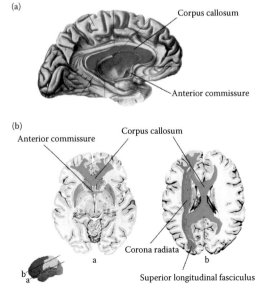

(a)

Corpus callosum

Anterior commissure

(b)

Anterior commissure

Corpus callosum

a

b

Corona radiata

b
a

Superior longitudinal fasciculus

FIGURE 1.14   (See caption in text.)

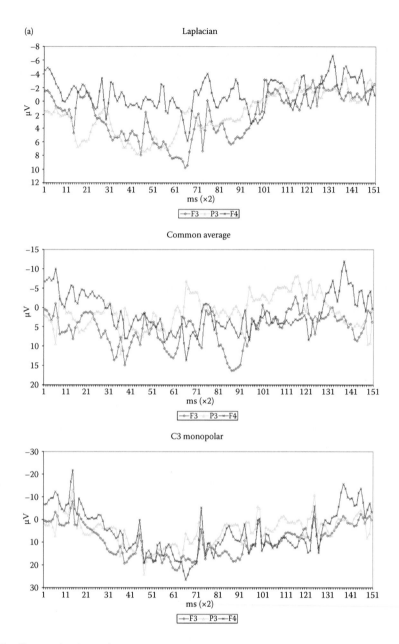

FIGURE 2.8 (See caption in text.)

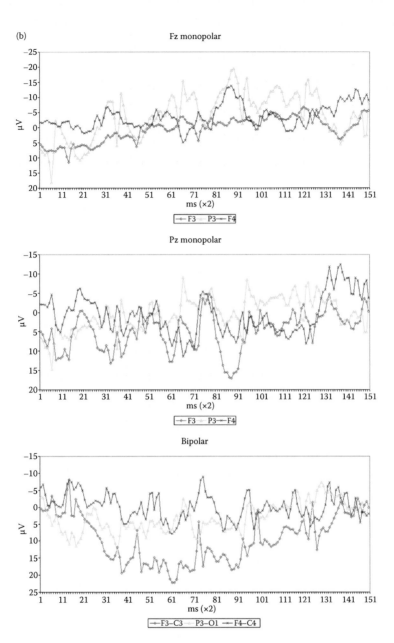

FIGURE 2.8 (continued)

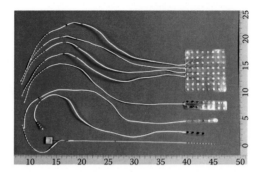

FIGURE 4.5   (See caption in text.)

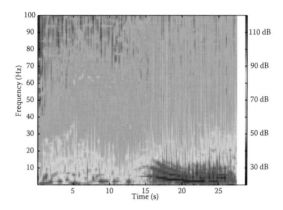

FIGURE 7.10   (See caption in text.)

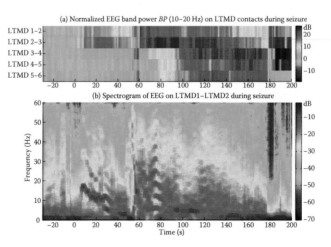

FIGURE 10.8   (See caption in text.)

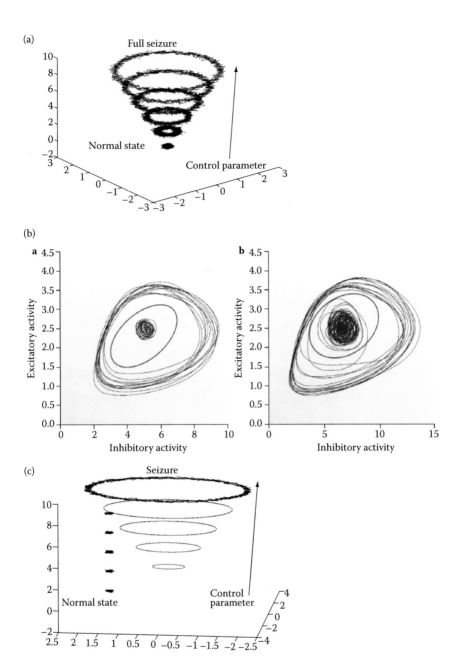

FIGURE 12.1 (See caption in text.)

(a)

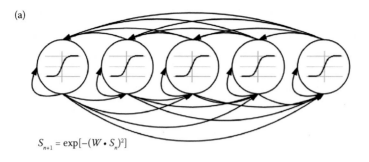

$$S_{n+1} = \exp[-(W \cdot S_n)^2]$$

(b)

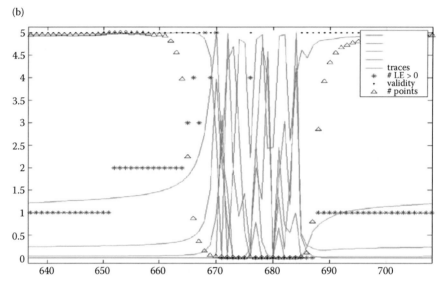

FIGURE 12.2 (See caption in text.)

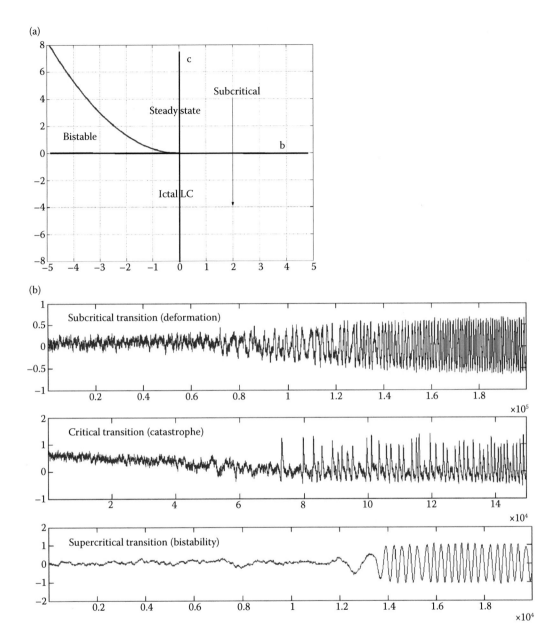

FIGURE 12.3 (See caption in text.)

(a)

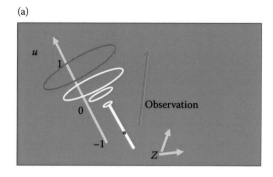

(b)

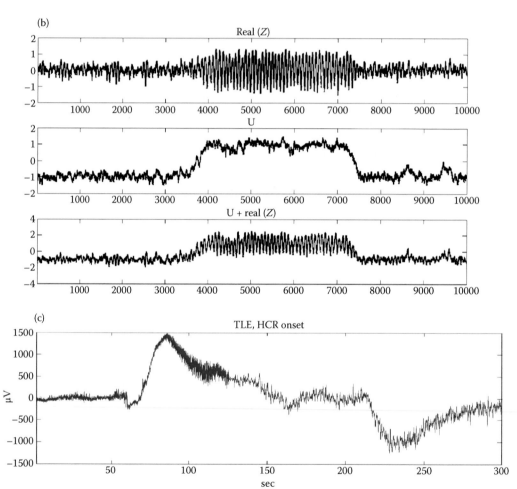

FIGURE 12.4 (See caption in text.)

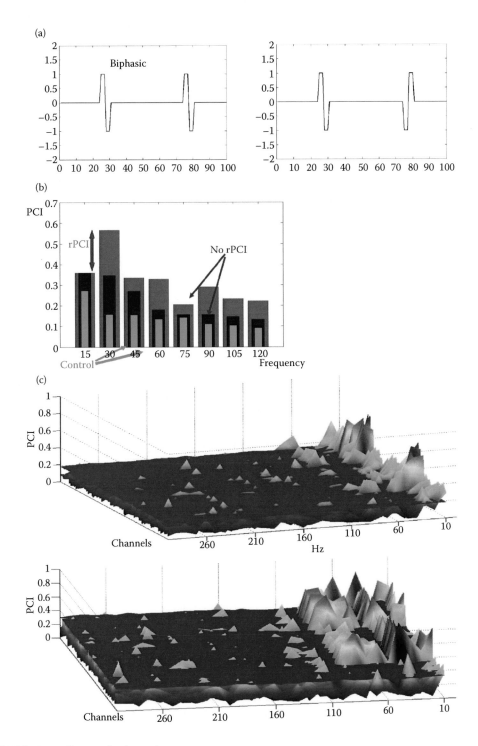

(a)

Biphasic

(b)

PCI

rPCI   No rPCI

15   30   45   60   75   90   105   120
Control   Frequency

(c)

PCI

Channels   260   210   160   110   60   10
                        Hz

PCI

Channels   260   210   160   110   60   10

FIGURE 12.6   (See caption in text.)

(a)

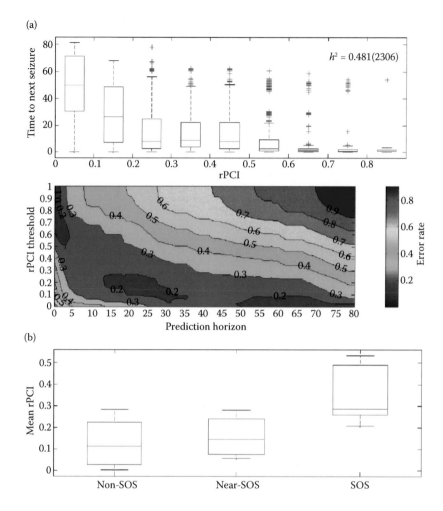

(b)

FIGURE 12.7 (See caption in text.)

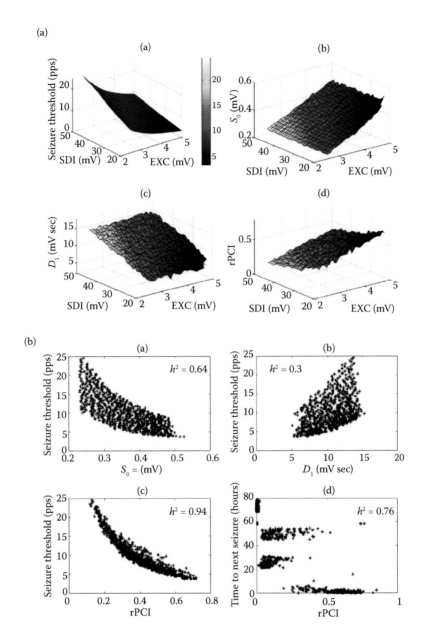

FIGURE 12.8 (See caption in text.)

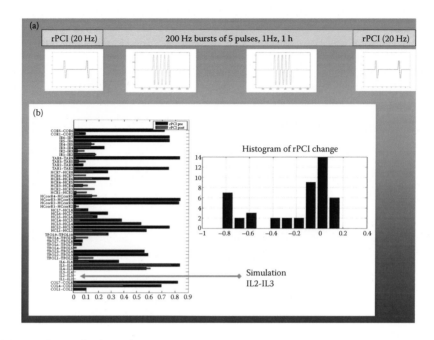

FIGURE 12.9 (See caption in text.)

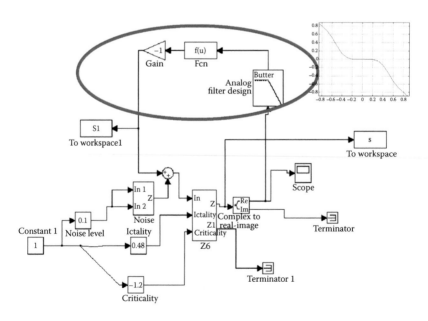

FIGURE 12.10 (See caption in text.)

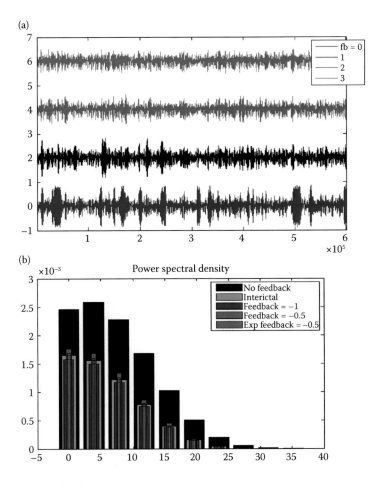

FIGURE 12.11 (See caption in text.)

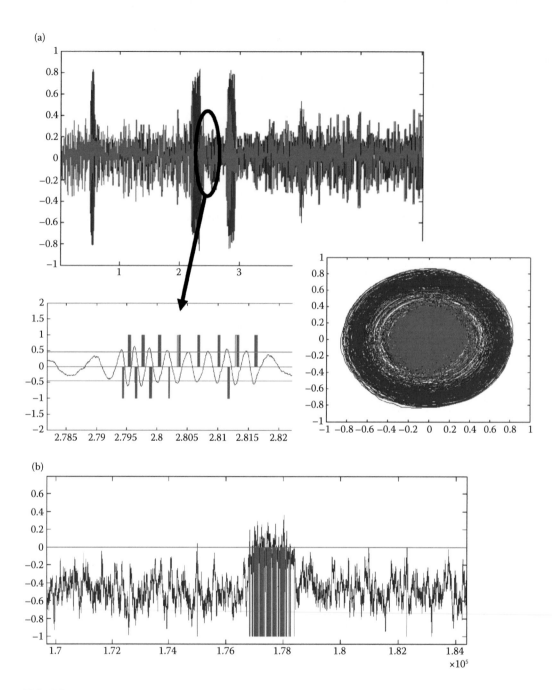

FIGURE 12.12 (See caption in text.)

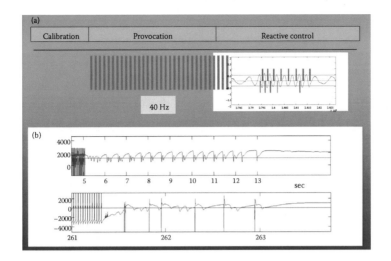

FIGURE 12.13 (See caption in text.)

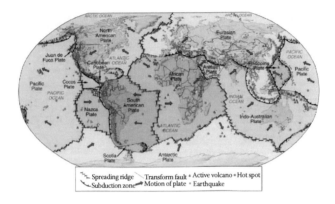

FIGURE 13.3 (See caption in text.)

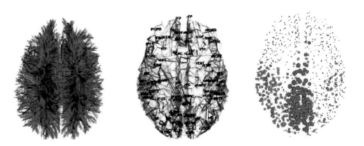

FIGURE 13.5 (See caption in text.)

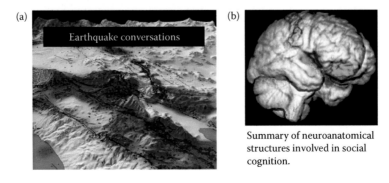

Summary of neuroanatomical structures involved in social cognition.

FIGURE 13.6 (See caption in text.)

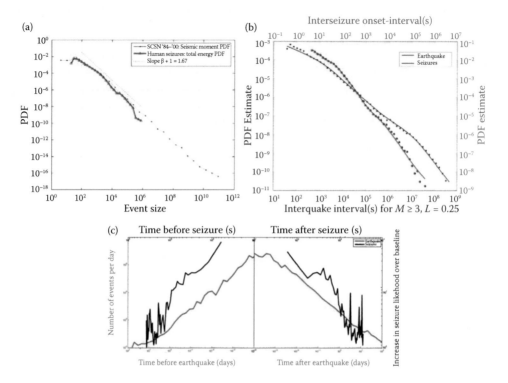

FIGURE 13.10   (See caption in text.)

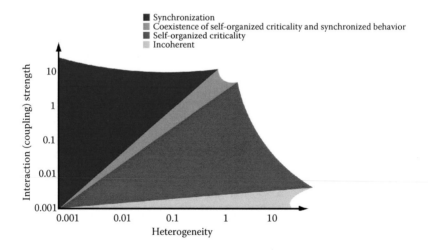

FIGURE 13.11   (See caption in text.)

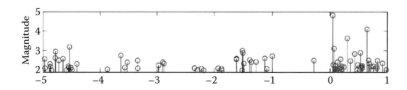

FIGURE 13.13   (See caption in text.)

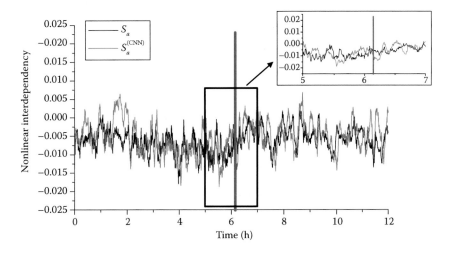

FIGURE 17.1 (See caption in text.)

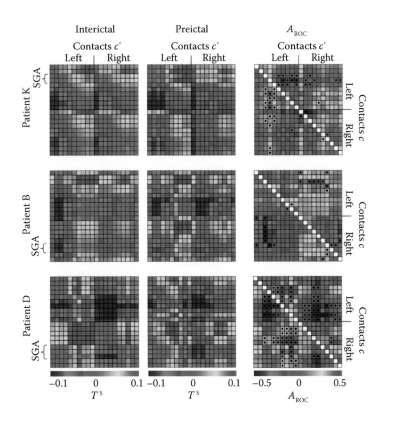

FIGURE 17.2 (See caption in text.)

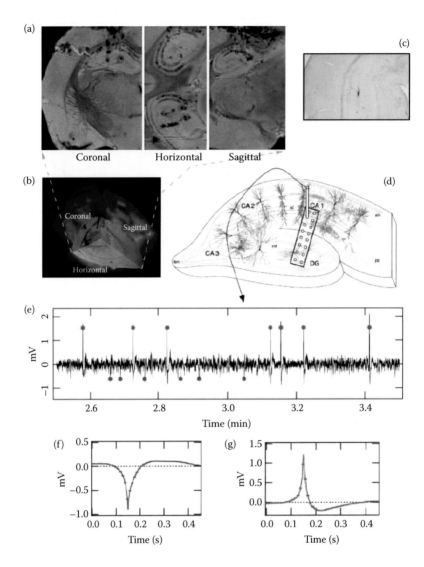

FIGURE 19.1 (See caption in text.)

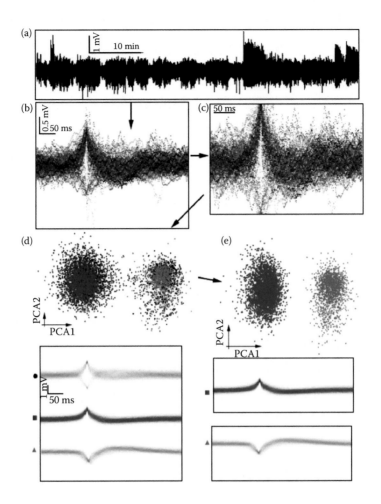

FIGURE 19.2 (See caption in text.)

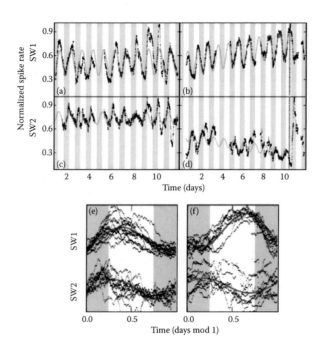

FIGURE 19.3 (See caption in text.)

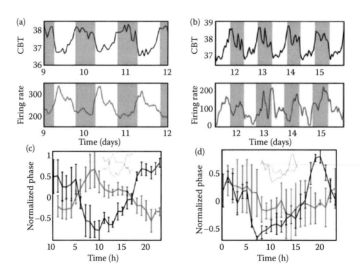

FIGURE 19.5 (See caption in text.)

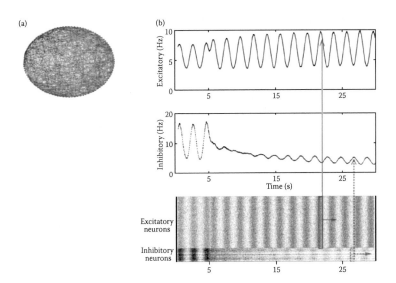

FIGURE 19.7 (See caption in text.)

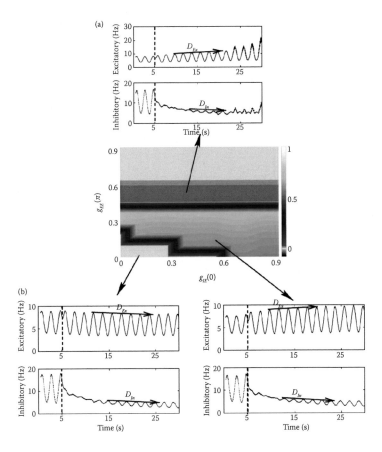

FIGURE 19.8 (See caption in text.)

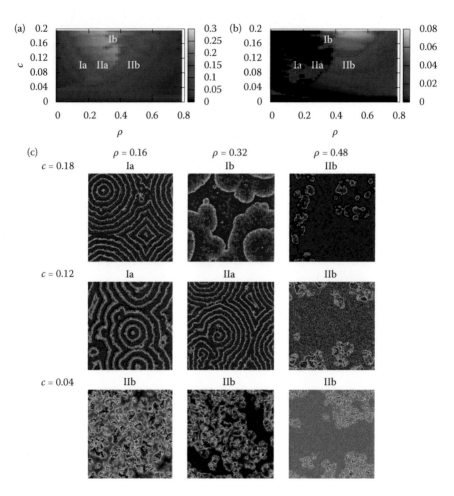

FIGURE 23.2 (See caption in text.)

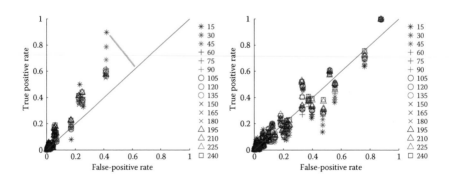

FIGURE 24.1 (See caption in text.)

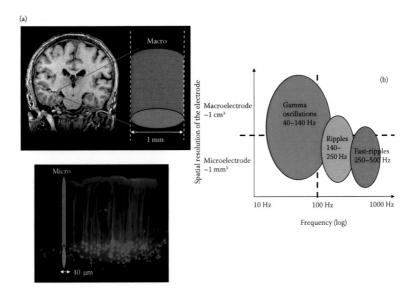

FIGURE 26.1 (See caption in text.)

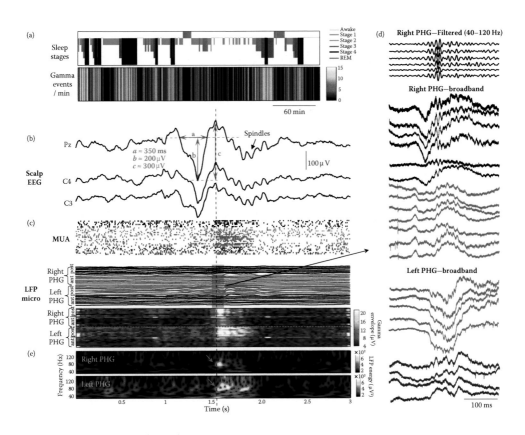

FIGURE 26.2 (See caption in text.)

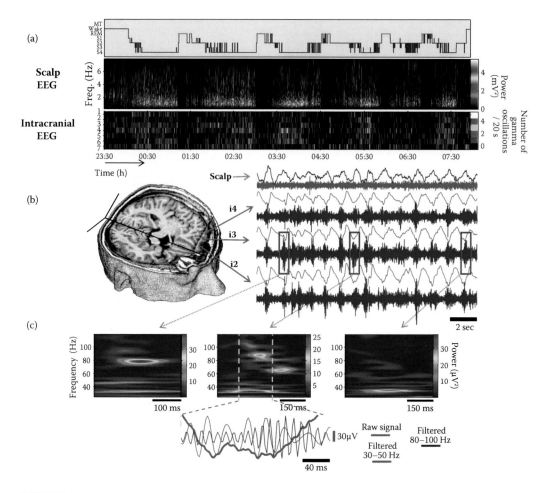

FIGURE 26.3 (See caption in text.)

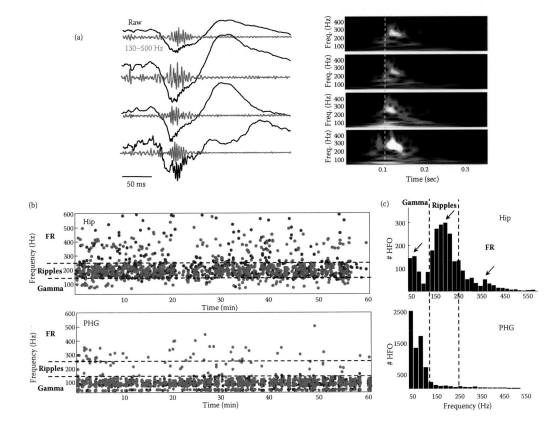

FIGURE 26.4 (See caption in text.)

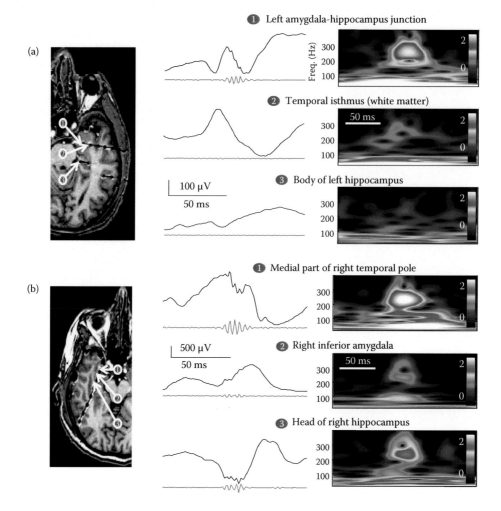

FIGURE 26.5 (See caption in text.)

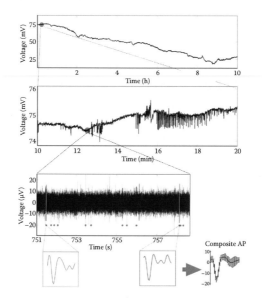

FIGURE 29.1 (See caption in text.)

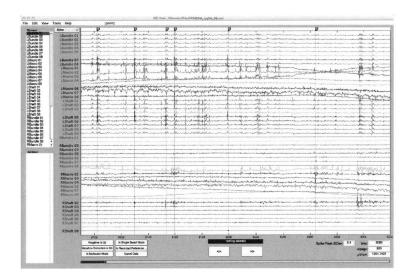

FIGURE 29.3   (See caption in text.)

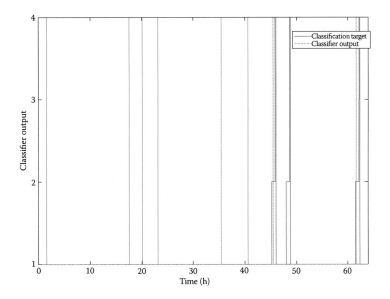

FIGURE 30.3   (See caption in text.)

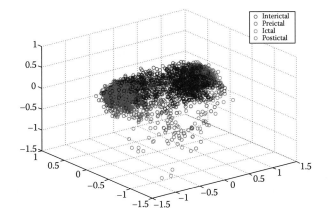

FIGURE 30.4   (See caption in text.)

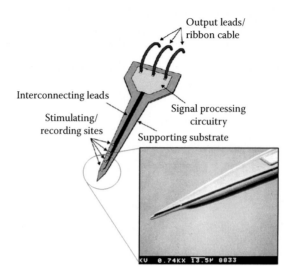

FIGURE 31.2 (See caption in text.)

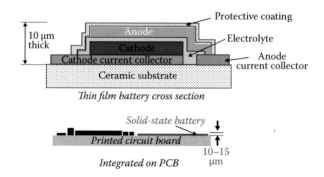

FIGURE 31.4 (See caption in text.)

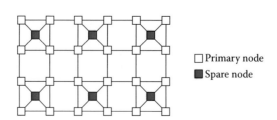

FIGURE 31.5 (See caption in text.)

17 Preictal Directed Interactions in Epileptic Brain Networks

Klaus Lehnertz, Dieter Krug, Matthäus Staniek,
Dennis Glüsenkamp, and Christian E. Elger

CONTENTS

17.1 INTRODUCTION

Synchronization plays an important role in brain function and dysfunction (Schnitzler and Gross 2005; Uhlhaas and Singer 2006; Buzsáki 2006). A prominent example for pathophysiologic neuronal synchronization is epilepsy together with its main symptom—the epileptic seizure, which reflects the clinical signs of an excessive and hypersynchronous activity of extended neuron networks in cortex. Gaining deeper insights into the complex spatial-temporal dynamics of seizure generation, spread, and termination calls for analysis techniques that allow one to characterize both strength and direction of interactions between brain regions involved in the epileptic process. Bivariate time series analysis techniques that were developed over the past years (Pikovsky et al. 2001; Boccaletti et al. 2002; Pereda et al. 2005; Stam 2005; Hlaváčková-Schindler et al. 2007; Lehnertz et al. 2009) can be classified into two different groups by which dynamical aspect—strength or direction of interaction—they try to characterize. The investigation of interaction strength, whose application dominated in former seizure prediction studies (Mormann et al. 2007; Osterhage et al. 2008), is mainly concerned with the question of whether there is an interaction between brain regions and how strong it is. Interestingly, these studies indicate seizure precursors in more remote and, in some cases, even contralateral brain areas, thus underlining the importance of brain regions outside the epileptic focus but within an epileptic network that might be responsible for ictogenesis (Lehnertz et al. 2007). More recent developments allow one to characterize the direction of interactions in order to infer possible causal relationships (in a driver-responder sense) between brain regions. We present here two approaches from the latter class of bivariate time series analysis techniques. The first approach makes use of our measures for nonlinear interdependence (Arnhold et al. 1999; Quian Quiroga et al. 2002; Chicharro and Andrzejak 2009). Since a characterization of directed interactions in multichannel EEG time series requires computational resources that grow quadratically with the number of recording sites, we approximate nonlinear interdependence with so-called cellular neural networks (CNN) (Krug et al. 2007; Chernihovskyi et al. 2008). The second approach combines the information-theoretic concept of transfer entropy (Schreiber 2000) with concepts from the theory of symbolic dynamics (Hao 1989; Daw et al. 2003) and allows a fast and robust characterization of directed interactions from EEG recordings with a measure called symbolic transfer entropy (Staniek and Lehnertz 2008). We show that symbolic transfer entropy can be calculated analytically with CNN.

17.2 MEASURING DIRECTED INTERACTIONS IN THE EPILEPTIC BRAIN

17.2.1 NONLINEAR INTERDEPENDENCE

The concept of generalized synchronization (Pikovsky et al. 2001) assumes some functional relationship between the states of two dynamical systems in the sense that states of one system can be calculated from states of the other system. Properties of this functional relationship (e.g., smoothness, differentiability, etc.) provide information about the systems' synchronization state. Unfortunately, the application of this concept to real world systems is difficult since exact properties of the relationship are usually unknown and can, in general, not be determined analytically. In a previous study (Arnhold et al. 1999), we proposed the concept of nonlinear interdependence, which allows one to estimate the strength and the direction of interactions and thus to detect driver-responder relationships.

Consider two dynamical systems X and Y and their reconstructed state space vectors $\vec{x}_n = \left(x_n, \ldots, x_{n-(m-1)\tau}\right)$ and $\vec{y}_n = \left(y_n, \ldots, y_{n-(m-1)\tau}\right)$, where m denotes the embedding dimension and τ the time delay (Takens 1981). Let $r_{n,j}$ and $s_{n,j}, j = 1, \ldots, k$ denote the time indices of the k nearest neighbors of $\vec{x}_n$ and $\vec{y}_n$, respectively. The mean squared Euclidean distance between vector $\vec{x}_n$ and its k nearest neighbors in the reconstructed state space of system X can be defined as:

$$R_n^{(k)}(X) = \frac{1}{k} \sum_{j=1}^{k} \left(\vec{x}_n - \vec{x}_{r_{n,j}}\right)^2. \tag{17.1}$$

By replacing the time indices $r_{n,j}$ of the nearest neighbors with those found in the state space of system Y, we define the Y-conditioned mean squared Euclidean distance

$$R_n^{(k)}(X \mid Y) = \frac{1}{k} \sum_{j=1}^{k} \left(\vec{x}_n - \vec{x}_{s_{n,j}}\right)^2, \tag{17.2}$$

and derive the nonlinear interdependence S as

$$S^{(k)}(X \mid Y) = \frac{1}{K} \sum_{n=1}^{K} \frac{R_n^{(k)}(X)}{R_n^{(k)}(X \mid Y)}, \tag{17.3}$$

where K denotes the total number of state space vectors. By using other averages, the nonlinear interdependences H (Arnhold et al. 1999) and N (Quian Quiroga et al. 2002) are defined as

$$H^{(k)}(X \mid Y) = \frac{1}{K} \sum_{n=1}^{K} \log \frac{R_n(X)}{R_n^{(k)}(X \mid Y)}, \tag{17.4}$$

$$N^{(k)}(X \mid Y) = \frac{1}{K} \sum_{n=1}^{K} \frac{R_n(X) - R_n^{(k)}(X \mid Y)}{R_n(X)}, \tag{17.5}$$

where $R_n(X)$ denotes the mean distance of a vector to all other state space vectors:

$$R_n(X) = \frac{1}{K-1} \sum_{j \neq n} \left(\vec{x}_n - \vec{x}_j \right)^2. \tag{17.6}$$

Note that $S^{(k)}(X|Y) \neq S^{(k)}(Y|X)$ (the same holds true for H and N). This inequality can be exploited to distinguish effectively driving and responding elements and to detect asymmetry in the interaction of subsystems by defining an asymmetric measure for nonlinear interdependence

$$S_a = \frac{S(X \mid Y) - S(Y \mid X)}{2}. \tag{17.7}$$

H_a and N_a are defined similarly based on H and N. These asymmetric measures can be used to identify the direction of an interaction between a *more active* and a *more passive* system (Arnhold et al. 1999; Quian Quiroga et al. 2002; Pereda et al. 2005; Ansari-Asl et al. 2006; Osterhage et al. 2007).

Estimating either a symmetric or an asymmetric measure for nonlinear interdependence from long-term, multichannel EEG recordings requires high computational resources that limit the use of this analysis technique, particularly with respect to real-time EEG analyses. In order to overcome this limitation, we have proposed cellular neural networks (CNN) (Chua 1998) as an alternative computational tool for EEG analyses (Sowa et al. 2005; Chernihovskyi et al. 2008). In our previous work (Krug et al. 2006, 2007, 2008) we have showed that CNN allow one to approximate a symmetric measure for nonlinear interdependence (to characterize the strength of an interaction) with an accuracy of about 90%. We here show that a similar approximation accuracy can also be achieved for an asymmetric measure for nonlinear interdependence.

In Figure 17.1 we show, as an example, the temporal evolution of S_a and its CNN-based approximation $S_a^{(CNN)}$ calculated from EEG data recorded for 12 h from within the epileptic focus and from a contralateral homologous site. We used a moving-window technique with nonoverlapping segments

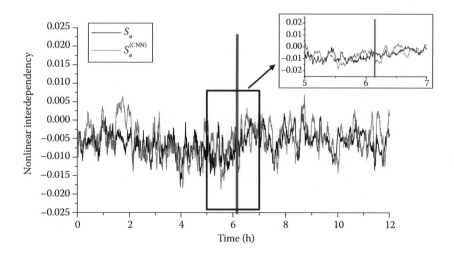

FIGURE 17.1 **(See color insert.)** Exemplary temporal evolution of nonlinear interdependence S_a and its CNN-based approximation $S_a^{(CNN)}$. Intrahippocampal EEG recordings from within the epileptic focus and from a contralateral homologous site from a patient with drug-resistant mesial temporal lobe epilepsy (200-Hz sampling frequency; 16-bit analog-to-digital converter; bandpass filter settings: 0.85–85 Hz (12 dB/oct.)). Negative values indicate the epileptic focus to be more active. The black vertical line marks time of electrical seizure onset. The inset shows a 2-h periictal period.

of 20.48 s (corresponding to 4096 data points) to estimate S_a in a time-resolved manner (embedding dimension $m = 10$, time delay $\tau = 25$, $k = 10$ nearest neighbors). With a supervised learning strategy we trained the CNN with only a few minutes of EEG data (see Krug et al. 2007 for details) and achieved an average deviation between S_a and $S_a^{(CNN)}$ of only 9%. Thus, both strength and direction of an interaction between EEG data can be approximated with CNN with high accuracy.

17.2.2 SYMBOLIC TRANSFER ENTROPY

Schreiber introduced the concept of transfer entropy (TE) to quantify the flow of information between two time series and thus to determine driving and responding systems (Schreiber 2000). Let us consider time series x_i and y_i, $i = 1,\ldots,N$ as realizations of dynamical systems X and Y. TE is closely related to the concept of Granger causality (Granger 1969) and specifies the deviation from the generalized Markov property $p(x_{i+1} \mid x_{i+1}^{(k)}, y_i^{(l)}) = p(x_{i+1} \mid x_{i+1}^{(k)})$, where k and l denote the number of previous states of X and Y, respectively. Unfortunately, TE is difficult to estimate from field data, and analysis techniques that had been proposed previously to estimate it require fine-tuning of parameters, make great demands on the data, and are highly sensitive to noise contributions (Kaiser and Schreiber 2002; Verdes 2005; Lungarella et al. 2007). In order to overcome some of these limitations, we recently proposed a modified version of TE, the so-called symbolic transfer entropy (STE) (Staniek and Lehnertz 2008) making use of a symbolization technique (Bandt and Pompe 2002). We convert a time series x_i in a symbol sequence $\hat{x}_i$ by reordering m amplitude values $X_i = \{x(i),x(i + \tau),\ldots,x(i + (m - 1)\tau)\}$ in an increasing order $\{x(i + (j_1 - 1)\tau) \leq x(i + (j_2 - 1) \leq \ldots \leq x(i + (j_m - 1)\tau)\}$, where τ denotes the time delay, and m the embedding dimension. Thus, every X_i is uniquely mapped onto one of the $m!$ possible permutations and a symbol can be defined as $\hat{x}_i \equiv (j_1, j_2, \ldots, j_m)$. With symbol sequences $\hat{x}_i$ and $\hat{y}_i$ joint and conditional probabilities can be estimated using the relative frequency of symbols, and STE is defined as

$$T^S(Y,X) = \sum p\left(\hat{x}_{i+1}, \hat{x}_i^{(k)}, \hat{y}_i^{(l)}\right) \log \frac{p\left(\hat{x}_{i+1} \mid \hat{x}_i^{(k)}, \hat{y}_i^{(l)}\right)}{p\left(\hat{x}_{i+1} \mid \hat{x}_i^{(k)}\right)}. \tag{17.8}$$

The index $T^S(X, Y)$ is defined in complete analogy. We use $k = l = 1$ in the remainder, i.e., only a single previous state of systems X and Y is incorporated in the probabilities of transition to the next state of X. With the directionality index $T^S = T^S(X, Y) - T^S(X, Y)$, we quantify the preferred direction of information flow. For unidirectional couplings with X as the driver, positive values of T^S are expected. On the contrary, negative values are expected, if system Y drives X. For symmetric bidirectional couplings $T^S \approx 0$ holds.

We now present our findings obtained from a time-resolved analysis of the preferred direction of information flow between EEG recordings from the seizure–generating area and from other brain regions. We retrospectively analyzed long-term (average duration: 156 h, range: 47–286 h) EEG data that had been recorded intrahippocampally (total number of electrode contacts: $L = 20$) from 15 patients with drug-resistant temporal lobe epilepsy during presurgical evaluation (200-Hz sampling frequency; 16 bit analog-to-digital converter; bandpass filter settings: 0.85–85 Hz (12 dB/ oct.)). EEG recordings captured 79 seizures (mean: 5.2, range: 3–10 per patient). All patients signed informed consent forms that their clinical data might be used and published for research purposes. All patients achieved complete seizure control after surgery, thus the seizure generating area can be assumed to be contained within the resected area.

We used a moving-window technique (window duration: 20.48 s ($N = 4096$ data points); no overlap between consecutive segments) to calculate the directionality index T^S for all combinations of pairs of electrode contacts (c, c') in a time-resolved manner. We hypothesized that a preictal state is accompanied by a transition in the preferred direction of information flow between different brain regions. Assuming that a preictal period of 4-h duration exists, we compared the frequency distributions of values of T^S from the preictal periods with those from interictal periods (all data that were recorded at least 4 h prior to and 1 h after a seizure) using an evaluation scheme based on receiver operating characteristic (ROC) statistics (Mormann et al. 2005). As a quantifying criterion for the difference between preictal and interictal frequency distributions, we used the deviation of the ROC area from 0.5 (i.e., identical distributions), A_{ROC} (c, c'), and assessed its statistical significance with seizure time surrogates (Andrzejak et al. 2003).

In Figure 17.2 we show exemplary matrices representing the average preferred direction of information flow between different brain regions (sampled from pairs of electrode contacts (c, c')) during interictal and preictal periods in three patients together with a matrix that comprises A_{ROC} values for each pair of contacts. In patient K, there was a preferred direction of information flow from the seizure-generating area to all other ipsilateral and contralateral brain regions during both interictal and preictal periods. The corresponding A_{ROC} values indicated that this focal driving decreased significantly during the preictal period. On the contrary, in Patient B we observed a preferred direction of information flow from ipsi- and contralateral brain regions to the seizure-generating area during both interictal and preictal periods. Preictally, the driving of contralateral brain regions by ipsilateral brain regions was significantly decreased. In Patient D, there was a preferred direction of information flow from all ipsilateral but nonfocal brain regions to contralateral brain regions both interictally and preictally, but during the latter period, the driving from ipsilateral to contralateral brain regions was significantly enhanced.

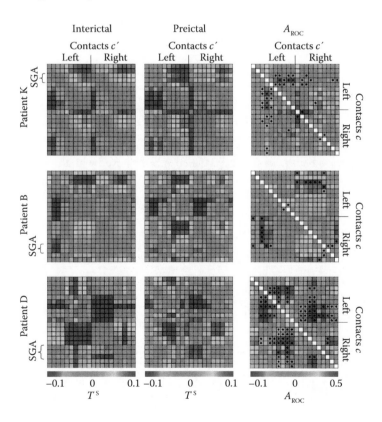

FIGURE 17.2 **(See color insert.)** Average preferred direction of information flow between different brain regions during interictal (left) and preictal (middle) periods and corresponding A_{ROC} values (right) for three patients. Positive values of the directionality index T^S indicate that the brain region sampled at contact c ($R(c)$) drives the brain region sampled at contact c' ($R(c')$). SGA marks electrode contacts that cover the seizure-generating area. Negative A_{ROC} values denote a decreased influence of $R(c)$ on $R(c')$ during preictal periods. Black dots denote significant ($p < 0.05$) alterations after validation with seizure times surrogates.

In addition to these spatially resolved analyses of preictal directed interactions in the epileptic brain, we also investigated global preictal interactions between all ipsilateral (I) and contralateral (C) recording sites using

$$\bar{A} = \frac{1}{(L/2)^2} \sum_{c \in I; c' \in C} A_{\mathrm{ROC}}(c, c')$$

(17.9)

as a quantifying measure. By definition, $A_{\mathrm{ROC}}(c, c')$ will be zero for identical distributions or can attain negative or positive values. Thus, $\bar{A} < 0$ indicates a global decrease of the preferred direction of information flow from all ipsilateral to all contralateral recording sites during preictal periods. We observed $\bar{A} < 0$ in 11 out of 15 patients (see Figure 17.3) and concluded that a decreased global influence of the seizure-generating area in an epileptic brain network appears to reflect a preictal state, at least in the majority of patients.

We note that symbolic transfer entropy can be calculated analytically with CNN. With the relationship $p(x \mid y) = \frac{p(x,y)}{p(y)}$ for the conditional probability and using algebraic identities and laws for the logarithm, we can rewrite Equation 17.8 as a sum of Shannon entropies (Glüsenkamp 2009):

$$\begin{aligned}
T^S(Y, X) &= \sum p(\hat{x}_{i+1}, \hat{x}_i^{(k)}, \hat{y}_i^{(l)}) \log\left(p(\hat{x}_{i+1}, \hat{x}_i^{(k)}, \hat{y}_i^{(l)})\right) \\
&+ \sum p(\hat{x}_i^{(k)}) \log\left(p(\hat{x}_i^{(k)})\right) \\
&- \sum p(\hat{x}_i^{(k)}, \hat{y}_i^{(l)}) \log\left(p(\hat{x}_i^{(k)}, \hat{y}_i^{(l)})\right) \\
&- \sum p(\hat{x}_{i+1}, \hat{x}_i^{(k)}) \log\left(p(\hat{x}_{i+1}, \hat{x}_i^{(k)})\right).
\end{aligned}$$

(17.10)

An efficient CNN-based binarization of EEG data can be achieved with the so-called *thresholding template* and the Shannon entropy can be approximated by counting the frequency of occurrence of each symbol with the so-called *symbol-matching template* (Wang 2007). Both templates are from

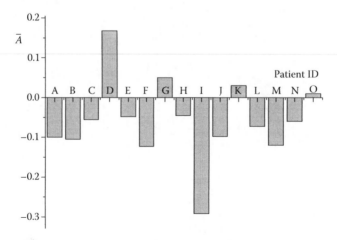

FIGURE 17.3 Alterations of the preferred direction of information flow from the ipsilateral to the contralateral hemisphere during the preictal state in 15 patients. Negative values of $\bar{A}$ indicate a global decrease of the influence of the ipsilateral on the contralateral hemisphere.

the standard CNN-template library (Chua 1998) and can be executed directly on existing CNN software and hardware platforms.

17.3 CONCLUSION

We have presented two bivariate EEG analysis approaches that allow one to characterize the direction of interactions between brain regions. The first approach exploits geometric properties in the reconstructed state spaces, and nonlinear interdependence quantifies both strength and direction of an interaction. The second approach quantifies the preferred direction of information flow between two time series with symbolic transfer entropy. We have shown that both measures for directed interactions can be efficiently estimated with CNN. Although an unambiguous inference of directed interaction from EEG recordings is far from being resolved, our findings indicate that a decreased global influence of the focus in an epileptic brain network appears to reflect a preictal state, at least in the majority of patients. Future studies should include analyses of EEG data from a larger group of patients and a comparison with other approaches to assess directed interactions in epileptic brain networks.

ACKNOWLEDGMENT

This work was supported by the Deutsche Forschungsgemeinschaft.

REFERENCES

Andrzejak, R. G., F. Mormann, T. Kreuz, C. Rieke, A. Kraskov, C. E. Elger, and K. Lehnertz. 2003. Testing the null hypothesis of the nonexistence of a preseizure state. *Phys. Rev. E* 67:010901(R).

Ansari-Asl, K., L. Senhadji, J.-J. Bellanger, and F. Wendling. 2006. Quantitative evaluation of linear and nonlinear methods characterizing interdependencies between brain signals. *Phys. Rev. E* 74:031916.

Arnhold, J., P. Grassberger, K. Lehnertz, and C. E. Elger. 1999. A robust method for detecting interdependences: Application to intracranially recorded EEG. *Physica D* 134:419–430.

Bandt, C., and B. Pompe. 2002. Permutation entropy: A natural complexity measure for time series. *Phys. Rev. Lett.* 88:174102.

Boccaletti, S., J. Kurths, G. Osipov, D. L. Valladares, and C. S. Zhou. 2002. The synchronization of chaotic systems. *Phys. Rep.* 366:1–101.

Buzsáki, G. 2006. *Rhythms of the brain.* New York: Oxford University Press, USA.

Chernihovskyi, A., D. Krug, C. E. Elger, and K. Lehnertz. 2008. Time series analysis with cellular neural networks. In *Seizure prediction in epilepsy. From basic mechanisms to clinical applications*, B. Schelter, J. Timmer, and A. Schulze-Bonhage, eds., 131–148. New York: Wiley-VCH.

Chicharro, D., and R. G. Andrzejak. 2009. Reliable detection of directional couplings using rank statistics. *Phys. Rev. E* 80:026217.

Chua, L. O. 1998. *CNN: A paradigm for complexity.* Singapore: World Scientific.

Daw, C., C. Finney, and E. Tracy. 2003. A review of symbolic analysis of experimental data. *Rev. Sci. Instrum.* 74:915–930.

Glüsenkamp, D. 2009. *Messung direktionaler Abhängigkeiten in dynamischen Systemen mit Zellularen Nichtlinearen Netzen (in German)*, Diploma thesis, Dept. of Epileptology and Helmholtz-Institute for Radiation and Nuclear Physics, University of Bonn.

Granger, C. 1969. Investigating causal relations by econometric models and cross–spectral methods. *Econometrica* 37:424–438.

Hao, B. L. 1989. *Elementary symbolic dynamics and chaos in dissipative systems.* Singapore: World Scientific.

Hlaváčková-Schindler, K., M. Paluš, M. Vejmelka, and J. Bhattacharya. 2007. Causality detection based on information-theoretic approaches in time series analysis. *Phys. Rep.* 441:1–46.

Kaiser, A., and T. Schreiber. 2002. Information transfer in continuous processes. *Physica D* 166:43–62.

Krug, D., A. Chernihovsky, H. Osterhage, C. E. Elger, and K. Lehnertz. 2006. Estimating generalized synchronization in brain electrical activity from epilepsy patients with cellular nonlinear networks. In *Proceedings of the 2006 10th IEEE International Workshop on Cellular Neural Networks and their Applications.* V. Tavsanoglu and S. Arik, eds. Piscataway, NJ: IEEE Press.

Krug, D., C. E. Elger, and K. Lehnertz. 2008. A CNN-based synchronization analysis for epileptic seizure prediction: Inter- and intraindividual generalization properties. In *Proceedings of the 2008 11th International Workshop on Cellular Neural Networks and their Applications.* D. L. Vilarino, D. C. Ferrer, and V. M. B. Sanchez, eds. Piscataway, NJ: IEEE Press.

Krug, D., H. Osterhage, C. E. Elger, and K. Lehnertz. 2007. Estimating nonlinear interdependences in dynamical systems using cellular nonlinear networks. *Phys. Rev. E* 76:041916.

Lehnertz, K., S. Bialonski, M.-T. Horstmann, D. Krug, A. Rothkegel, M. Staniek, and T. Wagner. 2009. Synchronization phenomena in human epileptic brain networks. *J. Neurosci. Methods.* 183:42–48.

Lehnertz, K., M. Le Van Quyen, and B. Litt. 2007. Seizure prediction. In *Epilepsy: A comprehensive textbook, 2nd ed.* J. Engel Jr. and T. A. Pedley, eds., 1011–1024. Philadelphia: Lippincott, Williams & Wilkins.

Lungarella, M., A. Pitti, and Y. Kuniyoshi. 2007. Information transfer at multiple scales. *Phys. Rev. E* 76:056117.

Mormann, F., R. Andrzejak, C. E. Elger, and K. Lehnertz. 2007. Seizure prediction: The long and winding road. *Brain* 130:314–333.

Mormann, F., T. Kreuz, C. Rieke, R. G. Andrzejak, A. Kraskov, P. David, C. E. Elger, and K. Lehnertz. 2005. On the predictability of epileptic seizures. *Clin. Neurophysiol.* 116:569–587.

Osterhage, H., S. Bialonski, M. Staniek, K. Schindler, T. Wagner, C. E. Elger, and K. Lehnertz. 2008. Bivariate and multivariate time series analysis techniques and their potential impact for seizure prediction. In *Seizure prediction in epilepsy. From basic mechanisms to clinical applications.* B. Schelter, J. Timmer, and A. Schulze-Bonhage, eds., 189–208. Weinheim: Wiley–VCH.

Osterhage, H., F. Mormann, T. Wagner, and K. Lehnertz. 2007. Measuring the directionality of coupling: Phase versus state space dynamics and application to EEG time series. *Int. J. Neural. Syst.* 17:139–148.

Pereda, E., R. Quian Quiroga, and J. Bhattacharya. 2005. Nonlinear multivariate analysis of neurophysiological signals. *Prog. Neurobiol.* 77:1–37.

Pikovsky, A. S., M. G. Rosenblum, and J. Kurths. 2001. *Synchronization: A universal concept in nonlinear sciences.* Cambridge: Cambridge University Press.

Quian Quiroga, R., A. Kraskov, T. Kreuz, and P. Grassberger. 2002. Performance of different synchronization measures in real data: A case study on electroencephalographic signals, *Phys. Rev. E* 65:041903.

Schnitzler, A., and J. Gross. 2005. Normal and pathological oscillatory communication in the brain. *Nat. Rev. Neurosci.* 6:285–296.

Schreiber, T. 2000. Measuring information transfer. *Phys. Rev. Lett.* 85:461–464.

Sowa, R., A. Chernihovskyi, F. Mormann, and K. Lehnertz. 2005. Estimating phase synchronization in dynamical systems using cellular nonlinear networks. *Phys. Rev. E* 71:061926.

Stam, C. J. 2005. Nonlinear dynamical analysis of EEG and MEG: Review of an emerging field. *Clin. Neurophysiol.* 116:2266–2301.

Staniek, M., and K. Lehnertz. 2008. Symbolic transfer entropy. *Phys. Rev. Lett.* 100:158101.

Takens, F. 1981. Detecting strange attractors in turbulence. In *Dynamical systems and turbulence (Warwick 1980),* Vol. 898 of *Lecture notes in mathematics,* D. A. Rand and L.-S. Young, eds., 366–381. Berlin: Springer-Verlag.

Uhlhaas, P. J., and W. Singer, 2006. Neural synchrony in brain disorders: Relevance for cognitive dysfunctions and pathophysiology. *Neuron* 52:155–168.

Verdes, P. F. 2005. Assessing causality from multivariate time series. *Phys. Rev. E* 72:026222.

Wang, L. 2007. *Detection of complexity changes in dynamical systems with symbolic analysis and cellular neural networks.* Diploma thesis, Dept. of Epileptology and Helmholtz-Institute for Radiation and Nuclear Physics, University of Bonn.

18 Seizure Prediction and Observability of EEG Sources

Elma O'Sullivan-Greene, Levin Kuhlmann,
Andrea Varsavsky, David B. Grayden,
Anthony N. Burkitt, and Iven M. Y. Mareels

CONTENTS

18.1 INTRODUCTION

On the topic of obstacles to seizure prediction, from a summary of informal discussions at the Third International Seizure Prediction Workshop in Freiburg, Zaveri, Frei, and Osario noted that, "put bluntly,... it is not necessarily possible to predict states of a dynamical system on the basis of a limited amount of measurements" (Zaveri et al. 2008). This chapter seeks to formally investigate what it is we can expect to glean from the electroencephalogram (EEG) toward the effort of seizure prediction.

Epileptic seizures are commonly associated with "abnormally excessive or synchronous neuronal activity in the brain" (Fisher et al. 2005). Visual inspection of the EEG in people with epilepsy indicates that transitions from nonseizure to seizure states often occur with a synchronization of the recorded voltages on several electrodes. This has led to the hypothesis that seizures correspond to a pathological synchronization of neural activity across different brain regions.

It follows that much of the seizure prediction work to date has attempted to track synchrony across brain regions through the application of linear and nonlinear techniques to both scalp and intracranial EEG data. Linear methods applied have included crosscorrelation (Mormann et al. 2005) and phase analysis based on the Hilbert transform (Mormann et al. 2000; Le Van Quyen et al. 2005). Several nonlinear techniques have utilized the Takens/Aeyels embedding theorem (Takens 1981; Aeyels 1981) in an effort to reconstruct the state space, with embedding dimensions of 5–16. Measures including correlation dimension (Babloyantz and Destexhe 1986; Lehnertz and Elger 1998) and Lyapunov exponents (Iasemidis et al. 1990) were applied to these reconstructed state

273

spaces to investigate if there was a predictable transition to a lower-dimensional (more synchronized) brain state. All these approaches to EEG signal analysis, both linear and nonlinear, have failed to reliably predict seizures better than a random predictor (Mormann et al. 2007).

In light of this failure to predict seizures as the onset of mass synchrony in the brain, we ask the question of what may we actually expect from a low-dimensional signal—multichannel EEG—extracted from a high-dimensional system—the brain—using finite precision measurements. We abstract the problem to the study of a simple network of linear oscillators with linear interconnection. Such a model neglects the complexities of biologically realistic neuron-dynamics and instead formulates the problem as one of a generic network of oscillators upon which an EEG-like measurement is made. In this context, the problem constraints are scale (i.e., the number of individual oscillators in the system) and measurement resolution (i.e., accuracy of the EEG measurement). Through simulation of this observability problem we show that the EEG is but a poor reflection of the actual underlying brain activity and may be of itself insufficient to allow us to infer the nature of the dynamics for the purpose of seizure prediction.

To complement the observability analysis, we also present an investigation of synchrony-based seizure prediction for prediction horizons less than 15 minutes. While longer prediction horizons may be informative, we have focused on those less than 15 minutes because we are primarily interested in prediction for an implantable seizure control device. This timeframe should be adequate for intervention through, for example, local drug delivery or electrical stimulation. The results demonstrate that synchrony-based seizure prediction for short prediction horizons performs similarly to random prediction. These findings are consistent with the observability results and taken together they highlight specific shortcomings of the EEG signal.

18.2 OBSERVABILITY IN A COUPLED OSCILLATOR MODEL FOR EEG

We propose to model brain dynamics as a coupled oscillator model, where seizure dynamics are represented as synchronization events. The brain is simulated as a matrix of N coupled oscillators, each of which represents a region of brain tissue that is oscillating at a particular frequency. The EEG is modeled as the weighted sum of the activity generated by all these oscillations. The region represented by each oscillation is arbitrary—it can represent one neuron, a small group of neurons, or medium/large-scale network activity. This generic model is scalable from microelectrode recordings, where each oscillator represents a small group of neurons to intracranial EEG from macroelectrodes, where each oscillator represents an area of cortex.

A very similar coupled-oscillator model was described by Wright et al. in 1985 to model state changes in the brain (Wright 1985). Their motivation was that the existing brain models of the time were constrained by "simplified neuronal relationships and laws of interaction," whose "ideas are difficult to put to critical test, and each [model] necessarily ignores certain problems treated in the others." Mathematical models have developed considerably since this work; however, Wright's comments still stand. These newer models are not yet at a level of sophistication suitable for parameter fitting to EEG for the specific purpose of seizure prediction.

While synchrony of coupled oscillators within the brain is undoubtedly a nonlinear process, here we limit a first analysis to the linear case and investigate observability using linear methods. If the extent of information that an EEG-like output can reveal is limited in the linear case, nonlinear efforts are unlikely to be more productive.

In our model, each oscillator is described by the well-known equation describing the movement of a pendulum clock,

$$\frac{d^2\theta_i(t)}{dt^2} + \omega_i^2\theta_i(t) = F_i(t),$$ (18.1)

where $i = 1, 2, \ldots, N$ is the oscillator number, $\theta_i(t)$, is the angular position (in radians) of the pendulum i at time t, ω_i is the natural frequency of oscillation of the ith oscillator and $F_i(t)$ is the forcing term for the ith oscillator.

An example system with $N = 4$ interconnected oscillators is shown in Figure 18.1. The coupling between oscillator i and oscillator j is represented by the parameter α_{ij}. We assume that coupling is positive ($\alpha_{ij} \geq 0$) and symmetric ($\alpha_{ij} = \alpha_{ji}$). The forcing term $F_i(t)$ of each oscillator is described by the sum of its inputs, so that

$$F_i(t) = F_{in} + \sum_{j=1}^{N} \alpha_{ij}(\theta_j(t) - \theta_i(t)). \tag{18.2}$$

$F_{in}(t)$ is the external input to each oscillator and can be assumed to be a sinusoid of frequency ω_{in}, noise, zero, or any other realistic function to describe input from outside the area being modeled.

A model of the generated EEG waveform is the weighted sum of all the oscillations, where each oscillation is described by the solution to Equation 18.1 with N networked oscillators or clocks. To find this solution, we derive a linear state-space equation,

$$\frac{d\boldsymbol{\theta}(t)}{dt} = \mathbf{A}\boldsymbol{\theta}(t) + \mathbf{B}u(t); \tag{18.3}$$

$u(t)$ is the input signal, $\mathbf{A}$ and $\mathbf{B}$ are the system and input matrices, respectively, and $\boldsymbol{\theta}(t)$ is a vector that defines all the states in the system.

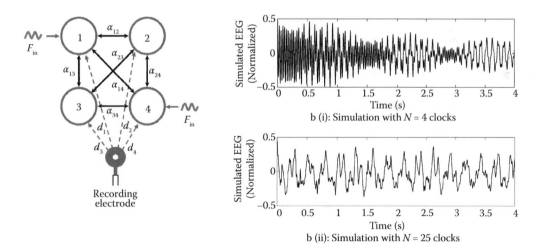

(a) Model of coupled oscillator

b (i): Simulation with $N = 4$ clocks

b (ii): Simulation with $N = 25$ clocks

(b) Example of model's EEG electrode recordings

FIGURE 18.1 (a) An example of coupled oscillator model with $N = 4$ clocks. Each clock generated a waveform according to Equation 18.1 and contributes to the behavior of all the other clocks through constant coupling parameters α_{ij}, as shown. F_{in} is a forcing term representing the possibility of an input forcing signal. The EEG is modeled as the measurement taken from a sensor at distance d_i to each clock, and is computed as in Equation 18.6. (b) EEG simulations using the coupled oscillator model illustrated in (a) show waveforms that are consistent with typical EEG sequences, even with relatively few clocks in the network.

Equation 18.1 can be described by two states, so that a system with $N = 2$ clocks has a total of 4 states. When $N = 2$, these states are defined as

$$\boldsymbol{\theta}(t) = \begin{bmatrix} \theta_{11}(t) \\ \theta_{12}(t) \\ \theta_{21}(t) \\ \theta_{22}(t) \end{bmatrix} = \begin{bmatrix} \theta_1(t) \\ d\theta_1/dt \\ \theta_2(t) \\ d\theta_2/dt \end{bmatrix} \tag{18.4}$$

and

$$\frac{d\boldsymbol{\theta}}{dt} = \begin{bmatrix} 0 & 1 & 0 & 0 \\ -\omega_1^2 - \alpha_{12} & 0 & +\alpha_{12} & 0 \\ 0 & 0 & 0 & 1 \\ +\alpha_{21} & 0 & -\omega_2^2 - \alpha_{21} & 0 \end{bmatrix} \boldsymbol{\theta}(t) + \begin{bmatrix} 0 \\ 1 \end{bmatrix} F_{in}(t). \tag{18.5}$$

This can be arbitrarily extended to an N-clock system by adding two states per clock and ensuring that all the coupling parameters are included in the matrix $\mathbf{A}$. An EEG waveform $y(t)$ is modeled as the linear combination of the states $\boldsymbol{\theta}(t)$,

$$y(t) = \mathbf{C}\boldsymbol{\theta}(t) \tag{18.6}$$

$\mathbf{C}$ is a projection matrix that normalizes the spectrum of the transfer function to unit energy and weights the contributions of $\theta_i(t)$ according to the distance of the oscillator from the electrode. In the case of $N = 2$,

$$\mathbf{C} = \begin{bmatrix} 0 & \dfrac{1}{d_1} & 0 & \dfrac{1}{d_2} \end{bmatrix}, \tag{18.7}$$

where d_i is the constant that scales $d\theta_i/dt$ according to how far the clock i is from the electrode in unit distance.

Finally, a model of the *measured* EEG signal, as opposed to the actual signal, is $y(t)$ sampled at discrete times $n = 1,2,3,\ldots$. Each sample from the digitized signal $y[n]$ is stored to a resolution of K bits, so that a measurement can represent a maximum of 2^K values.

This model is a very simple and idealized description of the EEG because the physiology that determines the oscillations is not considered and it describes regional brain activity as oscillations at a single frequency. Nevertheless, simulations shown in Figure 18.1b reveal behavior that could easily be part of a real EEG sequence. Thus, although limited, this model can replicate realistic behavior. In the next section, we investigate what information about the states $\boldsymbol{\theta}(t)$ can be recovered from the EEG signal $y[n]$.

18.2.1 A Look at the Observability of This System

To predict seizures through synchrony, we would need to be able to track the behavior of our oscillators over time. This essentially formulates the prediction problem as one of observability of the underlying generators or sources of EEG. In this section, we ask the question, "How much can the EEG measurement tell us about the underlying generators?" The coupled oscillator model is appropriate to address this question because we know exactly what the generators are, however unrealistic these may be. There is a well-developed theory about the observability of a model in the form of Equations 18.3 and 18.6. In control theory, observability is a measure that gives an estimate of whether the states in $\theta(t)$ can be reconstructed from a particular measurement, in this case the EEG $y[n]$. If our system is observable then, given an EEG record, we can compute the amplitudes and phases of each of the N coupled oscillators. If our system is not observable, there is no way of knowing the activity of all the oscillators.

In the continuous sense, a system is observable if the matrix, ϑ, is full rank, where

$$\vartheta = \begin{bmatrix} \mathbf{C} & \mathbf{CA} & \mathbf{CA}^2 & \ldots & \mathbf{CA}^{N-1} \end{bmatrix}'. \tag{18.8}$$

Here, $'$ denotes transpose and full rank implies ϑ contains N independent directions as its column (or, equally, row) vectors. Full rank essentially means that we can construct N simultaneous equations to solve for the initial state from our series of output measurements. As each clock in our model has two states, we would need the rank N to be double the number of clocks. With an infinite precision EEG signal, observability theory tells us that the observability matrix for our system of coupled oscillator clocks is full rank (O'Sullivan-Greene et al. 2009b), however, when the EEG is measured to a finite resolution, this is not the case. A matrix, ϑ, can be decomposed into its directions through singular value decomposition (Antoulas 2005). By examining the magnitude of the singular-values, we can determine the percentage of states that are observable. Given, K bits in our EEG measurement, we have 2^K levels with which to represent the range of values from largest to smallest singular value. To make our system as well posed as possible, in the sense that we wish to consider the limit of observability under the best possible conditions, we consider our observability matrix in the digital sense where $\mathbf{A}$ is transformed to $e^{\mathbf{A}T}$, with T being the digital sample time in seconds.

$$\vartheta = \begin{bmatrix} \mathbf{C} & \mathbf{C}e^{\mathbf{A}T} & \mathbf{C}e^{2\mathbf{A}T} & \ldots & \mathbf{C}e^{(N-1)\mathbf{A}T} \end{bmatrix}' \tag{18.9}$$

This transformation, together with the normalization of clock frequencies, ω, ensures that the magnitude of the singular values of the ϑ matrix do not blow up to infinity or become arbitrarily small for large N (since the magnitude of $e^{(N-1)\mathbf{A}T}$ remains unity for all N). The normalization also ensures that the magnitude of the largest singular value is of the order of unity. It follows that the percentage of states observable is the percentage of singular values with magnitudes that lie above $1/2^K$, where K is the number of bits. For example, with 10 bit sampling and the largest singular value of magnitude 1, the smallest singular value we can distinguish is one with a magnitude above $1/2^{10} \approx 10^{-3}$. In general terms, the visible oscillators from the EEG are those whose state pairs have corresponding singular value with magnitude greater than $1/2^K$. For the interested reader, a more complete treatment of observability theory can be found in Kailath's text on Linear Systems (Kailath 1980).

In Figure 18.2, the percentage of observable states as a function of increasing coupling strength, α, is shown for the clock network with normalized frequencies. The oscillator frequencies were chosen at random from a uniform distribution in the range (0.0001, 1). We simulated 50 trials of frequency distribution and the plotted observability percentage is the mean value over these trials. We find almost full observability for $\alpha = 1$. Observability is almost, but not exactly, 100% due to the stochastic nature of the frequency distribution. Coupling was chosen here to have a geometrically decaying structure with strongest coupling strength to the nearest neighboring oscillator. Oscillator i is coupled to its nearest neighbors (oscillator $i \pm 1$) with strength α and to oscillators $i \pm n$ with strength α^n where $\alpha \leq 1$. Coupling strength in the brain between neurons and between groups of neurons is certainly much less than one. One crude estimate to illustrate the likely order

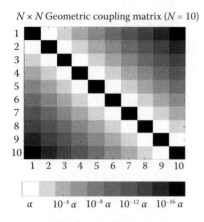

$N \times N$ Geometric coupling matrix ($N = 10$)

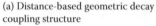

α $10^{-4}\alpha$ $10^{-8}\alpha$ $10^{-12}\alpha$ $10^{-16}\alpha$

(a) Distance-based geometric decay coupling structure

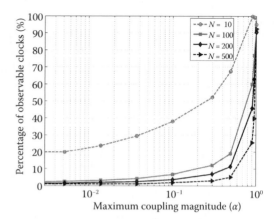

(b) Observability as a function of coupling strength for the coupling structure in (a)

FIGURE 18.2 (a) An image showing the structure of the coupling matrix for a 2-D network of 10 clocks. The white boxes indicate that each clock is coupled to its nearest neighbors with the maximum coupling magnitude of α. Coupling strength to other clocks is geometrically decaying by distance of clock i to clock j as indicated by the accompanying grayscale bar. The black boxes indicate that each clock is not coupled to itself. (b) The percentage of observable clocks as a function of coupling strength, for various networks of N clocks. For large coupling strengths, α approaching 1, the system is quite observable; however, realistic strength of connection between brain areas is on the order of 10^{-3} where observability is poor.

of coupling strength is based on numbers of connections. Take for example a minicolumn of cortical tissue (0.03 mm radius, 3 mm height), one of the approximate 2×10^8 minicolumn volumes in the whole cortex, each containing about 100 neurons with around 10,000 synapses per neuron (Nunez 2006). Arbitrarily assume we require five synapses to establish a connection between minicolumns. Then, the number of available corticocortical connections that emanate from a single minicolumn is 200,000 (10,000 $\times$ 100 ÷ 5). Given this limit, we can find the ratio of available connections to the total number of connections for a network of minicolumns that was fully interconnected (in the sense of each minicolumn connecting to all the others) (Johansson and Lanser 2004). That connection ratio is just 1/1000.

$$\frac{\text{Available Connections}}{\text{Total Possible Connections}} = \frac{200,000 \times 2.10^8}{(2.10^8)^2} = \frac{1}{1000}. \tag{18.10}$$

An $\alpha = 1/1000$ corresponds to a very low level of observability as shown on the left-hand side of the plot in Figure 18.2b.

We have shown how few underlying oscillators we can see given the small coupling strengths we expect in the real brain. However, even for a reasonably large coupling strength of $\alpha = 0.6$, we can show how rapidly observability is lost as the size of our network increases. Figure 18.3 shows the percentage of observable states as a function of N, the number of clocks, when a single EEG measurement with resolutions $K = 10$, 14, and 24 bits is used. This plot shows that observability rapidly falls off as the number of clocks in the network increases with approximately a $1/N$ scaling law. When $K = 14$, like in many typical EEG machines (Compumedics 2002), less than 5% of all states can be reconstructed, even in a relatively small system with $N = 1000$ oscillators (2000 states). This is very limiting given that, in a system such as the brain, where the number of neurons is on the order of 10^{11}, we expect to find a much larger number of oscillators than 1000. Thus, the EEG

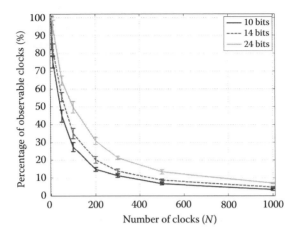

FIGURE 18.3   The percentage of observable clocks is plotted as a function of increasing network order, N (where N is the number of clocks). The network was coupled using a geometric distance-based coupling as illustrated in Figure 18.2, with a maximum coupling strength of 0.6, corresponding to the far right-hand side of Figure 2b. The error bars show mean and standard deviation. The amount of oscillators or brain activity we can see dramatically falls off with increasing network order; it behaves with a $1/N$ scaling law. Furthermore, the effect of increasing quantization bits in the EEG has little real effect.

does not unambiguously describe the generators of neural activity—even in an ideal situation where there is no noise, more than 95% of the oscillations cannot be identified.

If the EEG was measured to infinite precision then the system would indeed be observable (O'Sullivan-Greene et al. 2009b). However, it is shown in Figure 18.3 that even with equipment that is sophisticated by today's standards ($K = 24$) the number of observable states only increases to just under 7% (Compumedics 2009). Again this is an upper limit that declines significantly if noise is introduced into the model. One can simulate the effect of noise as a reduction of the number of effective bits available. The plots shown also include the case of 2^{10} to very conservatively model this case of reduced available quantization levels in the presence of noise. Having considered the effect of a finite resolution measurement, we look at the time sample constraint in the next section.

18.2.2   A QUESTION OF TIME

Time considerations for EEG analysis are often discussed in the context of seizure prediction because the EEG is nonstationary (Reike et al. 2003; Ebersole 2005). Certainly, any attempt to track the activity or dynamics of a system over time, based on a set of measurements, requires the underlying laws of nature not to change significantly over that analysis window. Here, we wish to look at suitable time window durations from the point of view of bits of information.

This interpretation of the observation problem can be reached by purely counting what we actually measure. The latter is the measurement rate of the system with, say, 10^{11} neurons (Nunez 2006). For a standard EEG machine with 10 bits per sample and 512 samples per second, the measurement rate is 5120 bits per second (bits/s) assuming the information is independent from sample to sample (Compumedics 2002). If we had 64 electrodes or channels, then the rate would increase by 5120 × 64 bits/s. Should we wish to observe just 1 bit of information about each neuron (i.e., 2 possible levels of activity per neuron, firing or not firing), then we would require 84 h of EEG data. With estimates of EEG dynamical stationarity on the order of 10 s (Dikanev et al. 2005), this is clearly unrealistic. The fact that seizures are at all detectable from the EEG implies that they must correspond to a vastly simplified mode of behavior in the brain.

We can also examine the reverse argument. Should we have 10 s of stationary data, and we use this to build a dynamical model for the current behavior of the brain for the purpose of predicting its future behavior, how many states can we detect? We use the term state in the dynamical systems sense, where each state variable describes the rate of change of a particular parameter. So a system with N states can be described by N first-order differential equations (or, equally, a single Nth order differential equation). By considering epilepsy to be a global network phenomenon, perhaps our concern is to determine how many states we need to model the brain network involved. Of the 10^{11} neurons in the brain, 10^{10} reside in the cortex. Each neuron makes 10^4 connections (Nunez 2006), so the number of connections or network nodes in the brain is of the order 10^{14}. With our information rate of 5120×64 bits/s we would have $5120 \times 64 \times 10$ bits in a 10 s quasistationary period. If we wish to describe the activity of each node over time (with only a single state) to the accuracy of 10 bits (2^{10} discrete levels) then we could, at most, detect 5120×64 states. This is an upper bound as the effects of the inevitable noise and sample-to-sample dependence are not considered here. While this may seem a considerable number of first order differential equations, it is sobering to think that this is just 0.0000003% of the total network. Can the brain activity network be adequately described by a model whose order is reduced by such a factor? Are the network dynamics of our brain indeed that simple? The prospect seems unlikely to us. Let us now consider how practical studies have fared.

18.3 ANALYSIS OF REAL EEG DATA FOR SYNCHRONY-BASED SEIZURE PREDICTION

As described in the Introduction, this chapter looks at the consistencies between theoretical results of observability of simulated EEG signals and the ability to predict seizures based on synchrony measures calculated from intracranial EEG (iEEG) data. Here we summarize our approach and results for synchrony-based seizure prediction and relate these back to the theoretical analysis of observability.

The prediction analysis addresses five major aspects of proper evaluation of synchrony-based seizure prediction (Mormann et al. 2005) applied to individuals by:

1. Analyzing synchrony between all iEEG channel pairs to find the channel pairs that provide the best synchrony-based seizure prediction performance for a given patient
2. Performing the analysis on long-term continuous iEEG data instead of discontinuous segments of data
3. Analyzing the different times over which preictal changes in synchrony could take place within prediction horizons less than 15 minutes before a seizure
4. Determining whether or not decreases or increases in synchrony are relevant to seizure prediction
5. Comparing the performance of a synchrony-based predictor with a random predictor

18.3.1 INTRACRANIAL EEG DATA

This analysis involves continuous long-term iEEG recordings from three patients recorded at the Epilepsy Center of the University Medical Center, Freiburg, Germany. This data was provided to participants in the seizure prediction contest of the Third and Fouth International Workshops on Seizure Prediction (http://www.iwsp4.org), and will likely be incorporated into the seizure prediction contest of the Fifth International Workshop on Seizure Prediction. All patients suffered from pharmacoresistant focal epilepsy. Invasive iEEG recordings were done via stereotactically implanted depth, subdural strip, and grid electrodes.

The data was obtained using a Neurofile NT™ digital video-EEG system with 128 channels at a sampling rate of 512 Hz and a 16 bit A/D converter. An integrated high and low pass filter restricted EEG frequency content to 0.032-97 Hz. The data were visually inspected by board-certified epileptologists who marked clinical and electroencephalographic events. Prediction was evaluated based on the times of electroencephalographic seizures.

18.3.2 THE SEIZURE PREDICTION ALGORITHM

The mean phase coherence (MPC) was applied to the iEEG data as a measure of phase synchrony between two channels. The calculation of the MPC follows that described by Kuhlmann et al. (2009) and Mormann et al. (2003). The MPC sequence for a given channel pair was determined using a sliding, nonoverlapping window of 20 s duration. A 4 minute median filter was applied to reduce noise. Two separate modes of prediction were evaluated: decreases and increases in synchrony. For decreases in synchrony, a prediction was made if the MPC sequence dropped below a threshold. For increases in synchrony, a prediction was made if the MPC went above a threshold. A full range of thresholds were analyzed to generate ROC (Receiver Operating Characteristic) curves of sensitivity versus False Positive Rate (FPR). The overall performance of a channel pair and method, R, was defined to be the area above the ROC curve and below the line sensitivity equal to one. Thus, R values close to zero indicated good performance. Channel pairs and methods were ranked based on their R value to find the best performing channel pairs.

Two different approaches to thresholding were investigated. The first approach is dynamic threshold (DT), where for a given window of data, the threshold was defined as

$$TH = \mu \pm k\sigma, \tag{18.11}$$

where μ and σ are the mean and standard deviation, respectively, of the MPC time series over the previous 4 minutes, and k is a factor that determines the height of the threshold relative to the mean. Addition and subtraction in Equation 18.11 are used for detecting increases and decreases in synchrony, respectively. The dynamic threshold was employed to account for nonstationarity of the EEG. The second approach is DTL: dynamic threshold with lumped crossings, where a threshold crossing was not called a prediction if the previous window crossed the threshold. As a result, a prediction was called only at the beginning of a contiguous set of threshold crossings. This lumping of crossings was adopted as a way to reduce the FPR.

The seizure prediction performance was evaluated by defining the prediction horizon to be a period starting some fixed time before a seizure and ending at the seizure onset. If a prediction/threshold crossing occurred within the prediction horizon, it was considered a true prediction; otherwise, it was considered to be a false prediction. The performance of synchrony-based seizure prediction was evaluated for prediction horizon durations of 1, 2, 3, 5, and 15 minutes. A similar investigation to that presented here has also been performed using the evaluation method referred to as the seizure prediction characteristic (Schelter et al. 2006; Winterhalder et al. 2003; Maiwald et al. 2004; Kuhlmann et al. 2009).

In summary, seizure prediction performance was evaluated for each channel pair for increases or decreases in synchrony, for different prediction horizon parameter values, for a large set of thresholds, and for each of the thresholding methods. All channel pairs were analyzed for each patient. Channel pairs and methods were ranked using the overall performance metric, R.

The seizure prediction performance of each channel pair for each patient was compared to the performance of a random predictor based on a Poisson process (Schelter et al. 2006; Schindler 2007; Winterhalder et al. 2003; Snyder et al. 2008), for which the interval between two consecutive alarms of the random predictor is exponentially distributed. The probability of raising an alarm in a given time interval of fixed width is (Schelter et al. 2006)

$$P_{Poisson} = \frac{FP}{N_w}. \tag{18.12}$$

The variable *FP* is the number of false positives occurring in the periods outside of the seizure prediction horizons and N_w is the total number of windows points during these periods. The variable $P_{Poisson}$ is the probability of raising an alarm at any single point of the time series with N_w windows. The probability of raising at least one alarm during a period *T* windows in length is given by (Schelter et al. 2006)

$$P_T = 1 - (1 - P_{Poisson})^T. \tag{18.13}$$

The parameter *T* was the duration of a particular prediction horizon in seconds divided by the window duration, which means that P_T gives the probability of predicting a single seizure with a particular prediction horizon.

A binomial distribution with probability P_T can be used to determine the probability of randomly predicting at least *l* of *L* seizures for a given patient. It is given by (Schelter et al. 2006)

$$P_{binom}(l; L; P_T) = \sum_{j \geq l} \binom{L}{j} P_T^j (1 - P_T)^{L-j}. \tag{18.14}$$

For a given parameter, $\alpha \in [0,1]$, the sensitivity $S_r(\alpha)$ of an unspecific random prediction is defined as (Schad et al. 2008)

$$S_r(\alpha) = \frac{1}{L} \max\{l \mid P_{binom}(l; L; P_T) > \alpha\}. \tag{18.15}$$

For a given value of α, the resulting sensitivity will be an upper bound on the sensitivity obtained by the bottom $(1 - \alpha) \times 100\%$ of random predictor trials, and a lower bound on the sensitivity obtained by the top $\alpha \times 100\%$ of random predictor trials. A caveat with this random predictor formulation is that it does not take into account the fact that the EEG signals are dependent on their spatial location and the signals have various dependencies (Schelter et al. 2006). It is difficult to estimate these dependencies. If performance of synchrony-based prediction for all channel pairs falls within the performance boundaries of the random predictor presented here, then it could indicate that dependencies between channels are weak and synchrony-based prediction is random. Bootstrapping techniques provide an alternative validation tool to random prediction but are not considered here (Andrzejak et al. 2009).

18.3.3 Seizure Prediction Results

The performance of the best performing channel pair and method for each patient is given in Table 18.1 for both decreases and increases in synchrony. Channel pairs and methods were ranked based on the *R* measure, and the optimal threshold was picked based on visual inspection of sensitivity versus FPR curves of the best pair and the random predictor. In particular, the optimal threshold was taken to be the threshold corresponding to the greatest difference between the sensitivities obtained with the best pair and the random predictor at the same FPR. For all patients, for both increases and decreases in synchrony, the best channel generally performs above random prediction for $\alpha = 1/N_p$ (where N_p is the total number of channel pairs for a given patient) at reasonable FPRs. This does not necessarily mean that performance is better than random prediction because, for $\alpha = 1/N_p$, the corresponding random predictor sensitivity is a lower bound on the performance for the top $1/N_p \times 100\%$ of random predictor trials.

Ideally, one would like FPR < 1 hr^{-1} for a seizure control device although FPR < 5 hr^{-1} could potentially be tolerated in the case of electrical stimulation, depending on the amount of charge

TABLE 18.1

Best Channel Pair Performance Based on Preictal Changes in Synchrony for Best Method and Parameter Settings

Patient	No. Seizures	No. Channels	Recording Duration (hr)	Method	PH (min.)	Electrode Pair	S	S_r	FPR (hr⁻¹)	Prediction Time (min.)
				Decreases in Synchrony						
1	6	60	100.1	DT	1	37,49	0.5	0.33	1.57	0.8 ± 0.4
2	10	44	30.2	DTL	10	14,36	0.7	0.6	1.27	5.7 ± 3.3
3	5	22	101.9	DTL	10	5,17	1.0	0.6	0.82	7.3 ± 2.0
				Increases in Synchrony						
1	6	60	100.1	DTL	15	24,60	0.83	0.83	1.31	7.8 ± 4.4
2	10	44	30.2	DTL	15	5,31	0.7	0.4	0.31	5.1 ± 5.3
3	5	22	101.9	DTL	10	9,22	0.8	0.6	0.95	5.8 ± 3.5

Note: Methods: DT for dynamic threshold, DTL for dynamic threshold and lumped crossings. MPC decreases or increases were evaluated for prediction horizons (PH) of 1, 3, 5, 10, 15 min.; S: synchrony-based sensitivity; S_r: random predictor sensitivity for $\alpha = 1/N_p$; FPR: false positive rate.

delivered by a particular method. Looking across the patients, slightly different prediction horizon durations gave the best performance for each patient, which is consistent with the notion that different epileptic brains follow different trajectories toward seizures. The mean prediction times (i.e., the average time between a true prediction and the seizure onset) were also different across patients and they showed high variability, as evidenced by the magnitudes of the standard deviations relative to the mean prediction times.

A clearer idea of performance of the best channels compared to random prediction can be obtained from Figure 18.4. Here, the sensitivity versus FPR curve of the best performing channel pair and method given in Table 18.1 (solid black line) is shown for each patient for decreases (top three plots) and increases (bottom three plots) in synchrony. In addition, the top 5% performing channel pairs for the same parameters are displayed (gray solid line) along with the curve at the fifth percentile of performance (dashed black line). The curves for the random predictor performance are given for $\alpha = 1/N_p$ (dash-dot line) and $\alpha = 0.05$ (dotted line). These curves correspond to lower bounds on the sensitivity of the top $1/N_p \times 100\%$ and top 5% of expected random predictor performance, respectively. The lumping of adjacent threshold crossings into single predictions had an unpredictable effect on sensitivity for high FPRs as can be seen for patient 1 for increases in synchrony. This results from the nonstationarity of the EEG and the variable times over which the signal remains below or above threshold. It can be seen that the performance of the top 5% of channel pairs fell largely in between the random predictor curves, indicating that synchrony-based prediction appears to be due to random noise. For the best performing channel pairs, there is a small set of thresholds for which synchrony-based prediction sensitivity is above random prediction sensitivity for $\alpha = 1/N_p$, however, this is only a lower bound on the sensitivity for the top $1/N_p \times 100\%$ of random predictor trials.

Figure 18.5 demonstrates the relationship between the seizure onset channels and seizure prediction performance for all patients for preictal decreases (top three plots) and increases (bottom three plots) in synchrony, respectively. These results correspond to those produced by the same parameters given in Table 18.1. In each subfigure, a given pixel corresponds to a particular channel pair for that patient. The brighter the pixel, the better the performance of that channel pair. Performance is represented as $-\log(R + 1)$ for visualization. In each figure, gray arrows indicate the seizure onset channels for that patient. It can be seen that, for both decreases and increases in synchrony, performance appears randomly distributed across space (or channel pairs). In some cases, the best

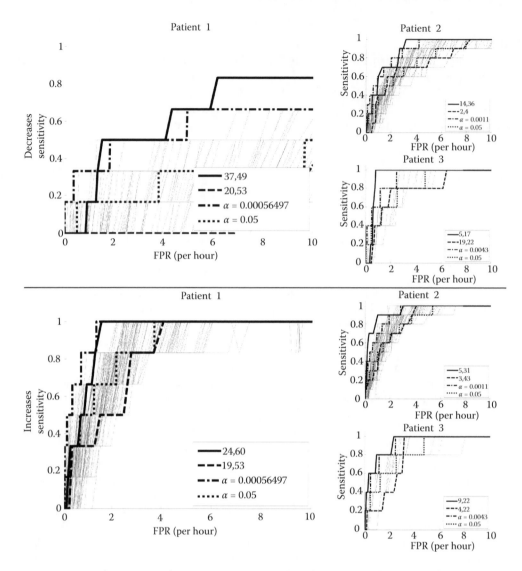

FIGURE 18.4 Performance of preseizure decreases (top three plots) or increases (bottom three plots) in synchrony compared to random prediction for each patient. Results for patient 1 have been scaled for clarity. Sensitivity versus FPR curves are plotted for the top performing 5% of EEG channel pairs for each patient; the method and parameters given in Table 18.1. In each figure for each patient, the legend indicates from top to bottom: (1) the best performing channel pair (solid black curve); (2) the channel pair at the fifth percentile of performance (dashed black curve); the performance of a random predictor for (3) $\alpha = 1/N_p$ (dash-dot curve); and (4) $\alpha = 0.05$ (dotted curve). The remainder of the top 5% of channel pairs are plotted in light gray.

performing (i.e., brighter) channel pairs were seizure onset channels or close to them and, in other cases, the best pairs were away from the seizure focus.

18.4 DISCUSSION

We have shown that in the most ideal of situations, a very limited number of oscillators are observable from an EEG-like measure. We discussed an idealized situation without measurement noise and artifacts in the EEG. The presence of both of these will further hinder practical observability in the EEG record. With relatively few states observable, the large-scale measurements of EEG on

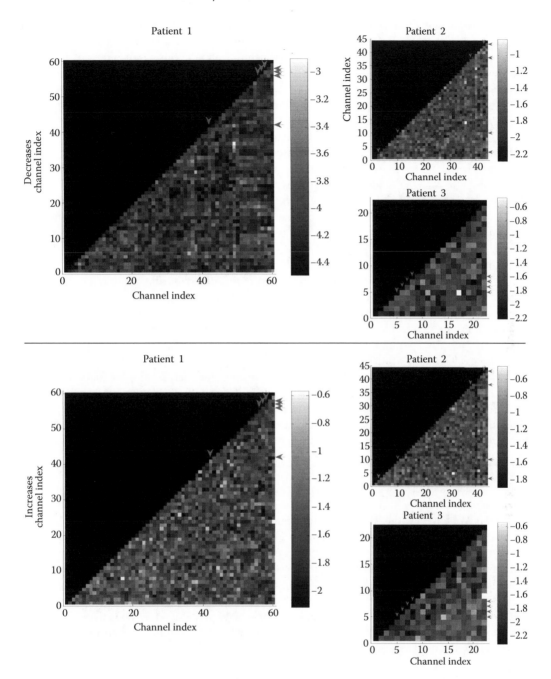

FIGURE 18.5 Spatial distribution of performance of preseizure decreases (top three plots) or increases (bottom three plots) in synchrony for each patient and the same parameters as in Table 18.1 and Figure 18.4. Results for patient 1 are enlarged for clarity. Brighter pixels indicate better performance. Gray arrows indicate seizure onset channels.

both scalp and cortical surface are unlikely to reveal any underlying synchrony of oscillations until the brain is in such an advanced state of synchrony that the seizure has already entrained large areas of cortex.

While full observability requires more information than is strictly needed to measure synchronization, the issue here is the extent to which the number of underlying oscillators is higher than the subset that is actually influencing the measure. Even with a network of the modest order of 1000 clocks, the percentage of observable states is just 5% (for 14 bit measurement resolution and a coupling magnitude of 0.6). An attempt to measure synchronization in this case would only guarantee success if over 95% of the network were synchronized. For the brain, 95% synchronized constitutes a global seizure and synchronization measures would flag it too late for clinical usefulness. It may be the case that when 5% of the system synchronizes, it aligns with the 5% visible through the EEG, allowing early seizure detection. However, this is unlikely to be the case for large-scale measures of EEG. This is highly suggestive that seizure prediction derived through synchrony increases would be better enabled via localized measurement, for example with smaller scale electrodes in cortex or in deep brain structures, where the network may begin to synchronize. Our extensive analysis of synchrony in real EEG data prior to seizure onset supports this theory. The results show that the performance of the top 5% of channel pairs corresponds well with the performance of the top 5% of random predictor trials. If the performances of some of these top channels were better than random prediction, it would allow for the conclusion that these few channels were miraculously well placed.

In that case, why are seizures and spikes even detectable with large-scale EEG measures on the cortex and scalp? We do not contradict the existence of well-performing seizure detection algorithms. The results in this chapter imply the brain must be in a very advanced state of synchronization across huge areas of cortex before we can see it in the EEG. By the time a sufficient number of oscillators have synchronized enough to be observed in the EEG it is most likely that the seizure would have already commenced. This may indicate why attempts at seizure detection have had more success with large-scale EEG measures than seizure prediction (Varsavsky 2010).

From both the theoretical and practical results, we conclude that tracking the build-up of synchronization in the brain from scalp or cortical scale EEG is an ill-posed problem. The brain is not going to reveal its secrets by looking at large-scale EEG measures. More localized electrophysical recordings (Worrell et al. 2008) may prove a more useful measurement and open the possibility of intervention to prevent seizures; however, this presents a paradox in requiring precise knowledge of where to locate the electrodes.

We would strongly advocate that a worthy way forward is through electrically evoked potentials, where safe levels of electrical stimulation via intracranial EEG electrodes are delivered to probe the state of the brain as it heads toward seizure (Suffczynski et al. 2008; O'Sullivan-Greene et al. 2009a). We envisage that the lack of information in the EEG signal can be overcome through electrical probing for two reasons. First, probing with an input signal is a way to raise EEG signals above the noise level by the ability to average over evoked potential trials. Second, through use of a certain input signal (with perhaps a certain frequency spectral content) we expect the brain response to be specific to that probing signal. So when the EEG only lets us see a small fraction of the underlying dynamics, we can tune our input signal to ensure that the fraction of dynamics visible in the EEG gives us the specific information necessary for seizure prediction. We hope that the injection of current into a region of brain tissue to estimate the level of excitability or synchronizability in that region will prove a reliable preictal marker towards successful clinical seizure prediction in the future.

ACKNOWLEDGMENTS

This work was supported by an Australian Research Council Linkage Project grant (LP0560684), The Bionic Ear Institute, St. Vincent's Hospital Melbourne, Perpetual Trust Cassidy Bequest Gift Fund, and National ICT Australia (NICTA). We wish to sincerely thank Michelle Chong, Mark Cook, Dean Freestone, Karen Fuller, Matthieu Gilson, Colin Hales, Alan Lai, Stephan Lau, Dragan Nesic, Andre Peterson, and Simon Vogrin for their time and insight during many discussions surrounding this work. We are grateful to the Freiburg Seizure Prediction Group at the Epilepsy Center of the University Medical Center, Freiburg, Germany, for providing data from three patients.

This data is available online (https://epilepsy.uni-freiburg.de/seizure-prediction-workshop-2007/prediction-contest).

REFERENCES

Aeyels, D. 1981. Generic observability of differentiable systems. *SIAM J. Control Optim.* 19:595–603.

Andrzejak, R. G., D. Chicharo, C. E. Elger, and F. Mormann. 2009. Seizure prediction: Any better than chance? *Clin. Neurophysiol.* 120:1465–1478.

Antoulas, A. C. 2005. *Approximation of large-scale dynamical systems.* Philadelphia: Society for Industrial and Applied Mathematics.

Babloyantz, A., and A. Destexhe. 1986. Low-dimensional chaos in an instance of epilepsy. *Proceedings of the National Academy of Sciences of the United States of America* 83:3513.

Compumedics. 2002. *Profusion EEG user guide.* Abbotsford, Victoria, UK: ©Compumedics Limited.

Compumedics. 2009. ™*SynAmps2 Specifications.* Charlotte, NC: ©Compumedics Neuroscan.

Dikanev, T., D. Smirnov, R. Wennberg, J. L. Perez Valazquez, and B. Bezruchko. 2005. EEG nonstationarity during intracranially recorded seizures: Statistical and dynamical analysis. *Clin. Neurophysiol.* 116:1796–1807.

Ebersole, J. S. 2005. In search of seizure prediction: A critique. *Clin. Neurophysiol.* 116:489–492.

Fisher, R. S., W. Van Emde Boas, W. Blume, et al. 2005. Epileptic seizures and epilepsy: Definitions proposed by the International League against Epilepsy (ILAE) and the International Bureau for Epilepsy (IBE). *Epilepsia* 46:470–472.

Iasemidis, L. D., J. C. Sackellares, H. P. Zaveri, and W. J. Williams. 1990. Phase space topography and the Lyapunov exponent of electrocorticograms in partial seizures. *Brain Topogr.* 2:187.

Johansson, A. C., and A. Lanser. 2004. Towards cortex sized attractor ANN. In *Biologically inspired approaches to advanced information technology.* A. J. Ijspeert, M. Masayuki and Wakamiya, N., eds. New York: Springer.

Kailath, T., ed. 1980. *Linear systems.* Upper Saddle River, NJ: Prentice-Hall.

Kuhlmann, L., D. Freestone, A. Lai, A. N. Burkitt, M. J. Cook, K. Fuller, D. B. Grayden, L. Seiderer, S. Vogrin, I. M. Mareels, and M. J. Cook. 2009. Patient-specific bivariate-synchrony-based seizure prediction for short prediction horizons. *Epilepsy Res.* 91(2–3):214–231.

Le Van Quyen, M., J. Soss, V. Navarro, R. Robertson, M. Chavez, M. Baulac, and J. Martinerie. 2005. Preictal state identification by synchronization changes in long-term intracranial EEG recordings. *Clin. Neurophysiol.* 116:559–568.

Lehnertz, K., and C. E. Elger. 1998. Can epileptic seizures be predicted? Evidence from nonlinear Time Series Analysis of brain electrical activity. *Phys. Rev. Lett.* 80:5019.

Maiwald, T., M. Winterhalder, R. Aschenbrenner-Scheibe, H. U. Voss, A. Schulze-Bonhage, and J. Timmer. 2004. Comparison of three nonlinear seizure prediction methods by means of the seizure prediction characteristic. *Physica D* 194:357–368.

Mormann, F., R. G. Andrzejak, C. E. Elger, and K. Lehnertz. 2007. Seizure prediction: The long and winding road. *Brain* 130:314–333.

Mormann, F., T. Kreuz, R. G. Andrzejak, P. David, K. Lehnertz, and C. E. Elger. 2003. Epileptic seizures are preceded by a decrease in synchronization. *Epilepsy Res.* 53:173–185.

Mormann, F., T. Kreuz, C. Rieke, R. G. Andrzejak, A. Kraskovc, P. David, C. E. Elger, and K. Lehnertz. 2005. On the predictability of epileptic seizures. *Clin. Neurophysiol.* 116:569–587.

Mormann, F., K. Lehnertz, P. David, and C. E. Elger. 2000. Mean phase coherence as a measure for phase synchronisation and its application to the EEG of epileptic patients. *Physica D* 114:358–369.

Nunez, P. L., and R. Srinivasan. 2006. *Electric Fields of the Brain—The neurophysics of EEG.* Oxford: Oxford University Press.

O'Sullivan-Greene, E., I. Mareels, D. Freestone, L. Kuhlmann, and A. Burkitt. 2009a. A paradigm for epileptic seizure prediction using a coupled oscillator model of the brain. *Proceedings of the Annual International Conference of the IEEE Engineering in Medicine and Biology Society (EMBC)* 2009:6428–6431.

O'Sullivan-Greene, E., I. Mareels, L. Kuhlmann, and A. N. Burkitt. 2009b. Observability issues in networked clocks with applications to epilepsy. *Proceeding of the Annual International IEEE Control & Decision Conference (CDC)* 2009:3527–3532.

Reike, C., F. Mormann, R. G. Andrzejak, T. Kreuz, P. David, C. E. Elger, and K. Lehnertz. 2003. Discerning nonstationarity from nonlinearity in seizure-free and preseizure EEG recordings from epilepsy patients. *IEEE Trans. Biomed. Eng.* 50:634–639.

Schad, A., K. Schindler, B. Schelter, T. Maiwald, A. Brandt, J. Timmer, and A. Schulze-Bonhage. 2008. Application of a multivariate seizure detection and prediction method to non-invasive and intracranial long-term EEG recordings. *Clin. Neurophysiol.* 119:197–211.

Schelter, B., M. Winterhalter, T. Maiwald, A. Brandt, A. Schad, A. Schulze-Bonhage, and J. Timmer. 2006. Testing statistical significance of multivariate time series analysis techniques for epileptic seizure prediction. *Chaos* 16:013108.

Schindler, K., C. E. Elger, and K. Lehnertz. 2007. Changes of EEG synchronization during low-frequency electric stimulation of the seizure onset zone. *Epilepsy Res.* 77:108–119.

Snyder, D. E., J. Echauz, D. B. Grimes, and B. Litt. 2008. The statistics of a practical seizure warning system. *J Neural Eng.* 5:392–401.

Suffczynski, P., S. Kalitzin, F. L. Da Silva, J. Parra, D. Velis, and F. Wendling. 2008. Active paradigms of seizure anticipation: Computer model evidence for necessity of stimulation. *Physical Review E* 78(5 Pt 1): 051917.

Takens, F. 1981. Detecting strange attractors in turbulence. *Lecture notes in mathematics: Dynamical systems and turbulence.* 898:366–381.

Varsavsky, A., I. Mareels, and M. Cook. 2010. *Epileptic seizures and the EEG: Measurement, models, detection and prediction.* Boca Raton, FL: CRC Press.

Winterhalder, M., T. Maiwald, H. U. Voss, R. Aschenbrenner-Scheibe, J. Timmer, and A. Schulze-Bonhage. 2003. The seizure prediction characteristic: A general framework to assess and compare seizure prediction methods. *Epilepsy Behav.* 4:318–325.

Worrell, G. A., A. B. Gardner, S. M. Stead, S. Hu, S. Goerss, G. J. Cascino, F. B. Meyer, R. Marsh, and B. Litt. 2008. High-frequency oscillations in human temporal lobe: Simultaneous microwire and clinical macro-electrode recordings. *Brain* 131:928–937.

Wright, J. J., R. R. Kydd, and G. J. Lees. 1985. State-changes in the brain viewed as linear steady-states and non-linear transitions between steady-states. *Biol. Cybern.* 53:11–17.

Zaveri, H. P., M. G. Frei, and I. Osario. 2008. State of Seizure Prediction. In *Seizure prediction in epilepsy.* Schelter, B., Timmer, J. and Schulze-Bonhage, A., eds. New York: Wiley-VCH.

19 Circadian Regulation of Neural Excitability in Temporal Lobe Epilepsy

Paul R. Carney, Sachin S. Talathi,
Dong-Uk Hwang, and William Ditto

CONTENTS

19.1 INTRODUCTION

Epilepsy is the single most common serious brain disorder in every country of the world and is also one of the most universal of all medical disorders, affecting all ages, races, social classes, and nations (Janca et al. 1997). Among the 40 epileptic clinical syndromes, temporal lobe epilepsy (TLE) is the most common chronic partial epilepsy, affecting nearly 50 million people worldwide. TLE is a heterogeneous disorder and is thought to develop via a cascade of dynamic biological events that alter the balance between excitation and inhibition in limbic neural networks. Although it is not yet clear which mechanisms are necessary or sufficient for the development of epilepsy, a number of recent studies have provided evidence for circadian rhythmicity in the occurrence of seizures, both in humans and in animal models of chronic epilepsy (Arida et al. 1999; Quigg et al. 1998, 2000; Herman et al. 2001; Hofstra and de Weerd 2009; Hofstra et al. 2009). Several factors are thought to contribute to the pattern of seizure recurrence, such as state-dependent changes in neuronal excitability associated with the sleep–wake cycle, daily rhythms of hormone release, and body temperature (Quigg et al. 1998, 2001; Herman et al. 2001).

This work is motivated by the fact that very limited information is available to correlate in vivo limbic neural excitability changes with circadian factors in epilepsy. Preliminary experiments coupled with computational modeling have generated novel hypotheses for circadian control of neural excitability in epilepsy (Talathi et al. 2009). We focus our attention on CA1 because its cell properties have been so extensively studied in normal and epileptic brain; CA1 cells are often referred to as the "model" CNS neuron. Spontaneous sharp wave (SW) CA1 activity is employed to evaluate large-scale hippocampus synaptic activity and neural excitability. These studies are then evaluated in the setting of the circadian cycle. Forced desynchrony experiments are employed in order to establish a role for the circadian timekeeper with regard to CA1 neural excitability.

Our results suggest that the acutely induced seizures perturb the phase relationship of SW activity within the CA1 region with respect to the circadian rhythm. The resulting perturbation in phase appears to produce an imbalance in the firing dynamics, such that the network within the hippocampus becomes increasingly excitable, eventually leading to spontaneous epileptic limbic seizures. Based on these findings, we postulate that understanding these phase relationships and subsequent dynamics and network reorganizations with respect to the circadian influence may provide fundamental and profound new directions in understanding how healthy and injured neuronal networks interact.

19.2   RELATIONSHIP BETWEEN EPILEPSY AND CIRCADIAN RHYTHM

Circadian rhythm is any physiological activity that regularly oscillates, has period of oscillations of ~24-h, and persists in environments lacking time cues. Large numbers of mammalian physiologic functions show circadian rhythmicity, including basic intracellular processes, enzyme activities, hormonal secretions, and more complex regulated functions such as core body temperature, sleep and wakefulness, and rest and activity cycle (Beersma and Gordijn 2007). One function of the circadian timekeeping system is to maintain appropriate phase relationships among these rhythms. Chronic disorders of the central nervous system have been shown to alter circadian rhythmicity and the timing of sleep and wakefulness through alterations of the suprachiasmatic nucleus (SCN), retinohypothalamic tract (RHT), and limbic networks (Morton et al. 2005; Maywood et al. 2006). However, despite these observations, it is remarkable that the role and influence of circadian factors in the generation and maintenance of seizures has been so little explored. One explanation for the paucity of research in this area stems from the fact that TLE develops over an extended interval with effects that continue to progress over a long period of time. Consequently, clinical studies of epilepsy development are particularly difficult because the latent evolution into epilepsy may take years, and longitudinal datasets of electrical brain activity are rarely available for analysis. In addition, since sleep deprivation is known to enhance seizures, the controlled experimental protocol that is required to expose endogenous circadian rhythm (Duffy and Dijk 2002) is not usually performed on epileptic patients. These limitations present a serious challenge for the development of a more complete understanding of the interaction between circadian rhythm and epilepsy. Alternatively, animal models of TLE represent excellent test beds for advancing our understanding of in vivo limbic network neuronal dynamics and circadian factors in response to injury. To this end, several animal models of chronic TLE have been developed in recent years. These models differ from acute seizure models in that the experimenter does not induce seizures, but rather the seizures occur spontaneously (for a review, refer to the works by Loscher 1997, 2002; Sarkisian 2001). The utility of any given animal model depends on the extent to which the model shares the underlying pathophysiology and mechanisms involved. The better an animal model resembles the type of human epilepsy that it attempts to replicate, the more useful it will be at understanding and curing a given subclass of epilepsy (Stables et al. 2002). The self-sustaining limbic status epilepticus (SSLSE) rat model, initially described by Lothman and Bertram (1990), has many of the features associated with human chronic TLE, including similar electrophysiological correlates, etiology, pathological changes in the limbic system, and seizure-induced behavioral manifestations (Bertram and Cornett 1993; Quigg et al. 1998; Sanchez et al. 2006). The seizures in this TLE model are recurrent, spontaneous, and chronic in nature.

The SSLSE rat model is developed using two-month old male Harlan Sprague-Dawley rats by direct electrical stimulation of the ventral hippocampal region until sustained electrographic and behavioral seizures are observed. High-amplitude population spikes corresponding to the field activity of excitatory postsynaptic potentials are observed immediately after status epilepticus from CA1, CA3, and dentate gryus. Over time, rats show loss of cells in the hippocampus formation, parahippocampal gyrus, mossy fiber sprouting, atrophy, and gliosis (Bertram 1997). Also, after a period of 2 to 4 weeks, rats begin to have spontaneous Racine scale (Racine et al. 1972; Racine 1972) seizures, with normal life spans. The spontaneous behavioral and electrographic seizures may persist for many months and are typically bilaterally synchronous in the hippocampi. The mechanisms by means of which spontaneous seizures occur in the status epilepticus model are not yet clear. However, it has been shown that chronic hippocampal stimulation causes a diminution of GABA-A mediated inhibition in CA1 and dentate gyrus areas (Mangan and Bertram 1997). Importantly, the SSSLE rat model shares many of the characteristics associated with human TLE, including similar electrophysiological correlates, etiology, pathological changes in the limbic system, and seizure-induced behavioral manifestations and disturbance in sleep cycle (Quigg et al. 1999, 2000).

19.3 CIRCADIAN AND SLEEP–WAKE DEPENDENT MODULATION IN TEMPORAL LOBE EPILEPSY

In epilepsy, a number of factors are thought to contribute to the pattern of seizure recurrence, such as state-dependent changes in neuronal excitability associated with the sleep–wake cycle and daily rhythms of hormone release and core body temperature (Quigg et al. 1998, 2001; Herman et al. 2001). Although it is not yet clear which mechanisms are necessary or sufficient for the development of epilepsy, a number of recent studies have provided new insights into the role played by the circadian timing system (Carney et al. 2005). To this end, circadian periodicity in the occurrence of seizures has been recognized for more than a century (Gowers 1881) and is well documented for human TLE (Arida et al. 1999; Quigg et al. 1998; Quigg 2000; Herman et al. 2001). Disturbances of circadian organization can have a direct effect on physiological functions and disease. It is conceivable that permanent alterations in neuronal excitability or structural lesions associated with the development of epilepsy might alter rhythm expression. Some epileptic patients were reported to manifest altered melatonin and cortisol rhythms, perhaps related in part to seizure episodes (Schapel et al. 1995; Bazil et al. 2000). Patients suffering from frequent epileptic attacks exhibited irregularities in the daily rhythm of body temperature (Laakso et al. 1993). Likewise, kindled seizures in rats altered the complexity of temperature rhythms and either advanced or delayed the acrophase or rhythm peak, depending on the time of day the seizures were evoked (Quigg et al. 1998, 2001). In acute status epilepticus models of TLE, marked changes in the daily activity patterns of rats were found up to 12 weeks after pilocarpine-induced status epilepticus (Leite et al. 1990). Interestingly, the changes in activity patterns occurred without SCN cell loss, suggesting that abnormal rhythmicity may result from functional, seizure-induced disruptions of SCN activity rather than physical damage to the circadian timing system itself. In another study, pilocarpine-induced status epilepticus rats exhibited an increase in spontaneous locomotor activity during both light and dark phases 1-week after status epilepticus (Steward and Leung 2001, 2003). When the nucleus accumbens was ablated, these rats did not exhibit changes in spontaneous activity, suggesting that the increase in spontaneous locomotor activity may be due in part to the hippocampal CA1 hyperexcitability driving the descending limbic–motor pathway via the nucleus accumbens. However, despite the increase in both spontaneous locomotor activity and CA1 region interictal spikes, epileptic rats maintained a near 24-h circadian rhythm. Indeed, several other lines of evidence indicate that CA1 neurons are under circadian influence. For example, in vitro experiments have shown diurnal variations in long-term potentiation in CA1 pyramidal cells and interneurons (Harris and Teyler 1984; Raghavan et al. 1999) and long-term potentiation is easily induced in CA1 slices prepared during the day as compared to night time (Harris and Teyler 1984). These observations of rhythmicity in CA1 neurons (Brewis et al. 1966; Barnes et al. 1977) are particularly intriguing and may suggest, among other factors, a relationship between the circadian influence and seizures.

Chronic in vivo experiments using a model of human TLE show a relation between neural excitability, circadian rhythm, and spontaneous limbic seizures. Chronic hippocampus CA1 region local field potential (LFP) microwire recordings (Williams et al. 1999) were obtained from seven Harlan Sprague-Dawley SSSE rats and two age-matched sham controls. Continuous time-locked video-EEG data was obtained and screened for Racine spontaneous seizures and each dataset was scored for wake and sleep states (Johns et al. 1977; Neuhaus and Borbely 1978; Clark and Radulovacki 1988). Rats were monitored for an average of 5 weeks and characteristics of the dataset are summarized in Table 19.1.

At the end of the recording session, SSSE rats were sacrificed and their intact brains were excised. Each brain was imaged with high-field magnetic resonance microscopy to confirm the location of the electrode placement as is shown in Figure 19.1a and b, along with an associated brain histological slice with iron staining shown in Figure 19.1c. The University of Florida Institutional Animal Care and Use Committee approved all the protocols and procedures for the experiments.

TABLE 19.1
Mean Duration of Epileptogenic Phase (Time Interval from Status Epilepticus to the Observation of First Spontaneous Seizure) and the Racine Seizure Grade for All the Epileptogenic Rats

Rat	Duration of Recording (Days)	Time to First Spontaneous Seizure (Days)	Racine Grade of First Spontaneous Seizure
E1	91	21	4
E2	50	13	3
E3	60	14	4
E4	14	5	1
E5	18	NA	NA
E6	39	NA	NA
E7	28	NA	NA

LFPs recorded from microwires in CA1 were analyzed to extract spontaneous SW activity. SWs were qualitatively identified as high-amplitude, short–time duration (100–200-ms), spatially localized patterns of spontaneous electrical activity that were monophasic, with either positive or negative excursion in their relative electrical activity (Niedermyer and Silva 2004; White et al. 2010). SWs were also quantitatively identified and classified by dividing LFP CA1 region microwire recordings into 1-h nonoverlapping windows in order to ensure a lack of drift in the amplitude of the EEG signal. Putative SWs were selected as those events whose amplitude exceeds a statistical threshold value of $T_{th} = m\sigma$, where σ is the second moment estimated from the median value of the EEG data over a 1-h time window. The choice of median value is motivated by the fact that the dominant signal content of the EEG is low-amplitude background activity. It has been our observation that the EEG data is relatively stationary over any random sampling of 1-h time window (Talathi et al. 2008). The multiple m of σ was determined as representing the T_{th} value where the numbers of candidate events selected first plateau before decreasing again to zero from a random sampling of 1-h windows over the entire duration of the experiment. SWs are, therefore, those events whose absolute amplitude always exceeds m standard deviations of the background EEG data at any given instant in time. The peak of each event detected was determined from a second-order polynomial fit of a region around the maximum amplitude of the data in a window of 1.5 s duration centered on the threshold crossing. Data from the temporal window that best captured the entire profile of the candidate SW were extracted. This consisted of a time window of 0.45 s centered on the event including 0.15 s before the fitted peak to 0.3 s after the fitted peak. These peak-aligned candidate events were then merged into a larger set of candidate SW over a period of 24 h and sorted using a modification of a well-established clustering algorithm (Fee et al. 1996). A representative example of SWs extracted from raw LFP data is shown in Figure 19.1e–g. In Figure 19.2, we demonstrate the flow chart for the extraction and classification of SW events.

The time evolution of the normalized firing rates (defined as the number of SW observed per unit time) of SW1 and SW2 from both an epileptogenic TLE and control rat is shown in Figure 19.3a–d. Key points from Figure 19.3 are the following:

1. Approximately 24-h circadian modulation in the firing rate of the SW, both during the control and the epileptogenic time periods.
2. Lack of an observed drift in the firing rate of the SW in the data obtained from healthy control rats (Figure 19.3a and c).
3. Marked upward drift in the firing rate of SW1 and a corresponding marked downward drift in the firing rate of the SW2 over several weeks (Figure 19.3b and d).

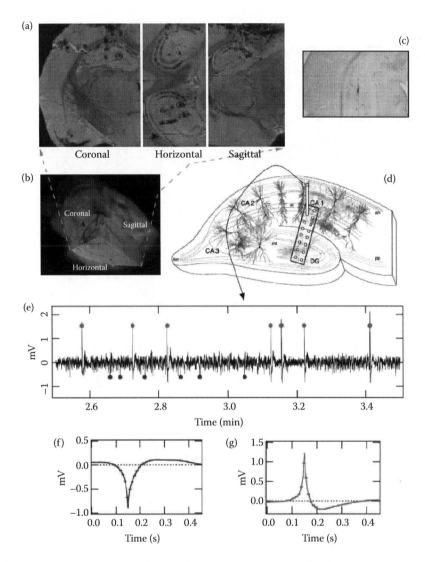

FIGURE 19.1 (See color insert.) A three-dimensional, gradient-echo MR image, acquired at 17.6 T (750 MHz) with a Bruker Advance system (Bruker NMR Instruments, Billerica, MA) is shown in (a) and (b). (a) Orthogonal slices with the tip location of right-side microarray electrode 2 (shown as the black spot within the red circles) terminating in CA1. (b) The complete three-dimensional image volume illustrating an electrode tract (shown as a black line) within the brain. The electrode tract appears visible in the MR image because iron, which accumulated around the site of electrode insertion, shortens the MR transverse relaxation time. (c) Using another rat brain treated in the same way, the presence of iron is confirmed in the histological slice, which was prepared using Perl strain with diamine-benzidine-tetrahydrochloride. In this slice, the iron surrounding a tract is visible as the black region in the middle of the slice. (d) A schematic diagram of the hippocampus and the location of the implanted electrode array (blue box and circles in the box). (e) A sample trace of 1-min-duration trace from extracellular recording with microwire electrode 2. The time of the SW2 event is denoted in green and the time of the SW1 event, detected through the automated clustering algorithm, is represented in red (Fee et al. 1996). The mean amplitude profile of the SWs (with standard error representing 95% confidence level) obtained from a total of (f) 40,000 SW1 and (g) 24,500 SW2 events over the entire epileptogenic period of recordings from rat E3. (Adapted from Talathi et al., *Neuroscience Letter*, Elsevier, 455, 145–149, 2009.)

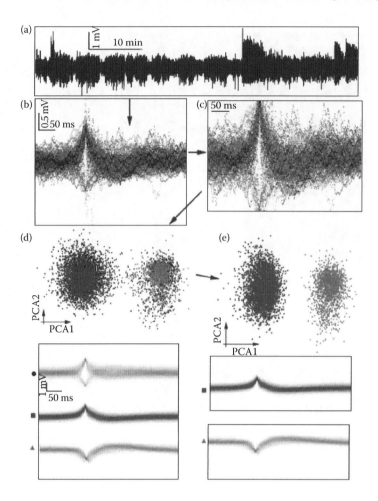

FIGURE 19.2 (See color insert.) (a) Sample extracellular recordings trace of 1-h in duration from a microwire electrode implanted in the CA1 region of an epileptogenic rat. (b) Raster plot of the spike events pooled over a 24-h period of the continuous extracellular recordings. The pooled spike events represent the set of spike events selected from a non-overlapping time window of 1 h of the recorded data, whose amplitude exceeds a threshold of $5s$, where s is the standard deviation computed for each hour of the extracellular recorded data. (c) Pooled spike events are then normalized in amplitude and peak-adjusted to discount for any large amplitude fluctuations and jitter in timing of the raster spike events before feeding them into the automated clustering algorithm. (d) A typical output of the clustering procedure resulting in three separate clusters (shown in black, blue, and red) with the corresponding probability-density plots of the spike patterns is shown. (e) The two primary clusters representing the two population spike patterns (SW1 in red and SW2 in blue) and the corresponding shape profiles represented in the probability density plots are shown. The final two spike patterns are selected by determining the cross-correlation of the events in the third cluster (black) with the mean shape profiles of the events in the two primary clusters. Only those events in the cluster (shown in black) whose cross-correlations are >75% are included in one of the two final primary spike clusters. (Adapted from Talathi et al., *Neuroscience Letter*, Elsevier, 455, 145–149, 2009.)

4. Circadian modulation (~24 h) of firing rates of SW locked in phase in the control period (Figure 19.33e); during the interitctal period, the two SW oscillate antiphase with respect to each other with a marked shift in the rhythmic activity of SW1 (Figure 19.3f).

In Figure 19.4, we summarize the results of the phase shift in the circadian-like firing activity of the SW and the imbalance in their firing rates during epileptogenesis from the SW data obtained

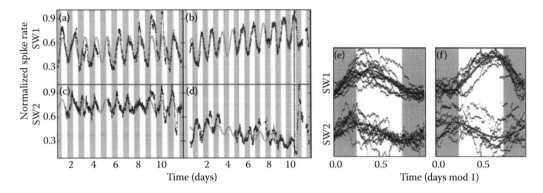

FIGURE 19.3 **(See color insert.)** (a) through (d) show the firing rates of the SWs recorded from a healthy control rat and an epileptogenic rat. Red dotted lines represent the least-squares fit of the firing rate data to a function $f(t) = at + b\sin(wt + c)$, where $w = 7.2722 \times 10^{-5}$ Hz. The fitted line is shown as a guide for the eye to follow the circadian pattern in the firing activity of PS. The gaps in the firing rate data (around day 5 in both control and epileptogenic period) reflect the absence of recordings on those days due to technical problems). (e) and (f) The phase of circadian oscillations of NPS and PPS from the healthy control rat and the epileptogenic rat is shown in (e) and (f), respectively. The red line is a least squares fit to the phase data with a function $f(t) = a\sin(wt + b)$, where $w = 2p$. The diurnal day-night cycle is shown in the background. The dotted line in (b) and (d) shows the time of occurrence of the first spontaneous epileptic seizure. (f) and (g) Epileptogenic latent period and the average number of PPS and NPS events observed per hour during the control time period. (Adapted from Talathi et al., *Neuroscience Letter*, Elsevier, 455, 145–149, 2009.)

from three epileptogenic rats and two healthy age-matched controls. The imbalance in the firing rates is quantified by estimating the drift $\left\langle D = \dfrac{df}{dt} \right\rangle$ (f is the firing rate) in the firing activity of SW through least-squares fit of the drift in the baseline firing rate to a straight line $\Delta f = D\Delta t + c$. In Figure 19.4a, we plot the mean value of D (with error bars representing the standard error corresponding to a 95% confidence interval). From Figure 19.4a, we see that, while the firing rates are in balance (D^0) in healthy controls, $D > 0$ during the latent period in epileptogenic rats ($p \approx$ 0.0044, two sample t test). This implies an evolving imbalance in the firing activity of the SW. The phase relationship between the circadian-like firing activity of the SW is quantified through a least squares-fit of the detrended modulo 24-h firing rate data (detrending implies the removal of the drift in the baseline of the circadian-like rhythm of the firing rate) with a sinusoidal function $f(t) =$ $a\sin(wt + b)$, with $\omega = 7.2722 \times 10^{-5}$ Hz. The phase is associated with the time T_X ($X = $ SW1, SW2) of maximum value obtained by $f(t)$ and is given as $\phi_X = \dfrac{2\pi T_X}{24}$. The mean value of phase for the SW (with standard error corresponding to 95% confidence interval) is shown in Figure 19.3b. The relative phase difference is quantified as $\Delta\phi = |\phi_{SW1} - \phi_{SW2}|$. In-phase firing activity of the SW occurs when $\Delta\phi < \pi/2$. We see that during the control period, SWs are phase-locked with a lag of around $\pi/4$ radians; however, following status the phase-lag increases to approximately $3\pi/4$ radians. It is, therefore, highly likely that the observed rhythms in the SW firing rates in our preliminary studies and their relative phase shift observed following injury in the epileptogenesis latent period are highly correlated to the sleep–wake cycles of the rat. In order to determine whether the rhythm of SW firing rate is truly circadian, it is essential to discount for the masking effects of the sleep–wake cycle on the SW firing activity within the hippocampus.

Sleep–wake cycles are known to have strong reciprocal influence on epilepsy (Bazil et al. 2000). Sleep disorders have been intrinsically linked to TLE. For example, interitctal paroxysmal activity in patients with TLE has been shown to increase in slow wave sleep (Malow et al. 1999). In addition, the localization of the primary epileptogenic areas has been shown to be more reliable in

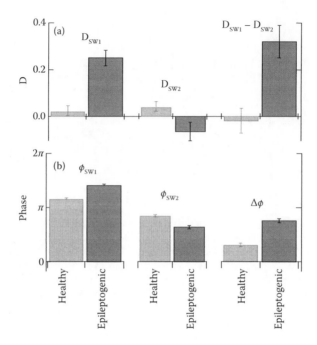

FIGURE 19.4 (a) The mean amplitude of the drift in the firing rate, D. (b) The circadian phase shift in the relative firing activity of the SWs during the control and the latent periods. (Adapted from Talathi et al., *Neuroscience Letter*, Elsevier, 455, 145–149, 2009.)

paradoxical sleep (Peraita-Adrados 2004). The *forced* desynchrony protocol is an extremely useful tool for evaluating the influence of the circadian pacemaker on physiological variables because it allows separation of the confounding effect of the sleep–wake schedule from the output of the endogenous circadian pacemaker (Angles-Pujolras et al. 2007; Cambras et al. 2007; Horacio et al. 2008). This imposed desynchrony between the sleep–wake schedule and the output of the circadian pacemaker driving the temperature rhythm occurs under conditions in which the non-24-h zeitgeber (external cue) is outside the range of entrainment or range of capture of the circadian system. Experimental exposure of human subjects to rest–activity cycles and dim-light light–dark cycles with periods ~24-h, but outside the range of entrainment for several circadian rhythms induces internal desynchronization. In such studies, the rhythms of core body temperature (CBT) and other physiological variables including plasma melatonin and cortisol, sleep propensity, and paradoxical sleep oscillate out of synchrony with the imposed rest–activity cycle with a period near 24 h (Czeisler and Dijk 2001; Saper et al. 2005; Saper et al. 2005). However, it was not known whether internal desynchronization of physiological and behavioral rhythms represented the activity of two independent oscillators of distinct anatomical origin until a rat model of forced desynchrony was developed in which rats were exposed to 22-h light–dark cycles (11 h of light: 11 h of dark) and exhibited two stable locomotor activity rhythms with different period lengths in individual animals (Campuzano et al. 1998). Importantly, this study demonstrated that one rhythm with a period of 22 h and entrained to the light–dark cycle is associated with the expression of clock genes in the ventrolateral SCN. The other rhythm with a period longer than 24 h and not entrained to the light–dark cycle was associated with gene expression in the dorsomedial SNC (de la Iglesia et al. 2004). These findings suggested that the uncoupling of subpopulations of neuronal oscillators within the SCN could lead to the desynchronization of specific circadian physiological and behavioral processes, similar to that observed in humans (Cambras et al. 2007). Coupling within the SCN may be due in part to gap junctions and a GABA-dependent mechanism (Colwell 2000; Liu and Reppert 2000;

Albus et al. 2005; Carpentieri et al. 2006). Indeed, the degree of coupling has been suggested to be a key factor in determining the circadian rhythm pattern of physiological rhythms both in terms of seizure occurrence (Quigg et al. 1999, 2000) and CA1 neural excitability in epilepsy (Talathi et al. 2009). A better understanding of the coupling mechanism is essential to understanding the role of the SCN and to developing new treatments for circadian rhythm sleep disorders for patients with chronic epilepsy.

We are currently employing the technique of *forced desynchronization* in order to determine the relative dependence of SW rhythm on the sleep–wake cycle and the endogenous circadian master clock. We postulate that the range of entrainment (ROE) of CBT will be near 23 to 27 h with zeitgebers composed of regular light–dark cycles as previously demonstrated (Zulley et al. 1981; Wever 1983). A forced desynchronization is expected to occur when the period of zeitgeber is beyond the ROE of the temperature rhythm. Conversely, the sleep–wake activity rhythm is postulated to remain synchronized to the zeitgeber. Until recently, lack of a robust animal model for circadian forced desynchronization had hampered the progress in our understanding of the influence of circadian disorders on neurophysiologic diseases. However, recent work (Cambras et al. 2007; Horacio et al. 2008) has shown that circadian rhythms can be internally desynchronized in a predictable manner in rats when these animals are maintained in an artificially shortened 22-h light–dark cycle. It is interesting that forced desynchronization has been associated with the stable dissociation of rhythmic clock gene expression within the ventrolateral SCN and the dorsolateral SCN (Dallmann et al. 2007; Van Dongen et al. 1999). Accordingly, animals are maintained on a symmetric light–dark cycle of period $T = 22$ h to induce internal desynchronization of their circadian rhythms. Light, consisting of cool white light (100–300 lux) for $T/2$ h, and darkness, consisting of dim red light (<1 lux) for $T/2$ h, is used to create the light–dark cycles (Cambras et al. 2007). Rat activity and the EMG are measured as a marker for sleep–wake cycle. Animals are fed according to the standard feeding schedule during the light cycle, and have access to the exercise wheel only during the typical light cycle (Dallmann and Mrosovsky 2006; Dallmann et al. 2007). In order to maintain the multifactorial zeitgeber environment, the feeding schedule and the exercise access is advanced by the same $T/2$-h shift. In Figure 19.5a and b, we show a sample raw time trace of the CBT rhythm and the corresponding rhythm of SW extracted from LFP recordings during the baseline pre–status epilepticus electrical stimulation. As expected for nocturnal animals, the circadian rhythm of CBT is high during the subjective-night and achieves trough at around 6–10 h of subjective-day. The circadian rhythm of SW from CA1 is in exact anticorrelation with the rhythm of CBT with peak firing activity observed during the early phase of subjective day. However, following status epilepticus and after the rat has entered the chronic epilepsy phase at 2–4 weeks after injury, the phase of circadian rhythm of CA1 SW exhibits a positive correlation with respect to the CBT endogenous rhythm (Figure 19.5). We chose a time epoch in the chronic phase when the rat did not exhibit spontaneous seizures, since seizures have been shown to transiently perturb the phase of circadian oscillations (Quigg et al. 2001). We observe that during this time interval, no significant shift in the phase of CBT rhythm was observed. However, a significant shift in the phase of CA1 SW rhythm was observed, which corresponds with our preliminary analysis. From these results, we hypothesized that this shift in phase of CA1 circadian activity is associated with perturbations in the circadian drive onto the CA1 network, which in turn triggers a hyperexcitable brain state, thereby resulting in the occurrence of spontaneous epileptic seizures. The circadian control hypothesis suggests that the evolving imbalance in the firing rates of SW1 and SW2 is the result of an abrupt shift in the phase of their activity relative to the circadian cycle, which in turn triggers homeostatic mechanisms (Davis 2006). This results in changes in the interaction strength between subpopulations of hippocampal neurons, which in turn serves to modulate their firing activity. Together, the net result is that of a hyperexcitable brain state. Finally, it is conceivable that the changes in CA1 neural activity may reflect both circadian regulation of hippocampal long-term potentiation (Chaudhary et al. 2005) and molecular changes and homeostatic control of neural activity (Davis 2006).

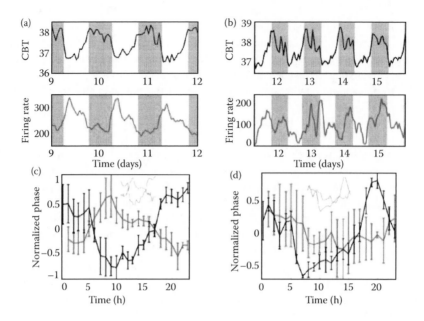

FIGURE 19.5 **(See color insert.)** The circadian rhythms of core body temperature (CBT) in healthy and epileptic states (during a continuous time epoch when no spontaneous seizures were observed) are shown in the top panels and the corresponding circadian rhythms of hippocampal sharp waves (of type SW2) are shown in the bottom panels of (a) and (b), respectively (diurnals light–dark rhythm is shown in the background). In (c) and (d), we characterize the phase relationship between the rhythms of CBT (shown in black) and SW2 (shown in red) in healthy and epileptic states, respectively. Inset shows the mean profile of the phase activity of these rhythms. (Adapted from Talathi et al., *Neuroscience Letter*, Elsevier, 455, 145–149, 2009.)

Our computational study is derived from the following three key experimental observations drawn from Figure 19.3 and 19.4:

a. The phase shift between the firing rates of SW1 and SW2 is almost instantaneous compared to the time scale of the emergence of the imbalance.
b. The phase shift persists throughout the entire duration of the latent period when the brain gradually evolves toward imbalance.
c. The circadian-like oscillation is preserved in healthy brain.

These observations lead us to believe that the experimentally observed phase shift is caused by an external source rather than emerging from dynamics from the network. Although emergent dynamics through changes within the network could cause the phase shift, it is highly unlikely that this would happen instantaneously as is observed experimentally. Moreover, the phase shift is greater than $\pi/2$ (i.e., multiple hours in real time). Such a large shift in the phase of oscillations cannot occur through changes in the internal dynamics of the CA1 network, since the time scale of oscillations of neurons in the CA1 network is a few tens of seconds at most (width of population spike events). Finally, since the circadian-like oscillations persist both before and after injury, it is more likely that the experimentally observed phase shift results from changes in the external forcing on the CA1 network. Accordingly, we hypothesize that sustained brain injury such as status epilepticus alters the target firing rate of the hippocampal network. Homeostatic synaptic plasticity (HSP) then regulates the synaptic strength of excitatory and inhibitory neurons to force the firing rate to approach the modified target firing rate. The circadian control hypothesis then suggests a model for HSP where the synaptic strengths are regulated in a phase-dependent manner. We have investigated the implications of our circadian control hypothesis on a computational model of the CA1 network both using a simple two-dimensional Wilson–Cowan model (Talathi et al. 2009; Wilson and Cowan 1972), and a detailed biophysical network model of interacting neuron oscillators (Fisher et al. 2009). Here, we briefly summarize the key findings from our computational studies. In both simple mean field model and detailed biophysical network model, we assume the SW1 represents firing activity of populations of excitatory neurons and SW2 represents firing activity of populations of inhibitory neurons.

The mean-field model allows the study of the interaction between populations of excitatory and inhibitory neurons in the CA1. In Figure 19.6a and b, we show the schematic diagram of the interactions between the two populations in the healthy controls and the epileptic animals. Accordingly, if $X(t)$ represents the firing rate of excitatory neurons and $Y(t)$ represents the firing rate of inhibitory neurons, then using the framework of the Wilson–Cowan model, the coupled pair of ordinary differential equations (ODEs) governing the evolution of X and Y are given as:

$$\tau_x \frac{dX}{dt} = -X + S(A\sin(\omega t + \Delta\phi) + c_1 X - c_2 Y + P)$$

$$\tau_y \frac{dY}{dt} = -Y + S(B\sin(\omega t) + c_3 X - c_4 Y + Q)$$

(19.1)

where $\tau_x \ll \omega^{-1}$ and $\tau_y \ll \omega^{-1}$. $S(x) = (1 + \exp(-\alpha x))^{-1}$ is the subpopulation response function (Wilson and Cowan 1972). S gives the expected proportion of cells in a subpopulation, which would respond to a given level of excitation if none of them were initially in the absolute refractory state. The strength of local interactions between the populations of excitatory and inhibitory neurons in the CA1 area is represented by c_j ($j = 1 \ldots 4$) represent the strength of local interactions between the populations of excitatory and inhibitory neurons in the CA1 area. Since sparse recurrent connections between CA1 excitatory neurons exist in normal CA1, in our modeling of the interaction between the excitatory and inhibitory neurons, we set $c_1 = 0$ for the healthy control paradigm (Figure 19.6a). Figure 19.6a and b represents the strength of external circadian drive with respect to the firing activity of the excitatory and inhibitory neurons, respectively. $\Delta\phi$ represents the phase difference with regard to the timing of the circadian input on the two-neuron population, which are modulated in epilepsy according to our experimental results. P represents the excitatory input from the hippocampal CA3 Schaffer collateral-commissural projections onto the CA1 excitatory neurons (Anderson et al. 1971). Q represents the excitatory input onto the CA1 interneurons via the temporoammonic pathway from layer II of the entorhinal cortex (Lacaille and Schwartzkroin 1988). Finally, according to our proposed circadian control hypothesis, the asymptotic strength of interaction between the firing rates of populations of neurons in the CA1 c_j^∞ is considered to be dependent on the phase difference $\Delta\phi$ of the circadian input. We model c_j^∞ through linear dependence on $\Delta\phi$. The evolution of c_j, following a perturbation in phase $\Delta\phi$ modeled through a linear first-order ODE given by: $\frac{dc_j}{dt} = \frac{c_j^\infty - c_j}{\tau_L}$.

Figure 19.6c–f demonstrates the output of the Wilson–Cowan model. Results simulate the conditions from in vivo experiments in both epileptic and age-matched control rats (Figures 19.3 and 19.4). $\Delta\phi = \pi/4$ represents the condition observed in healthy controls. Accordingly, the firing rates of CA1 neurons exhibit slow modulations at the frequency of the external sinusoidal input w (Figure 19.6c). In addition, there is maintenance of balance in the relative firing rates of the network activity. Following injury, there is a sudden shift in the phase of circadian drive to the two populations of interacting neurons resulting in $\Delta\phi = 3\pi/4$. This results in modulation of the interaction terms c_j through the homeostatic plasticity update equations described previously. In addition, it is known that status epilepticus results in selective loss of neurons in layer III of the entorhinal cortex (Du and Schwarcz 1992). Therefore, in our computational model, we set $Q = 0$ (Figure 19.6b). This results in a sudden decrease in the firing rate of inhibitory neurons, mimicking our observation of a decrease in the firing rate of SW2. The nonlinear interaction between the firing rates of populations

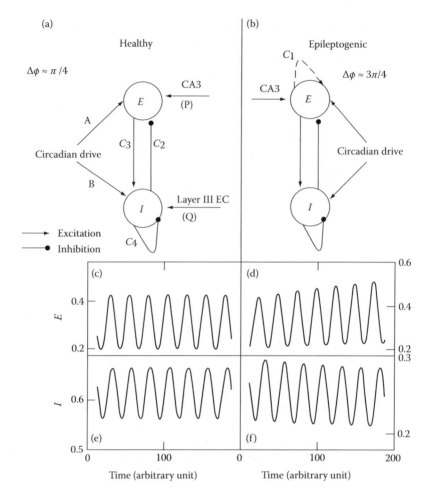

FIGURE 19.6 (a and b) The schematic diagram of the interaction between the excitatory (E) and the inhibitory (I) populations of neurons in a healthy brain and an epileptic brain. (c through f) The firing activity generated by the model (Equation 19.1) for the healthy and epileptic periods. The model parameters are, $A = 0.5$, $B = 0.25$, $\omega = 2\pi/25$, $\tau_x = \tau_y = 1$, $\tau_L = 200$, $P = 0.025$ for both the healthy control and the latent time periods. $Q = 1.9$ for the healthy brain and $Q = 0$ for the epileptic brain. (Adapted from Talathi et al., *Neuroscience Letter*, Elsevier, 455, 145–149, 2009.)

of CA1 neurons through Equation 19.1 then results in the further decrease in firing activity of the inhibitory population and a corresponding increase in the firing rate of the excitatory population (Figure 19.66d–f). Thus, using the constraints imposed through the known anatomical connectivity patterns within the hippocampus and the proposed circadian control hypothesis, with the simple two-dimensional models, we are able replicate our preliminary experimental findings of an evolving imbalance in the firing rates of CA1 SW following status epilepticus.

The detailed biophysical CA1 network model is constructed using 1000 neurons, 800 of which are excitatory and 200 of which are inhibitory in a random network architecture. Any given neuron is connected to another neuron in the network (independent of the type of neuron) with a probability of 10% drawn from a uniform random distribution. The schematic of the resulting architecture is shown in Figure 19.7a. This figure demonstrates an architecture of 100 neurons for clarity; however, the ratios of inhibitory to excitatory neurons and synaptic connections are equivalent. Inhibitory neurons are represented by the white circles and excitatory neurons are represented by the black circles. Synaptic connections in which the presynaptic neuron is excitatory are shown in red, and synaptic connections in which the presynaptic neuron is inhibitory are shown in blue. Before the phase shift, we make excitatory to excitatory synaptic connections relatively weak in comparison to the excitatory to inhibitory synaptic connections in an effort to mimic the sparse recurrent CA1 excitatory connections seen in the hippocampus. After the phase shift, we increase the relative weight of the excitatory to excitatory synaptic connections when compared to excitatory to inhibitory connections in a manner representative of the model of the brain during the latent period. Neuronal dynamics is modeled using a simple two-dimensional parameterized set of ordinary differential equations as given by (Fisher et al. 2009).

$$\left.\begin{aligned} \frac{dv_j}{dt} &= 0.04v_j^2 + 5v_j + 140 - u_j + I \\ \frac{du_j}{dt} &= a(bv_j - u_j) \end{aligned}\right\} \text{if } v_j \geq 30, \text{ then } \begin{cases} v_j \leftarrow c \\ u_j \leftarrow u_j + d \\ v_i \leftarrow v_i + g_{ij}, \forall i \end{cases} \tag{19.2}$$

Here, v mimics fast neuronal spiking dynamics, and u mimics the slow channel kinetics of neuron membrane. Once the spike crosses the threshold of 30 mV, the membrane potential and recovery variable are reset according to Fisher et al. (2009) for this neuron; the membrane potential of all neurons that receive input from the spiking neuron are immediately increased by the synaptic weight, g_{ij}. Here, i models the synaptic current. Each neuron also receives sinusoidal input current representing the circadian drive as given by

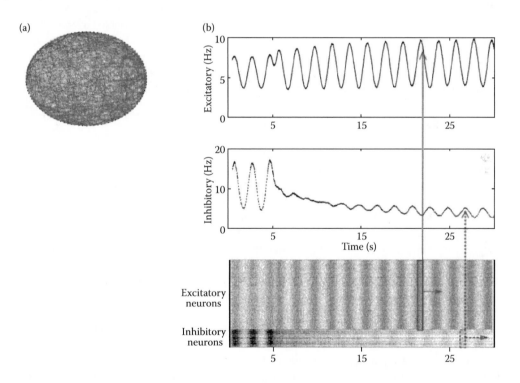

FIGURE 19.7 **(See color insert.)** Schematic diagram of the neuronal network used in our simulations. The figure has been scaled down to 100 neurons for clarity; however, the proportions of inhibitory to excitatory neurons as well as the connection ratio remain the same. The inhibitory neurons are white and the excitatory neurons are black. Synaptic connections in which the presynaptic neuron is inhibitory are shown in blue. Synaptic connections in which the presynaptic neuron is excitatory are shown in red. (b) Spike rate evolution to represent the networks ability to reproduce the experimentally observed drift in firing rates of the two populations of neurons. (Adapted from Fisher et al., *Biol Cybern*, Springer, 102, 427–440, 2009.)

$$I_E(t) = \alpha_E + C_E \sin(\omega t + \Delta\phi)$$

$$I_I(t) = \alpha_I + C_I \sin(\omega t)$$

(19.3)

Again, as in the case for the mean field model and, based on our proposed circadian control hypothesis, the synaptic weights are assumed to be modulated in a circadian phase–dependent manner following the update rule given by

$$\frac{dg_{ij}}{dt} = (g_{ij}^\infty - g_{ij})/\tau_L, \text{ where } g_{ij}^\infty = \alpha_{ij}\Delta\phi + \beta_{ij}.$$

(19.4)

The network model was analyzed to determine the set of parameters that mimics the same dynamics as experimentally observed in Figure 19.3. The results are presented in Figure 19.7b. The status injury resulting in instantaneous phase shift $\Delta\phi$ happens at time $t = 5$ s. The average firing rate $f_X(t)$ ($CX = Ex, In$) at any given time t is computed as follows. We count the number of spikes in the window ranging from $(t - 0.5)$ s to t s for the neuron population in question. We then normalize this count by the number of neurons in that population and divide the normalized value by the window size to get a spike rate, which is in hertz. The drift in firing rates is then computed by estimating $D_X = <df_X/dt>$, as described above. We have analyzed the difference in the drift $\Delta D = D_{Ex} - D_{IN}$ as a function of the initial and final asymptotic synaptic weights for the various synapse types. The initial asymptotic synaptic weights are defined as $g_{ij}^\infty = g_{ij}^\infty(0)$ when $\Delta\phi = 0$ prior to status injury and the final asymptotic synaptic weights are $g_{ij}^\infty = g_{ij}^\infty(\pi)$ when $\Delta\phi = \pi$ following status injury, as in Figure 19.3. We first identified a set of asymptotic synaptic weights that yielded dynamics similar to the experimental results, as shown in Figure 19.7b. For each pairwise combination of synapse types ($g_{EE}, g_{EI}, g_{IE}, g_{II}$), we would alter the initial asymptotic synaptic weight and the final asymptotic synaptic weight while keeping the other asymptotic synaptic weights fixed. We then reran the simulations and calculated ΔD with the new asymptotic synaptic weights to see how changing the asymptotic synaptic weights of a given type of synapse affected the outcome of the experiment. In Figure 19.8, we present an example plot of this analysis when we vary $g_{IE}(0)$ and $g_{EE}(\pi)$. This means the synaptic weights of all inhibitory to excitatory connections before the shift and the synaptic weights of all excitatory to excitatory connections after the shift are modified, while the remaining synaptic weights are kept constant. Any experiment that produces spike rates that yield $D_{Ex} < 0$ or $D_{In} > 0$ signify a set of weights that yielded invalid drifts and are shown in yellow. Any experiments that do not produce an excitatory spike rate shifted by π in relation to the inhibitory spike rate are shown in red. The remaining colors indicate results that are representative of the dynamics observed experimentally. Also shown in the plot are the excitatory (solid line) and inhibitory (dashed line) spike rates as a function of time for three different pairs of asymptotic synaptic weights. A detailed analysis such as presented in Figure 19.8 gives insight into which type of synapse has the most influence and the role of the synapse type in the evolution of an epileptic seizure as per the circadian control hypothesis. As summarized in Fisher et al. (2009), we found that synaptic connections that involve excitatory neurons, particularly those in which the presynaptic neuron is excitatory, have more significant effects on the drift as opposed to synaptic junctions with inhibitory neurons. The most likely reason for this is that the excitatory neurons outnumber the inhibitory neurons four to one. This analysis of the synapse type agrees with previous results, which also show the importance of excitatory neurons and the ineffectiveness of inhibitory neurons when it comes to HSP (Turrigiano and Nelson 2004; Frolich et al. 2008).

19.4 CONCLUSIONS

In summary, these animal model experimental and computational results suggest an evolving imbalance in hippocampal CA1 excitability following status epilepticus as exhibited by the firing activity of

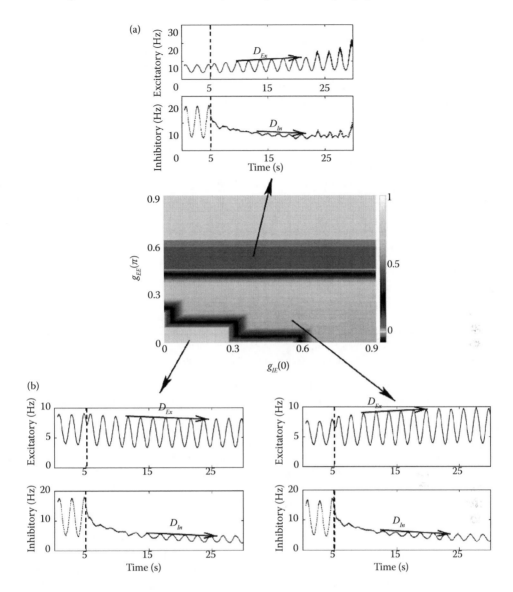

FIGURE 19.8 **(See color insert.)** A plot of ΔD as you vary $g_{IE}(0)$ and $g_{EE}(\pi)$ while the remaining asymptotic synaptic weights are kept constant. $g_{IE}(0)$ and $g_{EE}(\pi)$ are the asymptotic synaptic weight values for the inhibitory to excitatory connections, before and after the phase shift $t = 5$ s, respectively. Red and yellow indicate a set of asymptotic synaptic weights, which yields results that are not similar to experimental results in Figure 19.3. Red depicts those simulations which produced inhibitory and excitatory spike rates that oscillated in the same phase after the phase shift as opposed to antiphase, as was seen experimentally. Yellow depicts those simulations, which either yielded excitatory spike rates with a negative drift or inhibitory spike rates with a positive drift. Also shown are three examples of the spike rate evolution for the chosen pairs of asymptotic synaptic weights. The shift in the excitatory driving current occurs at time $t = 5$ s, shown by the dotted line. The increase or decrease in the spike rates is made apparent with the use of arrows indicating the drift for the respective spike rate. (Adapted from Fisher et al., *Biol Cybern*, Springer, 102, 427–440, 2009.)

two distinct classes of SWs. Moreover, the phase of SW rhythm in CA1 is shifted relative to the endogenous circadian rhythm as demonstrated by the measurement of the CBT constant under controlled 24-h light–dark conditions. These results lead us to propose a circadian control mechanism for the circadian phase-induced imbalance in CA1 SW firing activity. The implication of our circadian control hypothesis was tested by employing a two-dimensional Wilson–Cowan model for CA1. The model consists of two, data-driven key assumptions, namely, that the two subclasses of CA1 SW represent synergistic interaction between excitatory and inhibitory neuronal population, and that SW firing rate activity is under ~24-h circadian control. Importantly, the results of the model allowed us to replicate the observed temporal dynamics of firing activity of populations of neurons in the CA1 network and provided support for our circadian control hypothesis. Future in vivo animal model experiments combined with the computational model are expected to have long-term impacts and realizable clinical applications in epilepsy and associated sleep disorders neuroscience. For instance, identification of circadian factors in seizure occurrence can lead to novel treatments. Indeed, a better understanding about the limbic and SCN coupling mechanisms is essential to understanding the role of circadian rhythm and seizures, and to developing new treatment strategies for individuals with chronic epilepsy.

REFERENCES

Albus, H., M. J. Vansteensel, S. Michel, G. D. Block, and J. H. Meijer. 2005. A GABAergic mechanism is necssary for coupling dissociable ventral and dorsal regional oscillators within the circadian clock. *Curr. Biol.* 15(10):886–893.

Anderson, P., T. V. Bliss, and K. K. Skrede. 1971. Lamellar organization of hippocampal pathways. *Exp. Brain Res.* 13(2):222–238.

Angles-Pujolras, M., A. Diez-Noguera, and T. Cambras. 2007. Exposure to T-cycles of 22 and 23 h during lactation modifies the later dissociation of motor activity and temperature circadian rhythms in rats. *Chronobiol. Int.* 24(6):1049–1064.

Arida, R. M., F. A. Scorza, C. A. Peres, and E. A. Cavalheiro. 1999. The course of untreated seizures in the pilocarpine model of epilepsy. *Epilepsy Res.* 34(2–3):99–107.

Barnes, C. A., B. L. McNaughton, G. V. Goddard, R. M. Douglas, and R. Adamec. 1977. Circadian rhythm of synaptic excitability in rat and monkey central nervous system. *Science* 197(4298):91–92.

Bazil, C. W., D. Short, D. Crispin, and W. Zheng. 2000. Patients with intractable epilepsy have low melatonin, which increases following seizures. *Neurology* 55(11):1746–1748.

Beersma, D. G., and M. C. Gordijn. 2007. Circadian control of the sleep-wake cycle. *Physiol. Behav.* 90(2–3):190–195.

Bertram, E. H. 1997. Functional anatomy of spontaneous seizures in a rat model of limbic epilepsy. *Epilepsia* 38(1):95–105.

Bertram, E. H., and J. Cornett. 1993. The ontogeny of seizures in a rat model of limbic epilepsy: Evidence for a kindling process in the development of chronic spontaneous seizures. *Brain Res.* 625(2):295–300.

Brewis, M., D. C. Poskanzer, C. Rolland, and H. Miller. 1966. Neurological disease in an English city. *Acta. Neurol. Scand.* 42(Suppl 24):1–89.

Cambras, T., J. R. Weller, M. Angles-Pujoras, M. L. Lee, A. Christopher, A. Diez-Noguera, J. M. Krueger, and H. O. de la Iglesia. 2007. Circadian desynchronization of core body temperature and sleep stages in the rat. *Proc. Natl. Acad. Sci. U. S. A.* 104(18):7634–7639.

Campuzano, A., J. Vilaplana, T. Cambras, and A. Diez-Noguera. 1998. Dissociation of the rat motor activity rhythm under T cycles shorter than 24 hours. *Physiol. Behav.* 63:171–176.

Carney, P. R., R. B. Berry, and J. D. Geyer. 2005. Introduction to sleep and sleep monitoring: The basics. In *Clinical sleep disorders*. Geyer, J. D., Carney, P. R., and Berry, R. B., eds. Philadelphia: Lippincott, Williams & Wilkins.

Carpentieri, A. R., M. A. Pujolras, J. J. Chiesa, A. D. Noguera, and T. Cambras. 2006. Effect of melatonin and diazepam on the dissociated circadian rhythm in rats. *J. Pineal. Res.* 40(4):318–325.

Chaudhary, D., L. M. Wang, and C. S. Colwell. 2005. Circadian regulation of hippocampal long term potentiation. *J. Biol. Rhythms* 20:225–236.

Clark, F. M., and M. Radulovacki. 1988. An inexpensive sleep-wake state analyzer for the rat. *Physiol. Behav.* 43(5):681–683.

Colwell, C. S. 2000. Rhythmic coupling among cells in the suprachiasmatic nucleus. *J. Neurobiol.* 43(4):379–388.

Czeisler, C. A., and D. J. Dijk. 2001. *Human circadian physiology and sleep-wake regulation*. New York: Kluwer.

Dallmann, R., G. Lemm, and N. Mrosovsky. 2007. Toward easier methods of studying nonphotic behavioral entrainment in mice. *J. Biol. Rhythms* 22(5):458–461.

Dallmann, R., and N. Mrosovsky. 2006. Scheduled wheel access during daytime: A method for studying conflicting zeitgebers. *Physiol. Behav.* 88(4–5):459–465.

Davis, G. W. 2006. Homeostatic control of neural activity: From phenomenology to molecular design. *Annu. Rev. Neurosci.* 29:307–323.

de la Iglesia, H. O., T. Cambras, W. J. Schwartz, and A. Diez-Noguera. 2004. Forced desynchronization of dual circadian oscillators within the rat suprachiasmatic nucleus. *Curr. Biol.* 14(9):796–800.

Du, F., and R. Schwarcz. 1992. Aminooxyacetic acid causes selective neuronal loss in layer III of the rat medial entorhinal cortex. *Neurosci. Lett.* 147(2):185–188.

Duffy, J. F., and D. J. Dijk. 2002. Getting through to circadian oscillators: Why use constant routines? *J. Biol. Rhythms* 17(1):4–13.

Fee, M. S., P. P. Mitra, and D. Kleinfeld. 1996. Automatic sorting of multiple unit neuronal signals in the presence of anisotropic and non-Gaussian variability. *J. Neurosci. Methods* 69(2):175–188.

Fisher, N., S. S. Talathi, W. Ditto, and P. R. Carney. 2009. Effect of phase on homeostatic spike rates. *Biol. Cybern.* 102(5):427–440.

Frohlich, F., M. Bazhenov, and T. J. Sejnowski. 2008. Pathological effects of homeostatic synaptic scaling on network dynamics in diseases of the cortex. *J. Neurosci.* 28(7): 1709–1720.

Gowers, W. R. 1881. *Epilepsy and other chronic convulsive diseases: Their causes, symptoms and treatment*. London: J. & A. Churchill.

Harris, K. M., and T. J. Teyler. 1984. Developmental onset of long-term potentiation in area CA1 of the rat hippocampus. *J. Physiol.* 346(1):27–48.

Herman, S. T., T. S. Walczak, and C. W. Bazil. 2001. Distribution of partial seizures during the sleep–wake cycle: Differences by seizure onset site. *Neurology* 56(11):1453–1459.

Hofstra, W. A., and A. W. de Weerd. 2009. The circadian rhythm and its interaction with human epilepsy: A review of literature. *Sleep Med. Rev.* 3(6):413–420.

Hofstra, W. A., B. E. Grootemarsink, R. Dieker, J. van der Palen, and A. W. de Weerd. 2009. Temporal distribution of clinical seizures over the 24-h day: A retrospective observational study in a tertiary epilepsy clinic. *Epilepsia* 50(9):2019–2026.

Horacio, O. D. L. I., T. Cambras, and A. D. Noguera. 2008. Circadian internal desynchronization: Lessons from a rat. *Sleep Biol. Rhythms* 6:76–83.

Janca, A., L. Prilipko, and J. A. Costa e Silva. 1997. The World Health Organization's global initiative on neurology and public health. *J. Neurol. Sci.* 145(1):1–2.

Johns, T. G., et al. 1977. Automated analysis of sleep in the rat. *Electroencephalogr. Clin. Neurophysiol.* 43(1):103–105.

Laakso, M. L., L. Leinonen, T. Hatonen, A. Alila, and H. Heiskala. 1993. Melatonin, cortisol and body temperature rhythms in Lennox-Gastaut patients with or without circadian rhythm sleep disorders. *J. Neurol.* 240(7):410–416.

Lacaille, J. C., and P. A. Schwartzkroin. 1988. Stratum lacunosum-moleculare interneurons of hippocampal CA1 region. II. Intrasomatic and intradendritic recordings of local circuit synaptic interactions. *J. Neurosci.* 8(4):1411–1424.

Leite, J. P., Z. A. Bortolotto, and E. A. Cavalheiro. 1990. Spontaneous recurrent seizures in rats: An experimental model of partial epilepsy. *Neurosci. Biobehav. Rev.* 14(4):511–517.

Liu, C., and S. M. Reppert. 2000. GABA synchronizes clock cells within the suprachiasmatic circadian clock. *Neuron* 25(1):123–128.

Loscher, W. 1997. Animal models of intractable epilepsy. *Prog. Neurobiol.* 53(2):239–258.

Loscher, W. 2002. Animal models of drug-resistant epilepsy. *Novartis Found. Symp.* 243:149–159; Discussion 159–166, 180–185.

Lothman, E. W., E. H. Bertram, J. Kapur, and J. L. Stringer. 1990. Recurrent spontaneous hippocampal seizures in the rat as a chronic sequela to limbic status epilepticus. *Epilepsy Res.* 6(2):110–118.

Malow, B. A., L. M. Selwa, D. Ross, and M. S. Aldrich. 1999. Lateralizing value of interictal spikes on overnight sleep-EEG studies in temporal lobe epilepsy. *Epilepsia* 40:1587–1592.

Mangan, P. S., and E. H. Bertram, III. 1997. Shortened-duration GABA(A) receptor-mediated synaptic potentials underlie enhanced CA1 excitability in a chronic model of temporal lobe epilepsy. *Neuroscience* 80(4):1101–1111.

Maywood, E. S., J. O'Neill, G. K. Wong, A. B. Reddy, and M. H. Hastings. 2006. Circadian timing in health and disease. *Prog. Brain Res.* 153:253–269.

Morton, A. J., N. I. Wood, M. H. Hastings, C. Hurelbrink, R. A. Barker, and E. S. Maywood. 2005. Disintegration of the sleep-wake cycle and circadian timing in Huntington's disease. *J. Neurosci.* 25(1):157–163.

Neuhaus, H. U., and A. A. Borbely. 1978. Sleep telemetry in the rat. II. Automatic identification and recording of vigilance states. *Electroencephalogr. Clin. Neurophysiol.* 44(1):115–119.

Niedermyer, E., and L. Silva. 2004. *Electroencephalography: Basic principles, clinical applications and related field*, 5th ed. Baltimore: Williams & Wilkins.

Peraita-Adrados, R. 2004. Epilepsy and sleep–wake cycle. *Rev. Neurol.* 38:173–175.

Quigg, M. 2000. Circadian rhythms: Interactions with seizures and epilepsy. *Epilepsy Res.* 42(1):43–55.

Quigg, M., H. Clayburn, M. Straume, M. Menaker, and E. H. Bertram. 1999. Hypothalamic neuronal loss and altered circadian rhythm of temperature in a rat model of mesial temporal lobe epilepsy. *Epilepsia* 40(12):1688–1696.

Quigg, M., H. Clayburn, M. Straume, M. Menaker, and E. H. Bertram. 2000. Effects of circadian regulation and rest-activity state on spontaneous seizures in a rat model of limbic epilepsy. *Epilepsia* 41(5):502–509.

Quigg, M., M. Straume, M. Menaker, and E. H. Bertram. 1998. Temporal distribution of partial seizures: Comparison of an animal model with human partial epilepsy. *Ann. Neurol.* 43(6):748–755.

Quigg, M., M. Straume, T. Smith, M. Menaker, and E. H. Bertram. 2001. Seizures induce phase shifts of rat circadian rhythms. *Brain Res.* 913(2):165–169.

Racine, R., V. Okujava, and S. Chipashvilli. 1972. Modification of seizure activity by electrical stimulation. 3. Mechanisms. *Electroencephalogr. Clin. Neurophysiol.* 32(3):295–299.

Racine, R. J. 1972. Modification of seizure activity by electrical stimulation. II. Motor seizure. *Electro-Encephalogr. Clin. Neurophysiol.* 32(3):281–294.

Raghavan, A. V., J. M. Horowitz, and A. C. Fuller. 1999. Diurnal modulation of long-term potentiation in the hamster hippocampal slice. *Brain Res.* 833(2):311–314.

Sanchez, J. C., T. H. Mareci, W. M. Norman, J. C. Principe, W. L. Ditto, and P. R. Carney. 2006. Evolving into epilepsy: Multiscale electrophysiological analysis and imaging in an animal model. *Exp. Neurol.* 198(1):31–47.

Saper, C. B., G. Cano, and T. E. Scammell. 2005. Homeostatic, circadian, and emotional regulation of sleep. *J. Comp. Neurol.* 493(1):92–98.

Saper, C. B., T. E. Scammell, T. C. Chou, and J. Gooley. 2005. Hypothalamic regulation of sleep and circadian rhythms. *Nature* 437(7063):1257–1263.

Sarkisian, M. R. 2001. Overview of the current animal models for human seizure and epileptic disorders. *Epilepsy Behav.* 2(3):201–216.

Schapel, G. J., R. G. Beran, D. L. Kennaway, J. McLoughney, and C. D. Matthews. 1995. Melatonin response in active epilepsy. *Epilepsia* 36(1):75–78.

Stables, J. P., E. H. Bertram, H. S. White, D. A. Coulter, M. A. Dichter, M. P. Jacobs, W. Loscher, D. H. Lowenstein, S. L. Moshe, J. L. Noebels, and M. Davis. 2002. Models for epilepsy and epileptogenesis: Report from the NIH workshop, Bethesda, Maryland. *Epilepsia* 43(11):1410–1420.

Steward, L. S., and L. S. Leung 2001. Diurnal variation in pilocarpine-induced generalized tonic–clonic seizure activity. *Epilepsy Res.* 44:207–212.

Steward, L. S., and L. S. Leung. 2003. Temporal lobe seizures alter the amplitude and timing of rat behavioral rhythm. *Epilepsy Behav.* 4:153–160.

Talathi, S. S., D. U. Hwang, W. L. Ditto, T. Mareci, H. Sepulveda, M. Spano, and P. R. Carney. 2009. Circadian control of neural excitability in an animal model of temporal lobe epilepsy. *Neurosci. Lett.* 455(2):145–149.

Talathi, S. S., D. U. Hwang, M. L. Spano, J. Simonotto, M. D. Furman, S. M. Myers, J. T. Winters, W. L. Ditto, and P. R. Carney. 2008. Non-parametric early seizure detection in an animal model of temporal lobe epilepsy. *J. Neural. Eng.* 5(1):85–98.

Turrigiano, G. G., and S. B. Nelson. 2004. Homeostatic plasticity in developing nervous system. *Nat. Rev. Neurosci.* 5(2): 97–107.

Van Dongen, H. P., et al. 1999. Searching for biological rhythms: Peak detection in the periodogram of unequally spaced data. *J. Biol. Rhythms* 14(6):617–620.

Wever, R. A. 1983. Fractional desynchronization of human circadian rhythms. A method for evaluating entrainment limits and functional interdependencies. *Pflugers Arch.* 396:128–137.

White, A., P. A. Williams, J. L. Hellier, S. Clark, E. F. Dudek, and K. J. Staley. 2010. EEG spike activity precedes epilepsy after kainate-induced status epilepticus. *Epilepsia* 51:371–383.

Wilson, H. R., and J. D. Cowan. 1972. Excitatory and inhibitory interactions in localized populations of model neurons. *Biophys. J.* 12(1):1–24.

Zulley, J., R. Wever, and J. Aschoff. 1981. The dependence of onset and duration of sleep on the circadian rhythm of rectal temperature. *Pflugers Arch.* 391:314–318.

20 Use of Dynamical Measures in Prediction and Control of Focal and Generalized Epilepsy

Shivkumar Sabesan, Leon Iasemidis, Konstantinos Tsakalis, David M. Treiman, and Joseph Sirven

CONTENTS

20.1 INTRODUCTION

Epileptic seizures are manifestations of epilepsy, a common dynamical disorder second only to stroke and Alzheimer's disease among neurological disorders. Of the world's 50 million people with epilepsy, about one-third have seizures that are not controlled by antiepileptic drugs (AEDs) (Engel et al. 2007). One of the most disabling aspects of epilepsy is the seemingly unpredictable nature of seizures. If seizures cannot be controlled, the patient experiences major limitations in family, social, educational, and vocational activities. In addition, status epilepticus (SE), a life-threatening condition in which seizure activity occurs continuously, is often successfully treated only with extreme intervention (Alldredge et al. 2007; Treiman and Walker 2006). Until recently, the general belief in the medical community was that epileptic seizures could not be anticipated. Seizures were assumed to be abrupt transitions that occurred randomly over time. However, theories based on reports from clinical practice and scientific intuition, like the "reservoir theory" postulated by Lennox, have pointed to the possibility of seizure predictability (Lennox 1946). Various feelings of auras (patients' reports of sensations of an upcoming seizure) also are common in the medical literature.

The ability to predict epileptic seizures significantly before their occurrence (onset) may lead to novel diagnostic tools and treatments for epilepsy. Electromagnetic stimulation and/or administration of anti-epileptic drugs during the preictal period may disrupt the observed dynamical entrainment of normal brain sites with the epileptogenic focus, and thus lead to a significant reduction of epileptic seizures. Evaluation of antiepileptic drugs and protocols, within the context of a patient's seizure susceptibility period and/or preictal period—as detected by seizure prediction algorithms—may lead to the design of new, more effective drugs with fewer side effects (Theodore and Fisher 2004).

20.2 BACKGROUND

20.2.1 Historical Developments in Seizure Prediction

The 1980s saw the emergence of new signal processing methodologies, based on the mathematical theory of nonlinear dynamics, for the study of spontaneous formation of organized spatial, temporal, or spatiotemporal patterns in physical, chemical, and biological systems (Degn et al. 1987; Frank et al. 1990; Freeman 1987). These methodologies quantify the complexity and randomness of the signal from the perspective of dynamical invariants (e.g., dimensionality of the attractor through correlation dimension, or entropy rate through Kolmogorov entropy), and represent a drastic departure from the signal processing techniques based on the linear model (e.g., Fourier analysis). The dynamical hypothesis changed some long-held beliefs about seizures. Iasemidis, Sackellares et al. (1994) reported the first evidence that the transition to epileptic seizures may be consistent with a deterministic process and that the ictal electroencephalogram (EEG) can be better modeled as an output of a nonlinear rather than a linear system (Iasemidis et al. 1988, 1990). The existence of long-term preictal periods (order of minutes) was initially shown using nonlinear dynamical analysis of EEG from subdural arrays over the temporal lobe (Iasemidis et al. 1988; Iasemidis and Sackellares 1991), and later from scalp (Iasemidis et al. 1996; Sackellares et al. 2000) and depth EEG (Casdagli et al. 1996; Iasemidis and Sackellares 1996; Sackellares et al. 2000); this demonstration raised the feasibility of developing seizure prediction algorithms by monitoring the temporal evolution of the short-term Lyapunov exponents (STL_{max}), that is, measures of chaos and stability of a system. The possibility of focus localization and seizure detection with these techniques was also reported (Iasemidis et al. 1997a, 1997b).

The basis of these findings was the discovery of a progressive preictal increase of spatiotemporal entrainment (dynamical synchronization) between critical sites of the brain as a precursor to epileptic seizures. The algorithm that first showed this preictal entrainment was based on the statistical convergence of STL_{max} among critical electrode sites adaptively selected over time and has been successfully implemented for prospective prediction of epileptic seizures. Global optimization techniques were applied for selecting the critical groups of electrode sites (Iasemidis et al. 2001, 2003c). The adaptive seizure prediction algorithm (ASPA) was tested in continuous (0.76 to 5.84 days) intracranial EEG recordings from a group of five patients with refractory temporal lobe epilepsy. For a 3-h seizure prediction horizon, the algorithm predicted 80% of seizures with a false prediction rate of 0.16/h. Seizure warnings occurred on an average of 71.7 min before ictal onset. These results indicated that ASPA could be used for diagnostic and therapeutic purposes. A skepticism regarding application of STL_{max} to predict seizures (Lai et al. 2003) was addressed in Iasemidis et al. (2005). In later reports, the algorithm's performance was compared with ones of statistically-based naive prediction schemes that did not utilize information from the EEG (Sackellares et al. 2006; Yang et al. 2004). An ongoing debate exists as to which statistical comparisons are optimal, and on the standards for a good seizure prediction algorithm performance (Andrzejak et al. 2009; Chaovalitwongse et al. 2006). The success of ASPA triggered a number of studies (unfortunately mostly retrospective ones) with a substantially poorer predictive performance than expected for measures other than STL_{max}, like the correlation dimension (Aschenbrenner-Scheibe et al. 2003), the similarity index (Winterhalder et al. 2003), and accumulated energy (Maiwald et al. 2004). In

2005, several groups published a series of studies that were carried out on a set of five continuous multiday EEG recordings provided by different national and international epilepsy centers for the First International Collaborative Workshop on Seizure Prediction (Lehnertz and Litt 2005). One of the aims of this workshop was to have different groups test and compare their methods on identical EEG datasets. Results from the different groups for the most part showed a poor performance of univariate measures (D'Alessandro et al. 2005; Mormann et al. 2005). A better performance was reported for bivariate and multivariate measures (Iasemidis et al. 2005; Le Van Quyen et al. 2005), with Iasemidis et al. (2005) STL_{max} method being the only prospective one. In addition, this method exhibited the best performance sensitivity and specificity on the common datasets we analyzed. For further review of the literature on seizure prediction see the works by Iasemidis (2003), Lehnertz et al. (2007), and Sackellares (2008).

20.2.2 RESETTING OF BRAIN DYNAMICS

A byproduct of this line of analysis was the counterintuitive finding that the majority (81%) of analyzed seizures in patients with temporal lobe epilepsy (TLE) irreversibly reset (disentrained) postictally the observed preictal dynamical entrainment (Iasemidis et al. 2004). This finding, combined with the observed nonresetting of brain dynamics during the preictal periods, implied that seizures occur when there is a real need to reset the pathologically synchronized brain's dynamics (Iasemidis et al. 2004; Sabesan et al. 2009a). Recent reports (Mormann et al. 2003) have claimed that what characterizes the preictal period is disentrainment rather than entrainment of brain dynamics. However, we have found that this reverse concept of preictal disentrainment is either not present or not consistent across seizures in the same patient and/or across patients (Sabesan et al. 2007a).

20.2.3 SEIZURE CONTROL

Employing neuronal population models that are capable of exhibiting seizure-like behavior, we have shown that entrainment (disentrainment) of the populations' STL_{max}, with increased (decreased) coupling between populations, resembles the observed preictal dynamical entrainment and postictal disentrainment of STL_{max} at critical sites in the epileptic brain (Iasemidis et al. 2003b; Tsakalis et al. 2006). In agreement with burst phenomena in adaptive systems, seizures in these models occur if the existing (internal to the models) feedback loops are pathological so that they lack the ability to compensate fast enough for excessive increases in the network coupling. This situation eventually leads to seizure-like transitions (Iasemidis et al. 2003b; Prasad et al. 2007). Motivated by these findings, using a control-oriented approach, we developed a functional model for an external seizure controller. During periods of abnormally high synchrony, the developed control scheme provides appropriate desynchronizing feedback to maintain normal synchronization levels between neural populations (homeostasis of dynamics) (Tsakalis and Iasemidis 2006). This feedback control view of epileptic seizures and the developed seizure control strategies we called feedback decoupling were validated on coupled chaotic oscillator models (Tsakalis et al. 2006) and biologically plausible neurophysiologic models (Chakravarthy et al. 2007, 2009).

20.3 METHODS

The brain is inherently a nonlinear system. Therefore, it makes sense to move one step forward from linear methods of analysis (methods like Fourier transform and linear models), that can accurately analyze and characterize only linear systems, to methods that can analyze and correctly characterize linear and nonlinear systems. This move has progressively produced surprisingly refreshing results for epilepsy. Under certain conditions, through the method of delays (Packard et al. 1980; Takens 1981), sampling of a single variable of a nonlinear system over time can determine the state variables of the system that are related to it. In the case of the EEG, this method can be used to reconstruct a multidimensional state space of the brain's electrical activity at a corresponding brain site. A dimension of seven has typically been used to reconstruct the state space from EEG in epilepsy. This selection of dimension for the states (vectors) in the state space, together with a corresponding time delay of 20 ms between their components, has been justified in the literature (Iasemidis et al. 2000; Sackellares et al. 2000).

Among the important measures of the dynamics of a linear or nonlinear system are the Lyapunov exponents that measure the average information flow (bits/s) a system produces along local eigen directions through its movement in its state space. Positive Lyapunov exponents denote generation of information, while negative exponents denote destruction of information. A chaotic nonlinear system possesses at least one positive Lyapunov exponent, and it is because of this feature that its behavior looks random, even though as a system it is deterministic. Methods for calculating these measures of dynamics from experimental data have been published (Wolf et al. 1985; Rosenstein et al. 1993; Iasemidis and Sackellares 1991; Kantz 1994). The estimation of the largest Lyapunov exponent from a chaotic system has been shown to be more reliable and reproducible than the estimation of the remaining exponents, especially when the actual value of the dimension of the state space is not known and changes over time (Vastano and Kostelich 1985). The brain, being nonstationary, is never in a steady state in the strictly dynamical sense, at any location. We have shown that, in the case of a nonstationary system with transients like epileptic spikes, using the short-term maximum Lyapunov exponent (STL_{max}) (Iasemidis and Sackellares 1991; Iasemidis et al. 2000) is a more accurate characterization of the average information flow than the one using the regular maximum Lyapunov exponent (Wolf et al. 1985; Kantz 1994; Rosenstein et al. 1993). STL_{max} is estimated from sequential EEG segments of 10 s in duration per recording site to create a set of STL_{max} profiles over time for all recording sites.

A statistical measure of entrainment between two brain sites, i and j, with respect to a measure of their dynamics (e.g., STL_{max}), has been developed in the past (Iasemidis and Sackellares 1996; Iasemidis et al. 2000). Specifically, the T_{ij} between measures at electrode sites i and j and at time t is defined as

$$T_{ij}^t = \frac{\left|\bar{D}_{ij}^t\right|}{\hat{\sigma}_{ij}^t/\sqrt{m}}, \tag{20.1}$$

where $\bar{D}_{ij}^t$ and $\hat{\sigma}_{ij}^t$ denote the sample mean and standard deviation, respectively, of all the m differences between the STL_{max} values (one STL_{max} value is produced per 10-s EEG segment) at electrodes i and j, within a moving window $w_t = [t, t - m*10.24 \text{ s}]$. If the true mean μ_{ij}^t of the differences D_{ij}^t is equal to zero, and σ_{ij}^t are independent and normally distributed, T_{ij}^t is asymptotically distributed as the t-distribution with $(m - 1)$ degrees of freedom. We have shown that these independence and normality conditions are satisfied in EEG (Iasemidis et al. 2003c). Therefore, we define disentrainment (dynamical desynchronization) between electrode sites i and j when T_{ij} is significantly different from zero at a significance level α. The desynchronization condition between the electrode sites i and j, as detected by the paired t-test, is

$$T_{ij}^t > t_{\alpha/2,m-1} = T_{th}, \tag{20.2}$$

where $t_{\alpha/2, m-1} = T_{th}$ is the $100(1 - \alpha/2)$ critical value of the t-distribution with $m - 1$ degrees of freedom. If $T_{ij}^t \leq t_{\alpha/2,m-1}$ (which means that we do not have satisfactory statistical evidence at the α level that the differences of values of a measure between electrode sites i and j within the time window $w(t)$ are not zero), we consider that sites i and j are synchronized with each other (with respect to the utilized measure of synchronization) at time t. Using $\alpha = 0.01$ and $m = 60$, that is, using a window

$w(t)$ of 10 min in duration, the threshold $T_{th} = 2.662$. It is noteworthy that similar STL_{max} values at two electrode sites (and therefore low values of the T-index) do not necessarily mean that these sites even interact. However, when there is a progressive convergence (synchronization) over time of the measures at these sites, the probability that they are unrelated diminishes.

This is exactly what occurs before seizures. We observe significantly low values of T-index (i.e., convergence of the STL_{max} profiles of critical sites) long (ranges between 30–120 min) before seizure onset. Moreover, the trend toward low values of the T-index starts long before the seizure's onset; as the seizure approaches, the critical pairs of sites become progressively entrained and stay entrained thereafter, until the seizure's onset. This observation of preictal synchronization was also tested using other measures of brain dynamics, namely, (a) amplitude, (b) phase, (c) Shannon entropy, and (d) Kullback-Leibler entropy. Results from such an analysis of 43 seizures recorded from two patients with temporal lobe epilepsy showed that critical sites selected on the basis of STL_{max} have longer and more consistent preictal trends before a majority of seizures than the ones selected from the other four measures of synchronization (Iasemidis et al. 2003a; Sabesan et al. 2003b, 2007a).

After seizure's onset, we observe a trend toward disentrainment of the preictally entrained pairs of sites (T-index moves rapidly toward higher values above T_{th}). We have called this reversal of dynamics brain resetting (Iasemidis et al. 2004; Sabesan et al. 2009a). We have shown in the past that the observed dynamical resetting is specific to seizures at the $\alpha = 0.05$ statistical significance level. This observation may reflect a passive mechanism (e.g., high electrical activity during a seizure might deplete critical neurotransmitters and thus deactivate critical neuroreceptors in the entrained neuronal network). An alternative explanation is that release of neuropeptides in the brain, due to seizure activity, may contribute to the temporary repair of the pathological feedback that allowed the dynamical entrainment to last for a long time prior to a seizure's onset.

20.4 RECENT RESULTS

20.4.1 Status Epilepticus—A Failed Case of Brain Resetting

Given the association of the phenomenon of resetting of brain dynamics with seizures as described above, one would predict that status epilepticus (SE) is associated with failure of seizures to reset the brain. That is what we actually have observed in EEG records analyzed from SE patients in the emergency room (ER), the intensive care unit (ICU), and the epilepsy monitoring unit (EMU). The correspondence of the T-index values to the changing medical condition of the patient over time was remarkable, as was the effect (success or failure) of the different AEDs on SE. It was observed that AEDs can reset the pathological brain dynamics. High T-index values were observed for successful resetting of the brain dynamics by an AED, and low T-index values when the effect of the AED was temporary (and therefore unsuccessful) (Good et al. 2004; Faith et al. 2009). Moreover, the long-term recovery trend in the T-index over days indicated that this measure could be used for detection of a patient's seizure susceptibility; that is, to identify periods when a patient is more vulnerable to seizure recurrences (Iasemidis et al. 2009).

20.4.2 Seizure Control-Modeling Studies

While seizures can currently be anticipated with relatively good sensitivity and specificity, the question is whether we can effectively intervene to change the pathological brain dynamics (long-term entrainment) in a timely manner and prevent a seizure from occurring. During the preictal dynamical transition described in the previous section, multiple regions of the brain progressively (minutes to hours) approach a similar dynamical state. These observations are in agreement with observations in systems of coupled neural population models with internal (i.e., part of the overall system) pathological feedback. We have shown that by increasing the interpopulation coupling and detuning their internal feedback mechanism, these systems move from spatiotemporal chaotic states into spatially synchronized, quasiperiodic, seizure-like states (Tsakalis and Iasemidis 2006). These transitions also are characterized by spatial convergence of the maximum Lyapunov exponents of the individual populations (Iasemidis et al 2003b; Tsakalis et al. 2006). Such results directly address the process of ictogenesis and appear to support the hypotheses that: a) seizures may result from the inability of internal feedback mechanisms to provide timely compensation/regulation of coupling between brain sites, and b) seizure control can be achieved by feedback decoupling of the pathological sites via externally provided appropriate stimuli. These stimuli are functions of both the EEG and the coupling of the respective brain sites. We have called such a closed-loop control scheme feedback decoupling (Tsakalis and Iasemidis 2006; Tsakalis et al. 2006). Thus, the observed long-term dynamical entrainment prior to seizures can be interpreted as an indicator of pathology in the internal feedback of the network. This rationale can, of course, easily be applied to reflex epilepsies, too.

20.4.3 SEIZURE CONTROL-ANIMAL STUDIES

Motivated by our modeling studies, we first showed that impulsive electrical stimulation can temporally disentrain the dynamics of a rat's epileptic brain (larger T-index values poststimulus rather than prestimulus). Second, we utilized our adaptive seizure prediction algorithm in the feedback branch of a closed-loop electrical stimulation scheme for seizure control. In the baseline mode (no stimulation, just spontaneous occurrence of seizures), the sensitivity and specificity of the seizure prediction algorithm on the epileptic rats were tested, and we found very similar values to the ones we have reported from patients with focal epilepsy. In the control mode, the seizure prediction program was used to analyze the recorded EEG from the rat's brain (continuous real-time estimation of STL_{max} and T-index every 10 s), issue seizure warnings and trigger the stimulator to stimulate the brain at each warning. We called this scheme a just-in-time seizure control scheme. In that study (Good et al. 2009), the lithium-pilocarpine (LP) model of acute SE in rats was chosen as the animal model for chronic epilepsy due to the similarities it exhibits with human TLE. Male Sprague-Dawley rats were used for the study. Three to four weeks after the induced SE, rats were implanted with a six-microwire monopolar electrode array targeted to four cortical and two hippocampal locations; two Teflon-coated tungsten bipolar twisted stimulating electrodes were implanted in the centromedial thalamic nucleus. A stimulator/seizure prediction program was developed in-house to control an A-M Systems Model 2300 stimulator unit (Calsborg, WA). At a seizure warning, a train of square pulses of amplitude 600 µA, pulse-width 100 µsec, and frequency 130 Hz was administered across the centromedial thalamic electrodes. The train of pulses was 1 min in duration and was delivered in a charge balanced bipolar cathodic fashion. In one rat, we found that the seizure rate was reduced sharply within 1 day following the onset of the closed-loop control scheme described above. The rat became seizure-free during the third and fourth day, but seizures slowly started to reappear thereafter. However, it was found that the seizure-free state was attained only when resetting of the brain dynamics by the externally provided stimulus was achieved. Subsequent open-loop periodic stimulation (that is, without the parallel running of the seizure prediction software) in the same rat, over 10 days with the same stimulus parameters, and rate of stimulation equal to the mean rate of closed-loop stimulation, failed to reduce the rat's seizure rate and/or reset the brain dynamics. This result implies that a) exact timing, and not period, of stimulation is critical for seizure control, b) entrained brain dynamics increase the probability of seizure occurrence (and thus the seizure rate), whereas disentrained dynamics do not (and thus decrease the seizure rate), and c) in addition to the timely stimulation, a new type of stimulus should be incorporated into an advanced and successful closed-loop control. Our feedback decoupling scheme for seizure control offers some guidance in this quest (Tsakalis and Iasemidis 2006; Tsakalis et al. 2006). Its application to biologically plausible simulation models with seizures has generated results that appear to out-perform both the currently in use open- and closed-loop control schemes and may explain their successes and failures.

20.4.4 Epileptogenic Focus Localization via Information Flow Analysis

In recent years, experimental and theoretical advancements in the study of synchronization of chaotic systems have forged an unprecedented growth in using measures of synchronization to study brain dynamics. Commonly used tools for the estimation of synchronization between data series are the linear cross-correlation in the time domain and the cross-coherence in the frequency domain (Priestly 1981). A mathematically more general (nonparametric, model-free) and statistically rigorous approach for the detection of linear and nonlinear interdependences between time series is the mutual information (MI). However, MI, being a symmetrical function with respect to its arguments, cannot detect directional flow of information and causal relationships unless one of the time series is time delayed. Although the introduction of a time delay in MI is an important improvement toward the goal of causality detection, the resulting measure presents a set of difficulties. One important issue is that its estimation requires a relatively large amount of high quality data (noise-free, stationary, that is, characteristics rarely met in real world data where experimental signals are typically short, nonstationary, and likely to be masked at least by measurement noise), especially if the causal effect under investigation is weak. A second issue is the poor sensitivity and specificity of MI to existing and/or nonexisting time delays in the causal structure of the analyzed time series respectively (Schiff et al. 1996).

To study the directional aspect of interactions, several other approaches have been employed (Chen et al. 2004; Sabesan et al. 2003b, 2007b; Schiff et al. 1999; Schreiber 2000). One of these approaches is based on the improvement of the prediction of a series' future values by incorporating information from another time series. Such an approach was originally proposed by Wiener (1956) and later formalized by Granger in the context of linear regression models of stochastic processes. Granger causality was then extended to nonlinear systems by a) utilizing local linear models in reduced neighborhoods, estimating the resulting statistical quantity and then averaging it over the entire dataset (Schiff et al. 1996), or b) considering an error reduction that is triggered by added variables in global nonlinear models (Chen et al. 2004).

Despite the relative success of the above approaches in detecting the direction of interactions in special cases, these approaches essentially are model-based (parametric) methods (linear or nonlinear). That is, they either make assumptions about the structure of the interacting systems or the nature of their interactions, and as such, they may suffer from all the shortcomings associated with modeling systems and signals of unknown structure. For a detailed review of parametric and nonparametric (linear or nonlinear) measures of causality, we refer the reader to the work by Hlaváčková-Schindler et al. (2007). To overcome this problem, an information theoretic approach to identify the direction of information flow and quantify the strength of coupling between complex systems/signals has recently been suggested (Schreiber 2000). This method was based on the study of transitional probabilities of the states of systems or signals under consideration. The resulted measure was termed transfer entropy (TE).

We have shown (Sabesan et al. 2003a, 2007b) that the direct application of TE as proposed by Schreiber (2000) may not always give the expected results, and that tuning of certain parameters involved in the TE estimation plays a critical role in detecting the correct direction of the information flow between time series. We extended the application of the improved TE to the epileptic brain for the study of localization of the epileptogenic focus (Sabesan et al. 2009b).

Results from such an information flow-based analysis of long-term EEG data, recorded from four patients with temporal lobe epilepsy at two different epilepsy centers, showed a 100% lateralization and localization of their epileptogenic focus. Furthermore, the average probability of the focus being active over hours of continuous EEG, across all unifocal patients, was 0.69. This suggests that an epileptogenic focus may be well active in the interictal period, in addition to the preictal and ictal periods. Detecting the time periods when the focus is active could also help us understand the underlying mechanisms of ictogenesis (seizure generation), as well as design and implement timely interventions for seizure control even in the interictal periods (Iasemidis et al. 1997b; Tsakalis and Iasemidis 2006).

20.4.5 Improvements in Seizure Prediction

More recently, we have developed an automated, reference-independent seizure prediction scheme, based on spatial synchronization of the available recording electrodes. This practical seizure prediction scheme (PSPS) was tested prospectively on continuous, long-term intracranial EEG recordings from ten patients with refractory temporal lobe epilepsy (Sabesan et al. 2010). We have shown that, for a 120-min seizure prediction horizon, the PSPS predicts 88% of all seizures across ten patients with an average false prediction rate of 0.12/h. Seizure warnings were issued by PSPS on an average of 42.7 min before ictal onset. These results are a significant improvement over the ones from our first-generation prospective prediction algorithm (ASPA), especially when taking into consideration the additional features of PSPS: robustness (e.g., independence from when the algorithm starts the analysis), no need for training (the algorithm does not need any training datasets to estimate its parameters). Furthermore, PSPS was statistically validated by testing its performance against a random and a periodic prediction scheme (Andrzejak et al. 2009; Yang et al. 2004). It was shown that PSPS outperforms the periodic and random prediction schemes (Sabesan et al. 2010). Therefore, PSPS represents the first of its kind practical and feasible seizure prediction scheme and can either be embedded into existing neurostimulation devices or used as a stand-alone epilepsy monitoring device. In the next section, we show the application of this new prediction algorithm to EEG recorded from two patients with refractory primary generalized epilepsy.

20.4.6 Dynamical Changes in Primary Generalized Epilepsy

The term primary generalized epilepsy normally encompasses disorders characterized by attacks of absence, myoclonic jerks, and generalized convulsions that may all occur in the same individual. In

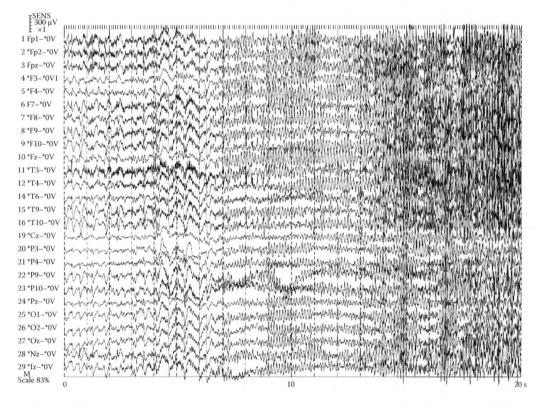

FIGURE 20.1 A 20-s EEG segment at the onset of a primary generalized JME seizure recorded from Patient 1 using a standard 10–20 referential montage. The ictal discharge begins as a series of high amplitude sharp and slow wave complexes simultaneously in all electrodes. The seizure lasted for 150 s (the full duration of this seizure is not shown in this figure).

contrast to focal (partial) epilepsy, in which abnormal electrographic changes originate from a small brain region, primary generalized epilepsy manifests abnormal electrographic changes occurring simultaneously from a large, diffuse region of the brain. Juvenile myoclonic epilepsy (JME) is a type of primary generalized epilepsy that affects approximately 7% of adolescent and adult epilepsy patients. JME is characterized by myoclonic seizures, combined with generalized tonic-clonic seizures or absence seizures. Seizures are precipitated by sudden awakening, sleep deprivation, photic stimulation, or alcohol consumption. Figure 20.1 depicts a typical ictal EEG recording, centered about the time of the onset of an epileptic seizure in Patient 1.

To test the existence of a detectable preictal state prior to generalized seizures, we evaluated our new seizure prediction PSPS algorithm on scalp EEG recorded from two patients with JME. These two patients identified for analysis were the only ones with JME in a database of 125 patients with long-term EMU recordings collected between 03/09 and 09/09 at Mayo Clinic, Scottsdale, Arizona. Each of the two patients had two and three seizures respectively during their stay at EMU. We let PSPS run prospectively from the beginning of each recording. Figure 20.2 shows the warnings issued by PSPS (Sabesan et al. 2010) in both patients. Black vertical lines denote seizures. The warnings were retrospectively classified as true or false warnings using a seizure prediction horizon (SPH) of 2 h. The true and false warnings, evaluated retrospectively with a seizure prediction horizon SPH = 120 min, are shown in Figure 20.1 by solid and dotted arrows, respectively. It is interesting to note that in

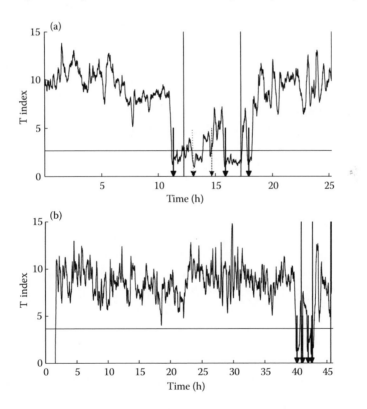

FIGURE 20.2 The dynamical transitions (warnings) of PSPS in two patients with juvenile myoclonic epilepsy. The seizures in Patient 1 (top panel) and Patient 2 (bottom panel) were predictable using PSPS. The solid arrows represent true positives and the dotted arrows false positives, determined using SPH = 2 h. The black horizontal line denotes the seizure prediction threshold, $T_{0.001,60} = 3.662$. A T-index below $T_{0.001,60}$ triggers a warning. The black vertical lines are the recorded seizures in these patients. PSPS was not trained in these data sets. The specificity of the algorithm is noteworthy: PSPS does not give any warnings for a long time (10 h in Patient 1, 40 h in Patient 2) before seizures. The average seizure prediction time across seizures was 1 h for Patient 1, and 40 min for Patient 2.

both patients, no warnings occurred during the long seizure-free interval prior to their first seizure. It is also important to note that the average T-index values fall below the threshold (denoting synchronization) 1 h and 40 min before the first seizure in Patients 1 and 2, respectively. Since subsequent seizures in both patients occurred in quick succession, the T-index values remained relatively low after the first seizure. Following the last seizure, the T-index curve shows an upward trend toward higher T-index values (denoting desynchronization). This observation is consistent across both patients. These results imply that, contrary to current belief, a long-term preictal synchronization of dynamics may also exist prior to certain types of generalized seizures, rendering them predictable.

20.5 FUTURE GOALS

In summary, our group's past and ongoing research on the EEG dynamics and epilepsy suggests the following three central concepts that can be used for seizure prediction and control. First, we have shown that seizures of focal and/or generalized origin are manifestations of progressive recruitment of brain sites in an abnormal hypersynchronization of their dynamics. The onset of such recruitment occurs long before the electrographical onset of a seizure and progressively culminates in the seizure (on the order of minutes for seizure prediction, on the order of hours to days for seizure susceptibility). We now have evidence that this entrainment occurs in focal and at least one type of generalized epilepsy. Second, time-irreversible resetting of the observed preictal dynamical recruitment occurs postictally via hysteresis. Preictal and postictal periods could be mathematically (statistically) defined and detected from dynamical analysis of the EEG. Complete or partial resetting of the preictal entrainment of the epileptic brain after a seizure may affect the route to the subsequent seizure (examples of seizures not resetting this pathology of dynamics have been demonstrated in status epilepticus, with human and animal subjects). The partial resetting of the pathology of dynamics may explain the observed apparently nonstationary nature (e.g., clustering) of seizure occurrences. Third, through control-oriented modeling, we have postulated a feedback control view of epileptic seizures, wherein epileptic seizures are hypothesized to be a result of the inability of the internal feedback/regulatory mechanisms of the brain to track excessive synchronization changes between the epileptogenic focus and other brain areas.

All the above concepts are interrelated, and have constituted the basis for development of seizure prediction algorithms, epileptogenic focus localization algorithms, and first-generation closed-loop seizure control algorithms. These concepts may provide the key ingredients for the development of robust brain pacemakers for the treatment of epilepsy in the near future. In addition, these concepts can be used for the evaluation of the efficacy of existing AEDs for seizure control, as well as for the design of novel drugs that would specifically target the resetting of the pathology of the dynamics in the epileptic brain. Finally, the developed measures and methodology, incorporated in medical devices for the evaluation of long-term seizure susceptibility, either to provide the proper dose of medication or external intervention or just warn for the need of it.

ACKNOWLEDGMENTS

We would like to acknowledge the support from U.S. National Institutes of Health (EB002089 BRP Grant on Brain Dynamics), the Barrow Neurological Foundation, the Epilepsy Research Foundation of America and Ali Paris Fund for LKS Research, the National Science Foundation (Grant No. 0601740), and the Arizona Science Foundation (a Competitive Advantage Award grant, 2008).

REFERENCES

Alldredge, B. K., D. M. Treiman, T. P. Bleck, and S. D. Shorvon. 2007. Treatment of status epilepticus. In *Epilepsy: A comprehensive textbook*. J. Engel and T. A. Pedley, eds., 1357. Philadelphia: Lippincott Williams & Wilkins.

Andrzejak, R. G., D. Chicharro, C. E. Elger, and F. Mormann. 2009. Seizure prediction: Any better than chance? *Clin. Neurophysiol.* 120(8):1465–1478.

Aschenbrenner-Scheibe, R., T. Maiwald, M. Winterhalder, H. U. Voss, J. Timmer, and A. Schulze-Bonhage. 2003. How well can epileptic seizures be predicted? An evaluation of a nonlinear method. *Brain* 126(12):2616–2626.

Casdagli, M. C., L. D. Iasemidis, J. C. Sackellares, S. N. Roper, R. L. Gilmore, and R. S. Savit. 1996. Characterizing nonlinearity in invasive EEG recordings from temporal lobe epilepsy. *Physica D* 99(2–3):381–399.

Chakravarthy, N., S. Sabesan, L. D. Iasemidis, and K. Tsakalis. 2007. Controlling synchronization in a neuron-level population model. *Int. J. Neural Syst.* 17(2):123–138.

Chakravarthy, N., K. Tsakalis, S. Sabesan, and L. Iasemidis. 2009. Homeostasis of brain dynamics in epilepsy: A feedback control systems perspective of seizures. *Ann. Biomed. Eng.* 37(3):565–585.

Chaovalitwongse, W. A., L. D. Iasemidis, P. M. Pardalos, P. R. Carney, D. S. Shiau, and J. C. Sackellares. 2006. Reply to comments on performance of a seizure warning algorithm based on the dynamics of intracranial EEG by Mormann, F., Elger, C. E., and Lehnertz, K. *Epilepsy Res.* 72(1):85–87.

Chen, Y., G. Rangarajan, J. Feng, and M. Ding. 2004. Analyzing multiple nonlinear time series with extended Granger causality. *Physics Letters A* 324(1):26–35.

D'Alessandro, M., G. Vachtsevanos, R. Esteller, J. Echauz, S. Cranstoun, G. Worrell, L. Parish, and B. Litt. 2005. A multi-feature and multi-channel univariate selection process for seizure prediction. *Clin. Neurophysiol.* 116(3):506–516.

Degn, H., L. F. Olsen, and A. V. Holden. 1987. *Chaos in biological systems*. New York: Springer.

Engel, J., T. A. Pedley, J. Aicardi, M. A. Dichter, and S. Moshé. 2007. *Epilepsy: A comprehensive textbook*. Philadelphia: Lippincott Williams & Wilkins.

Faith, A., S. Sabesan, S. Pati, J. Drazkowski, K. Noe, L. Tapsell, J. Sirven, and L. Iasemidis. 2009. Dynamical analysis of the EEG in the treatment of human status epilepticus by antiepileptic drugs. *Epilepsia* 49(Suppl. 7):112.

Frank, G. W., T. Lookman, M. A. H. Nerenberg, C. Essex, J. Lemieux, and W. Blume. 1990. Chaotic time series analyses of epileptic seizures. *Physica D* 46(3):427–438.

Freeman, W. J. 1987. Simulation of chaotic EEG patterns with a dynamic model of the olfactory system. *Biol. Cybern.* 56(2):139–150.

Good, L. B., S. Sabesan, L. D. Iasemidis, K. Tsakalis, and D. M. Treiman. 2004. Brain dynamical disentrainment by anti-epileptic drugs in rat and human status epilepticus. In *Proceedings of the (IEEE) 26th Annual International Conference of the Engineering in Medicine and Biology Society* (1):1–4.

Good, L. B., S. Sabesan, S. T. Marsh, K. Tsakalis, D. Treiman, and L. Iasemidis. 2009. Control of synchronization of brain dynamics leads to control of epileptic seizures in rodents. *Int. J. Neural Syst.* 19(3):173–196.

Hlaváčková-Schindler, K., M. Paluš, M. Vejmelka, and J. Bhattacharya. 2007. Causality detection based on information-theoretic approaches in time series analysis. *Phys. Rep.* 441(1):1–46.

Iasemidis, L. D. 2003. Epileptic seizure prediction and control. *IEEE Trans. Biomed. Eng.* 50(5):549–558.

Iasemidis, L. D., J. Chris Sackellares, H. P. Zaveri, and W. J. Williams. 1990. Phase space topography and the Lyapunov exponent of electrocorticograms in partial seizures. *Brain Topogr.* 2(3):187–201.

Iasemidis, L. D., R. L. Gilmore, S. N. Roper, and J. C. Sackellares. 1997a. Preictal-postictal versus postictal analysis for epileptogenic focus localization. *J. Clin. Neurophysiol.* 14:144.

Iasemidis, L. D., R. L. Gilmore, S. N. Roper, J. C. Sackellares, et al. 1997b. Epileptogenic focus localization by dynamical analysis of interictal periods of EEG in patients with temporal lobe epilepsy. *Epilepsia* 38(Suppl. 8):213.

Iasemidis, L. D., Olson, R. S. Savit, and J. C. Sackellares. 1994. Time dependencies in the occurrences of epileptic seizures. *Epilepsy Res.* 17(1):81–94.

Iasemidis, L. D., P. M. Pardalos, J. C. Sackellares, and D. S. Shiau. 2001. Quadratic binary programming and dynamical system approach to determine the predictability of epileptic seizures. *J. Comb. Optim.* 5(1):9–26.

Iasemidis, L. D., P. M. Pardalos, D. S. Shiau, W. Chaovalitwongse, K. Narayanan, S. Kumar, P. R. Carney, and J. C. Sackellares. 2003a. Prediction of human epileptic seizures based on optimization and phase changes of brain electrical activity. *Optim. Methods Softw.* 18(1):81–104.

Iasemidis, L. D., A. Prasad, J. C. Sackellares, P. M. Pardalos, and D. S. Shiau. 2003b. On the prediction of seizures, hysteresis and resetting of the epileptic brain: Insights from models of coupled chaotic oscillators. In *Order and Chaos*, Vol. 8, eds. T. Bountis and S. Pneumatikos, 283–305. Thessaloniki, Greece: Publishing House of K. Sfakianakis.

Iasemidis, L. D., J. C. Principe, and J. C. Sackellares. 1996. Spatiotemporal dynamics of human epileptic seizures. In *3rd Experimental Chaos Conference*, eds. R. G. Harrison, L. Weiping, W. Ditto, L. Pecora, and S. Vohra, 26–30. Singapore: World Scientific.

Iasemidis, L. D., J. C. Principe, and J. C. Sackellares. 2000. Measurement and quantification of spatio-temporal dynamics of human epileptic seizures. *Nonlinear Biomedical Signal Processing* 2:294–318.

Iasemidis, L. D., S. Sabesan, L. Good, N. Chakravarthy, D. M. Treiman, J. Sirven, and K. Tsakalis. 2009. A new look into epilepsy as a dynamical disorder: Seizure prediction, resetting and control. In *Encyclopedia of Basic Epilepsy Research*, Philip Schwartzkroin, Vol. 3, 1295–1302. Amsterdam: Elsevier.

Iasemidis, L. D., and J. C. Sackellares. 1991. The temporal evolution of the largest Lyapunov exponent on the human epileptic cortex. In *Measuring chaos in the human brain*, 49–82. Singapore: World Scientific.

Iasemidis, L. D., and J. C. Sackellares. 1996. Chaos theory and epilepsy. *Neuroscientist* 2(2):118–125.

Iasemidis, L. D., D. S. Shiau, W. Chaovalitwongse, J. C. Sackellares, P. M. Pardalos, J. C. Principe, P. R. Carney, A. Prasad, B. Veeramani, and K. Tsakalis. 2003c. Adaptive epileptic seizure prediction system. *IEEE Trans. Biomed. Eng.* 50(5):616–627.

Iasemidis, L. D., D. S. Shiau, P. M. Pardalos, W. Chaovalitwongse, K. Narayanan, A. Prasad, K. Tsakalis, P. R. Carney, and J. C. Sackellares. 2005. Long-term prospective on-line real-time seizure prediction. *Clin. Neurophysiol.* 116(3):532–544.

Iasemidis, L. D., D. S. Shiau, J. C. Sackellares, P. M. Pardalos, and A. Prasad. 2004. Dynamical resetting of the human brain at epileptic seizures: Application of nonlinear dynamics and global optimization techniques. *IEEE Trans. Biomed. Eng.* 51(3):493–506.

Iasemidis, L. D., K. Tsakalis, J. C. Sackellares, and P. M. Pardalos. 2005. Comment on inability of Lyapunov exponents to predict epileptic seizures. *Physical Review Letters* 94(1):19801.

Iasemidis, L. D., H. P. Zaveri, J. C. Sackellares, and W. J. Williams. 1988a. Modelling of ECoG in temporal lobe epilepsy. In *Proceedings of the 25th Annual Rocky Mountains Bioengineering Symposium*, Vol. 24, 187–193.

Iasemidis, L. D., H. P. Zaveri, J. C. Sackellares, W. J. Williams, and T. W. Hood. 1988b. Nonlinear dynamics of electrocorticographic data. *J. Clin. Neurophysiol.* 5:339.

Kantz, H. 1994. A robust method to estimate the maximal Lyapunov exponent of a time series. *Physics Letters A* 185:77–87.

Lai, Y. C., M. A. F. Harrison, M. G. Frei, and I. Osorio. 2003. Inability of Lyapunov exponents to predict epileptic seizures. *Phys. Rev. Lett.* 91(6):68102–05.

Le Van Quyen, M., J. Soss, V. Navarro, R. Robertson, M. Chavez, M. Baulac, and J. Martinerie. 2005. Preictal state identification by synchronization changes in long-term intracranial EEG recordings. *Clin. Neurophysiol.* 116(3):559–568.

Lehnertz, K., and B. Litt. 2005. The first international collaborative workshop on seizure prediction: Summary and data description. *Clin. Neurophysiol.* 116(3):493–505.

Lehnertz, K., F. Mormann, H. Osterhage, A. Muller, J. Prusseit, A. Chernihovskyi, M. Staniek, D. Krug, S. Bialonski, and C. E. Elger. 2007. State-of-the-art of seizure prediction. *J. Clin. Neurophysiol.* 24(2):147–153.

Lennox, W. G. 1946. *Science and seizures: New light on epilepsy and migraine.* New York: Harper.

Maiwald, T., M. Winterhalder, R. Aschenbrenner-Scheibe, H. U. Voss, A. Schulze-Bonhage, and J. Timmer. 2004. Comparison of three nonlinear seizure prediction methods by means of the seizure prediction characteristic. *Physica D* 194(3–4):357–368.

Mormann, F., T. Kreuz, R. G. Andrzejak, P. David, K. Lehnertz, and C. E. Elger. 2003. Epileptic seizures are preceded by a decrease in synchronization. *Epilepsy Res.* 53(3):173–185.

Mormann, F., T. Kreuz, C. Rieke, R. G. Andrzejak, A. Kraskov, P. David, C. E. Elger, and K. Lehnertz. 2005. On the predictability of epileptic seizures. *Clin. Neurophysiol.* 116(3):569–587.

Packard, N. H., J. P. Crutchfield, J. D. Farmer, and R. S. Shaw. 1980. Geometry from a time series. *Phys. Rev. Lett.* 45(9):712–716.

Prasad, A., L. D. Iasemidis, S. Sabesan, and K. Tsakalis. 2005. Dynamical hysteresis and spatial synchronization in coupled non-identical chaotic oscillators. *Pramana* 64(4):513–523.

Priestly, M. B. 1981. *Spectral analysis and time series*, Vol. 1 of the *Univariate Series* and Vol. 2 of the *Multivariate Series*, Prediction and Control. San Diego: Academic Press.

Rosenstein, M. T., J. J. Collins, C. J. De Luca, et al. 1993. A practical method for calculating largest Lyapunov exponents from small data sets. *Physica D* 65(1–2):117–134.

Sabesan, S., N. Chakravarthy, K. Tsakalis, P. Pardalos, and L. Iasemidis. 2009a. Measuring resetting of brain dynamics at epileptic seizures: Application of global optimization and spatial synchronization techniques. *Journal of Combinatorial Optimization* 17(1):74–97.

Sabesan, S., L. Good, N. Chakravarthy, K. Tsakalis, P. M. Pardalos, and L. Iasemidis. 2007a. Global optimization and spatial synchronization changes prior to epileptic seizures. In *Optimization in biomedicine, Springer optimization and its applications series*, eds. P. Pardalos, C. Alves, and L. Vicente, 103–125. New York: Springer.

Sabesan, S., L. Good, K. Tsakalis, A. Spanias, D. Treiman, and L. Iasemidis. 2009b. Information flow and application to epileptogenic focus localization from intracranial EEG. *IEEE transactions on neural systems and rehabilitation engineering* 17(3):244–253.

Sabesan, S., L. D. Iasemidis, K. Tsakalis, D. M. Treiman, and J. S. Sirven. Towards a practical seizure prediction algorithm: Performance evaluation using intracranial EEG. *International Journal of Neural Systems*, 2010, forthcoming.

Sabesan, S., K. Narayanan, A. Prasad, A. Spanias, and L. D. Iasemidis. 2003a. Improved measure of information flow in coupled nonlinear systems. In *Proceeding of the International Conference on Modeling and Simulation*, 24–26.

Sabesan, S., K. Narayanan, A. Prasad, A. Spanias, J. C. Sackellares, and L. D. Iasemidis. 2003b. Predictability of epileptic seizures: A comparative study using Lyapunov exponent and entropy based measures. *Biomedical Sciences Instrumentation* 39:129–135.

Sabesan, S., K. Narayanan, A. Prasad, K. Tsakalis, A. Spanias, and L. Iasemidis. 2007b. Information flow in coupled nonlinear systems: Application to the epileptic human brain. In *Data mining in biomedicine, Springer optimization and its applications series*, P. Pardalos, V. Boginski, and A. Vazacopoulos, eds., 483–504. New York: Springer.

Sackellares, J. C. 2008. Seizure prediction. *Epilepsy Currents* 8(3):55–59.

Sackellares, J. C., L. D. Iasemidis, D. S. Shiau, R. L. Gilmore, and S. N. Roper. 2000. Epilepsy—when chaos fails. In *Chaos in the brain*. K. Lehnertz, J. Arnhold, P. Grassberger, and C. E. Elger, eds., 12–133. Singapore: World Scientific.

Sackellares, J. C., D. S. Shiau, J. C. Principe, M. C. K. Yang, L. K. Dance, W. Suharitdamrong, W. Chaovalitwongse, P. M. Pardalos, and L. D. Iasemidis. 2006. Predictability analysis for an automated seizure prediction algorithm. *J. Clin. Neurophysiol.* 23(6):509–520.

Schiff, S. J., P. So, T. Chang, R. E. Burke, and T. Sauer. 1996. Detecting dynamical interdependence and generalized synchrony through mutual prediction in a neural ensemble. *Physical Review E* 54(6):6708–6724.

Schreiber, T. 2000. Measuring information transfer. *Phys. Rev. Lett.* 85(2):461–464.

Takens, F. 1981. Detecting strange attractors in turbulence. *Lecture notes in mathematics* 898(1):366–381.

Theodore, W. H., and R. S. Fisher. 2004. Brain stimulation for epilepsy. *Lancet Neurol.* 3(2):111–118.

Treiman, D. M., and M. C. Walker. 2006. Treatment of seizure emergencies: Convulsive and non-convulsive status epilepticus. *Epilepsy Res.* 68:77–82.

Tsakalis, K., N. Chakravarthy, S. Sabesan, L. D. Iasemidis, and P. M. Pardalos. 2006. A feedback control systems view of epileptic seizures. *Cybern. Syst. Anal.* 42(4):483–495.

Tsakalis, K., and L. D. Iasemidis. 2006. Control aspects of a theoretical model for epileptic seizures. *Int. J. Bifurcat. Chaos* 16(7):2013–2027.

Vastano, J. A., and E. J. Kostelich. 1985 Comparison of algorithms for determining Lyapunov exponents from experimental data. Presented at International Conference on Dimensions and Entropies in Chaotic Systems, ed. G. Mayer-Kress, Pecos River, NM, September 11.

Wiener, N. 1956. *Modern mathematics for the engineers [Z]*. Series 1. New York: McGraw-Hill.

Winterhalder, M., T. Maiwald, H. U. Voss, R. Aschenbrenner-Scheibe, J. Timmer, and A. Schulze-Bonhage. 2003. The seizure prediction characteristic: A general framework to assess and compare seizure prediction methods. *Epilepsy Behav.* 4(3):318–325.

Wolf, A., J. B. Swift, H. L. Swinney, and J. A. Vastano. 1985. Determining Lyapunov exponents from a time series. *Physica D* 16(3):285–317.

Yang, M. C. K., D. S. Shiau, and J. C. Sackellares. 2004. Testing whether a prediction scheme is better than guess. *Quantitative neuroscience: Models, algorithms, diagnostics, and therapeutic applications*, 251–266.

21 Time-Series-Based Real-Time Seizure Prediction

Pooja Rajdev and Pedro P. Irazoqui

CONTENTS

21.1 INTRODUCTION

About 50 million people in the world are suffering from epilepsy, 30% of whom are resistant to all medical therapies (Kwan and Brodie 2000). Although the drugs and devices pertinent to this disease have advanced over time, the percentage of patients who experience a decrease in seizure frequency has not corresponded proportionally. Only about 60% of the patients on medication and 35% of the patients using the vagal nerve stimulator have experienced a decrease in seizure frequency. One of the prominent treatment measures for patients with severe refractory seizures is surgery to cut away brain tissue that contains the seizure focus (resection) or disconnecting the corpus callosum to prevent the spread of seizures (callosotomy). A working algorithm that can predict seizures holds immense promise for such patients, as it will allow them to take appropriate precautions keeping the risk of a seizure attack to a minimum.

A number of in vitro and in vivo animal experiments have identified areas in the brain that, when stimulated, decrease seizure activity (Durand and Warman 1994; Psatta 1983; Takano et al. 1995). Preliminary results from well-controlled studies of the responsive neural stimulation (RNS) and vagus nerve stimulation (VNS) also demonstrate that electrical stimulation of the epileptogenic zone and the vagal nerve, respectively, can terminate epileptiform afterdischarges in select populations of epilepsy patients (Salanova and Worth 2007). Given the recent success obtained with these studies and deep brain stimulation for movement disorders, a renewed effort to develop a closed-loop prosthesis that can intervene before the clinical onset of a seizure has been pursued.

Clinicians, engineers, and neuroscientists have proposed many new methodologies for the implementation of an automated, implantable seizure-control device that can successfully predict and abort the epileptic activity prior to clinical onset. These proposed devices can be of two major types: a) blind seizure-control devices that do not respond to any physiological activity, and b) intelligent devices that are triggered by a seizure-predicting algorithm. The intelligent devices are preferred over the blind stimulators, as responsive stimulation provides for temporal specificity and avoids the possibility of tissue damage caused by excessive exposure to stimulation. It also allows us to optimize the stimulation therapy necessary based on feedback provided from the properties of the electrographic state of the brain during the seizure. A major contingency for the success of these intelligent devices, however, is the effective implementation of seizure-prediction algorithms.

21.1.1 History of Prediction Algorithms

With the aim of automating visual analysis of brain activity, researchers have been trying to extract features predictive of an impending seizure from continuous EEG recordings of the brain. A question of particular interest has been whether the transition to the seizure (ictal) state is an abrupt phenomenon or if it evolves from a preictal state with characteristic features. Until relatively recently, the general assumption was that the occurrence of seizures was a random process that could not be predicted (Milton et al. 1987). However, the advent of various techniques such as single photon emission computed tomography (SPECT), functional magnetic resonance imaging (fMRI) and electroencephalography (EEG), has led researchers to try to better understand the physiological events that give rise to seizures. Motivated groups began working on mathematical models and algorithms to predict seizures (Mormann et al. 2005).

The earliest works on the predictability of seizures date back to the 1970s (Viglione and Walsh 1975). Based on the hypothesis that the seizures follow a pattern in their development, Viglione and Walsh used pattern recognition techniques to predict grand mal seizures. In the 1980s, many theories on neuronal recruitment and hypersynchrony were developed (Traub and Wong 1982) and it was believed that the neurons were capable of mutual excitation leading to synchronization of many neurons in the preictal period of a seizure (Mormann et al. 2003). Intensive research was also done to analyze epileptic spikes and to test the hypothesis that the rate of spikes would increase in the preictal period (Lange et al. 1983; Gotman 1985).

The emergence of a new physical–mathematical theory of nonlinear dynamics in the 1980s brought about new tools to apply within the field of seizure prediction. Studies based on the concept of phase-space topography (Iasemidis et al. 1990), correlation dimension (Lehnertz and Elger 1995), and accumulated signal energy (Litt et al. 2001) claimed very encouraging results. However, most of them only quantified certain epileptic brain dynamics and identified measures that seemed to have some success in differentiating the pre-ictal states from the ictal states. However, one significant problem with most of these nonlinear measures and associated prediction algorithms is their computational complexity, which has hindered their real-time implementation in a warning device. An online real-time seizure-detection or early prediction system is quintessential to automated warning and responsive therapy delivery systems. This topic has become of increasing interest over the past several years (Osorio et al. 1998; Qu and Gotman 1997; Shoeb et al. 2004).

21.1.2 Criteria for Successful Implementation of Prediction Algorithms

In addition to having an online automated algorithm, the criteria for successful implementation of prediction algorithms also includes a need for adaptive architectures to account for inter- and intraindividual variability (Sackellares et al. 2006; Haas et al. 2007). Haas et al. have reiterated these criteria by their adaptation of seizure-detection algorithms to an individual seizure's fingerprint by using both seizure and nonseizure training segments.

Current seizure-monitoring methods severely restrict patient mobility since they require the subject to be constantly connected to video and computerized EEG equipment for extended periods, even days at a time. Furthermore, this inconvenient approach is expensive and its execution requires trained professionals. To effectively reach and apply to a broader base of patients necessitates a portable, wearable monitoring and predicting device that is less cumbersome and will likely assist in reducing hospitalization costs and patient discomfort. Our eventual goal is the creation of an implantable intelligent seizure-control device. However, our success in this endeavor is contingent upon a working real-time seizure prediction or early detection algorithm that can be miniaturized into a handheld or implantable device upon its hardware implementation.

Ambulatory brain recording and data collection devices that can monitor seizure activity will have a significant impact on the efficiency of seizure-prediction algorithms. Scalp EEG data can be considered to be simply a compilation of the summed potentials of millions of neurons. However, seizure activity can only be recognized in an EEG after the seizure has advanced to the cerebral cortex. Using local field potentials can be far more effective for early detection and possibly enable the prevention of the proliferation of a seizure beyond the focus. In the case of seizures of focal onset, electric discharges are likely to develop in the epileptogenic zone (most likely the hippocampus in the case of temporal lobe epilepsy) and expand further to the cortex (Iasemidis et al. 2005; Rogowski et al. 1981). It has been reported that changes in the firing patterns of neurons in the preseizure period are not reflected in the averaged field potentials that we examine in the EEG data (Litt et al. 2001). Distortion of cortical data is inevitable in the case of scalp EEG's due to the filtration and attenuation caused by layers of cerebrospinal fluid, skull, tissue, and scalp. Therefore, conventional methods that are derived from scalp EEG recordings are harder to correlate with the associated neurophysiology of focal seizures. Capturing intracranial neural data directly from the focus of the seizure is theorized to be more advantageous with respect to the prediction of seizure activity since the epileptic spikes can be recognized prior to the expansion of such activity to the cerebral cortex (Iasemidis et al. 2005; Rogowski et al. 1981). Additional advantages include a higher signal-to-noise ratio and better spatial resolution. However, that comes at the cost of an invasive surgery to implant the deep brain electrodes. Thus, to successfully record and process LFPs, miniaturized implantable devices will be necessary.

21.2 SIGNAL MODELING AND ADAPTIVE PREDICTION

Signal modeling is a broad class of processing techniques where a stochastic signal of interest is modeled as a certain type of process, such as an autoregressive moving average (ARMA) process. It is clear that if one is successful in developing a parametric model for the behavior of some signal, then that model can also be used for predicting future values of the signal. The focus in this chapter will be on modeling a signal as an autoregressive (AR) process, also known as an all-pole model, which is a special class of the more general ARMA process model. Using the AR model, the location of multiple peaks within the frequency spectrum can be obtained. The AR model can also be used within the framework of the Wiener filter, a powerful adaptive filter that is employed in a wide variety of applications.

The data analysis and prediction scheme is a four-step process. The first step enhances the signal by attenuating the low-frequency noise and the high-frequency artifacts that would lead to false positive detection of seizures (Legatt et al. 1980). The next step is the adaptive autoregressive modeling and prediction, which is explained below and in the work by Rajdev et al. (2010). This is followed by envelope detection and a binomial decision rule. Figure 21.1 presents a block diagram of the algorithm implemented in real-time.

In this section, we discuss an adaptive Wiener prediction algorithm that has been optimized to run in real time. The Wiener coefficients are reiteratively calculated to model the LFP data, thus allowing the current data to be compared with baseline data in the recent past. This algorithm also has the advantage of low computational complexity, making its real-time implementation on a digital signal processor (DSP) feasible. This feature makes this algorithm, should it prove capable of accurate seizure prediction or early detection in a prospective large scale validation, a definitive step toward development of a miniaturized portable/wearable or implantable warning device.

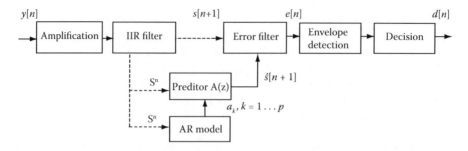

FIGURE 21.1 Block diagram of the real time algorithm. The LFP signal y[n] is amplified by the TDT pre-amp and passed through an IIR band pass filter (10–500Hz). A best fit AR model is then determined and a future value of the signal s[n + 1] is estimated. The error filter then computes the error between the observed and the estimated value. The envelope of the error is used by the decision rule to make a binary decision. (Reproduced with permission from Rajdev et al., *Computers in Biology and Medicine*, 40(1), 97–108.)

21.2.1 WIENER PREDICTION

The main objectives of time-series prediction are description, modeling, and forecasting. The description step describes the current series using statistics and feature extraction (such as correlation matrices), the modeling step uses the description of the past and current values of the series to identify suitable models and corresponding model parameters, and the forecasting step then uses the model to estimate future values of the series. The inherent assumption in this is that we expect the future to be much like the past. The prediction error can be computed and used for detection of changes in how well the model describes the recent data by comparing errors between forecasted values and future values (Chatfield 2000).

The linear prediction model considers a signal $s[n]$ to be the output of some system with a white noise input $u[n]$, such that the following relation holds:

$$s[n] = \sum_{k=1}^{p} a[k]s[n-k] + G * u[n], \tag{21.1}$$

where $a[k]$, $1 \le k \le p$ and the gain G are the parameters of the system, and $S^n = (s[n-1],\ldots,s[n-N])$ are the previous observations of the signal. This equation says that the output $s[n]$ is a linear function of the past outputs. This equation when transformed to the frequency domain is of the form:

$$H(z) = \frac{1}{1 + \sum_{k=0}^{p} a[k]z^{-k}} = \frac{1}{A(z)}, \tag{21.2}$$

thus forming an all-pole model, also known as the autoregressive (AR) model. This is a widely used model, and was further developed by Norbert Wiener for the filtering and prediction of stationary time series.

In the Wiener prediction model, the future is computed as a linear combination of the p past values of the signal. The notation $\hat{s}[n]$ refers to the estimate of $s[n]$, which is obtained as the output of a predictor filter, as given below, where $a[k]$ are the AR coefficients.

$$\hat{s}[n] = \sum_{k=1}^{p} a[k]s[n-k]. \tag{21.3}$$

The error between the estimated and observed values $\varepsilon[n]$, also known as the residual, is given by

$$\varepsilon[n] = (s[n] - \hat{s}[n])$$

$$= \left(s[n] - \sum_{k=1}^{p} a[k]s[n-k] \right). \tag{21.4}$$

In order to obtain the optimal model parameters $a[k]$, the least squares method was used (Poor 1994). In this method, we minimize the expected value of the square of the error E by setting

$$\frac{\partial E}{\partial a_i} = 0, \qquad 0 \le i \le p, \tag{21.5}$$

where

$$E = \mathrm{E}(\varepsilon^2[n]) = (s[n] - \hat{s}[n])^2$$

$$= \left(s[n] - \sum_{k=1}^{p} a[k]s[n-k] \right)^2. \tag{21.6}$$

Solving Equation 21.5 for all n, we obtain the normal equations given below:

$$\sum_{k=1}^{p} a_k \mathrm{E}[s_{n-k}s_{n-i}] = -\mathrm{E}[s_n s_{n-i}], \qquad 0 \le i \le p. \tag{21.7}$$

For a stationary process $s[n]$,

$$E[s_{n-k}s_{n-i}] = R_{ss}[i-k], \tag{21.8}$$

where $R_{ss}[i]$ is the autocorrelation of the process.

An estimate of the autocorrelation function $\hat{R}_{ss}$ is obtained by

$$\hat{R}_{ss}[k] = \frac{1}{N} \sum_{n=0}^{N-1-|k|} s[n]s[n+|k|], \qquad |k| \le N-1. \tag{21.9}$$

Substituting Equation 21.7, we obtain the Wiener–Holf equations given below:

$$\sum_{l=0}^{p} \hat{a}[k]\hat{R}_{ss}[k-l] = -\hat{R}_{ss}[k], \qquad k = 0, 2, \ldots, p. \tag{21.10}$$

These equations can be solved by inverting the autocorrelation matrix $\hat{R}_{ss}$ but this is computationally very demanding. Levinson derived an elegant and efficient recursive procedure for solving such types of equations to determine the coefficients that involves solving a symmetric Toeplitz-type equation (Poor 1994). Details of estimating and validating the model are discussed elsewhere (Rajdev et al. 2010).

In the frequency domain, the residual term can be obtained as

$$E(z) = \left[1 + \sum_{k=1}^{p} a[k]z^{-k} \right] S(z) = A(z)S(z). \tag{21.11}$$

Minimizing $E(z)$ in the frequency domain, we obtain an equation to calculate the autocorrelation from the signal spectrum by an inverse Fourier transform:

$$R(i) = \frac{1}{2\pi} \int_{-\pi}^{\pi} P(\omega)\cos(i\omega) d\omega. \tag{21.12}$$

In this way, we can approximate any spectrum arbitrarily closely by an all-pole model, and the least squares error criterion in the time domain translates into a spectral matching criterion in the frequency domain. Due to the duality of information between autocorrelation and power spectral density, this algorithm may be interpreted as detecting any significant change in the PSD of a signal away from the long-term ergodic average described by Equation 21.4, what may typically amount to the interictal background PSD.

Using the method described above, the autoregressive coefficients reiteratively model the statistical characteristics of the neural LFP data and attempt to estimate the future values of the signal. This adaptive nature of the algorithm detects the preseizure period by tracking the changes in the model. In order for the algorithm to make a binary decision to predict seizures, thresholds on the prediction error are placed. Instead of a fixed present value, a scaled version of the mean energy of the prediction error signal of the baseline signal in the recent past is used as the threshold to predict seizures.

21.2.2 RESULTS

The prediction algorithm was developed in Matlab® and then a real-time implementation was implemented on a floating point Texas Instruments TMS320C6713T DSP (Rajdev et al. 2010). The algorithm was tested on in vivo recordings from kainite-treated rats, an animal model for temporal lobe epilepsy. Bipolar stainless steel electrodes were stereotaxically implanted into the dorsal dentate gyrus of Long Evans rats. After a postsurgical recovery period, the unrestrained awake animals were injected intraperitoneally with doses of kainite to induce status epilepticus. Seizure activity was carefully monitored and LFPs were amplified, band-pass filtered, and recorded at 1526 Hz using a TDT System 3 (Tucker-Davis Technologies, Alachua, FL, USA). Recorded signals were then resampled and sent into the TMS320C6713DSK digital signal processor where the sampling frequency was set at its lowest rate of 8 KHz. The signal was digitized using a 32-bit ADC and passed through the prediction and decision algorithm. Complete details of the animal model and recording setup are discussed elsewhere (Rajdev et al. 2010).

In order to characterize the performance of the algorithm, the Wiener algorithm was compared with other naive prediction schemes including the random and periodic predictor (Winterhalder et al. 2003; Sackellares et al. 2006). The ROC curve is a commonly used curve that characterizes the algorithm by displaying the relation between sensitivity and false positive rate (over a range of λ). The values of λ were varied from 1 to 2 for the Wiener algorithm (where λ is a multiplicative factor for the prediction threshold), 1 to 20 min for the mean of exponential distribution, and from

1 to 20 min for the periodic prediction scheme. A standard metric to evaluate ROC curves is the area under the ROC curve. However, in this case, the horizontal axis is the FPR of the algorithm, which is unbounded. Thus we used a metric proposed by Sackellares et al. (2006) that computes the area above the ROC curve (AAC). The AAC gives a measure of the performance of the algorithm, with smaller AAC values indicating better performance. The mean AAC of the Wiener algorithm is 4.44, while the random and periodic schemes are 38.63 and 28.5, respectively (t-test $p < 0.05$). The predictability power (PP) of the algorithm, which is defined as the difference in AAC of the algorithm and the naive algorithm, normalized by the area of the naive predictor is another method to characterize algorithms. The PP of the Wiener algorithm when compared to the naive schemes was greater than 0.5, suggesting superior predicting capabilities (Figure 21.2).

In the Matlab analysis, it was observed that the prediction error started building up a few seconds before the seizure electrographic onset. This reflects a change in the underlying AR model that the Wiener coefficients could not capture. ANOVA test suggests that the means of the squared prediction error in the baseline, preictal, and ictal regions are significantly different (Figure 21.3) and a good thresholding algorithm should be able to demarcate one state from another.

Testing of this algorithm on kainite-treated rats resulted in prediction of seizures 27 ± 17 s before clinical onset, with 94% sensitivity and a false positive rate of 0.009 per min (as shown in Table 21.1a) (Rajdev et al. 2010).

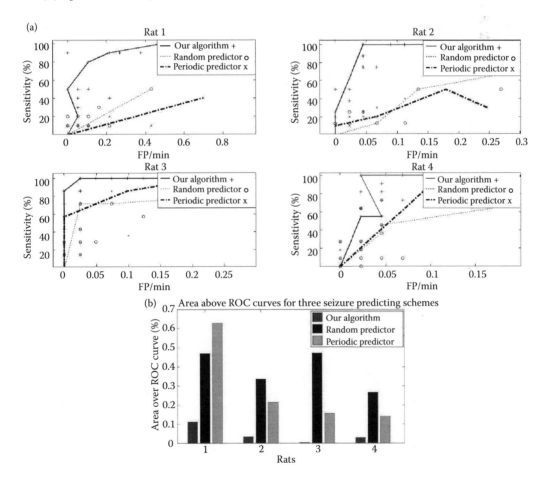

FIGURE 21.2 (a) ROC curves of the Wiener algorithm, random predictor, and periodic predictor for the four rats. (b) Area above ROC curves for the three prediction schemes for all four rats. (Reproduced with permission from Rajdev et al., *Computers in Biology and Medicine*, 40(1), 97–108.)

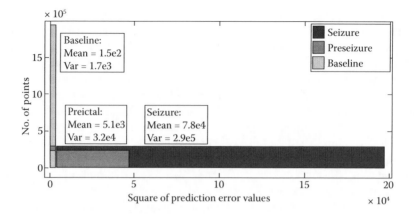

FIGURE 21.3 Values of prediction error squared in the baseline, preictal, and ictal regions. ANOVA test suggests that the means are significantly different ($p < 0.05$). (Reproduced with permission from Rajdev et al., *Computers in Biology and Medicine*, 40(1), 97–108.)

Real-time performance was also evaluated by coding the algorithm in C and loading it on the DSP 6713. Real-time testing resulted in prediction of seizures 6.7 ± 5.6 s before clinical onset, with 92% sensitivity and a false positive rate of 0.08 per min (as shown in Table 21.1b) (Rajdev et al. 2010). Some seizures were not predicted, but only detected in their early phase by the real-time implementation. The reason for this might be the frame-based processing approach used in the implementation.

A major limitation of the method presented here is the use of a frame-based approach, which causes an inherent delay in the real-time prediction of the seizure. This is especially prevalent when the algorithm is implemented on the DSP.

The Wiener prediction algorithm has many advantages. Because of the adaptive nature of the algorithm, supervised training is not required. Also, the proposed method is not computationally demanding and is easily implementable on a DSP.

21.3 CONCLUSIONS

Electrical stimulation of the brain has the potential to be a safe therapy for epilepsy. However, in order for the development of intelligent stimulators that combine seizure prediction with the ability

TABLE 21.1a
Performance Results Using Matlab

ID	# of Seizures	Sensitivity (%)	FP/Min	Selectivity (%)	Median Latency (sec)	Mean Latency (sec)	Std of Latency (sec)
1	32	96.87	0.0064	96.87	19.96	26.02	20.84
2	27	96.29	0.0095	92.85	34.79	33.29	18.12
3	25	88	0.0063	95.65	29.185	31.34	19.61
4	25	96	0.0143	92.30	13.824	15.51	10.50

Source: Rajdev et al., *Computers in Biology and Medicine*, 40(1), 97–108.

TABLE 21.1b
Performance Results Using TI DSP

ID	# of Seizures	Sensitivity (%)	FP/Min	Selectivity (%)	Median Latency (sec)	Mean Latency (sec)	Std of Latency (sec)
1	14	92.85	0.115	76.47	6.33	5.63	4.29
2	14	92.85	0.077	81.25	7.68	8.35	7.63
3	24	91.67	0.074	88	5.18	6.35	4.87
4	18	88.89	0.080	88.89	6.91	7.24	5.34

Source: Rajdev et al., *Computers in Biology and Medicine*, 40(1), 97–108.

to deliver therapeutic stimulation, efficient real-time prediction algorithms need to be developed. It is proposed that development of seizure-prediction/early detection algorithms that perform in real time—operating on single unit data or local field potentials received through an implantable device that records focal neural activity in the brain of animal models—would be the most effective.

REFERENCES

Chatfield, C. 2000. *Time series forecasting*. Boca Raton, FL: Chapman & Hall/CRC.

Durand, D. M., and E. N. Warman. 1994. Desynchronization of epileptiform activity by extracellular current pulses in rat hippocampal slices. *J. Physiol.* 480(Pt 3):527–537.

Gotman, J. 1985. Automatic recognition of interictal spikes. *Electroencephalogr. Clin. Neurophysiol. Suppl.* 37:93–114.

Haas, S. M., M. G. Frei, and I. Osorio. 2007. Strategies for adapting automated seizure detection algorithms. *Med. Eng. Phys.* 29(8):895–909.

Iasemidis, L. D., J. C. Sackellares, H. P. Zaveri, and W. J. Williams. 1990. Phase space topography and the Lyapunov exponent of electrocorticograms in partial seizures. *Brain Topogr.* 2(3):187–201.

Iasemidis, L. D., D. S. Shiau, P. M. Pardalos, W. Chaovalitwongse, K. Narayanan, A. Prasad, K. Tsakalis, P. R. Carney, and J. C. Sackellares. 2005. Long-term prospective on-line real-time seizure prediction. *Clin. Neurophysiol.* 116 (3):532–544.

Kwan, P., and M. J. Brodie. 2000. Early identification of refractory epilepsy. *N. Engl. J. Med.* 342(5):314–319.

Lange, H. H., J. P. Lieb, J. Engel, Jr., and P. H. Crandall. 1983. Temporo-spatial patterns of pre-ictal spike activity in human temporal lobe epilepsy. *Electroencephalogr. Clin. Neurophysiol.* 56(6):543–555.

Lehnertz, K., and C. E. Elger. 1995. Spatio-temporal dynamics of the primary epileptogenic area in temporal lobe epilepsy characterized by neuronal complexity loss. *Electroencephalogr. Clin. Neurophysiol.* 95(2):108–117.

Litt, B., R. Esteller, J. Echauz, M. D'Alessandro, R. Shor, T. Henry, P. Pennell, C. Epstein, R. Bakay, M. Dichter, and G. Vachtsevanos. 2001. Epileptic seizures may begin hours in advance of clinical onset: A report of five patients. *Neuron* 30(1):51–64.

Milton, J. G., J. Gotman, G. M. Remillard, and F. Andermann. 1987. Timing of seizure recurrence in adult epileptic patients: A statistical analysis. *Epilepsia* 28(5):471–478.

Mormann, F., T. Kreuz, R. G. Andrzejak, P. David, K. Lehnertz, and C. E. Elger. 2003. Epileptic seizures are preceded by a decrease in synchronization. *Epilepsy Res.* 53(3):173–185.

Mormann, F., T. Kreuz, C. Rieke, R. G. Andrzejak, A. Kraskov, P. David, K. Lehnertz, and C. E. Elger 2005. On the predictability of epileptic seizures. *Clin. Neurophysiol.* 116:569–587.

Osorio, I., M. G. Frei, and S. B. Wilkinson. 1998. Real-time automated detection and quantitative analysis of seizures and short-term prediction of clinical onset. *Epilepsia* 39(6):615–627.

Poor, H. V. 1994. *An introduction to signal detection and estimation*, 2nd ed., 398. New York: Springer, New York.

Psatta, D. M. 1983. Control of chronic experimental focal epilepsy by feedback caudatum stimulations. *Epilepsia* 24(4):444–454.

Qu, H., and J. Gotman. 1997. A patient-specific algorithm for the detection of seizure onset in long-term EEG monitoring: Possible use as a warning device. *IEEE Trans. Biomed. Eng.* 44(2):115–122.

Rajdev, P., M. P. Ward, J. Rickus, R. Worth, and P. P. Irazoqui. 2010. Real-time seizure prediction from local field potentials using an adaptive Wiener algorithm. *Comput. Biol. Med.* 40(1):97–108.

Rogowski, Z., I. Gath, and E. Bental. 1981. On the prediction of epileptic seizures. *Biol. Cybern.* 42(1):9–15.

Sackellares, J. C., D. S. Shiau, J. C. Principe, M. C. Yang, L. K. Dance, W. Suharitdamrong, W. Chaovalitwongse, P. M. Pardalos, and L. D. Iasemidis. 2006. Predictability analysis for an automated seizure prediction algorithm. *J. Clin. Neurophysiol.* 23(6):509–520.

Salanova, V., and R. Worth. 2007. Neurostimulators in epilepsy. *Curr. Neurol. Neurosci. Rep.* 7(4):315–319.

Shoeb, A., H. Edwards, J. Connolly, B. Bourgeois, T. Treves, and J. Guttag. 2004. Patient-specific seizure onset detection. *Conf. Proc. IEEE Eng. Med. Biol. Soc.* 1:419–422.

Takano, K., T. Tanaka, T. Fujita, H. Nakai, and Y. Yonemasu. 1995. Zonisamide: Electrophysiological and metabolic changes in kainic acid-induced limbic seizures in rats. *Epilepsia* 36(7):644–648.

Traub, R. D., and R. K. Wong. 1982. Cellular mechanism of neuronal synchronization in epilepsy. *Science* 216(4547):745–747.

Viglione, S. S., and G. O. Walsh. 1975. Proceedings: Epileptic seizure prediction. *Electroencephalogr. Clin. Neurophysiol.* 39(4):435–436.

Winterhalder, M., T. Maiwald, H. U. Voss, R. Aschenbrenner-Scheibe, J. Timmer, and A. Schulze-Bonhage. 2003. The seizure prediction characteristic: A general framework to assess and compare seizure prediction methods. *Epilepsy Behav.* 4(3):318–325.

22 Optimizing Seizure Detection Algorithms toward the Development of Implantable Epilepsy Prostheses

Shriram Raghunathan and Pedro P. Irazoqui

CONTENTS

22.1 INTRODUCTION

Technological advances allow for miniaturization of devices that were inconceivable just a decade ago. The advancement of integrated circuit fabrication combined with research on wireless powering strategies and biocompatible packaging advancements now allows for the development of portable devices to better manage disease state and even help with the treatment of several chronic disorders. The cardiac pacemaker is a classic example of the integration of medical science and engineering technology that is enhancing quality of life for millions of patients across the world. The application of cutting-edge engineering solutions for the treatment of epilepsy would lead to a significant improvement in the quality of life of over 50 million epileptics all over the world (Begley et al. 1994). Epilepsy therapy today is predominantly pharmaceutical, with the exception of surgical resection in certain cases if the patient's seizure focus is accurately identified and meets a set of eligibility criteria. Alternate therapies mainly involving electrical stimulation have sparked recent interest, especially with successful seizure control sometimes achieved with the use of vagus-nerve stimulators (Labar et al. 1999). Deep brain stimulation and even direct stimulation of the epileptogenic focus have shown promising results from animal and some preliminary human studies (Graves and Fisher 2005). However, a majority of the devices employ continuous or periodic stimulation protocols (either intentionally or as a result of the challenges in implementing complex signal-processing circuitry on a battery-powered platform) (Osorio et al. 2005; Peters et al. 2001; Lesser et al. 1999). Closed-loop studies that trigger stimulation upon detection or prediction of an upcoming seizure are few and not thoroughly documented (Osorio et al. 2001; Politsky et al. 2005). The experiments that validate this concept mostly employ bedside computers to implement the detection algorithms or use in vitro animal studies to evaluate the concept (Peters et al. 2001; Kossoff et al.

2004). The Neuropace responsive stimulation device is the only known implantable platform to attempt to integrate these signal-processing techniques onto a portable electrical stimulator and is currently under FDA investigation for efficacy (Sun et al. 2008).

The effects of continuous electrical stimulation, although not clearly understood, point toward an increased stimulation threshold for the activation of neuron populations with time (Agnew and McCreery 1990; McCreery et al. 1997). It is also believed that constant stimulation, even if charge balanced, contributes to certain irreversible electrochemical changes in the electrode–tissue interface that would cause this therapy to become less effective with time (McCreery et al. 1997). Over the past few decades, there have been a number of seizure detection and prediction algorithms reported to work with varying levels of efficacy using both animal and human data (Gotman 1982; Gotman et al. 1997; Iasemidis et al. 2003). Integrating these algorithms to trigger electrical stimulation, or a therapeutic intervention such as focal drug delivery, could greatly enhance the efficacy of the therapy by limiting the stimulus to the immediate preictal period, decreasing overall stimulus delivery over time and thus the likelihood of neuronal desensitization and damage. Furthermore, fewer triggers of electrical stimulus would increase the battery life significantly because stimulus currents often tend to be a significant contributor to battery drain.

Rapid developments in CMOS integrated-circuit fabrication allow for ultra-low-power, custom, application-specific integrated circuits (ASICs) to be designed to implement moderately complex signal-processing functions, while still maintaining low average power consumption from the battery source. Ultra-low-power digital circuit design allows for CMOS circuits to be operated below or near threshold-supply voltages at low throughput frequencies. Biomedical applications such as epilepsy prostheses may not require GHz clock operation, and can tolerate a low frequency of operation, especially if it comes with significant power reduction (Hyung-Il and Roy 2001). We have developed seizure detection techniques that tradeoff computational complexity for low-power hardware operation while still maintaining an acceptable detection efficacy (Raghunathan et al. 2009a, 2009b). We also discuss the potential of such a technique to enhance the detection efficacy of any other detection algorithm when used in combination. Finally, new technologies to wirelessly power and operate these devices are briefly discussed with the intent of introducing the next generation of implantable medical devices to treat epilepsy.

22.2 SEIZURE DETECTION

22.2.1 INTERTHRESHOLD CROSSING INTERVAL (ITCI)-BASED DETECTION

The threshold-crossing-interval-based detection algorithm that was developed by our group is based on the basic features used to visually identify electrographic seizures (Raghunathan et al. 2009a). It attempts to trade computational complexity for low-power hardware implementation. For instance, extraction of dominant frequency from a window or section of data may require its Fourier-transform computation along with digitization, and in some cases some memory or storage elements as well. On the other hand, an estimate of the dominant frequency may be obtained by measuring the time interval between amplitude-filtered "events" or threshold crossings. The inverse of this interthreshold crossing interval (ITCI) reflects an approximate estimate of the dominant frequency of the signal. Although different than the traditional fast fourier transform (FFT) the proposed ITCI technique can be easily implemented in hardware with the use of basic digital computation blocks such as a binary counter and combinational gates and a clock. These tradeoffs, while approximating the desired feature to be extracted, allow for real-time implementation using ultra-low-power circuits, thereby allowing for a long battery life when implemented in an epilepsy prosthesis. Furthermore, with the advent of multichannel recording systems that require detection algorithms to simultaneously screen multiple channels for electrographic seizures, it is extremely important that the hardware power consumption per channel of recording be minimized. The proposed algorithm is intended for integration with implantable microchips designed for neural recording systems, such as those now commonly reported in the literature (Jochum et al. 2009; Harrison 2008). The designed algorithm interfaces with such implantable systems and utilizes band-pass-filtered data between 10 and 500 Hz for processing.

The ITCI-based detection approach utilizes a combination of amplitude and frequency along with a third component related to rhythmicity of the signal to classify seizures. The algorithm is intended to be used with field potential data, which are typically obtained at higher rates of sampling to include a much broader bandwidth than traditional clinical EEG. The recorded data is screened for "events," which are defined as an amplitude-threshold crossing when compared to a reference that is proportional to the baseline average amplitude with time. This baseline signal is constantly updated by sampling the background at rates less than 10 Hz to avoid false events due to a DC drift in the recording data. Outliers in the individual samples from baseline may be prevented from contaminating the data by a second averaging performed on the sampled data where excessive baseline drifts are observed. (The present algorithm did not employ background averaging on-chip.) Any activity over the marked threshold is time stamped and marked as a threshold crossing. The second processing stage then calculates the time interval between two successive threshold-crossing time stamps to determine if the interval is less than a preset threshold. The interval between two successive threshold crossings is termed the interthreshold-crossing interval (ITCI), an approximate estimate of the dominant frequency of the amplitude-filtered data. The second stage is designed to identify any high-frequency activity with an ITCI less than the preset threshold. After a pair of amplitude-screened threshold crossings is classified as falling below the ITCI threshold ($ITCI_{thresh}$), a master counter is incremented to keep a count of the number of consecutive occurrences of ITCI threshold crossings. This increases the selectivity of the algorithm with minimal additive computation cost, as the third stage would reject detections that could be caused due to brief spiking or short-term artifacts that could trigger wrong threshold-crossing timestamps. After a pair of threshold crossings is detected that does not fit the ITCI threshold criterion, this master counter is reset, thereby resetting the queue of consecutive threshold-crossing pairs that reflect high-frequency activity. In other words, the detection algorithm screens the data for high-amplitude, rhythmic, and sustained high-frequency activity before flagging detection. The algorithm can be programmed to detect a seizure when the queue in the master counter reaches a programmed value (N_{Stage}). Like any detection algorithm, choosing optimal thresholds for threshold crossing (K_{amp}), interthreshold-crossing-interval ($ITCI_{thresh}$), and consecutive detections (N_{Stage}) is critical to the performance of the algorithm. The methods used to quantitatively set these thresholds based on a sample data set are documented by Raghunathan et al. (2009a). Figure 22.1 depicts a flowchart that captures the discussed decision tree of the threshold-crossing-based, seizure-detection algorithm.

22.2.2 HARDWARE DEVELOPMENT

The key advantage to trading off complex computational functions in the algorithm is its easy hardware implementation. Figure 22.2 describes a block-level schematic of the digital circuit implementation of the discussed seizure-detection algorithm. The recorded neural data is conditioned (amplified by a factor of 1000× and filtered between 10- and 500-Hz bandwidth) by a standard neural-recording headstage (Tucker-Davis Technologies). With the advent of a number of wireless neural-recording headstages that are also low power, the integrated-circuit solution to this stage is more commonly documented in the literature (Harrison et al. 2009; Jochum et al. 2009). The comparator flags threshold crossings to start and stop a clocked binary counter. The output of the binary counter between start and stop reflects the ITCI between the two marked threshold crossings. This ITCI is then compared with the programmed threshold ($ITCI_{thresh}$) before deciding to increment or reset the master counter queue. This keeps track of the number of consecutive pairs of threshold-crossings that meet the ITCI criterion. Finally, the queue length is constantly compared to the desired threshold (N_{Stage}) to flag a seizure detection when a certain number of consecutive

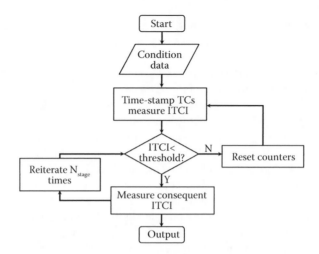

FIGURE 22.1 Flowchart depicting the operation of the ITCI-based seizure-detection algorithm. (Reproduced from Raghunathan et al., *Journal of Neural Engineering*, 6(5), 056005, 2009. With permission.)

threshold-crossing pairs have been detected that also meet the ITCI criterion. Figure 22.3 shows a snapshot of the recorded data with inter-threshold-crossing-intervals marked using dots in the bottom half corresponding to the same timescale as the data.

The hardware marked out threshold crossings and calculated the interthreshold-crossing intervals corresponding to consecutive crossings, which was then plotted on the same time scale as the data to obtain Figure 22.3. The distribution of IEI during a seizure episode is considerably different than that during normal baseline activity as shown in the Figure 22.3. The vertical arrow marks the point of electrographic onset of the seizure, as marked by a combination of data and video observation by a clinical epileptologist. The data obtained from the designed hardware, which was prototyped on a printed circuit board using commercially available digital circuit components, was averaged and plotted on the same time scale to obtain the plot. The ITCI measured from the PCB implementation was recorded using a data-acquisition system (National Instruments DAQ) and processed in software to obtain the fitting curve that shows the ITCI trend with progression of

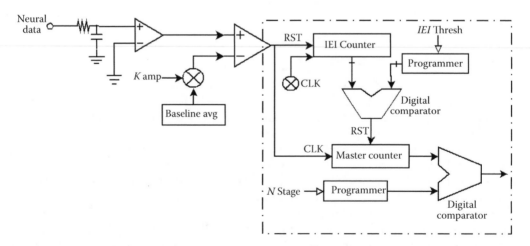

FIGURE 22.2 Block diagram of digital circuit schematic implementing the detection algorithm. (Reproduced from Raghunathan et al., *Journal of Neural Engineering*, 6(5), 056005, 2009. With permission.)

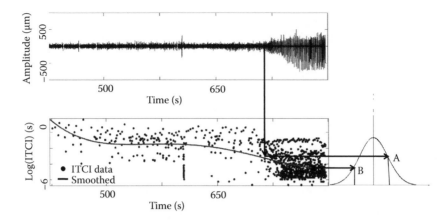

FIGURE 22.3 Progression of ITCI from baseline into seizure states. The normal plot represents the distribution of ITCI during a seizure episode with points A and B marked out to indicate their temporal location on the recorded data. (Reproduced from Raghunathan et al., *Journal of Neural Engineering*, 6(5), 056005, 2009. With permission.)

data from baseline to seizure. The solid line represents the windowed median of the recorded ITCI values obtained from the system.

22.2.3 ALGORITHM EVALUATION

The proposed detection algorithm was tested on in-vivo data recorded from kainate treated rats, used to model human temporal lobe epilepsy. Two-channel stainless steel microwire electrodes were implanted in the hippocampus close to the dentate gyrus to record field potential data (band filtered from 10 to 500 Hz) at a sampling rate of 1.5 kHz. A commercial neural-recording headstage (Tucker Davis Technologies) was used to acquire data, which included two stages of amplification (totaling 1000× gain) and a 60-Hz notch filter on hardware. The surgery and data-acquisition methods are documented elsewhere (Raghunathan et al. 2009a). The recorded data was then streamed into the designed hardware prototype at the same rate of recording to mimic an in-vivo setting. The algorithm was quantified based on detections that were validated through visual inspection of data and video by an expert epileptologist at the Indiana University School of Medicine as a gold standard. The seizures were classified on a Racine scale and no distinction was made between electrographic and clinical seizures for the purposes of this classification. The algorithm is designed to detect electrographic seizures, with the understanding that not all electrographic seizures may have clinical symptoms. Detections were quantified purely on the basis of number of electrographic seizures observed, detected, and missed. The algorithm was quantified in terms of its false positives, average time taken to detect a seizure after its electrographic onset (EO), and number of seizure misses. Table 22.1 documents the results obtained from six animals that were used to evaluate the algorithm.

The K_{amp} factor depicts the threshold used in proportion to the number of times over baseline average amplitude; FP, FN, SEN, and SEL represent false positives, false negatives, selectivity, and sensitivity, respectively. The thresholds (K_{amp}, $ITCH_{thresh}$, and N_{Stage}) were chosen based on a training data set specific to each animal. A statistical training model has been proposed that takes part of baseline and part of seizure-activity samples from each animal in the study and minimizes a cost function to solve for threshold values. The algorithm was then evaluated on a different testing data set from the same animal to quantify its performance, as documented in Table 22.1. An elaborate discussion on the methods used to train the algorithm, including the statistical model proposed, is provided by Raghunathan et al. (2009a). Data was divided into segments to stream into the hardware

TABLE 22.1
Results from Hardware Implementation of Detection Algorithm Evaluated Using Kainite-Treated Rats

ID	K_{amp}	Total Seizures	Number of Segments	Avg. Duration (min)	Detections	FP	FN	SEL	SEN
1	5	18	4	22	16	2	2	0.889	0.889
2	4	22	5	18	21	3	1	0.875	0.955
3	6	15	8	16	14	2	1	0.875	0.934
4	4	18	8	20	17	1	1	0.944	0.944
5	4	17	6	12	17	3	0	0.850	1
6	5	19	12	13	19	2	0	0.904	1

Source: Raghunathan et al., *Journal of Neural Engineering*, 6(5), 056005, 2009. With permission.

of varying average durations listed in the table. The segments were selected to mimic continuous operation, only omitting parts where there was an unnatural recording artifact, such as cable touch while unplugging the animal or amplifier saturation during stages of continuous seizing during an episode of status epilepticus. The segments were chosen to ensure comparable if not longer interictal baseline recording as compared to seizure durations. The detection delay was also quantified for each seizure and averaged across animals to obtain an average detection delay of 8.5 s [5.97, 11.04] with a standard deviation of 6.85 s from electrographic onset. The large standard deviation was due to differences in quality of data and type of seizure progressions observed from animal to animal (high-frequency onset versus lower-frequency onset). Detection delay was measured from the point of electrographic onset to the point of first detection by the hardware. The N_{Stage} queue on the PCB was incremented with successive pairs of threshold crossings that met the $ITCI_{thresh}$ criteria and was reset every time a single pair of threshold crossings did not meet this threshold. A detection was flagged by hardware when the N_{Stage} count crossed the programmed N_{Stage} threshold, indicating that there were N_{Stage} number of consecutive threshold crossings in amplitude that were spaced closer than $ITCI_{thresh}$ seconds. False positives were counted for every detection that was raised during what was termed nonseizure or baseline data. The false positives were observed mainly due to sustained artifacts in data usually caused by excessive movement. The N_{Stage} threshold had a strong check on artifacts caused by short bursts of spiking activity. For the purposes of quantification, every detection was counted as a false positive if there were multiple detections observed for a long burst of spikes that was termed baseline by inspection. The inherent design of the hardware does not allow for rapid multiple detections because a detection is only asserted if there is a sustained number of threshold crossings that meet both amplitude and frequency criteria.

22.3 INCREASING SPECIFICITY OF DETECTION ALGORITHMS IN COMBINATION

One approach to increasing the specificity—defined by the ability of the algorithm to measure the proportion of true positives that are correctly classified as such—is by its combination with other detection algorithms so that the chances of missing a seizure are reduced. The effect of "AND-ing" multiple algorithms makes detections more restrictive than any of the individual algorithms used, which can be used to attempt to reduce FPs but do so at the risk of missing seizures that are detected by some but not all of the individual algorithms. As a further caution, it must be understood that "AND-ing" may have a significant negative impact on detection speed because resulting detections cannot occur until the latest detection time of the individual algorithms, and may even be later (or missed altogether) if the individual detections do not properly overlap in time.

On the other hand, "OR-ing" can be used to attempt to improve sensitivity and speed and is especially useful when the individual algorithms accurately detect different types of signal changes of interest. However, "OR-ing" individual algorithms may also reduce specificity because the resulting combination will now detect the superset of all FPs that occur for any of the individual algorithms. An important criterion to validate the feasibility of such multialgorithm approaches is the feasibility in hardware (measured in terms of power consumption, memory requirements, and real-time feasibility, amongst other criteria). An incremental improvement in overall algorithm performance may come at a significant hardware cost, thereby reducing battery life of the implant. The proposed seizure-detection technique may be used in combination with other algorithms to increase the overall detection efficacy at a negligible hardware cost. To illustrate this, we evaluated the performance of previously published detection features and quantified the false positive rates on untrained data based on thresholds set using a training data set. Coastline, Hjorth variance, nonlinear energy, and wavelet energy were independently evaluated as detection features on data recorded from five

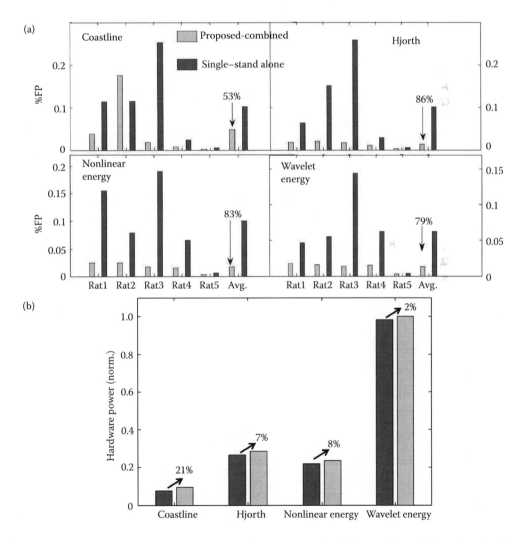

FIGURE 22.4 (a) The effect of the combination of the ITCI-based algorithm with standalone detection features on false positive rates for each of five animals in the study. (b) The effect of the proposed combination on the total power consumption, indicating the percent increase for each of the four detection features compared in the study.

animals treated with Kainic acid. These features have been used in seizure-detection algorithms and described in detail elsewhere (White et al. 2006; Hjorth 1975; Pfurtscheller and Lopes da Silva 1999; D'Allesandro et al. 2003; Khan and Gotman 2003).

The combination of the proposed detection algorithm resulted in a 53%, 86%, 83%, and 79% decrease in the average false positive rate when compared to using the coastline, Hjorth variance, nonlinear-energy, and wavelet-energy detection features, respectively, as a stand-alone detection algorithm. The hardware power consumption of each of the four detection features was obtained by implementing the feature using standard VHDL synthesis techniques (180-nm CMOS models) at a supply voltage of 0.8 V when clocked at 1 MHz. The threshold-crossing-based detection tool was then added to the hardware to recalculate the new hardware power estimate of the combination. The significant decrease in false positive rate increases the specificity, given that no decrease in true positives was observed. The proposed combination uses the detection output from the stand-alone measure (such as coastline) and uses them as threshold crossings with identical processing as the original threshold-crossing interval algorithm after that. These results are summarized in Figure 22.4. This example illustrates the power of the proposed algorithm in working with multiple other algorithms in combination in an implantable epilepsy prosthesis application.

22.4 FUTURE DIRECTIONS

The advent of several implantable neural recording systems on application-specific integrated circuits (ASICs) opens up avenues for implementing signal-processing schemes that combine some of these optimized seizure-detection algorithms on the same board (Harrison 2008; Harrison et al. 2009). The main advantage of integrating detection algorithms with neural conditioning and amplification is the elimination of the need to transmit recorded or digitized data to a signal-processing block. Usually, wireless data-transmission schemes are preferred both from a patient safety and a patient mobility aspect. However, multichannel recording systems are often bottlenecked by the data-transmission bandwidth supported by a low-power transmission scheme limited to operation in the medical implant communication service (MICS) band (Jan 2003). Integrating detection algorithms onboard could also reduce the amount of data to be transmitted to just time stamps of detections, as opposed to full-bandwidth raw data, thus allowing for multichannel recording schemes. Figure 22.5 shows an illustration of an integrated epilepsy prosthesis and also a screenshot of a complete seizure-detection system integrated with a neural-recording amplifier that is currently under fabrication. Wireless powering schemes would facilitate easy recharging of the batteries for such implanted devices, and in some cases may be able to eliminate the need for a battery at all. There have been published reports of wireless powering and recharging in medical implants, and our research group at Purdue University is also actively developing novel radio-frequency (RF) powered schemes that would facilitate continuous powering of such low-power systems (Ghovanloo and Atluri 2007; Harrison et al. 2009).

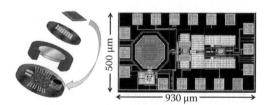

FIGURE 22.5 (left) Top-down illustration of integrated epilepsy prosthesis with recording electrode array on top; batteries and inductive powering coil are integrated with the microchip to form a cylindrical package. (right) Layout screen capture of seizure detection system integrated with neural recording amplifier and wireless transmission capabilities on a 0.13-μm CMOS process (5M, 1P).

The integration of ASIC design methodologies to adapt seizure-detection algorithms for this new application space would facilitate the development of more efficient therapeutic interventions to treat epilepsy. Multichannel seizure monitoring and tracking devices could also be a powerful tool in epilepsy research, helping us better understand the temporal dynamics or accurately identify the epileptogenic focus to develop new treatment paradigms for this chronic neural disorder.

REFERENCES

Federal Communications Commission (FCC). 2003. MICS band plan. In *FCC rules and regulations*, Part 95, January.

Agnew, W., and D. B. McCreery. 1990. Considerations for safety with chronically implanted nerve electrodes. *Epilepsia* 31:S27–S32.

Begley, C. E., J. F. Annegers, D. R. Lairson, T. F. Reynolds, and W. A. Hauser. 1994. Cost of epilepsy in the United States: A model based on incidence and prognosis. *Epilepsia* 35:1230–1243.

D'Allesandro M., R. Esteller, G. Vachtsevanos, A. Hinson, J. Echauz, and B. Litt. 2003. Epileptic seizure prediction using hybrid feature selection over multiple intracranial EEG electrode contacts: A report of four patients. *IEEE Trans. Biomed. Eng.* 50:603–613.

Ghovanloo, M., and S. Atluri. 2007. A wideband power-efficient inductive wireless link for implantable microelectronic devices using multiple carriers. *IEEE Transactions on Circuits and Systems-I* 54:2211–2221.

Gotman, J. 1982. Automatic recognition of epileptic seizures in the EEG. *Electroencephalogr. Clin. Neurophysiol.* 54:530–540.

Gotman, J., D. Flanagan, and J. Zhang. 1997. Automatic seizure detection in the newborn: Methods and initial evaluation. *Electroencephalogr. Clin. Neurophysiol.* 103:356–362.

Graves, N. M., and R. S. Fisher. 2005. Neurostimulation for epilepsy, including a pilot study of anterior nucleus stimulation. *Clin. Neurosurg.* 52:127–134.

Harrison, R. R. 2008. The design of integrated circuits to observe brain activity. *Proceedings of the IEEE* 96:1203–1216.

Harrison, R. R., R. J. Kier, C. A. Chestek, V. Gilja, P. Nuyujikian, S. Ryu, B. Greger, F. Solzbacher, and K. V. Shenoy. 2009. Wireless neural recording with single low-powered integrated circuit. *IEEE Trans. on Neural Systems and Rehab. Eng.* 17:322–329.

Hjorth, B. 1975. An on-line transformation of EEG scalp potentials into orthogonal source derivations. *Electroencephalogr. Clin. Neurophysiol.* 39:526–530.

Hyung-Il, K., and K. Roy. 2001. Ultra-low power DLMS adaptive filter for hearing aid applications. In *Proceedings of ISLPED* 2001:352–357.

Iasemidis, L. D., D.-S. Shiau, W. Chaovalitwongse, J. C. Sackellares, P. M. Pardalos, J. C. Principe, P. R. Carney, A. Prasad, and B. Veeramani. 2003. Adaptive epileptic seizure prediction system. *IEEE Trans. Biomed. Eng.* 50:616–627.

Jochum, T., T. Denison, and P. Wolf. 2009. Integrated circuit amplifiers for multi-electrode intracortical recording. *J. Neural. Eng.* 6:012001.

Khan, Y. U., and J. Gotman. 2003. Wavelet based automatic seizure detection in intracerebral electroencephalogram. *Clin. Neurophysiol.* 114:898–908.

Kossoff, E. H., E. K. Ritzl, J. M. Politsky, A. M. Murro, J. R. Smith, R. B. Duckrow, D. D. Spencer, and G. K. Bergey. 2004. Effect of an external responsive neurostimulator on seizures and electrographic discharges during subdural electrode monitoring. *Epilepsia* 45:1560–1567.

Labar, D., J. Murphy, and E. Tecoma. 1999. Vagus nerve stimulation for medication-resistant generalized epilepsy. *Neurology* 52:1510–1512.

Lesser, R. P., S. H. Kim, L. Beyderman, D. L. Miglioretti, W. R. S. Webber, M. Bare, B. Cysyk, G. Krauss, and B. Gordon. 1999. Brief bursts of pulse stimulation terminate afterdischarges caused by cortical stimulation. *Neurology* 53:2073–2081.

McCreery, D. B., T. G. H. Yuen, W. Agnew, and L. A. Bullara. 1997. A characterization of the effects on neuronal excitability due to prolonged microstimulation with chronically implanted microelectrodes. *IEEE Trans. Biomed. Eng.* 44:931–939.

Osorio, I., M. G. Frei, B. F. J. Manly, S. Sunderam, N. C. Bhavaraju, and S. B. Wilkinson. 2001. An introduction to contingent (closed-loop) brain electrical stimulation for seizure blockage, to ultra-short-term clinical trials, and to multidimensional statistical analysis of therapeutic efficacy. *J. Clin. Neurophysiol.* 18:533–544.

Osorio, I., M. G. Frei, and S. Sunderam. 2005. Automated seizure abatement in humans using electrical stimulation. *Ann. Neurol.* 57:258–268.

Peters, T. E., N. C. Bhavaraju, M. G. Frei, and I. Osorio. 2001. Network system for automated seizure detection and contingent delivery of therapy. *J. Clin. Neurophysiol.* 18:545–549.

Pfurtscheller, G., and F. H. Lopes da Silva. 1999. Event-related EEG/MEG synchronization and desynchronization: Basic principles. *Clin. Neurophysiol.* 110:1842–1857.

Politsky, J. M., R. Estellar, A. M. Murro, J. R. Smith, P. Ray, Y. D. Park, and M. J. Morrell. 2005. Effects of electrical stimulation paradigm on seizure frequency in medically intractable partial seizure patients with a cranially implanted responsive cortical neurostimulator. In *Proceedings of the Annual Meeting of American Epilepsy Society (AES),* Washington, DC, December 2–6, 2005. Washington, DC: AES.

Raghunathan, S., S. K. Gupta, M. P. Ward, R. M. Worth, K. Roy, and P. P. Irazoqui. 2009a. The design and hardware implementation of a low-power real-time seizure detection algorithm. *J. Neural Eng.* 6(5):056005.

Raghunathan, S., M. P. Ward, K. Roy, and P. P. Irazoqui. 2009b. A low-power implantable event-based seizure detection algorithm. In *Proceedings of the 4th International IEEE EMBS Conference on Neural Engineering,* Antalya, Turkey, April 29–May 2, 2009, 151–154. Washington, DC: IEEE.

Sun, F. T., M. J. Morrell, and R. E. Wharen. 2008. Responsive cortical stimulation for the treatment of epilepsy. *Neurotherapeutics* 5:68–74.

White, A. M., P. A. Williams, D. J. Ferraro, S. Clark, S. D. Kadam, F. E. Dudek, and K. J. Staley. 2006. Efficient unsupervised algorithms for the detection of seizures in continuous EEG recordings from rats after brain injury. *J. Neurosci. Method* 152:255–266.

23 Initiation and Termination of Seizure-Like Activity in Small-World Neural Networks

Alexander Rothkegel, Christian E. Elger, and Klaus Lehnertz

CONTENTS

23.1 INTRODUCTION

It is an unchallenged viewpoint that the collective behavior of a complex dynamical network like the brain emerges from both the local dynamics of the constituents and from the structure of the network that connects them (Strogatz 2001). Epileptic seizures are generally considered as a short episode of overly synchronized firing of neurons. It could advance our understanding of ictogenesis to identify combinations of neuron and network properties that lead to synchronized seizure-like firing, to asynchronous firing, or to a seizure-prone configuration with intermittent changes between these two dynamic states (Wendling 2005; Soltesz and Staley 2008; Lytton 2008). We aim at developing and studying such neuron network models that can spontaneously (i.e., without change of parameters) switch between asynchronous and synchronous firing. Previously proposed models (Suffczynski et al. 2004; Breakspear et al. 2006) that can achieve this use strong noise influences that allow the system to change spontaneously between attractors corresponding to seizure-free and seizure-like activity. The stochastic nature of these transitions leads to exponentially distributed seizure durations and interseizure intervals. Although this may be applicable to absence seizures, it lacks a concrete mechanism for seizure termination and possibly cannot account for dynamics observed with focal seizures.

As properties of single neurons have been extensively investigated, there is a plethora of neuron models available to choose from (Izhikevich 2007). Moreover, studies in animal models of epilepsy and in resected tissue from epilepsy patients provide us with a detailed knowledge about altered physiological properties of neurons from epileptic tissue (McCormick and Contreras 2001; Jefferys 2003). Despite these experimental successes, the question of which of these alterations are actually involved in ictogenesis (and which may be considered as byproducts) is still not satisfactorily answered. Moreover, our knowledge about synaptic networks and their pathophysiological alterations is still limited, and meaningful notions with which to describe brain networks are still

341

few. As a simple model for the synaptic wiring, we here consider small-world networks, similar to those introduced by Watts and Strogatz (1998). These networks have a spatial structure with some tunable amount of long-range connections that are also believed to be present in neural tissue (Bullmore and Sporns 2009). We study the collective dynamical behavior of neural networks of this type and investigate the possibility of intermittent changes between asynchronous and synchronous firing.

23.2 METHODS

We investigate networks of n excitatory, identical, nonleaky, integrate-and-fire neurons with constant input currents I, time delays τ, and refractory periods ϑ. (Burkitt 2006a, 2006b). When a neuron n is not refractory, we describe its state by the membrane potential $x_n(t) \in [0,1]$, whose time evolution in absence of excitations is given by $\frac{d}{dt} x_n(t) = I$. Additionally, we introduce, at times t_i, membrane potential jumps of size c due to the firing of presynaptic neurons at time $t_i - \tau$:

$$x_n(t_i^+) = x_n(t_i) + c.$$

If the membrane potential exceeds 1 due to such an excitation (or due to the linear charging), the neuron fires and stays refractory for a time period of ϑ, thereby ignoring all incoming excitations. Subsequently, the neuron reenters the excitatory state with $x_n = 0$. For the sake of simplicity, we normalize the intrinsic frequency to 1 by setting $I := \frac{1}{1-\vartheta}$. As the considered neurons are oscillatory, we can define for every neuron n a phase ϕ_n such that in absence of excitations $\frac{d}{dt}\phi_n = 1$. Phases in $[0,\vartheta)$ then indicate refractory neurons, while phases $\phi_n \in [\vartheta,1)$ indicate a membrane potential of $x_n = \frac{\phi_n - \vartheta}{1-\vartheta}$. Using this notion, we define an order parameter r using circular statistics (Mardia and Jupp 2000) that reflects the amount of synchrony in the network at a given time t:

$$r(t) = \frac{1}{N}\left|\sum_{n \in N} e^{2\pi i \phi_n(t)}\right|.$$

While a value of 1 is reached for complete synchrony of all neurons only, 0 is attained for any balanced distribution of the phases of neurons. The introduction of phases also allows for a natural definition of a randomized initial condition by choosing homogeneously distributed phases [0,1] for all neurons. Note that in this way, some of the neurons are refractory at the beginning of a simulation. For the findings presented here, we chose a refractory period of $\vartheta = 0.05$ and a time delay of $\tau = 0.01$.

As a synaptic network connecting these neurons, we choose small-world networks that contain long-range and short-range connections. We start the network construction with a lattice of 300 × 300 neurons and connect each neuron bidirectionally to its 36 nearest neighbors. Afterward, we replace with probability ρ each directed connection by a unidirectional connection between two randomly chosen neurons, excluding self-connections and multiple connections between the same neurons. To characterize the collective dynamical behavior, we choose a (small) sampling interval δ and measure—in addition to the order parameter r—the mean firing rate f of neurons. We are interested in whether f shows interchanging periods of high and low activity that we assume to mimic seizure-like activity (see Figure 23.1).

Depending on the size of the coupling strength c, either the constant current or the synaptic coupling between neurons will dominate. We will investigate both scenarios separately and investigate the possibility for intermittent behavior of periods with and without collective firing.

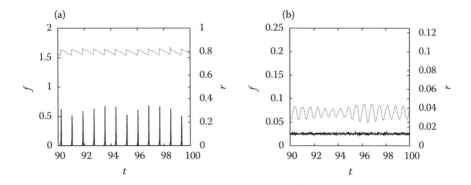

FIGURE 23.1 Exemplary time series of the order parameter r (dashed, right ordinate) and the mean firing rate f (solid, left ordinate) for a network with (a) and without (b) collective firing.

23.3 RESULTS

23.3.1 LARGE COUPLING STRENGTH

We start by investigating the case of large synaptic coupling. As we consider both time delay and refractoriness in networks with spatial structure, we can observe wave phenomena like cyclic waves, spiral waves, and more complicated turbulent-like wave pattern. As we already investigated in an earlier study (Rothkegel and Lehnertz 2009), these patterns may or may not coincide with the collective firing of neurons. When studying the dynamical behavior in dependence on neuron or network parameters, we can relate parameter regimes to different dynamical behavior that evolves from a given initial condition.

Figure 23.2 shows mean and variance of the firing rate f dependent on the rewiring probability ρ and the coupling strength c. Statistics were estimated for $t \in [50, 100]$. The complementing snapshots were taken at $t = 100$.

For parameter settings well inside these regimes, the network rapidly attains the corresponding dynamical behavior. At boundaries, however, we observe spontaneous changes even after long periods of seemingly stationary behavior. Figure 23.3 shows an exemplary time series of the mean firing rate f. Initially, f exhibits periodic behavior with interchanging periods of high and low activity. The network dynamics demonstrates cyclic waves and collective firing (regime I-B in Figure 23.2). The transition at $t = 120$ is due to a spontaneous creation of a spiral wave that soon dominates the network dynamics. Compared to the initial state, the spiral wave shows a high degree of phase-locking between neurons. Therefore, the network does not switch back to collective firing. These changes are possible from Ib to IIa, from Ib to Ia and from IIb to IIa (using the notation for dynamical behaviors introduced in the caption of Figure 23.2). Transitions to states with collective firing were not observed. When starting from distributed phases with parameters such as those in Figure 23.3, there is a high probability that the spiral wave is generated instantly. If not, the network begins generating cyclic waves and settles to the semistable state Ib. To investigate the nature of the transition from Ib to Ia, we estimated the distribution of times a network (with a specific rewiring probability ρ) remains in Ib before switching to IIa. To ensure that a network started in state Ib, we only evaluated realizations for which we were able to observe oscillatory behavior after $t = 20$. Figure 23.4 shows a histogram of the distribution of lifetimes of state Ib, which is consistent with an exponential law. This result can be regarded as an indication that the transition is in some way random and may not be predicted. Note that this randomness is generated although the considered dynamical system is purely deterministic.

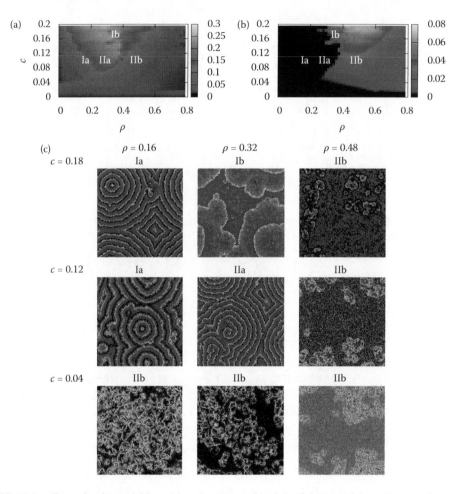

FIGURE 23.2 **(See color insert.)** Mean (a) and variance (b) of the firing rate f dependent on the rewiring probability ρ and the coupling strength c. Labels mark regimes corresponding to different dynamical behaviors. (Ia) cyclic waves and no collective firing; (Ib) cyclic waves and collective firing; (IIa) spiral waves, turbulent-like pattern and no collective firing; (IIb) spiral waves, turbulent-like pattern and collective firing. (c) Spatial distribution of membrane potentials for different rewiring probabilities ρ and coupling strengths c. Red points: neurons which have fired but their excitations have not reached postsynaptic neurons yet (indicating wave fronts), blue points: refractory neurons whose excitations already reached postsynaptic neurons, gray points: charging neurons with lightness encoding the membrane potential $0 < x_n < 1$.

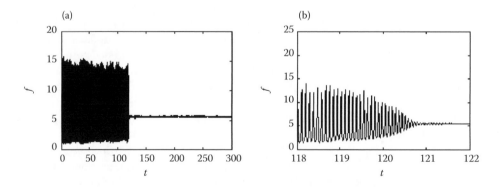

FIGURE 23.3 (a) Exemplary time series of the mean firing rate f of 300×300 neurons on a small-world network with rewiring probability $\rho = 0.35$ and coupling strength $c = 0.15$. (b) Detailed view of the period $t \in [118,122]$.

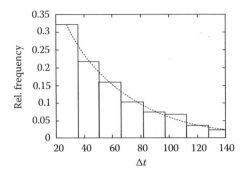

FIGURE 23.4 Histogram of the distribution of periods Δt, small-world networks of 300×300 neurons ($\rho = 0.33$, $c = 0.16$) stay in state Ia before switching to state Ib. 626 transitions were observed. The dashed line shows an exponential fit.

23.3.2 SMALL COUPLING STRENGTH

Next, we investigate the case of small coupling strengths, where the input currents are only slightly modulated by the synaptic coupling. Here, we observe no waves but less structured asynchronous patterns instead, which change form and position slowly and form regions that may be embedded in regions of synchronously firing neurons. In order to indicate changes in the amount of synchrony in the network, we use the order parameter r. Note that r itself attains a periodic behavior that reflects the collective oscillation of the neurons. While the collective firing of the synchronous part increases r, it drops continuously between the collective firing. Therefore, we consider the series of local maxima $R(t)$ of $r(t)$.

In Figure 23.5, we present the dependence of the variance of $R(t)$ on the rewiring probability ρ and the coupling strength c. As before, the variance is calculated for $t \in [50, 100]$. For a certain domain in the (ρ, c)-plane centered around $\rho = 0.45$ and $c = 0.02$, we observe large fluctuating values for $R(t)$ with no convergence to a state of fixed amount of synchrony. Depending on parameters, these fluctuations take the form of regular oscillations, irregular oscillations, or of short randomly occurring episodes of synchronous firing separated by asynchronous firing. Figure 23.6 (top) shows an exemplary time series of $R(t)$ with the latter behavior. In order to investigate the nature of the transitions into and out of the episodes of synchronous firing, we measured times of transitions for a single network of 100×100 neurons that was observed for $t \in [0,200000]$, starting from 10 different sets of homogeneous distributed initial conditions. In these simulations, a total of 13,989 episodes of synchronous firing were observed. In Figure 23.6 (bottom), we show histograms of the durations of the asynchronous firing between episodes and of the durations of the episodes. While

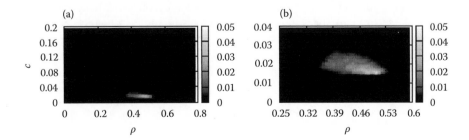

FIGURE 23.5 (a) Variance of $R(t)$ (the series of local maxima of the order parameter $r(t)$) as a function of the rewiring probability ρ and the coupling strength c. (b) Detailed view of $(\rho, c) \in [0.25,0.6] \times [0, 0.04]$.

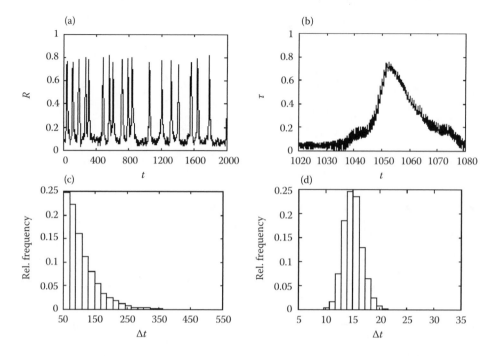

FIGURE 23.6 (a) Local maxima $R(t)$ of the order parameter $r(t)$ for a network with $\rho = 0.5$ and $c = 0.0145$. The dynamics shows short episodes of synchronous firing, which are separated by asynchronous firing. (b) A closer inspection of $r(t)$ for the first episode. (c) Histogram of intervals between episodes for a simulation of 100×100 neurons with $\rho = 0.5$ and $c = 0.0145$. (d) Histogram of the duration of episodes for the same simulation.

the distribution of times between episodes again resembles an exponential function (at least for the right tail), the distribution of the duration of episodes is peaked. This can possibly be regarded as an indication of a mechanism that terminates the synchronous firing.

23.4 CONCLUSION

We have studied small-world networks of nonleaky integrate-and-fire neurons driven by constant input currents. We investigated the possibility for spontaneous changes between synchronous and asynchronous states. For large coupling strengths where the synaptic coupling dominates over the input currents, we observed different wave patterns that may or may not occur concurrently with collective firing of neurons. For specific parameter settings, the networks may change from a state with collective firing and some wave patterns, to a state without collective firing and wave patterns. The lifetimes of synchronous states appeared to be exponentially distributed. For small coupling strengths, we observed transitions both into and out of synchronous firing. The periods between synchronous firings again appeared to be exponentially distributed, and we observed a peaked distribution for the duration of the synchronous periods. The transitions are possible even without noise influences and without the change of external control parameters.

ACKNOWLEDGMENT

This work was supported by the Deutsche Forschungsgemeinschaft.

REFERENCES

Breakspear, M., J. A. Roberts, J. R. Terry, S. Rodrigues, N. Mahant, and P. A. Robinson. 2006. A unifying explanation of primary generalized seizures through nonlinear brain modeling and bifurcation analysis. *Cereb. Cortex* 16:1296–1313.

Bullmore, E., and O. Sporns. 2009. Complex brain networks: Graph theoretical analysis of structural and functional systems. *Nat. Rev. Neurosci.* 10:186–198.

Burkitt, A. N. 2006a. A review of the integrate-and-fire neuron model: I. Homogeneous synaptic input. *Biol. Cybern.* 95:1–19.

Burkitt, A. N. 2006b. A review of the integrate-and-fire neuron model: II. In-homogeneous synaptic input and network properties. *Biol. Cybern.* 95:97–112.

Izhikevich, E. M. 2007. *Dynamical systems in neuroscience: The geometry of excitability and bursting.* Cambridge, MA: MIT Press.

Jefferys, J. G. R. 2003. Models and mechanisms of experimental epilepsies. *Epilepsia* 44(Suppl. 12):44–50.

Lytton, W. W. 2008. Computer modelling of epilepsy. *Nat. Rev. Neurosci.* 9:626–637.

Mardia, K. V., and P. E. Jupp. 2000. *Directional statistics.* New York: Wiley.

McCormick, D. A., and D. Contreras. 2001. On the cellular and network bases of epileptic seizures. *Annu. Rev. Physiol.* 63:815–846.

Rothkegel, A., and K. Lehnertz. 2009. Multistability, local pattern formation, and global collective firing in a small-world network of non-leaky integrate-and-fire neurons. *Chaos* 19:015109.

Soltesz, I., and K. Staley. 2008. *Computational neuroscience in epilepsy.* San Diego: Academic Press.

Strogatz, S. H. 2001. Exploring complex networks. *Nature* 410:268–276.

Suffczynski, P., S. Kalitzin, and F. H. Lopes da Silva. 2004. Dynamics of non-convulsive epileptic phenomena modeled by a bistable neuronal network. *Neuroscience* 126:467–484.

Watts, D. J., and S. H. Strogatz. 1998. Collective dynamics of "small-world" networks. *Nature* 393:440–442.

Wendling, F. 2005. Neurocomputational models in the study of epileptic phenomena. *J. Clin. Neurophysiol.* 22:285–356.

24 Are Interaction Clusters in Epileptic Networks Predictive of Seizures?

Stephan Bialonski, Christian E. Elger, and Klaus Lehnertz

CONTENTS

24.1 INTRODUCTION

Analysis techniques designed to reliably predict epileptic seizures from recordings of brain electrical activity could make a considerable contribution to improving the quality of life of our patients by providing newer and more effective treatments and expanding our knowledge of how seizures are generated in the brain. While research in the 1990s mainly focused on a characterization of seizure precursors using univariate time series analysis techniques, recent years have seen a shift toward a characterization of interdependencies between different brain regions during the preictal period. For an overview, see the works by Mormann et al. (2007) and Schelter et al. (2008). In addition to bivariate time series analysis techniques that aim at estimating the strength and direction of an interaction (Lehnertz et al. 2009), the success of network theory in physics, biology, and other scientific fields (Boccaletti et al. 2006; Arenas et al. 2008) has inspired the development of multivariate analysis techniques. These methods aim at characterizing the interdependence structure between arbitrary numbers of interacting dynamical systems.

Here we employ a multivariate analysis approach by estimating interaction networks from multichannel EEG data recorded intracranially from epilepsy patients during presurgical evaluation. Nodes of an interaction network represent recording sites while the weights of the links between nodes are estimated using a bivariate measure of signal interdependence. We present a method to identify clusters in interaction networks, i.e., groups of strongly interconnected nodes with low-weight links to nodes that are not members of the group. We study the temporal evolution of interaction clusters and address the question of whether interaction clusters undergo specific changes prior to epileptic seizures that may guide the identification of a preictal state.

24.2 IDENTIFYING CLUSTERS IN INTERACTION NETWORKS

Cluster identification techniques have been developed—sometimes in parallel—in different scientific fields such as physics (Fortunato 2010), mathematics, or engineering (Theodoridis and Koutroumbas 2003). Most methods deal with two challenges, namely, identification of clusters and estimation of an appropriate number of clusters justified by the data. While hierarchical approaches usually produce a series of partitions with a varying number of clusters from which one has to choose, nonhierarchical methods typically require the number of clusters to be chosen a priori. Various quality functions have been introduced, for example, in works by Milligan and Cooper (1985), Newman and Girvan (2004), and Rummel et al. (2008), which evaluate a given partition with respect to the data and are supposed to achieve an extremum for the partition with an optimum number of clusters.

The time-resolved identification of clusters in interaction networks derived from multiday, multichannel EEG data calls for methods that are computationally feasible, while at the same time being robust with respect to noise contamination. For this reason, we chose a method that met both requirements in simulation studies (Allefeld and Bialonski 2007). Let $\mathbf{R}$ denote a matrix with entries R_{ij}, which represent values of some interdependence between time series x_i and x_j, $i, j, \in \{1, \ldots, N\}$. We assume $\mathbf{R}$ to be symmetric and $R_{ij} \geq 0$ for all i and j. $\mathbf{R}$ can be interpreted as an adjacency matrix of a weighted interaction network with nodes that are associated with sensors sampling the signals and with link weights that correspond to the strength of interdependence between signals. In order to identify clusters within such a network, we consider a random walk on the link structure where the probability P_{ij} of a transition from node j to node i is proportional to R_{ij}. More specifically, we define the transition probability matrix $\mathbf{P}$ of a Markov chain, $\mathbf{P} = \mathbf{R}\mathbf{D}^{-1}$, where $\mathbf{D}$ is a diagonal matrix with entries $d_{jj} = \sum_{l=1}^{N} R_{lj}$. The key idea is that nodes should belong to clusters within which the stochastic process stays long and only rarely jumps from nodes of one cluster to nodes of another cluster. Thus, an intuitive choice for quantifying the distance between nodes i and j in terms of transition probabilities would be to consider the vector distance between the ith and j^{th} column vectors of $\mathbf{P}$. Additionally, we exploit the time evolution of the stochastic process to explore the connectivity structure of nodes from a local to a global perspective by considering powers of $\mathbf{P}$. Note that $(\mathbf{P}^{\tau})_{ij}$ with $\tau \geq 0$ represents the probability of a transition from node j to i in τ steps. We now define the distance d^2 between nodes as a weighted distance,

$$d^2(i, j) := \sum_{k=1}^{N} c_k \left| (\mathbf{P}^{\tau})_{ki} - (\mathbf{P}^{\tau})_{kj} \right|^2$$

$$= \sum_{k=1}^{N} |\lambda_k|^{2\tau} (A_{ki} - A_{kj})^2, \tag{24.1}$$

where the weights c_k are defined as $c_k := \Sigma_{i,j} R_{ij} / \Sigma_j \mathbf{R}_{kj}$, and A_{ki} denotes the ith component of the kth normalized left eigenvector of $\mathbf{P}$ $\left(\sum_i A_{ki}^2 / c_i = 1 \right)$. λ_k is the associated eigenvalue with $\lambda_1 = 1 > |\lambda_2| \geq |\lambda_3| \geq \cdots \geq |\lambda_N|$, see the work by Allefeld and Bialonski (2007) for further details. d^2 is also known as *diffusion distance* (Nadler et al. 2006; Coifman and Lafon 2006; Lafon and Lee 2006). Varying τ allows us to explore the cluster structure on different scales, where $\tau \to \infty$ lets d^2 vanish for all pairs of nodes ($|\lambda_k|^{\tau} \to 0$ for $k > 1$, $A_{1i} = 1 \forall i$) and belongs to the perspective of all nodes belonging to a single cluster, whereas $\tau = 0$ makes $\mathbf{P}^{\tau}$ the identity matrix ($|\lambda_k|^{\tau} = 1 \forall k$), increases d^2 for all pairs of nodes, and belongs to the perspective of a network disintegrating into as many clusters as there are nodes. Thus, in order to identify the number q of clusters, we determine the corresponding time scale $\tau = \tau(q)$ by requiring the $(q + 1)$st eigenvalue to vanish, $|\lambda_{(q+1)}|^{\tau} = \xi$, where $0 < \xi \ll 1$ is a small number (e.g., $\xi = 0.01$), leading to $\tau(q) = \ln \xi / \ln |\lambda_{(q+1)}|$. d^2 is also known as *diffusion distance* between position vectors $\vec{o}(j) = \left(|\lambda_k|^{\tau} A_{kj} \right)$, $k = 1, \ldots, N$, associated with nodes j, which leads us to a remarkable result: Since all contributions from terms with $k > q$ can be neglected for appropriately chosen $\tau(q)$

and are zero for $k = 1$ ($A_{1i} = 1 \forall i$), we obtain an effective dimensionality reduction by considering Euclidean distances between vectors

$$\vec{o}_{\text{red}}(j) = \left(\left| \lambda_k \right|^\tau A_{kj} \right), k = 2, \ldots, q \qquad (24.2)$$

in ($q - 1$)-dimensional space only. We obtain clusters of nodes within this space by employing a standard k-means clustering algorithm (MacQueen 1967), which we initialize with rough estimates of the cluster locations (Allefeld and Bialonski 2007). Note that this approach resembles spectral clustering methods (Shi and Malik 2000; von Luxburg 2007; Chung 1997), which perform a k-means clustering on the eigenvectors of the q largest eigenvalues of $\mathbf{P}^{\mathrm{T}}$ (T here denotes transposition). We let the method determine clusters for varying numbers $q = 1, \ldots, N$ and choose the clustering that maximizes the so-called *modularity* (Newman and Girvan 2004), a quality function that has been successfully applied in various contexts and whose limitations have been studied thoroughly (Fortunato and Barthélemy 2007).

24.3   DATA ANALYSIS AND RESULTS

We retrospectively analyzed multiday, multichannel EEG data (total recording time: 90 days, mean: 154 h/patient, range: 45–267 h; average number of recording sites: 63, range: 32–76) recorded intracranially from 14 patients (patients A–N) who underwent presurgical evaluation of pharmacoresistant focal epilepsies. Data had been sampled at 200 Hz (16 bit analog-to-digital converter) within the frequency band 0.3–70 Hz using a referential montage, and the recordings captured a total number of 119 seizures (mean: 8.5, range: 6–14 seizures/patient). All patients signed informed consent that their clinical data might be used and published for research purposes.

We identified clusters in interaction networks from EEG data in a time-resolved and frequency-specific manner. For this purpose, we split the data into consecutive nonoverlapping windows of 20.48 s duration ($T = 4096$ sampling points) and band-pass filtered the data in the δ, ϑ, α, β_1, and β_2 frequency bands. We then extracted phase time series $\phi_i(t)$, $\phi_j(t)$ from the windowed and filtered EEG time series $x_i(t)$, $x_j(t)$ ($i, j \in \{1, \ldots, N\}$, N denotes the number of recording sites (nodes)) by utilizing the Hilbert transform and estimating the mean phase coherence (Mormann et al. 2000)

$$R_{ij} = \left| \frac{1}{N} \sum_{t=1}^{T} e^{i(\phi_i(t) - \phi_j(t))} \right|, \qquad (24.3)$$

as a measure for interdependence for all nonredundant combinations of recording sites i and j. R_{ij} takes on values between 0 (no phase synchronization) and 1 (perfect phase synchronization). In this way we obtained, for each window and each frequency band, a matrix $\mathbf{R}$, which we used to identify the clustering (i.e., a set of clusters) as described in Section 24.2.

We then investigated whether the clustering exhibits changes prior to epileptic seizures that may guide the identification of a preictal state. We assumed that a preictal period exists and lasts for a certain amount of time T_p. If data from a preictal period amounted to less than 70% of T_p (e.g., due to recording gaps or due to seizure clustering), it was excluded from subsequent analyses. Likewise, we discarded data from recording periods within 60 minutes after the onset of a seizure in order to exclude effects from the ictal and postictal periods. We varied T_p from 15 to 240 minutes (in steps of 15 minutes) and determined the number n_i of windows during interictal periods, as well as the number n_p of windows during preictal periods. For all clusters contained in the $n_i + n_p$ clusterings, we determined their occurrences in all windows, and for a given cluster c we obtained the number $n_p^{(c)}$ of occurrences in windows from the preictal periods as well as the number $n_i^{(c)}$ of occurrences

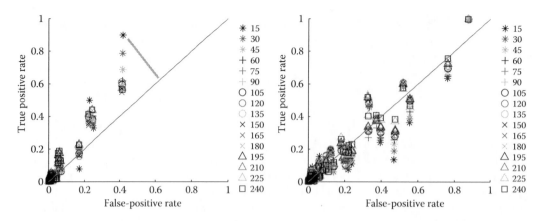

FIGURE 24.1 (See color insert.) Exemplary ROC spaces. Each point in space represents a cluster, together with the presumed duration T_p of a preictal period (color- and symbol-coded, see legend). Durations are given in minutes. Left: ROC space for data in the ϑ-band from Patient A. The gray line indicates the distance (V) to the diagonal for an exemplary cluster in ROC space; the larger V, the higher the predictive power of the respective cluster. Right: Same as the left figure, but for data in the β_2-band from Patient E.

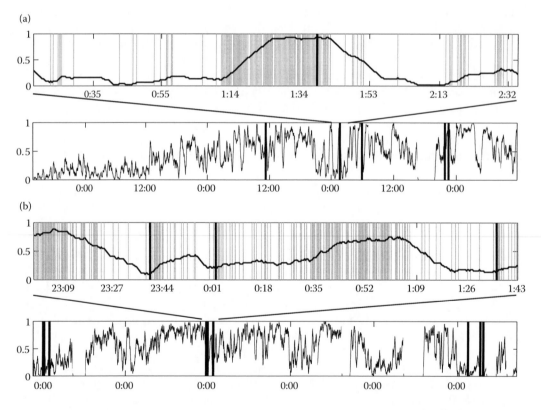

FIGURE 24.2 Time courses of occurrences (indicated as vertical gray lines; 15 minutes moving-average smoothing as black line) of the most predictive cluster (Figure 24.1). Seizures are marked by vertical black lines, and white areas indicate recording gaps. (a) Top: Enlarged view of a periictal recording from patient A (ϑ-band). Bottom: Complete recording. (b) Same as (a) but for Patient E (β_2-band).

TABLE 24.1
Summary of Cluster Characteristics with Highest Predictive Values W

ID	SOZ	Outcome	Loc	W	N^*	T_p	f
A	NC	1A	ipsi HC	0.48	10	15	ϑ
B	NC	1A	near SOZ	0.45	1	15	α
C	NC	1A	near SOZ	0.44	2	15	β_1
D	HC+NC[a]	2B	ipsi HC	0.36	10	15	ϑ
E	HC	1A	near SOZ	0.33	1	15	β_2
F	NC	1A	SOZ	0.32	2	105	α
G	HC	n. s.	ipsi NC	0.30	1	15	β_1
H	HC	1A	contra NC[c]	0.28	2	60	β_2
I	HC+NC[b]	2A	NC	0.27	8	75	β_2
J	NC	1A	contra NC	0.25	2	30	δ
K	HC	1A	contra HC	0.24	8	210	β_2
L	HC	1A	SOZ	0.23	4	15	δ
M	NC	1A	near SOZ	0.20	2	15	ϑ
N	HC	1A	contra NC	0.14	1	15	β_2

Note: Summary includes all investigated patients. ID = patient ID; SOZ = seizure onset zone, according to presurgical work-up; HC = hippocampus; NC = neocortex; outcome = outcome according to Engel classification; n.s. = no surgery; Loc = location of cluster; N^* = the number of recording sites in most predictive cluster; T_p = presumed duration of preictal phase in minutes; f = frequency band; [a] = a neocortical lesion; [b] = multifocal seizures; and [c] = near nonepileptogenic lesion.

in windows from the interictal periods. Thus, the true positive rate $TPR^{(c)} = n_p^{(c)}/n_p$ and false positive rate $FPR^{(c)} = n_i^{(c)}/n_i$ can be defined for each cluster c, for each duration T_p of a preictal period, as well as for each considered frequency band. Figure 24.1 shows exemplary ROC spaces, which are defined by FPR and TPR as the x and y axes, respectively (Provost and Fawcett 2001). Each point in ROC space corresponds to a cluster c with its $TPR^{(c)}$ and $FPR^{(c)}$ values for a given duration T_p of the preictal period. Since the diagonal of the ROC space constitutes the set of points obtained for a random predictor, we are interested in points deviating from the diagonal. Points above the diagonal represent clusters whose frequency of occurrence is higher during the presumed preictal period than during the interictal periods, while the opposite is true for points below the diagonal. Note that a cluster with $TPR = 1$ and $FPR = 0$ (or $TPR = 0$ and $FPR = 1$) would perfectly indicate a preictal state due to its unequivocal occurrence (or absence) in the preictal period only. For each cluster c, we quantify its predictive power $W^{(c)}$ as a deviation from the diagonal in ROC space $V^{(c)} = W^{(c)}/\sqrt{2}$, where $W^{(c)} := |TPR^{(c)} - FPR^{(c)}| \in [0,1]$ with $W^{(c)} = 1$ ($W^{(c)} = 0$), indicating the cluster to perfectly indicate (or not to indicate) a preictal state.

We observed sets of points (cascades) in ROC space with varying TPR values but almost the same FPR value. Interestingly, points within each cascade reflected the very same cluster but for different T_p values. For a cluster within the cascades, we observed the TPR value increase (Figure 24.1, left) with decreasing duration T_p of the presumed preictal period. The opposite could also be observed (Figure 24.1, right), where the TPR values of a cluster within a cascade decreased with decreasing T_p (i.e., the frequency of occurrence of such a cluster diminished when approaching a seizure). In both cases, W increased with decreasing T_p possibly indicating a gradual build-up of some process prior to an impending seizure that is reflected in a gradual reorganization of the cluster structure in interaction networks. We highlight these observations in Figure 24.2a and b (top), where we show time courses of occurrences of clusters with the largest W values (Figure 24.1) for two patients. On a longer time scale (Figure 24.2a and b, bottom), we observed that the frequency of occurrence of clusters (shown as a moving-average of 15-minutes duration of the discrete cluster occurrences) varied strongly in time. It remains to be shown whether this variability can be related to the specific nature of the epileptic process or whether it also reflects other influencing factors such as alterations of the antiepileptic medication, physiologic activities, or the sleep–wake cycle. We summarize our findings for all investigated patients in Table 24.1. In 9 out of 14 patients, we observed the highest predictive power W of interaction clusters for a preictal period of 15 minutes duration and for interactions in the α- and β-frequency bands. The size of the most predictive cluster (i.e., the number of involved recording sites) varied largely from 1 to 10, and quite often we observed these clusters to be located near the seizure onset zone or in ipsilateral brain regions. Nevertheless, in four patients the most predictive cluster was located in contralateral brain regions.

24.4 CONCLUSION

We have presented a method that allows one to identify—in a time-resolved manner—interaction clusters within functional networks derived from EEG recordings. We retrospectively analyzed multiday, multichannel intracranial EEG data from 14 epilepsy patients and studied whether a change in the frequency of occurrence of different clusters may indicate an impending seizure. We observed in many cases the frequency of occurrence of clusters to continuously increase or to continuously decrease when approaching a seizure, and the highest predictive power of these alterations could be observed for a presumed preictal period lasting 15 minutes. This may indicate that a preictal state may be regarded as a gradual build-up of a process or a continuing reorganization of an epileptic network on a time scale of minutes prior to seizures. However, we also observed the frequency of occurrence of clusters to vary largely in time, which might indicate that cluster occurrences may also reflect physiologic activity (Bialonski and Lehnertz 2006). Taken together, it is not yet clear whether specific aspects of the dynamics of interaction clusters can be regarded as predictive of an impending seizure. Nevertheless, studying interaction networks with concepts from network theory can be regarded as a promising approach to gain deeper insights into the complex spatial-temporal dynamics of the ictogenic process in epileptic networks.

REFERENCES

Allefeld, C., and S. Bialonski. 2007. Detecting synchronization clusters in multivariate time series via coarse-graining of Markov chains. *Phys. Rev. E* 76:066207.

Arenas, A., A. Diaz-Guilera, J. Kurths, Y. Moreno, and C. Zhou. 2008. Synchronization in complex networks. *Phys. Rep.* 469:93–153.

Bialonski, S., and K. Lehnertz. 2006. Identifying phase synchronization clusters in spatially extended dynamical systems. *Phys. Rev. E* 74:051909.

Boccaletti, S., V. Latora, Y. Moreno, M. Chavez, and D.-U. Hwang. 2006. Complex networks: Structure and dynamics. *Phys. Rep.* 424:175–308.

Chung, F. R. K. 1997. *Spectral graph theory,* Vol. 92. In *CBMS regional conference series in mathematics.* Providence, RI: American Mathematical Society.

Coifman, R. R., and S. Lafon. 2006. Diffusion maps. *Appl. Comput. Harmon. Anal.* 21:5–30.

Fortunato, S. 2010. Community detection in graphs. *Phys. Rep.* 486:75–174.

Fortunato, S., and M. Barthélemy. 2007. Resolution limit in community detection. *Proc. Natl. Acad. Sci. USA* 104:36–41.

Lafon, S., and A. B. Lee. 2006. Diffusion maps and coarse-graining: A unified framework for dimensionality reduction, graph partitioning, and data set parameterization. *IEEE T Pattern Anal.* 28:1393–1403.

Lehnertz, K., S. Bialonski, M.-T. Horstmann, D. Krug, A. Rothkegel, M. Staniek, and T. Wagner. 2009. Synchronization phenomena in human epileptic brain networks. *J. Neurosci. Methods* 183:42–48.

MacQueen, J. B. 1967. Some methods for classification and analysis of multivariate observations. In *Proceedings of 5th Berkeley Symposium on Mathematical Statistics and Probability*, eds. M. L. Cam and J. Neyman, 281–297.

Milligan, G. W., and M. C. Cooper. 1985. An examination of procedures for determining the number of clusters in a data set. *Psychometrika* 50:159–179.

Mormann, F., R. Andrzejak, C. E. Elger, and K. Lehnertz. 2007. Seizure prediction: The long and winding road. *Brain* 130:314–333.

Mormann, F., K. Lehnertz, P. David, and C. E. Elger. 2000. Mean phase coherence as a measure for phase synchronization and its application to the EEG of epilepsy patients. *Physica D* 144:358–369.

Nadler, B., S. Lafon, R. R. Coifman, and I. Kevrekidis. 2006. Diffusion maps, spectral clustering and eigenfunctions of Fokker-Planck operators. In *Advances in neural information processing systems*, 955–962. Cambridge, MA: MIT Press.

Newman, M. E. J., and M. Girvan. 2004. Finding and evaluating community structure in networks. *Phys. Rev. E* 69:026113.

Provost, F., and T. Fawcett. 2001. Robust classification for imprecise environments. *Mach. Learn.* 42: 203–231.

Rummel, C., M. Müller, and K. Schindler. 2008. Data-driven estimates of the number of clusters in multivariate time series. *Phys. Rev. E* 78:066703.

Schelter, B., J. Timmer, and A. Schulze-Bonhage (eds). 2008. *Seizure prediction in epilepsy: From basic mechanisms to clinical applications*. Berlin: Wiley-VCH.

Shi, J., and J. Malik. 2000. Normalized cuts and image segmentation. *IEEE T Pattern Anal.* 22:888–905.

Theodoridis, S., and K. Koutroumbas. 2003. *Pattern recognition*, 2nd ed., San Diego: Elsevier.

Von Luxburg, U. 2007. A tutorial on spectral clustering. *Stat. Comput.* 17:395–416.

25 Preictal Spikes in the Hippocampus of Patients with Mesial Temporal Lobe Epilepsy

C. Alvarado-Rojas, M. Valderrama,
G. Huberfeld, and M. Le Van Quyen

CONTENTS

25.1 INTRODUCTION

Epileptic seizures are not randomly occurring events, but are instead the product of nonlinear dynamics in brain circuits, and are expected to be detectable with some antecedence (Le Van Quyen et al. 2001). Indeed, it has long been observed that the transition from the interictal state (far from seizures) to the ictal state (seizure) is not sudden and may be preceded from minutes to hours by pre-ictal clinical, metabolic, or electrical changes (Lehnertz et al. 2007). In recent years, new answers to this question have begun to emerge from quantitative analyses of the electroencephalogram (EEG). Most current approaches use quantities that draw inferences about the level of EEG complexity, such as an effective correlation dimension (Lehnertz and Elger 1998), correlation density (Le Van Quyen et al. 2001), or Lyapunov exponents (Sackellares et al. 1999). More recently introduced, bivariate measures that estimate dynamical interactions between two time series of two EEG channels such as phase synchronization or other measures for generalized synchronization were especially promising for seizure prediction (Mormann et al. 2003; Le Van Quyen et al. 2005). Even if the existence of a preictal period is not completely confirmed, these observations strongly suggest that distinct electrical characteristics can be determined between preictal and interictal periods. Nevertheless, until now, a physiological interpretation of these changes remains unknown. A few studies have hypothesized that changes in interictal epileptiform spikes can anticipate the pathophysiological recruitments that give rise to a seizure. However, in animal models of focal epilepsy, no consistent preictal changes were demonstrated in interspike interval distributions (Angeleri 1982) or in the occurrence of optimal spike rates (Elazar and Blum 1974). In humans, conflicting findings were reported: Wieser (1989) reported a decrease in spiking prior to seizure. Lange et al. (1983) observed that the spatial organization of spike patterns changes in a systematic fashion several minutes prior to temporal lobe seizures, whereas others (Lieb et al. 1978; Katz et al. 1991) found no change in spiking prior to seizures. Gotman et al. (1982) utilized prolonged telemetry recordings (11–16 days) in six patients whose medication levels were stable; spike rates were sometimes found to increase seconds prior to seizures, but the main finding was that repeated seizures caused a build-up in spike rate. It is apparent from the above finding that merely quantifying spike rates is unlikely to yield information capable of heralding seizure onset.

The principal objective of our study here is to investigate the appearance of certain spike morphologies in intracranial EEG signals from patients with mesial temporal lobe epilepsy (MTLE), caused by hippocampal sclerosis (HS). This injury, strongly associated with insults such as febril seizures in childhood, is characterized by a neuronal loss together with an astrogliosis in regions CA1 and CA3 of the hippocampus (Blümcke et al. 2002). Seizures that are poorly controlled by current pharmacotherapy require surgical extraction of the region that generates them, even if these areas are involved in episodic memory processes (Wieser 2004). In the case of HS, it is not known how circuit reorganizations or intrinsic abnormalities of neurons trigger the seizures. In a recent in vitro study in postoperative human epileptic tissue associated with HS, it has been shown that electrical epileptic activity is generated by the neuronal networks some hours after extraction (Huberfeld et al. 2008). The transition from the interictal state to seizure can be characterized by the progressive emergence of population events, the preictal spikes, that are distinct from interictal discharges (Huberfeld et al. 2009). These preictal events with different cellular properties have larger voltage amplitude and are generated at a single focus and spread faster, compared to interictal spikes. In particular, it has been demonstrated that they depend only on glutamatergic transmission and are preceded by pyramidal cell firing whereas interictal events are preceded by interneuron firing and depend on both glutamatergic and depolarizing GABAergic mechanisms (Huberfeld et al. 2009).

In our study, we examine whether specific spike morphologies can also be detected in vivo and whether they can be useful as electrophysiological markers of preictal changes. Hippocampal spikes are automatically detected first in long-term intracranial EEG signals using a singularity detection method based on the value of wavelet transform coefficients and their local maxima (Mallat and Hwang 1992). Once discontinuities are detected, different morphologic characteristics are evaluated including the initial and final time, the peak-to-peak voltage amplitude, the time duration, and the slope between the two peaks of each event. Values of amplitude and duration confirm the existence of a distinct population immediately before seizure onsets. In order to examine their behaviors over long periods of time, we develop a semiautomatic classification by selecting one preictal spike for reference and comparing it to others, using crosscorrelation estimation. We report that preictal spikes are not unique to the period immediately before the seizure onset, but they can appear up to 10 h before the onset. Furthermore, results suggested an evolution in amplitude in burst of preictal spikes, which presumably originate from the same source.

25.2 MATERIALS AND METHODS

25.2.1 EEG DATABASE AND PREPROCESSING

The algorithms are applied to intracranial electroencephalographic signals from five patients with MTLE. They all present at least one complex partial seizure during a recording period of 1 or 2 weeks prior to surgery. The recording system allows us to have a sampling rate of 400Hz and up to 64 depth electrodes implanted in different structures of the temporal lobe such as amygdala and hippocampus. In fact, the last region is an interesting zone because it is the most common focus of epileptic activity in MTLE. Additionally, bipolar montage is used in order to reduce the influence of the reference electrode. Electrodes at the hippocampus present higher electrical activity, so we discard those electrodes at different locations.

Because noise needs to be removed from the signal, a process of prefiltering is applied. A high-pass filter of 0.5 Hz and a notch filter of 50 Hz are used here. Frequency sampling-based filters of finite impulse response (FIR) with a zero-phase delay are preferred to avoid phase distortion.

25.2.2 Detection of Epileptic Spikes

25.2.2.1 Wavelet Transform

In order to represent the signal $f(t)$ in a two-dimensional space, we compute its real wavelet transform $Wf(u,s)$. This important mathematical tool measures time-frequency variations of the signal by correlating it with a zero-average function $\psi(t)$, called the mother wavelet (Mallat 1999). Real transform guarantees energy conservation if the wavelet satisfies the admissibility condition; it means that $\widehat{\psi}(0) = 0$ and $\widehat{\psi}(\omega)$ is continuously differentiable. A family of functions can be obtained by varying the scale s and position u of the function:

$$\psi_{u,s}(t) = \frac{1}{\sqrt{|s|}} \psi\left(\frac{t-u}{s}\right). \tag{25.1}$$

As the mother function, we select the second derivative of Gaussian function, also known as the Mexican hat, since its shape is similar to that of a spike. The normalized expression for the Mexican hat function and its Fourier transform is:

$$\psi(t) = \frac{2}{\sqrt{3\sigma} \pi^{1/4}} \left(\frac{t^2}{\sigma^2} - 1\right) \exp\left(\frac{-t^2}{2\sigma^2}\right) \tag{25.2}$$

$$\widehat{\psi}(\omega) = \frac{-\sqrt{8}\sigma^{5/2}\pi^{1/4}}{\sqrt{3}} \omega^2 \exp\left(\frac{-\sigma^2\omega^2}{2}\right). \tag{25.3}$$

The wavelet transform can be written as the inner product in $L^2(\mathbb{R})$ of a function $f(t)$ and the mother function $\psi(t)$:

$$Wf(u,s) = \langle f, \psi_{u,s}\rangle = \frac{1}{\sqrt{s}} \int_{-\infty}^{\infty} f(t)\psi^*\left(\frac{t-u}{s}\right) dt. \tag{25.4}$$

Since $\psi(t)$ is a real-valued even function, it means that $\psi^*(t) = \psi(-t)$, the above integral can be seen as a convolution operation (Chui 1997). In order to avoid redundancy, we use a modification of this transform—the dyadic wavelet transform—obtained as a discrete version with a scale sequence $s = \{2^j\}_{(j\in\mathbb{Z})}$ (Mallat 1999). Since the range of discrete scales covers all the frequency values, dyadic transform defines a complete representation of the signal.

$$Wf(u,2^j) = \langle f, \psi_{u,2^j}\rangle = \frac{1}{\sqrt{2^j}} \int_{-\infty}^{\infty} f(t)\psi^*\left(\frac{t-u}{2^j}\right) dt. \tag{25.5}$$

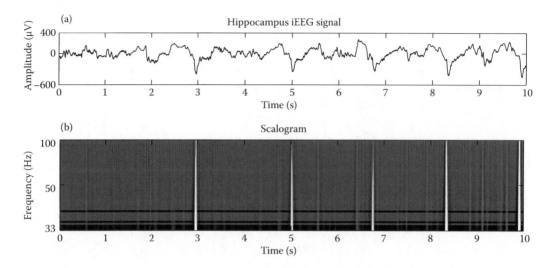

FIGURE 25.1 A segment of intracranial EEG signal recorded from a sclerotic hippocampus and the corresponding energy density in the time-frequency domain. Epileptic spikes are sharper than the background, so the wavelet transform coefficients are higher across different scales at these events (white color).

Furthermore, calculation of $|Wf(u,s)|^2$ corresponds to the energy on each Heisenberg box, determined by the region where the energy of $\psi_{u,s}(t)$ and $\hat{\psi}_{u,s}(\omega)$ is concentrated (Mallat 1999). This energy density, called a scalogram, can be visualized as a time-frequency plot (Figure 25.1).

25.2.2.2 Modulus Maxima

According to the Lipschitz regularity theorem, higher values of wavelet coefficients characterize sharper transitions of the signal across different scales. Localization and characterization of discontinuities can be achieved by finding the local maxima of its wavelet transform, the so-called wavelet transform modulus maxima (WTMM) (Mallat and Hwang 1992). Over a 30-s windowed transform of the original signal, we find each local maximum with the following expression (Mallat 1999):

$$\frac{\partial Wf(u_0,s_0)}{\partial u} = 0. \tag{25.6}$$

Since one spike represents a singularity, all are detected following the wavelet transform modulus maxima at fine scales. The threshold of the maxima value is computed by receiver operating characteristics (ROC) curves, used commonly in decision-making for finding a tradeoff between sensitivity and specificity of the detection method (Fawcett 2006). From each patient, 10 random segments are selected. Sensitivity and specificity are computed for 50 different thresholds, selecting the optimum point closer to a sensitivity and specificity of 1.

25.2.3 FEATURE EXTRACTION

We compute some relevant features of the spikes, which characterize the waveforms obtained by the detection algorithm, so that a classification process can be performed. Maximum and minimum amplitudes are first computed. Then, the time instants corresponding to the beginning and end of each pattern are calculated, with the first zero-crossing point before and after the two maximum magnitudes, respectively. Duration and peak-to-peak slope can also be extracted from these measures (Figure 25.2). As we regard the distribution of the features extracted, different waveforms are evident. Specifically, a distinct population of spikes with higher amplitude appears immediately before the seizure onset.

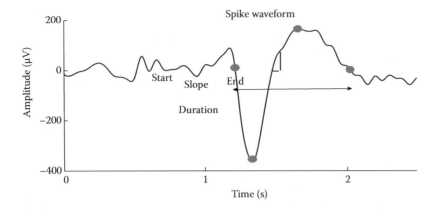

FIGURE 25.2   Standard epileptic spike waveform generated by abnormal electrical activity of a neuronal network. Characteristics extracted in order to perform the segmentation, peak amplitude, time duration, and sharpness suggest two different types of patterns labeled here as preictal and interictal.

25.2.4   SEMIAUTOMATIC CLASSIFICATION OF PREICTAL SPIKES WITH CROSSCORRELATION

In order to segment those preictal spikes, we develop a semiautomatic method that applies a similarity criterion with one of them as the reference waveform. Since all epileptiform patterns are detected and a distinct preictal waveform is evidenced, we select one of these spikes as reference and we compare it to others by crosscorrelation of the waveforms. Crosscorrelation is defined as follows:

$$f * g = \int_{-\infty}^{+\infty} f^*(-\tau)g(t-\tau)d\tau . \tag{25.7}$$

As a convolution operation, this integral makes it easier to compare two waveforms for different time-lags and it additionally avoids the effect of nonsignificant time delays for similar patterns (Figure 25.3). To decrease the number of variables, the maximum value of crosscorrelation is

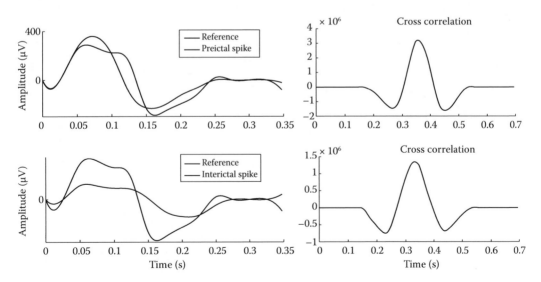

FIGURE 25.3   Comparison of the waveform of a preictal spike, previously selected as the reference by visual inspection, to the detected patterns by crosscorrelation operation. As expected, the maximum value of crosscorrelation is higher for similar patterns, preictal spikes, than for interictal ones.

TABLE 25.1
Contingency or Confusion Matrix for a Binary Classifier Showing Possible Results of the Detection Output

	Positive Condition Spike	Negative Condition Nonspike
Positive Detection	TP	FP
Negative Detection	FN	TN

Note: TP, true-positive; FP, false-positive; FN, false-negative; TN, true-negative.

selected as a reliable factor of decision. We fit the probability density function *t*-student to the data and label as preictal all patterns that appear to be far enough from the distribution mean (a probability of 0.05 is chosen).

25.3 RESULTS

Using the validation of an expert electroencephalographer (G. Huberfeld), we define a metric to evaluate the performance of the detection algorithm, which has a single output indicating whether the analyzed segment of the signal contains a spike or not. The pathological condition (the presence of one spike) and the result of the detection method form a contingency matrix (two-by-two) that shows all the possible combinations between them (Table 25.1). The elements along the major diagonal represent the correct decision. Furthermore, the true positive rate (the so-called sensitivity) can be seen as the relation between true spikes and positive detection of one event as an epileptic spike. Similarly, the specificity defines the relation between artifacts (nonepileptic activity) and nondetected events (Fawcett 2006), which is the capacity of the method to reject false events. In addition, we applied a ROC analysis to determine the optimal parameter. This technique consists of setting different thresholds and computing the true positive rate and false positive rate (1 − specificity). Thus, curves are constructed when plotting true versus false positive rates. To select the best threshold, it is necessary to find the nearest point near to the value (0,1) in the graph (Figure 25.4). Over five patients, we obtained optimal values for sensitivity of 85.3% and specificity of 90.4%.

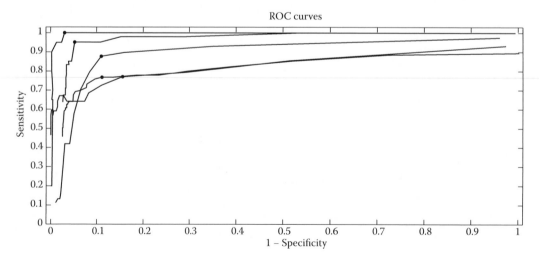

FIGURE 25.4 Receiver operating characteristic (ROC) curve showing the assessment metrics for the detection method, sensitivity, and specificity. For different thresholds, these values are calculated and the optimal threshold is then selected.

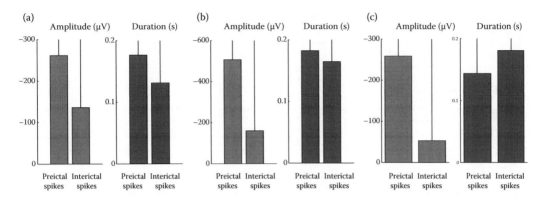

FIGURE 25.5 Bars representing (for three different patients) the voltage amplitude and time duration for the two kinds of patterns—preictal and interictal spikes. Higher amplitude is evident in preictal patterns and duration also appears to be longer for some patients.

Using these optimal parameters, our detection algorithm is first applied to short time periods (30 min) before seizures. The characteristics previously described are extracted for each detected spike and suggest the existence of a distinct morphology immediately before seizure onsets (in 9/9 seizures from 5/5 patients, Figure 25.5). Although the morphologies of these preictal spikes are different between patients, each individual patient presents very similar characteristics. The most relevant features for the distinction of different spike morphologies are the maximal and minimal amplitude of the spike and its duration. Using a criterion for separating preictal spikes from all other detected spikes based on the crosscorrelation of the waveforms, we then analyze long periods of time (10 h before seizures). A probability density function (t-student) is fitted to the data and a probability value of 0.05 is selected as the threshold. All spikes considered as preictal by the classification process are isolated and superposed (Figure 25.6a). We first observe that assumed preictal spikes are not unique to the period immediately previous to the seizure onset, but they could appear several hours before (Figure 25.7). In particular, preictal spikes are particularly expressed during the sleep periods. Furthermore, we observe that the amplitude of preictal spikes often evolve slowly in time. In particular, high-amplitude preictal spikes appear to have predecessors with the same

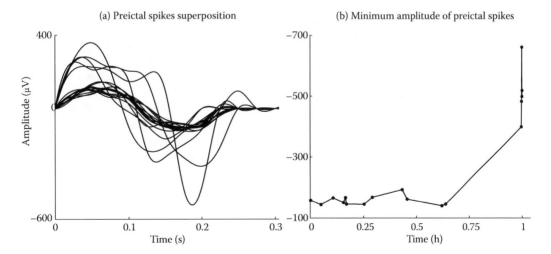

FIGURE 25.6 (a) Superposition of epileptic spikes identified as preictal after crosscorrelation classification is applied. (b) A different population of higher amplitude spikes is generally found to be preceded by similar preictal events of lower amplitude and suggests a time evolution.

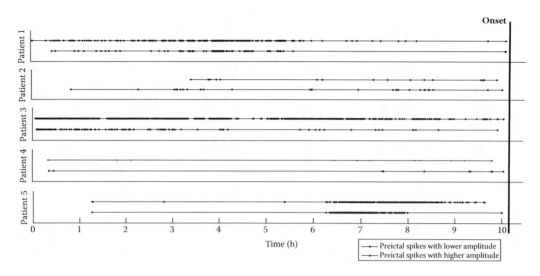

FIGURE 25.7 Timeline showing the distribution of preictal spikes classified by methods over a period of 10 h. Patterns of lower amplitude (possible predecessors) are discriminated from those of higher amplitude (dotted lines).

morphology but with lower amplitudes (Figure 25.6b). Although limited by the relatively small sample size, these observations are strengthened by the replicability of results across subjects.

25.4 DISCUSSION AND CONCLUSIONS

Following a statement attributed to Rasmussen, there is a general impression that there are specific red spikes, which are generated within epileptogenic tissue, and nonspecific green spikes, which are generated in more normal tissue (Engel et al. 2010). Recently, new in vitro data confirm that ictal-like activities generated by the human epileptic subiculum are preceded by novel preictal population activities (Huberfeld et al. 2009). In particular, these preictal spikes have specific cellular properties and depend on glutamatergic rather than mixed GABA/glutamatergic mechanisms. This novel population activity appears to be a crucial feature of ictogenesis in the subiculum of patients with temporal lobe epilepsy associated with HS. We ask here whether similar events could be detected in vivo in EEG recordings from the epileptic hippocampus of these patients. On the basis of five patients, we report that seizures from patients with temporal lobe epilepsies are reliably preceded by a distinct preictal form of epileptic spikes, interspersed with other interictal discharges. In particular, even if preictal spikes are strongly expressed at the beginning of the seizure, we observe that they are not unique to the onset period and can be present several hours before. Furthermore, it is suggested that sleep cycles have some influence on the presence of these patterns, as well as on the evolution of voltage amplitude, until preictal spikes appear. Therefore, we believe that the investigations of spike morphologies are of potential importance not only because they could reveal fundamental mechanisms responsible for epileptogenesis, but because these transients have potential clinical value as putative biomarkers of preseizure states in combination with other electrographic patterns (e.g., high-frequency oscillations).

REFERENCES

Angeleri, F., S. Giaquinto, and G. F. Marchesi. 1982. Temporal distribution of interictal and ictal discharges form penicillin foci in cats. In *Synchronization of EEG activity in epilepsies*, eds. H. Petsche and M. A. B. Brazier, 221–234. New York: Springer.

Blümcke, I., M. Thom, and O. D. Wiestler. 2002. Ammon's horn sclerosis: A maldevelopmental disorder associated with temporal lobe epilepsy. *Brain Pathol.* 12:199–211.

Chui, C. K. 1997. *Wavelets: A mathematical tool for signal processing.* Philadelphia: SIAM.

Elazar, Z., and B. Blum. 1974. Interictal discharges in tungsten foci and EEG seizure activity. *Epilepsia* 15:599–610.

Engel, J., A. Bragin, R. Staba, and I. Mody. 2009. High-frequency oscillations: What is normal and what is not? *Epilepsia* 50:598–604.

Fawcett, T. 2006. An introduction to ROC analysis. *Pattern Recognit. Lett.* 27:861–874.

Gotman, J., J. Ives, P. Gloor, A. Olivier, and L. F. Quesney. 1982. Changes in interictal EEG spiking and seizure occurrence in humans. *Epilepsia* 23:432–433.

Huberfeld, G., L. Menendez de la Prida, S. Clemenceau, J. Pallud, I. Cohen, M. Le Van Quyen, M. Baulac, and R. Miles. 2009. Pyramidal cells and interneurons interactions in the initiation of ictal discharges in human epileptic tissue in vitro. Paper presented at the 28th International Epilepsy Congress, June 28–July 2, 2009, Budapest, Hungary.

Katz, A., D. A. Marks, G. McCarthy, and S. S. Spencer. 1991. Does interictal spiking rate change prior to seizure? *Electroenceph. Clin. Neurophysiol.* 79:153–156.

Lange, H. H., J. P. Lien, J. Engel, and P. H. Crandall. 1983. Temporo-spatial patterns of preictal spike activity in human temporal lobe epilepsy. *Electroenceph. Clin. Neurophysiol.* 56:543–555.

Le Van Quyen, M., J. Martinerie, V. Navarro, et al. 2001. Anticipation of epileptic seizures from standard EEG recordings. *The Lancet* 357:183–188.

Le Van Quyen, M., J. Soss, V. Navarro, et al. 2005. Preictal state identification by synchronization changes in long-term intracranial EEG recordings. *Clin. Neurophysiol.* 116:559–568.

Lehnertz, K., and C. E. Elger. 1998. Can epileptic seizures be predicted? Evidence from nonlinear time series analysis of brain electrical activity. *Phys. Rev. Lett.* 80:5019–5022.

Lehnertz, K., M. Le Van Quyen, and B. Litt. 2007. Seizure prediction. In *Epilepsy: A comprehensive textbook,* 2nd ed., eds. J. Engel and T. A. Pedley, 1011–1024. Philadelphia: Lippincott Williams & Wilkins.

Lieb, J. P., S. C. Woods, A. Siccardi, P. Crandall, D. O. Walter, and B. Leake. 1978. Quantitative analysis of depth spiking in relation to seizure foci in patients with temporal lobe epilepsy. *Electroenceph. Clin. Neurophysiol.* 44:641–663.

Mallat, S., and W. Hwang. 1992. Singularity detection and processing with wavelets. *IEEE Trans. Inf. Theory* 38(2):617–643.

Mallat, S. 1999. *A wavelet tour of signal processing.* San Diego: Academic Press, Inc.

Mormann, F., R. G. Andrzejak, T. Kreuz, et al. 2003. Automated detection of a pre-seizure state based on a decrease in synchronization in intracranial EEG recordings from epilepsy patient. *Phys. Rev. E, Statistical, Nonlinear, and Soft Matter Physics* 67:21912.1–21912.10.

Sackellares, J. C., L. D. Iasemidis, D. S. Shiau, R. L. Gilmore, and S. N. Roper. 1999. Epilepsy—when chaos fails. In *Chaos in the brain?* eds. K. Lehnertz, and C. E. Elger, 10–12. Singapore: World Scientific.

Wieser, H. G. 1989. Preictal EEG findings. *Epilepsia* 30:669.

Wieser, H. G. 2004. ILAE Commission Report. Mesial temporal lobe epilepsy with hippocampal sclerosis. *Epilepsia* 45:695–714.

Section V

The State of Seizure Prediction:
Seizure Generation

26 Microanalysis and Macroanalysis of High-Frequency Oscillations in the Human Brain

B. Crépon, M. Valderrama, C. Alvarado-Rojas, V. Navarro, and M. Le Van Quyen

CONTENTS

26.1 INTRODUCTION: DR. JEKYLL OR MR. HYDE?

High-frequency oscillations (HFOs) greater than 40 Hz are thought to underlie several physiological functions of the normal brain (Buzsaki and Draguhn 2004) but are also a characteristic feature of the epileptic brain and may, for example, play a causal role in the initiation of focal epileptic seizure (Köhling et al. 2000; Traub et al. 2001; Bragin et al. 1999, 2002a; Grenier et al. 2003) or in epileptogenesis (Bragin et al. 2004; Khalilov et al. 2005). Thus it is important to determine the differences as well as the common basic underlying mechanisms involved in the production of such normal and pathological activities (Le Van Quyen et al. 2006).

In vitro and in vivo studies and recordings in humans have revealed the presence of several different types of HFOs in different frequency bands (Figure 26.1). One subgroup is gamma oscillations (40–120 Hz). In recent years, gamma oscillations have been extensively investigated because they are hypothesized to be important in a range of waking functions and reflect an increased alertness, such as sensory binding (Singer and Gray 1995; Lachaux et al. 2005), attention (Fries et al. 2001), and encoding and retrieval of memory traces (Montgomery and Buzsáki 2007). For all these functions, gamma activities are thought to provide a temporal structure relative to which the activities of individual neurons are organized in a millisecond timescale across distributed neural networks (Singer and Gray 1995). In particular, gamma (40–100 Hz) oscillations can be observed in the CA1 and CA3 regions of the hippocampus in rats, especially during exploratory behavior (Bragin et al. 1995). Furthermore, researchers have shown that the gamma oscillations in the entorhinal cortex are coupled to the oscillations in the dentate hilar regions and that this coupling provides for effective neuronal communication and synaptic plasticity in the perforant pathway (Chrobak and Buzsaki 1998). Networks of fast-spiking inhibitory interneurons connected by gap junctions are known to

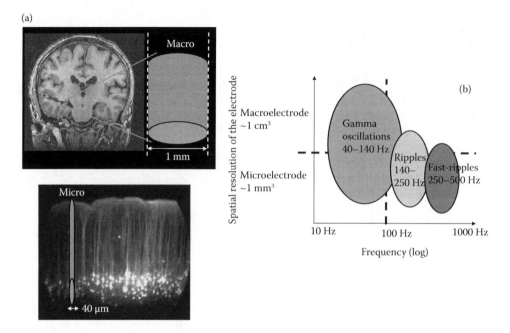

FIGURE 26.1   (See color insert.) Macrorecording and microrecordings of HFOs in the human brain. (a) Conventional intracranial depth macroelectrodes (e.g., from Ad-Tech Medical Instruments, Racine, WI) have a contact area around 1 mm², which presumably record the activity of several cubic centimeters of cortex of brain tissue. Microelectrodes are capable of probing the fine structure of cerebral cortex and have a typical size of 30–40 µm in diameter (e.g., Ad-Tech Medical Instrument). (b) The three groups of HFOs (gamma, ripples, and fast ripples) can be recorded with both micro- and macroelectrodes.

be crucial for the genesis of these oscillations by providing high-frequency synchronous inhibitory postsynaptic potentials (IPSPs) to local excitatory neurons (Bartos et al. 2007; Cardin et al. 2010).

In addition to normal brain function, oscillations in the gamma range are also often seen in intracranial EEG recordings at seizure onset of partial seizures in patients with temporal and extratemporal epilepsy and are highly localized in the seizure onset zone. The reported frequencies on intracranial EEG recordings varied and included 20–80 Hz (Alarcon et al. 1995), 40–120 Hz (Fisher et al. 1992), 60–100 Hz (Worrell et al. 2004), 70–90 Hz (Traub et al. 2001), and 80–110 Hz (Allen et al. 1992). Furthermore, it has been shown that removal of the onset zone from which gamma activities were localized resulted in good seizure control for localization-related epilepsy patients (Alarcon et al. 1995). Similar results were reported by another group who showed that 64% of infantile spasms are characterized by rhythmic gamma activity of 50–100 Hz (Kobayashi et al. 2004). Additionally, in neocortical epilepsy, gamma oscillations were also intermittently present throughout the interictal period and a significant increase in interictal gamma was reported several minutes prior to onset of an epileptic seizure (Worrell et al. 2004), suggesting that gamma may be useful for identifying periods of increased predisposition to clinical seizures.

One other subgroup of HFOs is the ripple (140–200Hz), which was originally described as physiological activity in the hippocampus. Hippocampal ripples are usually associated with sharp waves (Buzsáki et al. 1992) that occur in rodent hippocampal–entorhinal networks during immobility or slow-wave sleep (O'Keefe and Nadel 1978; Buzsáki et al. 1992, 2003). Sharp-wave ripple complexes originate from the synchronized firing of CA3 cells and spread downstream as spatially coherent oscillations along the CA1-subicular-entorhinal axis (Chrobak and Buzsáki 1996). This coactivation of hippocampal and neocortical pathways are thought to be crucial for memory consolidation processes, during which memories are gradually translated from short-term hippocampal to longer-term neocortical stores (Buzsáki 1989; Lee and Wilson 1992; Wilson and McNaughton 1994). On the cellular level, ripples reflect the summed activity of pyramidal cells and interneurons coupled to each oscillatory cycle. In support of this, recent work has indicated that both pyramidal neurons and various classes of interneurons display their highest discharge probability around the negative peak of the ripple cycle (Ylinen et al. 1995; Klausberger et al. 2003; Le Van Quyen et al. 2008). In particular, the currents responsible for ripples—maximal at the cell body layer—are believed to correspond to synchronized IPSPs on the somata of pyramidal cells (Ylinen et al. 1995).

Closely related to ripples, the fast ripple (FR) is a third subgroup with a frequency of 250–500 Hz. FRs were first recorded during interictal periods with microelectrodes from the hippocampus and entorhinal cortex of patients with mesial temporal lobe (MTL) epilepsy (Bragin et al. 1999, 2002b; Staba et al. 2004). Like ripples, FRs are more strongly expressed during quiet wakefulness and slow-wave sleep. Similar oscillations are detected in rodent models of mesial temporal lobe epilepsy (MTLE) (Bragin et al. 2002a; Foffani et al. 2007). Unlike ripples, they are recorded exclusively from epileptogenic areas of the lesioned hippocampus and occur at higher frequencies in the epileptogenic MTL containing atrophic hippocampi and amygdala than in contralateral MTL (Staba et al. 2004). Therefore, they provide a promising diagnostic marker for epileptogenic areas associated with hippocampal sclerosis and consequent synaptic reorganization. FR differ from ripples not only in their frequency range but by the fact that they can be recorded from dentate gyrus where normal ripples never occur. They appear to be localized exclusively in areas capable of generating spontaneous seizures and they are generated by small discrete clusters of neurons, whereas generation of ripples is much more diffuse (Bragin et al. 2000, 2002a). In contrast to ripples, which are dependent on inhibitory mechanisms, fast ripples appear to reflect summated action potentials of bursting principal neurons (Bragin et al. 2002a), possibly coupled by gap junctions (Roopun et al. 2010). It is therefore possible that loss of critical inhibitory influences in these localized epileptogenic regions could mediate the transition of normal ripples to pathological fast ripples. Nevertheless, the distinction between ripples and FR is not as clear as previously suggested (Engel et al. 2009). For example, a study tracing the development of fast ripples after intrahippocampal kainate injection in rats documented the appearance of ripple frequency oscillations in dentate gyrus along with fast ripple frequency oscillations in this structure and elsewhere in hippocampus and parahippocampal regions (Bragin et al. 2004). It was concluded that the dentate ripples were also pathological markers of epileptogenesis, presumably reflecting the same mechanisms as FR, rather than that of normal ripples. This finding raises the question of whether at least some ripple oscillations outside dentate gyrus in patients with MTLE and animal models of this disorder might also be pathological.

26.2 MICRORECORDING AND MACRORECORDING OF SLEEP-RELATED GAMMA OSCILLATIONS IN THE HUMAN CORTEX

Gamma oscillations, usually associated with waking functions, are strongly expressed in cortical networks during slow-wave sleep (SWS). In particular, in vivo (Steriade et al. 1996; Grenier et al. 2001; Isomura et al. 2006; Mukovski et al. 2007; Mena-Segovia et al. 2008) and in vitro (Dickson et al. 2003; Compte et al. 2008) recordings in the neocortex indicate that gamma oscillations occur during "UP" states, i.e., rhythmic cycles of suprathreshold membrane potential depolarizations occurring synchronously in large neuronal populations and reflected on electroencephalography (EEG) recordings as large-amplitude slow waves (Steriade et al. 1993). Network dynamics during UP states have been proposed to be equivalent to those observed during the waking state (Destexhe et al. 2007), and this has provided useful and testable predictions about the mechanisms underlying rapid changes in functional connectivity across cortical networks during waking behavior (Luczak et al. 2007; Haider and McCormick 2009). In particular, during these highly dynamic states, gamma frequency fluctuations in inhibitory and excitatory synaptic potentials determine the precise probability and timing of action potential generation, even at the millisecond level (Hasenstaub et al. 2005). Nevertheless, the full details of their large-scale coordination across multiple cortical networks are still unknown. Furthermore, it is not known whether oscillations with similar characteristics are also present in the human brain. In a recent study, made in collaboration with the Department of Neurology at UCLA (R. Staba, A. Bragin, and J. Engel), we have examined the existence of gamma oscillations during seizure-free and polysomnographically defined sleep-wake states using microelectrodes (Figure 26.1a), recorded in parallel with macroscopic scalp EEG (Le Van Quyen et al. 2010). In a group of nine patients, we observed that gamma was present across all stages of vigilance, but that gamma was less frequent during wakefulness, stage 1, and REM, and more prominent during stage 2, and that high rates occurred during sleep stages 3 and 4 (Figure 26.2a). The highest rates of gamma episodes were recorded in the parahippocampal gyrus (origin of the perforant path input to the hippocampal formation in the medial temporal lobe) with 13.3 ± 6.4 gamma oscillations per minute during SWS. Figure 26.2b through d illustrates typical patterns of detected gamma activities recorded across 30 microelectrodes in the right and left posterior parahippocampal gyri during a sleep stage 2. In either the raw signals or those filtered between 40 and 120 Hz, large-amplitude fast sinusoidal waves appeared in many channels as discrete events that were clearly distinguishable from background activity (Figure 26.2b). As best seen in the envelope amplitude of the filtered signals in the gamma range (Figure 26.2c, bottom), these occurrences of gamma activity appeared simultaneously between homotopic sites, forming large spatiotemporally coherent patterns of increased activity separated by equally coherent periods of little activity. The wavelet transformed-energy scalogram, which represents the spectral energy with respect to time and frequency, more clearly illustrates gamma activity in homotopic parahippocampal sites between hemispheres (Figure 26.2c). Prominent gamma frequency oscillations, with distinct narrow band peaks centered around 70–80 Hz, were observed and lasted a few hundred milliseconds. These patterns were correlated with positive peaks of EEG slow oscillations (here a K-complex, fulfilling standard polysomnographic criteria of duration/amplitude and followed by standard sleep spindles around 12 Hz) and marked increases in local cellular discharges (Figure 26.2b), which suggests that they were associated with cortical UP states. Over our group of patients, we confirmed that distinct narrow band oscillations in the low (40–80 Hz) and high (80–120 Hz) gamma ranges were strongly expressed during seizure-free SWS and were correlated with positive peaks of EEG slow oscillations and increases in local cellular discharges. These data confirm and extend earlier animal studies reporting that gamma oscillations are transiently expressed during normal sleep UP states.

In another study on 20 patients (Valderrama et al. 2010), we reported that gamma oscillations can also be recorded with conventional intracranial macroelectrodes (contact area around 1 mm², Figure 26.1a). Using sleep staging from simultaneous recorded scalp EEG signals, we confirmed that gamma oscillations (30–120 Hz) were prominent during SWS and can be observed in the majority of analyzed intracranial contacts (53%, 390 of a total of 740). In Figure 26.3a, the simultaneous evolution of the EEG power and frequency rate of gamma activities is presented for one entire night of a representative patient. A clear increment of the power in the slow oscillation frequency range (~ 0.3–2 Hz) can be seen as the subject enters into the deepest stages of sleep (stages 2–4). This increased rate of slow waves is in turn accompanied by an increment on the density of gamma events, measured as the number of detected events per time interval (20 s), presented here for seven intracranial contacts (Figure 26.3a, bottom). For our group of patients, the mean density of gamma events during SWS periods was 1.156 ± 0.46. To illustrate the relationship between the gamma activity and the presence of slow waves, Figure 26.3b presents a segment during sleep stage 4. A predominant slow activity can be seen in raw signals for both surface and depth electrodes. When regarding the evolution of the gamma-range filtered signals, near-periodic bursts of increased amplitude can be observed in all intracranial contacts (but not in the scalp one) at the depth-negative phase (respectively, surface-positive phase) of the slow oscillations in the raw signals. In either of the raw signals (without any previous filtering or preprocessing), or in those filtered between 40 and 120 Hz, large-amplitude fast sinusoidal waves appeared in all channels as discrete events that were clearly distinguishable from background activity (Figure 26.3c). For an individual recording

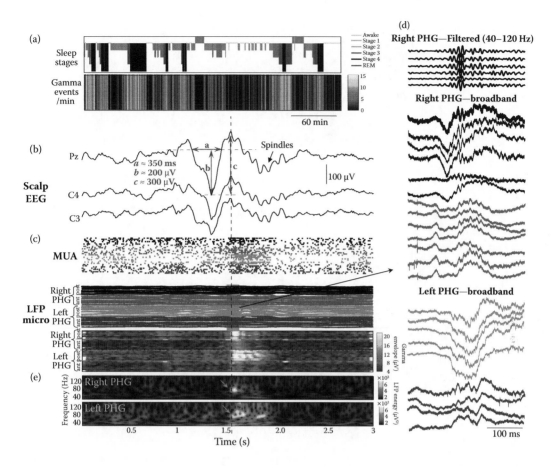

FIGURE 26.2   (See color insert.) Gamma oscillations recorded with microelectrodes during sleep. (a) All-night detections of gamma patterns (here recorded in the posterior parahippocampal gyrus) during all physiological stages from quiet wakefulness, sleep stage 1–4, and REM. The number of detected gamma patterns per minute was indicated in colors as function of sleep stages. (b) Gamma patterns recorded during a K-complex in sleep stage 2. (top) A K-complex, fulfilling standard polysomnographic criteria of duration (*a*) and amplitude (*b–c*), was recorded from several scalp EEG channels (here C3, C4, and Pz). (c) Corresponding recording of 30 microelectrodes in the field potentials (LFP) of the right and left posterior parahippocampal gyri (PHG, *ant*: anterior part and *post*: posterior part) indicates that gamma episodes in the depth were temporally correlated with positive peaks of EEG slow waves and appeared simultaneously between homotopic sites. (d) Display of a single gamma episode appearing simultaneously, in either the raw signals or those filtered between 40 and 120 Hz (top), in the right and left posterior parahippocampal gyri. (e) Corresponding wavelet transforms of two homotopic sites revealing nearly simultaneous gamma oscillations with distinct narrow band frequencies around 70–80 Hz (green arrows).

contact, gamma oscillations appeared on subsequent cycles of the slow oscillation and were composed of pure oscillatory bursts in narrow frequency bands in the low (~ 30–50 Hz) gamma range, high (~ 60–120 Hz) gamma range, or in broad-band events composed of a few mixed frequencies (Figure 26.3c).

What can be the physiological function of these gamma oscillations during sleep? Slow-wave sleep, and particularly the cortical slow oscillation, is important for consolidating memory traces acquired during waking (Huber et al. 2004; Marshall et al. 2006). Gamma oscillations during sleep might also support these consolidation processes (Molle et al. 2004; Sejnowski and Destexhe 2000), thus complementing the roles of gamma oscillations in encoding and retrieval of memory traces

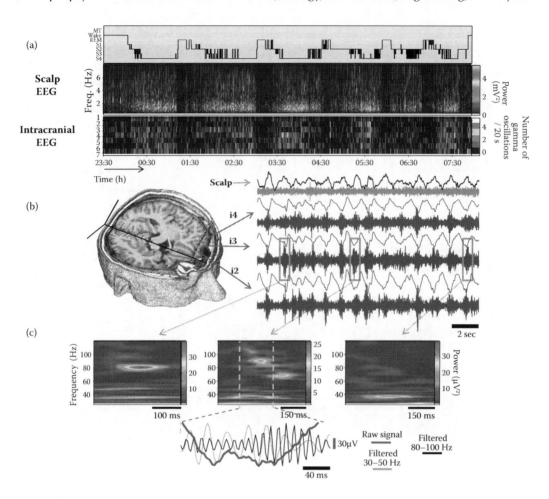

FIGURE 26.3 **(See color insert.)** Gamma oscillations recorded with macroelectrodes during sleep. (a) All-night hypnogram of a single subject (top), evolution of the power spectrum of one scalp electrode (middle), density of gamma events (30–120 Hz) detected on seven intracranial contacts located in the frontal and temporal cortex (bottom). (b) 3D MRI reconstruction of the patient's brain presenting one depth electrode and its corresponding six intracranial contacts in the left orbitofrontal gyrus (left). Simultaneous recordings during deep sleep (S4 stage) of slow waves at the scalp level and gamma activities (30–120 Hz, filtered signal presented below each raw one) of three depth contacts (right). (c) Time-frequency representations of the three bursts of gamma activity illustrating examples of pure oscillations with a narrow frequency in the high gamma range (left), in the low gamma range (right), and complex oscillations composed of a few mixed low and high frequencies (middle), as also shown in the filtered signals (bottom).

during wakefulness (Montgomery and Buzsáki 2007; Sejnowski and Paulsen 2006). Furthermore, consistent with unit studies in animals (Isomura et al. 2006; Wolansky et al. 2006) or functional neuroimaging in humans (Dang-Vu et al. 2008), we also reported that gamma patterns have their strongest rates of detection and earliest activations in the parahippocampal gyrus. As a major relay between the hippocampus and the neocortex, the parahippocampal cortex may play a critical role in sleep-associated consolidation of memory by transferring information between hippocampus and neocortex (Clemens et al. 2007).

We have reported here the presence of neocortical gamma oscillations during seizure-free periods and have shown that these oscillations coincide with nonepileptic activities, such as slow sleep oscillations. Nevertheless, gamma oscillations in the same frequency range are often seen in intracranial EEG recordings throughout the interictal period and at seizure onset of neocortical epilepsy; furthermore, they are highly localized in the seizure onset zone (Worrell et al. 2004). In addition to these observations, experimental models have led to the suggestion that they could be involved in the initiation of seizures (Grenier et al. 2003). This raises again the question about the relations between physiological and epileptic gamma oscillations. As suggested by others (Worrell et al. 2004), we also believe that the increased high-frequency activity seen at the onset of some neocortical seizures is the result of an aberration of the same physiological mechanisms underlying network activations in SWS. Indeed, as already reported in patients with frontal-lobe seizures (Della Marca et al. 2007), epileptic gamma oscillations appear to be closely related to sleep, especially in the prefrontal and orbitofrontal cortex. Strong gamma oscillations may be driven in the epileptic cortex by transient paroxysmal neuronal depolarization, facilitated by the strong potential fluctuations that occur during the slow oscillation and which are known to evolve into seizures (Bragin et al. 2007). Indeed, epileptic processes are well known to be characterized by a general tendency of hypersynchronization and exaggeration of normal oscillations, as seen, for example, in the epileptic facilitation of sleep spindles (Steriade et al. 1994) or ripples (Clemens et al. 2007). This aberrant expression implies the same physiological mechanisms underlying memory consolidation, leading to a progressive reinforcement of local epileptogenic circuits (Worrell et al. 2004).

26.3 MICRORECORDING AND MACRORECORDING OF HFOS IN THE HUMAN MEDIAL TEMPORAL LOBE

Discrete ripples (120–250 Hz) and FR (250–500 Hz) were first described in humans with microelectrodes (40-μm diameter) that record a local field potential from a volume of around 1 mm^3 (Bragin et al. 1999). The ripples were more prominent during NREM sleep (Staba et al. 2004). FR occurrences are significantly associated with the seizure onset zone (SOZ) (Bragin et al. 2002; Engel et al. 2009). Figure 26.4a illustrates an example of FR with a peak power at 250 Hz recorded in the epileptic entorhinal cortex (Bragin et al. 2002b). Note, as observed in a large percentage of cases (~85%), this FR is associated with a sharp wave. To better subclassify HFOs in different frequency bands and in region-specific patterns, in collaboration with the Department of Neurology at UCLA (R. Staba, A. Bragin, and J. Engel), we have systematically detected all HFOs occurring during SWS between 40 and 600 Hz in the hippocampus and adjacent parahippocampal gyri, ipsilateral to side of seizure onset, $n = 5$ patients (Le Van Quyen et al. 2010). Figure 26.4b through c illustrates the occurrence times for a single patient and the detection histogram in our group of patients. In the hippocampus, analysis of frequencies at maximum power showed a multimodal distribution in the high-frequency range with three main peaks around 50 Hz, 190 Hz, and 360 Hz. These data confirm that three groups of independent HFOs co-exist: gamma events between 40 and 130Hz, ripples between 130 and 250 Hz, and FR between 250 and 600 Hz. Both ripples and FR were strongly expressed in the epileptogenic hippocampus during SWS (ripple rate of 2.8 ± 2.3/min; FR rate of 1.3 ± 1.1/min). At a slow rate (0.27 ± 0.12/min), oscillations in the low gamma range around 50 Hz can also be identified in the hippocampus of all investigated patients. In contrast, in the parahippocampal gyri of all investigated patients, only one single frequency band was dominant in the gamma range between 40 and 130 Hz (Figure 26.5a through b). In two of five cases and consistent with the spatial extent of SOZ, ripples and FR occurred at a low rate (1.13 ± 0.58/min) at discrete locations involving a few channels; these were often temporally correlated with the occurrences of gamma oscillations. These data therefore indicated differences in HFO expressions between the hippocampus and surrounding medial-temporal cortices. Histologically, the cortical structure of the parahippocampal gyrus is six-layered (like the neocortex), but the hippocampus is three-layered (archicortex). Therefore, although all three groups of HFOs can be generated in the hippocampus or parhippocampal gyrus, our observations suggest that, during SWS, ripples and FR are more strongly generated by hippocampal networks that have a common archaic phylogenetic origin. In contrast, gamma oscillations are preferentially expressed by cortical networks.

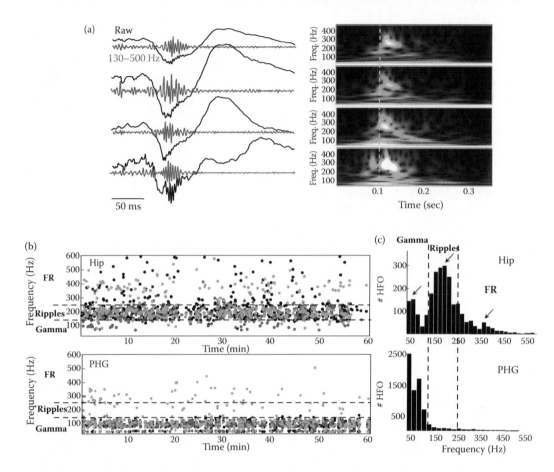

FIGURE 26.4 **(See color insert.)** High-frequency oscillations with microelectrodes in the epileptic medial temporal lobe. (a) FR recorded with 4 bundles of microelectrodes in the epileptic entorhinal cortex (left) with their corresponding time-frequency representations (right). (b) Detection times and mean frequencies of high-frequency oscillations (gamma, ripples, and FR) between 40 and 600 Hz (three individual channels are depicted by different colors) for a single patient in the epileptic hippocampus (Hip) and adjacent parahippocampal gyrus (PHG) and (c) frequency histogram over five subjects. Note, in the hippocampus, a multimodal distribution in the high-frequency range with three main peaks in the gamma, ripples, and fast ripple range. In contrast, in the parahippocampal gyrus, only one single frequency band was dominant in the gamma range between 40 and 130 Hz.

Recently, Urrestarazu et al. (2007) convincingly showed that interictal HFOs may be detected and used to localize the SOZ in intracranial records obtained from broad-band depth electrodes of small size (surface area 0.9 mm²). Using hybrid depth electrodes containing microwires and macroelectrodes, Worrell et al. (2008) confirmed that HFOs can be recorded in the hippocampal SOZ using standard clinical macroelectrodes of contact area greater than 1 mm²; these electrodes presumably record the activity of ~20 mm² of brain tissue (Worrell et al. 2008). They reported that the distribution of HFO frequencies recorded with macroelectrodes falls off more rapidly than in records made with microwires; they also observed HFO > 200 Hz only rarely. Due to the low rate of detection of HFOs using clinical macroelectrodes, the authors could not draw reliable conclusions on the spatial extent of HFOs in epileptic patients. In a recent study (Crépon et al. 2010), we confirmed that HFOs could be detected when signals from conventional intracranial EEG macroelectrodes were sampled at 1024 Hz. The mean frequency of all selected events was 261 ± 53 Hz, the mean amplitude was 11.9 ± 6.7 μV, and mean duration was 22.7 ± 11.6 ms. They occurred at a

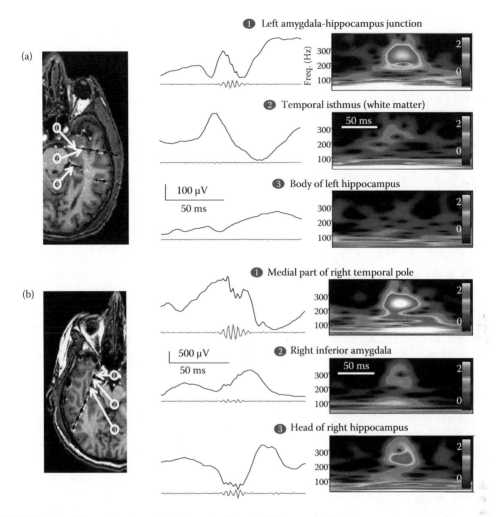

FIGURE 26.5 **(See color insert.)** (a) A typical local HFO recorded from one macroelectrode at the left amygdalo-hippocampal junction (top). Note the HFO was temporally correlated with a sharp wave which can also be seen in the two adjacent contacts without an HFO (middle bottom). (b) A representative example of an HFO recorded from three macroelectrodes contacts. Note that the HFO amplitude is minimal at contact 2 situated between contacts 1 and 3, suggesting a synchronization phenomenon rather than simple diffusion.

rate of 1.6 per min. About 95% of HFOs were nested within a spike or a sharp wave, typically just after the maximal deflection. Therefore, these HFOs have similar properties in terms of frequency, amplitude, and occurrence to those recorded with microelectrodes; they are also coincident with sharp waves and their localizations are similar. Most HFO patterns (80% of the cases) were limited to a single recording contact (Figure 26.5a). Nevertheless, in 20% of the cases, HFOs were recorded from two or three contacts (with a separation of 10 mm between adjacent contacts on the same electrode, Figure 26.5b). These data suggest that networks that generate HFOs, or those through which they propagate, can extend over volumes of cm^3.

Concerning the correlation between HFOs and the seizure onset zone, for each of the nine patients suffering from MTLE, HFOs were generated in the SOZ or in the earliest propagation zone. In particular, HFOs can be recorded in the archeocortex of amygdala or hippocampus as well as the mesocortex of temporal pole or para hippocampal gyrus. In contrast, they were never found outside the mesiotemporal structures or in the healthy amygdala or hippocampus. Specifically HFOs were not detected in patients suffering from neocortical epilepsy. These results imply that neocortical epileptic networks do not generate HFOs, which contradicts other recent studies (Jacobs et al. 2008, 2009). A reason for these differences may result from the nature of the epileptic syndrome. In our patients, no lesions were detected by MRI, whereas a heterotopia or dysplasia were identified in the previously mentioned studies. Another reason may lie in the electrode size. HFOs were first recorded with microelectrodes, then with electrodes with an active surface area less than 1 mm^2; HFOs were, however, recorded less easily with larger contacts (Worrell et al. 2008).

Taken together, conventional intracranial EEG can effectively detect and characterize HFOs at frequencies above 200 Hz. On the basis of these observations, we postulate that HFOs recorded by macroelectrodes in the medial temporal lobe reflect the same local, highly synchronous oscillations as those recorded with microelectrodes. This conclusion is supported by the co-occurrence of inter-ictal FR and HFOs in simultaneous records made with microwires and macroelectrodes (Worrell et al. 2008).

26.4 CONCLUSION

At large scale, EEG signals represent the superposition of various neural activations, both physiological and pathophysiological, constituting a dynamic system that is additionally influenced by external and internal factors such as sensory input, circadian rhythms, or pharmacological therapy. An additional problem is the signal strength of these components. For any activity to be directly visible in surface EEG, synchronization of a large neural network is necessary. For example, an epileptic transient is only visible when at least 6–10 cm^2 of cortex or even larger neuron populations are synchronously activated (Tao et al. 2005). Microelectrode observations in epileptic patients have suggested HFOs may initiate in microdomains 1 mm^3 or even smaller, and similar observations have been made in animal models. Although the spatial sampling of intracranial EEG is probably not sufficient to resolve the individual neural elements of the assemblies generating these submillimeter HFOs, several recent observations suggested that some of these high-frequency events can be recorded with macroelectrodes. These HFOs may represent the summation of multiple assemblies oscillating in phase at broadly tuned frequencies in the high-frequency range. The observations made by our group reinforce this by demonstrating that a standard macroelectrode can record gamma oscillations (Valderrama et al. 2010) or $FR > 200$ Hz (Crépon et al. 2010). Furthermore, these data suggest that HFOs may sometimes be recorded from two or three contacts (with a separation of 10 mm between adjacent contacts on the same electrode) and may result from a neuronal synchrony manifest over volumes of 1 cm^3. This is particularly the case for patients with mesial TLE. These findings suggest that events specific to MTLE produce neural networks that can generate hypersynchronous, high-frequency activity (>200Hz). They may involve the synaptic reorganization that is known to occur during secondary epileptogenic processes (Bragin et al. 2004; Khalilov et al. 2005; Le Van Quyen et al. 2006). HFOs of a similar frequency content, around 240 Hz, have also been identified in patients just before seizure onset at sites close to seizure initiation (Jirsch et al. 2006) as well as in experimental models of hippocampal seizures (Traub et al. 2001; Khosravani et al. 2005). In contrast, seizures arising from neocortical structures have been shown to begin with lower-frequency fast oscillations in the gamma range of 30–100 Hz (Allen et al. 1992; Alarcon et al. 1995; Worrell et al. 2004). This suggests that neuronal networks specific to an epileptic mesial temporal cortex produce hypersynchronous events at frequencies above 200 Hz not only between seizures, but also at seizure onset. Thus, the possibility to record HFOs at frequencies greater than 200 Hz with macroelectrodes has important clinical implications, notably in identifying periods of increased predisposition to clinical seizures in MTL epilepsy. From a practical point of view, although important scientific information on HFOs can be gained from in vivo microelectrode recordings of patients with epilepsy, the experimental methods used are technically complex and require a highly skilled intraoperative team. There are also significant experimental time limitations, as well as constraints on the brain regions that may be safely studied. Compared to microelectrodes, macroelectrodes may permit easier characterization of HFOs within a larger brain volume and therefore provide a better understanding of the propagation of this high-frequency epileptic activity. Furthermore, the use of standard intracranial electrodes permitted a systematic, simultaneous mapping of multiple cortical sites, thereby providing a large-scale analysis of HFOs during the interictal and preictal period.

REFERENCES

Alarcon, G., C. D. Binnie, R. D. Elwes, and C. E. Polkey. 1995. Power spectrum and intracranial EEG patterns at seizure onset in partial epilepsy. *Electroencephalogr. Clin. Neurophysiol.* 94:326–337.

Allen, P. J., D. R. Fish, and S. J. Smith. 1992. Very high-frequency rhythmic activity during SEEG suppression in frontal lobe epilepsy. *Electroencephalogr. Clin. Neurophysiol.* 82:155–162.

Bartos, M., I. Vida, and P. Jonas. 2007. Synaptic mechanisms of synchronized gamma oscillations in inhibitory interneuron networks. *Nat. Rev. Neurosci.* 8:45–56.

Bragin, A., G. Jandó, Z. Nádasdy, J. Hetke, K. Wise, and G. Buzsáki. 1995. Gamma (40–100 Hz) oscillation in the hippocampus of the behaving rat. *J. Neurosci.* 15:47–60.

Bragin, A., J. Engel, C. L. Wilson, I. Fried, and G. W. Mathern. 1999. Hippocampal and entorhinal cortex high-frequency oscillations (100–500Hz) in human epileptic brain and in kainic acid-treated rats with chronic seizures. *Epilepsia* 40:127–137.

Bragin, A., C. L. Wilson, and J. Engel. 2000. Chronic epileptogenesis requires development of a network of pathologically interconnected neuron clusters: A hypothesis. *Epilepsia* 41S:144–152.

Bragin, A., I. Mody, C. Wilson, and J. Engel. 2002a. Local generation of fast ripples in epileptic brain. *J. Neurosci.* 22:2012–2021.

Bragin, A., C. L. Wilson, R. J. Staba, M. Reddick, I. Fried, and J. Engel Jr. 2002b. Interictal high-frequency oscillations (80–500 Hz) in the human epileptic brain: Entorhinal cortex. *Ann. Neurol.* 52:407–415.

Bragin, A., C. L. Wilson, J. Almajano, I. Mody, and J. Engel. 2004. High-frequency oscillations after status epilepticus: Epileptogenesis and seizure genesis. *Epilepsia* 45:1017–1023.

Bragin, A., P. Claeys, K. Vonck, D. Van Roost, C. Wilson, P. Boon, and J. Engel. 2007. Analysis of initial slow waves (ISWs) at the seizure onset in patients with drug resistant temporal lobe epilepsy. *Epilepsia* 48:1883–1894.

Buzsáki, G. 1989. Two-stage model of memory-trace formation: A role for "noisy" brain states. *Neuroscience* 31:551–570.

Buzsáki, G., Z. Horvath, R. Urioste, J. Hetke, and K. Wise. 1992. High-frequency network oscillation in the hippocampus. *Science* 256:1025–1027.

Buzsáki, G., D. L. Buhl, K. D. Harris, J. Csicsvari, B. Czeh, and A. Morozov. 2003. Hippocampal network patterns of activity in the mouse. *Neuroscience* 116:201–211.

Buzsaki, G., and A. Draguhn. 2004. Neuronal oscillations in cortical networks. *Science* 304:1926–1929.

Cardin, J. A., M. Carlen, K. Meletis, U. Knoblich, F. Zhang, K. Deisseroth, L. H. Tsai, and C. I. Moore. 2009. Driving fast-spiking cells induces gamma rhythm and controls sensory responses. *Nature* 459:663–667.

Chrobak, J. J., and G. Buzsáki. 1996. High-frequency oscillations in the output networks of the hippocampal-entorhinal axis of the freely behaving rat. *J. Neurosci.* 16:3056–3066.

Chrobak, J. J., and G. Buzsaki. 1998. Gamma oscillations in the entorhinal cortex of the freely behaving rat. *J. Neurosci.* 18:388–398.

Clemens, Z., M. Mölle, L. Eross, P. Barsi, P. Halász, and J. Born. 2007. Temporal coupling of parahippocampal ripples, sleep spindles and slow oscillations in humans. *Brain* 130:2868–2878.

Compte, A., R. Reig, V. F. Descalzo, M. A. Harvey, G. D. Puccini, and M. V. Sanchez-Vives. 2008. Spontaneous high-frequency (10–80 Hz) oscillations during UP states in the cerebral cortex in vitro. *J. Neurosci.* 28:13828–13844.

Crépon, B., V. Navarro, D. Hasboun, S. Clemenceau, J. Martinerie, M. Baulac, C. Adam, and M. Le Van Quyen. 2010. Mapping interictal oscillations greater than 200 Hz recorded with intracranial macroelectrodes in human epilepsy. *Brain* 133:33–45.

Dang-Vu, T. T., M. Schabus, M. Desseilles, G. Albouy, M. Boly, A. Darsaud, S. Gais, G. Rauchs, V. Sterpenich, G. Vandewalle, J. Carrier, et al. 2008. Spontaneous neural activity during human slow wave sleep. *Proc. Natl. Acad. Sci. U.S.A.* 105:15160–15165.

Della Marca, G., C. Catello Vollono, C. Barba, M. Fuggetta, D. Restuccia, and G. Colicchio. 2007. High-frequency ECoG oscillations in the site of onset of epileptic seizures during sleep. *Sleep Medicine* 8:96–97.

Destexhe, A., S. W. Hughes, M. Rudolph, and V. Crunelli. 2007. Are corticothalamic 'UP' states fragments of wakefulness? *Trends Neurosci.* 30:334–342.

Dickson, C. T., G. Biella, and M. de Curtis. 2003. Slow periodic events and their transition to gamma oscillations in the entorhinal cortex of the isolated guinea pig brain. *J. Neurophysiol.* 90:39–46.

Engel, J., A. Bragin, R. Staba, and I. Mody. 2009. High-frequency oscillations: What is normal and what is not? *Epilepsia* 50:598–604.

Fisher, R. S., W. R. Webber, R. P. Leeser, S. Arroyo, and S. Uematsu. 1992. High frequency EEG activity at the start of seizures. *J. Clin. Neurophysiol.* 9:441–448.

Foffani, G., Y. G. Uzcategui, B. Gal, and L. Menendez de la Prida. 2007. Reduced spike-timing reliability correlates with the emergence of fast ripples in the rat epileptic hippocampus. *Neuron* 55:930–941.

Fries, P., J. H. Reynolds, A. E. Rorie, and R. Desimone. 2001. Modulation of oscillatory neuronal synchronization by selective visual attention. *Science* 291:1560–1563.

Grenier, F., I. Timofeev, and M. Steriade. 2001. Focal synchronization of ripples (80–200 Hz) in neocortex and their neuronal correlates. *J. Neurophysiol.* 86:1884–1898.

Grenier, F., I. Timofeev, and M. Steriade. 2003. Neocortical very fast oscillations (ripples, 80–200 Hz) during seizures: Intracellular correlates. *J. Neurophysiol.* 89:841–852.

Haider, B., and D. A. McCormick. 2009. Rapid neocortical dynamics: Cellular and network mechanisms. *Neuron* 62:171–189.

Hasenstaub, A., Y. Shu, B. Haider, U. Kraushaar, A. Duque, and D. A. McCormick. 2005. Inhibitory postsynaptic potentials carry synchronized frequency information in active cortical networks. *Neuron* 47:423–435.

Huber, R., M. F. Ghilardi, M. Massimini, and G. Tononi. 2004. Local sleep and learning. *Nature* 430:78–81.

Isomura, Y., A. Sirota, S. Ozen, S. Montgomery, K. Mizuseki, D. A. Henze, and G. Buzsáki. 2006. Integration and segregation of activity in entorhinal-hippocampal subregions by neocortical slow oscillations. *Neuron* 52:871–882.

Jacobs, J., P. LeVan, R. Chander, J. Hall, F. Dubeau, and J. Gotman. 2008. Interictal high-frequency oscillations (80–500 Hz) are an indicator of seizure onset areas independent of spikes in the human epileptic brain. *Epilepsia* 49:1893–1907.

Jacobs, J., P. Levan, C.-É. Châtillon, A. Olivier, F. Dubeau, and J. Gotman 2009. High frequency oscillations in intracranial EEGs mark epileptogenicity rather than lesion type. *Brain* 132:1022–1037.

Jirsch, J. D., E. Urrestarazu, P. LeVan, A. Olivier, F. Dubeau, and J. Gotman. 2006. High-frequency oscillations during human focal seizures. *Brain* 129:1593–1608.

Khalilov, I., M. Le Van Quyen, H. Golzan, and Y. Ben-Ari. 2005. Epileptogenic actions of GABA and fast oscillations in the developing hippocampus. *Neuron* 48:787–796.

Khosravani, H., C. R. Pinnegar, J. R. Mitchell, B. L. Bardakjian, P. Federico, and P. L. Carlen. 2005. Increased high-frequency oscillations precede in vitro low-Mg seizures. *Epilepsia* 46:1188–1197.

Klausberger, T., P. J. Magill, L. F. Marton, J. D. Roberts, P. M. Cobden, G. Buzsaki, and P. Somogyi. 2003. Brain-state- and cell-type-specific firing of hippocampal interneurons in vivo. *Nature* 421:844–848.

Kobayashi, K., M. Oka, T. Akiyama, T. Inoue, K. Abiru, T. Ogino, H. Yoshinaga, Y. Ohtsuka, and E. Oka. 2004. Very fast rhythmic activity on scalp EEG associated with epileptic spasms. *Epilepsia* 45:488–496.

Köhling, R., M. Vreugdenhil, E. Bracci, and J. G. Jefferys. 2000. Ictal epileptiform activity is facilitated by GABAA receptor-mediated oscillations. *J. Neurosci.* 20:6820–6829.

Lachaux, J. P., N. George, C. Tallon-Baudry, J. Martinerie, L. Hugueville, L. Minotti, P. Kahane, and B. Renault. 2005. The many faces of the gamma band response to complex visual stimuli. *Neuroimage* 25:491–501.

Lee, A. K., and M. A. Wilson. 2002. Memory of sequential experience in the hippocampus during slow wave sleep. *Neuron* 36:1183–1194.

Le Van Quyen, M., I. Khalilov, and Y. Ben-Ari. 2006. The dark side of high-frequency oscillations in the developing brain. *Trends Neurosci.* 29:419–427.

Le Van Quyen, M., A. Bragin, R. Staba, B. Crépon, C. L. Wilson, and J. Engel. 2008. Cell type-specific firing during ripple oscillations in the hippocampal formation of humans. *J. Neurosci.* 28:6104–6110.

Le Van Quyen, M., R. Staba, A. Bragin, C. Dickson, M. Valderrama, I. Fried, and J. Engel. 2010. Large-scale microelectrode recordings of high frequency gamma oscillations in human cortex during sleep. *J. Neurosci.* 30:7770–7782.

Luczak, A., P. Bartho, S. L. Marguet, G. Buzsáki, and K. D. Harris. 2007. Sequential structure of neocortical spontaneous activity in vivo. *Proc. Natl. Acad. Sci. U.S.A.* 104:347–352.

Marshall, L., H. Helgadottir, M. Mölle, and J. Born. 2006. Boosting slow oscillations during sleep potentiates memory. *Nature* 444:610–613.

Mena-Segovia, J., H. M. Sims, P. J. Magill, and J. P. Bolam. 2008. Cholinergic brainstem neurons modulate cortical gamma activity during slow oscillations. *J. Physiol.* 586:2947–2960.

Molle, M., L. Marshall, S. Gais, and J. Born. 2004. Learning increases human electroencephalographic coherence during subsequent slow sleep oscillations. *Proc. Natl. Acad. Sci. U.S.A.* 101:13963–13968.

Montgomery, S. M., and G. Buzsáki. 2007. Gamma oscillations dynamically couple hippocampal CA3 and CA1 regions during memory task performance. *Proc. Natl. Acad. Sci. U.S.A.* 104:14495–14500.

Mukovski, M., S. Chauvette, I. Timofeev, and M. Volgushev. 2007. Detection of active and silent states in neocortical neurons from the field potential signal during slow-wave sleep. *Cereb. Cortex* 17:400–414.

O'Keefe, J., and L. Nadel. 1978. *The hippocampus as a cognitive map*. Oxford: Oxford University Press.

Roopun, A. K., J. D. Simonotto, M. L. Pierce, A. Jenkin, C. Nicholson, I. S. Schofield, R. G. Whittaker, M. Kaiser, M. A. Whittington, R. D. Traub, and M. O. Cunningham. 2010. A nonsynaptic mechanism underlying interictal discharges in human epileptic neocortex. *Proc. Natl. Acad. Sci. U.S.A.* 107:338–343.

Sejnowski, T. J., and A. Destexhe. 2000. Why do we sleep? *Brain Res.* 886:208–223.

Sejnowski, T. J., and O. Paulsen. 2006. Network oscillations: Emerging computational principles. *J. Neurosci.* 26:1673–1676.

Singer, W., and C. M. Gray. 1995. Visual feature integration and the temporal correlation hypothesis. *Annu. Rev. Neurosci.* 18:555–586.

Staba, R. J., C. L. Wilson, A. Bragin, D. Jhung, I. Fried, and J. Engel. 2004. High-frequency oscillations recorded in human medial temporal lobe during sleep. *Ann. Neurol.* 56:108–115.

Steriade, M., A. Nunez, and F. Amzica. 1993. A novel slow (< 1 Hz) oscillation of neocortical neurons in vivo: Depolarizing and hyperpolarizing components. *J. Neurosci.* 13:3252–3265.

Steriade, M., D. Contreras, and F. Amzica. 1994. Synchronized sleep oscillations and their paroxysmal developments. *Trends Neurosci.* 17:199–208.

Steriade, M., F. Amzica, and D. Contreras. 1996. Synchronization of fast (30–40 Hz) spontaneous cortical rhythms during brain activation. *J. Neurosci.* 16:392–417.

Tao, J. X., A. Ray, S. Hawes-Ebersole, and J. S. Ebersole. 2005. Intracranial EEG substrates of scalp EEG interictal spikes. *Epilepsia* 46:669–676.

Traub, R., M. A. Whittington, E. H. Buhl, F. LeBeau, A. Bibbig, S. Boyd, H. Cross, and T. Baldeweg. 2001. A possible role for Gap junctions in generation of very fast EEG oscillations preceding the onset of, and perhaps initiating, seizures. *Epilepsia* 42:153–170.

Urrestarazu, E., R. Chander, F. Dubeau, and J. Gotman. 2007. Interictal high-frequency oscillations (100–500 Hz) in the intracerebral EEG of epileptic patients. *Brain* 130:2354–2366.

Valderrama, M., B. Crépon, V. Botella, C. Adam, V. Navarrro, and M. Le Van Quyen. 2010. Cortical mapping of gamma oscillations during human slow wave sleep. Submitted.

Ylinen, A., A. Bragin, Z. Nadasdy, G. Jando, I. Szabo, A. Sik, and G. Buzsáki. 1995. Sharp wave-associated high-frequency oscillation (200 Hz) in the intact hippocampus: Network and intracellular mechanisms. *J. Neurosci.* 15:30–46.

Wilson, M. A., and B. L. McNaughton. 1994. Reactivation of hippocampal ensemble memories during sleep. *Science* 265:676–679.

Worrell, G. A., L. Parish, S. D. Cranstoun, R. Jonas, G. Baltuch, and B. Litt. 2004. High-frequency oscillations and seizure generation in neocortical epilepsy. *Brain* 127:1496–1506.

Worrell, G. A., A. B. Gardner, S. M. Stead, S. Hu, S. Goerss, G. J. Cascino, F. B. Meyer, R. Marsh, and B. Litt. 2008. High-frequency oscillations in human temporal lobe: Simultaneous microwire and clinical macroelectrode recordings. *Brain* 131:928–937.

Section VI

The State of Seizure Prediction: Seizure Control

27 Vagus Nerve Stimulation Triggered by Machine Learning–Based Seizure Detection

Initial Implementation and Evaluation

Ali Shoeb, Trudy Pang, John V. Guttag, and Steven C. Schachter

CONTENTS

27.1 INTRODUCTION

For most of the history of epilepsy, pharmacologic, surgical, and diet-based treatments were the only therapeutic options available to patients with refractory epilepsy. In 1997, the US Food and Drug Administration (FDA) approved the use of vagus nerve stimulation (VNS) as an adjunctive therapy for the treatment of refractory seizures; the vagus nerve stimulator became the first implantable device for the treatment of epilepsy (Ben-Menachem 2002; Schachter 1998). The VNS pulse generator is implanted in an infraclavicular, subcutaneous pocket and delivers stimuli to the mid-cervical portion of the left vagus nerve.

VNS therapy is delivered in two modes. In automatic mode, the implanted pulse generator automatically delivers vagus nerve stimuli at programmed intervals. In on-demand mode, the patient or their caregiver initiates VNS in response to symptoms of a seizure. Initiating on-demand stimulation requires holding a permanent magnet for 1–2 seconds over the implanted pulse generator.

Several clinical trials have been conducted to assess the therapeutic efficacy of automatic-mode VNS. The E03 and E05 trials demonstrated that automatic-mode VNS reduces seizure frequency by more than 50% in more than 20% of patients within 3 months of device implantation (Handforth 1998; VNS Study Group 1995). The XE5 trial demonstrated that the efficacy of automatic VNS is long lasting and improves significantly with time (DeGiorgio 2000). Based on these and other studies, VNS was deemed a safe and effective adjunctive therapy for intractable epilepsy.

Preclinical studies involving rat (Woodbury 1990) and canine (Zabara 1992) models of seizures suggested that initiating VNS during a seizure could terminate it or lessen its severity. Clinical studies designed to quantify the therapeutic impact of on-demand-mode VNS have also been conducted. Hammond (1992) recorded an EEG tracing from an adult that illustrated the abrupt termination of an electrographic seizure following the initiation of on-demand VNS. Boon (2001) concluded, through a review of patient seizure diaries, that two-thirds of patients receiving on-demand stimulation were able to interrupt seizures. Similarly, Morris (2003) analyzed seizure diaries from the E04 trial (Labar 1999) and noted that 53% of patients capable of receiving on-demand stimulation experienced seizure termination or diminution. Finally, Major (2008) noted that 50% (8/16) of patients with tuberous sclerosis complex (TSC) and intractable epilepsy reported that on-demand mode VNS aborted or attenuated their seizures. One difficulty with these human studies is that they are based on speculation by patients or caregivers about what would have happened had the VNS not been engaged.

A significant proportion of VNS therapy patients depend on others to initiate on-demand stimulation because the physical or cognitive symptoms of a seizure leave them unable to do so themselves (Boon 2001). Depending on a caregiver to initiate on-demand stimulation has two consequences: (1) it denies patients the potential therapeutic benefit of on-demand stimulation in the absence of caregivers, and (2) it potentially reduces stimulation efficacy because caregivers may not always be able to initiate on-demand stimulation immediately upon the clinical (let alone the electrographic) onset of a seizure. There is already some evidence suggesting that the likelihood of affecting seizure progression decreases with longer delays between seizure onset and the start of stimulation (Hammond 1992).

A system that automatically initiates on-demand-mode VNS following computerized seizure onset detection could, in theory, relieve caregivers, increase a patient's sense of independence and security, and more frequently terminate or diminish seizure symptoms in individuals with seizures that respond to on-demand stimulation. The system may also benefit patients who never realized they were on-demand responders because neither they nor their caregivers were able to initiate stimulation following the onset of a seizure. Finally, the system could provide EEG tracings that allow physicians to study brain-wave correlates of a patient's sense of seizure termination or diminution following on-demand VNS.

In this chapter, we describe the design and clinical testing of a noninvasive computerized system that automatically initiates on-demand VNS following the detection of a seizure. The computerized system detects seizures through a machine learning–based, patient-specific detection algorithm that combines analysis of the scalp electroencephalogram (EEG) and surface electrocardiogram (ECG). Our system is an example of a closed-loop neurostimulator that delivers antiseizure therapy contingent on the detection of a seizure using noninvasive physiologic signals. Other seizure detection and treatment systems described in the literature (Sun 2008; Kossoff 2004; Peters 2001) deliver therapy based on analysis of the intracranial electroencephalogram (IEEG), an invasive physiologic signal.

In Section 27.2, we present a block diagram of our seizure-triggered VNS system. In Section 27.3, we discuss the machine learning–based, patient-specific algorithm used by the system to detect seizures. In Section 27.4, we outline the protocol used in our pilot clinical tests of the system. In Section 27.5, we present a case study illustrating the feasibility of initiating on-demand VNS following real-time, noninvasive seizure onset detection. Section 27.5 also illustrates the improvement in seizure detection performance made possible through the fusion of information from multiple physiologic signals.

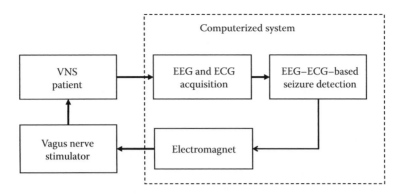

FIGURE 27.1   Block diagram of seizure-triggered vagus nerve stimulation system (VNS).

27.2   SEIZURE-TRIGGERED VNS SYSTEM OVERVIEW

Figure 27.1 shows a block diagram of the computerized system; all the components of the computerized system are external to the patient. The computerized system is composed of a commercial acquisition system (Digitrace 1800 Plus from SleepMed, Inc.) that collects the EEG and ECG of a patient, a computer that analyzes the EEG and ECG in real time using the algorithm described in Section 27.3, and an electromagnet that is worn by the patient and positioned so that it rests over the implanted VNS pulse generator.

When the computerized system detects the onset of a seizure, it energizes the electromagnet worn by the patient for a period of 1.5 seconds. The magnetic field produced by the electromagnet triggers the implanted generator to initiate on-demand stimulation of the vagus nerve. The electromagnet initiates on-demand stimulation through the FDA-approved mechanism that is triggered when a permanent magnet is passed over the implanted generator.

A future, seizure-triggered vagus nerve stimulator would not rely on an electromagnet to initiate on-demand stimulation. An electromagnet capable of generating the magnetic field necessary to trigger the VNS pulse generator is large and heavy and therefore impractical for everyday wear. Furthermore, a shift in the electromagnet's position relative to the VNS pulse generator can abolish the electromagnet's ability to trigger on-demand VNS. Nevertheless, we chose to use an electromagnet to link seizure detection and VNS because it enabled us to work with VNS therapy patients without redesigning the VNS pulse generator.

An ambulatory version of the proposed closed-loop VNS system might be implemented as follows. The ECG and EEG required by the system can be captured using active, dry electrodes (Taheri 1994) positioned over the chest and region of the head demonstrating the earliest seizure-associated EEG change. Salient features of the signals can be locally extracted and wirelessly transmitted to a digital signal processor (DSP) that determines whether or not seizure activity is ongoing (Verma 2009). Finally, detection of seizure activity by the DSP will trigger the VNS generator through a dedicated DSP–VNS radio link.

27.3   MACHINE LEARNING–BASED SEIZURE ONSET DETECTION

To detect a seizure rapidly and reliably, we adopted a patient-specific approach that utilizes machine learning techniques. Although patient-nonspecific algorithms can exhibit impressive detection performance on certain seizure types (Meier 2008; Saab 2005), commercially available patient-nonspecific algorithms (Wilson 2004) tend to exhibit worse sensitivity, specificity, and longer detection delays when compared to patient-specific algorithms (Shoeb 2007; Wilson 2005). This performance disparity is primarily due to the large variability in the character of seizure and nonseizure EEG that exists across patients, and the relative consistency of seizure and nonseizure EEG for a given patient.

Our patient-specific approach infers the onset of a seizure by analyzing the EEG alone, or in concert with other physiologic signals. Factoring information extracted from a second physiologic signal into the seizure detection process is useful for detecting seizures associated with subtle or nonspecific changes within the EEG. The second signal and the EEG complement each other if the changes in each signal that suggest the presence of a seizure rarely coincide during nonseizure states and often coincide at the time of an actual seizure. The surface electrocardiogram (ECG), a noninvasive measure of the electrical activity of the heart, is a good candidate for a complementary signal. It is easy to acquire, and there is significant evidence that seizures for many patients are associated with changes in heart rate (Zijlman 2002).

In our patient-specific approach, a classifier determines whether an observed epoch of synchronized EEG and ECG more closely resembles epochs from an individual's seizures or epochs from awake, sleep, and active periods. The classifier accomplishes this task by comparing EEG and ECG features extracted from the observed epoch with features extracted from training seizure and nonseizure epochs; the classifier then assigns the observation to the class (seizure or nonseizure) whose features most resemble those of the observation.

The block diagram in Figure 27.2 illustrates more concretely the processing stages that comprise our patient-specific detector. The detector passes a L_1-second epoch of N EEG channels (typically $N = 18$) and a L_2-second epoch of a lead-II ECG channel through two feature extractors. In general, $L_1 \neq L_2$ because EEG and ECG features may not necessarily be computed over the same timescales. However, to simplify the implementation and deployment of our algorithm, we set $L_1 = L_2 = 5$ seconds.

The EEG feature extractor characterizes the spectral and spatial structure of the EEG within the input epoch because seizures are often associated with changes in these properties. To characterize the spectral structure of an EEG epoch, a filterbank composed of $M = 8$ filters measures the energy falling within the passband of each filter. The filters collectively span the frequency range 0–24 Hz because most seizure and nonseizure EEG activity falls within this range (Ebersole 2003). For EEG channel k, the energy measured by filter i is denoted by the feature $x_{i,k}$ as shown in Figure 27.2. The eight features extracted from each of the N channels are then concatenated to

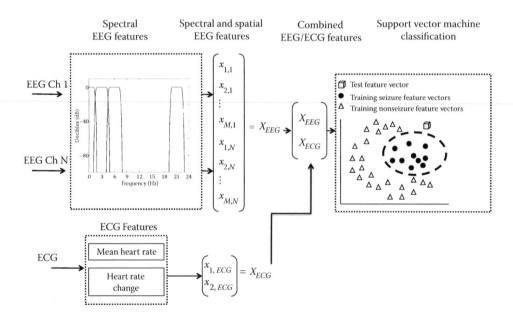

FIGURE 27.2 Block diagram of patient-specific seizure onset detection algorithm.

form an $8N$-dimensional feature vector $\mathbf{X}_{EEG}$ that captures spatial and spectral information within the 5-second epoch.

The ECG feature extractor attempts to characterize heart-rate dynamics within the observed epoch through two features. The first feature, $\mathbf{X}_{1,ECG}$, is the mean heart rate within the observed epoch. The second feature, $\mathbf{X}_{2,ECG}$, is the difference between the terminal and initial instantaneous heart-rate measurements within the observed epoch. The mean heart rate measured by $\mathbf{X}_{1,ECG}$ and the heart-rate change measured by $\mathbf{X}_{2,ECG}$ can deviate significantly from a patient's baseline following the onset of seizure-associated tachycardia or bradycardia. These features are concatenated to form the vector $\mathbf{X}_{ECG}$. Finally, the vectors $\mathbf{X}_{EEG}$ and $\mathbf{X}_{ECG}$ are combined to form a single feature vector that automatically captures the relationship between spectral and spatial information extracted from the EEG with heart-rate and heart-rate-change information extracted from the ECG.

The combined EEG and ECG feature vector is assigned to the seizure or nonseizure class using a support vector machine (SVM) classifier (Shawe-Taylor 2000). The SVM determines the class membership of the observed feature vector by noting on which side of a decision boundary the feature vector falls. The decision boundary, which divides the feature space into seizure and nonseizure regions, is determined by training the SVM on feature vectors from multiple seizures and at least 24 h of nonseizure EEG. When ECG is not available or not used, the SVM only uses the information within the feature vector $\mathbf{X}_{EEG}$ for the purpose of seizure onset detection.

27.4 CLINICAL STUDY DESIGN

The primary purpose of our pilot clinical study was to evaluate the feasibility of initiating on-demand VNS following real-time, noninvasive, computerized seizure onset detection. The study followed a protocol approved by the committee on clinical investigations at the Beth Israel Deaconess Medical Center (BIDMC) in Boston, Massachusetts, USA. Study participants were adult, long-term users of VNS who reported no adverse effects from on-demand VNS. Participants were admitted to the BIDMC General Clinical Research Center (GCRC) for a period lasting up to 5 days. Admission to the GCRC took place after study staff obtained informed consent. No changes to a study participant's anti-epileptic drug (AED) regimen or VNS stimulation parameters were made during the study period.

During the admission period, the participant's video, EEG, and ECG signals were recorded using the Digitrace 1800 Plus Recorder and Digital Video System (SleepMed, Inc.). Furthermore, a continuous signal that reflects whether the vagus nerve stimulator is on or off was collected using electrodes placed over the left side of the participant's neck. After the first 24–48 h of the study, an electroencephalographer manually searched the entirety of the collected EEG and video for clinical seizures, which are events associated with changes in both an individual's EEG and behavior. Research staff facilitated this search by highlighting periods of time during which study participants exhibited behavior consistent with previous accounts of their seizures. The electroencephalographers also identified abnormal EEG activity not associated with behavioral correlates. The abnormal activity, which we refer to as epileptiform discharges, includes prominent spikes or short bursts of abnormal waveforms.

Next, the computerized system was trained to detect the marked clinical seizures using both the EEG gathered in the first 24–48 h of the study, and prerecorded EEG from the study participant available on record at BIDMC. In the absence of clinical seizures, the computerized system was trained to detect a type of epileptiform discharge identified by the electroencephalographer. On subsequent days of the study, the computerized system was set to automatically activate the participant's VNS therapy system whenever real-time detection of a clinical seizure or epileptiform discharge occurred.

At the conclusion of the study we evaluated the sensitivity, specificity, and delay with which the computerized system initiated on-demand VNS. Using the data collected during the first 24–48 h of the study as a baseline, we also examined how on-demand VNS altered the characteristics of clinical seizures or epileptiform discharges. For epileptiform discharges, we evaluated whether on-demand VNS altered the duration of the associated EEG change. For clinical seizures, we examined whether on-demand VNS altered the EEG or behavior associated with the seizure (ictal) and post-seizure (postictal) periods.

27.4.1 SUMMARY OF CLINICAL STUDY FINDINGS

We have previously reported (Shoeb 2009) the performance of our system on five patients (A–E). For Patients A and B, the computerized system successfully initiated on-demand VNS in response to seizures. For Patients C and D, no clinical seizures were recorded during the study period, but the system successfully initiated on-demand VNS in response to epileptiform discharges. For Patient E, neither seizures nor epileptiform discharges were observed during the evaluation period.

During clinical testing of the system on Patient A, which lasted 81 consecutive hours, the computerized system detected 5/5 daytime seizures and initiated VNS within 5 seconds of the appearance of seizure activity within the EEG. Relative to the four seizures used to train the system, on-demand VNS did not appear to alter the electrographic or behavioral characteristics of Patient A's seizures. During the same testing session, the computerized system initiated false stimulations at the rate of 1 false stimulus every 2.5 h while the subject was at rest and not ambulating. Furthermore, false detections primarily occurred while the subject was awake as opposed to asleep. During periods of ambulation around the research floor, study staff allowed the system to log detections but disabled computerized VNS activation because ambulation triggered false detections. When periods of ambulation are included, the computerized system's false detection rate became 2.5 false detections per hour; false detections are not uniformly distributed in time, but concentrated within periods of ambulation.

During clinical testing of the system on Patient B, which lasted for 26 consecutive hours, the computerized system declared 0 false detections, and detected 1/1 seizures. The system initiated on-demand VNS within 16 seconds of the appearance of seizure activity within the EEG. Relative to the single seizure used to train the system, on-demand VNS did not appear to alter the duration of ictal or postictal EEG. However, VNS did appear to decrease Patient B's anxiety and increase her awareness during the postictal period.

27.5 ON-DEMAND VNS USING A DETECTOR THAT COMBINES EEG AND ECG: A CASE STUDY

Patient A participated twice in our clinical study. During Patient A's first admission (Shoeb 2009), EEG features were used alone to detect seizures. In contrast, during Patient A's second admission, which occurred approximately 6 months after the first, a single classifier examined features from both the EEG and ECG to establish the presence or absence of seizure activity. This case study illustrates the performance gain made possible by incorporating information from a second physiologic signal into the seizure detection process.

27.5.1 MEDICAL HISTORY AND SEIZURE CHARACTERISTICS

Patient A is a 39-year-old woman with a long history of refractory complex partial seizures. At the time of admission to our study, she was experiencing 30–40 seizures per month. Clinically, Patient A's seizures last for 1–2 min and consist of repeatedly asking questions and blank stares; her seizures are not accompanied by automatisms or tonic-clonic movements. Following a seizure, Patient A is confused for a few minutes and requires reorientation by friends or family. Patient A is a VNS therapy patient; however, because she does not experience a warning prior to the onset of a seizure she is unable to use the on-demand mode of VNS.

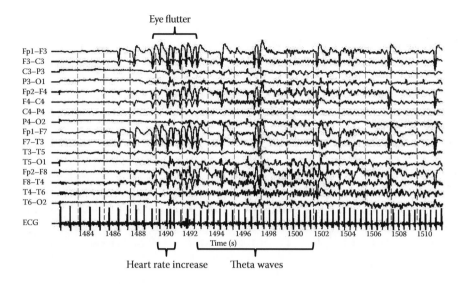

FIGURE 27.3 Typical EEG and ECG characteristics associated with Patient A's seizures.

Figure 27.3 illustrates the typical EEG signature of Patient A's seizures. The onset of the seizure, at 1486 seconds, involves rapid eye blinking (eye flutter), which manifests in the EEG as high-amplitude deflections on the channels FP1–F7, FP1–F3, FP2–F8, FP2–F4. Coincident with the onset of eye flutter, the patient's heart rate rapidly increases, as can be seen in the ECG channel in Figure 27.3 and in the heart-rate profile shown in the left panel of Figure 27.4. Finally, at 1492 seconds, 3–4 Hz theta waves appear on the right-sided EEG channel T4–T6, while the patient's heart rate remains elevated. Electrographic signs of the seizure end around 1510 seconds.

27.5.2 EEG–ECG–BASED ACTIVATION OF ON-DEMAND VNS

At the start of Patient A's second admission, a total of 9 prerecorded seizures and 28 h of EEG containing both awake and sleep epochs were used to train a detector. The awake and sleep EEG

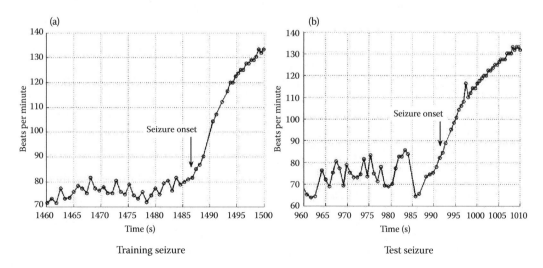

FIGURE 27.4 Heart-rate profile associated with the training seizure in Figure 27.3 and the test seizure in Figure 27.5.

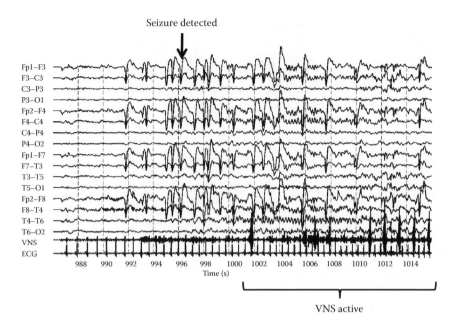

FIGURE 27.5 Seizure detected by computerized system during Patient A's second admission. The seizure, which began at 992 seconds, was detected at 996 seconds and on-demand VNS was initiated in response. The VNS pulse generator became active at 1002 seconds.

and 5/9 of the training seizures were collected during Patient A's first admission; the remaining four training seizures were collected prior to the first admission. Note that no preprocessing was done to account for the fact that much of the training data was recorded 6 months before the second admission.

Figure 27.5 illustrates the only seizure Patient A experienced during her second admission. Seizure onset begins with rapid eye blinking at 992 seconds. The detector declared seizure onset at 996 seconds, which is 4 seconds after the onset of eye flutter and 2 seconds before the onset of theta wave activity, and then energized the electromagnet for 1.5 seconds. The VNS pulse generator needs to be exposed to a magnetic field lasting at least 1 second to initiate a cycle of on-demand stimulation. Beginning at 1002 seconds, a spike train appears on the VNS channel, confirming the initiation of VNS. The VNS generator activity does not appear immediately on the VNS channel following the end of the magnetic pulse because the generator gradually increases the stimulation strength. The heart-rate change that accompanied the eye flutter and triggered the detector is shown in the right panel of Figure 27.4. Note the similarity of both the seizure EEG shown in Figure 27.5 and the heart-rate profile shown in the right panel of Figure 27.4, to the training EEG (Figure 27.3) and ECG (left panel of Figure 27.4) recorded 6 months earlier.

As was the case in Patient A's first admission, on-demand VNS did not seem to alter the electrographic or behavioral characteristics of the seizure observed during the second admission. Three of the possible explanations for this include: (1) the on-demand stimulus was not initiated early enough in the course of the seizure, (2) the on-demand VNS stimulus parameters (pulse current, frequency, and duration) were not set appropriately, and (3) Patient A is not a responder to on-demand-mode VNS.

27.5.3 COMPARING EEG AND EEG-ECG BASED SEIZURE DETECTION PERFORMANCE

In this section, we compare the performance of the seizure detection approaches used in the first and second admissions of Patient A on seizure and nonseizure data from both admissions. The data consists of 70 h of nonseizure data and six seizures (five from the first admission and one from the second admission). The seizure detector used in the second admission combined EEG and ECG features ($\mathbf{X_{EEG}}$ and $\mathbf{X_{ECG}}$ in Section 27.3) to detect the earliest manifestation of Patient A's seizures, the coincidence of eye flutter, and heart-rate change. We refer to this detector as the "flutter EEG+ECG" detector. The detector used in the first admission relied only on EEG features ($\mathbf{X_{EEG}}$ in Section 27.3) to detect a later manifestation of Patient A's seizures, the onset of 3–4 Hz, right-side-predominant theta waves. We refer to this detector as the "theta EEG" detector. The detector used in the first admission was not taught to detect the EEG correlate of the earliest manifestation of Patient A's seizure (eye flutter) because that would result in a detector with poor specificity. This will be illustrated by examining the performance of a detector that uses EEG features to detect the EEG correlate of eye flutter. We refer to this detector as the "flutter EEG" detector. It is important to note that these detectors automatically learned to recognize their target activity using the support-vector machine algorithm and feature vectors derived from training seizures. These detectors are not hand-coded, special-purpose algorithms for detecting theta-wave activity, eye flutter, or heart-rate change.

The left panel of Figure 27.6 shows the false-alarm rates associated with each detector. Because eye flutter is an activity that is frequently observed in the nonseizure state, the flutter EEG detector exhibited the highest false detection rate (~11 false detections per 24-h period). The theta EEG detector generated false detections at a lower rate (~7 false detections per 24-h period), which is consistent with right-side-predominant theta-wave activity having a greater association with Patient A's seizure state. The flutter EEG+ECG detector had the lowest false alarm rate (~5 false detections per 24-h period) because the coincidence of eye flutter and heart-rate change most often occurred within the seizure state.

The right panel of Figure 27.6 illustrates the delay with which each detector declares the onset of a seizure. The onset of a seizure is taken to be the onset of eye flutter within the context of heart-rate change (Section 27.5.1), and not the onset of theta waves as observed in a previous report (Shoeb 2009). The flutter EEG+ECG detector consistently declared seizure onset before the other two detectors. In summary, for Patient A, the detector that combined EEG and ECG could detect the earliest stage of Patient A's seizures with a latency and false detection rate lower than that possible using EEG alone.

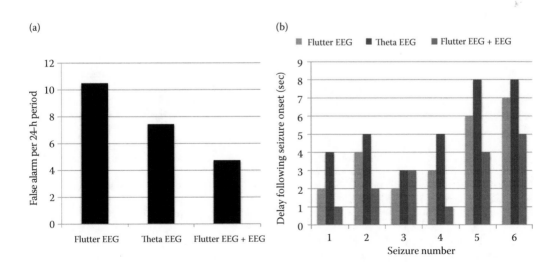

FIGURE 27.6   The fusion of EEG and ECG features results in a detector (flutter EEG+ECG) that detects seizures with a latency and false detection rate lower than those achieved by detectors that only use EEG features (flutter EEG and theta EEG).

27.6 CONCLUSIONS AND FUTURE WORK

In this chapter, we reported on the feasibility of initiating on-demand VNS following real-time, noninvasive seizure onset detection. Furthermore, we demonstrated that seizure detection performance can be enhanced through the fusion of information extracted from multiple physiologic signals whose dynamics are influenced by the seizure state. In addition to being used to initiate VNS in response to seizure onset, the system we developed could be used to trigger other extracranial, noninvasive brain stimulation modalities such as repetitive transcranial magnetic stimulation (rTMS), transcranial direct current stimulation (tDCS), and trigeminal nerve stimulation (TNS) (Fregni 2007; DeGiorgio 2006). Future clinical studies of our system will be designed to shed light on the impact of on-demand VNS on the electrographic and behavioral changes associated with the seizure and postseizure states.

ACKNOWLEDGMENTS

We would like to thank the individuals that participated in this research study as well as the nurses of the Beth Israel Deaconess Medical Center General Clinical Research Center for their dedication. The project described in this report was supported by Grant Number UL1 RR025758 and M01-RR-01032 Harvard Clinical and Translational Science Center, from the National Center for Research Resources. The content is solely the responsibility of the authors and does not necessarily represent the official views of the National Center for Research Resources or the National Institutes of Health.

REFERENCES

Ben-Menachem, E. 2002. Vagus nerve stimulation for the treatment of epilepsy. *Lancet Neurol.* 1(8):477–482.
Boon, P., K. Vonck, P. Van Walleghem, et al. 2001. Programmed and magnet-induced vagus nerve stimulation for refractory epilepsy. *J. Clin. Neurophysiol.* 18(5):402–407.
DeGiorgio, C. M., S. C. Shcahcter, A. Handforth, M. Salinsky, J. Thompson, B. Uthman, et al. 2000. Prospective long-term study of vagus nerve stimulation for the treatment of refractory seizures. *Epilepsia* 41(9):1195–1200.
DeGiorgio, C. M., A. Shewmon, D. Murray, and T. Whitehurst. 2006. Pilot study of trigeminal nerve stimulation (TNS) for epilepsy: A proof-of-concept trial. *Epilepsia* 47(7):1213–1215.
Ebersole, J., and T. Pedley. 2003. *Current practice of clinical electroencephalography*. Philadelphia: Lippincott Williams & Wilkins.
Fregni, F., and A. Pascual-Leone. 2007. Technology insight: Noninvasive brain stimulation in neurology-perspectives on the therapeutic potential of rTMS and tDCS. *Nat. Clin. Pract. Neurol.* 3(7):383–393.
Hammond, E. J., B. M. Uthman, S. A. Reid, and B. J. Wilder. 1992. Electrophysiological studies of cervical vagus nerve stimulation in humans: EEG effects. *Epilepsia*, 33(6):1013–1020.
Handforth, A., C. M. DeGiorgio, S. C. Schachter, B. M. Uthman, D. K. Naritoku, E. S. Tecoma, et al. 1998. Vagus nerve stimulation therapy for partial onset seizures: A randomized active control trial. *Neurology* 51(1):48–55.
Kossoff, E. H., E. K. Ritzl, J. M. Politsky, A. M. Murro, J. R. Smith, R. B. Duckrow, et al. 2004. Effect of an external responsive neurostimulator on seizures and electrographic discharges during subdural electrode monitoring. *Epilepsia* 45(12):1560–1567.
Labar, D., J. Murphy, and E. Tecoma. 1999. Vagus nerve stimulation for medication-resistant generalized epilepsy. *Neurology* 52(7):1510–1512.
Major, P., and E. A. Thiele. 2008. Vagus nerve stimulation for intractable epilepsy in tuberous sclerosis complex. *Epilepsy Behav.* 13(2):357–360.
Meier, R., H. Dittrich, A. Schulze-Bonhage, and A. Aertsen. 2008. Detecting epileptic seizures in long-term human EEG: a new approach to automatic online and real-time detection and classification of polymorphic seizure patterns. *J. Clin. Neurophysiol.* 25(3):119–131.
Morris, G. L. III. 2003. A retrospective analysis of the effects of magnet-activated stimulation in conjuction with vagus nerve stimulation therapy. *Epilepsy Behav.* 4(6):740–745.

Peters, T. E., N. C. Bhavaraju, M. G. Frei, and I. Osorio. 2001. Network system for automated seizure detection and contingent delivery of therapy. *J. Clin. Neurophysiol.* 8(6):545–549.

Saab, M. E., and J. Gotman. 2005. A system to detect the onset of epileptic seizures in scalp EEG. *Clin. Neurophysiol.* 116(2):427–442.

Schachter, S. C., and C. Saper. 1998. Vagus nerve stimulation. *Epilepsia* 39(7):677–686.

Shawe-Taylor, J., and N. Cristianini. 2000. *Support vector machines and other kernel-based learning methods.* Cambridge: Cambridge University Press.

Shoeb, A., B. Bourgeois, S. Treves, S. C. Schachter, and J. V. Guttag. 2007. Impact of patient-specificity on seizure onset detection performance. *Conference Proceedings of IEEE Engineering in Medicine and Biology Society* 2007:4110–4114.

Shoeb, A., T. Pang, J. V. Guttag, and S. C. Schachter. 2009. Non-invasive computerized system for automatically initiating vagus nerve stimulation following patient-specific detection of seizures or epileptiform discharges. *Int. J. Neural Syst.* 19(3):157–172.

Sun, F. T., M. J. Morell, and R. E. Wharen Jr. 2008. Responsive cortical stimulation for the treatment of epilepsy. *Neurotherapeutics* 5(1):68–74.

Taheri, B. A., R. T. Knight, and R. L. Smith. 1994. A dry electrode for EEG recording. *Electroencephalogr. Clin. Neurophysiol.* 90(5):376–383.

The Vagus Nerve Stimulation Study Group. 1995. A randomized controlled trial of chronic vagus nerve stimulation for treatment of medically intractable seizures. *Neurology* 45(2):224–230.

Verma, N., A. Shoeb, J. V. Guttag, and A. P. Chandrakasan. 2009. A micro-power EEG acquisition SoC with integrated seizure detection processor for continuous patient monitoring, *Symposium on VLSI Circuits* 2009:62–63.

Wilson, S. B., M. L. Scheuer, R. G. Emerson, and A. J. Gabor. 2004. Seizure detection: Evaluation of the reveal algorithm. *Clin. Neurophysiol.* 115(10):2280–2291.

Wilson, S. B. 2005 A neural network method for automatic and incremental learning applied to patient-dependent seizure detection. *Clin. Neurophysiol.* 116:1785–1795.

Woodbury, D., and J. W. Woodbury. 1990. Effects of vagal stimulation on experimentally induced seizures in rats. *Epilepsia* 31(Suppl. 2):S7–S19.

Zabara, J. 1992. Inhibition of experimental seizures in canines by repetitive vagal stimulation. *Epilepsia* 33(6):1005–1012.

Zijlman, M., D. Flanagan, and J. Gotman. 2002. Heart-rate changes and ECG abnormalities during epileptic seizures: Prevalence and definition of an objective clinical sign. *Epilepsia* 43(8):847–854.

28 Low-Frequency Stimulation as a Therapy for Epilepsy

Jeffrey H. Goodman

CONTENTS

28.1 INTRODUCTION

One of the goals of improved seizure detection is to expand the window for therapeutic intervention (Litt et al. 2001). A new therapeutic approach that can be combined with seizure detection is the use of responsive deep brain electrical stimulation (DBS) (Sun et al. 2008). Anecdotal evidence dating back to the 1930s suggested that direct electrical stimulation of different brain regions can decrease seizure frequency and/or severity (Walker 1938). Unfortunately, many of these early studies consisted of isolated, uncontrolled case reports of conflicting results (Cooper et al. 1973; Cooper and Upton 1978; Wright et al. 1978). The first controlled clinical study that examined the effect of centromedian thalamic stimulation in patients with intractable epilepsy did not observe a significant decrease in the number of seizures (Fisher et al. 1992). The mixed results of these early studies led to a decreased interest in DBS as a therapy for epilepsy. However, the recent success obtained in the treatment of Parkinson's disease and other movement disorders with DBS (Benabid et al. 2009) as well as the decrease in seizure frequency observed in some patients treated with vagus nerve stimulation (VNS) (Schmidt 2001) has led to a reexamination of DBS as a therapy for epilepsy. The basic hypothesis is that if stimulation of a peripheral nerve can decrease seizure activity, then delivering the stimulus in the brain, closer to the site of seizure onset or into a seizure-gating network, should result in improved efficacy.

Before DBS can become a viable addition to pharmacotherapy and surgical resection as a treatment for epilepsy, there are several fundamental questions that need to be addressed. The questions of where and how to stimulate and how to insure that the therapeutic stimulation is safe remain unresolved (Goodman 2004, 2009). In terms of where to stimulate, two approaches have been taken. The first is the activation of seizure-gating networks (Gale 1992) by stimulating a number of structures including, but not limited to, the caudate (Sramka and Chkenkeli 1990, 1997), the anterior thalamus (Hodaie et al. 2002; Lockman and Fisher 2009), and the subthalamic nucleus (Benabid et al. 2002). The second is the direct electrical stimulation of an identified seizure focus (Boon et al. 2007, Lesser et al. 1999; Osorio et al. 2001, 2005; Velasco et al. 2000, 2007). In each of these studies, the results were variable suggesting that—in addition to location—the stimulation parameters may also be critical, with stimulation frequency being particularly important (Lado et al. 2003; Mirski et al. 1997).

The majority of studies that have examined the efficacy of DBS applied high-frequency stimulation (HFS) as a result of its proven efficacy in the treatment of movement disorders. However, while HFS is effective in some patients, given the number of patients who do not respond, alternative stimulation paradigms may capture additional members of the patient population. Durand and Bikson (2001) hypothesized that low-frequency stimulation (LFS) would have a greater potential for clinical benefit because its effect outlasts the duration of the stimulation. In support of this hypothesis, several clinical studies have reported that LFS can effectively decrease interictal as well as ictal activity. Yamamoto et al. (2002, 2006), Tkesheleshvili et al. (2003), and Carrington et al. (2007) have reported that a single presentation of 1-Hz sine wave stimulation can elevate kindled afterdischarge threshold for days. Earlier animal studies by Gaito and his colleagues (Gaito 1980; Gaito et al. 1980) provided the first evidence that low-frequency sine wave stimulation (LFSWS) could interfere with kindled seizures, but these studies were poorly designed and difficult to interpret. The idea that a low-frequency DC stimulus could quench the kindled seizure was put forward by Weiss et al. (1998). A later study by Velisek et al. (2002) demonstrated that 1-Hz pulsatile stimulation could delay kindling acquisition in immature rats. In vitro slice studies have also demonstrated that LFS can block epileptiform activity (Albensi et al. 2004; Barbarosie and Avoli 1997; Khosravani et al. 2003). My laboratory has focused on building upon these earlier studies to develop low-frequency stimulation as a therapy for epilepsy (Goodman et al. 2005). The ultimate goal of these studies is to combine safe, low-frequency stimulus paradigms with early seizure detection in a responsive implantable stimulator.

28.2 METHODS

28.2.1 KINDLING SEIZURE MODEL

The efficacy of low-frequency sine wave (LFSWS) and low-frequency pulsatile stimulation (LFS) were tested in the kindling seizure model. The kindling model is well suited for an examination of the efficacy of therapeutic stimulation paradigms because the investigator has control over when the seizures occur and the duration of the electrographic seizure and the severity of the behavior convulsion can be readily measured. In the kindling model, seizures are initiated with an electrical stimulus delivered to a limbic or cortical structure. The kindling process, first described by Goddard et al. (1969), is characterized by the repeated, spaced presentation of a short duration stimulus that initially is subconvulsive but over time induces a permanent change in brain function such that the initially subconvulsive stimulus consistently elicits a generalized convulsion. Initially, the electrographic response to this stimulus consists of a hypersynchronous increase in excitability called an afterdischarge (AD). Repeated daily stimulation results in an increase in the duration and complexity of the AD. The growth of the AD is accompanied by the development of behavioral convulsions that can be scored on a 1–5 scale according to severity (Racine 1972). Seizures with behavioral scores of stage 1–2 are considered equivalent to partial seizures, while behavioral seizure scores of stage 3–5 are generalized. Once an animal exhibits three consecutive stage 5 seizures the animal is considered to be fully kindled and the kindling process is complete. Fully kindled animals will consistently exhibit a stage 5 seizure in response to the kindling stimulus that was initially subconvulsant. The efficacy of a given therapy during kindling acquisition can be assessed by detecting changes in AD threshold, AD duration, and the number of stimulations required for the animal to become fully kindled. Similar features except for the time to kindle can be used to assess efficacy of preemptive low-frequency stimulation in fully kindled animals. By examining the effect of preemptive low-frequency stimulation during kindling acquisition and in fully kindled animals the potential antiepileptogenic as well as the antiepileptic efficacy of this stimulation paradigm can be assessed.

28.2.1.1 Electrode Implantation

Bipolar, Teflon-coated, stainless steel electrodes (127 μm diameter wire) were implanted bilaterally into the basolateral amygdalae or dorsal hippocampi of adult male Sprague-Dawley rats (275–325 g), as described by Goodman et al. (2005). Briefly, each animal was anesthetized with a combination of ketamine and xylazine (1 mL/kg; 80 mg/kg ketamine, 12 mg/kg xylazine, i.p.) and placed in a stereotaxic apparatus. A midline incision was made and the skull surface exposed. Burr holes were drilled for the placement of anchor screws and depth electrodes. The coordinates measured from bregma for electrode placement in the amygdala were: A–P, 2.3 mm; M–L, ±4.5 mm from midline; and depth, 8.5 mm from skull surface and, for placement in the hippocampus: A–P, –3.0 mm; M–L, ±3.7 mm from midline; and depth 3.2 mm from skull surface (Paxinos and Watson 1998). Each electrode was lowered to its appropriate stereotaxic coordinate, cemented in place with dental acrylic, and connected to a socket. The wound was treated with disinfectant and closed with sterile wound clips.

28.2.1.2 AD Threshold Testing

A minimum of one week after electrode implantation, the AD threshold for each electrode was tested using the following kindling stimulus: 60 Hz, 1ms pulse duration, sine wave for 1–2 s. AD testing was initiated with a current intensity of 25 μA and was increased in 25-μA steps until an AD was observed. The electrode with the lower threshold and/or better recording quality was selected for kindling and low-frequency stimulation. Rats with an AD threshold above 250 μA were excluded from the study.

28.2.1.3 Effect of Preemptive Low-Frequency Stimulation on Kindling Acquisition in Amygdala-Kindled Rats

The rats were randomly assigned to the control or experimental group. Control rats were stimulated twice daily with the following kindling stimulus: 60 Hz, 1-ms pulse, 400-μA sine wave for 1–2s. The experimental rats received the same stimulus; however, immediately before the delivery of the kindling stimulus, the experimental rats received the following preemptive LFSWS: 1 Hz, 50-μA sine wave (100 μA peak to peak) for 30 s. AD duration and behavior seizure score, as defined by Racine (1972), were measured for each animal after each stimulation. The number of stimulations required to reach the first stage 5 seizure and the number of stimulations required for the animal to become fully kindled were also measured. Data were collected for the first 20 stimulations, which is greater than what is normally required to fully kindle an amygdala-kindled rat (Goddard et al. 1969) in case the LFSWS delayed the time to kindle.

28.2.1.4 Effect of Preemptive Low-Frequency Stimulation on the Kindled State

In these experiments, a LFSWS (1 Hz, 50 μA) was delivered for 30 s immediately before delivery of the kindling stimulus (60 Hz, 1-ms pulse, 400 μA for 1–2 s) to rats fully kindled in the basolateral amygdala or dorsal hippocampus. The incidence of stage 5 seizures and AD duration were measured for each animal. The effect of preemptive low-frequency pulsatile stimulation (LFS) was also examined in rats fully kindled in the dorsal hippocampus. For this experiment, a long term depression (LTD)-like stimulus (1 Hz, 1 ms biphasic pulse, 50 μA square wave) was delivered for 15 min immediately before delivery of the kindling stimulus.

28.2.1.5 Data Analysis

To assess the effect of preemptive LFSWS on kindling acquisition, the values for AD duration and behavioral seizure score were averaged after each stimulation. Statistical significance was determined using a two-way ANOVA with one repeated measure followed by the Hotelling T^2 test for AD duration and the Mann-Whitney U test for comparison of behavioral seizure score. Comparison of AD duration or seizure incidence between two groups was made using Fisher's exact test. The threshold for significant was set at $p < 0.05$.

28.2.2 SAFETY

At the conclusion of each experiment, each animal was perfusion-fixed with 4% paraformaldehyde and the brain was removed and sectioned on a vibratome (50 μm). Slide-mounted sections were Nissl-stained and assessed for damage. In a separate group of animals, the LFSWS was delivered continuously for 6 min to determine if prolonged exposure to this stimulus caused damage. This tissue was examined by Nissl stain and Fluoro-jade.

28.3 RESULTS

28.3.1 EFFECT OF PREEMPTIVE LOW-FREQUENCY STIMULATION ON KINDLING ACQUISITION

Kindling acquisition in the amygdala is characterized by a growth in AD duration and complexity accompanied by an increase in behavioral seizure score. It usually takes 9–12 stimulations for an amygdala-kindled rat to become fully kindled. In this experiment, the coupling of the LFSWS to the kindling stimulus did not significantly alter the number of stimulations required for the experimental animals to become fully kindled. However, preemptive LFSWS did significantly decrease AD duration ($p < 0.01$, Figure 28.1a) and behavioral seizure score ($p < 0.001$, Figure 28.1b). More importantly, preemptive LFSWS significantly decreased the incidence of AD in response to the kindling stimulus by 30% (Figure 28.2, $p < 0.001$), which is significant since the kindling stimulus rarely fails to elicit an AD.

28.3.2 EFFECT OF PREEMPTIVE LOW-FREQUENCY STIMULATION IN FULLY KINDLED ANIMALS

Baseline seizure frequency was determined for rats fully kindled in the amygdala or dorsal hippocampus. The incidence of stage 5 seizures in these animals in response to the kindling stimulus ranged from 98%–100%. When LFSWS was delivered immediately before the kindling stimulus, the incidence of stage 5 seizures significantly dropped from 98% to 42% (Figure 28.3, $p < 0.0001$) in amygdala-kindled rats and from 100% to 70% ($p < 0.0001$) in hippocampal-kindled rats (data not shown). It is important to recognize that when these animals failed to exhibit a stage 5 seizure, the failure was complete with no AD and no lower scored behavioral convulsions. When seizures did occur there was no effect on AD duration. In a separate experiment, a LTD-like LFS (1 Hz, 15 min) was delivered immediately before the kindling stimulus in rats fully kindled in the hippocampus. This LFS had no effect on the incidence of stage 5 seizures or on AD duration in these animals (data not shown).

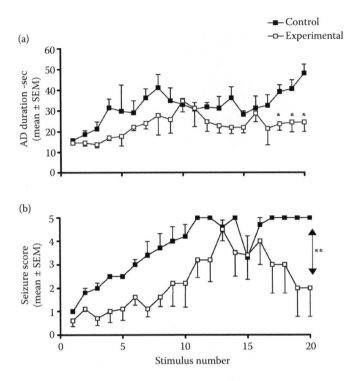

FIGURE 28.1 Effect of preemptive LFSWS on kindling acquisition. LFSWS significantly decreased AD duration ($p > 0.01$) (a) and the behavioral seizure score ($p < 0.001$) (b). (From Goodman, J. H. et al., *Epilepsia*, 46(1), 1–7, 2005. With permission.)

28.3.3 SAFETY

The electrode location for each animal was identified in Nissl-stained tissue. We did not detect evidence of damage at any of the electrode sites where LFSWS was delivered. In naive animals that received 6 min of continuous LFSWS there was no evidence of damage in Nissl-stained and Fluoro-jade-stained tissue. These results suggest that the decrease in seizure incidence observed after LFSWS was not due to obvious tissue damage.

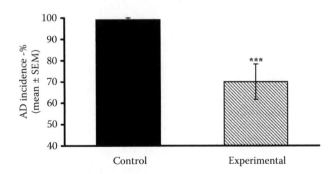

FIGURE 28.2 Preemptive LFSWS significantly decreased the incidence of AD by 30% during kindling acquisition ($p < 0.0001$). (From Goodman, J. H. et al., *Epilepsia*, 46(1), 1–7, 2005. With permission.)

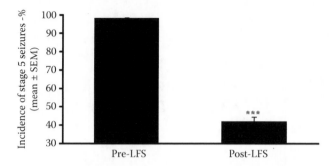

FIGURE 28.3 Preemptive LFSWS significantly decreased the incidence of stage 5 kindled seizures in rats fully kindled in the amygdala from 98% in the control rats to 42% in experimental rats ($p < 0.0001$). (From Goodman, J. H. et al., *Epilepsia*, 46(1), 1–7, 2005. With permission.)

28.4 DISCUSSION

The ability to combine therapeutic intervention with seizure detention should lead to improved outcomes in patients with epilepsy. DBS is an exciting new therapy for epilepsy that can readily be combined with seizure detection technology in an implantable device. However, the development of DBS for epilepsy is still in its infancy, with issues of where and how to stimulate yet to be resolved. Stimulation frequency is likely a key parameter in determining the potential efficacy of a given stimulus. There are many examples of the animal (Hablitz and Rea 1976; Lado et al. 2003; Lockhard et al. 1979; Vercuiel et al. 1998; Wyckhuys et al. 2009) and clinical studies (Benabid et al. 2002; Boon et al. 2007; Hodaie et al. 2002; Lockman and Fisher 2009; Osorio et al. 2001, 2005; Velasco et al. 2000, 2007) that have assessed the efficacy of electrical stimulation utilizing HFS. While HFS is effective in some patients, the utilization of the alternative LFS may capture additional patients. There is clinical evidence that LFS can be effective (Tkesheleshvili et al. 2003; Yamamoto et al. 2007) but further basic research is required to properly characterize this stimulation paradigm to optimize efficacy as well as to insure safety. Low-frequency 1 Hz pulsatile stimulation effectively decreased interictal and ictal-like activity in several in vitro preparations (Albensi et al. 2004; Barbarosie and Avoli 1997; Khosravani et al. 2003) and has been shown to be effective in several in vivo animal studies (Velisek et al. 2002; Wu et al. 2008; Zhang et al. 2009; Zhu-Ge et al. 2007). LFSWS has also been shown to be effective by Gaito and colleagues (1980), but their observations were largely ignored. In the present study, we examined the effect of preemptive LFSWS during kindling acquisition and in fully kindled rats.

28.4.1 Effect of Preemptive LFSWS on Kindling Acquisition

While LFSWS had no effect on the number of stimulus required for the animals to become fully kindled, it did significantly decrease AD duration and behavioral seizure score. However, the observation that LFSWS decreased the incidence of AD in response to the kindling stimulus is more significant since, in control animals, a kindling stimulus rarely fails to elicit an AD. The failure to elicit an AD can only be attributed to the LFSWS.

28.4.2 Effect of Preemptive 1-Hz Sine Wave and 1-Hz Pulsatile Stimulation in Fully Kindled Rats

Preemptive LFSWS significantly decreased the incidence of stage 5 seizures in animals fully kindled in the amygdala and hippocampus by 30%–60%. It is significant that when the kindling stimulus failed to elicit a stage 5 seizure, the seizure appeared to be completely blocked. There was no

AD or behavioral seizure. This can only be attributed to the LFSWS. The observation that LFSWS was effective in amygdala-kindled as well as hippocampal-kindled rats indicates that effectiveness of LFSWS is not structure-dependent. In contrast, a preemptive LTD-like 1-Hz pulsatile stimulus had no effect on the incidence of stage 5 seizures in rats fully kindled in the hippocampus.

28.4.3 Why Was LFSWS Effective?

The mechanism underlying the anticonvulsant effect of LFSWS is unclear. In many ways, 1-Hz stimulation mimics the frequency of interictal and postictal spikes. It has been hypothesized that the interictal spike may be inhibitory and that an increase in interictal activity may decrease the likelihood of an ictal event. The observation by Carrington et al. (2007) that the same LFSWS used in this study elevated the AD threshold for a period of days could explain why it was more difficult to elicit a kindled seizure in these animals. It is not clear why, in this study, the LFSWS was effective while the LFS was not, since LFS has been reported to be effective in other studies (Velisek et al. 2002; Wu et al. 2008; Zhang et al. 2009; Zhu-Ge et al. 2007). In this study, the LFS was tested using a fixed current intensity. It is possible that if the intensity of the LFS were increased it would become effective. However, it is important to recognize that the total charge delivered by a 1-s continuous sine wave is much greater that what is delivered by a LTD-like stimulus of the same amplitude using a 1-ms pulse duration. There may also be something unique about sinusoidal stimulation. There are reports of sinusoidal field stimulation suppressing epileptiform activity (Bikson et al. 2001; Lian et al. 2003; Sunderam et al. 2009) and axonal conduction (Jensen and Durand 2007); however, these stimulations were delivered at a much higher frequency.

28.4.4 Future Goals

Since kindling is a seizure model and not a true model of epilepsy, it will be important to determine whether LFSWS is also effective in experimental models that produce spontaneous seizures. The injection of tetanus toxin into the neocortex or hippocampus creates a nonlesional epileptic focus that leads to spontaneous seizure activity similar to what is observed in human partial epilepsy (Benke and Swann 2004; Nilsen et al. 2005). Alternatively, pilocarpine-induced status epilepticus creates a hippocampal lesion similar to what is seen in human mesial temporal sclerosis that is accompanied by spontaneous seizures (Goodman 1998). A demonstration that LFSWS can decrease seizure activity in these models will increase the likelihood that this type of DBS can be effective in human epilepsy. The ultimate goal is to combine LFSWS with improved seizure detection in an implantable responsive stimulator.

REFERENCES

Albensi, B. C., G. Ata, E. Schmidt, J. D. Waterman, and D. Janigro. 2004. Activation of long-term synaptic plasticity causes suppression of epileptiform activity in rat hippocampal slices. *Brain Res.* 13:56–64.

Barbarosie, M., and M. Avoli. 1997. Ca3-driven hippocampal-entorhinal loop controls rather than sustains in vitro limbic seizures. *J. Neurosci.* 17:9308–9314.

Benabid, A. L., S. Chabardes, J. Mitrofanis, and P. Pollak. 2009. Deep brain stimulation of the subthalamic nucleus for the treatment of Parkinson's disease. *Lancet Neurol.* 8:67–81.

Benabid, A. L., L. Minotti, A. Koudsie, A. de Saint Martin, and E. Hirsch. 2002. Antiepileptic effect of high-frequency stimulation of the subthalamic nucleus (Corpus Luysi) in a case of medically intractable epilepsy caused by focal dysplasia: A 30-month follow-up: Technical case report. *Neurosurgery* 50:1385–1392.

Benke, T. A., and J. Swann. 2004. The tetanus toxin model of chronic epilepsy. *Adv. Exp. Med. Biol.* 548:226–238.

Bikson, M., J. Lian, P. J. Hahn, W. C. Stacey, C. Sciortino, and D. M. Durand. 2001. Suppression of epileptiform by high frequency sinusoidal fields in rat hippocampal slices. *J. Physiol.* 531:181–191.

Boon, P., K. Vonck, V. De Herdt, A. Van Dycke, M. Goethals, L. Goossens, M. Van Zandijcke, T. De Smedt, I. Dewaele, R. Achten, W. Wadman, F. Dewaele, J. Caemaert, and D. Van Roost. 2007. Deep brain stimulation in patients with refractory temporal lobe epilepsy. *Epilepsia* 48:1551–1560.

Carrington, C. A., K. L. Gilby, and D. C. McIntyre. 2007. Effect of low-frequency stimulation on amygdala-kindled afterdischarge thresholds and seizure profiles in fast- and slow-kindled rat strains. *Epilepsia* 48:1604–1611.

Chkhenkeli, S. A., and I. S. Chkhenkeli. 1997. Effects of therapeutic stimulation of nucleus caudatus on epileptic electrical activity of brain in patients with intractable epilepsy. *Stereotact. Funct. Neurosurg.* 69:221–224.

Cooper, I. S., I. Amin, and S. Gilman. 1973. The effect of chronic cerebellar stimulation on epilepsy in man. *Trans. Am. Neurol. Assoc.* 98:192–196.

Cooper, I. S., and A. R. Upton. 1978. Effects of chronic cerebellar stimulation on epilepsy, the EEG and cerebral palsy in man. *Electroencephalogr. Clin. Neurophysiol.* S34:349–354.

Durand, D. M., and M. Bikson. 2001. Suppression and control of epileptiform activity by electrical stimulation: A review. *Proc. IEEE* 89:1065–1082.

Fisher, R. S., S. Uematsu, G. L. Krauss, B. J. Cysyk, R. McPherson, R. P. Lesser, B. Gordon, P. Schwerdt, and M. Rise. 1992. Placebo-controlled pilot study of centromedian thalamic stimulation in treatment of intractable seizures. *Epilepsia* 33:841–851.

Gaito, J. 1980. The effect of variable duration one hertz interference on kindling. *Can. J. Neurol. Sci.* 7: 59–64.

Gaito, J., J. N. Nobrega, and S. T. Gaito. 1980. Interference effect of 3 Hz brain stimulation on kindling behavior induced by 60 Hz stimulation. *Epilepsia* 21:73–84.

Gale, K. 1992. Subcortical structures and pathways involved in convulsive seizure generation. *J. Clin. Neurophysiol.* 9:264–277.

Goddard, G. V., D. C. McIntyre, and C. K. Leech. 1969. A permanent change in brain function resulting from daily electrical stimulation. *Exp. Neurol.* 25:295–330.

Goodman, J. H. 1998. Experimental models of status epilepticus. In *Neuropharmacology methods in epilepsy research*, eds. S. L. Peterson and T. E. Albertson, 97–125. New York: CRC Press.

Goodman, J. H. 2004. Brain stimulation as a therapy for epilepsy. *Adv. Exp. Med. Biol.* 548:239–247.

Goodman, J. H. 2005. Low frequency sine wave stimulation decreases the incidence of kindled seizures. In *Kindling*, 6th ed., eds. M. Corcoran and S. Moshe, 343–353. New York: Springer.

Goodman, J. H. 2009. Deep brain and peripheral nerve stimulation as potential therapies for epilepsy. In *Encyclopedia of basic epilepsy research,* Vol. 3, ed. P. A. Schwartzkroin, 1421–1426. Oxford: Academic Press.

Goodman, J. H., R. E. Berger, and T. K. Tcheng. 2005. Preemptive low-frequency stimulation decreases the incidence of amygdale-kindled seizures. *Epilepsia* 46:1–7.

Hablitz, J. J., and G. Rea. 1976. Cerebellar nuclear stimulation in generalized penicillin epilepsy. *Brain Res. Bull.* 1:599–601.

Hodaie, M., R. A. Wennberg, J. O. Dostrovsky, and A. M. Lozano. 2002. Chronic anterior thalamus stimulation for intractable epilepsy. *Epilepsia* 43:603–608.

Jensen, A. L., and D. M. Durand. 2007. Suppression of axonal conduction by sinusoidal stimulation in rat hippocampus in vitro. *J. Neural Eng.* 4:1–16.

Khosravani, H., P. L. Carlen, and J. L. PerezVelazquez. 2003. The control of seizure-like activity in the hippocampal slice. *Biophys. J.* 84:687–695.

Lado, F. A., L. Velisek, and S. L. Moshe. 2003. The effect of subthalamic stimulation of the subthalamic nucleus on seizures is frequency dependent. *Epilepsia* 44:157–164.

Lesser, R. P., S. H. Kim, L. Beyderman, D. L. Miglioretti, W. R. Webber, M. Bare, B. Cysyk, G. Krauss, and B. Gordon. 1999. Brief bursts of pulse stimulation terminate afterdischarges caused by cortical stimulation. *Neurology* 53:2073–2081.

Lian, J., M. Bikson, C. Sciortino, W. C. Stacey, and D. M. Durand. 2003. Local suppression of epileptiform activity by electrical stimulation in rat hippocampus in vitro. *J. Physiol.* 547:427–434.

Litt, B., R. Esteller, J. Eschauz, M. D'Alessandro, R Shor, T. Henry, P. Pennell, C. Epstein, R. Bakay, M. Dichter, and G. Vachtsevanos. 2001. Epileptic seizures may begin hours in advance of clinical onset: A report of five patients. *Neuron* 30:51–64.

Lockard, J. S., G. A. Ojemann, W. C. Condon, and L. L. DuCharme. 1979. Cerebellar stimulation in alumina-gel monkey model: Inverse relationship between clinical seizures and EEG interictal bursts. *Epilepsia* 20:223–234.

Lockman, J., and R. S. Fisher. 2009. Therapeutic brain stimulation for epilepsy. *Neurol. Clin.* 27:1031–1040.

Mirski, M. A., L. A. Rossell, J. B. Terry, and R. S. Fisher. 1997. Anticonvulsant effect of anterior thalamic high frequency electrical stimulation in the rat. *Epilepsy Res.* 28:89–100.

Nilsen, K. E., M. C. Walker, and H. R. Cock. 2005. Characterization of the tetanus toxin model of refractory focal neocortical epilepsy in the rat. *Epilepsia* 46:179–187.

Osorio, I., M. G. Frei, B. F. Manly, S. Sunderam, N. C. Bhavaraju, and S. B. Wilkinson. 2001. An introduction to contingent (closed-loop) brain electrical stimulation for seizure blockage, to ultra-short term clinical trials, and to multidimensional statistical analysis of therapeutic efficacy. *J. Clin. Neurophysiol.* 18(6):533–544.

Osorio, I., M. G. Frei, S. Sunderam, J. Giftakis, N. C. Bhavaraju, S. F. Schaffner, and S. B. Wilkinson. 2005. Automated seizure abatement in humans using electrical stimulation. *Ann. Neurol.* 57(2):258–268.

Paxinos, G., and C. Watson. 1998. *The rat brain in stereotaxic coordinates*, 4th ed. San Diego: Academic Press.

Racine, R. J. 1972. Modification of seizure activity by electrical stimulation, II: Motor seizure. *Electroencephalogr. Clin. Neurophysiol.* 32:281–294.

Schimdt, D. 2001. Vagus nerve stimulation for the treatment of epilepsy. *Epilepsy Behav.* 2:S1–S5.

Sramka, M., and S. A. Chkhenkeli. 1990. Clinical experience in intraoperative determination of brain inhibitory structures and application of implanted neurostimulators in epilepsy. *Stereotact. Funct. Neurosurg.* 54–55:56–59.

Sun, F. T., M. J. Morrell, and R. E. Wharen Jr. 2008. Responsive cortical stimulation for the treatment of epilepsy. *Neurotherapeutics* 5:68–74.

Sunderam, S., N. Chernyy, N. Peixoto, J. P. Mason, S. L. Weinstein, S. J. Schiff, and B. J. Gluckman. 2009. Seizure entrainment with polarizing low-requency electric fields in a chronic animal epilepsy model. *J. Neurol. Eng.* 6:046009.

Tkesheleshvili, D., H. Zaveri, J. Krystal, S. Spencer, D. D. Spencer, and I. Cavus. 2003. Effects of 1-Hz intracranial stimulation on interictal spiking and EEG power in humans. *Epilepsia* 44:246.

Velasco, A. L., F. Velasco, M. Velasco, D. Trejo, G. Castro, and J. D. Carrillo-Ruiz. 2007. Electrical stimulation of hippocampal epileptic foci for seizure control: A double-blind, long-term follow-up study. *Epilepsia* 48:1895–1903.

Velasco, M., F. Velasco, A. L. Velasco, B. Boleaga, F. Jimenez, F. Brito, and I. Marquez. 2000. Subacute electrical stimulation of the hippocampus blocks intractable temporal lobe seizures and paroxysmal EEG activities. *Epilepsia* 41:158–169.

Velisek, L., J. Veliskova, and P. K. Stanton. 2002. Low-frequency stimulation of the kindled focus delays basolateral amygdala kindling in immature rats. *Neurosci. Lett.* 326:61–63.

Vercuiel, L., A. Benazzouz, C. Deransart, K. Bressand, C. Marescaux, A. Depaulis, and A. L. Benabid. 1998. High-frequency stimulation of the subthalamic nucleus suppresses absence seizures in the rat: Comparison with neurotoxin lesion. *Epilepsy Res.* 31:39–46.

Walker, A. E. 1938. An oscillographic study of the cerebello-cerebral relationship. *J. Neurophysiol.* 1:16–23.

Weiss, S. R., A. Eidsath, X. L. Li, T. Heynen, and R. M. Post. 1998. Quenching revisited: Low level direct current inhibits amygdala-kindled seizures. *Exp. Neurol.* 154:185–192.

Wright, G. D., D. L. McLellan, and J. G. Brice. 1984. A double-blind trial of chronic cerebellar stimulation in twelve patients with severe epilepsy. *J. Neurol. Neurosurg. Psychiatr.* 47:769–774.

Wu, D. C., Z. H. Xu, S. Wang, Q. Fang, D. Q. Hu, Q. Li, H. L. Sun, S. H. Zhang, and Z. Chen. 2008. Time-dependent effect of low-frequency stimulation on amygdaloid-kindled seizures in rats. *Neurobiol. Dis.* 31:74–79.

Wyckhuys, T., R. Raedt, K. Vonck, W. Wadman, and P. Boon. 2010. Comparison of hippocampal Deep Brain Stimulation with high (130Hz) and low frequency (5Hz) on afterdischarges in kindled rats. *Epilepsy Res.* 88:239–246.

Yamamoto, J., A. Ikeda, T. Satow, K. Takeshita, M. Takayama, M. Matsuhashi, R. Matsumoto, S. Ohara, N. Mikuni, J. Takahashi, S. Miyamoto, W. Taki, N. Hashimoto, J. C. Rothwell, and H. Shibasaki. 2002. Low-frequency electrical stimulation has an inhibitory effect on epileptic focus in mesial temporal lobe epilepsy. *Epilepsia* 43:491–495.

Yamamoto, J., A. Ikeda, M. Kinoshita, R. Matsumoto, T. Satow, K. Takeshita, M. Matsuhashi, N. Mikuni, S. Miyamoto, N. Hashimoto, and H. Shibasaki. 2006. Low-frequency electric cortical stimulation decrease interictal and ictal activity in human epilepsy. *Seizure* 15:520–527.

Zhang, S. H., H. L. Sun, Q. Fang, K. Zhong, D. C. Wu, S. Wang, and Z. Chen. 2009. Low-frequency stimulation of the hippocampal CA3 subfield is antiepileptic and anti-ictogenic in rat amygdaloid kindling model of epilepsy. *Neurosci. Lett.* 455:51–55.

Zhu-Ge, Z. B., Y. Y. Zhu, D. C. Wu, S. Wang, L. Y. Liu, W. W. Hu, and Z. Chen. 2007. Unilateral low-frequency stimulation of the central piriform cortex inhibits amygdaloid-kindled seizures in Sprague-Dawley rats. *Neuroscience* 146:901–906.

Section VII

The State of Seizure Prediction: Technology

29 Large-Scale Electrophysiology
Acquisition, Storage, and Analysis

*Matt Stead, Mark R. Bower, Benjamin H. Brinkmann,
Christopher Warren, and Gregory A. Worrell*

CONTENTS

29.1 LARGE-SCALE ELECTROPHYSIOLOGY

Recent advances in recording technology have enabled investigators to record the full range of extracellular brain activity—spanning single neuron action potentials, microdomain local field oscillations, to high-amplitude ultraslow field potential oscillations originating from macroscale neuronal ensembles (Brinkmann 2009). Spatiotemporal multiscale recordings make possible the study of the range of spatial and temporal scales involved in seizure generation, i.e., ictogenesis (Figure 29.1). There is increasing evidence that improved spatiotemporal resolution in intracranial EEG (iEEG) recordings using microwire and macroelectrode arrays has the potential to play an important role in the understanding of normal and pathological neurophysiology in focal epilepsy (Bragin 2002; Worrell 2004; Urrestarazu 2007; Schevon 2008). Evidence acquired from long-term iEEG monitoring of both interictal and ictal states is used clinically to delineate the epileptogenic zone for surgical resection. Newly identified interictal microdomain markers of epileptogenic activity (Worrell 2008; Stead 2010; Schevon 2008) illustrate the potential of expanding the spatial and temporal bandwidth of clinical iEEG recording.

29.1.1 ACQUISITION OF WIDE-BANDWIDTH INTRACRANIAL EEG

Our laboratory acquires wide-bandwidth iEEG data in epilepsy patients undergoing long-term electrographic monitoring for presurgical evaluation using hybrid microwire and clinical macroelectrode arrays under a Mayo Clinic institutional review board (IRB)-approved protocol. These microelectrode arrays, consisting of clusters of 0.5–1.0-mm–spaced, 40-micron diameter, platinum-iridium microwires interspersed between clinical macroelectrodes, are manufactured by Adtech Medical Instrument Corporation, Racine, WI, and PMT, Chanhassen, MN under FDA 510k clearance. The impedance of the microwires ranges between 200 and 500 kOhms. Hybrid subdural grids, as well as hybrid depth electrodes are used in order to permit adequate coverage and access to medial temporal and subcortical areas. (Van Gompel 2008a, 2008b; Worrell 2008). The subdural

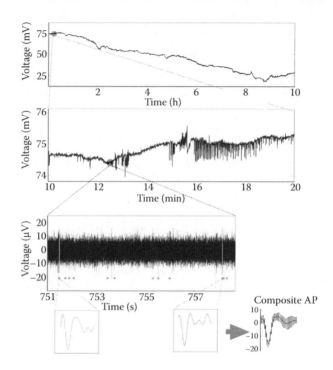

FIGURE 29.1 **(See color insert.)** Multiscale electrophysiology allows data analysis across a wide range of temporal and voltage scales.

grid macroelectrodes provide broad coverage of the potential ictal onset zone, while the surface microwires are concentrated near the center of the grid to provide more detailed microdomain information at the most likely location of the ictal onset zone. The hybrid subdural grid electrodes do not penetrate the pial surface, and do not increase the invasiveness of the intracranial EEG monitoring procedure.

Electrophysiology recordings are performed with a scalable (32–320 channel) system (Figure 29.2) developed in conjunction with Neuralynx (Neuralynx Inc., Bozeman, Montana). Signals are acquired using a single 24-bit DC-coupled low-noise A/D converter for each channel, sampling at 32 kHz with a 9 kHz antialiasing filter. This design permits signals to be captured with a net dynamic range of 264 mV at 1-μV resolution. The 24-bit A/D converter output represents 18 bits of information in each recorded sample, with the lowest 6 bits of information produced by the A/D converters consisting of random noise with an inverse-frequency spectral power correlation. Recorded samples are packetized and transferred via a fiber optic connection at 600 Mbit/sec to a dedicated PC running the Neuralynx Cheetah software system (http://www.neuralynx.com/).

Data are streamed from the Neuralynx Cheetah system across a dedicated high-capacity ethernet connection to a dedicated acquisition server, which stores the data stream on a storage area network (SAN). The SAN is managed by Apple's XSan 2.2 (Apple computer, Cupertino, CA) software with two dedicated metadata controllers. The system has 104 Terabytes of RAID storage consisting of both Apple XServe and Promise VTrak E-series (Promise Technology, Milpitas, CA) RAID arrays. The RAID units are connected via a QLogic fiber channel network, consisting of four high-speed switches (QLogic Corp., Aliso Viejo, CA). A separate metadata ethernet network connects the metadata controllers and RAID units for the purpose of maintaining an accurate catalog of the locations and extents of files stored on the RAID units. Data are accessed by SAN client computers, which are also connected to the fiber channel fabric, as well as the local ethernet network.

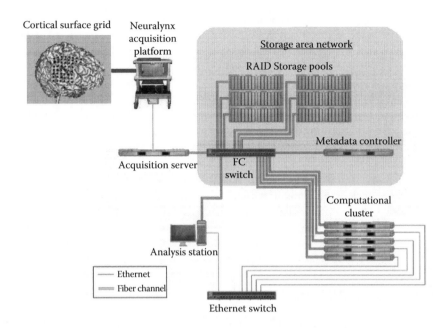

FIGURE 29.2 Schematic describing the wide-bandwidth electrographic data acquisition system and the network and data storage infrastructure needed to support it.

Large-scale wide-bandwidth recording is challenging due to the practical challenges presented by the large amounts of recorded data generated. Often these challenges have been addressed by reducing the duration, number of channels, sampling rate, dynamic range, or resolution of recordings in order to limit acquired data to manageable quantities. Our recordings are routinely performed for presurgical evaluation of epilepsy patients using hybrid arrays with over 100 channels, with the capability available to acquire up to 320 channels. Continuous recordings performed at a sampling frequency of 32 kHz at 18-bit data resolution stored uncompressed on disk would fill 250 Mb per channel per hour, assuming samples were stored contiguously with no delimiters. Presurgical epilepsy recordings are commonly performed for up to 7 days, which, for a typical recording with 100 channels, would fill 4 Tb of disk space, while a full recording session of 320 channels for this duration would fill 13 Tb. Commonly used file formats store samples as either 16-bit integers (requiring data loss from our recordings) or 32-bit integers (increasing the practical storage needs further).

In addition to the raw disk space required to store our recorded data, as the size of data files increases, so does the probability of disk storage errors within the file. Keeping online backup copies of files this large is not practical, and data backup is limited to a tape copy of each file. Detection and repair or isolation of any data errors that may occur is crucial. An additional concern regarding patient data is ensuring confidentiality, either during transmission of data across unsecured networks, or for collaboration with other medical or academic institutions. The United States Health Insurance Portability and Accountability Act (HIPAA) requires any patient-protected health information transmitted over a nonsecure network to be encrypted with a minimum 112-bit symmetric encryption (Federal Register 2003).

29.1.2 Multiscale Electrophysiology Format

These challenges have been overcome with the development of Multiscale Electrophysiology Format (MEF) (Brinkmann 2009) by our laboratory. The MEF format compresses samples up to 24-bits in length using range-encoding (Martin 1979; Bodden 2002), a form of integer arithmetic encoding that employs byte-wise scaling to improve the speed of encoding and decoding. Range-encoded difference (RED) compression begins by calculating the differences between sequential samples in the data block. The range and relative frequencies of difference values are then used to range encode the group of data samples. Calculating difference values in time-series data reduces the data's variance, and enables the range encoding algorithm to achieve greater compression than it could on the raw sample values alone. Principal among the RED algorithm's advantages are its lossless compression rate and its computational speed, as well as the algorithm's ability to handle statistical variations in the recorded data, which is particularly beneficial when dealing with nonstationary signals such as iEEG (Cranstoun 2002). This property of the algorithm also results in improved compression ratios in filtered or slowly varying data, without modification to the algorithm or its parameters.

Recorded data are stored in a series of independent compressed data blocks, each of which begins with a block header containing important information about that block, including the number of samples encoded, the maximum and minimum sample values in the block, and the time index of the first sample in the block. A 32-bit cyclically redundant checksum (CRC) value (Peterson 1961) is calculated from each compressed block using the Koopman polynomial (Koopman 2002), which has a Hamming distance of 4 for data streams up to 114,663 bits in length. The CRC value is the first entry in each block's header and it enables detection of data errors within the compressed block from disk or network transmission errors. The use of multiple independent data blocks has the effect of limiting the potential damage data errors can cause. If a particular data block were corrupted beyond repair, it could be removed without compromising the remaining data in the file. In some compression schemes, notably difference encoding, a single data error can propagate throughout the remaining data in the file. The compressed data blocks are stored maintaining 8-byte alignment in order to enable direct access to block header fields and to assist file recovery.

In order to permit direct access to any of the variable-length compressed data blocks within the file, a catalog of the starting recorded sample index, microsecond uUTC block start time, and offset within the file to the data block is calculated and stored. This catalog is stored as a series of 8-byte integer triplets at the end of the MEF file. This catalog allows direct data block access based on a recorded sample's time stamp or recording sample number. These time stamps are stored as microsecond coordinated universal time (uUTC), which is defined by the number of microseconds since midnight January 1, 1970, GMT. Eight-byte integer microseconds provide adequate resolution for EEG recordings, and avoid the use of floating point data types and the associated errors that can arise from truncation of the least significant bits.

Each MEF file begins with a fixed-length one kilobyte file header with optional 128-bit AES encryption (NIST 2001) of patient identifying information, as well as separate optional encryption of the file's technical parameters and data. The MEF header begins with an unencrypted section describing the byte order of integers in the file, the encryption algorithm used and which areas of the file are encrypted, the file's version (in anticipation of future MEF versions), the length of the header, the institution where the file originated, and a general-use unencrypted text field. This is followed by a subject section containing patient name and identification number fields. The subject section of the header is encrypted by a separate password, and can be encrypted or decrypted independently from the rest of the file's header. Next comes a session section, containing a wide range of information regarding the data acquisition. The session section of the header includes the number of recorded samples in the file, the data sampling frequency, the channel name, the start and end uUTC times of the recording session, the file offset to the block index offset catalog, a scale factor to convert from recorded sample values to microvolt units, the size of the largest compressed block in the file, and the maximum and minimum recorded values. This section of the file header can optionally be encrypted with a separate password, which, if used, is stored in the subject-encrypted section of the header, allowing the subject password to serve as a sort of master key to the entire file. The session password can also optionally be used to encrypt the leading coefficients of the range encoding model in each of the data block headers, rendering the data itself impossible to read without the encryption key. The use of a three-tiered encrypted format allows flexibility in what information is available to the end user. For example, if a research collaboration required patient data to be transferred to an academic institution, the original MEF file could be shared with only the session password provided, making patient-identifying information inaccessible, but leaving the technical details of the recording and data unobscured.

MEF is an open-source file format, and a file format specification document, posix-standard C source code, and MATLAB functions to generate and read MEF files have been made freely available, along with example multi-scale EEG data and annotations, on our laboratory's Internet site (http://mayoresearch.mayo.edu/mayo/research/msel/). In addition, Neuralynx is working to implement support for saving recordings directly in MEF format from its recording equipment.

29.1.3 MULTISCALE ANNOTATION FORMAT (MAF)

Labels, descriptors, or specifications attached to portions of raw data are called metadata or annotations. In a sense, they can be thought of as data about data. The metadata and annotations describing multiscale data clearly fit the criteria of "Big Data" (Howe 2008), if only because of the sheer number of events that may be observed across different temporal and spatial scales in a multiscale data set. In addition, the communication, storage, and analysis of multiscale data annotations in regards to clinical applications poses additional problems, such as issues regarding patient privacy. Fortunately, each of these problem domains has produced general, well-tested solutions that have been packaged into broadly available software packages, many of which exist as open-source software. XML describes a family of related formats that allow the communication of application-specific information between users and programs. Databases are widely used to store and query data in an organized manner. Object-oriented programming methodologies underlie a number of analysis tools written in modern programming languages, including Java and Objective-C. When integrated, these approaches have led to a bioinformatics revolution that has permanently altered many fields of biology (e.g., genomics with GenBank) with a few notable exceptions, such as systems neurophysiology.

The Multiscale Annotation Format (MAF) developed by our group provides an integrated software package and data format tailored to the practical problems of generating, storing, and communicating systems neurophysiology annotations and metadata. One foundation upon which MAF was based is the XML-based Clinical Data Exchange (XCEDE) schema, developed by the Biomedical Informatics Research Network (BIRN, http://www.nbirn.net). A particular strength of this schema is the ability to represent many-to-many relationships between groups of data, even in linear, text-based representations of data, such as XML. For example, a given patient may see many different doctors, while a given doctor may see many different patients, creating a many-to-many relationship. Describing the complete set of patient-doctor relationships for a given clinic could be done easily in a linear text file if replication of information was not considered a problem: all of the information related to a given doctor could be replicated under the heading of each patient seen by that doctor, and all of the information related to each patient could be replicated under the heading of each doctor caring for those patients. Such an approach, however, leads to descriptive files that are needlessly long, and the replication of data increases the opportunities for error and confusion. XCEDE solves this problem by attaching a unique ID to each data element of a given type (e.g., person, hospital) and then using this ID to describe arbitrary relationships between elements without replicating detailed information. Because XCEDE focuses on issues related to clinical data, the representation of research data required the construction of several new XML tag types, which we have labeled as the XREDE (XML-based Research Data Exchange) format.

A unique identification value for each element of data lends itself to solutions beyond the communication of data provided by XML: a unique ID is required for description of an element in a database, and is helpful for packaging related data as elements in an instantiation of an object-oriented class. Persistence of multiscale data in a database provides several benefits over flat file storage in operating system folders, including error-checking during data entry, increased search capability, and flexible linkage of data elements. This allows users to find new relationships within existing datasets through a process that has come to be called data-mining. MAF assigns each XML element type to a unique database table, creating an intuitive mapping between a single XML tag or element and a single database entry. In a similar fashion, each database table can be associated with a class in an object-oriented programming language, such as Java. The information in an individual XML element can then be represented as the data within an instantiation, or object, of that class. As an example, MAF includes an XML tag type, a relational database table and an object-oriented class that are each called "Subject," which contains all of the information related to a single person, along with software tools for flexibly converting that information between XML elements, software objects, and relational database entries. This class-tag-table format provides a solution to the problem of organizing big data sets across multiple working groups by standardizing data communication, analysis, and storage according to accepted, widely used, and robust software packages: JDOM for XML, Java for the object-oriented language, and MySQL for the relational database.

We have written a MATLAB-based EEG viewer, eeg_view, that is capable of identifying and annotating features in multiscale EEG data, and uses MAF to load and save annotation data, such as the times associated with seizure onsets, artifacts in data sets, and the occurrence of interictal, epileptiform events (Figure 29.3). Annotations may be created manually by a user selecting time points of interest using a mouse or keyboard, or they may be created by automated analysis programs, the results of which can subsequently be reviewed and refined by an expert. Annotations can then be saved to an XML file that stores information related to the subject, EEG channels, analysis task, and individual annotation events that can subsequently be stored to a database or loaded into data analysis software. Previous annotations can be reloaded into the EEG viewer by simply selecting the appropriate XML file, which allows the original EEG channels and events to be reloaded into the

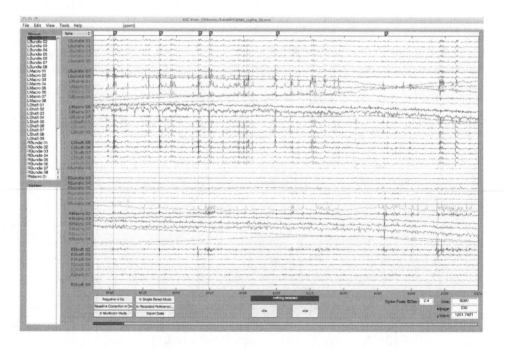

FIGURE 29.3   (See color insert.) Multiscale EEG viewer showing recorded macroelectrode and microwire channels acquired in an epilepsy patient. Interictal spikes were marked using a semiautomated spike peak detector and are shown by flags at the top of the screen and vertical lines through the EEG traces. The spike annotations are stored in an XML file using MAF, allowing the user to close and reload the annotation session as desired.

data viewer. The success of the class-tag-table approach to the problem of viewing and annotating EEG data suggests that this approach will be applicable to a wide range of data analysis tasks.

29.2 CLINICAL APPLICATION AND RESULTS

To date, our lab has recorded wide-bandwidth electrophysiology data from over 45 patients using clinical macroelectrodes and hybrid microwire arrays, representing over 7000 h of wide-bandwidth recordings. The majority of these patients suffer from temporal or neocortical epilepsy, but a few represent patients without epilepsy implanted with subdural electrodes over motor cortex and receiving cortical stimulation to treat medically intractable facial pain. All patient recordings were performed under a clinical research protocol approved by our institution's IRB. Our findings support the hypothesis that human epileptic brain is composed of pathological microcircuits generating microseizures and HFO. We have recently demonstrated that these microdomain epileptiform events are electrographic signatures (biomarkers) of an epileptic brain, and may be involved in the generation of seizures.

29.3 SUMMARY

Systems electrophysiology recorded across a wide bandwidth from large numbers of electrodes and over a wide dynamic range has the potential to expand our understanding of human and animal neurophysiology. Here we present our innovative approach to clinical electrophysiological recording and the solutions to some of the numerous technical challenges to this process that must be overcome in order for this potential to be realized.

ACKNOWLEDGMENTS

The authors acknowledge the contributions of Karla Crockett, Andrew Gardner, and Cindy Nelson. This work was supported by the National Institutes of Health (Grant K23 NS47495 and R01-NS039069), the Minnesota Partnership for Biotechnology, and by an Epilepsy Therapy Development Project grant from the Epilepsy Foundation of America.

REFERENCES

Bodden, E., M. Clasen, and J. Kneis. 2002. Arithmetic coding in a nutshell. In *Proseminar Datenkompression 2001*. Aachen: University of Technology.

Bragin, A., I. Mody, C. L. Wilson, and J. Engel Jr. 2002. Local generation of fast ripples in epileptic brain. *J. Neurosci.* 22(5):2012–2021.

Brinkmann, B. H., M. R. Bower, K. A. Stengel, G. A. Worrell, and M. Stead. 2009. Large-scale electrophysiology: Acquisition, compression, encryption, and storage of big data. *J. Neurosci. Methods* 180(1):185–192.

Centers for Medicare & Medicaid Services (CSM). 2003. HHS. Health insurance reform: Security standards; Final rule. *Fed. Regist.* 68(34):8334–8381.

Cranstoun, S. D., H. C. Ombao, R. von Sachs, W. Guo, and B. Litt. 2002. Time-frequency spectral estimation of multichannel EEG using the auto-SLEX method. *IEEE Trans. Biomed. Eng.* 49(9):988–996.

Gardner, A., G. Worrell, E. Marsh, D. J. Dlugos, and B. Litt. 2007. Human and automated detection of high-frequency oscillation in clinical intracranial EEG recordings. *J. Clin. Neurophysiol.* 118:1134–1143.

Howe, D., M. Costanzo, P. Fey, T. Gojobori, L. Hannick, W. Hide, D. Hill, R. Kania, M. Schaeffer, S. St Pierre, S. Twigger, O. White, and S. Rhee. 2008. Big data: The future of biocuration. *Nature* 455(7209):47–50.

Koopman, P. 2002, June. 32-bit cyclic redundancy codes for internet applications. The International Conference on Dependable Systems and Networks. 459.

Martin, G. N. N. 1979. Range encoding: An algorithm for removing redundancy from a digitised message. Video & Data Recoding Conference, Southampton.

NIST. 2001, November. Federal Information Processing Standards Publication 197. *Announcing the Advanced Encryption Standard (AES)*. Springfield, VA: NTIS.

Peterson, W. W., and D. T. Brown. 1961. Cyclic codes for error detection. *Proceedings of the IRE* 49: 228.

Schevon C. A., S. K. Ng, J. Cappell, R. R. Goodman, G. McKhann, A. Waziri, A. Branner, A. Sosunov, C. E. Schroeder, and R. G. Emerson. 2008. Microphysiology of epileptiform activity in human neocortex. *J. Clin. Neurophysiol.* 25(6):321–330.

Stead, S. M., M. R. Bower, B. H. Brinkmann, K. Lee, W. R. Marsh, F. B. Meyer, B. Litt, J. J. VanGompel, and G. A. Worrell. 2010. Microseizures and the spatiotemporal scales underlying the genesis of focal seizures in human partial epilepsy. *Brain* 133(9):2789–2797.

Urrestarazu, E., R. Chander, F. Dubeau, and J. Gotman. 2007. Interictal high-frequency oscillations (100–500 Hz) in the intracerebral EEG of epileptic patients. *Brain* 130(Pt 9):2354–2366.

Van Gompel, J. J., G. A. Worrell, M. L. Bell, T. A. Patrick, G. D. Cascino, C. Raffel, W. R. Marsh, and F. B. Meyer. 2008a. Intracranial electroencephalography with subdural grid electrodes: Techniques, complications, and outcomes. *Neurosurgery* 63(3):498–505.

Van Gompel, J. J., S. M. Stead, C. Giannini, F. B. Meyer, W. R. Marsh, T. Fountain, E. So, A. Cohen-Gadol, K. Lee, and G. A. Worrell. 2008b. Phase I trial: Safety and feasibility of intracranial electroencephalography using hybrid subdural electrodes containing macro- and microelectrode arrays. *Neurosurg. Focus* 25(3):E23.

Worrell, G. A., A. B. Gardner, S. M. Stead, S. Hu, S. Goerss, G. J. Cascino, F. B. Meyer, W. R. Marsh, and B. Litt. 2008. High-frequency oscillations in human temporal lobe: Simultaneous microwire and clinical macroelectrode recordings. *Brain* 131(Pt 4):928–937.

Worrell, G. A., L. Parish, S. D. Cranstoun, R. Jonas, G. Baltuch, and B. Litt. 2004. High-frequency oscillations and seizure generation in neocortical epilepsy. *Brain* 127:1496–1506.

30 EPILAB: A MATLAB® Platform for Multifeature and Multialgorithm Seizure Prediction

B. Direito, R. P. Costa, H. Feldwisch-Drentrup,
M. Valderrama, S. Nikolopoulos, B. Schelter,
M. Jachan, C. A. Teixeira, L. Aires, J. Timmer,
M. Le Van Quyen, and A. Dourado

CONTENTS

30.1 INTRODUCTION

During the past several decades, an enormous effort has been devoted by the scientific and clinical communities to seizure prediction in medically refractory epilepsy. Several papers have been published in journals and conference proceedings proposing different predictors to meet this challenge. These studies have evolved from straightforward descriptions of seizure precursors to controlled studies applying novel prediction algorithms to long-term EEG recordings; for an extensive review, see the work by Mormann et al. (2007). However, most studies are designed based on selected data subsets and when tested prospectively on extended, out-of-sample data, fail in most cases. In addition, the vast majority of these studies reflect the trade off between sensitivity (the ability to predict seizures) and specificity, usually described by the false-positive rate, with none achieving clinical applicability.

The feeling that there is no predictor with performance statistically acceptable for all real-life scenarios is becoming a consensus in the research community. For these reasons, epilepsy research has focused on the uniqueness of each patient's brainwave patterns, suggesting that the studies should try a multiplicity of techniques and algorithms to adjust patient-specific parameters, and, hopefully, find a suitable one with clinical utility.

EPILAB is a computational framework that has been conceived and developed along these lines to allow a rapid design and training of several predictors for long-term EEG datasets and to compare their performance. This framework intends to provide a simple graphical user interface to simplify the use of the methodologies and computational tools available for nonspecialists.

Several open-source toolboxes have been developed for EEG multichannel processing, including MEA-Tools (Egert et al. 2002), EEGLAB (Delorme et al. 2004), BSMART (Cui et al. 2008), ERPWAVELAB (Morup et al. 2006), and MATS (Kugiumtzis et al. 2010), where MATLAB (The Mathworks, Inc.) has been a predominant choice as a platform. Despite their usefulness, these toolboxes are not suitable for seizure-predictor studies.

EPILAB includes a variety of feature-extraction methods and prediction algorithms that are complemented with statistical validation schemes and extensive graphical capabilities. From the software engineering perspective, it has been conceived as an open architecture where new features, algorithms, and predictors can be easily introduced. EPILAB accepts different data formats and new ones can be easily included. EPILAB is an application that will be made available to the scientific and clinical communities.

The development of EPILAB is part of the EPILEPSIAE Project* that aims to develop an intelligent alarming system, the "Brainatics," which will hopefully be able to predict seizures. Additionally, the system should be transportable and allow the patients to assess their actual risk in real time using appropriate algorithms and their efficient combinations.

This chapter describes EPILAB, a new software tool to ease the development, testing, and validation of seizure prediction algorithms encompassing within the framework univariate and bivariate feature-extraction methods, computational intelligence algorithms, algorithms based on thresholds (with or without consideration of circadian dependencies), and statistical methods for results assessment. Lastly, we reflect on possible future developments.

30.2   EPILAB SOFTWARE ARCHITECTURE

EPILAB represents one of the milestones proposed by the EPILEPSIAE project and has been developed by its research partners. The toolbox was designed using MATLAB version R2008a (v7.6) and has been tested up to release R2010a (v7.10). The framework was developed in Microsoft Windows (XP, Vista, and 7) and Linux (Debian and Ubuntu) environments, has been successfully tested on Mac OS X, and is expected to work on other operating systems that support MATLAB.

Recent releases of MATLAB have included the object-oriented programming (OOP) paradigm. Modularity and encapsulation enabled by OOP were two of the main specifications required in the development of EPILAB, especially to incorporate new methods and enable the continuous updating and development of a project by different users. Another important advantage of MATLAB is the possibility of establishing interfaces between EPILAB and previously developed toolboxes.

---

* EPILEPSIAE, Evolving Platform for Improving Living Expectations of Patients Suffering from IctAl Events, EU FP7 Grant 211713, http://www.epilepsiae.eu

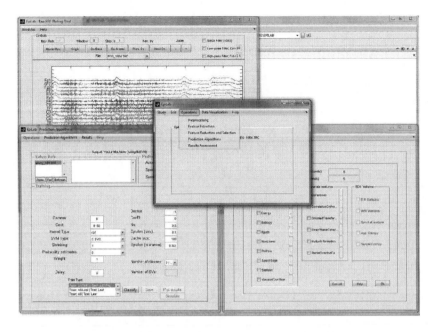

FIGURE 30.1 EPILAB main window and functionalities available; the user can use the GUI to interact with the functions and access data directly using the command line.

A collection of C/MATLAB functions for signal processing and data visualization is accessible through a user-friendly graphical user interface (GUI).

30.2.1 GUI STRUCTURE

The GUI was designed using a multiwindow approach with the following stages: data loading, preprocessing, feature extraction, prediction algorithms, assessment, and data visualization. This structure intends to ease the integration of custom scripts developed by more advanced users.

The main GUI presents an interface to the most important signal-processing and data-edition modules. The application interface presents four menus: *Study*, *Edit*, *Operations*, and *Data Visualization*. The GUI Study menu has two main items: *New displays* a dialog where the user can open an EEG file or an entire folder with a set of files, and *Open* presents an interface from which the user can load a previously created study allowing the user to further explore the facilities available. The *Edit* menu presented in the main window handles the shortcuts to metadata visualization and editing. The GUI Operations menu has five items. The first two items represent shortcuts to the feature extraction methods' and prediction algorithms' GUIs, from which the user can easily interact with the methods implemented in the toolbox. This menu also presents the shortcuts to the *Preprocessing*, *Feature Reduction and Selection*, and *Results Assessment* interfaces. Finally, the *Data Visualization* menu allows the user to access the different available options for data display (see Figure 30.1).

30.2.2 DATA STRUCTURES

EPILAB uses the object "*STUDY*" to store EEG and ECG file information, event information, metadata, and computed features. Different file formats can be accessed through plugins that can be integrated within the software. TRC files from Micromed S.p.A., binary files from Nicolet, EPILEPSIAE binary files format,* and MATLAB MAT-files plugins are currently implemented and new formats

---

* EPILEPSIAE binary format was developed by the EPILEPSIAE partners to increase the efficiency of data exchange.

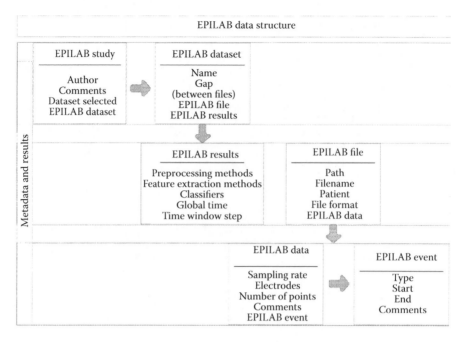

FIGURE 30.2 Data structure based on the MATLAB object-oriented programming.

can be added. Advanced users may access this information using the MATLAB command line. A schematic representation of the complete data structure is provided in Figure 30.2.

30.3 UNIVARIATE AND MULTIVARIATE FEATURES

30.3.1 DATA PREPROCESSING

One of the most needed developments for seizure-prediction algorithms is their prospective, out-of-sample testing based on continuous long-term recordings. To avoid memory limits due to the size of data files, EEG data is loaded only when and while it is required by the methods (data streaming). When a file (or a set of files) is selected during the creation of a new study, the file path is saved in the data structure and a pointer to the last loaded sample is defined. This pointer is then updated during the data streaming. The metadata existent in the different file formats is integrated into the EPILAB data structure.

Standard signal-preprocessing methods available in EPILAB include Finite Impulse Response filters and normalization. These simple methods allow basic preprocessing before the application of feature-extraction techniques.

30.3.2 FEATURE EXTRACTION AND FEATURE REDUCTION

The implemented methods follow a moving time-window approach. The size of this time window is parameterized by the user as well as its moving time step. If the window size exceeds the step size, the data segments will naturally overlap. EPILAB saves the computed features in the STUDY object as described above. Alternatively, the extracted features can be saved in binary files to improve data exchange with new studies using the same data. Table 30.1 summarizes the set of computable features incorporated in EPILAB.

TABLE 30.1
Feature Extraction Methods Implemented in the EPILAB Platform

	Group		Feature
EEG	Univariate		AR Modelling Predictive Error
			Decorrelation Time
			Energy
			Entropy
		Horjth	Mobility
			Complexity
		Nonlinear	Largest Lyapunov Exponent
			Effective Correlation Dimension
		Relative Power	Delta Band (0.1–4 Hz)
			Theta Band (4–8 Hz)
			Alpha Band (8–15 Hz)
			Beta Band (15–30 Hz)
			Gamma Band (30–2000 Hz)
		Spectral Edge	Power
			Frequency
		Statistics	1st Moment (Mean)
			2nd Moment (Variance)
			3rd Moment(Skewness)
			4th Moment (Kurtosis)
		Energy of Wavelet Transform Coef.	Several Decomposition Levels
	Multivariate		Coherence
			Correlation on the Prob. of Recurrence
			Directed Transfer Function
			Mean Phase Coherence
			Mutual Information
			Partial Directed Coherence
ECG		RR-Statistics	Mean
			Variance
			Skewness
			Kurtosis
		BPM-Statistics	Mean
			Variance
			Minimum
			Maximum
		Frequency domain	Very Low Freq. (<0.04 Hz)
			Low Freq (0.04–0.15 Hz)
			High Freq (0.15–0.4 Hz)
			Approximate Entropy

30.3.2.1 EEG Features

The attempts to extract seizure predictors from surface EEG recordings date back to the 1970s and these initial studies led several groups to use linear approaches for the seizure prediction problem (Mormann et al. 2007). Later, the rise of the theory of nonlinear systems in the 1980s originated novel approaches to the subject. During the past few decades, bivariate measures have been used to relate EEG patterns of preseizure periods with the dynamics observed in biological neural networks (Chávez et al. 2003). EPILAB integrates most of the features that have been used by the numerous researchers in this area. The current challenges that seizure prediction faces are related to the methodologies used and their statistical evaluation to obtain clinical applicability, which remains unachieved thus far.

30.3.2.1.1 Univariate Features

Here we give a brief introduction to the univariate methods presented in Table 30.1. The energy of the signal quantifies temporal patterns of the EEG (Litt et al. 2001) and the spectral analysis allows the description of specific patterns in certain frequency bands, such as the well-known delta, theta, alpha, beta, and gamma frequency bands. The autoregressive modelling error is a measure of the predictiveness of a given time series. Lyapunov exponents and correlation dimension model the nonlinear chaotic behavior of the brain (Iasemidis et al. 1990; Chaovalitwongse et al. 2006). The nonlinear system analysis in EPILAB is based upon the open-source toolbox TSTOOL (DPI Gottingen 1997).

30.3.2.1.2 Bivariate and Multivariate Features

Other significant studies on seizure prediction emphasize the importance of multivariate features, such as synchronization measures. These allow the differentiation of the ictal period from the interictal period based on synchronization patterns. Several bivariate measures associated with brainwave synchronization are available in EPILAB, such as coherence and mean phase coherence (Chávez et al. 2003).

30.3.2.2 ECG

Some properties of the ECG have been related to seizure precursors (Novak et al. 1999). Some of the properties that can be calculated are the R–R interval statistics (mean, variance, maximum, and minimum), the approximate entropy (descriptor of irregularity of the R–R interval), frequency domain features [divided in three bands: very low (<0.04 Hz), low (<0.15 Hz) and high frequency (<0.4 Hz)] and beats per minute (BPM) statistics (mean, variance, maximum, and minimum).

30.3.2.3 Features Selection and Reduction

As an example, the use of the features provided in Table 30.1 for a regular montage of 31 EEG electrodes plus ECG results in a high-dimensional feature space with more than 1,000 features. Using all these dimensions may lead to the curse of dimensionality (also known as the Hughes effect). Feature selection and reduction methods transform this high-dimensional feature space into a lower-dimensional space, either by selecting a subset of features (e.g., using search algorithms) or performing a data transformation. With these methods, computational complexity is substantially reduced while the loss of information is minimized.

EPILAB implements principal component analysis (PCA) and multidimensional scaling (MDS). Additionally, EPILAB has an interface to the toolbox VISRED (Dourado et al. 2007), which consists of user-friendly software for data reduction, classification, and clustering. VISRED includes several techniques for MDS and clustering, as well as extensive visualization techniques.

30.4 SEIZURE PREDICTION ALGORITHMS

30.4.1 Computational Intelligence Algorithms

Recently, some machine learning algorithms, such as artificial neural networks (ANN) and support vector machines (SVM) have been introduced to the seizure-prediction problem. In these approaches, the detection of the preictal phase is defined as a classification problem on pattern-by-pattern basis (see Figure 30.3). Using the supervised learning principle, a training set (in-sample) is selected and its classification target (class) is defined, followed by the training of a classifier. The number of classes can be two (preictal and nonpreictal) for the binary classification or four (ictal, preictal, interictal, and postictal) depending on the user's goal. The nonlinear boundaries given by

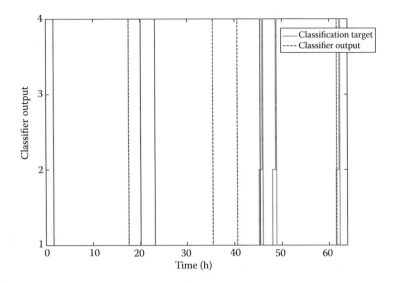

FIGURE 30.3 **(See color insert.)** Alarms generated by the output of an SVM model (red line), and the target classification levels (green line). The four classification levels correspond to (1) interictal, (2) preictal, (3) ictal, and (4) postictal. The plot displays results from a dataset of more than 63 h. The model is able to predict two seizures (three occurred within the testing set) with a false-positive rate of 0.09/h.

these classifiers tend to be a good fit for many nonlinear problems (Bishop 2006). The evaluation of the classifier is performed for the remaining dataset (out-of-sample).

The selection of the training set (in-sample) and of the evaluation set (out-of-sample) can be accomplished using different approaches: cross-validation, multiple evaluations with different combinations of the training and testing datasets, and hold out (i.e., 70% for training and 30% for testing). Manual selection of datasets using the implemented visualization routines is also possible. Unbalanced training sets, where one class dominates all others (notice that the extent of the interictal period generally exceeds by far the other periods), can also influence the quality of the classifier. In EPILAB, the biased datasets can be corrected by equalizing the number of samples of the preictal and nonpreictal periods.

To evaluate the results, five different metrics are used: (1) point-by-point sensitivity, (2) point-by-point specificity, (3) point-by-point accuracy, i.e., the proportion of correct classified points, (4) the false alarm rate per hour, and (5) sensitivity related to the number of seizures correctly predicted. These metrics are largely used in this field of research, easing the comparison process between different case studies (Costa et al. 2008; Mirowski et al. 2008). The similarity between evaluation schemes of the different prediction algorithms highlights another important EPILAB feature, i.e., the support for the incorporation of new algorithms developed by other users. This represents a great advance in the development of seizure-prediction algorithms, allowing a quick and transparent comparison among different algorithms for the same data.

30.4.1.1 Artificial Neural Networks

ANN are machine learning techniques inspired by the brain. They are capable of approximating any function, given the right training and data input (Bishop 1998). Different ANN architectures have been applied to the classification of ictal and preictal periods, using different features (Petrosian et al. 2000; Costa et al. 2008; Mirowski et al. 2008), with promising results.

Costa and coworkers (2008) suggested that the application of recurrent ANN with memory may achieve better results for seizure prediction when compared to other architectures. Convolutional neural networks (Mirowski et al. 2008) are a type of multilayer ANN with optimal and time-invariant learning of local features present in the input matrix, which is particularly relevant to this problem due to the high variability in the time domain.

EPILAB implements several ANN architectures, such as feed-forward multilayer networks, layer-recurrent networks, Elman networks, distributed time delay, and feed-forward input time delay as implemented in the neural networks MATLAB toolbox (Demuth et al. 2010). These networks can be fully parameterized using a friendly interface.

30.4.1.2 Support Vector Machines

Support vector machines are another important machine learning technique that has been shown to achieve, in some cases, better generalization than ANN (Cortes et al. 1995). Recent attempts have shown that SVM can obtain a high specificity and a low false alarm rate (Mirowski et al. 2008; Chisci et al. 2010). However, as with ANN, experiments with longer data sets (on the order of days) remain to be performed using accurate validation methods.

The open-source library libSVM (Chang et al. 2001) forms the base of the SVM classifiers in EPILAB. This library is implemented in C and is invoked from MATLAB, thereby allowing the SVM to achieve a very good computational performance. The full parameterization of SVM classifiers is available through a GUI, where it is possible to choose different kernel functions (linear, polynomial, radial basis function, or sigmoid) among other parameters. Both ANN and SVM models can be applied in the high-dimensional feature space as well as in the reduced feature space computed by MDS or other feature-selection methods.

30.4.1.2.1 Combination of MDS and SVM

The classical MDS method requires the processing of the entire dataset in a batch mode to obtain the data transformation. Such a configuration restricts real-time implementation. To overcome these restraints, preliminary studies were performed assuming the linearity of the MDS transformation.

In summary, a transformation matrix was determined using the training set and then used prospectively, sample by sample, in the test set. Each sample (represented in a lower dimensional space) was then classified using the SVM model on the reduced training set. The training and testing sets considered were similar in the MDS and non-MDS cases. The number of reduced variables was determined using 80% of the sum of the eigenvalues of the variables that resulted from the transformation. In a case study with 132 features (all univariate features available in EPILAB used for six scalp electrodes) and preictal period of 30 min, the MDS transformation resulted in a reduced nine-dimensional space. The best result (point-by-point analysis) using MDS and SVM has a 21% sensitivity, 80% specificity, and 75% accuracy, while the complete dataset (without MDS transformation) presents 31% sensitivity, 70% specificity, and 66% accuracy.

These and other preliminary results seem to suggest that the combination of methods can be an important step toward the reduction of the complexity presented by this problem, without major drawbacks in the classification results.

30.4.1.2.2 Results Visualization

Feature-reduction and feature-selection methods can also contribute to the analysis and selection of the training subsets. Using the first three dimensions obtained by MDS, it is possible to see the dispersion of data and, for example, to analyze the influence of different preictal periods (see Figure 30.4).

30.4.2 STATISTICAL VALIDATION OF PREDICTOR OUTPUTS

The seizure prediction characteristic (Winterhalder et al. 2003) evaluates sensitivity and false prediction rate using the concepts of seizure occurrence period (SOP, period during which the seizure is expected to occur) and seizure prediction horizon (SPH, minimum window of time between the alarm and the beginning of the SOP to render an intervention—therapeutic or behavioral). The authors

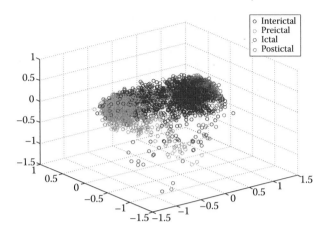

FIGURE 30.4   (See color insert.) Three-dimensional representation of a high-dimensional feature space using the three first variables resulting from MDS method. Here, the preictal period considered was 30 min and the total of number of features in the original dataset was 132.

suggest this general framework to evaluate and compare different predictors. Later, Schelter et al. (2006) stated that the performance of a good predictor has to be superior to the performance given by a random predictor. EPILAB implements these methods to assess the performance of predictors.

30.5   GRAPHICAL FUNCTIONALITIES

One of the facilities most requested (by clinicians, technicians, and developers) is the ability to visualize the data. Here we consider not only the raw data but also the computed features. The graphical representation of EEG data allows the users to identify outliers, noise-related artifacts, and other specific patterns, whereas feature visualization permits the user to identify variations in the feature pattern around specific events (e.g., preictal events). Another interesting possibility is the comparison of features (see Figure 30.5).

EPILAB allows the possibility of navigating through data using two approaches: a sliding window (time based) and an event-based window (displaying the data segments containing events). Visualization functions were also designed considering possible memory limitations.

EPILAB visualization routines include channel selection, zooming facilities, and window size selection. Filters were also considered and implemented (notch, low-pass, and high-pass filters). To ease the analysis of long-term recordings and respective features, a "movie-like" visualization is also possible.

30.6   FUTURE WORK

Prediction algorithms based on circadian concepts will be implemented in EPILAB to complement the computational intelligence algorithms with approaches that take into account physiological considerations.

The extraction of several features from dozens of EEG channels leads to an explosion in the number of features and possible combinations. To avoid overfitting and to improve the model performance, as well as to increase the classification speed, an appropriate feature selection is essential (Guyon et al. 2003). This selection has been done empirically, which is time consuming and prone to human error. To overcome this problem, a future improvement to EPILAB will be the addition of search algorithms developed for hard optimization problems, such as simulated annealing, tabu search, evolutionary algorithms, and ant-colony algorithms in single- or multicriteria frameworks.

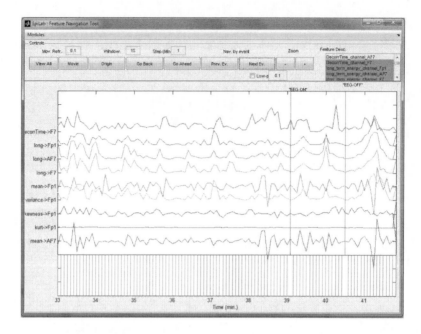

FIGURE 30.5 Feature navigation tool developed for EPILAB. Feature extraction methods were implemented in the EPILAB platform.

The continuous development of EPILAB will result in a public version that will be available at the end of the EPILEPSIAE project. In a longer time perspective, EPILAB will prioritize developments that will lead to an improved environment for data sharing and easier access to standardized, shared tools, as well as to shared environments for computational analysis.

30.6.1 New Contributions to EPILAB by Users

More experienced MATLAB users can integrate into the platform custom scripts, taking advantage of the different available features. Such users should be able to introduce new functions and interfaces that encapsulate different methods. The comparison between methods, the use of different classification strategies, and a correct statistical evaluation of the predictor performance are some of the advantages that the users can experience from EPILAB.

30.6.2 Database Connection

Another milestone of the EPILEPSIAE project is the creation of an epilepsy-based SQL database. The development of an interface between EPILAB and the database is also planned and represents one of the near-term aims of the framework. The interface should allow database querying facilities, which will eventually be useful in classification paradigms. Metadata information available in the database can provide additional information, which is important for seizure prediction (e.g., pharmacological treatments).

30.6.3 Summary

We have presented an overview of EPILAB, a computational framework designed for seizure-prediction studies. The project's main objective is to provide a platform that allows the design, training, testing, and evaluation of computational predictors and allow the community to contribute with their own methods. Preliminary work has allowed the design of hundreds of models with easy selection of the best methodologies and parameters with promising results.

It also represents an effort toward real-time studies in epilepsy, which are crucial for clinical applicability of seizure-prediction algorithms.

ACKNOWLEDGMENTS

We would like to thank Hitten Zaveri for all the valuable comments on the manuscript. This work was supported by the European Union (211713). HFD, MJ, BS, and JT would like to express their gratitude for additional support from the German Federal Ministry of Education and Research (01GQ0420), the German Science Foundation (Ti315/4-2, Sonderforschungsbereich-TR3), and the Excellence Initiative of the German Federal and State Governments. BD and RPC would like to acknowledge the support from the Portuguese Foundation for Science and Technology (SFRH / BD / 47177 / 2008).

REFERENCES

Andrezjak, R., D. Chicharro, C. Elger, and F. Mormann. 2009. Seizure prediction: Any better than chance? *Clin. Neurophysiol.* 120(8):1465–1478.

Andrezjak, R., F. Mormann, T. Kreuz, C. Rieke, A. Kraskov, C. E. Elger, and K. Lehnertz. 2003. Testing the null hypothesis of the non-existence of a preseizure state. *Phys. Rev. E* 67(1 Pt 1):1–4.

Bao, F. S., J.-M. Gao, J. Hu, D. Y.-C. Lie, Y. Zhang, and K. J. Oommen. 2009. Automated epilepsy diagnosis using interictal scalp EEG. In *31st Annual International Conference of the IEEE Engineering in Medicine and Biology Society.* Minneapolis.

Bettus, G., F. Wendling, M. Guye, L. Valton, J. Régis, P. Chauvel, and F. Bartolomei. 2008. Enhanced EEG functional connectivity in mesial temporal lobe epilepsy. *Epilepsy Res.* 81(1):58–68.

Bishop, C. M. 1998. *Neural networks and machine learning,* vol. 168. New York: Springer.

Bishop, C. M. 2006. *Pattern recognition and machine learning*, vol. 16, eds. M. Jordan, J. Kleinberg, and B. Scholkopf. New York: Springer.

Chang, C., and C. Lin. 2001. LIBSVM: A library for support vector machines.

Chaovalitwongse, W., O. Prokopyev, and P. Pardalos. 2006. Electroencephalogram (EEG) time series classification: Applications in epilepsy. *Ann. Oper. Res.* 148(1):227–250.

Chávez, M., M. Le Van Quyen, V. Navarro, M. Baulac, and J. Martinerie. 2003. Spatio-temporal dynamics prior to neocortical seizures: Amplitude versus phase coupling. *IEEE Trans. Biomed. Eng.* 50(5):571–583.

Chisci, L., A. Mavino, G. Perferi, M. Sciandrone, C. Anile, G. Collicchio, and F. Fugetta. 2010. Real time epileptic seizure prediction using AR models and support vector machines. *IEEE Trans. Biomed. Eng.* 57(5):1124–1132.

Cortes, C., and V. Vapnik. 1995 Support-vector networks. *Machine Learning* 20(3): 273–297.

Costa, R., P. Oliveira, G. Rodrigues, B. Leitão, and A. Dourado. 2008. Epileptic seizure classification using neural networks with 14 features. In *Proceedings of the 12th International Conference on Knowledge-Based Intelligent Information and Engineering, Part II.* Zagreb, Croatia, 281–288.

Cui, J., L. Xu, S. L. Bressler, M. Ding, and H. Liang. 2008. BSMART: A MATLAB/C toolbox for analysis of multichannel neural time series. *Neural Networks, Special Issue on Neuroinformatics* 21(8):1094–1104.

Delorme, A., and S. Makeig. 2004. EEGLAB: An open source toolbox for analysis of single-trial EEG dynamics including independent component analysis. *J. Neurosci. Methods* 134(1):9–21.

Demuth, H., M. Beale, and M. Hagan. 2010. User's Guide. *Neural network toolbox 6.* Natick, MA: MathWorks.

Direito, B., A. Dourado, M. Vieira, and F. Sales. 2008. Combining energy and wavelet transform for epileptic seizure prediction in an advanced computational system. In *Proceedings of the 2008 International Conference on Biomedical Engineering and Informatics.* Sanya, China, 380–385.

Dourado, A., E. Ferreira, and P. Barbeiro. 2007. VISRED—Numerical data mining with linear and nonlinear techniques. *Proceedings of the In Advances in Data Mining. Theoretical Aspects and Applications.* Leipzig, Germany, 93–106.

DPI Gottingen. 1997. *TSTOOL Home Page,* viewed April 2007, http://www.physik3.gwdg.de/tstool.

Egert, U., T. Knott, C. Schwarz, M. Nawrot, A. Brandt, S. Rotter, and M. Diesmann. 2002. MEA-Tools: An open source toolbox for the analysis of multielectrode data with MATLAB. *J. Neurosci. Methods* 117(1):33–42.

Guyon, I., and A. Elisseeff. 2003. An introduction to variable and feature selection. *J. Mach. Learn. Res.* 3:1157–1182.

Iasemidis, L. D., J. D. Sachellares, H. P. Zaveri, and W. J. Williams. 1990. Phase space topography and the Lyapunovs exponent of electrocorticograms in partial seizures. *Brain Topogr.* 2(3):187–201.

Kugiumtzis, D., and A. Tsimpiris. 2010. Measures of Analysis of Time Series (MATS): A MATLAB toolkit for computation of multiple measures on time series data bases. *J. Stat. Softw.* 33(5):1–28.

La Van Quyen, M., J. Martineria, V. Navarro, M. Baulac, and F. J. Varela. 2001. Characterizing neurodynamic changes before seizures. *J. Clin. Neurophysiol.* 18(3):191–208.

Litt, B., and J. Echauz. 2002. Prediction of epileptic seizures. *Lancet Neurol.* 1(1):22–30.

Litt, B., R. Esteller, J. Echauz, M. D'Alessandro, R. Shor, T. Henry, P. Pennell, C. Epstein, R. Bakay, M. Dichter, and G. Vachtsevanos. 2001. Epileptic seizures may begin hours in advance of clinical onset: A report of five patients. *Neuron* 30(1):51–64.

Mirowski, P. W., Y. LeCun, D. Madhavan, and R. Kuzniecky. 2008. Comparing SVM and convolutional networks for epileptic seizure prediction from intracranial EEG. *IEEE Machine Learning for Signal Processing.*

Mormann, F., T. Kreuz, C. Rieke, R. Andrzejak, A. Kraskov, P. David, C. E. Elger, and K. Lehnertz. 2005. On the predictability of epileptic seizures. *Clin. Neurophysiol.* 116(3):569–587.

Mormann, F., R. Andrzejak, C. E. Elger, and K. Lehnertz. 2007. Seizure prediction: The long and winding road. *Brain* 130(2):314–333.

Morup, M., L. K. Hansen, and S. M. Arnfred. 2007. ERPWAVELAB: A toolbox for multi-channel analysis of time-frequency transformed event related potentials. *J. Neurosci. Methods* 161:361–268.

Novak, V., A. L. Reeves, P. Novak, P. A. Low, and F. Sharbrough. 1999. Time-frequency mapping of R-R interval during complex partial seizures of temporal lobe origin. *J. Auton. Nerv. Syst.* 77:195–202.

Petrosian, A., D. Prokhorov, R. Homan, R. Daseiff, and D. Wunsch. 2000. Recurrent neural network based prediction of epileptic seizures in intra- and extracranial EEG. *Neurocomputing* 30(1):201–218.

Sackellares, J. C. 2008. Seizure prediction. *Epilepsy Curr.* 8(3):55–59.

Schelter, B., M. Winterhalder, T. Maiwald, A. Brandt, A. Schad, A. Schulze-Bonhage, and J. Timmer. 2006. Testing the statistical significance of multivariate time series analysis techniques for epileptic seizure prediction. *Chaos* 16(1):013108.

Valdés-Sosa, P. A., J. M. Sánchez-Bornot, A. Lage-Castellanos, M. Vega-Hernández, J. Bosch-Bayard, L. Melie-Garcia, and E. Canales-Rodriguez. 2005. Estimating brain functional connectivity with sparse multivariate autoregression. *Philos. Trans. R. Soc. B* 360(1457):969–981.

Winterhalder, M., T. Maiwald, H. U. Voss, R. Aschenbrenner-Scheibe, J. Timmer, and A. Schulze-Bonhage. 2003. The seizure prediction characteristic: A general framework to assess and compare seizure prediction methods. *Epilepsy Behav.* 4:318–325.

31 Emerging Technologies for Brain-Implantable Devices

Bruce Lanning, Bharat S. Joshi, Themis R. Kyriakides,
Dennis D. Spencer, and Hitten P. Zaveri

CONTENTS

31.1 INTRODUCTION

Information gathered from the direct measurement of the cortex and subcortical structures has been central to the current understanding of epilepsy and its treatment. As measurement of the brain improves with the advent of new sensor modalities and technologies, so does our understanding of epilepsy and options for its control. Direct measurement of the brain in epilepsy and the possibility of intervention to control the expression of this disorder has been a long-lasting interdisciplinary effort involving engineers, physicists, mathematicians, epileptologists, neuroscientists, and neurosurgeons. However, transitioning new and emerging technologies fostered in academic, research, medical, and industrial environments into a successful product presents unique challenges. This chapter reviews some of the emerging technologies for brain-implantable devices for epilepsy and the challenges for their manufacture and commercialization.

31.2 BACKGROUND

Research in epilepsy relevant to the topic explored here involves a two-pronged approach of monitoring and interpreting brain activity over multiple length and time scales in combination with intervention to prevent seizures. Historically, to monitor brain activity, investigators have employed noninvasive and invasive electrodes down to micrometer- and nanometer-length scales (Gesteland et al. 1959) with an end goal of measuring electrical signals at the cellular level in the central nervous system. Although it is possible to obtain information at the single-neuron level with a single microelectrode, understanding the nature of signal communication and processing in the complex neural networks of the brain has conventionally required an array of electrodes, including depth electrodes, deployed in subdural strips and grids, as well as sophisticated multimodal sensors in various proximities and sizes (length scales).

Alongside the need to track seizures with sophisticated multimodal sensor arrays in epileptic networks is the need to arrest or prevent clinical events. Motivated by the successful therapeutic approaches in implantable cardiac devices, similar types of intervention approaches can be applied to neurological disorders. For example, in some cases electric fields have been shown to prevent seizure onset in a variety of locations when applied to portions of epileptic networks. Focal cooling can take advantage of channel dynamics, slowing down their activity to make cells less excitable. Yet another intriguing avenue is the use of implantable devices to elute (deliver) antiepileptic drugs (AEDs) focally. An intervention approach could also be based on raising an alarm to allow the patient to seek safety and use a fast-acting AED.

Whether for monitoring or intervention, conventional measurement/sensing devices are proving to be inadequate and self limiting in terms of (1) size (mass and volume), (2) accuracy and sensitivity, (3) spectral bandwidth, (4) spatial and temporal multimodal resolution, and (5) material biocompatibility. There is an ongoing need, therefore, to build a more effective electric (and/or chemical) interface to the cellular world in addition to a self-powered, implantable device platform with signal processing, data management, and wireless telemetry. Fortunately, over the past 30–50 years, there has been tremendous advancement in the technology of transistors and integrated circuitry (IC). In the case of transistors, there has been a dramatic decrease in size (to dimensions in the tens of nanometers) along with a corresponding increase in complexity (or functionality) over time following the trend known as Moore's law, wherein the number of transistors on an integrated circuit doubles

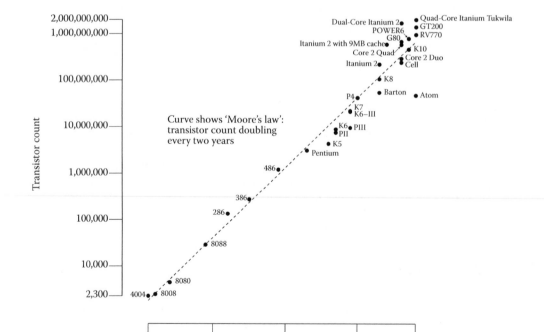

FIGURE 31.1 Graphical representation of Moore's law depicting the increased number of transistors per unit area. In the case of medical implants, this electronic paradigm has not only reduced the physical size of a processor, but increased the functionality and speed at a lower operating power requirement.

every 24 months (Figure 31.1) (Intel Corporation). As a result, the computation power of a state-of-the-art microprocessor, such as Intel's Pentium 4 with 42 million transistors per chip at ~75 watts of power, is able to process at a rate of one tenth to one twentieth the rate of the human brain at 20 million billion computations per second.

With respect to medical implants, the technological advancements presented above for the transistor, as per Moore's law, will directly help to reduce an implant's size and cost, as well as improve its embedded signal processing speed and functionality, sensitivity, dynamic range, and overall energy efficiency (power reduction). In the case of an implant or biosensor, there still remains the challenge of the body's response—specifically, tissue response—to a foreign object; i.e., does the interface or interaction between a foreign object and the body affect the measurement or function of the implanted device? Sensors, whether chemical/electrochemical, optoelectronic, or mechanical, must be compatible with the human body and not in themselves contribute to the measurement. It is fortunate that the primary backbone material in transistors and integrated circuitry is silicon, a material that has been shown to be biocompatible. Further development, however, is required to create a stable electronic and biological interface within the body.

Advancements in microelectronics have provided a tool set of emerging technologies, i.e., microprocessors (logic), sensors, power, telemetry, and biocompatible materials, that can be used in the development of the next generation of medical implantable devices. As discussed above, the computational speed of today's microprocessors nearly match that of the human brain, with dramatic increases in numbers of operations and functions on a per-unit-volume and weight basis, due to the development of transistors on a 20- to 30-nanometer length scale. An overview of how these emerging technologies can and are being incorporated into medical implant devices, in addition to the commercialization challenge, is discussed below.

31.3   EMERGING TECHNOLOGIES

31.3.1   MULTISCALE AND MULTIMODAL SENSING

As the need for more comprehensive intracranial monitoring increases, a larger number of electrode contacts is required to cover a larger amount of the neocortex with fine spatial. In addition to the need for studying seizures spatially, there is also a need to sample electrophysiological activity at higher frequencies as well; i.e., reports have linked high-frequency activity spatially to the seizure onset area and temporally to the time of seizure. Although new emerging technologies, as discussed below, could potentially permit characterization of both the spatial and temporal topology of seizures, it is important to establish what recording bandwidth will be necessary and sufficient for accurate seizure detection and prediction. In other words, as spatial resolution increases, bandwidth increases and, correspondingly, power consumption and heat dissipation also increase; the challenge then is one of balancing bandwidth with power.

The IC fabrication technologies of film growth, electronic doping, lithography, and etching (both anisotropic and isotropic etching, as well as boron etch stops in silicon) have enabled high-density, micron-scale implantable electrodes and probes. Through monolithic integration with signal-processing circuitry, the IC technology has enabled the single, miniaturized biocompatible platform shown schematically in Figure 31.2 (Wise et al. 2004). Examples of integrated silicon electrodes for field potentials from the University of Michigan and the University of Utah (Wise et al. 2004; Harrison et al. 2006) are shown in Figure 31.3. In comparison to traditional metal microelectrodes or micropipettes, a biocompatible silicon electrode provides a versatile interface to the electronic circuitry.

In addition to the intracranial electroencephalography (icEEG) measurement, there has been a desire to enhance real-time sensing of the extracellular environment by incorporating other sensing modalities such as pH, [K^+], [Ca^{2+}], and neurotransmitters, which may allow improved understanding of functional changes within the brain. Although traditional techniques for determining ion

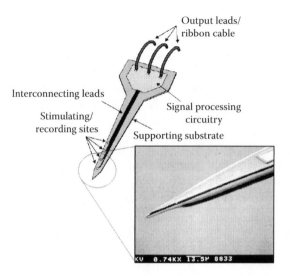

FIGURE 31.2 **(See color insert.)** Example of a silicon neural probe enabled by the microfabrication technologies of the integrated circuit/electronics industry. (Reproduced from Wise, K. D. et al., *Proceedings of the IEEE*, 92, 76–97. With permission.)

concentrations, such as microdialysis and ion-selective probes, have led to important advances in both epilepsy and trauma, there are considerable technical difficulties, such as time-consuming, labor-intensive offline analysis and inherent kinetic limitations, which have slowed the widespread adoption of these methods for clinical decision making. Integration of multiple modalities or sensors onto a micron-scale, silicon-based platform would help to overcome these difficulties through the creation of nanometer-scale structures, such as quantum confining channels or dots [i.e., a semiconductor whose excitons are confined in all three spatial dimensions (hence point or dot) and as a result have properties that are between those of bulk semiconductors and those of discrete molecules], that would increase temporal resolution and selectivity and, by their reduced overall feature size, increase spatial resolution. For example, investigators at the Mayo clinic (Kimble et al. 2009) have developed a wireless instantaneous neurotransmitter concentration sensing system (WINCS) to measure extracellular neurotransmitter concentration in vivo with nearly real-time graphical display on a nearby personal computer; again, this is an example of leveraging the advancements in electronics to advance functional neurosurgery.

Utah 3-D array
flip-chip bonding

8-channel
integrated system
(U. of Michigan)

FIGURE 31.3 Representation of a self-contained, implantable local-field-potential measurement system using silicon-based, monolithic integration methods from IC technology. (Reproduced from Wise, K. D. et al., *Proceedings of the IEEE*, 92, 76–97. With permission.)

Another approach to incorporating multiple modalities onto a single chip is to use the ubiquitous field effect transistor (FET) to transduce a biological signal into an electrical one. By replacing the doped semiconductor region between the source and drain electrodes of a traditional FET transistor with a one-dimensional wire or channel, and by coating an ion-specific dielectric or combined dielectric/bioreceptor layer to serve as the gate electrode, a nanometer-scale, ion-selective sensor is formed (Pijanowska et al. 2005). This type of "bioFET"-based sensor is small (less than 100 nm) and has high sensitivity (nanomole concentration levels) with inherently fast response times (millisecond); furthermore, it can be integrated with other types of sensors into dense array configurations for increased spatial resolution, and can be fabricated using conventional, low-cost complimentary metal-oxide semiconductor (CMOS) processing methods (CMOS is a technology for constructing integrated circuitry). This type of FET-based sensor architecture, derived from the integrated electronics industry, represents one way to provide integrated, addressable multimodal measurements on a single, miniaturized, biocompatible platform.

31.3.2 Power

Operating, controlling, and managing the aforementioned multimodal sensor array and its corresponding data management function requires power. Although brain monitoring in epilepsy is conventionally performed with passive sensing of electrical field potentials, the advantages of powered devices, which include signal conditioning, amplification, and digitization at the source, are multiple and include the recovery of very low-power biopotentials before they are corrupted by noise. In addition, by digitizing the signals at the point of measurement, power and serial data can be routed over a small set of wires, independent of the number of electrode contacts used to sense the brain. It is clear that future approches will employ active (powered) devices for sensing the brain. One option is to store power onboard the implant with stable, long-term storage devices, such as batteries or capacitors, with the future trend toward high-cycle-life rechargeable lithium batteries such as the thin film, solid-state lithium battery shown in Figure 31.4 (ITN Energy Systems). The top schematic in Figure 31.4 is an enlarged view of the thin-film battery construction in cross-section and the lower schematic is a representation of a battery mounted onto a circuit board in comparison to the discrete IC components (represented in grey). Alternative approaches attempt to harvest power from local energy sources, such as thermal gradients, mechanical vibrations, chemistry (fuels), fluid flow (microturbines), or electromagnetic fields (both active and passive, and including solar), in the surrounding environment, although the harvested energy is typically low and less reliable than a battery.

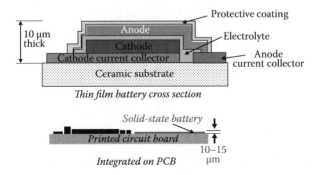

FIGURE 31.4 (See color insert.) All solid-state, thin-film rechargeable battery technology with long life (>40,000 recharge/discharge cycles) and high energy density (>300 Wh/kg). At less than 8-microns thick, this battery is not only flexible (i.e., conformable), but can be directly integrated into (or onto) the printed-circuit-board platform.

31.3.3 DATA AND CONTROL COMMUNICATIONS (TELEMETRY)

Using micro antennas, data/communication signals can be transferred wirelessly, in addition to harvesting energy (via induction), to power an implantable device or recharge an onboard battery. At frequencies between 1 and 10 MHz, electromagnetic energy is able to penetrate through the body with little energy loss and therefore can serve as both a two-way communication link (with limited bandwidth) and power source. An example of a technology in widespread use that utilizes an inductive link for combined power harvesting and two-way communication is the radio frequency identification (RFID) tag. Used for tracking and identification, the RFID tag provides a passive or active telemetry system with large memory capacity, wide reading range, and fast processing at a cost comparable to printed bar code technology, some features of which can be applied to implantable medical devices.

An inductive power/communication link consists of a receiving coil (antenna) in close proximity (near field) to a radiating (transmitting) coil to capture a portion of the radiated electromagnetic field, analogous to the circuit in a transformer (Soontornpipit et al. 2005). In the case of the body, embedding an antenna has its design challenges because the conductive medium of the body affects efficiency and performance; additionally, an antenna's physical size scales with frequency. For the range of typical communication frequencies, an optimum antenna size would be physically large, whereas for biocompatibility considerations the antenna should be as small as possible.

However, as the demand for more icEEG and multimodal monitoring channels increases, the typical data communication links in the MHz to GHz frequencies will be challenged in bandwidth and speed. Furthermore, a fundamental performance limitation exists due to electromagnetic interference issues. One way to overcome the limitations of the "low frequency" links is to use an infrared (IR) or optical data communication link. In the near IR, optical data can be transmitted without error up to distances of 45 mm in tissue and at rates that can handle the demand from increased numbers of multimodal sensors (Abita and Schneider 2004), although near IR has it limitations with respect to attenuation (absorption) and directionality.

31.3.4 FAULT TOLERANCE

As CMOS devices, such as transistors, continue to shrink in size, a gap is being created between device technology and its ability to deliver performance proportional to device density, principally due to thermal dissipation, thermal noise, and leakage current. The possibility that these factors may render Moore's law of scaling less applicable in the future has enhanced interest in semiconductor nanostructures, which offer unique advantages, such as extremely low power dissipation (Schwalke 2008). These emerging nanotechnologies, together with the scaling of CMOS devices, are expected to revolutionize electronic systems, including biomedical systems. However, as electronic devices get smaller and enter nanoscale levels, these nanodevices are expected to have higher manufacturing defects. Furthermore, due to the feature size of the devices, they will operate at reduced noise margins, which may result in a higher rate of transient faults. Furthermore, the need for continuous real-time monitoring of brain activity for extended periods of time can result in permanent faults due to aging and a loss of function due to the brain's foreign-body reaction.

We believe these factors will necessitate a need for fault-tolerant designs to assure system reliability. Fault-tolerant strategies will have to be incorporated into the architecture design to tolerate both defects introduced during manufacturing and transient and permanent faults. The architecture will have to be flexible to achieve the designed function and accommodate reconfiguration in the event of faults. We define the reliability of a system as the conditional probability that the system performs its tasks correctly throughout a given time interval $[t_0, t]$, given that the system was performing correctly at time t_0.

A possible sensing solution may consist of several sensor grids where each grid could be connected to its network of analog switches and bioinstrumentation amplifiers to amplify, condition,

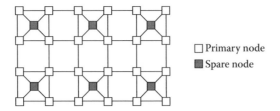

FIGURE 31.5 **(See color insert.)** A fault-tolerant scheme based on interstitial redundancy (1, 4) where a primary node can be replaced by one spare node and a spare node can replace one of the four primary nodes. A cluster, consisting of four primary nodes and the associated spare node, is considered functional if one of the failed primary nodes can be successfully replaced by the spare node or a failed spare node is successfully identified so that it is not used to replace a faulty primary node.

and digitize the signals close to the source of the signal. These sensor grids can communicate with each other through a network. To achieve a higher level of reliability, we propose a hierarchical fault-tolerant design could be used to accommodate failure of sensors, pathways, and switching circuits. For example, each sensor within a sensor network could be connected to a node consisting of switches and a bioinstrumentation amplifier, and hardware redundancy could be incorporated into the design at the node level. An approach based on interstitial redundancy (Singh 1988) can be used for the resultant network of sensors and nodes. In this approach, spare nodes are placed in interstitial sites within the network. Each spare node can replace any of the primary nodes that are adjacent to it. The main advantage of this scheme is that reconfiguration of the network in response to a detected fault is simple due to the locality of the redundant node. Figure 31.5 shows the configuration where a simple interstitial redundancy is considered. Fault recovery coverage, C, is the measure of a system's ability to recover from faults and maintain its operational state. A sensor network is considered to be functional if each cluster of sensors is working, i.e., at least four of the five nodes within each cluster are working. The reliability, R, of each cluster is given by

$$R_{\text{cluster}}(t) = p(t)^5 + C \binom{5}{1} p(t)^4 (1 - p(t)),$$

where $p(t)$ is the reliability of each switch and its associated electronics. It is assumed that when a failure occurs it is independent of other failures. The first term in the above equation is the probability that all five nodes are working at a given time t. The probability that four nodes are fault free and one node is faulty is given by $p(t)^4 (1 - p(t))$. Because any one of the five nodes (four primary and one spare nodes) can fail, this term is multiplied by $\binom{5}{1}$, read as "5 choose 1." If there are n primary nodes, then the reliability of the network is

$$R_{\text{network}}(t) = [R_{\text{cluster}}(t)]^{n/4}.$$

Figure 31.6 shows the reliability of the fault-tolerant system for $C = 1$ for different values of n and p, both with and without redundancy. Here, $C = 1$ implies that the implantable system completely recovers from faults. Clearly, redundancy improves the reliability of the network. It can also be observed that the impact of redundancy is higher when the survival probability, p, is lower.

31.3.5 Biocompatibility/Material Selection

In terms of fabrication methodology to produce miniature devices, the advances brought about by the IC industry discussed thus far have provided sophisticated electronic functions to interface

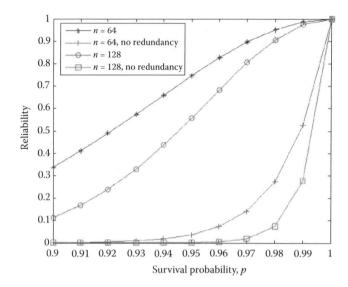

FIGURE 31.6 Estimated reliability of the configuration shown in Figure 31.5 with $n = 64$ and $n = 128$ contacts. There is a clear improvement in reliability with redundancy. It can also be observed that the impact of redundancy is higher when the survival probability, p, is lower.

with the nervous system. Unfortunately, the effectiveness of these devices can be severely limited by reaction of tissue to insertion and the continuous presence of the implant (i.e., a foreign body). In general, biomaterials, medical devices, and tissue-engineered constructs induce a foreign-body reaction (FBR) that is characterized by the formation of a collagenous capsule that aims to isolate the implant from the surrounding tissue (Mikos et al. 1998; Anderson 2007). In the brain, the FBR is characterized by the presence of both reactive astrocytes, which form a glial scar, and activated microglia (Kim et al. 2004); the FBR has shown to be the main cause of failure of brain electrode contacts (Biran et al. 2005; Polikov et al. 2005). Moreover, studies have shown that subdural electrodes induce a variety of responses, including hemorrhage and histopathology, consistent with a FBR (Stephan et al. 2001; Swartz 1996). Electrode-induced chronic inflammation has been shown to be associated with local neurodegeneration (McConnell et al. 2009). Specifically, axonal pathology in the form of hyperphosphorylation of the protein tau was observed 16 weeks following implantation in rat cortex.

Recently, one of the coauthors at Yale University obtained a detailed characterization of the FBR in the cortex of mice implanted with an inert biomaterial and observed prolonged leakage of the blood brain barrier (BBB) (Tian and Kyriakides 2009). This observation is consistent with the reported occurrence of hemorrhage in retrospective analysis of subdural implantation studies. Thus, the FBR in the brain appears associated with persistent inflammation, reactive gliosis, and disrepair of the BBB, which could all contribute to neuronal degeneration. The mouse brain cortex was chosen as model tissue and all the processing and fixation protocols were developed for the proper preservation of tissue for histological and immunohistochemical analysis (Tian and Kyriakides 2009). These included staining sections with histological stains such as hematoxylin and eosin and by immunohistochemistry with antibodies specific for inflammatory cell markers, astrocytes, and components of blood vessels. In addition, quantitative immunohistochemistry was performed to enumerate various parameters of the FBR. Utilization of this model also allowed the evaluation of the brain FBR toward novel materials, such as amorphous platinum-based bulk metallic glass, which is more compliant than the crystalline form of the alloy (Schroers et al. 2009). Specifically, it was shown that the increased elasticity of this material resulted in reduced gliosis.

Traditionally, and through years of evolution and experience, material selection for intracranial sensors and devices has been based on the inertness or immunity from biological or tissue response. For example, traditional biocompatible electrical conductors have been metals such as platinum, gold, iridium (iridium oxide), and tantalum, although more recently nonmetallic conductors such as polysilicon (doped silicon) and polymers (polyethylene, polysilane, etc.) are being used. In the case of dielectric and/or packaging and encapsulant materials for implantables, both polymer-based materials, such as silicone, epoxy, polyimide, polyurethane, and parylene, and ceramic materials such as Si_3N_4, SiO_2, and TiN have been identified.

Another factor to consider in biocompatibility is device size. As the scale of implantable sensors is reduced to micron and submicron scales, material defects and surfaces begin to dominate properties that can negatively affect performance; whereas, if the scale is reduced beyond submicron into the nanometer dimensions, there is the potential to harness positive types of effects, such as quantum confinement and the like, to enhance material properties. To address the issue of defects that are a byproduct of fabrication processes, low-cost, vacuum-based processes, such as physical vapor deposition (sputtering or evaporation) or chemical vapor deposition (low-pressure CVD, plasma-enhanced CVD, or atomic-layer deposition) can be used to fabricate thin-film materials with engineered properties, such as structure, density, strain, and chemistry. Along with this capability to engineer thin-film properties during formation, surfaces can also be modified with energetic surface treatments to enhance properties, such as hydrophobicity, adhesion, surface charge density (polarity), permeability, and biocompatibility, all of which will enable a new generation of materials for implantable devices.

Finally, another factor to consider in biocompatibility is the form factor or conformability of the device. Although silicon, with its unique combination of electronic and biocompatibility properties is able to serve multiple functions as both the active material in the device and as a readily micromachinable support material, it is rigid. A flexible form factor, on the other hand, would more easily adapt to the contours of the body and has the potential of reducing the mass and volume of a device. Fortunately, somewhat paralleling the advancements of integrated devices on silicon have been the advancements of integrated biocompatible devices on flexible substrates, such as thin polyimide film. Techniques such as "flip-chip" bonding and high-density interconnects (HDI) have been incorporated into flexible form factors as shown in Figure 31.7. Along with electronics, shown on the left in Figure 31.7, energy storage devices have also been integrated onto flexible substrates (Figure 31.7, right-hand side).

In terms of its flexibility, biocompatibility, and ability to withstand the relatively high processing temperatures for electronic integration, polyimide has traditionally been a good candidate for implantable devices, although alternative lightweight and biodegradable options are being developed. More recently, flexible biodegradable films have been considered, such as silk (Kim et al. 2010) and a cladded PEG/polyimide film wherein, as the biodegradable component of the film dissolves, the residual film and circuitry shrink to conform to the contours of the brain. The question then becomes one of compliance of the film and device; if the device is too compliant or flimsy the film may wrinkle or fold and lose interface control and electrode and sensor registry.

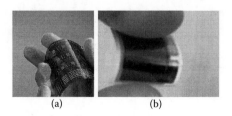

(a) (b)

FIGURE 31.7 Examples of integrated circuitry (a) and battery power storage (b) onto a flexible, biocompatible polyimide substrate. Direct integration onto thin plastic film enables circuitry and power at reduced volume and weight (mass).

In the case of a "cladded" PEG/polyimide composite structure, a polyimide-based film substrate is coated with a poly(ethylene glycol) hydrogel supplemented with poly-D-lysine (PEGDL), with higher Young's modulus, to increase the stiffness of the parent polyimide film. PEGDL, PEG, and many PEG derivatives have been investigated and widely applied for various purposes, such as drug delivery (Greenwald et al. 2003), glue for wounded nerves (Lore et al. 1999), and as a biomaterial for tissue engineering or cell culture (Riley et al. 2001; Mahoney and Anseth 2006). In addition, PEG hydrogel can be easily synthesized in situ from aqueous PEG-based precursor solution by photopolymerization. After PEG hydrogel degradation, the PEG matrix (or cladding in the case of polyimide) can remain at the original application site, as shown in a model of wounded nerve repair (Lore et al. 1999) and a therapeutic cardiomyocyte matrix transplant (Shapira-Schweitzer et al. 2007). Because of the biocompatibility and biodegradation of the PEG hydrogel, a solid cross-linked PEG hydrogel could be a desirable coating for electrodes to improve the material stiffness for placement and better conform to underlying cortical geometry after its degradation. Previously, a PEG polymer coating on an implant has been shown to provide sufficient mechanical stiffness in in vitro studies (O'Brien et al. 2001). The primary goal of the engineered biomaterial coating on the flexible electrode would be to provide a provisional stiffness for placement without compromising biocompatibility.

31.4 MANUFACTURABILITY AND COMMERCIALIZATION CHALLENGES

Advancements in electronics manufacturing/processing, such as wafer-level integration and packaging, have dramatically reduced the cost at both the discrete device level and microsystem level in addition to reducing the overall size of a package. All of the gains in electronics are directly applicable to the medical industry, particularly in the case of prosthetic implants. The challenge becomes one of how to manage development costs against the final prototype design and manufacturing. In other words, integration at an application-specific integrated circuit (ASIC) level in 0.18-μm CMOS technology can dramatically reduce the overall implant package size by up to 80% and improve onboard operating parameters such as power demand and heat dissipation in comparison with designs using conventional off-the-shelf (COTS) discrete hardware (as shown in Figure 31.8); however, overall costs are very high and increase dramatically with the typical iterative design steps that are required during development. Using low-cost COTS hardware in the design and proof-of-principle phase, however, will offer cost advantages as well as flexibility and versatility akin to existing, off-the-shelf hardware.

An alternative time and cost savings approach to COTS hardware is to use a field-programmable gate array (FPGA) or ASIC prototyping approach, i.e., electronic design automation tools, analysis, and prototyping. This approach enables a designer to verify design functionality without fabricating a chip, which can be very costly; it also ensures that the intellectual property (IP) technologies on

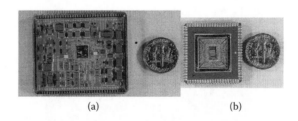

(a)   (b)

FIGURE 31.8 Representative examples of integrated circuitry using traditional off-the-shelf, discrete components soldered to circuit board (a) and chip-level ASIC (b). Monolithic integration at the ASIC level not only provides a smaller package, but greater functionality and speed, i.e., the implant in (a) is only a single-channel system whereas (b) is a 16-channel system.

TABLE 31.1
Comparison of Integrated Chip Level Features versus Discrete Package System Level Features

	ASIC (Chip Level)	Discrete Components (System Level)
Advantages	• Smaller size • Mixed signal integration • Efficient operation (low power/heat dissipation, high reliability) • Low production costs (at high volume) • IP protection (encryption)	• Cost-effective, more predictable development cycle • Low prototyping costs
Disadvantages	• High nonrecurring costs • Debugging software challenges • Significant development periods	• Large size

the system work at an acceptable level. Another advantage of this approach is that such prototyping can be used during a commercialization phase, for example when attracting investors, without investing substantial resources to fabricate ASIC chips.

Elimination of macroscale components through "chip-level" miniaturization cannot only minimize footprint and surface area, reduce power and heat dissipation, and minimize interaction with induced RF and magnetic fields, such as MRI, but ultimately reduce the cost of a device in a high-volume manufacturing scenario. A comparison of chip-level ASIC design features versus discrete off-the-shelf packaged design features is shown in Table 31.1. In some instances, discrete-circuit solutions may be used in conjunction with ASIC solutions to create a complete system with optimal performance and cost benefits.

Product commercialization requires a sound understanding of what is available and a vision of how to improve upon that. The process starts with dialogue between the marketplace (customer) and technology developer (i.e., working closely with the world's leading neuroscientists, as well as with leading research institutions and other medical device and technology companies) to ensure that a device will help advance treatment. To transition a prototype into product, rigorous design reviews, prototype testing, and postmanufacturing testing are required to ensure that meticulous design and performance criteria are met before new devices are offered to customers. After a promising technology has been developed and a product has been identified, its commercial and manufacturing viability can be demonstrated with internal and external (private or government) investment and then scaled up to manufacturing with investments from commercialization or strategic partners.

31.5 CONCLUSIONS—FUTURE PROSPECTS FOR BRAIN SENSING AND INTERVENTION

Development of an expandable implantable multimodal wireless microsystem that incorporates the emerging technologies discussed in this chapter will not only increase functionality and performance, but should also reduce risk and discomfort for patients. Additionally, a multimodal monitor, including icEEG, could provide the physician with invaluable data in NICU settings and during surgery—data that could improve decisions made for treatment of the brain. It has been widely understood that the work within the surgical treatment of epilepsy has influenced advancements in the treatments of other brain disorders and diseases. The development of a wireless multimodal brain sensing and intervention platform could open the door to further understanding and control of brain function in other disorders, diseases, and injuries.

In terms of functionality and performance, biocompatibility, miniaturization, and modularity, there are a number of emerging technologies—spurred by the electronics industry and discussed herein—that have and are being incorporated into the next generation of brain-implantable micro-systems for neurological applications. New brain implantable devices (Song et al. 2009; Irazoqui-Pastor et al. 2005) are featuring both wireless (transcutaneous) RF and IR links for power and high-speed data transfer, new materials that adapt to the physical and biological constraints of the environment, and nanometer-scale, addressable multimodal sensor architectures with enhanced sensitivity and spatial and temporal resolution—all packaged within a self-contained, miniaturized footprint that eliminates any cabling or physical connections outside the body.

Performance specifications of each implantable system vary, however, as a function of the number and type of channels and the expansion capacity of the system. For example, in the case of icEEG measurements, there are reported systems that have a large number of channels (>50) although they are only able to detect spike events as opposed to continuous, parallel acquisition of the full spectral band for all channels (Song et al. 2009). The advent of miniaturized, low-power, high-speed microprocessors and IR telemetry technologies, however, will provide solutions to the ever-expanding need for greater dynamic range (bandwidth) and sensitivity in the next generation of multimodal sensors.

Although the vision to create and incorporate a multimodal monitor into a wireless implant microsystem is at the forefront of emerging technology trends and is being investigated by many researchers, most of the activity remains embedded at the university level. Furthermore, the leading developments at universities are based on limited multimodal capabilities with poor prospects for expansion to meet user and market needs. In collaboration with strategic industrial partners, current products can be re-engineered to meet the growing need for greater numbers of sensors with better sensitivity and spatial and temporal resolution, with corresponding increases in bandwidth and data-transfer needs for wireless systems.

REFERENCES

Abita, J. L., and W. Schneider. 2004. Transdermal optical communications. *Johns Hopkins APL Technical Digest* 25:261–268.

Anderson, J. M., A. Rodriguez, and D. T. Chang. 2007. Foreign body reaction to biomaterials. *Semin. Immunol.* 20:86–100.

Biran, R., D. C. Martin, and P. A. Tresco. 2005. Neuronal cell loss accompanies the brain tissue response to chronically implanted silicon microelectrode arrays. *Exp. Neurol.* 195:115–126.

Gesteland, R. C., B. Howland, J. Y. Lettvin, and W. H. Pitt. 1959. Comments on microelectrodes. *Proceedings of the IRE* 47:1856–1862.

Greenwald, R. B., Y. H. Choe, J. McGuire, and C. D. Conover. 2003. Effective drug delivery by PEGylated drug conjugates. *Adv. Drug Deliv. Rev.* 55:217–250.

Harrison, R., P. Watkins, R. Kier, R. O. Lovejoy, D. J. Black, B. Greger, and F. Solzbacher. 2006. A low-power integrated circuit for a wireless 100-electrode neural recording system. *2006 IEEE International Solid-State Circuits Conference.* 2258–2267.

Intel Corporation, http://www.intel.com/museum/archive/history_docs/moore.htm.

Irazoqui-Pastor, P., I. Mody, and J. W. Judy. 2005. Recording brain activity wirelessly. *IEEE Eng. Med. Biol. Mag.* 24:48–54.

ITN Energy Systems, http://www.itnes.com/x.php?page=12.

Kim, D., J. Viventi, J. Amsden, J. Xiao, L. Vigeland, Y. S. Kim, J.A. Blanco, B. Panilaitis, E. S. Frechette, D. Contreras, and D. L. Kaplan. 2010. Dissolvable films of silk fibroin for ultrathin conformal bio-integrated electronics. *Nat. Mater.* 9(6):511–517.

Kim, Y. T., R. W. Hitchcock, M. J. Bridge, and P. A. Tresco. 2004. Chronic response of adult rat brain tissue to implants anchored to the skull. *Biomaterials* 25:2229–2237.

Kimble, C. J., D. Johnson, B. Winter, S. V. Whitlock, K. R. Kressin, A. E. Horne, J. C. Robinson, J. M. Bledsoe, S. J. Tye, S. Y. Chang, and F. Agnesi. 2009. Wireless instantaneous neurotransmitter concentration sensing system (WINCS) for intraoperative neurochemical monitoring. *31st Annual International Conference of the IEEE Engineering in Medicine and Biology Society, Proceedings*: 4856–4859.

Lore, A. B., J. A. Hubbell, D. S. Bobb Jr., M. L. Ballinger, K. L. Loftin, J. W. Smith, M. E. Smyers, H. D. Garcia, and G. D. Bittner. 1999. Rapid induction of functional and morphological continuity between severed ends of mammalian or earthworm myelinated axons. *J. Neurosci.* 19:2442–2454.

Mahoney, M. J., and K. S. Anseth. 2006. Three-dimensional growth and function of neural tissue in degradable polyethylene glycol hydrogels. *Biomaterials* 27:2265–2274.

McConnell, G. C., H. D. Rees, A. I. Levey, C. A. Gutekunst, R. E. Gross, and R. V. Bellamkonda. 2009. Implanted neural electrodes cause chronic, local inflammation that is correlated with local neurodegeneration. *J. Neural Eng.* 6:056003.

Mikos, A. G., L. V. McIntire, J. M. Anderson, and J. E. Babensee. 1998. Host response to tissue engineered devices. *Adv. Drug Deliv. Rev.* 33:111–139.

O'Brien, D. P., T. R. Nichols, and M. G. Allen. 2001. Flexible microelectrode arrays with integrated insertion devices. *14th IEEE International Conference on Micro Electro Mechanical Systems, Technical Digest*: 216–219.

Pijanowska, D. G., and W. Torbicz. 2005. Biosensors for bioanalytical applications. *Bull. Pol. Acad. Sci. (Tech. Sci.)* 53:251–260.

Polikov, V. S., P. A. Tresco, and W. M. Reichert. 2005. Response of brain tissue to chronically implanted neural electrodes. *J. Neurosci. Methods* 148:1–18.

Riley, S. L., S. Dutt, R. De La Torre, A. C. Chen, R. L. Sah, and A. Ratcliffe. 2001. Formulation of PEG-based hydrogels affects tissue-engineered cartilage construct characteristics. *J. Mater. Sci. Mater. Med.* 12:983–990.

Schroers, J., G. Kumar, T. M. Hodges, S. Chan, and T. R. Kyriakides. 2009. Bulk metallic glasses for biomedical application. *JOM Journal of Minerals, Metals and Materials Society* 61:21–29.

Schwalke, U. 2008. Nanotechnology: The power of small. *2008 International Conference on Signals, Circuits, and Systems*.

Shapira-Schweitzer, K., and D. Seliktar. 2007. Matrix stiffness affects spontaneous contraction of cardiomyocytes cultured within a PEGylated fibrinogen biomaterial. *Acta Biomater.* 3:33–41.

Singh, A. D. 1988. Interstitial Redundancy: An area efficient fault tolerance scheme for large area VLSI processor arrays. *IEEE Trans. Comput.* 37:1398–1410.

Song, Y. K., D. A. Borton, S. Park, W. R. Patterson, C. W. Bull, F. Laiwalla, J. Mislow, J. D. Simeral, J. P. Donoghue, and A. V. Nurmikko. 2009. Active microelectronic neurosensor arrays for implantable brain communications interfaces. *IEEE Trans. Neural Syst. Rehabil. Eng.* 17:339–345.

Soontornpipit, P., C. M. Furse, and Y. C. Chung. 2005. Miniatruized biocompatible microstrip antenna using genetic algorithm. *IEEE Trans. Antennas Propag.* 53:1939–1945.

Stephan, C. L., J. J. Kepes, K. Santa Cruz, S. B. Wilkinson, B. Fegley, and I. Osorio. 2001. Spectrum of clinical and histopathologic responses to intracranial electrodes: From multifocal aseptic meningitis to multifocal hypersensitivity-type meningovasculitis. *Epilepsia* 42:895–901.

Swartz, B. E., J. R. Rich, P. S. Dwan, A. DeSalles, M. H. Kaufman, G. O. Walsh, and A. V. Delgado-Escueta. 1996. The safety and efficacy of chronically implanted subdural electrodes: A prospective study. *Surg. Neurol.* 46:87–93.

Tian, W., and T. R. Kyriakides. 2009. Matrix metalloproteinase-9 deficiency leads to prolonged foreign body response in the brain associated with increased IL-1beta levels and leakage of the blood-brain barrier. *Matrix Biol.* 28:148–159.

Wise, K. D., D. J. Anderson, J. F. Hetke, D. R. Kipke, and K. Najafi. 2004. Wireless implantable microsystems: High density electronic interface to the nervous system. *Proceedings of the IEEE* 92:76–97.

Section VIII

Nocturnal Frontal Lobe Epilepsy:
A Paradigm for Seizure Prediction?

32 Familial and Sporadic Nocturnal Frontal Lobe Epilepsy (NFLE)— Electroclinical Features

Gholam K. Motamedi and Ronald P. Lesser

CONTENTS

32.1 INTRODUCTION

Prior to the establishment of nocturnal frontal lobe epilepsy (NFLE) as a form of epilepsy, a variety of clinical diagnoses had been suspected in these patients. These conditions included paroxysmal nocturnal dystonia (PND), paroxysmal arousal, episodic nocturnal wandering, non-REM parasomnias (e.g., sleep walking, night terrors, confusional arousals), primary psychiatric disorders, and epilepsy; in about 20% of cases no clear diagnosis could be made (Oldani et al. 1998b; Provini et al. 2000; Tinuper et al. 1990). The atypical presentation was one reason for the uncertainty about the true nature of NFLE, as was the lack of significant EEG findings during or between the episodes. Its name implies that this is a nocturnal disorder affecting the frontal lobe, and, clinically, is a form of epilepsy. This chapter reviews what is known about the disorder as a means of examining how well the manifestations of NFLE reflect its name.

32.2 CLINICAL PRESENTATION

Typical presentation of NFLE includes nocturnal, short-lasting, stereotyped movements of the limbs, axial muscles, and head, typically associated with arousal. The definition of motor behaviors in NFLE is somewhat complicated because many of the behaviors are minimal and minor, even though some are major and prolonged. The term "minimal" motor activity has been traditionally used to describe less-defined events, such as simple motor acts like body touching or scratching or rubbing the nose or head, grimacing, vocalization, moaning, simple body movements, limb flexion, or chewing. Minimal activities are usually brief, i.e., 3–10 seconds. The term minor motor activity is applied to those involving more body segments with movements that can be purposeful or semipurposeful, gross body movements, changes in position, and major sudden and abrupt or rhythmic body movements lasting for 10–30 seconds. The major motor behaviors include sudden and abrupt body or segmental movements, elevation of head and trunk, hyperextension of arms and trunk accompanied by dystonic or clonic movements, or fearful expression or panic sensation lasting for 5–60 seconds. Prolonged motor activities are typically complex motor behaviors with tonic–dystonic posture, bimanual and bipedal activity, axial movements, shouting, laughing, or deep breathing lasting for more than 1 min (Oldani et al. 1996, 1998b; Terzaghi et al. 2008). Oldani et al. (1998b) analyzed the motor activities recorded using video-EEG monitoring in 38 patients with NFLE. There were a total of 43.8 ± 16.1 motor activities/attacks per night, including 26.3 ± 10.8 minimal, 15.5 ± 6.2 minor, 1.6 ± 3.1 major, and only 0.4 ± 1.1 prolonged motor activities. The average seizure frequency was 6.5 ± 2.7/h. Twenty-two patients reported daytime sleepiness and 14 patients had daytime seizures (four generalized, six atonic, and four complex partial seizures).

Episodes often include nocturnal hypermotor activity, sudden arousal with dystonia/dyskinsia (42.1%), complex behaviors (13.2%), sleep-related violence such as verbal/gestural threat and obscenities (5.3%), episodic nocturnal wandering (≤3 min), and increased familial parasomnia (39%) (Oldani et al. 1998b; Tinuper 2007; Provini et al. 1999; Ryvlin et al. 2006; Derry et al. 2006; Vetrugno et al. 2005). Given these, NFLE is part of the differential diagnosis of nocturnal conditions with complex and violent behaviors, including REM sleep behavior disorder (RBD), or non-REM parasomnias.

Zucconi et al. (2000) studied sleep related problems in 16 patients with NFLE using Epworth sleepiness scale (ESS), difficulty waking, morning tiredness, and subjective poor sleep quality. They compared eight patients with daytime complaints (morning tiredness and/or excessive sleepiness) to eight patients without those complaints, matched for age, sex, number of diurnal and nocturnal seizures, and severity and frequency of seizures (ESS 11.5 ± 1.5 vs. 5.8 ± 1.9, $p = 0.001$). They also compared both groups to a group of eight healthy control patients based on their sleep microstructure. The results revealed an increase in sleep instability in patients with NFLE and daytime sleep complaints, indicating a relationship between sleep fragmentation, nocturnal motor seizures, and daytime symptoms. Although the exact mechanism of this finding remains elusive and while this difference might be attributed to differences in ictal and interictal EEG abnormalities, it suggests that NFLE might also be a sleep disorder.

32.3 DIFFERENTIAL DIAGNOSIS

There are clinical similarities between NFLE and slow sleep arousal disorders. The latter typically happens 90–120 min after sleep onset and patients do not have a clear recollection of the event. The EEG during these events may show delta slow-wave activity with no epileptiform discharges. On the contrary, NFLE commonly happens within 30 min of sleep onset, is associated with auras, and is associated with clear patient recollection of the event. The NFLE often happens during stage N2 sleep, although it can happen during slow (N3) sleep as well, or at other times of night. The ictal EEG shows epileptiform activity in less than 10% of patients (Derry et al. 2006; Tinuper et al. 2007). Although the limited number of EEG electrodes used in a polysomnogram (PSG) might lower the chances of recording epileptiform activity during NFLE, more extended electrode placement based on the 10–10 electrode system does not significantly increase the likelihood of recording seizure discharges.

Tinuper et al. (2007) have compared the characteristics of disorders of arousal, nightmares, REM sleep behavior disorders, and NFLE as follows: Disorders of arousal, such as night terrors, tend to start at 3–8 years of age and equally affect males and females, usually with a family history. The episodes are sporadic, occur in the first third of the night during slow (N3) sleep, are characterized by no patient recollection of the episode afterward, and tend to disappear with age. Disorders of arousal are usually triggered by sleep deprivation, fever, or stress. The episodes may last 1–10 min.

Nightmares occur within the first years of life with no gender preference and often with a family history. They are also sporadic, tend to occur in the last third of the night during REM sleep, and disappear with age. In contrast to disorders of arousal, such as sleep terror or confusional arousal, these patients can be easily awakened and seem alert with a clear recollection afterward. The episodes last 3–30 min.

In contrast, REM-sleep behavior disorder occurs in older individuals, usually older than 50 years of age, shows preference for males with no family history of similar episodes, and are unlikely to remit. The events occur more than 90 min after sleep onset with no particular triggers. The episodes typically last 1–2 min.

NFLE can occur at any age but typically starts in childhood (11.8 years, range 1–30), usually persisting throughout adult life (Oldani et al. 1998b), and is more common among men than women. They often have a family history of seizures. The seizures tend to increase in frequency over time. Episodes occur almost nightly during non-REM sleep, more often stage N2 sleep, and may happen at any time during the night. The episodes may or may not be affected by triggering factors. The seizures last from seconds to 3 min. Patients might not always have a clear recollection of the events. When compared to disorders of arousal, nightmares, and REM sleep behavior disorders, only the NFLE shows stereotypy in its clinical presentation.

32.4   EEG AND PSG FINDINGS

Of 45 nocturnal partial motor seizures in 6 patients with focal epilepsy, 43 seizures occurred during non-REM sleep. The seizures were significantly more frequent in stages 1 and 3 sleep ($p < 0.0001$) (Terzano et al. 1991). This study also reported decreased slow sleep (N3, formerly stages 3 and 4) and REM sleep; however, this finding is nonspecific and may be induced by primary sleep disorders, a variety of medications including antiepileptic drugs (AEDs), or frequent arousals and sleep deprivation caused by frequent nocturnal seizures.

Zucconi et al. (2000), in their study of sleep macro- and microstructure in ADNFLE, found no significant differences in stages of sleep between those with and without sleep-related complaints. However, there was increased sleep onset latency and decreased sleep efficiency between the two groups, with an even more significant difference between the two groups versus a control group. It remains unclear if these sleep changes are primarily part of the NFLE syndrome, are secondary to other underlying sleep disorders, or are due to the frequent arousals caused by the seizures.

Seizures in autosomal dominant partial epilepsies (ADPE) are mainly nocturnal. Picard et al. (2000) studied 71 patients with ADPE, including autosomal dominant nocturnal frontal lobe epilepsy (ADNFLE), familial temporal lobe epilepsy (FTLE), and ADPE with variable foci, as well as 33 nonepileptic at-risk family members. Seizures were mainly nocturnal in ADNFLE (26/29), FTLE (16/26), ADPE (10/16). Of the 71 patients, 62 also had EEG recordings. Interictal abnormalities were present in 46 of 62 patients, epileptiform discharges in 40 of 62, and focal slow activity in 21 of 46.

In a review of 23 patients with familial NFLE with an age of onset of 8.7 ± 4.8 years, all patients continued having seizures through adulthood, including generalized tonic–clonic, atonic, and complex partial seizures. Half of the patients had seizures while awake; about 60% of them had difficulty waking up in the morning and reported daytime tiredness (Oldani et al. 1998a). The same researchers performed an awake baseline EEG in a group of 23 patients with NFLE. The EEG was normal in all patients. In contrast, a 3-hour EEG following sleep deprivation showed interictal epileptiform discharges in 50% of the patients, and an overnight video-EEG monitoring showed epileptiform abnormalities in about 87% of the patients. Four patients had seizures while awake, along with ictal epileptiform activity during sleep (Oldani et al. 1998c). Therefore, longer EEG monitoring seems to have a significant advantage over regular EEG or short-term monitoring.

In another study, Oldani et al. (1998b) evaluated the association between the EEG and clinical presentation of 38 patients with ADNFLE from 30 unrelated families. The patients reported a variety of nocturnal events ranging from enuresis to violent behaviors during sleep, which suggested parasomnias. The clinical-electroencephalographic findings are summarized in Table 32.1.

TABLE 32.1
Clinical and EEG Findings in 38 Patients with ADNFLE

#	Sz Awake	Sz Asleep	Epil EEG
4	Y		I
4	Y		II
3	Y	maj/prol	I
1	Y		I
2	Y		
4		maj/prol	I/II
4		maj/prol	I
8			I/II
4		maj/prol	
4		brief	

Note: The first column indicates the number of patients with a given set of findings; Y in the second column indicates that patients in that group had seizures while awake; third column, maj/prol, indicates that major/prolonged attacks or brief motor seizures occurred during sleep; fourth column indicates when the EEG showed abnormalities that were ictal (I), interictal (II), or ictal and/or interictal (I/II). For further details, see the work by Oldani et al. (1998b).

About 12% of the patients had epileptiform discharges during a daytime EEG, 50% had interictal epileptiform activity either during the day or night, about 33% had frontal ictal epileptiform discharges, about 50% had frontal ictal rhythmic slow activity, about 10% had diffuse ictal EEG flattening, and about 25% had normal EEG. A review of the clinical events recorded on video showed episodes of threatening gestures and utterances, swearing during the episodes, elevation of the trunk with a fearful expression on the face, violently hitting the wall with a fist, sitting up with an astonished expression, and dystonic posturing.

A prominent EEG finding in patients with NFLE is cyclic alternating pattern (CAP), also known as periodic K-alpha; this is an endogenous non-REM sleep EEG pattern that is characterized by alternating transient discharges that are distinct from background EEG and occur periodically at 20- to 60-second intervals. The CAP may signify sleep instability, sleep disturbance, or both. During CAP intermittent periods with K complexes, delta slow waves, and arousals (phase A) interrupt the tonic theta/delta discharges of sleep (phase B). The hypothesis is that phase A of CAP reflects a condition of transient activation, whereas phase B represents a period of inhibition. Periodic, high-amplitude, delta slow-EEG discharges or K-complexes (CAP type A1) have been associated with sleep instability, i.e., brain attempts to preserve sleep. However, when they result in EEG arousal, CAP has been classified as type A2 or A3, depending on the degree of EEG desynchronization and the fast- and slow-wave components within the pattern (Terzano et al. 2001; Nobili et al. 2006). A study of sleep micro- and macrostructure showed a higher index of CAP in patients with sleep problems compared to those without (44.9 vs. 28/h). The mechanism of the proposed association between the CAP and sleepiness is unknown (Zucconi et al. 2000).

Terzaghi et al. (2008) investigated the arousal mechanism and its modulation of the expression of minor motor events (MME) and epileptiform discharges (ED) by scoring the CAP recorded through subdural electrodes. They found a close association between arousal and both MMEs and EDs, i.e., a high CAP index (per hour of non-REM sleep) of 37–84.9, along with an index of 11.6–73.8 for MME and 17.8–88 for ED. They hypothesized that ED might increase arousal level, which in turn would facilitate the occurrence of MMEs through cortical disinhibition of innate motor pattern generators. The variable expression of these could explain the heterogeneous expression of MME in NFLE (Terzaghi et al. 2008).

In addition to arousal instability, the EDs in NFLE may be associated with specific sleep-related movement syndromes, such as periodic leg movements of sleep (PLMS). A stereo-EEG (SEEG) recording in a 22-year-old female presenting with sleep-related symptoms and nocturnal seizures revealed ED in the left superior frontal gyrus. In this case, the PLMS seemed to be an epiphenomena of ED arising in nonmotor cortex, with the ED acting as an internal trigger to increase arousal instability and facilitate the occurrence of PLMS, which in turn caused excessive daytime sleepiness (Nobili et al. 2006).

Data obtained during invasive presurgical evaluation of three drug-resistant NFLE patients showed that major seizures started with similar MME. The SEEG showed ictal patterns at the start of seizures that were similar to those that occurred in association with MME. These discharges were seen in wide regions of the frontal lobe, including the supplementary sensori-motor motor area (SSMA), cingulate, and posterior middle frontal gyrus. The SEEG also revealed that all of the paroxysmal arousals (PA), even those resembling normal awakenings, were associated with polyspikes and slow activity followed by a low-voltage fast discharge over the dorsolateral superior frontal cortex, SSMA, and frontal cingulate (Nobili et al. 2005). These findings were consistent with the previous work by the same researchers that indicated the epileptic nature of MME and PA (Nobili et al. 2003).

On the other hand, invasive monitoring has demonstrated patients thought to have NFLE but whose seizures originated in extrafrontal cortex. One intracranial SEEG study demonstrated seizures originating from operculo-insular cortex (Picard et al. 2000). In other patients, invasive monitoring has shown that basal temporal cortex, amygdala, and hippocampus can give rise to seizures with typical frontal-lobe-like symptomatology. Ictal single photon emission computed tomography (SPECT) in another patient with NFLE showed hypoperfusion of the right temporal and frontal cortices that was hypothesized to contribute to the decreased awareness during the seizure and impaired recall upon awakening (Vetrugno et al. 2005). Cho et al. (2010) reported a young patient with frontal lobe epilepsy who presented with generalized, predominantly nocturnal, myoclonic seizures resembling juvenile myoclonic epilepsy (Cho et al. 2010).

Does the term NFLE appropriately define this group of patients? Some patients do have epilepsy. Others have a familial hypermotor nocturnal disorder with no EEG evidence of seizures or known genetic mutations. Rather than being a form of epilepsy, is NFLE a disorder of arousal or synchronization of the brain that at times happens to manifest itself as epilepsy, that can occur with or without motor phenomena, and that can occur both night and day? The existence of significant genetic and phenotypical heterogeneity makes it difficult to check each patient against uniform established diagnostic criteria. Invasive recordings might provide some answers to these questions when they are indicated in specific patients. Further developments in genetics, neuroimaging, and neurophysiology may eventually help determine why some patients have nocturnal and some have diurnal events, determine why some patients with nocturnal epilepsy have seizures of frontal-lobe origin while others do not, and delineate nocturnal sleep-related nonepileptic conditions from nocturnal epilepsy.

REFERENCES

Cho, Y. W., S. D. Yi, and G. K. Motamedi. 2010. Frontal lobe epilepsy may present as myoclonic seizures. *Epilepsy Behav.* 17(4):561–564.

Derry, C. P., J. S. Duncan, and S. F. Berkovic. 2006. Paroxysmal motor disorders of sleep: The clinical spectrum and differentiation from epilepsy. *Epilepsia* 47:1775–1791.

Nobili, L., S. Francione, R. Mai, L. Tassi, F. Cardinale, L. Castana, I. Sartori, G. Lo Russo, and M. Cossu. 2003. Nocturnal frontal lobe epilepsy: Intracerebral recordings of paroxysmal motor attacks with increasing complexity. *Sleep* 26:883–886.

Nobili, L., I. Sartori, M. Terzaghi, F. Stefano, R. Mai, L. Tassi, L. Parrino, M. Cossu, and G. Lo Russo. 2006. Relationship of epileptic discharges to arousal instability and periodic leg movements in a case of nocturnal frontal lobe epilepsy: A stereo-EEG study. *Sleep* 29:701–704.

Nobili, L., I. Sartori, M. Terzaghi, L. Tassi, R. Mai, S. Francione, M. Cossu, F. Cardinale, L. Castana, G. Lo Russo, and C. Munari. 2005. Intracerebral recordings of minor motor events, paroxysmal arousals and major seizures in nocturnal frontal lobe epilepsy. *Neurol. Sci.* 26(Suppl 3):s215–s219.

Oldani, A., L. Ferini-Strambi, and M. Zucconi. 1998a. Symptomatic nocturnal frontal lobe epilepsy. *Seizure* 7:341–343.

Oldani, A., M. Zucconi, R. Asselta, M. Modugno, M. T. Bonati, L. Dalprà, M. Malcovati, M. L. Tenchini, S. Smirne, and L. Ferini-Strambi. 1998b. Autosomal dominant nocturnal frontal lobe epilepsy. A video-polysomnographic and genetic appraisal of 40 patients and delineation of the epileptic syndrome. *Brain* 121(Pt 2):205–223.

Oldani, A., M. Zucconi, L. Ferini-Strambi, D. Bizzozero, and S. Smirne. 1996. Autosomal dominant nocturnal frontal lobe epilepsy: Electroclinical picture. *Epilepsia* 37:964–976.

Oldani, A., M. Zucconi, S. Smirne, and L. Ferini-Strambi. 1998c. The neurophysiological evaluation of nocturnal frontal lobe epilepsy. *Seizure* 7:317–320.

Picard, F., S. Baulac, P. Kahane, and L. Ferini-Strambi. 2000. Dominant partial epilepsies. A clinical, electrophysiological and genetic study of 19 European families. *Brain* 123(Pt 6):1247–1262.

Provini, F., G. Plazzi, and E. Lugaresi. 2000. From nocturnal paroxysmal dystonia to nocturnal frontal lobe epilepsy. *Clin. Neurophysiol.* 111(Suppl 2):S2–S8.

Provini, F., G. Plazzi, P. Tinuper, S. Vandi, E. Lugaresi, and P. Montagna. 1999. Nocturnal frontal lobe epilepsy. A clinical and polygraphic overview of 100 consecutive cases. *Brain* 122(Pt 6):1017–1031.

Ryvlin, P., L. Minotti, G. Demarquay, E. Hirsch, A. Arzimanoglou, D. Hoffman, M. Guénot, F. Picard, S. Rheims, and P. Kahane. 2006. Nocturnal hypermotor seizures, suggesting frontal lobe epilepsy, can originate in the insula. *Epilepsia* 47:755–765.

Terzaghi, M., I. Sartori, R. Mai, L. Tassi, S. Francione, F. Cardinale, L. Castana, M. Cossu, G. LoRusso, R. Manni, and L. Nobili. 2008. Coupling of minor motor events and epileptiform discharges with arousal fluctuations in NFLE. *Epilepsia* 49:670–676.

Terzano, M. G., L. Parrino, P. G. Garofalo, C. Durisotti, and C. Filati-Roso. 1991. Activation of partial seizures with motor signs during cyclic alternating pattern in human sleep. *Epilepsy Res.* 10:166–173.

Terzano, M. G., L. Parrino, A. Sherieri, R. Chervin, S. Chokroverty, C. Guilleminault, M. Hirshkowitz, M. Mahowald, H. Moldofsky, A. Rosa, R. Thomas, and A. Walters. 2001. Atlas, rules, and recording techniques for the scoring of cyclic alternating pattern (CAP) in human sleep. *Sleep Med.* 2:537–553.

Tinuper, P. 2007. Parasomnias versus epilepsy: Common grounds and a need to change the approach to the problem. *Epilepsia* 48:1033–1034.

Tinuper, P., A. Cerullo, F. Cirignotta, P. Cortelli, E. Lugaresi, and P. Montagna. 1990. Nocturnal paroxysmal dystonia with short-lasting attacks: Three cases with evidence for an epileptic frontal lobe origin of seizures. *Epilepsia* 31:549–556.

Tinuper, P., F. Provini, F. Bisulli, L. Vignatelli, G. Plazzi, R. Vetrugno, P. Montagna, and E. Lugaresi. 2007. Movement disorders in sleep: Guidelines for differentiating epileptic from non-epileptic motor phenomena arising from sleep. *Sleep Med. Rev.* 11:255–267.

Vetrugno, R., M. Mascalchi, A. Vella, R. Della Nave, F. Provini, G. Plazzi, D. Volterrani, P. Bertelli, A. Vattimo, E. Lugaresi, and P. Montagna. 2005. Paroxysmal arousal in epilepsy associated with cingulate hyperfusion. *Neurology* 64:356–358.

Zucconi, M., A. Oldani, S. Smirne, and L. Ferini-Strambi. 2000. The macrostructure and microstructure of sleep in patients with autosomal dominant nocturnal frontal lobe epilepsy. *J. Clin. Neurophysiol.* 17:77–86.

33 Nicotinic Acetylcholine Receptors in Circuit Excitability and Epilepsy
The Many Faces of Nocturnal Frontal Lobe Epilepsy

Ortrud K. Steinlein and Daniel Bertrand

CONTENTS

33.1 INTRODUCTION

More than 230 genes in the human genome are known to code for ion-channel subunits, and this number increases if auxiliary subunits are taken into account. Several highly conserved families of ion channels are known, including different classes of voltage-gated potassium, sodium, or chloride channels, and ligand-gated channels such as $GABA_A$ and acetylcholine receptors. Most ion-channel gene families have their origins in the earliest metazoans, but their largest increase in numbers took place in the vertebrate lineage where duplication events led to a multiplication of subunits in most families (Jegla et al. 2009). Ion channels are involved in nearly every physiological process and are the cause of many different rare and common disorders. Ion-channel mutations are mainly found in disorders that are caused by excitable cells, such as neurons, skeletal and smooth muscle, or heart muscle. Within the central nervous system, epilepsies constitute a major group of disorders associated with ion-channel mutations (Steinlein 2008). So far, at least 25 different ion-channel genes have been implicated with some form of epilepsy, and although many of these reports are still unconfirmed, several epilepsy genes have by now been firmly established. These include human epilepsy genes such as *SCN1A*, coding for the α1 subunit of the voltage-gated sodium channel (Escayg et al. 2000), *KCNQ2*, one of the voltage-gated potassium channel subunits responsible for the action-potential-controlling M-current (Biervert et al. 1998), as well as the nicotinic acetylcholine receptors (nAChR) discussed in the following sections.

33.2 NICOTINC ACETYLCHOLINE RECEPTORS

The nAChRs can be divided into muscular and neuronal types. Muscular nAChRs exist in only two different subtypes, with the fetal one that is expressed during prenatal life being almost completely regulated down around birth to give place to the adult subtype. Both the fetal and the adult muscular nAChR differ with respect to only one subunit, γ and ε, respectively. The neuronal nAChRs, however, can be described as a superfamily of ligand-gated ion channels with an unknown number of subtypes. Up to 12 genes (*CHRNA2–CHRNA10, CHRNB2–CHRNB4*) code for neuronal nAChR subunits (α2–α10, β2–β4) that are expressed in neuronal and nonneuronal tissues. They combine to build either α-bungarotoxin-sensitive receptors that are homomeric (α7, α8, and α9) or heteropentameric (α7α8, α9α10), or α-bungarotoxin-insensitive receptors that are heteromeric combinations of α and β subunits (α2–α6, β2–β4) (Figure 33.1). The neuronal nAChRs differ with respect to their numbers of acetylcholine binding sites, with five binding sites (one on each subunit interface) present on receptors composed only of α subunits and two binding sites (located at the interface between α and β subunits) on nAChRs built from a combination of α and β subunits. The latter nAChRs contribute mainly to the diversity of receptors; they can build either simple subtypes such as α2β2, α2β4, α3β2, α3β4, α6β2, and α6β4, or more complex subtypes that contain three different subunits (e.g., α4α6β2, α3α4β2, α3α4β4, α2α4β2). Each nAChR subtype has its own functional profile and many of them are only expressed in rather restricted areas in brain or in certain nonneuronal tissues. The mechanisms by which either one of these subtypes is chosen in a given cell are not known, but it is likely that they include epigenetic effects that control gene activity as well as direct interactions with chaperones and intrinsic affinities between pairs of subunits (Gotti et al. 2006).

Several different monogenic disorders that are caused by mutations in the muscular subtypes have already been linked to AChRs, including congenital myasthenia and multiple pterygia syndrome (Escobar syndrome) (Hoffmann et al. 2006; Morgan et al. 2006) and autosomal dominant nocturnal frontal lobe epilepsy (ADNFLE) that is due to mutations in at least two different nAChR subunits. ADNFLE belongs to the group of rare monogenic epilepsies that are caused by single gene mutations, whereas the majority of seizure disorders either have only a weak genetic background or are due to a combination of several genetic susceptibility factors. Like many other neurological disorders, monogenic epilepsies are known to be mostly caused by mutations in genes that code for either voltage-gated or ligand-gated ion channels. Within this group of epileptic channelopathies, ADNFLE had been the first seizure disorder for which a causative gene defect could be identified

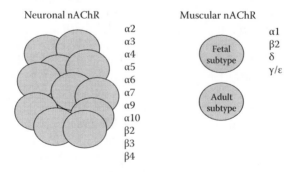

FIGURE 33.1 Schematic representation of possible subunits in human neuronal and muscular nAChRs.

(Steinlein et al. 1995). Our current state of knowledge about this fascinating disorder, the ion-channel subunits that cause it, as well as the implications of functional data gained on ADNFLE for future research will be presented and discussed in this chapter.

33.3   AUTOSOMAL DOMINANT NOCTURNAL FRONTAL LOBE EPILEPSY

ADNFLE was first described in 1995 and, compared to other neurological disorders, can therefore be considered as a rather young syndrome. However, it probably had been known much longer without having been recognized as an epileptic disorder. It is likely that at least some of the patients described to show nocturnal paroxysmal dystonia, paroxysmal arousals, or episodic nocturnal wanderings indeed had nocturnal frontal lobe epilepsy (Scheffer et al. 1995). In ADNFLE, the seizures usually start in the second decade of life but intra- as well as interfamilial variability is known to be considerable. Seizures often cluster and can cause significant disruption of the sleep pattern, followed by daytime fatigue. They arise from non-REM sleep and, as typical for frontal lobe seizures, are often hyperkinetic in nature. Some patients also tend to sleep walk during their nocturnal attacks. Ictal electroencephalography often does not reveal any epileptic activity, and video-polysomnography is a helpful tool to establish the diagnosis of frontal lobe seizures. The pattern of inheritance is autosomal dominant, meaning that mutation of only one of two homologous genes is required to cause the disorder. Offspring of ADNFLE patients have a 50% chance to inherit such a mutation, but, due to the reduced penetrance of most mutations, not every mutation carrier develops seizures. For most mutations, the penetrance rate is estimated to be about 70%, but mutations with much lower penetrance rates have been described (Leniger et al. 2003).

33.4   MUTATIONS IN NEURONAL nAChRs CAUSE ADNFLE

At least 2 of the 11 human neuronal nAChR subunit genes are known to cause ADNFLE. The first mutation identified in patients with this rare syndrome was an amino acid exchange in *CHRNA4*, the gene coding for the α4 subunit of the nAChR (Steinlein et al. 1995). The S248F mutation (or S280F, numbering according to reference sequence NP_000735.1) is located within the second transmembrane domain that contributes to the walls of the ion-channel pore. Four additional mutations that have been subsequently described affect the same functional domain within *CHRNA4* (Steinlein et al. 1997; Leniger et al. 2003; Hirose et al. 1999; Cho et al. 2003). A possible sixth *CHRNA4* mutation that had been recently found in a Chinese patient with sporadic nocturnal frontal lobe epilepsy is located in the extracellular loop between transmembrane domain 3 and 4 (Chen et al. 2009). The pathogenic nature of this mutation hasn't been verified by functional studies yet. The five known ADNFLE mutations in the β2 nAChR subunit gene, *CHRNB2*, are located either at the 3′ end of transmembrane domain 2 or within transmembrane domain 3 (De Fusco et al. 2000; McLellan et al. 2003; Hoda et al. 2008; Bertrand et al. 2005). In a third nAChR subunit, *CHRNA2*, a single mutation located in the first transmembrane domain has been described in an Italian family with a phenotype of sleep-related epilepsy that closely resembles ADNFLE (Table 33.1) (Aridon et al. 2006; Hoda et al. 2009).

The transmembrane domains represent functionally important parts of nAChR subunits but account for only a tiny portion of the complete coding regions in these genes. Furthermore, several ADNFLE mutations were found in families of different ethnic origins and from different countries and these mutations occurred more than once. The clustering of ADNFLE mutations, together with their repeated occurrence at the same amino acids, strongly suggests a position effect within the gene. It seems likely that only amino acids that have certain functions within the receptor are able to cause the phenotype of ADNFLE. So far it is not known exactly what the receptor functions critical for epileptogenesis are and which additional amino acids within these genes can be regarded as candidates for future mutation screening.

TABLE 33.1
Summary of Mutations in ADNFLE

Gene	Functional Domain	Mutation
CHRNA4	TM2	S248F (S280F)
		S252L (S284L)
		776ins3 (865–873insGCT)
		T265I (T293I)
	ICL2	R308H (R336H)[a]
CHRNB2	TM2	V287L
		V287M
	TM3	L301V
		V308A
		I312M
CHRNA2	TM1	I279N[b]

Note: TM, transmembrane region; ICL, intracellular loop.
[a] Not verified by functional testing yet.
[b] Sleep-related epilepsy that closely resembles ADNFLE.

33.5 THE CLINICAL VARIABILITY OF ADNFLE

The first ADNFLE mutation, α4-S248F, was associated with a phenotype of mostly pure epilepsy. The severity of the seizure disorder in the four families published thus far with this mutation (total of patients in these families: 67) varied considerably, but in most affected individuals the mutation α4-S248F was not associated with additional neurological features (Steinlein et al. 1995, 2000; Saenz et al. 1999; Magnusson et al. 2003; McLellan et al. 2003). Patients with the second mutation, α4-776ins3 (current nomenclature 865-873insGCT, according to reference sequence NP_000735.1), most often not only had epilepsy but also suffered from psychiatric problems such as schizophrenia or schizophrenia-like disorder (Steinlein et al. 1997; Magnusson et al. 2003). The comorbidity of epilepsy and psychiatric disorders could at first have been due to coincidence, especially because no second family with this mutation has been found so far for comparison. However, further ADNFLE families with other nAChR mutations came to attention in which a considerable percentage of patients showed other neurological problems in addition to epilepsy. Examples are the mutation α4-S252L (α4-S284L, numbering according to reference sequence NP_000735.1), which has so far been described in families of Japanese, Lebanese, and Korean origin. Of the 16 affected family members, 10 showed intellectual impairment in the range between low normal intelligence to mild or moderate mental retardation (Hirose et al. 1999; Phillips et al. 2000; Cho et al. 2003). Another interesting mutation is β2-I312M, which in families from England and Korea shows a strong association with significant memory deficits, especially in verbal memory (Bertrand et al. 2005; Cho et al. 2003). By now, enough families are known to indicate that ADNFLE mutations can be divided into two types: one that in most patients causes only epilepsy and one that often results in epilepsy accompanied by additional neurological features (Steinlein et al. in preparation). As shown in the next sections, these two groups of mutations not only differ with respect to their associated clinical phenotypes, but also in their pathofunctional profiles.

33.6 FUNCTIONAL CONSEQUENCES OF ADNFLE MUTATIONS

Evaluation of the functional consequences caused by the mutations in the genes encoding for the α4 or β2 subunits of the nicotinic acetylcholine receptors requires establishing a model to allow for physiological or behavioral evaluations. Over the years, two models have been established with the heterologous expression of the human genes and, more recently, the development of genetically modified rodents (mice and rats).

33.6.1 ADNFLE Mutations Cause a Gain of Function

Heterologous expression of genes encoding for the neuronal nicotinic acetylcholine receptors was initially carried out in *Xenopus* oocytes by injection of a mixture of the α4 and β2 mRNAs or cDNAs—reviewed in the work by Bertrand et al. (1991). This well-established method allows a faithful expression of many different membrane proteins, including ligand-gated channels; the method also allows investigation with the voltage-clamp technique. Initial results obtained by comparison of the electrophysiological properties of the α4β2 and α4S248Fβ2 mutant pointed out a series of modifications caused by this single-point mutation (Bertrand et al. 1998). A decisive point that must, however, be taken into account is that ADFNLE patients are heterozygous for the mutation and, in absence of evidence for an unbalance in gene expression, the two alleles must be equally expressed. To account for this genetic fact it is therefore necessary to examine properties of heterozygous expression of the normal and mutated alleles in the recombinant system. Functional studies carried out using "heterozygous-like" expression of the α4 + α4S248F and β2 confirmed the dominant nature of the mutation with an increased sensitivity to acetylcholine or, in other words, a gain of function (Bertrand et al. 2002).

Following these initial findings, the discovery of additional mutations in the α4 subunit (such as S252L) and later mutations in the β2 subunit (V287M, V287L) allowed further evaluation of the functional consequences of these mutations. Although it would be beyond the scope of this work to review in detail the effects of each of the mutations, it is important to recall that all the mutations identified so far cause a gain of function by increasing the acetylcholine sensitivity—reviewed in the work by Steinlein and Bertrand (2009).

33.6.2 NFLE Mutations in the α2 Gene

Although an important part of the understanding of the role of the nicotinic receptor in brain function comes from studies carried out in rodents, the correlation between these findings and human brain function may be limited. Analysis of the receptor expression using autoradiography and in situ expression revealed significant differences between rodent and monkey brains (Han et al. 2000). Namely, these studies pointed out the wide expression of the α2 subunit in the prefrontal cortex of the monkey brain, whereas expression of this subunit in rat brain is more limited. These data indicate that great care should be used when extrapolating data obtained from rodent models and that the α2 subunit may play an important role in cognitive function.

The identification of a mutation (I279N) in the gene encoding for the α2 subunit and its association with a form of nocturnal epilepsy confirmed the importance of this subunit in brain function (Aridon et al. 2006). Although for technical limitations these authors restricted their analysis to the expression of the α2 subunit with the β4 subunit, which does not correlate with the known receptor structure, the finding of a gain of function would be in agreement with data obtained at the α4β2 receptors. Subsequent experiments carried out in *Xenopus* oocytes confirmed that mutation in the α2 subunit could be expressed with β2 and that it conferred a gain of function (Hoda et al. 2009). Moreover, these studies confirmed the dominant character of the α2-I279N mutation when expressed in "heterozygous" manner. All together, these data illustrate that α2-associated epilepsies must be caused by a higher sensitivity of these receptors to acetylcholine.

33.7 WHAT GENETICALLY MODIFIED ANIMAL MODELS CAN TEACH US

Attempts to produce animal models reproducing the dysfunction observed in ADNFLE patients have been made by different laboratories. These strategies include the generation of knock-in mice, in which the gene encoding for the α4 subunit was mutated to cause a marked gain of function. The retained modification was the substitution of the leucine L9′ into a threonine in the channel domain—a mutation known to cause a profound increase in the acetylcholine sensitivity (Revah et al. 1991; Fonck et al. 2005). Although mice harboring this mutation showed distinct physiological and pharmacological profiles, they did not show spontaneous epileptic seizures (Teper et al. 2007). Alternative attempts made in mice include the introduction of the α4 subunit with the S252F, the leucine insertion 776ins3, or the S248F ADNFLE mutation (Klaassen et al. 2006; Tepper et al. 2007; Cannata et al. 2009). These mice show abnormal EEG features, increased sensitivity to nicotine-induced seizures, as well as additional neurological dysfunction.

Results obtained from these models suggest that the gain of function introduced in the nicotinic acetylcholine receptors causes excessive GABA release and, indirectly, neuronal hypersynchronization. This hypothesis was further tested in brain slices from normal and genetically engineered mice (Mann and Mody 2008).

A closer model to ADNFLE was obtained in rats by introducing the α4 mutation S284L (Hirose 1999). Rats harboring this mutation showed no major biological abnormalities, but had the characteristic epileptic phenotypes during slow-wave sleep (SWS) (Zhu et al. 2008). The use of such animal models is expected to provide new and valuable insights for the understanding of NFLE.

33.8 POSSIBLE ASSOCIATION BETWEEN MUTATIONS AND PHARMACOLOGICAL SENSITIVITY

Clinical data indicate that most patients suffering from ADNFLE associated with the α4S248F mutation were efficiently treated with low concentrations of the antiepileptic drug carbamazepine. However, differences in the response to carbamazepine treatment were noticed for different mutations, which suggests a possible differential pharmacological profile. Although the action of carbamazepine is mainly thought to be mediated by the inhibition of voltage-dependent sodium channels, evaluation of the putative effects of this compound at the nicotinic acetylcholine receptor revealed that it markedly inhibits the α4β2 receptors at concentrations comparable to those found in the cerebrospinal fluid of patients (Picard et al. 1999). Moreover, the α4S248F mutant displayed a higher sensitivity to carbamazepine than the wild type. This first observation suggested that carbamazepine may be efficiently counterbalancing the gain of function of the α4β2 receptor, thereby preventing the triggering of epileptic seizures. Determination of the IC_{50}, the dose required to produce fifty percent inhibition of the acetylcholine-evoked current, at different mutants revealed important differences in sensitivity for different mutations (Hogg and Bertrand 2004). This would suggest that patients harboring distinct mutation might differentially respond to the carbamazepine treatment. Although clinical data tend to support this hypothesis, the inability to establish a proper dose response curve in patients and to compare the efficacy of alternative antiepileptic treatment precludes definitive conclusions. Further documentation of the properties of additional mutations and possibly of clinical data are therefore needed before reaching a conclusion. Nonetheless, it could be suggested that treatment of patients suffering from ADNFLE, or the α2 NFLE, would consist of reducing the hypercholinergic activity of the nicotinic acetylcholine receptors. Analysis of the current pharmacopoeia indicates that mecamylamine, the active molecule of the antihypertensive drug inversine, which is a broad-spectrum noncompetitive inhibitor of nicotinic receptors, may be of interest and should require further scrutiny.

33.9 FUTURE DIRECTIONS

Future studies on the genetic and pathofunctional concepts underlying ADNFLE as a model disorder for human epilepsy should focus on different topics, extending our knowledge about the mutational spectrum of *CHRNA4* and *CHRNB2*, analyzing the possible involvement of other nAChRs subunits, and investigating the interaction between the clinical phenotype and naturally occurring genetic variation in these subunits. Routine screening for ADNFLE mutations mostly focuses on the transmembrane regions that carry all known mutations. However, it would be plausible that mutations in other, functionally important parts of *CHRNA4* and *CHRNB2* cause epilepsy, too, although not necessarily ADNFLE. Furthermore, it would be interesting to find out if mutations occur in some of the many other nAChR subunit genes and what kind of clinical phenotype they cause. Wherever additional mutations will be found, they do not act alone but are part of an unbelievably complex functional system based on naturally occurring variation in our genomes. Numerous single-nucleotide polymorphisms (SNPs) within and outside of genes can cause subtle changes in gene expression and protein function. Many such SNPs are known to be located in nAChR subunit genes, where they can be expected to modulate the effects of epilepsy mutations on receptor function. This complex web between rare mutations and naturally occurring genetic variation has to be kept in mind when interpreting functional data gathered from less complex experimental systems such as the *Xenopus* oocyte expression model.

REFERENCES

Aridon, P., C. Marini, C. Di Resta, E. Brilli, M. De Fusco, F. Politi, E. Parrini, I. Manfredi, T. Pisano, D. Pruna, G. Curia, et al. 2006. Increased sensitivity of the neuronal nicotinic receptor alpha 2 subunit causes familial epilepsy with nocturnal wandering and ictal fear. *Am. J. Hum. Genet.* 79:342–530.

Bertrand, D., E. Cooper, S. Valera, D. Rungger, and M. Ballivet. 1991. Electrophysiology of neuronal nicotinic acetylcholine receptors expressed in *Xenopus* oocytes, following nuclear injection of genes or cDNAs. In *Methods in neurosciences*, ed. M. Conn, 174–193. New York: Academic Press, Inc.

Bertrand, D., F. Picard, S. Le Hellard, S. Weiland, I. Favre, H. Phillips, S. Bertrand, S. F. Berkovic, A. Malafosse, J. Mulley. 2002. How mutations in the nAChRs can cause ADNFLE epilepsy. *Epilepsia* 43(Suppl 5):112–122.

Bertrand, D., F. Elmslie, E. Hughes, J. Trounce, T. Sander, S. Bertrand, O. K. Steinlein, et al. 2005. The CHRNB2 mutation I312M is associated with epilepsy and distinct memory deficits. *Neurobiol. Dis.* 20:799–804.

Bertrand, S., S. Weiland, S. F. Berkovic, O. K. Steinlein, and D. Bertrand. 1998. Properties of neuronal nicotinic acetylcholine receptor mutants from humans suffering from autosomal dominant nocturnal frontal lobe epilepsy. *Br. J. Pharmacol.* 125:751–760.

Biervert, C., B. C. Schroeder, C. Kubisch, S. F. Berkovic, P. Propping, T. J. Jentsch, and O. K. Steinlein. 1998. A potassium channel mutation in neonatal human epilepsy. *Science* 279:403–406.

Cannata, D. J., D. I. Finkelstein, and I. Gantois. 2009. Altered fast- and slow-twitch muscle fibre characteristics in female mice with a (S248F) knock-in mutation of the brain neuronal nicotinic acetylcholine receptor. *J. Muscle Res. Cell Motil.* 30:73–83.

Chen, Y., L. Wu, Y. Fang, Z. He, B. Peng, Y. Shen, and Q. Xu. 2009. A novel mutation of the nicotinic acetylcholine receptor gene CHRNA4 in sporadic nocturnal frontal lobe epilepsy. *Epilepsy Res.* 83:152–156.

Cho, Y. W., G. K. Motamedi, I. Laufenberg, S. I. Sohn, J. G. Lim, H. Lee, S. D. Yi, J. H. Lee, D. K. Kim, R. Reba, W. D. Gaillard, et al. 2003. A Korean kindred with autosomal dominant nocturnal frontal lobe epilepsy and mental retardation. *Arch. Neurol.* 60:1625–1632.

De Fusco, M., A. Becchetti, A. Patrignani, G. Annesi, A. Gambardella, A. Quattrone, A. Ballabio, E. Wanke, G. Casari, et al., 2000. The nicotinic receptor beta 2 subunit is mutant in nocturnal frontal lobe epilepsy. *Nat. Genet.* 26:275–276.

Escayg, A., B. T. MacDonald, M. H. Meisler, S. Baulac, G. Huberfeld, I. An-Gourfinkel, A. Brice, E. LeGuern, B. Moulard, D. Chaigne, C. Buresi, and A. Malafosse. 2000. Mutations of SCN1A, encoding a neuronal sodium channel, in two families with GEFS+2. *Nat. Genet.* 24:343–345.

Gotti, C., M. Zoli, and F. Clementi. 2006. Brain nicotinic acetylcholine receptors: Native subtypes and their relevance. *Trends Pharmacol. Sci.* 27:482–491.

Fonck, C., B. N. Cohen, R. Nashmi, P. Whiteaker, D. A. Wagenaar, N. Rodrigues-Pinguet, P. Deshpande, S. McKinney, S. Kwoh, J. Munoz, C. Labarca, et al. 2005. Novel seizure phenotype and sleep disruptions in knock-in mice with hypersensitive alpha 4* nicotinic receptors. *J. Neurosci.* 25:11396–11411.

Han, Z. Y., N. Le Novere, M. Zoli, J. A. Hill Jr, N. Champtiaux, and J. P. Changeux. 2000. Localization of nAChR subunit mRNAs in the brain of Macaca mulatta. *Eur. J. Neurosci.* 12:3664–3674.

Hirose, S., H. Iwata, H. Akiyoshi, K. Kobayashi, M. Ito, K. Wada, S. Kaneko, and A. Mitsudome. 1999. A novel mutation of CHRNA4 responsible for autosomal dominant nocturnal frontal lobe epilepsy. *Neurology* 53:1749–1753.

Hoda, J. C., W. Gu, M. Friedli, H. A. Phillips, S. Bertrand, S. E. Antonarakis, D, Goudie, R. Roberts, I. E. Scheffer, C. Marini, J. Patel, et al. 2008. Human nocturnal frontal lobe epilepsy: Pharmocogenomic profiles of pathogenic nicotinic acetylcholine receptor beta-subunit mutations outside the ion channel pore. *Mol. Pharmacol.* 74:379–391.

Hoda, J. C., M. Wanischeck, D. Bertrand, and O. Steinlein. 2009. Pleiotropic functional effects of the first epilepsy-associated mutation in the human CHRNA2 gene. *FEBS Lett.* 583:1599–1604.

Hoffmann, K., J. S. Muller, S. Stricker, A. Megarbane, A. Rajab, T. H. Lindner, M. Cohen, E. Chouery, L. Adaimy, I. Ghanem, V. Delague, et al. 2006. Escobar syndrome is a prenatal myasthenia caused by disruption of the acetylcholine receptor fetal gamma subunit. *Am. J. Hum. Genet.* 79:303–312.

Hogg, R. C., and D. Bertrand. 2004. Neuronal nicotinic receptors and epilepsy, from genes to possible therapeutic compounds. *Bioorg. Med. Chem. Lett.* 14:1859–1861.

Jegla, T. J., C. M. Zmasek, S. Batalov, and S. K. Nayak. 2009. Evolution of the human ion channel set. *Comb. Chem. High Throughput Screen* 12:2–23.

Klaassen, A., J. Glykys, and J. Maguire. 2006. Seizures and enhanced cortical gabaergic inhibition in two mouse models of human autosomal dominant nocturnal frontal lobe epilepsy. *Proc. Natl. Acad. Sci. U. S. A.* 103:19152–19157.

Leniger, T., C. Kananura, A. Hufnagel, S. Bertrand, D. Bertrand, and O. K. Steinlein. 2003. A new Chrna4 mutation with low penetrance in nocturnal frontal lobe epilepsy. *Epilepsia* 44:981–985.

Lipovsek, M., P. Plazas, J. Savino, A. Klaassen, J. Boulter, A. B. Elgoyhen, and E. Katz. 2008. Properties of mutated murine alpha4beta2 nicotinic receptors linked to partial epilepsy. *Neurosci. Lett.* 434:165–169.

Mann, E. O., and I. Mody. 2008. The multifaceted role of inhibition in epilepsy: Seizure-genesis through excessive gabaergic inhibition in autosomal dominant nocturnal frontal lobe epilepsy. *Curr. Opin. Neurol.* 212:155–160.

Magnusson, A., E. Stordal, E. Brodtkorb, and O. Steinlein. 2003. Schizophrenia, psychotic illness, and other psychiatric symptoms in families with autosomal dominant nocturnal frontal lobe epilepsy caused by different mutations. *Psychiatr. Genet.* 13:91–95.

McLellan, A., H. A. Phillips, C. Rittey, M. Kirkpatrick, J. C. Mulley, D. Goudie, J. B. Stephenson, J. Tolmie, I. E. Scheffer, S. F. Berkovic, and S. M. Zuberi. 2003. Phenotypic comparison of two Scottish families with mutations in different genes causing autosomal dominant nocturnal frontal lobe epilepsy. *Epilepsia* 44:613–617.

Morgan, N. V., L. A. Brueton, P. Cox, M. T. Greally, J. Tolmie, S. Pasha, I. A. Aligianis, H. van Bokhoven, T. Marton, L. Al-Gazali, J. E. Morton, et al. 2006. Mutations in the embryonal subunit of the acetylcholine receptor (CHRNG) cause lethal and Escobar variants of multiple pterygium syndrome. *Am. J. Hum. Genet.* 79:390–395.

Phillips, H. A., C. Marini, I. E. Scheffer, G. R. Sutherland, J. C. Mulley, and S. F. Berkovic. 2000. A de novo mutation in sporadic nocturnal frontal lobe epilepsy. *Ann. Neurol.* 48:264–267.

Picard, F., S. Bertrand, O. K. Steinlein, and D. Bertrand. 1999. Mutated nicotinic receptors responsible for autosomal dominant nocturnal frontal lobe epilepsy are more sensitive to carbamazepine. *Epilepsia* 40:1198–1209.

Revah, F., D. Bertrand, J. L. Galzi, A. Devillers-Thiéry, C. Mulle, N. Hussy, S. Bertrand, M. Ballivet, and J. P. Changeux. 1991. Mutations in the channel domain alter desensitization of a neuronal nicotinic receptor. *Nature* 353:846–849.

Sáenz, A., J. Galán, C. Caloustian, F. Lorenzo, C. Márquez, N. Rodríguez, M. D. Jiménez, J. J. Poza, A. M. Cobo, D. Grid, J. F. Prud'homme, and A. López de Munain. 1999. Autosomal dominant nocturnal frontal lobe epilepsy in a Spanish family with a Ser252Phe mutation in the CHRNA4 gene. *Arch. Neurol.* 56:1004–1009.

Scheffer, I. E., K. P. Bhatia, I. Lopes-Cendes, D. R. Fish, C. D. Marsden, E. Andermann, F. Andermann, R. Desbiens, D. Keene, and F. Cendes. 1995. Autosomal dominant nocturnal frontal lobe epilepsy. A distinctive clinical disorder. *Brain* 118:61–73.

Steinlein, O. K. 2008. Genetics and epilepsy. *Dialogues Clin. Neurosci.* 10:29–38.

Steinlein, O. K., and D. Bertrand. 2009. Nicotinic receptor channelopathies and epilepsy. *Pflugers Arch.* 460(2):495–503.

Steinlein, O. K., J. Stoodt, J. Mulley, S. Berkovic, I. E. Scheffer, and E. Brodtkorb. 2000. Independent occurrence of the CHRNA4 Ser248Phe mutation in a Norwegian family with nocturnal frontal lobe epilepsy. *Epilepsia* 41:529–535.

Steinlein, O. K., A. Magnusson, J. Stoodt, S. Bertrand, S. Weiland, S. F. Berkovic, K. O. Nakken, P. Propping, and D. Bertrand. 1997. An insertion mutation of the CHRNA4 gene in a family with autosomal dominant nocturnal frontal lobe epilepsy. *Hum. Mol. Genet.* 6:943–947.

Steinlein, O. K., J. C. Mulley, P. Propping, R. H. Wallace, H. A. Phillips, G. R. Sutherland, I. E. Scheffer, and S. F. Berkovic. 1995. A missense mutation in the neuronal nicotinic acetylcholine receptor alpha 4 sub-unit is associated with autosomal dominant nocturnal frontal lobe epilepsy. *Nat. Genet.* 11:201–203.

Teper, Y., D. Whyte, E. Cahir, H. A. Lester, S. R. Grady, M. J. Marks, B. N. Cohen, C. Fonck, T. McClure-Begley, J. M. McIntosh, C. Labarca, A. Lawrence, F. Chen, I. Gantois, P. J. Davies, S. Petrou, M. Murphy, J. Waddington, M. K. Horne, S. F. Berkovic, and J. Drago. 2007. Nicotine-induced dystonic arousal complex in a mouse line harboring a human autosomal-dominant nocturnal frontal lobe epilepsy mutation. *J. Neurosci.* 27:1012.

Zhu, G., M. Okada, S. Yoshida, S. Ueno, F. Mori, T. Takahara, R. Saito, Y. Miura, A. Kishi, M. Tomiyama, A. Sato, T. Kojima, G. Fukuma, K. Wakabayashi, K. Hase, H. Ohno, H. Kijima, Y. Takano, A. Mitsudome, S. Kaneko, and S. Hirose. 2008. Rats harboring S284L Chrna4 mutation show attenuation of synaptic and extrasynaptic GABAergic transmission and exhibit the nocturnal frontal lobe epilepsy phenotype. *J. Neurosci.* 28:12465–12476.

34 Channelopathies in Epileptology

*Frank Lehmann-Horn, Yvonne Weber, Snezana Maljevic,
Karin Jurkat-Rott, and Holger Lerche*

CONTENTS

34.1 INTRODUCTION

Epilepsy is one of the most common neurological disorders affecting ~3% of the world's population during their lifetime (Hauser et al. 1996), of which at least a third is primarily genetically determined. About 2% of the hereditary epilepsies are monogenic, of which most of the identified defects are related to pathologic channel function. Using a total prevalence of 0.9% for all forms of epilepsies, the prevalence of channel-related epilepsies is estimated to be fewer than 15:100,000 of the population, thus fulfilling the criterion for a chronic rare disease. In addition, ion channel genes may be involved as susceptibility genes in the pathogenesis of polygenic epilepsies.

Epilepsies are characterized by recurring seizures resulting from synchronized electrical discharges of neurons within the central nervous system. With regard to the complicated nature and the many different functions of the brain, there are a number of clinically differentiable types of both seizure and epilepsies. Epileptic semiology can include not only mild sensations of the patient himself that are not visible for other individuals (such as seen with an epigastric aura), but also transient blackouts (such as known for absence or complex–partial/dyscognitive seizures) or severe generalized tonic–clonic convulsions. The most important features used to classify epileptic seizures and epileptic syndromes are (1) the origin of the seizures, which can be focal or generalized in the surface EEG, and (2) the underlying cause, which can be symptomatic/structural–metabolic (e.g., due to cortical malformations, brain tumors, stroke) or idiopathic/genetic.

In this chapter, idiopathic epilepsy syndromes caused by malfunction or altered regulation of ion channel proteins are described. Although hereditary channel epilepsies are rare (fewer than 10% of all epilepsies), they are important for the elucidation of the pathogenesis of epileptic seizures in general. As more than 35% of the marketed drugs target ion channels, they also provide model disorders for therapeutic strategies of more frequent epilepsies. Because the genetic defect is permanent, the altered gene may only be expressed in a specific phase of life (neonatal, infancy, childhood, or adulthood), and in this specific phase of life, additional compensatory mechanisms may prevent seizures most of the time. Also, circadian regulation of channel properties occurs, which may be decisive for the manifestation or suppression of seizures (Ko et al. 2009). As in other channelopathies (Lehmann-Horn and Jurkat-Rott 1999), we will focus on (1) whether the mutations in a single-channel complex cause one, two, or more clinical phenotypes, (2) whether the mutations exert gain or loss of function on the channel and cellular level and if they can explain the clinical feature, and (3) if provocative factors and therapeutic strategies are available for the various clinical phenotypes.

34.2 AUTOSOMAL DOMINANT NOCTURNAL FRONTAL LOBE EPILEPSY

Autosomal dominant nocturnal frontal lobe epilepsy (ADNFLE, see Table 34.1) includes frequent brief seizures occurring in childhood with proximal hyperkinetic or distal dystonic manifestations, typically in clusters at night in sleep phase 2. Ictal video-electroencephalographic studies have revealed partial seizures originating from the frontal lobe and also from parts of the insula, suggesting a defect of a broader network. The age of onset is during puberty, and the disease usually worsens during adulthood. The penetrance of the disease is estimated at approximately 70% to 80%. A mutation was identified in the gene *CHRNA4* encoding the α4-subunit of a neuronal nicotinic acetylcholine receptor as the first ion-channel mutation found in an inherited form of epilepsy.

TABLE 34.1
Overview of the Most Important Channelopathies in Epileptology

Epilepsy	Acronym	Gene	Locus	Ion	Protein
ADNFLE	EFNL1	CHRNA4	20q13.3	Cations,	nAChRα4
	EFNL3	CHRNB2	1q21	particular	nAChRβ2
	EFNL4	CHRNA2	8p21	calcium	nAChRα2
BFNS	BFNS1	KCNQ2	20q13.3	Potassium	Kv7.2
	BFNS2	KCNQ3	8q24.22-24.3		Kv7.3
BFNIS	BFNIS	SCN2A	2q24.3	Sodium	Nav1.2
GEFS+	GEFS1	SCN1B	19q13.1	Sodium	Navβ1
	GEFS2	SCN1A	2q24		Nav1.1
	GEFS7	SCN9A	2q24		Nav1.7
	GEFS4	GABRG2	5q31.1-33.1	Chloride	GABA$_A$γ2
	GEFS5	GABRD	1p36.3		GABA$_A$δ
Dravet syndrome (SMEI)	SMEI	SCN1A	2q24	Sodium	Nav1.1
		SCN1B	19q13.1		Navβ1
CAE	ECA2	GABRG2	5q31.1-33.1	Chloride	GABA$_A$γ2
	ECA4	GABRA1	5q34-35		GABA$_A$α1
Susceptibility to CAE	ECA6	CACNA1H	16p13.3	Calcium	Cav3.2
JME	EJM5	GABRA1	5q34-35	Chloride	GABAA
	EJM6	CACNB4	2q22-23	Calcium	Cavβ4

Note: Diseases or susceptibilities are listed in column 1, their acronyms, genes, and chromosomal locations are given in columns 2 through 4, and the type of ions that are conducted by the corresponding channels and their specific protein names are provided in columns 5 and 6.

Several mutations in *CHRNA4* and in *CHRNB2*, which encodes the β2-subunit of neuronal nicotinic acetylcholine receptor, have been reported (Steinlein 2004). Recently, a mutation in *CHRNA2*, encoding the α2-subunit, has been detected in a slightly different phenotype (Aridon et al. 2006). All mutations in nAChRs reside in the pore-forming M2 (or M3) transmembrane segments. They increase the window calcium current either by shifting the activation curve to more negative potentials or by shifting the desensitization curve to less negative potentials (Steinlein and Bertrand 2008). The nicotinic acetylcholine receptor complex of the brain is distributed presynaptically and postsynaptically and along axons as well. If this complex is considered as a single-channel complex despite the various pentameric compositions, then it might be concluded that it is responsible for one clinical phenotype that shows a relatively broad spectrum. The mutations exert a gain of function. An important provocative factor is sleep phase 2. Carbamazepine, lamotrigine, and other sodium-channel blockers are beneficial. Carbamazepine may also be effective as an inhibitor of the neuronal nicotinic acetylcholine receptor.

34.3 BENIGN FAMILIAL NEONATAL OR INFANTILE SEIZURES

Benign familial neonatal seizures (BFNS) are dominantly inherited, with a penetrance of 85%. The seizures typically manifest within the first days of life and disappear spontaneously after weeks to months. Ictal EEGs indicate a partial seizure onset with frequent generalization, which is not always visible clinically: often, patients show hemitonic or hemiclonic symptoms, or apnea, but sometimes seizures appear as primarily generalized. Interictal EEGs are mostly normal. The risk of seizures recurring in adulthood is ~15%. Although psychomotor development is usually normal, an increasing number of cases with learning disabilities have recently been described (Borgatti et al. 2004; Steinlein et al. 2007).

Mutations have been identified in Kv7.2 and Kv7.3 potassium channels, which interact with each other and conduct the so-called M-current—an important current in the regulation of the firing rate of neurons (reviewed by Delmas and Brown 2005; Maljevic et al. 2008). As the M-current has slow kinetics, its main effect maybe regulating the subthreshold membrane potential and the termination of an action potential burst. Coexpression of heteromeric wild-type and mutant Kv7.2/Kv7.3 channels usually reveals a reduction in the resulting potassium current of ~20%–30%, which is apparently sufficient to cause BFNS. Even subtle changes in channel gating restricted to subthreshold voltages of an action potential are sufficient to cause BFNS, proving the physiological importance of this voltage range for the action of M-channels in a human disease model (Maljevic et al. 2008). From the genetic point of view, almost all mutations cause a haploinsufficiency. Treatment is only needed transiently and available drugs work usually well (for example, phenobarbital or valproate).

Clinically similar epilepsy syndromes that are genetically different from BFNS are benign familial neonatal/infantile seizures (BFNIS) and benign familial infantile seizures (BFIS). These phenotypes also display partial epileptic seizures with or without secondary generalization, but they occur between the age of 3 and 12 months (BFIS) or range from the neonatal to the infantile period (BFNIS) (Specchio and Vigevano 2006). BFIS can be associated with other neurological disorders, such as paroxysmal dyskinesia or migraine. Mutations in the *SCN2A* gene encoding one of the α-subunit $Na_V1.2$ of voltage-gated sodium channels expressed in the mammalian brain have been identified in BFNIS (reviewed by Reid et al. 2009). Functional investigations revealed a small persistent sodium current resulting in a gain-of-function effect predicting increased neuronal excitability. The age dependence of this syndrome could be explained by a transient expression of the respective Nav1.2 channels in the axon initial segments (AIS) of principal neurons in cortex and hippocampus during development and replacement later in life by Nav1.6 at these sites (Liao et al. 2010). A few *SCN2A* mutations with severe functional defects such as truncated proteins have been described in patients with intractable epilepsy and mental retardation (Mantegazza et al. 2010; Reid et al. 2009).

34.4 GENETIC EPILEPSY WITH FEBRILE SEIZURES PLUS, SEVERE MYOCLONIC EPILEPSY OF INFANCY (DRAVET SYNDROME), AND OTHER *SCN1A*-RELATED SYNDROMES

Genetic (formerly generalized) epilepsy with febrile seizures plus (GEFS+) is a childhood-onset syndrome featuring febrile convulsions and a variety of afebrile epileptic seizure types, often within the same pedigree. The penetrance is ~60%. Two-thirds of affected individuals were diagnosed as having febrile seizures (FS), which may be combined with either FS persisting after the sixth year of life, or with afebrile generalized tonic–clonic seizures (FS+). Additional seizure types such as absences, atonic, myoclonic–astatic, or focal seizures, particularly originating from the temporal lobe, may occur. Numerous additional mutations were subsequently identified in GEFS+ patients, accounting for 10% of cases. GEFS+ is caused by missense mutations in α and β1 subunits of the neuronal sodium channel, encoded by *SCN1A* and *SCN1B*, respectively. Some mutations may increase persistent sodium current, but a loss-of-function has been observed for most mutations. Reduced channel function is considered to be more significant than gain-of-function changes and leads to an overall loss-of-function phenotype (see severe myoclonic epilepsy of infancy [SMEI]) (Mantegazza et al. 2010).

Next to *SCN1A* and *SCN1B*, GEFS+ has been proposed to be associated with mutations in the homologous sodium channel α subunit genes encoded by *SCN2A* in a single family (Sugawara et al. 2001) and by *SCN9A* in potentially up to 5% of the patients with FS (Singh et al. 2009). The latter show a high penetrance of 95%. Functional expression has not yet been performed. Finally, several mutations in the gene *GABRG2* coding for a GABA-A receptor subunit, and one potential mutation in *GABRD*, have been identified. Dominant *GABRG2* mutations produce a decrease of GABA-activated chloride currents, thus reducing inhibitory currents, which in turn results in hyperexcitability (Eugene et al. 2007; Reid et al. 2009).

Severe myoclonic epilepsy of infancy (SMEI) or Dravet syndrome is characterized by clonic or tonic–clonic seizures in the first year of life that are often prolonged and associated with fever. During the course of the disease, patients develop afebrile generalized myoclonic, absence, or tonic–clonic seizures, but simple and complex partial seizures also occur. Cognitive deterioration appears in early childhood. Treatment is difficult in SMEI and related syndromes. Useful medications can be valproate (VPA), topiramate, or stiripentol (combined with clobazam and VPA). Since patients with SMEI sometimes have a family history of febrile or afebrile seizures, and in some families GEFS+ and SMEI overlap, SMEI may be regarded as the most severe phenotype of the GEFS+ spectrum (Reid et al. 2009; Ottman et al. 2010).

Somewhat milder forms compared with SMEI have been also described, such as borderline SMEI (SMEB) and intractable childhood epilepsy presents with generalized tonic–clonic seizures (ICEGTC). Families with some instances of ICEGTC in other family members affected by GEFS+ have been described. Therefore, we may conclude that the GEFS+ spectrum extends from simple FS to a variety of severe epileptic syndromes of childhood such as ICEGTC and SMEI, as also confirmed by genetic results described below (Weber and Lerche 2008; Mantegazza et al. 2010).

For SMEI and related forms, mutations in *SCN1A* encoding Nav1.1 have been identified (reviewed by Reid et al. 2009; Mantegazza et al. 2010). Together with GEFS+, more than 200 *SCN1A* mutations have been identified, accounting for 70% of cases (Meisler and Kearney 2005; Mantegazza et al. 2010). Mutation hotspots, such as sites of CpG deamination, account for 25% of de novo mutations (Kearney et al. 2006). Genetic screening for *SCN1A* is increasingly integrated in clinical diagnostics in Dravet's and related syndromes, as it confirms diagnosis, avoids other complicated, expensive and strenuous diagnostic procedures, and is an important factor for prognosis and genetic counseling (Ottman et al. 2010). One recessive *SCN1B* mutation has also been described in an SMEI patient (Patino et al. 2009).

Of the more than 200 $Na_V1.1$ mutations that have been described, most cause loss of function, with complete loss of function of one allele in Dravet's syndrome (Mantegazza et al. 2010). At first, it seems paradoxical that a loss of sodium channel function would produce epileptic seizures. However, knock-out mouse models have revealed that $Na_V 1.1$ seem to be the main sodium channel in inhibitory interneurons (Ogiwara et al. 2007; Yu et al. 2006), which may explain the pathophysiology of neuronal network hyperexcitability. Furthermore, it has been clinically observed that sodium channel blocking drugs aggravate epileptic seizures in patients with $Na_V 1.1$ mutations (Guerrini et al. 1998). This observation could be explained by a block of the remaining functional sodium channels that are generated by the healthy allele in inhibitory interneurons.

In a knock-in mouse model heterozygous for the GEFS+ associated mutation in the *SCN1B* gene, coding for the β_1 Na^+ channel subunit, the mutated proteins were absent from the AIS of pyramidal neurons where the WT subunits normally reside. AIS activity was increased, particularly at higher temperatures, which may explain the increased threshold for FS susceptibility observed in both patients and animals carrying the mutation (Wimmer et al. 2010).

34.5 IDIOPATHIC/GENETIC GENERALIZED EPILEPSY

Mutations were also identified rarely in families with classical idiopathic generalized epilepsies, namely childhood absence epilepsy (CAE) or juvenile absence epilepsy (JAE), juvenile myoclonic epilepsy (JME), and epilepsy with generalized tonic–clonic seizures on awakening (EGTCA). Absence seizures in CAE manifest typically around the sixth year of life and are of short duration, ~10 s, and typically occur in clusters of up to 100 seizures a day. In adolescence, generalized tonic–clonic seizures can occur. Myoclonic jerks are the clinical hallmark of JME, particularly of the upper extremities, which appear without loss of consciousness. They can be clinically subtle and escape recognition. JME also manifests during puberty, with seizures typically developing after awakening and being provoked by alcohol and sleep deprivation. Generalized tonic–clonic seizures also occur in adolescence in about 75% of the patients. The idiopathic generalized epilepsies may overlap within individuals and pedigrees and are typically associated with generalized spike-wave or poly–spike-wave discharges on EEG. Brain imaging is unremarkable.

A mutation in *GABRA1*, the gene encoding the α1-subunit of the GABA-A receptor, was identified in a family with JME (Cossette et al. 2002). The mutation leads to a loss of function of the GABA-A receptor, i.e., a decrease of inhibitory chloride currents, and thus leads to hyperexcitability (Cossette et al. 2002). Larger studies suggest that GABA-A receptor mutations are extremely rare (Dibbens et al. 2009). Two variants associated with JME have been described in the calcium-channel β subunit gene *CACNB4*, but they were not examined functionally and not much can be deduced about prevalence in the small population studied (Escayg et al. 2000).

For CAE with FS, a mutation in the γ2 subunit of the GABA-A receptor encoded by *GABRG2* has been described (Wallace et al. 2001) that also decreased GABA-activated chloride currents, which has been confirmed later on (Reid et al. 2009). A knock-in mouse model carrying this mutation exhibited frequent absence-like seizures. It was shown that cortical inhibition was reduced, whereas thalamic inhibition was not changed in this model (Tan et al. 2007). A *GABRA1* mutation associated with absence epilepsy revealed a loss of trafficking and membrane current. Functional coexpression of the wild type suggested that haploinsufficiency is the pathogenetic mechanism (Maljevic et al. 2006). Finally, variants in CAE and other subtypes have been described in *CACNA1H* encoding a neuronal voltage-gated T-type calcium channel. They were suggestive of a subtle gain of function by several different alterations in channel gating, which may explain a neuronal hyperexcitability (reviewed by Reid et al. 2009). Nevertheless, the significance of such findings with many mutations has to be validated by studying the mutation rate of this gene in the normal population.

34.6 CONCLUSIONS

Ion channels regulate neuronal excitability and synaptic transmission. Monogenetic ion-channel defects play an important role in the pathophysiology of several epileptic syndromes. Mutations in a single-channel complex can cause several different clinical phenotypes, and vice versa, the same clinical phenotype can be caused by mutations in distinct ion-channel complexes. As ion channels constitute one of the protein families that allow functional examination at the molecular level, expression studies of putative mutations have become standard in supporting the disease-causing nature of mutations. Functional expression can reveal both gain and loss of function of channel properties, and the decisive alteration is not always obvious. Several mechanisms on the channel and cell level can cause dominant diseases such as the channel epilepsies: simple loss of channel function (haploinsufficiency), dominant-negative effects on channel function, and gain of channel function. Although this functional study is quite helpful, overinterpretation of functional changes should be avoided, in particular, since most of these changes have been described in heterologous systems, i.e., mammalian nonneuronal cells or oocytes. Only few studies have reported changes in neuronal function on the cellular, network, or whole-animal level, and even in genetically altered animal models, we just start to understand underlying pathogenic mechanisms.

REFERENCES

Aridon, P., C. Marini, C. Di Resta, E. Brilli, M. De Fusco, F. Politi, E. Parrini, I. Manfredi, T. Pisano, D. Pruna, G. Curia, C. Cianchetti, M. Pasqualetti, A. Becchetti, R. Guerrini, and G. Casari. 2006. Increased sensitivity of the neuronal nicotinic receptor alpha 2 subunit causes familial epilepsy with nocturnal wandering and ictal fear. *Am J Hum Genet* 79(2):342–350.

Borgatti, R., C. Zucca, A. Cavallini, M. Ferrario, Panzeri, P. Castaldo, M. V. Soldovieri, C. Baschirotto, N. Bresolin, B. Dalla Bernardina, M. Taglialatela, and M. T. Bassi. 2004. A novel mutation in KCNQ2 associated with BFNS, drug resistant epilepsy, and mental retardation. *Neurology* 63:57–65.

Cossette, P., L. Liu, K. Brisebois, H. Dong, A. Lortie, M. Vanasse, J. M. Saint-Hilaire, L. Carmant, A. Verner, W. Y. Lu, Y. T. Wang, and G. A. Rouleau. 2002. Mutation of GABRA1 in an autosomal dominant form of juvenile myoclonic epilepsy. *Nat Genet* 31:184–189.

Dibbens, L. M., L. A. Harkin, M. Richards, B. L. Hodgson, A. L. Clarke, S. Petrou, I. E. Scheffer, S. F. Berkovic, and J. C. Mulley. 2009. The role of neuronal GABA(A) receptor subunit mutations in idiopathic generalized epilepsies. *Neurosci Lett* 453:162–165.

Escayg, A., M. De Waard, D. D. Lee, D. Bichet, P. Wolf, T. Mayer, J. Johnston, R. Baloh, T. Sander, and M. H. Meisler. 2000. Coding and noncoding variation of the human calcium-channel beta4-subunit gene CACNB4 in patients with idiopathic generalized epilepsy and episodic ataxia. *Am J Hum Genet* 66:1531–1539.

Eugène, E., C. Depienne, S. Baulac, M. Baulac, J. M. Fritschy, E. Le Guern, R. Miles, and J. C. Poncer. 2007. GABA(A) receptor gamma 2 subunit mutations linked to human epileptic syndromes differentially affect phasic and tonic inhibition. *J Neurosci* 27(51):14108–14116.

Guerrini, R., C. Dravet, P. Genton, A. Belmonte, A. Kaminska, and O. Dulac. 1998. Lamotrigine and seizure aggravation in severe myoclonic epilepsy. *Epilepsia* 39:508–512.

Hauser, W. A., J. F. Annegers, and W. A. Rocca. 1996. Descriptive epidemiology of epilepsy: Contributions of population-based studies from Rochester, Minnesota. *Mayo Clin Proc* 71:576–586.

Kearney, J. A., A. K. Wiste, U. Stephani, M. M. Trudeau, A. Siegel, R. RamachandranNair, R. D. Elterman, H. Muhle, J. Reinsdorf, W. D. Shields, M. H. Meisler, and A. Escayg. 2006. Recurrent de novo mutations of SCN1A in severe myoclonic epilepsy of infancy. *Pediatr Neurol* 34:116–120.

Ko, G. Y., L. Shi, and M. L. Ko. 2009. Circadian regulation of ion channels and their functions. *J Neurochem* 110:1150–1169.

Lehmann-Horn, F., and K. Jurkat-Rott. 1999. Voltage-gated ion channels and hereditary disease. *Physiol Rev* 79:1317–1371.

Lerche, H., Y. G. Weber, K. Jurkat-Rott, and F. Lehmann-Horn. 2005. Ion channel defects in idiopathic epilepsies. *Curr Pharm Design* 11:2737–2752.

Liao, Y., L. Deprez, S. Maljevic, J. Pitsch, L. Claes, D. Hristova, A. Jordanova, S. Ala-Mello, A. Bellan-Koch, D. Blazevic, S. Schubert, E. A. Thomas, S. Petrou, A. J. Becker, P. De Jonghe, and H. Lerche. 2010. Molecular correlates of age-dependent seizures in an inherited neonatal–infantile epilepsy. *Brain* 133:1403–1414.

Maljevic, S., K. Krampfl, J. Cobilanschi, N. Tilgen, S. Beyer, Y. G. Weber, F. Schlesinger, D. Ursu, W. Melzer, P. Cossette, J. Bufler, H. Lerche, and A. Heils. 2006. A mutation in the GABA(A) receptor alpha(1)-subunit is associated with absence epilepsy. *Ann Neurol* 59:983–987.

Maljevic, S., T. V. Wuttke, and H. Lerche. 2008. Nervous system Kv7 disorders: Breakdown of a subthreshold brake. *J Physiol* 586:1791–1801.

Mantegazza, M., G. Curia, G. Biagini, D. S. Ragsdale, and M. Avoli. 2010. Voltage-gated sodium channels as therapeutic targets in epilepsy and other neurological disorders. *Lancet Neurol* 9:413–424.

Meisler, M. H., and J. A. Kearney. 2005. Sodium channel mutations in epilepsy and other neurological disorders. *J Clin Invest* 115:2010–2017.

Ogiwara, I., H. Miyamoto, N. Morita, N. Atapour, E. Mazaki, I. Inoue, T. Takeuchi, S. Itohara, Y. Yanagawa, K. Obata, T. Furuichi, T. K. Hensch, and K. Yamakawa. 2007. Na(v)1.1 localizes to axons of parvalbumin-positive inhibitory interneurons: a circuit basis for epileptic seizures in mice carrying an Scn1a gene mutation. *J Neurosci* 27:5903–5914.

Ottman, R., S. Hirose, S. Jain, H. Lerche, I. Lopes-Cendes, J. L. Noebels, J. Serratosa, F. Zara, and I. E. Scheffer. 2010. Genetic testing in the epilepsies—Report of the ILAE Genetics Commission. *Epilepsia* DOI: 10.1111/J.1528–167.

Patino, G. A., L. R. Claes, L. F. Lopez-Santiago, E. A. Slat, R. S. Dondeti, C. Chen, H. A. O'Malley, C. B. Gray, H. Miyazaki, N. Nukina, F. Oyama, P. De Jonghe, and L. L. Isom. 2009. A functional null mutation of SCN1B in a patient with Dravet syndrome. *J Neurosci* Aug 26;29(34):10764–10778.

Reid, C. A., S. F. Berkovic, and S. Petrou. 2009. Mechanisms of human inherited epilepsies. *Prog Neurobiol* 87:41–57.

Singh, N. A., C. Pappas, E. J. Dahle, L. R. Claes, T. H. Pruess, P. De Jonghe, J. Thompson, M. Dixon, C. Gurnett, A. Peiffer, H. S. White, F. Filloux, and M. F. Leppert. 2009. A role of SCN9A in human epilepsies, as a cause of febrile seizures and as a potential modifier of Dravet syndrome. *PLoS Genet* 5:e1000649.

Specchio, N., and F. Vigevano. 2006 The spectrum of benign infantile seizures. *Epilepsy Res* 70 Suppl 1: S156–S167.

Steinlein, O. K., and D. Bertrand. 2008 Neuronal nicotinic acetylcholine receptors: From the genetic analysis to neurological diseases. *Biochem Pharmacol* 76(10):1175–1183.

Steinlein, O. K., C. Conrad, and B. Weidner. 2007. Benign familial neonatal convulsions: Always benign? *Epilepsy Res.* 73:245–249.

Steinlein, O. K. 2004. Genetic mechanisms that underlie epilepsy. *Nat Rev Neurosci* 5:443–448.

Sugawara, T., Y. Tsurubuchi, K. L. Agarwala, M. Ito, G. Fukuma, E. Mazaki-Miyazaki, H. Nagafuji, M. Noda, K. Imoto, K. Wada, A. Mitsudome, S. Kaneko, M. Montal, K. Nagata, S. Hirose, and K. Yamakawa. 2001. A missense mutation of the Na+ channel alpha II subunit gene Na(v)1.2 in a patient with febrile and afebrile seizures causes channel dysfunction. *Proc Natl Acad Sci U S A* 98:6384–6389. Erratum in 2001: *Proc Natl Acad Sci U S A* 98:10515.

Tan, H. O., C. A. Reid, F. N. Single, P. J. Davies, C. Chiu, S. Murphy, A. L. Clarke, L. Dibbens, H. Krestel, J. C. Mulley, M. V. Jones, P. H. Seeburg, B. Sakmann, S. F. Berkovic, R. Sprengel, and S. Petrou. 2007. Reduced cortical inhibition in a mouse model of familial childhood absence epilepsy. *Proc Natl Acad Sci U S A* 104:17536–17541.

Wallace, R. H., C. Marini, S. Petrou, L. A. Harkin, D. N. Bowser, R. G. Panchal, D. A. Williams, G. R. Sutherland, J. C. Mulley, I. E. Scheffer, and S. F. Berkovic. 2001. Mutant GABA(A) receptor gamma2-subunit in childhood absence epilepsy and febrile seizures. *Nat Genet* 28:49–52.

Weber, Y. G., and H. Lerche. 2008. Genetic mechanisms in idiopathic epilepsies. *Dev Med Child Neurology* 50:648–654.

Wimmer, V. C., C. A. Reid, S. Mitchell, K. L. Richards, B. B. Scaf, B. T. Leaw, E. L. Hill, M. Royeck, M. T. Horstmann, B. A. Cromer, P. J. Davies, R. Xu, H. Lerche, S. F. Berkovic, H. Beck, and S. Petrou. 2010. Axon initial segment dysfunction in a mouse model of genetic epilepsy with febrile seizures plus. *J Clin Invest* 120:2661–2671.

Yu, F. H., M. Mantegazza, R. E. Westenbroek, C. A. Robbins, F. Kalume, K. A. Burton, W. J. Spain, G. S. McKnight, T. Scheuer, and W. A. Catterall. 2006. Reduced sodium current in GABAergic interneurons in a mouse model of severe myoclonic epilepsy in infancy. *Nat Neurosci* 9:1142–1149.

35 Autosomal Dominant Nocturnal Frontal Lobe Epilepsy: Excessive Inhibition?

Molly N. Brown and Gregory C. Mathews

CONTENTS

Autosomal dominant nocturnal frontal lobe epilepsy (ADNFLE) was the first idiopathic epilepsy to be associated with a specific mutation that showed a single-gene inheritance pattern (Scheffer et al. 1995), and as such it became a model for understanding how a focal epilepsy could arise from a mutation in a widely expressed gene. The identified disease-causing mutations are in the α4 or β2 subunits of the nicotinic acetylcholine receptor (nAChR), a pentameric protein complex containing a cation-selective channel that produces a brief depolarizing potential upon activation. The known disease-causing mutations are all located near the proposed ion channel pore of the receptor complex (Steinlein and Bertrand 2008) and therefore affect the function of the receptor. nAChRs also possess variable permeability to calcium ions, conferring the potential to influence intracellular signaling pathways or directly modulate neurotransmitter release in addition to their depolarizing effects.

The frontal-lobe origin of seizures in ADNFLE and the association with mutant nAChRs suggest that cholinergic projections to the frontal cortex may somehow participate in the generation of the seizures. To understand how activation of mutant nAChRs can promote seizure activity, it is important to point out some basic concepts of cholinergic neurotransmission and the limitations of our knowledge in this area. Cholinergic neurons in the basal forebrain project widely throughout the frontal cortex, and their activity is correlated with the generation of fast (gamma frequency, 30–60 Hz) as well as theta frequency (4–8 Hz) cortical EEG activity in waking and REM sleep states (Jones 2005). The effects of acetylcholine (ACh) clearly involve both nAChR and muscarinic AChR activation in a complex interaction. The anatomical substrate of the rhythm generators and the precise role of ACh to facilitate these rhythms are incompletely understood.

35.1 FUNCTION OF NORMAL AND MUTANT α4β2 nAChRs

Acetylcholine may signal through two modes in the brain: A synaptic mode and a volume transmission mode. The brain nAChRs fall into two categories: lower-affinity synaptic receptors, primarily homomeric receptors composed of α7 subunits, and high-affinity receptors that are primarily localized to extrasynaptic sites. Neuronal cholinergic terminals rarely form clear synaptic junctions (Descarries et al. 1997), suggesting that volume transmission, in which released ACh activates receptors at distant, nonsynaptic sites, plays a major role in ACh's effects. Acetylcholine binds human α4β2-containing nAChRs with a very high affinity (Buisson and Bertrand 2001), and therefore α4β2 receptors would be able to generate signals from the low concentrations of ACh that are attained at extrasynaptic sites. After the α4β2 nAChR channel is opened in the presence of ACh, it enters a desensitized state more slowly than the lower-affinity α7-containing receptors, which makes this nAChR isoform particularly suitable to function in an environment with persistent low levels of ACh. Interestingly, although the identified ADNFLE mutations produced a variety of different effects on channel function, all mutations conferred enhanced sensitivity of α4β2-containing nAChRs to ACh (Bertrand et al. 2002). This gain-of-function alteration within the ADNFLE α4β2 nAChRs could thereby enhance the physiological response to low levels of ACh within the cortex.

Electrophysiological recordings have demonstrated that α4β2-containing nAChRs are presynaptically located on inhibitory axon terminals where they modulate GABA release. For example, rapid application of ACh (1 mM) evoked large inhibitory postsynaptic currents (IPSCs) measured in area CA1 pyramidal neurons from mouse hippocampal slices that were completely blocked by bicuculine, a GABA$_A$ receptor antagonist (Alkondon and Albuquerque 2001). Because ACh is an agonist at both homomeric α7- and α4β2-containing nAChRs, the authors demonstrated that a selective α4β2 receptor antagonist significantly decreased the size of these IPSCs. Tetrodotoxin (TTX), a blocker of voltage-gated sodium channels completely abolished the α4β2-evoked IPSCs, demonstrating that ACh triggered action-potential-dependent release of neurotransmitter GABA. In another study (Alkondon and Albuquerque 2004), these authors also investigated the role of α4β2-containing nAChRs in regulating GABA release from cortical interneurons by recording GABAergic postsynaptic currents (PSCs) from layer-V pyramidal neurons in rat brain slices. Acetylcholine (1 mM) significantly increased the amplitude and frequency of evoked GABAergic PSCs compared with choline, which selectively activates homomeric α7-containing nAChRs, indicating that α4β2-containing nAChRs play a predominant role in modulating GABA release at cortical inhibitory synapses.

35.2 ANIMAL MODELS OF ADNFLE

Because activation of α4β2-containing nAChRs strongly enhances GABA release in hippocampus and cortex, and because GABA is widely viewed as an inhibitory neurotransmitter that acts to dampen excitability and prevent seizures, it may seem paradoxical that a gain of function in these receptors could underlie a form of epilepsy. After multiple mutations associated with similar clinical manifestations had been identified, Steinlein recognized that nAChR channel properties alone would not sufficiently explain the pathophysiology of ADNFLE, but rather an understanding of the roles of nAChRs in cortical neuronal networks would be required (Steinlein 2004). Since that time, the generation of genetically altered animals harboring the human mutations have played an essential role in advancing our knowledge of the underlying pathophysiology of ADNFLE.

The recent generation of three ADNFLE mouse lines (Klaassen et al. 2006; Teper et al. 2007) has allowed investigators to explore the effect of these mutant nAChRs on GABA release and inhibitory circuitry both in vivo and in vitro. Klaassen and colleagues (2006) generated two knock-in mouse lines by creating two human ADNFLE-associated mutations in *Chrna4*, the mouse α4-nAChR subunit gene: *Chrna4^{S252F}* (an amino acid exchange) and *Chrna4^{+L264}* (an insertional mutation). To explore if these mutations in mice-reproduced key features of the ADNFLE syndrome in humans, the investigators recorded cortical EEGs in mice that were heterozygous for the mutation, mirroring the human genotype. Cortical EEG recordings in the two different heterozygous mutant strains revealed abnormal interictal EEG patterns with an increase in both δ (0.5–4 Hz) and θ (4–8 Hz) activity compared with their wild-type littermates. In addition, heterozygous mutant mice of both strains had seizures characterized by sudden onset of rhythmic high-voltage, low-frequency, and asymmetric spike-and-wave discharges. The spontaneous seizures occurred throughout the day were not associated with sleep, and were often characterized by hyperkinetic motor activity and a loss of balance.

In 2007, Teper and colleagues generated the third mouse line by genetically engineering a strain of mice harboring the human S248F mutation (a missense mutation) in *Chrna4* (the S252F mutation in the Klaassen line is the rat homologue to the human S248F mutation). In contrast to *Chrna4$^{S252F/wt}$* and *Chrna4$^{+L264/wt}$* mice, *Chrna4$^{S248F/wt}$* mutant mice did not exhibit spontaneous behavioral or electrographic seizures during overnight or daylight periods. However, after parenteral administration of nicotine (2 mg/kg), *Chrna4$^{S248F/wt}$* mice manifested a set of behaviors, collectively termed the dystonic arousal complex (DAC), that included saccadic behavior, forelimb dystonia behavior, and Straub tail (tail bent 180° from its natural position). Cortical EEG recordings during the DAC were normal in *Chrna4$^{S248F/wt}$* mice, raising the question of whether the DAC represents a seizure or other subcortically generated movement disorder. These *Chrna4$^{S248F/wt}$* mice did have seizures when challenged with a high dose (10 mg/kg) of nicotine. The phenotype displayed by these mice may be analogous to the nocturnal paroxysmal dystonia seen in some patients and should therefore be considered an important model with which to explore the spectrum of clinical manifestations in ADNFLE.

35.3 HOW DO ADNFLE MUTATIONS CAUSE EPILEPSY?

Is there any evidence of a link between these disease-causing mutations and altered regulation of GABA release? To investigate this question, Klaassen and colleagues (2006) performed whole-cell recordings in cortical pyramidal neurons in brain slices from adult wild-type and heterozygous mutant mice. Spontaneous excitatory or inhibitory postsynaptic currents (sEPSC or sIPSC, respectively) were not different in slices from wildtype and *Chrna$^{S252F/wt}$* mice. However, bath application of the nACh receptor agonist nicotine (1 μM) increased IPSC amplitude and frequency in *Chrna$^{S252F/wt}$*, which was quantified as the mean inhibitory current. Nicotine had no effect on EPSCs, suggesting that its only effects in the cortex were mediated through the GABAergic system. The nicotine-induced change in the mean inhibitory current in both *Chrna$^{S252F/wt}$* and *Chrna$^{+L264/wt}$* slices peaked at a 20-fold increase over baseline, then decreased during the continued presence of nicotine, presumably correlating with nAChR activation and desensitization. In contrast, nicotine increased the mean inhibitory current by only 2.5-fold in wild-type slices. This effect of nicotine was blocked by a selective α4β2 antagonist, but not by a selective homomeric α7 antagonist.

To determine the mechanism by which nicotine's activation of α4β2-containing nAChRs enhanced GABAergic neurotransmission, the authors applied nicotine to slices after blocking both voltage-gated sodium channels (with TTX) and calcium channels (with cadmium). The spontaneous miniature IPSCs (mIPSCs), or responses to release of single synaptic vesicles, observed were not different between wildtype and *Chrna4$^{S252F/wt}$* slices, and addition of nicotine did not alter the wild-type mIPSCs. However, nicotine increased the frequency and amplitude of mIPSCs in slices from *Chrna4$^{S252F/wt}$* mice, consistent with a direct activation of α4β2-containing nAChRs on presynaptic terminals of cortical interneurons to increase intraterminal calcium concentration and facilitate GABA release. However, the additional change in mIPSC amplitudes suggests that there may be additional effects of nicotine to potentiate GABAergic neurotransmission other than enhancing vesicle release.

These findings suggest that an exaggerated effect of ACh on presynaptic α4β2-containing nAChRs greatly enhances GABA release at cortical interneuron–pyramidal neuron synapses in ADFNLE mutant mice. To confirm the seemingly paradoxical finding that increased inhibitory output in the cortex could underlie the generation of seizures in *Chrna4$^{S252F/wt}$* mice, a GABA$_A$ receptor antagonist, picrotoxin, was administered in doses low enough to have no effect on wild-type mice. In *Chrna4$^{S252F/wt}$* mice, picrotoxin normalized the cortical EEG activity and abolished spontaneous seizure activity, strongly linking the increased inhibition with cortical seizure generation.

Teper and colleagues (2007) also investigated GABA release in the dystonic *Chrna4$^{S248F/wt}$* mouse line by evoking [^{3}H]GABA release with ACh in hippocampal synaptosomes from wild-type and mutant mice. The S248F mutation caused a leftward shift in the concentration-response curve for ACh-evoked release in the heterozygous mutant. This shift resulted in a reduced EC$_{50}$ for ACh-evoked GABA release, consistent with mutant α4β2-containing nAChRs conferring an increase in ACh sensitivity. Although Teper et al. argue that their mutant line displayed a dystonic behavior rather than seizures, they tested whether carbamazepine, a use-dependent sodium channel blocker that is particularly effective in the treatment of human ADNFLE, could prevent the nicotine-induced DAC in *Chrna4$^{S248F/wt}$* mice. Carbamazepine (40–60 mg/kg) was administered prior to nicotine (1 or 2 mg/kg) in both wild-type and *Chrna4$^{S248F/wt}$* mice. At the nicotine dose that induced the DAC (2 mg/kg, see above), carbamazepine did not inhibit the DAC. However, carbamazepine (60 mg/kg) abolished the forelimb dystonia component of the DAC induced by 1 mg/kg nicotine. Because carbamazepine also inhibits α4β2-containing nAChRs at the supraphysiological concentration administered (Picard 1999), the authors considered whether the blocking effect of carbamazepine could be due to an effect on receptor itself rather than on sodium channels. To investigate this question, they pre-administered a low dose of nicotine (0.1 mg/kg), which does not induce the DAC, to desensitize mutant nAChRs. This desensitization protocol abolished the subsequent induction of the DAC complex by high-dose nicotine in *Chrna4$^{S248F/wt}$* mice. These data suggest that carbamazepine may act directly on the mutant receptor to block the DAC rather than through an "anticonvulsant" effect due to sodium channel block. However, the study did not determine the pathophysiology of the DAC nor did it investigate further the role of GABA release and inhibitory circuits in this model.

Not all evidence from animal models implicates excessive inhibition in the pathophysiology of ADNFLE. Zhu and colleagues (2008) argued that a decrease, not an increase, in inhibition results from an ADNFLE mutation. These authors used transgenic technology to generate rats harboring the human S284L mutation in *Chrna4* (S248L-TG), a missense mutation distinct from the Klaassen S248F mouse strain. The S284L-TG rats manifested three distinct epileptic phenotypes during slow-wave sleep: paroxysmal arousals, paroxysmal dystonia, and epileptic wandering, each of which correlated with ictal discharges on EEG. Cortical EEG recordings also revealed abnormal interictal discharges originating in the sensorimotor cortex that appeared at 6 weeks of age. Therefore, in many respects the S284L-TG rats appear to mimic the human condition more closely than the three mouse lines discussed above. However, in contrast to the S248F and S252F mutations in α4β2-containing nAChRs, the S284L mutant did not show a nicotine-induced enhancement of inhibition that was observed in wild-type slices. The authors also observed a reduction in inhibitory synaptic transmission prior to seizure onset (experiments performed in cortical slices from 3-week-old rats). The authors propose that unlike *Chrna4$^{S248F/wt}$* and *Chrna4$^{S252F/wt}$* mice, the S284L mutation confers a loss of α4β2-containing nAChR function and thus results in a loss of the normal cholinergic enhancement of cortical inhibition. This study raises many questions, but there are several important technical differences between this model and the knock-in mouse models, besides the species difference. Transgenic technology does not replace the wild-type gene, but inserts the mutant gene elsewhere within the genome. Moreover, transgene transcription is driven by a non-native promoter, so the expression of the transgene is not subject to the same temporal and anatomical regulation as the native gene. Therefore, multiple differences in expression pattern of the mutant nAChR subunit could have influenced the findings. Nevertheless, it is important to consider that one phenotypical replica of human ADNFLE did not show evidence of altered inhibition in the frontal cortex.

35.4 A MODEL FOR EXCESSIVE INHIBITION PROMOTING SEIZURES

The two *Chrna4*-mutant mouse lines that share a common genotype (*Chrna4$^{S252F/wt}$* and *Chrna4$^{S248F/wt}$*) significantly advanced our understanding of a possible role for the GABA neurotransmitter system in the pathology that underlies ADNFLE. Mutations in both mouse lines represent a gain-of-function in α4β2-containing nAChRs that confer an increased sensitivity for ACh-evoked GABA release. Both studies also provided evidence to support the function of α4β2-containing nAChRs at GABAergic synapses to modulate vesicular GABA release. This effect of ACh is greatly enhanced in these ADNFLE mutant mice, resulting in an increase in vesicular GABA release and inhibitory output.

How could an increase in inhibitory output underlie the epileptic phenotype of ADNFLE mice? One explanation involves hypersynchronization of pyramidal neuron networks by GABAergic interneurons. Pyramidal neurons process cognitive information by firing at distinct frequencies that yield robust oscillatory activity. GABAergic interneurons that fire with these unique frequencies may provide the temporal structure required for this oscillatory activity. For example, hippocampal basket and chandelier cells are GABAergic interneurons that project their extensive axonal arbors to synapse exclusively with the cell body and proximal dendrites of CA1 pyramidal neurons. A single basket or chandelier cell forms synapses with hundreds of pyramidal neurons and thus is capable of synchronizing the spontaneous firing of multiple CA1 pyramidal neurons during theta oscillations (Cobb et al. 1995). In the hippocampus, the interneurons generating the α4β2-mediated effect on GABAergic transmission are those within and adjacent to the pyramidal cell layer (Albuquerque et al. 2009), most consistent with the basket and chandelier cells. In cortex, release of endogenous ACh increases the synchronization of local cortical interneuron networks (Bandyopadhyay et al. 2006), providing further support for the role of cholinergic inputs in network synchronization, although the anatomical localization of α4β2 receptors has not been determined.

In this model of ADNFLE, burst firing of cholinergic neurons results in pulsatile release of ACh, which diffuses to activate α4β2-containing nAChRs on GABAergic terminals. Because ADNFLE mutant receptors are abnormally sensitive to ACh, receptor activation occurs at levels of neurotransmitter that may not normally potentiate release, resulting in either stronger or more prolonged enhancement of GABAergic transmission than normal. This excessive inhibition causes a potentiated inhibition of a network of pyramidal neurons that, when relieved, results in hypersynchronization of the network output (Figure 35.1).

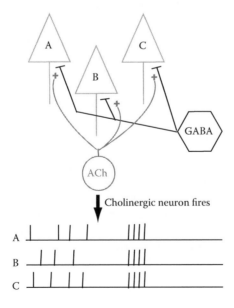

FIGURE 35.1   Model of enhanced inhibition resulting in hypersynchronization of excitatory networks. The model depicts three cortical pyramidal neurons (A,B,C) and their firing patterns (below). The pyramidal neurons are innervated by a single inhibitory interneuron (GABA). A cholinergic projection neuron releases acetylcholine (ACh) near the GABAergic presynaptic terminals. If release of ACh is sufficient to activate presynaptic nAChRs on GABAergic terminals, GABA release is enhanced and the pyramidal neurons are strongly inhibited. Release from the inhibition is followed by a period of hypersynchronization.

Alternatively, ACh may initiate cortical seizures by modulating other inhibitory circuits that synchronize cortical pyramidal neurons. For example, thalamic reticular neurons (TRNs) are exclusively GABAergic interneurons that exert strong inhibitory influence on glutamatergic thalamocortical (TC) neurons (Pinault 2004), and cholinergic inputs may modulate these circuits. These circuits have already been implicated in the seizures observed in absence epilepsy. Likewise, GABAergic projection neurons from the basal forebrain inhibit cortical pyramidal neurons and may synchronize their output; a role for $\alpha 4\beta 2$ receptors on these neurons has not been determined. Cholinergic afferents from the basal forebrain and brainstem nuclei that innervate the thalamus and cortex may modulate these inhibitory circuits to hypersynchronize cortical pyramidal neurons in ADNFLE mutant mice.

35.5 CONCLUSIONS

In humans, ADNFLE is characterized by a heterogeneous spectrum of clinical motor phenotypes, from paroxysmal arousals to episodes of paroxysmal dystonia to hyperkinetic seizures. This spectrum has called into question whether ADNFLE represents a true form of epilepsy, a movement disorder, or a combination of both. Although the ADNFLE mouse models do not precisely address this issue, they do recapitulate many features of the spectrum of human clinical motor phenotypes. On one end, $Chrna4^{S248F/wt}$ mice only display episodes of paroxysmal dystonia that correlate with normal scalp EEG activity. On the other end, $Chrna4^{S252F/wt}$ mice display hyperkinetic seizures that correlate with abnormal EEG activity. Therefore, the $Chrna4^{S248F/wt}$ and $Chrna4^{S252F/wt}$ mice likely represent a more mild and severe model of ADNFLE, respectively. Differences in the genetic backgrounds of the two mice strains may explain why a common mutation confers two distinct clinical manifestations. It will thus be important for investigators to explore the influence of modifier genes on behavioral phenotypes in these and future mouse models of ADNFLE. Also, in addition to the $\alpha 4$ subunit, studies have identified several mutations in the $\beta 2$ subunit gene ($Chrnb2$) of the $\alpha 4\beta 2$ nAChR that associate with ADNFLE (Steinlein and Bertrand 2008). Therefore, it will also be important for future investigations to generate mice that harbor mutations in the $\beta 2$ subunit gene to explore whether these mutations confer a similar motor phenotype as either the $Chrna4^{S248F/wt}$ or $Chrna4^{S252F/wt}$ mice.

Although the $Chrna4^{S248F/wt}$ and $Chrna4^{S252F/wt}$ mice exhibit the motor phenotypes associated with ADNFLE, the nocturnal pattern of seizure occurrence seen in human ADNFLE does not appear to be replicated in either mouse model. The paroxysmal manifestations of human ADNFLE occur primarily during the transition states of the sleep/wake cycle during non-REM sleep. The firing of cholinergic neurons is reduced during slow-wave sleep, and increased during REM sleep and waking. According to the working model of ADNFLE presented above, the rapid firing of cholinergic neurons during arousal could result in pulses of ACh release that transiently increase inhibitory output from specific interneuron populations and hypersynchronize pyramidal neuron networks. It is possible that such transient synchronizations occur during sleep state transitions or arousals and initiate seizures within frontal lobe circuits receiving numerous cholingeric inputs. However, the events observed in $Chrna4^{S252F/wt}$ mice and $Chrna4^{S252F/wt}$ mice did not correspond to certain sleep/wake states. The discrepancy may result from differences between mice and humans in the anatomy or physiology of the cholinergic projection system or in the expression pattern of nAChRs. If true, absence of the nocturnal phenotype in mouse models will likely hinder the use of mouse models to fully understand the cellular and circuit changes that underlie the generation of frontal lobe seizures in ADNFLE patients.

REFERENCES

Albuquerque, E. X., E. F. Pereira, M. Alkondon, and S. Rogers. 2009. Mammalian nictotinic acetylcholine receptors: From structure to function. *Physiol. Rev.* 89:72–120.

Alkondon, M., and E. X. Albuquerque. 2001. Nicotinic acetylcholine receptor alpha7 and alpha4beta2 subtypes differentially control GABAergic input to CA1 neurons in rat hippocampus. *J. Neurophysiol.* 86(6):3043–3055.

Alkondon, M, and E. X. Albuquerque. 2004. The nicotinic acetylcholine receptor subtypes and their function in the hippocampus and cerebral cortex. *Prog. Brain Res.* 145:109–120.

Bandyopadhyay, S., B. Sutor, and J. J. Hablitz. 2006. Endogenous acetylcholine enhances synchronized interneuron activity in rat neocortex. *J. Neurophysiol.* 95(3):1908–1916.

Bertrand, D., F. Picard, S. Le Hellard, S. Weiland, I. Favre, H. Phillips, S. Bertrand, S. F. Berkovic, A. Malafosse, and J. Mulley. 2002. How mutations in the nAChRs can cause ADNFLE epilepsy. *Epilepsia* 43(Suppl 5):112–122.

Buisson, B., and D. Bertrand. 2001. Chronic exposure to nicotine upregulates the human (alpha)4(beta)2 nicotinic acetylcholine receptor function. *J. Neurosci.* 21(6):1819–1829.

Cobb, S. R., E. H. Buhl, K. Halasy, O. Paulsen, and P. Somogyi. 1995. Synchronization of neuronal activity in hippocampus by individual GABAergic interneurons. *Nature* 378(6552):75–78.

Descarries, L., V. Gisiger, and M. Steriade. 1997. Diffuse transmission by acetylcholine in the CNS. *Prog. Neurobiol.* 53(5):603–625.

Jones, B. E. 2005. From waking to sleeping: Neuronal and chemical substrates. *Trends Pharmacol. Sci.* 26(11):578–586.

Klaassen, A., J. Glykys, J. Maguire, C. Labarca, I. Mody, and J. Boulter. 2006. Seizures and enhanced cortical GABAergic inhibition in two mouse models of human autosomal dominant nocturnal frontal lobe epilepsy. *Proc. Natl. Acad. Sci. U.S.A.* 103(50):19152–19157.

Pinault, D. 2004. The thalamic reticular nucleus: Structure, function and concept. *Brain Res. Rev.* 46(1):1–31.

Scheffer, I. E., K. P. Bhatia, I. Lopes-Cendes, D. R. Fish, C. D. Marsden, E. Andermann, F. Andermann, R. Desbiens, D. Keene, F. Cendes, J. I. Manson, J. E. C. Constantonou, A. McIntosh, and S. F. Berkovic. 1995. Autosomal dominant nocturnal frontal lobe epilepsy. A distinctive clinical disorder. *Brain* 118(Pt 1):61–73.

Steinlein, O. K. 2004. Genetic mechanisms that underlie epilepsy. *Nat. Rev. Neurosci.* 5(5):400–408.

Steinlein, O. K., and D. Bertrand. 2008. Neuronal nicotinic acetylcholine receptors: From the genetic analysis to neurological diseases. *Biochem. Pharmacol.* 76(10):1175–1183.

Teper, Y., D. Whyte, E. Cahir, H. A. Lester, S. R. Grady, M. J. Marks, B. N. Cohen, C. Fonck, T. McClure-Begley, J. M. McIntosh, C. Labarca, A. Lawrence, F. Chen, I. Gantois, P. J. Davies, S. Petrou, M. Murphy, J. Waddington, M. K. Horne, S. F. Berkovic, and J. Drago. 2007. Nicotine-induced dystonic arousal complex in a mouse line harboring a human autosomal-dominant nocturnal frontal lobe epilepsy mutation. *J. Neurosci.* 27(38):10128–10142.

Zhu, G., M. Okada, S. Yoshida, S. Ueno, F. Mori, T. Takahara, R. Saito, Y. Miura, A. Kishi, M. Tomiyama, A. Sato, T. Kojima, G. Fukuma, K. Wakabayashi, K. Hase, H. Ohno, H. Kijima, Y. Takano, A. Mitsudome, S. Kaneko, and S. Hirose. 2008. Rats harboring S284L Chrna4 mutation show attenuation of synaptic and extrasynaptic GABAergic transmission and exhibit the nocturnal frontal lobe epilepsy phenotype. *J. Neurosci.* 28(47):12465–12476.

36 How to Measure Circadian Rhythmicity in Humans

Wytske A. Hofstra and Al W. de Weerd

CONTENTS

In 1885, Gowers was the first to report that the occurrence of epileptic seizures is not entirely random. He classified patients into three groups based on the distribution of fits over the day: diurnal, nocturnal, and diffuse (Gowers 1885). Later, it was observed how diurnal seizures cluster at certain times of the day, i.e., upon awakening and in the late afternoon, and how nocturnal seizures tend to occur mainly at bedtime and the hours before awakening (Griffiths et al. 1938; Langdon-Down et al. 1929). These observations, that seizures may occur in patterns depending on the pathophysiology of the epileptic syndrome, have also been confirmed in more recent studies (Milton et al. 1987; Quigg et al. 1998b; Tauboll et al. 1991). The reason for these seizure patterns is not well understood. A hypothesis is that the circadian clock plays a significant role. The interaction between circadian rhythmicity and epilepsy has been studied; however, still relatively little is known (Hofstra et al. 2009).

Animal studies can be used to study the possible interaction between epilepsy and circadian rhythmicity in more detail. In an illustrative study by Quigg et al. (2000), epileptic rats were monitored with constant EEG registration and entrained to a 12-hour/12-hour light/dark cycle. Then, they were exposed to constant darkness. It was observed that during light/dark exposure, spontaneous limbic seizures occurred in statistically nonuniform patterns, with nearly twice as many seizures during the light. During constant darkness, seizures continued to occur in the same pattern observed during light/dark exposure when referenced to the circadian rhythm of these rats, suggesting that spontaneous limbic seizures recur in a true endogenously mediated circadian pattern.

More research is required to elucidate the possible interaction between the circadian rhythm and epilepsy. If this interaction indeed exists, it has important diagnostic and therapeutic consequences, such as improved control over epilepsy through administration of antiepileptic medication according to an individual's circadian rhythm.

36.1 CIRCADIAN RHYTHM AND OUTPUT

The mammalian biological clock consists of a hierarchy of oscillators, of which the master circadian pacemaker is found deep within the brain. This pacemaker is formed by the cells of the suprachiasmatic nuclei (SCN) that can be found within the anterior hypothalamus. The cells within these nuclei generate and maintain circadian rhythms in many physiological and psychological processes in the body, including the sleep/wake cycle, core body temperature, blood pressure, alertness, and the synthesis and secretion of several hormones, for example melatonin and cortisol (Hastings et al. 2007). The intrinsic period of the human biological clock is around 24.2 hours, with surprisingly small inter- and intraindividual ranges (Czeisler et al. 1999). This period is slightly longer than our 24-hour day, which is why it is named a circadian rhythm (with *circa* meaning *around* and *dies* meaning *day*). As in all rhythms, important features of the circadian clock are phase, amplitude, and period.

The circadian system has to be synchronized to the exact 24-hour day outside. This requires a daily adjustment of the SCN, which is termed entrainment. Entrainment is one of the key characteristics of the biological clock. It is accomplished by external cues, called Zeitgebers, which means *time givers* in German. Zeitgebers are, for instance, scheduled sleep, activity, temperature, and social external signals such as the clock and meals. However, by far the most important Zeitgeber is light or the solar light/dark cycle (Duffy et al. 2005). As light enters the eye, it is picked up by specialized photoreceptors in the ganglion cells of the retina (Berson et al. 2002). The signal travels from the retina to the SCN via the retinohypothalamic tract and there it resets the SCN. The extent of this resetting response depends on several factors, including wavelength, intensity, timing, number, pattern, and duration of exposure to light (Duffy et al. 2005).

The SCN exert their influence on the human body via projections throughout the hypothalamus, thalamus, and limbic system (Buijs et al. 2001). The pineal body is one of the main target organs of the SCN. This small brain structure is, amongst other things, responsible for the synthesis of the hormone melatonin from tryptophan. This production is highly rhythmic in all mammals examined thus far, with a characteristic low daytime level that ascends after the onset of darkness to a high output during the night (with its peak between 11:00 PM and 3:00 AM) and then falling sharply before the onset of light. Melatonin secretion follows a near-square-wave pattern that continues even in constant darkness (Ralph et al. 1971). As melatonin synthesis is suppressed by light, the circadian rhythm is similar in diurnal and nocturnal mammals (Lewy et al. 1980). The activity of the SCN is the other rare circadian rhythm occurring in phase in day-active and night-active species.

As mentioned, the master circadian clock is found in the SCN. Apart from this master pacemaker, there is convincing evidence that there are peripheral circadian oscillators in the human body that function more or less independently of the SCN. These peripheral oscillators are found in several organs, including liver, skeletal muscle, and testis and are all under the influence of the SCN (Plautz et al. 1997; Zylka et al. 1998).

36.2 INTERINDIVIDUAL DIFFERENCES

Circadian rhythmicity has been demonstrated in several processes, for example core body temperature (CBT) and secretion of several hormones. The phases of these rhythms vary between individuals. This variation has been attributed to factors such as gender, age, and especially "morningness/eveningness," which is the individual's preference in timing of sleep, wake, and activities (Vink et al. 2001). Morningness scores correlate with timing of the individual's circadian pacemaker (Duffy et al. 2001). Compared to evening types, morning types tend to schedule sleep earlier and experience earlier peaks of alertness and performance during the day (Andrade et al. 1992). Furthermore, markers of circadian rhythm (in the form of CBT, melatonin, and cortisol) show an earlier phase in morning types in comparison to evening types (Kerkhof et al. 1996; Duffy et al. 1999). However, morning and evening types paradoxically do not go to bed and wake up at the same circadian time (Duffy et al. 1999).

Various genes have been discovered that are, at least in part, responsible for the characteristic activity of the individual SCN and the differences between individuals (Cermakian et al. 2003). The SCN activity depends on the expression of auto regulatory translation–transcription feedback loops of genes, including the Period genes (Per1, Per2, Per3), the Clock gene, and two Cryptochrome genes (Cry 1, Cry2) (Van Gelder et al. 2003). In several animal studies it has been demonstrated that mutation or deletion of these genes leads to rhythms with abnormal periods or even arrhythmic phenotypes when tested under constant conditions (Cermakian et al. 2003; Ko et al. 2006). Therefore, it is no surprise that many researchers are studying whether there is a genetic basis for circadian rhythm disorders. This has been successful to a certain extent in a few circadian rhythm disorders (Toh et al. 2001; Xu et al. 2005). Furthermore, clock gene dysfunction might be important in the development of various diseases, including cancer (Lamont et al. 2007).

In this chapter, methods are described that are frequently used to measure circadian rhythm in humans. Protocols are discussed that can be used in research settings and are necessary in certain situations to ensure pure data collection. Also, frequently used biological phase markers are reviewed and sleep parameters, questionnaires, and actigraphy are also addressed.

36.3 MEASURING CIRCADIAN RHYTHM IN HUMANS

36.3.1 PROTOCOLS

It is necessary to rule out all influencing external factors (so-called masking factors) to study the underlying periodicity of the biological clock. Hereby, internal time can be desynchronized from external time, which is called forced desynchrony. Nathaniel Kleitman was the first to conduct such desynchronizing experiments. In 1938, he scheduled subjects to live on artificial day lengths in the Mammoth Cave in Kentucky (Cajochen et al. 2006). Under such conditions, the subjects' internal circadian clock could not entrain and, therefore, continued to oscillate with their own endogenous period. In the decades that followed, many researchers used the same principle to study circadian rhythmicity. For example, Dijk et al. scheduled subjects to live a 28-hour day, resulting in sleep episodes at all phases of the endogenous circadian cycle. Circadian and sleep/wake components are distinguished very well this way (Dijk et al. 1995). Unfortunately, this protocol is very labor intensive and also long lasting, as it takes a minimum of four weeks to complete. Therefore, other protocols have been proposed. For instance, Hiddinga et al. used a protocol of 120 hours to study body temperature in humans (Hiddinga et al. 1997). This protocol consists of six 20-hour days, demonstrating that a shorter forced desynchrony protocol can also be used to reliably differentiate the endogenous circadian rhythm.

In addition to the forced desynchrony protocols, the constant routine protocol has been developed to reveal unmasked circadian rhythms (Mills et al. 1978). A constant routine protocol minimizes or eliminates external factors that are known to obscure endogenous component of circadian rhythms. A constant routine works by keeping these external factors as constant as possible. During such a protocol, the patient is kept awake in a semirecumbent position and supplied equally distributed small meals for at least 24 hours. This is done in a room with constant temperature and humidity and under dim light conditions (Duffy et al. 2002). Although this constant-routine protocol can be very useful, it does not desynchronize the circadian pacemaker from the natural sleep/wake cycle. More important, because sleep is also a masking factor, subjects should be kept awake during the entire constant routine. This results in sleep deprivation, which may unfortunately influence the interpretation of results or modify the circadian patterns studied. Therefore, this protocol cannot be used in studies of diseases influenced by sleep deprivation, such as epilepsy (Hiddinga et al. 1997). To avoid buildup of sleep pressure, the multiple-nap protocol has been designed. This is a constant-routine protocol with various longer naps scheduled over a 24-hour day or longer and thereby accumulation of sleep pressure is prevented (Cajochen et al. 2001). This means that an important masking factor is strongly reduced and the circadian rhythm emerges very clearly. Moreover, it is a short protocol compared to the other long lasting forced desynchrony protocols.

Finally, the circadian rhythm can also be shifted by exposure to bright light when timing of the sleep/wake cycle is fixed. The extent and direction of the phase shift depend on the phase of the rhythm at the time of light exposure. Light late in the subjective night causes advances to an earlier phase and light early in the subjective night causes delay to a later phase (Voultsios et al. 1997).

36.4 MARKERS OF CIRCADIAN RHYTHM

Theoretically, all output rhythms driven by the circadian clock that can be measured can be used to assess the phase, period, and amplitude of the circadian rhythm. In practice, however, the most widely used measures are melatonin production, core body temperature, and cortisol production.

36.4.1 MELATONIN

Melatonin levels can be measured in plasma or saliva [with a concentration that is approximately three times lower than the serum concentration (Voultsios et al. 1997)] and its metabolite, 6-sulfatoxymela-tonin, can be measured in urine. Measuring the entire 24-hour melatonin rhythm is considered to be the most robust phase marker (Van Someren et al. 2007). However, this is a time-consuming method and also inconvenient for the subject. Therefore, this method is not frequently employed. More often, the moment at which the level of melatonin starts rising in the evening (the onset) is used as an indica-tion of the circadian phase. When using this method, it is very important to control light conditions. This is because ambient light intensities, as encountered outdoors (i.e., 3000–100,000 lx) can suppress the production of melatonin. Some authors report that even intensities as little as 100–180 lx (room light intensity) can produce phase shifts in the circadian rhythm (Boivin et al. 1998; Zeitzer et al. 2000). The degree of suppression by light seems to depend on the light history, i.e., to what intensity of light the subject was exposed prior to the study (Hebert et al. 2002; Smith et al. 2004). To avoid erroneous values during measurement, melatonin samples should be taken in dim light, which is less than 50 lx. Therefore, this procedure is termed the dim-light melatonin onset (DLMO) (Benloucif et al. 2008). As mentioned, the DLMO reflects the phase of the circadian rhythm and, if measured over more than one cycle, also the period of the endogenous circadian pacemaker. The DLMO is normally observed about 2 to 3 hours before habitual bedtime (1930–2200 hours) (Lewy et al. 1999).

Determining the DLMO from a partial melatonin profile (i.e., not over the entire 24 hours) can be done in various ways. First, an absolute threshold can be taken, for instance, in the range of 2 to 10 pg/mL (in serum). Although intraindividual differences in melatonin profiles are small, large differences in absolute production of melatonin between individuals exist. Some people are so-called low producers and produce very little melatonin. In these cases, the use of a threshold of 2 pg/mL is recommended (Benloucif et al. 2008). When measuring melatonin in saliva, an absolute threshold of 3 pg/ml can be used (Lewy et al. 1999). A second way of determining the DLMO is by calculating the threshold at two standard deviations above the average baseline samples (at least three samples needed prior to DLMO measurement) (Duffy et al. 2001). Finally, a visual estimate can be made of the point of change of the curve (baseline to rising).

The DLMO is considered to be one of the most reliable markers because it is minimally masked by exogenous factors. However, there is some evidence that several factors can mask the melatonin level to some degree. Masking factors include posture of the subject, exercise, sleep and sleep deprivation, caf-feine and certain drugs, e.g., NSAIDS and beta blockers (Deacon et al. 1994; Monteleone et al. 1990;

Murphy et al. 1996; Shilo et al. 2002; Stoschitzky et al. 1999; Zeitzer et al. 2007). When using the DLMO as phase marker, it is important to control these factors as much as possible to obtain reliable values.

Various studies have studied the relationship between age and secretion of melatonin. However, the results of these studies are inconsistent on whether this secretion is age related and whether melatonin levels decrease with increasing age (Munch et al. 2005; Zeitzer et al. 1999). Therefore, it remains to be elucidated whether this age-related reduction exists.

Melatonin is intensively studied with respect to epilepsy and seizures. Its precise value is still disputed because results are very inconsistent. For instance, Sandyk et al. (1992) reported that it can also have a proconvulsive effect in humans, although other authors found that melatonin has a depressive effect on brain excitability and can prevent seizures in several animal models (Champney et al. 1996; Lapin et al. 1998). Another inconsistency is found in the change of melatonin levels in epilepsy patients compared to people without epilepsy. Some have described low melatonin levels in epilepsy patients, whereas others have measured normal or elevated levels (Bazil et al. 2000; Laakso et al. 1993; Rao et al. 1989; Schapel et al. 1995). Likewise, in some studies levels were found to be elevated after complex partial seizures, whereas in other studies no changes were observed after complex partial and generalized tonic–clonic seizures (Bazil et al. 2000; Rao et al. 1989). In conclusion, further research is needed to define the precise effects of seizures on melatonin and its role in seizure prevention and epilepsy.

36.4.2 CORE BODY TEMPERATURE

In 1842 in Germany, it was found that core body temperature (CBT) possesses a circadian rhythm. Gierse showed that his own oral temperature reached a maximum in the early evening and a minimum in the early morning (Waterhouse et al. 2005). Over the years, various studies have confirmed this rhythm and have demonstrated that CBT is indeed one of the physiological processes regulated by the circadian pacemaker (Krauchi 2002).

The circadian rhythm of the CBT is a result of the combined action of heat production and loss. When heat loss exceeds heat production, CBT declines and vice versa (Waterhouse et al. 2005). The rhythm is characterized by a nocturnal decline, which is caused by greater heat loss and vasodilatation at distal skin regions (Krauchi et al. 1997). Human sleep is typically initiated when the decline is at its maximum rate (Campbell et al. 1994). After reaching its minimum, (the so-called *nadir*, reached at approximately 5:00 AM), temperature increases and reaches its maximum (at approximately 5:00 PM) during the day as a result of heat production surpassing heat loss.

It is relatively easy to collect continuous CBT data from a subject without much disturbance and the data can be analyzed immediately. Therefore, CBT is also a frequently used marker in studies on circadian rhythmicity. However, various factors are known to influence CBT, thereby masking the true endogenous signal. These factors include such behaviors as postural changes, physical activity, and meals and also external conditions, such as ambient temperature, sound, humidity, and bright light (Ancoli-Israel et al. 2003; Dauncey et al. 1983; Dijk et al. 1991; Gander et al. 1986; Krauchi et al. 1997; Krauchi et al. 2006; Moran et al. 1995). Earlier studies also hypothesized that sleep influences CBT; however, more recent studies contradict these findings (Krauchi et al. 2001). Another influencing factor is age. In the elderly, the CBT reaches its nadir earlier than in younger subjects (Czeisler et al. 1992). In a study by Quigg et al. (1998a), it was reported that the circadian rhythm of CBT in epileptic rats (models for mesial temporal lobe epilepsy) is more complex and polyphasic than that in normal rats. However, acute stimulated seizures did not affect this complexity.

CBT can be a good measure in circadian studies, but strict protocols have to be used. The constant-routine protocol is, as described earlier, designed to minimize the influence of masking factors by keeping these as constant as possible. Also, a forced-desynchrony protocol can be applied to distinguish the variable caused by masking factors from that related to the circadian pacemaker (Dijk et al. 1995; Hiddinga et al. 1997). Finally, it is also possible to minimize masking through the use of mathematical adjustments of the temperature rhythm (Waterhouse et al. 2000).

36.4.3 CORTISOL

Cortisol is a corticosteroid hormone produced by the zona fasciculata of the adrenal cortex. The circadian pacemaker also generates a rhythm in this hypothalamic–pituitary–adrenal secretion via a multisynaptic suprachiasmatic nucleus–adrenal pathway (Buijs et al. 1999). The secretion of cortisol is therefore highly rhythmic. The curve is characterized by declining levels throughout the day, a nocturnal period of quiescence, and a sharp rise in the second half of the night toward a morning maximum (the acrophase), particularly in the early morning, some hours before and just after waking. The amplitude of the secretory episodes declines throughout the morning and is minimal in the evening. The nadir is reached within approximately two hours after beginning sleep (Veldhuis et al. 1990).

Levels of cortisol can be measured in serum and saliva. Free cortisol in the serum diffuses freely into saliva and measurements of salivary cortisol reflect serum free cortisol concentrations more accurately than measurements of total cortisol in the serum (Umeda et al. 1981).

Several characteristics of rhythm in cortisol production can be used as markers of circadian rhythm, including the timing of the nadir or acrophase, the onset of the evening rise, and the start or end of the quiescent period (Van Cauter et al. 1996; Weibel et al. 2002). However, again, masking is important when measuring cortisol. Several factors are known to influence cortisol secretion. For example, physical and physiological stress also activates the hypothalamic–pituitary–adrenal axis, which results in bursts of secretory activity. Light is also an influencing factor because it raises the morning cortisol peak and can cause phase shifts (Benloucif et al. 2008; Leproult et al. 2001; Scheer et al. 1999). In addition, aging has its effects on the cortisol rhythm because, with increasing age, the cortisol rhythm shifts; the nadir and maximum of the curve are reached earlier and circadian amplitude is also reduced in elderly subjects (Van Cauter et al. 1996; van Coevorden et al. 1991). The sleep/wake cycle can influence the cortisol level in several ways. Deep sleep and sleep onset reduce or inhibit cortisol secretion, whereas sleep loss, light sleep, and awakenings result in elevated cortisol levels (Caufriez et al. 2002; Gronfier et al. 1998). High-protein meals can cause additional secretory episodes (Slag et al. 1981). With respect to epilepsy, several studies describe postictal elevations of cortisol levels (Culebras et al. 1987; Mehta et al. 1994; Rao et al. 1989). However, these changes were not observed in all studies (Molaie et al. 1987). With knowledge of all these masking factors, it is important to keep these as minimal or constant as possible when using cortisol levels to measure circadian rhythm.

36.4.4 COMBINATIONS AND CORRELATIONS

If studies are designed well using constant-routine or forced desynchrony protocols, the correlation between CBT and DLMO as phase markers is usually high (Sack et al. 2007; Shanahan et al. 1991). Therefore, this combination is frequently used in studies. DLMO and cortisol levels are also a well-known and frequently applied combination in circadian studies. For instance, Weibel et al. (2002) compared cortisol levels and the DLMO in a group of shift workers and found that the start of the quiescent period remained phase locked to the DLMO.

Recently, Klerman et al. (2002) compared the variability of all three markers discussed above. The conclusion was that methods using plasma melatonin as a marker may be considered more reliable than methods using CBT or cortisol as an indicator of circadian phase in humans.

Taken together, the DLMO is the most robust and widely used phase marker of these three. The data are easy to collect and easy to interpret, masking is relatively low, and masking factors can be controlled fairly simply.

36.4.5 SLEEP PARAMETERS

The sleep/wake cycle is a complex process in which several factors interact. According to the two-process model of sleep regulation, sleep timing and structure are determined by the interaction of two players: (1) the circadian pacemaker (process C), that promotes alertness during the subjective day and sleepiness during the subjective night and (2) a homeostatic increase in sleepiness (process S), which depends on the prior time awake (Borbely 1982; Daan et al. 1984). Other important factors are sleep inertia and temperature (Krauchi et al. 2005). As the circadian clock is only part of the whole process, merely a few sleep parameters are useful as phase markers of the circadian rhythm. The most reliable one with respect to sleep timing is the sleep midpoint (Wirz-Justice 2007). Studies have shown that the DLMO significantly correlates with sleep onset, sleep midpoint, and wake time in normal healthy young adults, but correlates most significantly with sleep midpoint (Martin et al. 2002). Also, the DLMO can be readily estimated in people whose sleep times are minimally influenced by external factors, such as family commitments, school, or work ("free" sleepers). On the other hand, this cannot be done in people whose habitual bedtimes are "fixed" by these commitments, such as those who have to get up early in the morning to work (Burgess et al. 2005).

In short, certain sleep parameters are being used in human circadian studies, but fall short in comparison to the other methods reviewed here.

36.4.6 QUESTIONNAIRES

Various questionnaires have been developed to study individual aspects of timing of daily activities and sleep. In 1976, the Morningness Eveningness Questionnaire (MEQ) was developed by Horne and Ostberg (1976). This is the most widely used questionnaire and differentiates morning and evening types. This differentiation, as explained earlier, does reflect some part of the circadian rhythm. The MEQ contains 19 items that determine when the subject is most active during the 24-hour day. Questions are, for example: "At what time in the evening do you feel tired and as a result in need of sleep?" (time scale) or "If you went to bed at 11 PM at what level of tiredness would you be?" (four-item answer). Most items are preferential, therefore determining the subject's preferences, not the actual timing. This questionnaire is also frequently used to investigate the correlation between these morningness–eveningness preferences (phenotypes) and genotypes (Katzenberg et al. 1998; Vink et al. 2001). A recent review concluded that, overall, the MEQ appears to be a fair predictor of the endogenous circadian phase or period (Sack et al. 2007).

Since then, several other questionnaires have been developed in an attempt to improve the MEQ and design questionnaires aimed at more specific groups or situations. For instance, the Circadian Type Questionnaire (CTQ) was originally developed to identify which individuals adjust readily to shift work. Later, this questionnaire was revised into the Circadian Type Inventory (CTI). However, very few studies on the correlation between circadian rhythm and this inventory have been published (Baehr et al. 2000). In 1980, the Diurnal Type Scale (DTS) of Torsvall and Åkerstedt (1980) was developed, also for use with shift workers. Nine years later, Smith et al. (1989) proposed an improved morningness scale (the Composite Scale of Morningness) composed of items from the MEQ and DTS. Fairly recently, the Munich Chronotype Questionnaire (MCTQ) was developed by Roenneberg and coworkers (2003) to determine the chronotype in a general population, just like the MEQ. Besides questions on employment and shift work, it focuses on actual timing of, for instance, going to bed, falling asleep, and waking up. Furthermore, it has the advantage of explicitly assessing sleep–wake patterns on working days and free days separately. In addition, the time spent outside during the day is assessed to correct for these factors when using timing of sleep as an indicator of the underlying chronotype (Zavada et al. 2005). Despite newer options, the MEQ is still the most widely used questionnaire.

36.4.7 ACTIGRAPHY

Actigraphy is an easy and noninvasive method of measuring the rest/activity cycle. A small actimetry sensor is worn by the subject, often on the nondominant wrist, to measure gross motor activity. It is based on the simple principle that there is more movement during wake periods and less during sleep.

The sensor (actimeter) has to be worn continuously for a minimum of 5 days to obtain reliable data on the subject's characteristics (Acebo et al. 1999). Recording activity is already a standard marker of circadian rhythm in animal studies. In human studies, however, it has not (yet) reached this status.

In comparison with the gold standard, which is polysomnography (PSG), actigraphy is a reliable and valid method for detecting sleep in a normal healthy adult population (Binnie et al. 1984; Jean-Louis et al. 1996; Van Someren et al. 2007). However, the method becomes less reliable when the subjects' sleep is more fragmented (Kushida et al. 2001). Actigraphy can be used as a good complementary assessment to determine sleep patterns in patients suspected of certain sleep disorders, such as insomnia and restless legs or periodic limb movement disorder (Hauri et al. 1992; Sforza et al. 1999). In addition, it can be applied in subjects with suspected circadian rhythm sleep disorders because actigraphy correlates well with sleep logs, PSG, and markers of the circadian phase in patients with these disorders (Morgenthaler et al. 2007; Nagtegaal et al. 1998). Sleep disturbance resulting from shift work or jet lag can also be detected by actigraphy (Ancoli-Israel et al. 2003). Finally, if PSG is cumbersome, actigraphy can be the best method to use because it is noninvasive, less expensive, can be done at home, allows long-term continuous recording, and can be used in populations in which PSG would be difficult (e.g., young children or patients with dementia or psychiatric disorders).

Although wrist activity appears to be a strong correlate of the entrained endogenous circadian phase, it is susceptible to masking and artefacts. As a consequence, actigraphy alone does not necessarily reflect the characteristics of the circadian clock. Therefore, in the study of human circadian rhythms, actimetry has certainly come to stay, but for now only as a good additional tool.

ACKNOWLEDGMENT

This work was financially supported by the Christelijke Vereniging voor de Verpleging van Lijders aan Epilepsie.

REFERENCES

Acebo, C., A. Sadeh, R. Seifer, O. Tzischinsky, A. R. Wolfson, A. Hafer, and M. A. Carskadon. 1999. Estimating sleep patterns with activity monitoring in children and adolescents: How many nights are necessary for reliable measures? *Sleep* 22(1):95–103.

Ancoli-Israel, S., R. Cole, C. Alessi, M. Chambers, W. Moorcroft, and C. P. Pollak. 2003. The role of actigraphy in the study of sleep and circadian rhythms. *Sleep* 26(3):342–392.

Andrade, M. M., A. A. Benedito-Silva, and L. Menna-Barreto. 1992. Correlations between morningness-eveningness character, sleep habits and temperature rhythm in adolescents. *Braz. J. Med. Biol. Res.* 25(8):835–839.

Baehr, E. K., W. Revelle, and C. I. Eastman. 2000. Individual differences in the phase and amplitude of the human circadian temperature rhythm: With an emphasis on morningness-eveningness. *J. Sleep Res.* 9(2):117–127.

Bazil, C. W., D. Short, D. Crispin, and W. Zheng. 2000. Patients with intractable epilepsy have low melatonin, which increases following seizures. *Neurology* 55(11):1746–1748.

Benloucif, S., H. J. Burgess, E. B. Klerman, A. J. Lewy, B. Middleton, P. J. Murphy, B. L. Parry, and V. L. Revell. 2008. Measuring melatonin in humans. *J. Clin. Sleep Med.* 4(1):66–69.

Berson, D. M., F. A. Dunn, and M. Takao. 2002. Phototransduction by retinal ganglion cells that set the circadian clock. *Science* 295(5557):1070–1073.

Binnie, C. D., J. H. Aarts, M. A. Houtkooper, R. Laxminarayan, S. A. Martin da Silva, H. Meinardi, N. Nagelkerke, and J. Overweg. 1984. Temporal characteristics of seizures and epileptiform discharges. *Electroencephalogr. Clin. Neurophysiol.* 58(6):498–505.

Boivin, D. B., and C. A. Czeisler. 1998. Resetting of circadian melatonin and cortisol rhythms in humans by ordinary room light. *Neuroreport* 9(5):779–782.

Borbely, A. A. 1982. A two process model of sleep regulation. *Hum. Neurobiol.* 1(3):195–204.

Buijs, R. M., and A. Kalsbeek. 2001. Hypothalamic integration of central and peripheral clocks. *Nat. Rev. Neurosci.* 2(7):521–526.

Buijs, R. M., J. Wortel, J. J. Van Heerikhuize, M. G. Feenstra, G. J. Ter Horst, H. J. Romijn, and A. Kalsbeek. 1999. Anatomical and functional demonstration of a multisynaptic suprachiasmatic nucleus adrenal (cortex) pathway. *Eur. J. Neurosci.* 11(5):1535–1544.

Burgess, H. J., and C. I. Eastman. 2005. The dim light melatonin onset following fixed and free sleep schedules. *J. Sleep Res.* 14(3):229–237.

Cajochen, C., V. Knoblauch, K. Krauchi, C. Renz, and A. Wirz-Justice. 2001. Dynamics of frontal EEG activity, sleepiness and body temperature under high and low sleep pressure. *Neuroreport* 12(10):2277–2281.

Cajochen, C., M. Munch, V. Knoblauch, K. Blatter, and A. Wirz-Justice. 2006. Age-related changes in the circadian and homeostatic regulation of human sleep. *Chronobiol. Int.* 23(1–2):461–474.

Campbell, S. S., and R. J. Broughton. 1994. Rapid decline in body temperature before sleep: Fluffing the physiological pillow? *Chronobiol. Int.* 11(2):126–131.

Caufriez, A., R. Moreno-Reyes, R. Leproult, F. Vertongen, C. E. Van Cauter, and G. Copinschi. 2002. Immediate effects of an 8-h advance shift of the rest-activity cycle on 24-h profiles of cortisol. *Am. J. Physiol. Endocrinol. Metab.* 282(5):E1147–E1153.

Cermakian, N., and D. B. Boivin. 2003. A molecular perspective of human circadian rhythm disorders. *Brain Res. Brain Res. Rev.* 42(3):204–220.

Champney, T. H., W. H. Hanneman, M. E. Legare, and K. Appel. 1996. Acute and chronic effects of melatonin as an anticonvulsant in male gerbils. *J. Pineal Res.* 20(2):79–83.

Culebras, A., M. Miller, L. Bertram, and J. Koch. 1987. Differential response of growth hormone, cortisol, and prolactin to seizures and to stress. *Epilepsia* 28(5):564–570.

Czeisler, C. A., J. F. Duffy, T. L. Shanahan, E. N. Brown, J. F. Mitchell, D. W. Rimmer, J. M. Ronda, E. J. Silva, J. S. Allan, J. S. Emens, D. J. Dijk, and R. E. Kronauer. 1999. Stability, precision, and near-24-hour period of the human circadian pacemaker. *Science* 284(5423):2177–2181.

Czeisler, C. A., M. Dumont, J. F. Duffy, J. D. Steinberg, G. S. Richardson, E. N. Brown, R. Sanchez, C. D. Rios, and J. M. Ronda. 1992. Association of sleep-wake habits in older people with changes in output of circadian pacemaker. *Lancet* 340(8825):933–936.

Daan, S., D. G. Beersma, and A. A. Borbely. 1984. Timing of human sleep: Recovery process gated by a circadian pacemaker. *Am. J. Physiol.* 246(2 Pt 2):R161–R183.

Dauncey, M. J., and S. A. Bingham. 1983. Dependence of 24 h energy expenditure in man on the composition of the nutrient intake. *Br. J. Nutr.* 50(1):1–13.

Deacon, S., and J. Arendt. Posture influences melatonin concentrations in plasma and saliva in humans. *Neurosci. Lett.* 167:191–194.

Dijk, D. J., C. Cajochen, and A. A. Borbely. 1991. Effect of a single 3-hour exposure to bright light on core body temperature and sleep in humans. *Neurosci. Lett.* 121(1–2):59–62.

Dijk, D. J., and C. A. Czeisler. 1995. Contribution of the circadian pacemaker and the sleep homeostat to sleep propensity, sleep structure, electroencephalographic slow waves, and sleep spindle activity in humans. *J. Neurosci.* 15(5 Pt 1):3526–3538.

Duffy, J. F., and D. J. Dijk. 2002. Getting through to circadian oscillators: Why use constant routines? *J. Biol. Rhythms* 17(1):4–13.

Duffy, J. F., D. J. Dijk, E. F. Hall, and C. A. Czeisler. 1999. Relationship of endogenous circadian melatonin and temperature rhythms to self-reported preference for morning or evening activity in young and older people. *J. Investig. Med.* 47(3):141–150.

Duffy, J. F., D. W. Rimmer, and C. A. Czeisler. 2001. Association of intrinsic circadian period with morningness-eveningness, usual wake time, and circadian phase. *Behav. Neurosci.* 115(4):895–899.

Duffy, J. F., and K. P. Wright Jr. 2005. Entrainment of the human circadian system by light. *J. Biol. Rhythms* 20(4):326–338.

Gander, P. H., L. J. Connell, and R. C. Graeber. 1986. Masking of the circadian rhythms of heart rate and core temperature by the rest-activity cycle in man. *J. Biol. Rhythms* 1(2):119–135.

Gowers, W. 1885, Course of epilepsy. In *Epilepsy and other chronic convulsive diseases: Their causes, symptoms and treatment*, ed. W. Gowers, 157–164. New York: William Wood.

Griffiths, G. M., and J. T. Fox. 1938. Rhythm in epilepsy. *Lancet* 2:409–416.

Gronfier, C., F. Chapotot, L. Weibel, C. Jouny, F. Piquard, and G. Brandenberger. 1998. Pulsatile cortisol secretion and EEG delta waves are controlled by two independent but synchronized generators. *Am. J. Physiol.* 275(1 Pt 1):E94–E100.

Hastings, M., J. S. O'Neill, and E. S. Maywood. 2007. Circadian clocks: Regulators of endocrine and metabolic rhythms. *J. Endocrinol.* 195(2):187–198.

Hauri, P. J., and J. Wisbey. 1992. Wrist actigraphy in insomnia. *Sleep* 15(4):293–301.

Hebert, M., S. K. Martin, C. Lee, and C. I. Eastman. 2002. The effects of prior light history on the suppression of melatonin by light in humans. *J. Pineal Res.* 33(4):198–203.

Hiddinga, A. E., D. G. Beersma, and R. H. Van den Hoofdakker. 1997. Endogenous and exogenous components in the circadian variation of core body temperature in humans. *J. Sleep Res.* 6(3):156–163.

Hofstra, W. A., and A. W. de Weerd. 2009. The circadian rhythm and its interaction with human epilepsy: A review of literature. *Sleep Med. Rev.* 13(6):413–420.

Horne, J. A., and O. Ostberg. 1976. A self-assessment questionnaire to determine morningness-eveningness in human circadian rhythms. *Int. J. Chronobiol.* 4(2):97–110.

Jean-Louis, G., H. von Gizycki, F. Zizi, J. Fookson, A. Spielman, J. Nunes, R. Fullilove, and H. Taub. 1996. Determination of sleep and wakefulness with the actigraph data analysis software (ADAS). *Sleep* 19(9):739–743.

Katzenberg, D., T. Young, L. Finn, L. Lin, D. P. King, J. S. Takahashi, and E. Mignot. 1998. A CLOCK polymorphism associated with human diurnal preference. *Sleep* 21(6):569–576.

Kerkhof, G. A., and H. P. Van Dongen. 1996. Morning-type and evening-type individuals differ in the phase position of their endogenous circadian oscillator. *Neurosci. Lett.* 218(3):153–156.

Klerman, E. B., H. B. Gershengorn, J. F. Duffy, and R. E. Kronauer. 2002. Comparisons of the variability of three markers of the human circadian pacemaker. *J. Biol. Rhythms* 17(2):181–193.

Ko, C. H., and J. S. Takahashi. 2006. Molecular components of the mammalian circadian clock. *Hum. Mol. Genet.* 15(Spec No 2):R271–R277.

Krauchi, K. 2002. How is the circadian rhythm of core body temperature regulated? *Clin. Auton. Res.* 12(3):147–149.

Krauchi, K., C. Cajochen, and A. Wirz-Justice. 1997. A relationship between heat loss and sleepiness: Effects of postural change and melatonin administration. *J. Appl. Physiol.* 83(1):134–139.

Krauchi, K., C. Cajochen, and A. Wirz-Justice. 2005. Thermophysiologic aspects of the three-process-model of sleepiness regulation. *Clin. Sports Med.* 24(2):287–300, ix.

Krauchi, K., V. Knoblauch, A. Wirz-Justice, and C. Cajochen. 2006. Challenging the sleep homeostat does not influence the thermoregulatory system in men: Evidence from a nap vs. sleep-deprivation study. *Am. J. Physiol. Regul. Integr. Comp. Physiol.* 290(4):R1052–R1061.

Krauchi, K., and A. Wirz-Justice. 2001. Circadian clues to sleep onset mechanisms. *Neuropsychopharmacology* 25(5 Suppl):S92–S96.

Kushida, C. A., A. Chang, C. Gadkary, C. Guilleminault, O. Carrillo, and W. C. Dement. 2001. Comparison of actigraphic, polysomnographic, and subjective assessment of sleep parameters in sleep-disordered patients. *Sleep Med.* 2(5):389–396.

Laakso, M. L., L. Leinonen, T. Hatonen, A. Alila, and H. Heiskala. 1993. Melatonin, cortisol and body temperature rhythms in Lennox-Gastaut patients with or without circadian rhythm sleep disorders. *J. Neurol.* 240(7):410–416.

Lamont, E. W., F. O. James, D. B. Boivin, and N. Cermakian. 2007. From circadian clock gene expression to pathologies. *Sleep Med.* 8(6):547–556.

Langdon-Down, M., and W. R. Brain. 1929. Time of day in relation to convulsions in epilepsy. *Lancet* 1:1029–1032.

Lapin, I. P., S. M. Mirzaev, I. V. Ryzov, and G. F. Oxenkrug. 1998. Anticonvulsant activity of melatonin against seizures induced by quinolinate, kainate, glutamate, NMDA, and pentylenetetrazole in mice. *J. Pineal Res.* 24(4):215–218.

Leproult, R., E. F. Colecchia, M. L'Hermite-Baleriaux, and C. E. Van Cauter. 2001. Transition from dim to bright light in the morning induces an immediate elevation of cortisol levels. *J. Clin. Endocrinol. Metab.* 86(1): 151–157.

Lewy, A. J., N. L. Cutler, and R. L. Sack. 1999. The endogenous melatonin profile as a marker for circadian phase position. *J. Biol. Rhythms* 14(3):227–236.

Lewy, A. J., T. A. Wehr, F. K. Goodwin, D. A. Newsome, and S. P. Markey. 1980. Light suppresses melatonin secretion in humans. *Science* 210(4475):1267–1269.

Martin, S. K., and C. I. Eastman. 2002. Sleep logs of young adults with self-selected sleep times predict the dim light melatonin onset. *Chronobiol. Int.* 19(4):695–707.

Mehta, S. R., S. K. Dham, A. I. Lazar, A. S. Narayanswamy, and G. S. Prasad. 1994. Prolactin and cortisol levels in seizure disorders. *J. Assoc. Physicians India* 42(9):709–712.

Mills, J. N., D. S. Minors, and J. M. Waterhouse. 1978. Adaptation to abrupt time shifts of the oscillator(s) controlling human circadian rhythms. *J. Physiol.* 285:455–470.

Milton, J. G., J. Gotman, G. M. Remillard, and F. Andermann. 1987. Timing of seizure recurrence in adult epileptic patients: A statistical analysis. *Epilepsia* 28(5):471–478.

Molaie, M., A. Culebras, and M. Miller. 1987. Nocturnal plasma prolactin and cortisol levels in epileptics with complex partial seizures and primary generalized seizures. *Arch. Neurol.* 44(7):699–702.

Monteleone, P., M. Maj, M. Fusco, C. Orazzo, and D. Kemali. 1990. Physical exercise at night blunts the nocturnal increase of plasma melatonin levels in healthy humans. *Life Sci.* 47(22):1989–1995.

Moran, D., Y. Shapiro, Y. Epstein, R. Burstein, L. Stroschein, and K. B. Pandolf. 1995. Validation and adjustment of the mathematical prediction model for human rectal temperature responses to outdoor environmental conditions. *Ergonomics* 38(5):1011–1018.

Morgenthaler, T., C. Alessi, L. Friedman, J. Owens, V. Kapur, B. Boehlecke, T. Brown, A. Chesson, Jr., J. Coleman, T. Lee-Chiong, J. Pancer, and T. J. Swick. 2007. Practice parameters for the use of actigraphy in the assessment of sleep and sleep disorders: An update for 2007. *Sleep* 30(4):519–529.

Munch, M., V. Knoblauch, K. Blatter, C. Schroder, C. Schnitzler, K. Krauchi, A. Wirz-Justice, and C. Cajochen. 2005. Age-related attenuation of the evening circadian arousal signal in humans. *Neurobiol. Aging* 26(9):1307–1319.

Murphy, P. J., B. L. Myers, and P. Badia. 1996. Nonsteroidal anti-inflammatory drugs alter body temperature and suppress melatonin in humans. *Physiol. Behav.* 59(1):133–139.

Nagtegaal, J. E., G. A. Kerkhof, M. G. Smits, A. C. Swart, and Y. G. Van Der Meer. 1998. Delayed sleep phase syndrome: A placebo-controlled cross-over study on the effects of melatonin administered five hours before the individual dim light melatonin onset. *J. Sleep Res.* 7(2):135–143.

Plautz, J. D., M. Kaneko, J. C. Hall, and S. A. Kay. 1997. Independent photoreceptive circadian clocks throughout Drosophila. *Science* 278(5343):1632–1635.

Quigg, M., H. Clayburn, M. Straume, M. Menaker, and E. H. Bertram. 1998a. Acute seizures may induce phase shifts in the circadian rhythm of body temperature in kindled rats. *Epilepsia* 39:136.

Quigg, M., H. Clayburn, M. Straume, M. Menaker, and E. H. Bertram, III. 2000. Effects of circadian regulation and rest-activity state on spontaneous seizures in a rat model of limbic epilepsy. *Epilepsia* 41(5):502–509.

Quigg, M., M. Straume, M. Menaker, and E. H. Bertram, III. 1998b. Temporal distribution of partial seizures: Comparison of an animal model with human partial epilepsy. *Ann. Neurol.* 43(6):748–755.

Ralph, C. L., D. Mull, H. J. Lynch, and L. Hedlund. 1971. A melatonin rhythm persists in rat pineals in darkness. *Endocrinology* 89(6):1361–1366.

Rao, M. L., H. Stefan, and J. Bauer. 1989. Epileptic but not psychogenic seizures are accompanied by simultaneous elevation of serum pituitary hormones and cortisol levels. *Neuroendocrinology* 49(1):33–39.

Roenneberg, T., A. Wirz-Justice, and M. Merrow. 2003. Life between clocks: Daily temporal patterns of human chronotypes. *J. Biol. Rhythms* 18(1):80–90.

Sack, R. L., D. Auckley, R. R. Auger, M. A. Carskadon, K. P. Wright, Jr., M. V. Vitiello, and I. V. Zhdanova. 2007. Circadian rhythm sleep disorders: Part I. Basic principles, shift work and jet lag disorders. An American Academy of Sleep Medicine review. *Sleep* 30(11):1460–1483.

Sandyk, R., N. Tsagas, and P. A. Anninos. 1992. Melatonin as a proconvulsive hormone in humans. *Int. J. Neurosci.* 63(1–2):125–135.

Schapel, G. J., R. G. Beran, D. L. Kennaway, J. McLoughney, and C. D. Matthews. 1995. Melatonin response in active epilepsy. *Epilepsia* 36(1):75–78.

Scheer, F. A., and R. M. Buijs. 1999. Light affects morning salivary cortisol in humans. *J. Clin. Endocrinol. Metab.* 84(9):3395–3398.

Sforza, E., M. Zamagni, C. Petiav, and J. Krieger. 1999. Actigraphy and leg movements during sleep: A validation study. *J. Clin. Neurophysiol.* 16(2):154–160.

Shanahan, T. L., and C. A. Czeisler. 1991. Light exposure induces equivalent phase shifts of the endogenous circadian rhythms of circulating plasma melatonin and core body temperature in men. *J. Clin. Endocrinol. Metab.* 73(2):227–235.

Shilo, L., H. Sabbah, R. Hadari, S. Kovatz, U. Weinberg, S. Dolev, Y. Dagan, and L. Shenkman. 2002. The effects of coffee consumption on sleep and melatonin secretion. *Sleep Med.* 3(3):271–273.

Slag, M. F., M. Ahmad, M. C. Gannon, and F. Q. Nuttall. 1981. Meal stimulation of cortisol secretion: A protein induced effect. *Metabolism* 30(11):1104–1108.

Smith, C. S., C. Reilly, and K. Midkiff. 1989. Evaluation of three circadian rhythm questionnaires with suggestions for an improved measure of morningness. *J. Appl. Psychol.* 74(5):728–738.

Smith, K. A., M. W. Schoen, and C. A. Czeisler. 2004. Adaptation of human pineal melatonin suppression by recent photic history. *J. Clin. Endocrinol. Metab.* 89(7):3610–3614.

Stoschitzky, K., A. Sakotnik, P. Lercher, R. Zweiker, R. Maier, P. Liebmann, and W. Lindner. 1999. Influence of beta-blockers on melatonin release. *Eur. J. Clin. Pharmacol.* 55(2)111–115.

Tauboll, E., A. Lundervold, and L. Gjerstad. 1991. Temporal distribution of seizures in epilepsy. *Epilepsy Res.* 8(2):153–165.

Toh, K. L., C. R. Jones, Y. He, E. J. Eide, W. A. Hinz, D. M. Virshup, L. J. Ptacek, and Y. H. Fu. 2001. An hPer2 phosphorylation site mutation in familial advanced sleep phase syndrome. *Science* 291(5506):1040–1043.

Torsvall, L., and T. Akerstedt. 1980. A diurnal type scale. Construction, consistency and validation in shift work. *Scand. J. Work Environ. Health* 6(4):283–290.

Umeda, T., R. Hiramatsu, T. Iwaoka, T. Shimada, F. Miura, and T. Sato. 1981. Use of saliva for monitoring unbound free cortisol levels in serum. *Clin. Chim. Acta.* 110(2–3):245–253.

Van Gelder, R. N., E. D. Herzog, W. J. Schwartz, and P. H. Taghert. 2003. Circadian rhythms: In the loop at last. *Science* 300(5625):1534–1535.

Van Someren, E. J., and E. Nagtegaal. 2007. Improving melatonin circadian phase estimates. *Sleep Med.* 8(6):590–601.

Van, C. E., R. Leproult, and D. J. Kupfer. 1996. Effects of gender and age on the levels and circadian rhythmicity of plasma cortisol. *J. Clin. Endocrinol. Metab.* 81(7):2468–2473.

van Coevorden, A., J. Mockel, E. Laurent, M. Kerkhofs, M. L'Hermite-Baleriaux, C. Decoster, P. Neve, and E. van Cauter. 1991. Neuroendocrine rhythms and sleep in aging men. *Am. J. Physiol.* 260(4 Pt 1):E651–E661.

Veldhuis, J. D., A. Iranmanesh, M. L. Johnson, and G. Lizarralde. 1990. Amplitude, but not frequency, modulation of adrenocorticotropin secretory bursts gives rise to the nyctohemeral rhythm of the corticotropic axis in man. *J. Clin. Endocrinol. Metab.* 71(2):452–463.

Vink, J. M., A. S. Groot, G. A. Kerkhof, and D. I. Boomsma. 2001. Genetic analysis of morningness and eveningness. *Chronobiol. Int.* 18(5):809–822.

Voultsios, A., D. J. Kennaway, and D. Dawson. 1997. Salivary melatonin as a circadian phase marker: Validation and comparison to plasma melatonin. *J. Biol. Rhythms* 12(5):457–466.

Waterhouse, J., B. Drust, D. Weinert, B. Edwards, W. Gregson, G. Atkinson, S. Kao, S. Aizawa, and T. Reilly. 2005. The circadian rhythm of core temperature: Origin and some implications for exercise performance. *Chronobiol. Int.* 22(2):207–225.

Waterhouse, J., D. Weinert, D. Minors, S. Folkard, D. Owens, G. Atkinson, I. Macdonald, N. Sytnik, P. Tucker, and T. Reilly. 2000. A comparison of some different methods for purifying core temperature data from humans. *Chronobiol. Int.* 17(4)539–566.

Weibel, L., and G. Brandenberger. 2002. The start of the quiescent period of cortisol remains phase locked to the melatonin onset despite circadian phase alterations in humans working the night schedule. *Neurosci. Lett.* 318(2)89–92.

Wirz-Justice, A. 2007. How to measure circadian rhythms in humans. *Medicographia* 29:84–90.

Xu, Y., Q. S. Padiath, R. E. Shapiro, C. R. Jones, S. C. Wu, N. Saigoh, K. Saigoh, L. J. Ptacek, and Y. H. Fu. 2005. Functional consequences of a CKI delta mutation causing familial advanced sleep phase syndrome. *Nature* 434(7033):640–644.

Zavada, A., M. C. Gordijn, D. G. Beersma, S. Daan, and T. Roenneberg. 2005. Comparison of the Munich chronotype questionnaire with the Horne-Ostberg's morningness-eveningness score. *Chronobiol. Int.* 22(2):267–278.

Zeitzer, J. M., J. E. Daniels, J. F. Duffy, E. B. Klerman, T. L. Shanahan, D. J. Dijk, and C. A. Czeisler. 1999. Do plasma melatonin concentrations decline with age? *Am. J. Med.* 107(5):432–436.

Zeitzer, J. M., D. J. Dijk, R. Kronauer, E. Brown, and C. Czeisler. 2000. Sensitivity of the human circadian pacemaker to nocturnal light: Melatonin phase resetting and suppression. *J. Physiol.* 526(Pt 3):695–702.

Zeitzer, J. M., J. F. Duffy, S. W. Lockley, D. J. Dijk, and C. A. Czeisler. 2007. Plasma melatonin rhythms in young and older humans during sleep, sleep deprivation, and wake. *Sleep* 30(11):1437–1443.

Zylka, M. J., L. P. Shearman, D. R. Weaver, and S. M. Reppert. 1998. Three period homologs in mammals: Differential light responses in the suprachiasmatic circadian clock and oscillating transcripts outside of brain. *Neuron* 20(6):1103–1110.

37 Seizure Prediction and the Circadian Rhythm

Tyler S. Durazzo and Hitten P. Zaveri

CONTENTS

37.1 INTRODUCTION

Seizure prediction through analysis of the electroencephalogram (EEG) remains a challenging and thus far unsuccessful task (Mormann et al. 2007). Algorithms employing measures such as accumulated energy and correlation density showed promise to predict seizures on select EEG segments, but have failed to show reproducibility when extended to unselected, long-term EEG (Lehnertz and Elger 1998; Litt et al. 2001). The general seizure prediction approach is based upon continuously calculating a parameter of interest using a moving window, with a seizure being predicted whenever the calculated metric moves outside a predefined range of values. This chapter discusses the circadian rhythm and its effects on seizure occurrence, and raises the question as to whether seizure prediction could be enhanced by taking the known information about the influence of the circadian rhythm into account. For instance, algorithms may benefit by simply varying—in circadian-like fashion—the threshold a measure must meet to predict a seizure. Recent research has provided evidence that seizures are more likely to occur at particular times of the day, depending on the region of the brain in which the epileptogenic focus is found (Durazzo et al. 2008; Hofstra et al. 2009; Pavlova, Shea, and Bromfield 2004; Quigg and Straume 2000; Quigg et al. 1998; Hofstra et al. 2009). Circadian-like distributions of seizures have been observed at both the individual and group levels, generating interest in the possibility of incorporating information of the circadian rhythm into seizure prediction algorithms.

37.2 CIRCADIAN RHYTHM

Many biological processes exhibit circadian fluctuations, including normal electrical activity of the brain and hormonal secretion driven by the hypothalamic–pituitary axis (Moore 1997). In turn, various pathological conditions have been described as showing circadian variation—both in severity of disease as well as the temporal likelihood of specific manifestations of the disease (Ingelsson et al. 2006; Manfredini et al. 2004; Manfredini et al. 2005; Michaud et al. 2005; Solomon 1992; Stephenson 2003).

The suprachiasmatic nuclus (SCN) of the hypothalamus serves as the dominant circadian pacemaker in mammals (Moore 1997). Remarkably, neurons of the SCN fire in a circadian rhythm even when isolated and maintained in tissue culture (Welsh et al. 1995). This firing pattern consists of a high discharge frequency during the day and a lower discharge frequency at night (Shibata and Moore 1988). Interestingly, this particular day-and-night variation in discharge frequency is similar across all mammalian species studied, regardless of whether the species is nocturnal or diurnal (Inouye 1996). All SCN neurons are gamma-aminobutyric acid (GABA) producing, and thus their output is inhibitory (Gao and Moore 1996).

The SCN sends the largest portion of its primary output to the hypothalamus, with a smaller portion sent to the basal forebrain and midline thalamus (Figure 37.1). The secondary output relayed from these structures is responsible for the circadian fluctuations observed in numerous biological processes. The basal forebrain and thalamus relay SCN rhythms to the neocortex, hippocampus, basal ganglia, and other forebrain structures. The hypothalamus relays SCN rhythms to the anterior pituitary, thereby providing circadian information to the endocrine system, including regulation of hormones that in turn can alter neuronal activity. The hypothalamus also relays SCN rhythms to the reticular formation and pineal gland, providing circadian variation to a number of processes, including the sleep–wake cycle and melatonin secretion. Circadian fluctuations in brain activity are detectable even from the scalp, as studies have revealed circadian variations in EEG parameters (Dijk 1999; Duckrow and Tcheng 2007; Cajochen et al. 2002).

Although it is apparent from the above discussion that circadian rhythms are manifestations of endogenous processes, it is important to note that the phase and period of the circadian pacemaker

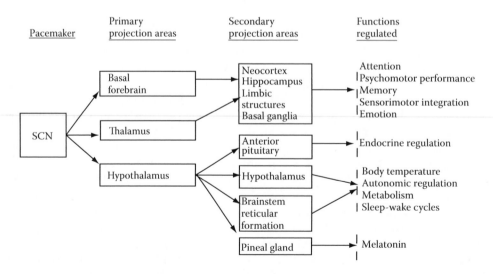

FIGURE 37.1 Overview of pathways involved in circadian rhythms. The SCN serves as the dominant circadian pacemaker in mammals, firing at high frequency during the day and lower frequency at night. All SCN neurons are GABA-producing, and thus their primary output is inhibitory. The functions regulated as a result of these pathways are shown in the last column. (Adapted from Moore, R. Y., *Annual Review of Medicine*, 48, 253–266, 1997.)

itself are set by external cues, referred to as *zeitgebers*, the most notable of which is light. In the absence of zeitgebers, such as during experimental isolation from light fluctuations, human circadian rhythms exhibit a periodicity of approximately 24.2 h (Czeisler et al. 1999). Zeitgebers encountered in normal daily living are responsible for entraining the SCN to a period of 24 h, thereby synchronizing endogenous circadian rhythms with the natural day-and-night cycle. Entrainment occurs principally through the retinohypothalamic tract (RHT)—a direct connection between photoreceptors in the retina and the SCN (Moore 1973). Complete transection of all retinal projections except the RHT results in blind animals lacking visual reflexes, but does not alter entrainment of circadian rhythms to any degree (Moore 1978). Nonphotic pathways for entrainment also exist, but exert a smaller degree of influence (Moore 1997).

37.3   CIRCADIAN INFLUENCE ON SEIZURE GENERATION

The tendency of seizures in some epileptic patients to occur at preferential periods of the day was first described over 100 years ago (Gowers 1885). Studies at that time found patients to have one of three different patterns of seizure occurrence: diurnal (daytime), nocturnal (nighttime), or diffuse (both day and night). Relatively recent work has extended and refined the notion that seizures do not occur randomly throughout the 24-h day; patients exhibit distinct times in which seizure generation is more probable (Durazzo et al. 2008; Hofstra and de Weerd 2009; Hofstra et al. 2009; Pavlova, Shea, and Bromfield 2004; Quigg and Straume 2000; Quigg et al. 1998; Hofstra et al. 2009). The particular period of the 24-h day in which a patient is most likely to generate a seizure appears to depend on the location of the epileptogenic focus and possibly the particular pathophysiology.

Quigg and Straume published a case report in which they describe a patient with two independent epileptogenic foci, each of which produced seizures with its own periodicity relative to the 24-h day (Quigg and Straume 2000) (Figure 37.2). Seizures in this patient originating from the right parietal lobe occurred predominantly at night (64% of these occurrences), with a peak incidence at 02:50. In contrast, seizures in this patient originating from the right temporal lobe occurred predominantly during the day (83% of these occurrences), with a peak incidence at 12:10. The temporal distributions of seizures in this patient were obtained using retrospective analysis of the patient's detailed seizure diary, which included 1009 seizures over a 5-year period. Each seizure in the diary was denoted as either consisting of pain in the right arm and hand (later confirmed to be of right

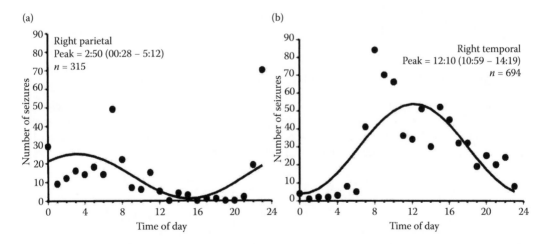

FIGURE 37.2 Case report of patient with two independent seizure foci. Analysis of 1009 seizures revealed that seizures generated from the patient's parietal lobe (a) were more likely to occur at night, whereas those from the temporal lobe (b) were more likely to occur during the day. (Adapted from Quigg, M., et al., *Annals of Neurology*, 43(6), 748–755, 1998.)

parietal origin), or the constellation of confusion, orofacial automatisms, and/or dystonic posturing of the left hands (later determined to be of right temporal origin). The epileptic foci in this patient were determined using imaging studies and intracranial EEG monitoring.

In addition to periodicity of seizure occurrence in individual patients, periodicity of seizure occurrence at the group level has also been described for patients grouped by pathophysiology. Quigg et al. (1998) pioneered the recent work in this area, with a retrospective study of 64 consecutive patients with mesial temporal lobe epilepsy (MTLE) and 26 with extratemporal lobe epilepsy (XTLE) admitted to the hospital for inpatient EEG monitoring. MTLE seizures ($n = 774$) were found to occur nonrandomly relative to the 24-h cycle, with 60% occurring during daylight hours. The overall distribution of these seizures closely fit a cosine function, with peak incidence at 15:00. Seizures grouped into the XTLE fraction ($n = 465$) were found to be randomly distributed relative to the 24-f cycle, with no statistical difference between daytime and nighttime occurrence.

In addition to human subjects, the work by Quigg et al. (1998) included evaluation of the time distributions of seizures generated from a rat model of MTLE, the postlimbic status epilepticus (PLS) model. EEG monitoring of 20 PLS rats yielded 547 recorded seizures. These seizures were found to occur in a pattern very similar to those found in human MTLE patients: 63% occurred during daytime hours (with lights on), with the overall distribution closely modeled by a cosine function with peak incidence at 16:48. The similarity between the seizure distributions of human MTLE with the rat model is remarkable because humans and rats have sleep/wake states 180 degrees out of phase relative to each other—rats, a nocturnal species, sleep during daylight hours and are active at night. These findings suggest that the sleep/wake state does not play a significant role in modulating seizure threshold in the mesial temporal lobe. However, considering the main circadian pacemaker, the SCN, fires in-phase between rats and humans (i.e., peak SCN activity occurs during daylight for both species), these findings suggest that other circadian rhythms driven by the SCN may play a major role in generating seizures in MTLE.

Other studies have attempted to look at the distribution of partial seizures arising from other regions of the brain in more detail (Table 37.1) (Durazzo et al. 2008; Hofstra and de Weerd 2009; Hofstra et al. 2009; Pavlova, Shea, and Bromfield 2004; Quigg and Straume 2000; Quigg et al. 1998; Hofstra et al. 2009). However, correctly determining periodicity in seizures generated from localized areas of the brain is not a simple task, with particular methodology required to generate data from which accurate conclusions may be drawn. The following section provides a point-by-point discussion of the components necessary to consider when studying and determining seizure distributions. Following that discussion, the findings from a rigorous study performed by the Epilepsy Program at Yale University is presented.

37.4 CONSIDERATIONS IN THE STUDY OF THE CIRCADIAN INFLUENCE ON SEIZURE GENERATION

A number of considerations must be taken into account when studying circadian seizure distributions, as differences in methodology can strongly influence the results and conclusions. The following section provides a detailed overview of the major components necessary to consider when attempting to determine circadian influence on seizure generation.

37.4.1 ENDOGENOUS VERSUS EXOGENOUS INFLUENCE ON RHYTHMS

A process truly influenced by circadian rhythms should be strictly dependent on the rhythms set in motion by the endogenous cyclic firing of the SCN. Therefore, to truly demonstrate circadian control, studies would need to be conducted with subjects in constant environmental conditions that lack exogenous factors which also cycle with 24-h periodicity. Examples of exogenous rhythms that could influence seizure threshold independently of circadian rhythms include: light/dark exposure, noise levels, meal times, social interactions (e.g., visiting hours), and administration of daily

TABLE 37.1

Summary of Studies Conducted over the Past 15 Years on Seizure Occurrence vs. Time of Day

Study	Methods	Epileptogenic Region	Number of Patients	Determined Distribution
Quigg et al. (1998)	(S, I) (a)	MTLE	64 (774)	Cosine with peak at 15:00
		XTLE	26 (465)	Random
Pavola et al. (2004)	(S, I) (a)	TLE	15 (41)	Nonrandom with peak at 15:00–19:00
		XTLE	11 (49)	Nonrandom with peak at 19:00–23:00
Durazzo et al. (2008)	(I) (a)	MTLE	45 (377)	Bimodal with primary peak at 16:00–19:00
		NTLE	34 (160)	Nonrandom with peak at 13:00–16:00
		FLE	23 (132)	Unimodal with peak at 4:00–7:00
		PLE	16 (77)	Gaussian with peak at 4:00–7:00
		OLE	13 (83)	Gaussian with peak at 16:00–19:00
Hofstra et al. (2009)	(S) (a)	TLE	65 (241)	Nonrandom with peak at 11:00–17:00
		XTLE	35 (171)	Random
Hofstra et al. (2009)	(I) (a, c)	MTLE	6 (85)	Nonrandom with peak at 11:00–17:00
		NTLE	8 (72)	Nonrandom with peak at 11:00–17:00
		FLE	14 (190)	Nonrandom with peak at 23:00–5:00
		PLE	4 (99)	Nonrandom with peak at 17:00–23:00
		OLE	1 (4)	Insufficient data

Note: Fourth column indicates the number of patients followed in parenthesis by the total number of seizures analyzed; S = scalp EEG; I = intracranial EEG; a = adults; c = children.

anticonvulsant medication. No study has controlled for the periodicity of these exogenous factors to determine their influence on seizure generation. Quigg's work using the rat PLS model of MTLE suggests that seizure distribution in human MTLE may be principally the result of endogenous circadian rhythms and not exogenous factors (Quigg et al. 1998). Similar work using animal models of other partial epilepsies would be beneficial to the field. A recent study separating sleep from the circadian rhythm using a forced desynchrony protocol in humans has suggested that interictal epileptiform discharges (IEDs) are independently influenced by both the circadian rhythm and sleep (Pavlova et al. 2009). Further studies using forced desynchrony in humans or animals models are required to determine if seizure generation in the various partial epilepsies are truly influenced by circadian rhythms.

37.4.2 Sleep/Wake State

The patient's state (sleep vs. wake), itself influenced by circadian rhythms, has been found to modulate seizure threshold (Bazil and Walczak 1997; Herman, Walczak, and Bazil 2001; Oldani et al. 1998). The degree of modulation varies according to epileptic syndrome and location of the seizure onset area. Sleep is well known to lower seizure threshold in some epileptic syndromes like autosomal dominant nocturnal frontal lobe epilepsy (ADNFLE) and Rolandic epilepsy of childhood (DallaBernardina 1991; Oldani et al. 1998). Its role in other syndromes, including partial epilepsy, is less understood. Sleep appears to increase the likelihood that temporal lobe seizures will secondarily generalize, yet has no effect on the likelihood that frontal lobe seizures will do the same (Bazil and Walczak 1997). Frontal lobe seizures themselves appear more likely to occur from sleep than wake, according to several studies (Bazil and Walczak 1997; Crespel, Baldy-Moulinier, and Coubes 1998; Herman, Walczak, and Bazil 2001). It should be noted, however, that none of these studies separated sleep from the nighttime hours, and thus sleep's individual influence on these phenomena apart from the other endogenous and exogenous rhythms of the night is unknown. The forced desynchrony study referenced in the previous section found that IEDs in patients with generalized epilepsy were 14 times more likely to occur during sleep than wake, independent of circadian rhythm; this strongly emphasizes the role of sleep in the generation of IEDs (Pavlova et al. 2009).

Within sleep, the different sleep stages appear to have opposing effects on IEDs and seizure threshold (Ferrillo et al. 2000; Herman, Walczak, and Bazil 2001; Malow et al. 1998; Minecan et al. 2002). Herman et al. (2001), who analyzed 264 seizures generated during sleep, found seizures more likely to begin during stage 1 (23%) and stage 2 (68%) than during slow-wave sleep (9%), with no seizures (0%) observed during REM sleep. In contrast, slow-wave sleep (stages 3 and 4) appears to activate IEDs, whereas REM sleep has been shown to decrease the occurrence and spread of IEDs (Malow et al. 1998).

37.4.3 GROUPING OF PATIENTS

When grouping patients, three factors must be accounted for: localization, pathology, and maturity of brain (adults/pediatrics). The various regions of the brain are differentially affected by the numerous rhythms set in motion by SCN discharges (Moore 1997). It is therefore necessary to study brain regions and structures separately if accurate circadian modulations of seizure threshold are to be determined. Only those patients whose seizures can confidently be localized as originating from one clearly defined focus should be included, and analysis should consist of grouping patients into groups that are as specific as possible. Furthermore, the pathophysiology of the patient's seizures should be taken into account. For instance, because the mechanism underlying seizure generation in a patient with posttraumatic epilepsy possibly differs from one with a mass lesion, it's not unreasonable to believe circadian rhythms may differentially influence the two. Along this same line of reasoning, because epileptogenesis in the immature brain differs from that of the mature brain (Rakhade and Jensen 2009; Rodriguez 2007), studies should separate adults and children.

37.4.4 SEIZURE DOCUMENTATION

Pivotal to studies on circadian variation in seizure occurrence is the ability to accurately and consistently detect seizures. Intracranial EEG is the gold standard in detecting seizure activity. Studies should not depend on clinical correlates to determine seizure occurrence because these depend on (1) the level of attention of clinical staff to the patient, and (2) the patient's baseline level of activity at the time of seizure. Both these factors make clinical correlates less likely to be observed at night—when there is typically less hospital staff and the patient is more likely to be sleeping. In essence, it should be recognized that studies using scalp-EEG and clinical correlates have a circadian-like variability in their ability to document seizures, with detection more likely to occur during daytime hours than nighttime hours.

37.4.5 SEIZURE CLUSTERING

The propensity of seizures to cluster in time is well known (Haut 2006). Following an initial primary seizure, a period of time may exist in which neural pathways are more vulnerable to generating and propagating additional seizures. These secondary seizures may be the result of increased excitation or impaired termination of the initial primary seizure. The period of time in which a primary seizure may increase the likelihood of additional seizures is unknown. One study of patients with bilateral foci found that seizures occurring within 8 h of each other were more likely to originate from the same focus than seizures further separated in time (Haut et al. 1997). Therefore, in studies of circadian modulation of seizure threshold, the possibility exists that some seizures mapped to one time were instead chiefly the result of a lower seizure threshold (and primary seizure) that occurred hours earlier. In addition, in studies with few patients, a single seizure cluster from one patient may be sufficient to erroneously produce a peak in the distribution for the group. It is important therefore to ensure that seizure clusters do not unduly impact the analysis.

37.4.6 IMPACT OF THE INDIVIDUAL PATIENT ON GROUP DISTRIBUTION

If circadian modulation of seizures generated from particular brain regions and pathophysiologies are to be determined, then analysis must be careful to not allow subjects with disproportionately high seizure counts to unduly influence the results. Patients undergoing inpatient intracranial EEG monitoring can have from 0 to more than 50 recorded seizures. To prevent excessive influence from those at the high end of the spectrum, the number of seizures each patient can contribute to the analysis should be limited. Without implementing a limit, a single patient could have undue influence on observations; for example, a rhythm unique to a single patient could be misinterpreted as the rhythm generated by a particular patient group (i.e., epileptogenic region or pathophyisology). Consequently, a sufficient number of seizures from a sufficient number of patients must be studied to allow the determination of seizure distribution as a function of time of day without undue impact from a single patient or a small subset of patients. The inclusion of a larger number of patients will also allow the determination of the seizure distribution with better temporal resolution (smaller bin size).

37.5 YALE UNIVERSITY STUDY OF THE CIRCADIAN INFLUENCE ON SEIZURE OCCURRENCE

A study on the distribution of seizures which incorporates many of the previous considerations was performed by the Yale University group (Durazzo et al. 2008). The study consisted of 131 consecutive adult patients who met strict inclusion criteria: (1) patients had only one seizure onset area, which was definitively localized by a team of neurologists and neurosurgeons to a degree of confidence high enough to warrant resective surgery; (2) patients were monitored with intracranial EEG only, thus allowing for the most accurate and uniform seizure detection possible, independent of the need for clinical correlates that could be partisan to daytime hours of increased baseline patient activity; (3) patients designated as MTLE were required to show pathology of hippocampal sclerosis on histology of resected tissue. Furthermore, data analysis in this study attempted to minimize the impact of seizure clusters or patients who experienced a high number of seizures (see discussion in Section 37.4).

The time distribution of seizures generated from various regions of the brain are summarized in Table 37.1 and depicted in Figure 37.3. Overall, through analysis of 669 seizures, this study was able to conclude that seizures do not occur randomly relative to the time of day, and that the pattern in which seizures occur relative to time is strongly dependent on the location of the epileptogenic focus. The study uncovered distinct circadian-like variation in seizure generation from each of the five different locations examined (mesial temporal, neocortical temporal, frontal, parietal, and occipital lobes). Most striking were the distributions observed from seizures generated by the parietal and occipital lobes. Both occur in Gaussian-like fashion, but precisely 180 degrees out of phase relative to each other: occipital seizures peaked between 16:00 and 19:00, whereas parietal seizures peaked between 4:00 and 7:00. Frontal seizures followed a unimodal pattern and peaked between 4:00 and 7:00. MTLE seizures occurred in a bimodal pattern, with a primary peak in the late afternoon between 16:00 and 19:00, and a secondary peak in the morning between 7:00 and 10:00. Seizures generated from the neocortical temporal lobe were unimodal and peaked slightly before the primary peak seen in MTLE.

In terms of patients with MTLE, both the Durazzo et al. (2008) and Quigg et al. (1998) studies agreed that seizures are most likely to be generated in the late afternoon. However, the distribution observed by Quigg et al. was cosine in character, whereas that reported by Durazzo et al. was

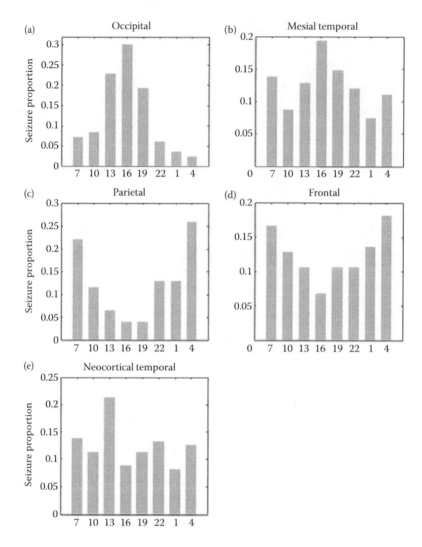

FIGURE 37.3 Seizure occurrence from various epileptogenic regions versus time of day. The study included 669 seizures recorded from 131 patients with definitively localized seizure foci. The 24-hour day is divided into eight 3-hours bins; the *x*-axis indicates the start time of each bin. (Adapted from Durazzo, T. S. et al., *Neurology*, 70(15), 1265–1271, 2008.)

bimodal; Durazzo et al. reported an additional peak of seizure occurrence in the late morning. Although these two studies are the only ones to look at true MTLE, i.e., patients with evidence of hippocampal sclerosis on pathology, they differed in methodology. The Durazzo et al. study only included patients monitored by intracranial EEG, whereas the Quigg et al. study included those monitored by either intracranial or scalp EEG. Intracranial EEG is the gold standard for localization and seizure detection. However, MTLE is unique in that, in current practice, the typical MTLE patient is usually localized without the need for intracranial studies. The majority of MTLE patients proceed directly to surgery without the need for intracranial EEG. Therefore, as stated in their conclusions, by restricting analysis to intracranial studies, Durazzo et al. in effect analyzed a group of atypical MTLE patients. Whether the secondary peak observed by Durazzo is due to greater sensitivity in seizure detection afforded by intracranial EEG or the result of a different patient base requires more study. No other study to date has adequately analyzed MTLE seizures. The group designated as MTLE by Hofstra et al. (2009) (Table 37.1) did not require the finding of hippocampal sclerosis on resected tissue, and furthermore, only three of their six patients had findings of temporal sclerosis on MRI.

In terms of seizures generated from regions outside the mesial temporal lobe, studies prior to Durazzo et al. had grouped all non-MTLE patients together into a single extratemporal lobe epilepsy (XTLE) group (Pavlova, Shea, and Bromfield 2004; Quigg et al. 1998). Using this grouping strategy, disagreement existed on whether partial seizures outside of the mesial temporal lobe occurred randomly or nonrandomly with respect to time of day (Hofstra, Spetgens et al. 2009; Pavlova, Shea, and Bromfield 2004; Quigg et al. 1998). The Durazzo et al. study found that when patients with seizures of extratemporal origin were considered to be a single group, the seizures appeared to occur uniformly throughout the day. However, by separately grouping seizures generated within the different lobes into different groups, Durazzo et al. uncovered that each region indeed showed its own characteristic circadian-like pattern of seizure generation. The observation that the regions had peaks and troughs at different times, some 180 degrees out of phase relative to each other, likely accounts for why statistically uniform distributions have been reported for extratemporal aggregates by other researchers.

Following the Durazzo et al. study, Hofstra et al. (2009) also studied the regions outside the temporal lobe individually. However, there were significant differences in methodology between these studies. The study by Durazzo et al. consisted of nearly four times the number of patients as the study by Hofstra et al., and patients in the Durazzo et al. study were required to pass strict inclusion criteria regarding localization. Furthermore, Hofstra et al. did not limit the number of seizures an individual patient could contribute to the group distribution and did not adequately account for the possibility of seizure clustering; they allowed up to 15 seizures from a single patient in a single time bin for a given day. This was particularly important given the low number of patients in the Hofstra et al. study; peaks reported for the groups in this study could potentially be the result of a seizure cluster in a single patient or unduly influenced by a subset of the patients. Other differences between the two studies exist, including the inclusion of children and the definition of MTLE in Hostra et al. As indicated above, Durazzo et al. required pathological verification of hippocampal sclerosis for a patient to be designated as MTLE, whereas in the Hofstra et al. study designation was only based on MRI findings. Despite these differences, peaks observed in their study have some overlap with peaks observed by Durazzo in MTLE, NTLE, and FLE.

37.6 DISCUSSION

We believe, in comparison to the other studies reported to date, the study by Durazzo et al. is the most rigorous in terms of the components outlined above; most notably the use of strict patient inclusion criteria that ensured correct localization, the use of intracranial EEG for seizure detection, and analysis of data to restrict the impact of individual patients and seizure clusters on the analysis.

As evident from Table 37.1, studies grouping seizures of different origins into the XTLE group produced results giving the impression that these seizures occurred randomly; however, by looking at the "extratemporal lobe" in higher resolution using strict localization criteria, Durazzo et al. revealed striking rhythmicity from these various regions.

Patients in the Durazzo et al. study had from 2–31 seizures during the monitoring period. To prevent the undue influence from those with a high number of seizures, Durazzo et al. implemented an algorithm to randomly select a maximum of eight seizures for analysis from each patient. No other study to date has attempted to correct for the potentially catastrophic misinterpretation of data that could occur if a given patient has an unduly high number of seizures. This is particularly important to consider in some of the other studies that built circadian distributions using as few as four patients, or less than 50 total seizures (Hofstra, Spetgens et al. 2009).

Future studies could improve on the Durazzo et al. study by taking sleep/wake state into account, performing a forced desynchrony study that records seizures, or by increasing the number of patients and seizures studied. The last of these could help improve the spatial and temporal resolutions of the seizure distributions by defining seizure distributions in a more specific manner than the lobe of seizure onset and with greater temporal resolution than the 3-h bins that were employed in that study. The possibility that sleep differentially affects seizure threshold depending on location of the epileptogenic focus requires more study, as does the role of sleep state in each syndrome.

37.7 CONCLUSIONS

Evidence indicating seizure occurrence relative to the time of day is not random continues to mount. Endogenous circadian rhythms and/or rhythmic exogenous factors likely play substantial roles in seizure occurrence. These roles appear to vary considerably according to the pathophysiology and location of the epileptogenic focus. Future work in this field could extend current knowledge by increasing the spatial resolution used to group patients based on epileptogenic focus; just as MTLE and NTLE produced seizures with different time distributions, studying the time of occurrence of seizures within structures and substructures within the other lobes could prove equally fruitful. Equally important would be to further group patients based on detailed pathophysiologies. Thus far the only pathophysiology investigated has been MTLE. Because seizure generation may differ based on not only location of focus, but also on underlying disease (e.g., cortical dysplasia vs. posttrauma), it is reasonable to believe circadian rhythms differentially influence the various pathophysiologies. Future work could also extend the current body of knowledge by performing analysis of the sleep/ wake state at the time of seizure occurrence.

As knowledge of circadian-like variation in seizure threshold emerges, questions arise as to the substrate responsible for these modulations. Apart from controlling rhythmic physiologic electrical activity, the SCN also generates rhythmicity in endocrine function, including the secretion of hormones with known influence on neuronal excitability (Moore 1997). Melatonin secretion peaks during the night, and was shown to increase hippocampal excitability (Musshoff et al. 2002; Stewart and Leung 2005), yet suppress activity in other regions (Munoz-Hoyos et al. 1998). Adenosine concentration increases in certain regions of the brain as duration of wakefulness increases (Porkka-Heiskanen, Strecker, and McCarley 2000), and has been found to suppress epileptic activity to a large degree (Boison 2005). The list continues: Histamine, dopamine, serotonin, norepinephrine, and the components of the hypothalamic–pituitary–adrenal axis are a few other substrates that both modify neuronal activity and vary in circadian fashion (Applegate, Burchfiel, and Konkol 1986; Applegate and Tecott 1998; Baram et al. 1996; Kole et al. 2001; McIntyre, Saari, and Pappas 1979; Tuomisto et al. 2001). Future studies, particularly those using animal models, would be helpful in narrowing down the list and determining the principle combinations contributing to circadian variation at each epileptogenic focus.

Incorporating the principle of circadian variation in seizure threshold by seizure-predication algorithms and chronotherapy would fortunately not need to wait for results from the aforementioned future experiments. The large body of evidence suggesting periodicity in at least some patients and pathophysiologies may be enough to begin the attempt of incorporating circadian rhythms into current algorithms. As mentioned in the introduction, this could easily be achieved by simply modulating—in circadian fashion—the algorithm's threshold value required to predict a seizure from any given measure. The precise phase of this threshold variation could be determined from one of two possible sources: (1) on an individual patient-by-patient basis, in which circadian variation is assessed through a "calibration" period, or through all previously recorded seizure records; or (2) predetermined circadian phases may be extrapolated from knowledge of the patient's pathophysiology and epileptogenic focus. At this stage of knowledge on circadian variation of seizures, the former source may be preferential, as it allows circadian phase to be determined for the precise pathophysiology and epileptogenic focus of the patient in question and it does not depend on the ability to match the patient to the broad categories thus far used to group patients' seizures. As future work in determining circadian variation at higher spatial and temporal resolution, aspects of both approaches could be incorporated within a seizure prediction algorithm.

REFERENCES

Applegate, C. D., J. L. Burchfiel, and R. J. Konkol. 1986. Kindling antagonism—Effects of norepinephrine depletion on kindled seizure supression after concurrent, alternate stimulation in rats. *Exp. Neurol.* 94(2):379–390.

Applegate, C. D., and L. H. Tecott. 1998. Global increases in seizure susceptibility in mice lacking 5-HT2C receptors: A behavioral analysis. *Exp. Neurol.* 154(2):522–530.

Baram, T. Z., W. G. Mitchell, A. Tournay, O. C. Snead, R. A. Hanson, and E. J. Horton. 1996. High-dose corticotropin (ACTH) versus prednisone for infantile spasms: A prospective, randomized, blinded study. *Pediatrics* 97(3):375–379.

Bazil, C. W., and T. S. Walczak. 1997. Effects of sleep and sleep stage on epileptic and nonepileptic seizures. *Epilepsia* 38(1):56–62.

Boison, D. 2005. Adenosine and epilepsy: From therapeutic rationale to new therapeutic strategies. *Neuroscientist* 11(1):25–36.

Cajochen, C., J. K. Wyatt, C. A. Czeisler, and D. J. Dijk. 2002. Separation of circadian and wake duration-dependent modulation of EEG activation during wakefulness. *Neuroscience* 114(4):1047–1060.

Crespel, A., M. Baldy-Moulinier, and P. Coubes. 1998. The relationship between sleep and epilepsy in frontal and temporal lobe epilepsies: Practical and physiopathologic considerations. *Epilepsia* 39(2):150–157.

Czeisler, C. A., J. F. Duffy, T. L. Shanahan, E. N. Brown, J. F. Mitchell, D. W. Rimmer, J. M. Ronda, E. J. Silva, J. S. Allan, J. S. Emens, D. J. Dijk, and R. E. Kronauer. 1999. Stability, precision, and near-24-hour period of the human circadian pacemaker. *Science* 284(5423):2177–2181.

DallaBernardina, B. 1991. Sleep and benign partial epilepsies of childhood: EEG and evoked potentials study. *Epilepsy Res. Suppl.* (2):83–96.

Dijk, D. J. 1999. Circadian variation of EEG power spectra in NREM and REM sleep in humans: Dissociation from body temperature. *J. Sleep Res.* 8(3):189–195.

Duckrow, R. B., and T. K. Tcheng. 2007. Daily variation in an intracranial EEG feature in humans detected by a responsive neurostimulator system. *Epilepsia* 48(8):1614–1620.

Durazzo, T. S., S. S. Spencer, R. B. Duckrow, E. J. Novotny, D. D. Spencer, and H. P. Zaveri. 2008. Temporal distributions of seizure occurrence from various epileptogenic regions. *Neurology* 70(15):1265–1271.

Ferrillo, F., M. Beelke, F. De Carli, M. Cossu, C. Munari, G. Rosadini, and L. Nobili. 2000. Sleep-EEG modulation of interictal epileptiform discharges in adult partial epilepsy: A spectral analysis study. *Clin. Neurophysiol.* 111(5):916–923.

Gao, B., and R. Y. Moore. 1996. Glutamic acid decarboxylase message isoforms in human suprachiasmatic nucleus. *J. Biol. Rhythms* 11(2):172–179.

Gowers, W. 1885. Course of epilepsy. In *Epilepsy and other chronic convulsive diseases: Their causes, symptoms and treatment.* New York: William Wood.

Haut, S. R. 2006. Seizure clustering. *Epilepsy Behav.* 8(1):50–55.

Haut, S. R., A. D. Legatt, C. Odell, S. L. Moshe, and S. Shinnar. 1997. Seizure lateralization during EEG monitoring in patients with bilateral foci: The cluster effect. *Epilepsia* 38(8):937–940.

Herman, S. T., T. S. Walczak, and C. W. Bazil. 2001. Distribution of partial seizures during the sleep-wake cycle—Differences by seizure onset site. *Neurology* 56(11):1453–1459.

Hofstra, W. A., and W. de Weerd. 2009. The circadian rhythm and its interaction with human epilepsy: A review of literature. *Sleep Med. Rev.* 13(6):413–420.

Hofstra, W. A., B. E. Grootemarsink, R. Dieker, J. van der Palen, and A. W. de Weerd. 2009. Temporal distribution of clinical seizures over the 24-h day: A retrospective observational study in a tertiary epilepsy clinic. *Epilepsia* 50(9):2019–2026.

Hofstra, W. A., W. P. J. Spetgens, F. S. S. Leijten, P. C. van Rijen, P. Gosselaar, J. van der Palen, and A. W. de Weerd. 2009. Diurnal rhythms in seizures detected by intracranial electrocorticographic monitoring: An observational study. *Epilepsy Behav.* 14(4):617–621.

Ingelsson, E., K. Bjorklund-Bodegard, L. Lind, J. Arnlov, and J. Sundstrom. 2006. Diurnal blood pressure pattern and risk of congestive heart failure. *JAMA* 295(24):2859–2866.

Inouye, S. I. T. 1996. Circadian rhythms of neuropeptides in the suprachiasmatic nucleus. In *Hypothalamic integration of circadian rhythms.* R. M. Buijs, A. Kalsbeek, H. J. Romijn, C. M. A. Pennartz, and M. Mirmiran, eds. Amsterdam: Elsevier.

Kole, M. H. P., J. M. Koolhaas, P. G. M. Luiten, and E. Fuchs. 2001. High-voltage-activated Ca2+ currents and the excitability of pyramidal neurons in the hippocampal CA3 subfield in rats depend on corticosterone and time of day. *Neurosci. Lett.* 307(1):53–56.

Lehnertz, K., and C. E. Elger. 1998. Can epileptic seizures be predicted? Evidence from nonlinear time series analysis of brain electrical activity. *Phys. Rev. Lett.* 80(22):5019–5022.

Litt, B., R. Esteller, J. Echauz, M. D'Alessandro, R. Shor, T. Henry, P. Pennell, C. Epstein, R. Bakay, M. Dichter, and G. Vachtsevanos. 2001. Epileptic seizures may begin hours in advance of clinical onset: A report of five patients. *Neuron* 30(1):51–64.

Malow, B. A., X. H. Lin, R. Kushwaha, and M. S. Aldrich. 1998. Interictal spiking increases with sleep depth in temporal lobe epilepsy. *Epilepsia* 39(12):1309–1316.

Manfredini, R., B. Boari, M. Gallerani, R. Salmi, E. Bossone, A. Distante, K. A. Eagle, and R. H. Mehta. 2004. Chronobiology of rupture and dissection of aortic aneurysms. *J. Vasc. Surg.* 40(2):382–388.

Manfredini, R., B. Boari, M. H. Smolensky, R. Salmi, O. la Cecilia, A. M. Malagoni, E. Haus, and F. Manfredini. 2005. Circadian variation in stroke onset: Identical temporal pattern in ischemic and hemorrhagic events. *Chronobiol. Int.* 22(3):417–453.

McIntyre, D. C., M. Saari, and B. A. Pappas. 1979. Potentiation of amygdala kindling in adult or infant rats by injections of 6-hydroxydopamine. *Exp. Neurol.* 63(3):527–544.

Michaud, M., M. Dumont, J. Paquet, A. Desautels, M. L. Fantini, and J. Montplaisir. 2005. Circadian variation of the effects of immobility on symptoms of restless legs syndrome. *Sleep* 28(7):843–846.

Minecan, D., A. Natarajan, M. Marzec, and B. Malow. 2002. Relationship of epileptic seizures to sleep stage and sleep depth. *Sleep* 25(8):899–904.

Moore, R. Y. 1973. Retinohypothalamic projection in mammals—Comparative study. *Brain Res.* 49(2):403–409.

Moore, R. Y. 1978. Neural control of pineal function in mammals and birds. *J. Neural Transm.* (13):47–58.

Moore, R. Y. 1997. Circadian rhythms: Basic neurobiology and clinical applications. *Annu. Rev. Med.* 48:253–266.

Mormann, F., R. G. Andrzejak, C. E. Elger, and K. Lehnertz. 2007. Seizure prediction: The long and winding road. *Brain* 130:314–333.

Munoz-Hoyos, A., M. Sanchez-Forte, A. Molina-Carballo, G. Escames, E. Martin-Medina, R. J. Reiter, J. A. Molina-Font, and D. Acuna-Castroviejo. 1998. Melatonin's role as an anticonvulsant and neuronal protector: Experimental and clinical evidence. *J. Child Neurol.* 13(10):501–509.

Musshoff, U., D. Riewenherm, E. Berger, J. D. Fauteck, and E. J. Speckmann. 2002. Melatonin receptors in rat hippocampus: Molecular and functional investigations. *Hippocampus* 12(2):165–173.

Oldani, A., M. Zucconi, R. Asselta, M. Modugno, M. T. Bonati, L. Dalpra, M. Malcovati, M. L. Tenchini, S. Smirne, and L. Ferini-Strambi. 1998. Autosomal dominant nocturnal frontal lobe epilepsy. A video-polysomnographic and genetic appraisal of 40 patients and delineation of the epileptic syndrome. *Brain* 121:205–223.

Pavlova, M. K., S. A. Shea, and E. B. Bromfield. 2004. Day/night patterns of focal seizures. *Epilepsy Behav.* 5(1):44–49.

Pavlova, M. K., S. A. Shea, F. Scheer, and E. B. Bromfield. 2009. Is there a circadian variation of epileptiform abnormalities in idiopathic generalized epilepsy? *Epilepsy Behav.* 16(3):461–467.

Porkka-Heiskanen, T., R. E. Strecker, and R. W. McCarley. 2000. Brain site-specificity of extracellular adenosine concentration changes during sleep deprivation and spontaneous sleep: An in vivo microdialysis study. *Neuroscience* 99(3):507–517.

Quigg, M., and M. Straume. 2000. Dual epileptic foci in a single patient express distinct temporal patterns dependent on limbic versus nonlimbic brain location. *Ann. Neurol.* 48(1):117–120.

Quigg, M., M. Straume, M. Menaker, and E. H. Bertram. 1998. Temporal distribution of partial seizures: Comparison of an animal model with human partial epilepsy. *Ann. Neurol.* 43(6):748–755.

Rakhade, S. N., and F. E. Jensen. 2009. Epileptogenesis in the immature brain: Emerging mechanisms. *Nat. Rev. Neurol.* 5(7):380–391.

Rodriguez, A. J. 2007. Pediatric sleep and epilepsy. *Curr. Neurol. Neurosci. Rep.* 7(4):342–347.

Shibata, S., and R. Y. Moore. 1988. Electrical and metabolic-activity of suprachiasmatic nucleus neurons in hamster hypothalamic slices. *Brain Res.* 438(1–2):374–378.

Solomon, G. D. 1992. Circadian-rhythms and migraine. *Cleve. Clin. J. Med.* 59(3):326–329.

Stephenson, R. 2003. Do circadian rhythms in respiratory control contribute to sleep-related breathing disorders? *Sleep Med. Rev.* 7(6):475–490.

Stewart, L. S., and L. S. Leung. 2005. Hippocampal melatonin receptors modulate seizure threshold. *Epilepsia* 46(4):473–480.

Tuomisto, L., V. Lozeva, A. Valjakka, and A. Lecklin. 2001. Modifying effects of histamine on circadian rhythms and neuronal excitability. *Behav. Brain Res.* 124(2):129–135.

Welsh, D. K., D. E. Logothetis, M. Meister, and S. M. Reppert. 1995. Individual neurons dissociated from rat suprachiasmatic nucleus express independently-phased circadian firing rhythms. *Neuron* 14(4):697–706.

38 Nocturnal Frontal Lobe Epilepsy
Metastability in a Dynamic Disease?

John G. Milton, Austin R. Quan, and Ivan Osorio

CONTENTS

The advantage of graphical approaches to complex dynamical systems, such as the brain, is that they can be used by neuroscientists to develop interpretations and key experiments even if the underlying mechanisms have not yet been fully identified. We illustrate this approach by showing that mathematical models of neural populations are particularly vulnerable for the production of paroxysmal transient events (possibly seizures) at times when changes in state occur as, for example, the transitions from wakefulness to sleep and through the sleep stages while sleeping. These arguments emphasize the importance of careful studies of the timing of seizure occurrences with respect to sleep stage transitions in patients with nocturnal frontal lobe epilepsy.

38.1 INTRODUCTION

The characteristic feature of nocturnal frontal lobe epilepsy (NFLE) is that seizures occur predominantly during sleep. Autosomal-dominant NFLE is associated with mutations in the alpha 4 subunit of neuronal acetylcholine receptors (Mann and Mody 2008). However, even in the same individual, seizures are highly variable with respect to their timing, semiology, and electroencephalographic features. These observations emphasize that the NFLE genetic mutation is a necessary, but not sufficient, condition for seizure occurrence in NFLE. In other words, there must be additional factors that, taken together with the receptor defect, determine when the seizure occurs. The identification of these additional factors may make it possible to develop methods based on seizure anticipation, such as delivery of well-timed electrical stimuli (Osorio and Frei 2009), to treat NFLE. In this way, investigations of NFLE can provide clues for the development of seizure prediction techniques in other types of epilepsy as well.

Here we consider NFLE from the perspective of a dynamic disease; for recent reviews see the works by Milton (2010a, 2010b). Since a seizure represents a paroxysmal and uncontrolled discharge of a large population of neurons, we first review three mechanisms for the occurrence of sudden events in dynamical systems. The observation that seizures in NFLE can occur during daytime naps suggests that the triggering events for seizures are more likely to be related to the neurophysiology of sleep rather than to circadian-rhythm-related phenomena. We use mathematical concepts to conjecture that seizures have a higher probability of occurrence at times when changes occur between macroscopic brain states.

The term macroscopic brain state refers to a collective, emergent dynamical property of the brain (Buzsáki 2006; Kelso 1999); specifically, wakefulness and each of the sleep stages represent different macroscopic brain states. At such transitions, there is an increased probability in mathematical models that transient, but relatively long-lived, oscillatory solutions can arise. We refer to these transient states as metastability and interpret them as epileptic seizures. Finally, since transitions between macroscopic brain states are expected to be associated with increased variability in key variables, it may be possible to develop measures to anticipate periods of increased seizure susceptibility.

38.2 SEIZURE ONSET

The mathematical description of a seizure involves a description of how variables change as a function of time, i.e., the dynamics of the neural population. By analogy with physical systems, the rate of change in a variable per unit of time is the quantity that most directly identified the underlying mechanisms. Thus, scientific hypotheses for seizure occurrence are formulated in terms of differential equations, e.g.,

$$\dot{x}(t) = f(x) \equiv \text{hypothesis}, \tag{38.1}$$

where x is a variable, $\dot{x} = dx/dt$, and $f(x)$ describes the hypothesis. The prediction corresponds to the solution of this equation and is the quantity to be compared with experimental observation. In order to determine a particular solution that corresponds to a given experimental condition, it is necessary to specify an initial condition, i.e., $x(t_0)$, the value of $x(t)$ at some instance in time, t_0. What is the nature of the hypothesis that can account for the sudden appearance of a qualitatively different dynamical behavior in a neural population that we recognize clinically as an epileptic seizure? We consider three different mechanisms that can produce a paroxysmal event in a dynamical system. We illustrate these mechanism using very simple examples. A fourth mechanism is briefly mentioned in Section 38.4.

38.2.1 SEIZURE ONSET DUE TO PARAMETER CHANGES

The first attempts to identify a mechanism for seizure onset were based on the distinction between a variable and a parameter; for a review, see Milton (2010b). A variable is anything that can be measured. A parameter is a variable that changes so slowly in comparison with other variables that, on an appropriately chosen time scale (e.g., milliseconds, seconds, minutes, hours, or days), it can be regarded to be constant. In other words, the concept of a dynamical disease introduced the notion of how variables evolving on many different time scales interact to produce the phenomena we observe at any instance in time. There is a direct relationship between parameter values and the qualitative nature of the dynamics produced by neurons and neural populations.

The relationship between parameters and dynamics can most easily be appreciated by choosing $f(x) = kx$ so that

$$\dot{x} = kx, \tag{38.2}$$

and the solution

$$x(t) = X_0 e^{kt}, \tag{38.3}$$

where k is a parameter, $t_0 = 0$, and $X_0 = (0)$. Figure 38.1 plots Equation 38.3 for different choices of k. There are two qualitatively different behaviors: when $k > 0$, all solutions $x(t)$ grow exponentially to infinity and when $k < 0$, all solutions decrease exponentially to zero. In parameter space, the switch between these two behaviors occurs abruptly; the transition point is $k = 0$. In mathematical terminology, the value of the parameter for which there is a qualitative change in dynamics (in this case $x(t)$ growing versus shrinking) is called the bifurcation point. A great deal of mathematical effort has been devoted to the demonstration that the number of possible different types of bifurcations is surprisingly small. These bifurcations differ with respect to the nature of the qualitative change in dynamics that occur; for example, bifurcations in which a stationary solution becomes an oscillating periodic solution, and vice versa, are called Hopf bifurcations. The key point is that sudden changes in dynamics occur as key parameters cross certain thresholds.

A possible example of seizure onset due to a parameter change is the triggering of a 3-Hz spike-and-wave absence seizure in a child by hyperventilation. If we imagine that the critical parameter is related to arterial pCO_2, then hyperventilation causes a pCO_2-related parameter to cross a bifurcation point causing a qualitative change in dynamics, leading to a seizure. Since the seizure typically causes respiration to slow, pCO_2 rises and, hence, we could imagine that the pCO_2-related parameter crosses the bifurcation point in the reverse direction and the seizure stops.

For the discussion that follows, it is convenient to introduce graphical approaches to the study of dynamics. These pictorial representations are particularly useful for attempts to translate observations made at the bedside to investigations that can be tested at the bench top. Here we introduce one such concept, namely the construction of a potential-like function, $\Phi(x)$, where

$$\Phi(x) = -\int_{x(t_0)}^{x(t)} f(s)\,ds. \tag{38.4}$$

Using $\Phi(x)$, we can imagine the dynamics of the system as a marble rolling over a complex surface that is described by $\Phi(x)$: the definition of $\Phi(x)$ ensures that the marble always rolls downhill (Hale and Kocek 1991). The potential surfaces corresponding to Equation 38.1 have the form of parabolas (see insets in Figure 38.1),

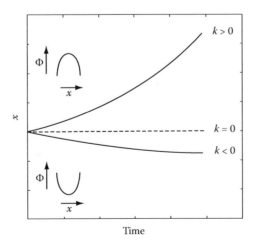

FIGURE 38.1 Comparison of the dynamics of Equation 38.2 for different choices of the parameter k. The insets show the potential $\Phi(x)$ for representative choices of k.

$$\Phi(x) = -\frac{kx^2}{2}. \tag{38.5}$$

When $k < 0$, the potential surface corresponds to a valley and, hence, the marble rolling on such a surface eventually comes to rest at the bottom of the valley. We say that such a steady state is stable to a perturbation. On the other hand, if $k > 0$, we have a hill and, when perturbed, the marble always rolls away from the top of the hill, no matter how small the perturbation may be. We say that such a steady state is unstable to a perturbation.

Equation 38.2 describes the special case when the steady state corresponds to a single value of x, namely $x = 0$. If we choose $x(t_0) = 0$, then $\dot{x} = 0$ and the dynamical system remains indefinitely at this point, therefore, it is referred to as a fixed-point. However, there are many other kinds of time varying steady states possible, including periodic and aperiodic oscillations and even chaotic rhythms. By analogy to a marble rolling on a surface, it has become common usage to replace the concept of a valley with the word attractor and a hill with the word repellor. Thus, we can have limit cycle attractors, chaotic attractors, and so on. Of course the mathematical construction of $\Phi(x)$ for these attractors and repellors is much more difficult than for the simple case of a fixed-point. However, the mental picture of a marble rolling over a complex surface composed of hills and valleys has proven useful for designing experiments. What must be kept in mind is that the dynamical object that exists within the valleys and on the hilltops can itself be very complex.

38.2.2 Seizure Onset Due to Changes in Initial Conditions

A second mechanism for generating a paroxysmal event arises in dynamical systems possessing two or more stable equilibria, i.e., dynamical systems for which $\Phi(x)$ contains two or more attractors. Such dynamical systems are said to exhibit multistability. Multistability is a common property of mathematical models of the nervous system (Hoppensteadt and Izhikevich 1997; Wilson and Cowan 1972). Indeed, the most common mode of oscillation onset is a subcritical Hopf bifurcation, which results in the coexistence of a fixed-point and limit cycle attractor. Moreover, a countably large number of coexistent attractors arise in models of delayed recurrent inhibitory loops (Foss and Milton 2000; Ma and Wu 2007). The key point is that multiple attractors exist even though all parameters are held constant; sudden changes in dynamics arise because of sudden changes in initial conditions that cause switches attractors.

We illustrate this phenomenon by choosing $f(x)$ to be a cubic nonlinearity so that Equation 38.1 becomes

$$\dot{x} = x - x^3, \tag{38.6}$$

And, hence,

$$\Phi(x) = -\frac{x^2}{2} + \frac{x^4}{4}. \tag{38.7}$$

It is well established that many of the properties of excitable systems such as neurons and cardiac cells are related to the presence of an underlying cubic nonlinearity (Milton 2010a). Figure 38.2 shows that $\Phi(x)$ is a double-well potential. There are three possible fixed-points determined by setting $\dot{x} = 0$; namely $x = 0$ and $x = \pm 1$. Using our marble rolling on a landscape analogy, it is easy to see that the $x = 0$ fixed-point is unstable (hill) and the $x = \pm 1$ fixed-points are stable (valleys); thus, Equation 38.6 describes a bistable dynamical system. Which fixed-point is observed at any instance in time depends on the choice of the initial condition, i.e., $x(t_0)$. The set of initial conditions

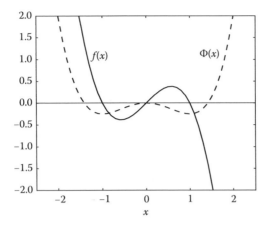

FIGURE 38.2 Comparison of $f(x)$ (solid line) and $\Phi(x)$ (dashed line) for a cubic nonlinearity. Note that the fixed-points of Equation 38.6 occur when $f(x) = 0$ and (B) correspond to either the bottom of a valley in $\Phi(x)$ (attractor) or to the top of a hill (repellor).

for which the $x = -1$ fixed-point is observed, i.e., all $x(t_0) < 0$, is called the basin of attraction for the $x = -1$ attractor. Similarly, all $x(t_0) > 0$ corresponds to the basin of attraction for the $x = 1$ attractor. Basins of attractions must necessarily be separated by an unstable object referred to as a separatrix; in this case, the separatrix corresponds to $x = 0$ and is a fixed-point repellor. These observations provided the motivation for neuroscientists to use brief electrical or sensory stimuli to change the qualitative nature of the dynamics of single neurons, isolated neural circuits, and epileptic neural populations; for more discussion see Milton (2010a).

38.2.3 METASTABILITY AND TIME DELAYS

Since neurons are typically separated by finite distances and axonal/dendritic conduction velocities are finite, it is clear that the effects of time delay, i.e., the time it takes for an action potential to travel from one neuron to another, must be taken into account. Consequently, mathematical models of realistic neural populations take the form of delay differential equations, e.g.,

$$\dot{x}(t) = f(x(t - \tau)), \tag{38.8}$$

where $x(t)$, $x(t - \tau)$ are, respectively, the values of the state variable at times t, $t - \tau$, and τ is the time delay. In contrast to ordinary differential equations, e.g., Equation 38.1, in order to obtain a particular solution of Equation 38.8, it is necessary to specify an initial function, $\Psi(x)$, on the interval $[t_0 - \tau, t_0]$. Consequently, Equation 38.8 describes an infinite dimensional dynamical system, since functions of finite length contain an infinite number of points. Here, we draw attention to the possibility that Equation 38.8 can give rise to metastable states. The occurrence of metastable states is intimately connected to the influence of $\Psi(x)$ on dynamics that are evolving in the close neighborhood of a separatrix. The key point is that, in contrast to multistability, sudden changes in dynamics are associated with the effects of unstable objects.

To illustrate metastability, we consider the effects of the initial function on the dynamics of two mutually inhibited neurons (Figure 38.3a)

$$\begin{aligned}
\dot{x}(t) &= -\tau x(t) - \tau S_2(y(t-1)) + \tau I_1 \\
\dot{y}(t) &= -\tau y(t) - \tau S_1(x(t-1)) + \tau I_2
\end{aligned}, \tag{38.9}$$

where

$$S_j = \frac{c_j u^{n_j}}{\theta_j^{n_j} + u^{n_j}} \quad j = 1, 2$$

describes sigmoidal functions representing the inhibitory influences, $\tau \equiv \tau_1 \equiv \tau_2$, and we have introduced the change of independent variable $t \rightarrow t/\tau$. This model has been extensively studied previously (Milton et al. 2010; Pakdaman et al. 1998). We focus on the case that there are three positive steady states: one unstable and two stable. This case corresponds to assuming that I_1 and I_2 are sufficiently large compared to the inhibitory effects. It can be shown that under these conditions, the stability of the equilibria is independent of τ. Thus, the observed dynamics depend solely on the choice of $\Psi(x, y)$.

Figure 38.3b, c, and d shows the effects of different choices of $\Psi(x, y)$ on the dynamics generated by Equation 38.9. For illustrative purposes, we consider $\Psi(x, y) = (X_0, Y_0)$, where X_0, Y_0 are positive constants. The unstable equilibrium point is $x = y = 0.2$. When $\Psi((x, y)$ is chosen sufficiently far to the left of the unstable fixed-point, dynamics quickly settle on a positive fixed-point (Figure 38.3b). A qualitatively similar behavior is observed when $\Psi(x, y)$ is chosen sufficiently far to the right of the unstable fixed-point in this case, quickly settling on a different positive fixed-point (Figure 38.3c). However, when $\Psi(x, y)$ is chosen sufficiently close to the unstable fixed-point, a dramatically different behavior is observed (Figure 38.3d). In particular, there occurs an oscillatory transient before the dynamics settle eventually onto one of the two stable fixed-points. This oscillatory transient corresponds to a metastable state.

There are three important features of this metastable state. First, the qualitative nature of the metastable dynamics is different than the coexisting attractors; in our example, it is oscillatory,

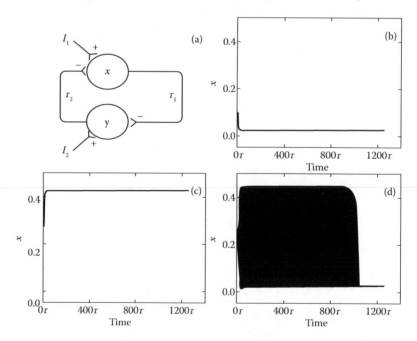

FIGURE 38.3 Dynamics of a neural network with two mutually inhibitory neurons. (a) schematic representation of the neural ciruit in which '–' indicates an inhibitory connection and '+' indicates an excitatory connection. (b), (c), and (d) show, respectively, the dynamics of Equation 38.9 for different choices of $\Psi(x, y)$: $x = 0.1, y = 0.3; x = 0.3, y = 0.1; x = y = 0.19$. Parameters in all simulations were: $c_1 = 0.4, c_2 = 0.6, I_1 = 0.5, I_2 = 0.4, n_1 = 2, n_2 = 2, \theta_1 = 0.2, \theta_2 = 0.2$ and $\tau_1 = \tau_2 = 8$. Only the activity of neuron x has been shown.

whereas the coexisting stable behaviors are fixed-point attractors. Second, the length of time that the metastable state persists is self-limited and depends on the choices of τ and $\Psi(x, y)$. In particular, the metastable state can persist for several orders of magnitude of time longer than τ. Thus, even if the neural delay is only a few ms, metastable events that last as long as typical clinical seizures, e.g., minutes, can readily be produced. Finally, numerical simulations suggest that the probability that a metastable state arises is likely to be very small, i.e., for most choices of $\Psi(x, y)$, metastability does not occur.

38.3   CHANGES IN BRAIN STATE AND SEIZURE SUSCEPTIBILITY

The observations in Section 38.2.3 imply that the probability of the occurrence of a metastable state (possibly a seizure) is highest at the times when a change in macroscopic brain state occurs. To make the connection to NFLE, we identify the macroscopic brain states as wakefulness and the various sleep stages. These considerations are justified by the fact that wakefulness and the various sleep stages are each characterized by well-defined electroencephalographic, physiological, and behavioral properties (Iber et al. 2007). Each macroscopic brain state corresponds to a distinct attractor, i.e., a valley in an appropriately constructed potential surface. Thus, the transition from wakefulness to drowsiness (N1 sleep) and through the successive stages of the sleep cycle can be conceptualized as the successive appearance of an attractor and its gradual replacement by another.

Figure 38.4 shows schematically a possible scenario for the replacement of one attractor by another due to a parameter (variable) change on a slower time scale. There are two key points. First, there is an intermediate stage in which two attractors coexist, i.e., bistability. Second, due to the presence of the slowly changing landscape, the initial function, Ψ, must at some point negotiate the region of the separatrix, particularly if random perturbations (noise) are present. It follows that there must be a finite probability that a metastable state, such as that shown in Figure 38.3c, occurs.

38.4   DISCUSSION

A fundamental limitation of theories that postulate seizure occurrence as a result of either a change in critical parameters or initial conditions is that one must hypothesize the occurrence of two unidentified events: one to trigger seizure onset, another to stop it. In contrast, the triggering process

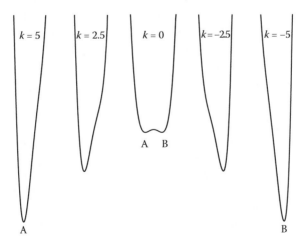

FIGURE 38.4   Schematic representation of the changes in $\Phi(x)$ as a parameter k changes slowly to cause one attractor (A) to disappear as another (B) appears. To illustrate this phenomenon, we took $\Phi(x) = k - \dfrac{x^2}{2} + \dfrac{x^4}{4}$ (Kelso 1999).

for a seizure related to metastability can be associated with a normal physiological process, i.e., the replacement of one macroscopic brain state by another during sleep and wakefulness. Moreover, by their very nature, metastable dynamical states are self-limited.

It is important to note that our use of the term metastability refers to a novel mechanism for producing long-lived transients that depends intimately on the role played by $\Psi(x, y)$ on a time-delayed dynamical system in the vicinity of a separatrix. Since $\Psi(x, y)$ can straddle the separatrix the marble on one side of the separatrix feels the pull of the other side, and vice versa, resulting in a transient confinement of trajectories to the neighborhood of the separatrix (Milton et al. 2008). In contrast to metastability in physical systems (Sornette 2004) and populations of neural oscillators (Freeman et al. 2006; Khosravani et al. 2003; Shanahan 2010), the metastability we introduce here is intimately associated with unstable states. When time delays are included in mathematical models of neural populations, complex metastable states can arise, even if neural spiking frequencies are low (Boustani and Destrexhe 2009; Timme and Wolf 2008).

Certainly relationships between sleep and epilepsy have long been recognized (Gloor et al. 1990; Steraide 2005). The cortical–thalamo circuit is a time-delayed recurrent inhibitory loop (Williams 1953) of which Equation 38.9 is an example. It is well established that multistability arises generically in delayed recurrent inhibitory loops (Foss and Milton 2000; Ma and Wu 2007). The model described by Equation 38.9 is clearly too simplistic to take into account the effects of the NFLE mutation in an acetylcholine receptor. However, this model does identify that a key issue for understanding seizure occurrence is to understand the connection between the properties of the basins of attraction (Lopes da Silva et al. 2003) and the probability that a metastable state arises—in other words, which factors increase and which factors decrease the probability that metastability arises. Thus, it will be important to study simple models in order to develop these understandings before taking on more physiologically detailed models.

Our observations emphasize the need for careful attention to the timing of seizure occurrence in patients with NFLE, with respect to the timing of sleep stage transitions. At first sight, the prediction is that seizures occur with increased probability at sleep stage transitions seems easy to test. However, the various criteria used to stage sleep are empirically-derived and, in particular, not easily related to the types of dynamical measures required to define a brain attractor; for discussion of this, see the works of Buzsáki (2006) and Kelso (1999). Moreover when parameters slowly change, the precise identification of the transition point is problematic when noise is present (Baer et al. 1989). On the other hand, it is known that the variance of fluctuations for a fixed-point attractor is directly proportional to the intensity of the noise and the magnitude of τ, and inversely proportional to the steepness of the walls of the basin of attraction (for a review see the work by Milton et al. (2004)). Indeed increased heart rate variability is associated with transitions between some sleep stages (Bunde et al. 1999; Telser et al. 2004). Thus, it may be possible to develop statistical techniques to identify time intervals during which transition between attractors are occurring as identifiers of time intervals of increased susceptibility for seizure occurrence.

The paradox of the study of rare human diseases attributed to single gene mutations is that investigations often provide more insights into how the healthy nervous system functions than into the design of effective treatment strategies. The lesson taught by NFLE is that it raises the possibility that triggering events for seizures can be the occurrence of a normal neurophysiological event associated with the sleeping brain. However, transitions between brain attractors also presumably arise in other brain activities, such as learning motor skills (Kelso 1999) and cognition (Buzsáki 2006). Although our speculations suggest that such transitions should also be associated with an increased risk of paroxysmal events it is not known why the overwhelming majority of such transitions, even in patients with epilepsy, are not associated with seizures. Nonetheless, it is clear that NFLE provides concrete reasons to believe that it may be possible to find the Holy Grail of epilepsy, i.e., prediction and prevention of seizures.

ACKNOWLEDGMENTS

We acknowledge support from the National Science Foundation (JM, NSF-0617072 and NSF-1028970) and NIH/NINDS (IO, 5R21NS056022).

REFERENCES

Baer, S. M., T. Erneux, and J. Rinzel. 1989. The slow passage through a Hopf bifurcation: Delay, memory effects, and resonance. *SIAM J. Appl. Math.* 49:55–71.

El Boustani, S., and A. Destrexhe. 2009. A master equation formalism for macroscopic modeling of asynchronous irregular activity states. *Neural Comp.* 21:46–100.

Bunde, A., S. Havlin, J. W. Kantekhardt, T. Pensel, J.-H. Peter, and K. Voight. 1999. Correlated and uncorrelated regions of heart-rate fluctuations during sleep. *Phys. Rev. Lett.* 85:3736–3739.

Buzsáki, G. 2006. *Rhythms of the brain.* New York: Oxford University Press.

Foss, J., and J. Milton. 2000. Multistability in recurrent neural loops arising from delay. *J. Neurophysiol.* 84:975–985.

Freeman, W. J., M. D. Holmes, G. A. West, and S. Vanhatalo. 2006. Dynamics of human neocortex that optimizes its stability and flexibility. *Int. J. Intell. Sys.* 21:881–901.

Gloor, P., M. Avoli, and G. Kostopoulos. 1990. Thalamocortical relationships in generalized epilepsy with bilaterally synchronous spike-and-wave discharge. In *Generalized epilepsy: Neurobiological approaches*, eds. M. Avoli, P. Gloor, G. Kostopoulos, and R. Naquet, 190–212. Boston: Birkhäuser.

Hale, J., and H. Kocek. 1991. *Dynamics and bifurcations.* New York: Springer-Verlag.

Hoppensteadt, F. C., and E. M. Izhikevich. 1997. *Weakly connected neural networks.* New York: Springer.

Iber, C., S. Ancoli-Israel, A. Chesson, and S. F. Quan. 2007. *The AASM manual for the scoring of sleep and associated events: Rules, terminology and technical specifications.* Westchester: American Academy of Sleep Medicine.

Kelso, J. A. S. 1999. *Dynamic patterns: The self-organization of brain and behavior.* Cambridge, MA: MIT Press.

Khosravani, H., P. L. Carlen, and J. L. Perez Velazquez. 2003. The control of seizure-like activity in the rat hippocampal slice. *Biophys. J.* 84:687–695.

Lopes da Silva, F. H., W. Blanes, S. Kalitizin, J. Gomez, P. Suffcyzynski, and F. J. Velis. 2003. Dynamical diseases of brain systems: Different routes to epileptic seizures. *IEEE Trans. Biomed. Eng.* 50:540–548.

Ma, J., and J. Wu. 2007. Multistability in spiking neuron models of delayed recurrent inhibitory loops. *Neural Comp.* 19:2124–2148.

Mann, E. O., and I. Mody. 2008. The multifaceted role of inhibition in epilepsy: Seizure-genesis through excessive GABAergic inhibition in autosomal dominant nocturnal frontal lobe epilepsy. *Curr. Opin. Neurol.* 21:155–160.

Milton, J., S. Small, and A. Solodkin. 2004. On the road to automatic: Dynamic aspects of skill acquisition. *J. Clin. Neurophysiol.* 21:134–143.

Milton, J., J. L. Cabrera, and T. Ohira. 2008. Unstable dynamical systems: Delays, noise and control. *EPL* 83:48001.

Milton, J., P. Naik, C. Chan, and S. A. Campbell. 2010. Indecision in neural decision making models. *Math. Model. Nat. Phenom.* 5:125–145.

Milton, J. G. 2010a. Neurodynamics and ion channels: A tutorial. In *Epilepsy: The intersection of neurosciences, biology, mathematics, engineering and physics.* I. Osorio, H. P. Zanvari, M. G. Frei, and S. Arthurs, eds., 111–124. Boca Raton, FL: CRC Press, Francis & Taylor.

Milton, J. G. 2010b. Epilepsy as a dynamic disease: A tutorial of the past with an eye to the future. *Epil. Behav.* 18(1–2):33–44.

Osorio, I., and M. G. Frei. 2009. Real-time detection, quantification, warning, and control of epileptic seizures: The foundations of a scientific epileptology. *Epilepsy Behav.* 16:391–396.

Pakdaman, K., C. Grotta-Ragazzo, C. P. Malta, O. Arino, and J.-F. Vobert. 1998. Effect of delay on the boundary of the basin of attraction in a system of two neurons. *Neural Netw.* 11:509–519.

Shanahan, M. 2010. Metastable chimera states in community-structured oscillator networks. *Chaos* 20:013108.

Sornette, D. 2004. *Critical phenomena in natural sciences: Chaos, fractals, self-organization and disorder: Concepts and tools.* New York: Springer.

Steraide, M. 2005. Sleep, epilepsy and thalamic reticular inhibitory neurons. *TINS* 28:317–324.

Telser, S., M. Staudacher, Y. Ploner, A. Amann, H. Hinterhuber, and M. Ritsch-Marte. 2004. Can one detect sleep stage transitions for on-line sleep scoring by monitoring the heart rate variability? *Somnologie* 8:33–41.

Timme, M., and F. Wolf. 2008. The simplest problem in the collective dynamics of neural networks: Is synchrony stable? *Nonlinearity* 21:1579–1599.

Williams, D. A. 1953. A study of thalamic and cortical rhythm in petit mal. *Brain* 76:50–59.

Wilson, H. R., and J. D. Cowan. 1972. Excitatory and inhibitory interactions in localized populations of model neurons. *Biophys. J.* 12:1–23.

Index

Page numbers followed by italicized f and t indicate figures and tables, respectively.